Springer Series in

MATERIALS SCIENCE 114

Springer Series in

MATERIALS SCIENCE

Editors: R. Hull R. M. Osgood, Jr. J. Parisi H. Warlimont

The Springer Series in Materials Science covers the complete spectrum of materials physics, including fundamental principles, physical properties, materials theory and design. Recognizing the increasing importance of materials science in future device technologies, the book titles in this series reflect the state-of-the-art in understanding and controlling the structure and properties of all important classes of materials.

99 **Self-Organized Morphology in Nanostructured Materials**
Editors: K. Al-Shamery, S.C. Müller, and J. Parisi

100 **Self Healing Materials**
An Alternative Approach to 20 Centuries of Materials Science
Editor: S. van der Zwaag

101 **New Organic Nanostructures for Next Generation Devices**
Editors: K. Al-Shamery, H.-G. Rubahn, and H. Sitter

102 **Photonic Crystal Fibers**
Properties and Applications
By F. Poli, A. Cucinotta, and S. Selleri

103 **Polarons in Advanced Materials**
Editor: A.S. Alexandrov

104 **Transparent Conductive Zinc Oxide**
Basics and Applications in Thin Film Solar Cells
Editors: K. Ellmer, A. Klein, and B. Rech

105 **Dilute III-V Nitride Semiconductors and Material Systems**
Physics and Technology
Editor: A. Erol

106 **Into The Nano Era**
Moore's Law Beyond Planar Silicon CMOS
Editor: H.R. Huff

107 **Organic Semiconductors in Sensor Applications**
Editors: D.A. Bernards, R.M. Ownes, and G.G. Malliaras

108 **Evolution of Thin-FilmMorphology**
Modeling and Simulations
By M. Pelliccione and T.-M. Lu

109 **Reactive Sputter Deposition**
Editors: D. Depla and S. Mahieu

110 **The Physics of Organic Superconductors and Conductors**
Editor: A. Lebed

111 **Molecular Catalysts for Energy Conversion**
Editors: T. Okada and M. Kaneko

112 **Atomistic and Continuum Modeling of Nanocrystalline Materials**
Deformation Mechanisms and Scale Transition
By M. Cherkaoui and L. Capolungo

113 **Crystallography and the World of Symmetry**
By S.K. Chatterjee

114 **Piezoelectricity**
Evolution and Future of a Technology
Editors: W. Heywang, K. Lubitz, and W. Wersing

Volumes 50–98 are listed at the end of the book.

Walter Heywang · Karl Lubitz
Wolfram Wersing

Piezoelectricity

Evolution and Future of a Technology

With 326 Figures

Springer

Prof. Dr. Walter Heywang
Schwabener Weg 9a
85630 Grasbrunn
Germany

Wolfram Wersing
Wagnerfeldweg 10
83346 Bergen
Germany

Dr. Karl Lubitz
Roentgenstr. 20
85521 Ottobrunn
Germany

Series Editors:

Professor Robert Hull
University of Virginia
Dept. of Materials Science and Engineering
Thornton Hall
Charlottesville, VA 22903-2442, USA

Professor Jürgen Parisi
Universität Oldenburg, Fachbereich Physik
Abt. Energie- und Halbleiterforschung
Carl-von-Ossietzky-Strasse 9–11
26129 Oldenburg, Germany

Professor R. M. Osgood, Jr.
Microelectronics Science Laboratory
Department of Electrical Engineering
Columbia University
Seeley W. Mudd Building
New York, NY 10027, USA

Professor Hans Warlimont
Institut für Festkörper-
und Werkstofforschung,
Helmholtzstrasse 20
01069 Dresden, Germany

Springer Series in Materials Science ISSN 0933-033x

ISBN 978-3-540-68680-4 e-ISBN 978-3-540-68683-5

Library of Congress Control Number: 2008930152

Typesetting: Data prepared by SPi using a Springer TEX macro package
Cover concept: eStudio Calamar Steinen
Cover production: WMX Design GmbH, Heidelberg

SPIN: 11555285 57/856/SPi
Printed on acid-free paper

9 8 7 6 5 4 3 2 1
springer.com

Foreword

When starting the European Materials Research Society (E-MRS) 25 years ago, the founders wanted to reestablish the famous tradition of the European materials community. The E-MRS conferences in Strasbourg and Warsow, which attract annually about 3,000 scientists from everywhere in the world, give proof of this successful initiative.

But in our global world it is not sufficient to be a top-player in science, market success must also be included: besides the well-known traditional pillars of Helmholtz, being research and education, innovation has to be included as a third column. Having this in mind, E-MRS has started "evolution and future of a technology" as a new book series, which is to describe exemplary cases of materials-based innovations.

After "Silicon," "Piezoelectricity" is the second volume of this series. Piezoelectricity itself is an excellent and instructive example: it was discovered as physical phenomenon by the Curie brothers in 1880, but has required new adapted materials and technologies. It was a long and sometimes troublesome way to go for making this initially minor effect a real and sometimes even superior competitor to the electro-dynamic respectively magnetic principal, which for a long time was the only way to transform electrical energy or signals into mechanical ones or vice versa effectively. Novel piezoelectric devices and novel design will open up means which avoid this roundabout in many cases.

The book "Piezoelectricity" is interesting not only for students, scientists, and engineers involved in this field, it is informative also for the entire innovation management in industry and politics. I hope that many books of this kind will follow in the new E-MRS series.

Strasbourg

August 2008 *Paul Siffert*

Preface

The volume as presented has originated from discussions with the European Materials Research Society. It is to fill a gap concerning the scientific and technical importance of piezoelectricity. We have to thank all who have supported this project, especially the publisher and all our authors for their excellent cooperation. We hope that we were able to meet the mentioned goal in spite of the broad diversity of the whole field.

München
August, 2008

Walter Heywang
Karl Lubitz
Wolfram Wersing

Contents

List of Contributors

E.K. Akdoğan
The Glenn Howatt Electronic
Ceramics Research Laboratory
Department of Materials Science
& Engineering
Rutgers-The State University
of New Jersey, Piscataway
NJ 08854, USA
eka@rci.rutgers.edu

S. Arnold
Physik Instrumente (PI) GmbH
& Co. KG
Auf der Roemerstrasse 1
D-76228 Karlsruhe, Germany
s.arnold@pi.ws

A. Ballato
US Army Communications-
Electronics, Research
Development and Engineering
Center Fort Monmouth
NJ 07703, USA
a.ballato@ieee.org

S. Bauer
Soft Matter Physics
Johannes Kepler University
Altenberger Str. 69
A-4040 Linz, Austria
sbauer@jk.uni-linz.ac.at

F. Bauer
Piézotech S.A. Rue de Colmar 9
F-68220 Hésingue
France
francois.bauer@piezotech.fr

H. Beige
Martin-Luther-University
Halle-Wittenberg
Department of Physics
Friedemann-Bach-Platz 6
D-06108 Halle
Germany
horst.beige@physik.uni-halle.de

D. Cathignol
Inserm
Institut national de la santé
et de la recherche médicale
151, cours Albert Thomas
F-69424 LYON Cedex 03
France
cathignol@lyon.inserm.fr

R.E. Cohen
Carnegie Institution
of Washington
Geophysical Laboratory
n5251 Broad Branch RD., N.W.
Washington, DC 20015, USA
cohen@gl.ciw.edu

L.E. Cross
Materials Research Institute
The Pennsylvania State University
187 MRL
University Park, PA 16802
USA
lec3@psu.edu

T. Fett
University of Karlsruhe
Institute for Ceramics
in Mechanical Engineering
Haid-und-Neustrasse 7
D-76131 Karlsruhe
Germany
theo.fett@ikm.uni-karlsruhe.de

G. Helke
CeramTec AG
Innovative Ceramic Engineering
Piezotechnology Lauf, retired
Luitpoldstr. 15
D-91207 Lauf
g.helke@t-online.de

W. Heywang
Siemens AG
Corporate Technology
München, Germany, retired
Schwabener Weg 9a
D-85630 Grasbrunn
Germany
walter.heywang@t-online.de

M.J. Hoffmann
University of Karlsruhe
Institute for Ceramics
in Mechanical Engineering
Haid-und-Neustrasse 7
D-76131 Karlsruhe
Germany
michael.hoffmann@ikm.uka.de

Y. Hosono
Corporate R&D Center
Toshiba Corp.
Kawasaki 212-8582
Japan
yasuharu.hosono@toshiba.co.jp

W. Kreher
Institut für Werkstoffwissenschaft
Technische Universität Dresden
D-01062 Dresden, Germany
wolfgang.kreher@tu-dresden.de

H. Kungl
University of Karlsruhe
Institute for Ceramics
in Mechanical Engineering
Haid-und-Neustrasse 7
D-76131 Karlsruhe, Germany
hans.kungl@ikm.uni-karlsruhe.de

R. Lerch
Department of Sensor Technology
University of Erlangen-Nuremberg
Paul-Gordan-Strasse 3/5
D-91052 Erlangen, Germany
reinhard.lerch@lse.eei.
uni-erlangen.de

K. Lubitz
Siemens AG
Corporate Technology
München, Germany, retired
Röntgenstrasse 20
D-85521 München, Germany
karllubitz@t-online.de

R. Mock
Siemens AG
Corporate Technology
Otto-Hahn-Ring 6
D-81730 München, Germany
randolf.mock@siemens.com

P. Muralt
Ceramics Laboratory
EPFL – Swiss Federal
Institute of Technology
(Ecole Polytechnique Fédérale
de Lausanne)
Lausanne 1015
Switzerland
paul.muralt@epfl.ch

P. Pertsch
PI Ceramic GmbH
Lindenstrasse
D-07589 Lederhose
Germany
p.pertsch@piceramic.de

G.A. Rossetti, Jr.
Materials Science and
Engineering Program
Institute of Materials Science
University of Connecticut
Storrs, CT 06269, USA
rossetti@ims.uconn.edu

A. Safari
The Glenn Howatt Electronic
Ceramics Research Laboratory
Department of Materials Science
& Engineering
Rutgers-The State University
of New Jersey, Piscataway
NJ 08854
USA
safari@rci.rutgers.edu

F. Seifert
University of Technology
Vienna Institute of
Sensor- and Actuatorsystems,
retired Gusshausstrasse 27-29
A-1040 Vienna
Austria
franz.seifert@tuwien.ac.at

N. Setter
Ceramics Laboratory
EPFL – Swiss Federal
Institute of Technology
(Ecole Polytechnique Fédérale
de Lausanne)
Lausanne 1015
Switzerland
nava.setter@epfl.ch

K. Spanner
Physik Instrumente (PI) GmbH
& Co. KG
Auf der Roemerstrasse 1
D-76228 Karlsruhe
Germany
k.spanner@pi.ws

T. Steinkopff
Siemens AG
Corporate Technology
Otto-Hahn-Ring 6
D-81730 München
Germany
thorsten.steinkopff@siemens.com

R. Theissmann
University of Duisburg-Essen
Faculty of Engineering
Institute for Nano Structures
and Technology (NST)
Bismarckstr. 81
D-47057 Duisburg
ralf.theissmann@uni-due.de

H. Thomann
Siemens AG
Corporate Technology
München, Germany, retired
Kössener Str. 15
D-81373 München
Germany

K. Uchino
International Center for Actuators and Transducers
Materials Research Institute
The Pennsylvania State University
University Park
PA 16802, USA
Micromechatronics Inc.
State College, PA 16803
USA
kenjiuchino@psu.edu

R. Veith
EPCOS PTE LTD
166 Kallang Way
Singapore 349249
richard.veith@epcos.com

K. Wagner
EPCOS AG
R&D/Surface Acoustic Wave Components
Anzinger Strasse 13
D-81671 Munich
Germany
karl.wagner@epcos.com

S. Wagner
University of Karlsruhe
Institute for Ceramics in Mechanical Engineering
Haid-und-Neustrasse 7
D-76131 Karlsruhe, Germany
susanne.wagner@ikm.uka.de

W. Wersing
Siemens AG
Corporate Technology
München, Germany, retired
Wagnerfeldweg 10
D-83346 Bergen, Germany
wolfram.wersing@t-online.de

A. Wolff
Siemens AG
Corporate Technology
Otto-Hahn-Ring 6
D-81730 München, Germany
andreas.wolff@siemens.com

Y. Yamashita
Toshiba Research Consulting Corp.
Kawasaki 212-8582, Japan
yohachi.yamashita@toshiba.co.jp

1

Introduction

L.E. Cross and W. Heywang

Barely noticed by the general public, piezoelectricity has become an innovation motor, which worldwide has initiated new markets with turnover of billions of dollars. The piezoelectric market covers a very wide range of technical applications; it is especially strong in the fields of information and communications, industrial automation, medical diagnostics, automation and traffic control, and in the defense industries.

All these innovations are materials based. Thus, on the one hand, the story of piezoelectric innovations cannot be understood as a simple market pull process. On the other hand, a simple science fiction of its inception with statements like "the recently invented class of piezoelectric substances offers many novel possibilities and by exploiting these opportunities an avalanche of novel products will develop changing our lives" does not hold either. Materials-based innovations are usually far more complex. There is mutual stimulation between advances in materials and processing technologies, on the one hand, and technical and economic requirements of new and improved applications on the other, i.e., both market pull and technology push are important intercorrelated factors.

Similarly, the way piezoelectricity entered substantially into the market place was most unusual. Piezoelectricity was first discovered by the Curie brothers in 1880, but no until one century later did we realize the current avalanche of products developing in this market. Interesting questions are what happened during the period of long delay on one hand, and what catalyzed the present avalanche of progress on the other.

From the perspective of solid-state physics, crystalline materials that become electrically polarized when subjected to mechanical stress and conversely change shape when under an applied electric field, namely piezoelectrics, certainly merit study. Technologically, however, the phenomenon only becomes interesting to the broader engineering community when it offers an effect large enough to usefully convert electrical power or signals into mechanical ones or vice-versa. This is the first essential condition, and there are also always a lot of additional needs like small conversion losses, temperature and long time stability, reproducibility, reliability, cost and other market needs. Out of the family of such requirement for a given device, and the measured material properties, a figure of merit can be formulated. Now we have to compare these figures of merit with the corresponding ones for already existing systems in current applications.

From general experience, it is known that a new novel technology will not succeed in application unless it offers at least a factor of three advantages. Otherwise, in the inertia born tendency for existing solutions to survive, the current approach will be improved by a factor of this order as soon as it is perceived to be endangered.

To effectively enter the market piezoelectrics always had to compete with well-known electromagnetic devices, as for example relays and electromagnetic motors. Only in the case of substantial advantage would they be accepted technically or commercially. Clearly this competition was the major reason for the above-mentioned long delay of the technical break-through to large scale applications.

First time a conceived major improvement in a practical device was achieved through piezoelectricity in 1921 with the development of the quartz crystal stabilized electrical oscillator. In this device, the weak piezoelectricity in quartz is used to excite resonant elastic vibrations where frequency is set by the ultra stable elastic properties of suitable crystal cuts. In 1924, the first crystal stabilized radio transmitter was placed into service, and by late 1930s all high frequency radio transmitters were under crystal control. The astounding success of this initially simple concept is attested by the fact that even now more than 80 years later, the quartz crystal controlled oscillator is still the secondary standard for timing and frequency control.

However, the avalanche of piezo technology starting in the market place was not possible until it was realized that the mixed oxide compound barium titanate ($BaTiO_3$) was a ferroelectric, and it was demonstrated in 1946 that $BaTiO_3$ ceramic, which can be easily fabricated and shaped at low price, can be made piezoelectric by an electrical poling process. First commercial piezoelectric devices based on $BaTiO_3$ ceramics were phonograph pickups and appeared on the market about 1947.

The piezoelectric effect in ferroelectric titanates is much stronger than in quartz because they exhibit different ferroelectric phases with only small energy differences. This can lead to the unusual combination of high dielectric permittivity in a strongly polar lattice in the broad vicinity of the phase boundaries. These conditions are essential for high piezoelectric constants and strong elasto-electric coupling. An advance of great practical importance was the discovery in 1954 of very strong piezoelectric effects in lead zirconate titanate solid solutions. In this material system, a morphotropic phase boundary exists between a tetragonal and rhombohedral ferroelectric phase, which is nearly independent of temperature. Thus, the anomalously high piezoelectric effect near to this boundary is almost constant over a wide temperature range, important for technical applications. However, one has to pay for the anomalously large piezoelectric effect at this phase boundary with the risk of higher instability of material properties. It was a long and arduous task to control these instabilities by proper adjustment of dopants and grain structure. This problem was eased by the development of two basic material classes adapted to quite different branches of applications: so called soft piezoelectric with as high as possible piezoelectric properties for actuator and ultrasonic applications and so-called hard piezoceramics with as high as possible stability for high power and precise frequency filter applications.

In addition to the material property, considerations on optimum design of device must consider the electrodes through which the field is to be applied. Electrodes must be robust and in very intimate contact with the high permittivity ceramic. Barely any other functional material exists with such a plethora of application-specific shaping requirements: plates, rods, disks, foils, tubes, all designed for either static, dynamic, or resonant applications. There are also composites made from combinations of piezoceramics and plastics, now very widely used in medical ultrasound and in hydrophone systems. Further more, to achieve enhanced elongation in actuators at smaller driving voltages, an inexpensive multilayer technology has been developed. In this way, it has become possible to achieve elongations on 0.1 mm in compact monolithic devices, a technology now used for modern fuel valves. These valves react ten times faster than conventional valves and are just revolutionizing motor management in the automotive industry. Thin film technologies are of developing interest for micro electronics and micro electro mechanical (MEMs), for high permittivity dielectric layers, in nonvolatile computer memories, pyroelectric thermal images, and biosensors.

It is very important to recall that beside *polycrystalline piezoelectrics*, ferroelectric piezo-crystals have also been developed. Today we have the stable ferroelectric phase single crystal lithium niobate ($LiNbO_3$) and lithium tantalate ($LiTaO_3$). Both are commercially available and widely used for surface acoustic wave devices, which play a major role in signal detection and processing in modern telecommunication systems.

Finally it should be mentioned that there are plastics that can be made ferro- respectively piezoelectric by stress and/or by electrical polarization because of an alignment of asymmetric molecules. They find use as inexpensive foils wherever cheapness or plasticity are relevant., e.g. in cheap microphones or shock gauges. Recent studies following analogous paths to electroceramics have demonstrated effective control of microstructure and properties, stepping stones in bridging to the growing field of biological technologies.

An over-all compilation of existing application of piezoelectricity is given in Table 1.1. The heavy-typed fields indicate where the number of installed piezoelectric samples have exceeded already several billions. Other fields are on the way to follow like the field of automotive applications, where piezoelectric sensors, valves, and injectors are key elements for clean and fuel saving motor

Table 1.1. Piezoelectricity, innovation fields, and important applications

Category	Innovation field	Materials and shaping	Main application
Frequency control and signal processing	**Frequency-/time standards**	Quartz single crystal plates	Precise frequency control
	Mechanical frequency filters	Ceramic plates of specifically tailored PZT	Inexpensive frequency control and filtering
	Surface acoustic wave (SAW) devices	$LiNbO_3$, $LiTaO_3$, Quartz single crystal substrates	Passive signal processing for wireless communication, identification, sensing, etc.
	Bulk acoustic wave (BAW) devices	Ceramic plates of hard PZT AlN, ZnO thin films	
Sound and ultrasound (US)	**Buzzer**	Ceramic tapes of soft PZT	Sonic alerts
	Microphones and speakers	Ceramic tapes of soft PZT PZT thin films	Telephone, blood pressure
	Ultrasonic (US) imaging	Diced plates of soft PZT or of PZNT single crystals PZT thin films	Medical diagnostics
	Hydrophonics	Hard PZT of various shapes soft PZT composites	Sources and detectors for sound location
	High power transducers and shock wave generation	Ceramic discs of hard PZT	Machining, US cleaning, lithotripsy
	Atomizer	Ceramic discs of soft PZT	Oil atomizers, humidifiers, aerosols
	Air ultrasound	Ceramic discs of soft PZT	Distance meter, intrusion alarm
Actuators and motors	**Printers**	Bars, tubes, multilayer ceramics of soft PZT PZT thin films	Needle drives and ink jet
	Motors and transformers	Rings, plates of hard PZT soft PZT multilayer ceramics PZT thin films	Miniaturized, compact motors and transformers
	Bimorph actuators	PZT multilayer ceramics	Pneumatics, micropumps, braille for the blind

	Multilayer actuators	Multilayer stacks of soft PZT	Fine positioning and optics
	Injection systems	Multilayer stacks of soft PZT	Automotive fuel valves
Sensors	Acceleration sensors	Rings, plates of soft PZT	Automotive, automation, medical
	Pressure and shock-wave sensors	$LiNbO_3$ substrates PVDF foils	
	Flow sensors	Soft PZT discs	
	Mass sensitive sensors	Quartz discs, Quartz substrates ZnO, AlN thin films	
Ignition	**Ignition**	Hard PZT cylinders	Gas and fuel ignition
Adaptronics	Adaptive devices	Various shapes of soft PZT, multilayer stacks of soft PZT	Active noise and vibration cancellation, adaptive control, airtail filter control

management. But also in fields where similar figures cannot be expected by nature, piezoelectricity has become a key technology, e.g., wherever ultrasound or mechanical sensing and positioning is concerned.

Furthermore, like every novel technology piezoelectricity has contributed to important scientific achievements or even made them possible. The most famous one is the tunnel microscope and the derived similar instruments for the investigation of atomistic structures. Generally spoken, the effective micro-coupling of mechanical and electric parameters has brought about a better insight into both atomistic as well as phonon-correlated cooperative phenomena.

Summarizing, concerning the multitude of piezoelectric materials and device applications existing already, today we may expect continuous future development and fascinating new novel application areas. Above all we can expect transfer of macrosystems into the micro and nano worlds, the search for lead free and environmentally more friendly highly active piezoelectrics will continue, and last but not least the piezoelectric effect in well textured or even single crystalline $Pb(ZrTi)O_3$ like systems could surmount the already known strong piezo effects by an order of magnitude enlarging once more the already very broad current field of applications.

Part I

Basics and Materials

2

Basic Material Quartz and Related Innovations

A. Ballato

2.1 General

2.1.1 Commercial and Technological Usages of Quartz

Although material quartz is of scientific interest in its own right, its volume of usage and variety of applications dictate its technological importance. The technological prominence of α-quartz stems largely from the presence of piezoelectricity, combined with extremely low acoustic loss. It was one of the minerals with which the Brothers Curie first established the piezoelectric effect in 1880. In the early 1920s, the quartz resonator was first used for frequency stabilization. Temperature-compensated orientations (the AT and BT shear cuts) were introduced in the 1930s, and assured the technology's success. By the late 1950s, growth of cultured bars became commercially viable, and in the early 1970s, cultured quartz use for electronic applications first exceeded that of the natural variety. The discovery of cuts that addressed compensation of stress and temperature transient effects occurred in the 1970s, and led to the introduction of compound cuts such as the SC, which has both a zero temperature coefficient of frequency, and is simultaneously stress-compensated [1–5]. Between 10^9 and 10^{10} quartz units per year were produced by 2000 at frequencies from below 1 kHz to above 10 GHz. Categories of application include resonators, filters, delay lines, transducers, sensors, signal processors, and actuators. Particularly noteworthy are the bulk- and surface-wave resonators; their uses span the gamut from disposable timepieces to highest precision oscillators for position-location, and picosecond timing applications. Stringent high-shock and high-pressure sensor operations are also enabled. Table 2.1 shows the major applications of quartz crystals. These applications are discussed subsequently in greater detail. For general background and historical developments, see [1, 6–11].

Crystal quartz for timekeeping currently has a global market of over 10^9 USD per year, and that for cellular communications exceeds 50×10^9 USD

Table 2.1. Major applications of quartz crystals

Military and aerospace	Research and metrology	Industrial	Consumer	Automotive
– Communications	– Atomic Clocks	– Communications	– Watches and clocks	– Engine control, stereo, clock
– Navigation/GPS	– Instruments	– Tele-communications	– Cellular and cordless phones, pagers	– Trip computer
– IFF	– Astronomy and geodesy	– Mobile/cellular/portable radio, telephone, and pager	– Radio and hi-fi equipment	– Navigation/GPS
– Radar	– Space tracking	– Aviation	– Color TV	
– Sensors	– Celestial navigation	– Marine	– Cable TV systems	
– Guidance systems		– Navigation	– Home computers	
– Fuzes		– Instrumentation	– VCR and video camera	
– Electronic warfare		– Computers	– CB and amateur radio	
– Sonobuoys		– Digital systems	– Pacemakers	
		– Displays	– Toys and games	
		– Disk drives		
		– Modems		
		– Tagging/identification		
		– Utilities		

per year. Useful rules-of-thumb for the temporal evolution of this technology are the following:

- Upper frequency limit: $f_0 = 10^{+r}$, where f_0 is nominal frequency, $r = 4 + 0.075\ (Y - 1920)$, and Y is year.
- Frequency accuracy (total fractional absolute frequency variations over all environmental ranges such as temperature, mechanical shock, and aging): $\Delta f/f_0 = 10^{-a}$, where $\Delta f = (f - f_0)$ is frequency shift, and $a = 3.5 + 0.05\ (Y - 1940)$.
- Aging: $<10^{-8}$/day for production units, and $<10^{-11}$/day for high precision units.
- Frequency stability: $\Delta f/f_0 = 10^{-s}$, where $s = 6 + 0.1(Y - X)$, with $X = 1920$ for laboratory versions, $X = 1940$ for commercial versions, and $X = 1960$ for large scale production models. Observation times are in the range 0.1–10 s. As with Moore's Law, a saturation of these exponential rates is predicted, but not yet observed.

2.1.2 Phases of the Silica System [12–15]

Silica (SiO_2) is the second most abundant molecule on earth, after H_2O. The phase diagram of the silica system is sketched in Fig. 2.1. This chapter discusses some of the properties and applications of the commercially significant α-quartz phase. Other phases appearing in the diagram are mentioned briefly below.

"Quartz" refers to crystalline α-SiO_2, arranged in point group symmetry 32 (Hermann-Mauguin)/D_3 (Schönflies). Figure 2.2 indicates its symmetry elements; z (x_3) is the axis of threefold symmetry. Three equivalent secondary axes x (x_1) are parallel to the twofold axes, and perpendicular to the x_3 axis; three equivalent y (x_2) axes are normal to x_3 and to the respective x_1 axes. The trigonal axis is denoted "optical" (x_3 a screw axis, rotating polarized light), the digonal (x_1) axes are denoted "electrical" and are piezoelectrically active; the x_2 axes are "mechanical" axes. Axial and sign conventions are given in various IEEE Standards and related papers [16–22].

The interlocking nature of the lattice bonds leads to a strong, rigid structure with scratch hardness of 0.667 (with alumina unity), and Moh hardness 7. Si–O bonds occur in pairs with lengths of 0.1598 and 0.1616 nm. Four oxygens coordinate each silicon, with each oxygen slightly off a line joining two neighboring silicons. Bonding is ∼0.6 covalent and 0.4 ionic, with average bond energies ∼4.85 eV/bond (∼0.468 MJ mol^{-1}), whereas a single Si–O bond has an energy of 3.82 eV [23]. The unit cell lattice constants are $a_o = 0.4913$ nm, and $c_o = 0.5404$ nm at room temperature [1, 14, 15, 24]. Quartz exhibits enantiomorphism, with chiral (left/right) pairs having space groups $P3_121/D_3{}^4$ and $P3_221/D_3{}^6$. Figure 2.3 portrays the external facet arrangement of a completely developed, right-handed specimen.

Upon heating, α-SiO_2 ("low quartz") undergoes a phase transition at 573.3°C, becoming β-SiO_2 ("high quartz"); in the transition, the silicon atoms

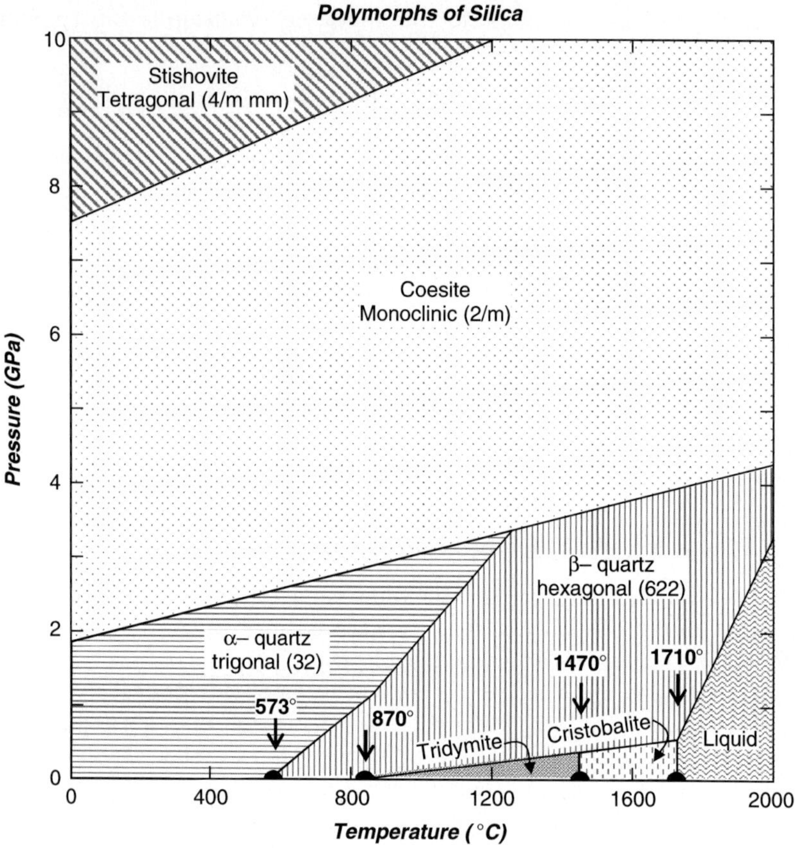

Fig. 2.1. Polymorphs of silica

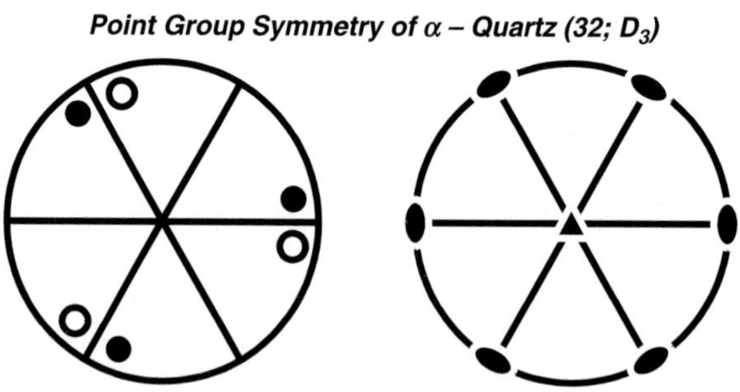

Fig. 2.2. Symmetry elements for α-quartz, point group 32

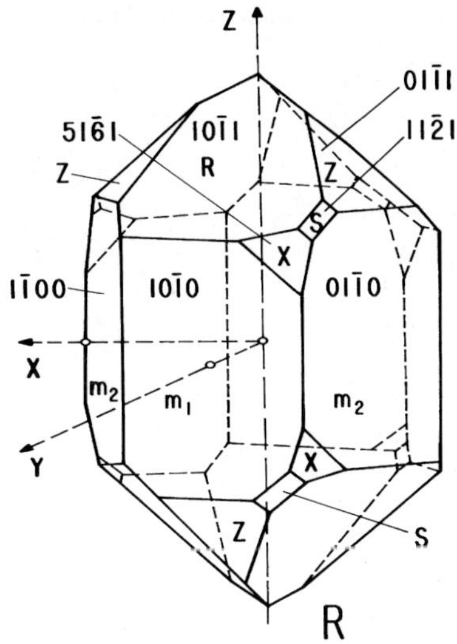

Fig. 2.3. Idealized, right-hand quartz, showing natural facets

move only 0.03 nm, changing the trigonal axis to hexagonal, and the point group symmetry to 622/D_6. Beta quartz is also enantiomorphic, composed of space groups $P6_222/D_6{}^4$ and $P6_422/D_6{}^5$; handedness is preserved in the transition. Enantiomorphism implies the presence of piezoelectricity. Beta quartz is not stable below 573°C. At 600°C, unit cell dimensions are $a_0 = 0.501$ and $c_0 = 0.547$ nm, and density $\rho = 2.533$. α-tridymite (point group uncertain; mmm or 2/m) is metastable below 870°C, to ~117°C; $\rho = 2.28$. β-tridymite (6/m mm) is stable between 870 and 1,470°C, and metastable down to ~163°C; $\rho = 2.24$. α-cristobalite (422) is metastable to below 200°C; $\rho = 2.32$. β-cristobalite (m3m) is stable between 1,470 and 1,710°C, and metastable down to 268°C; $\rho = 2.24$. Density units are Mg/m^3. Coesite is of point symmetry 2/m, that of stishovite is 4/m mm; both are metastable under ordinary conditions. With the exception of α-cristobalite, the other phases are not piezoelectric. The symmetry diagram for β-quartz is given in Fig. 2.4.

Transitions between different phases of a single polymorph (α-β transformations) involve only bond warping, and not bond breaking; these displacive transformations proceed rapidly. Cooling through the α-β quartz transition invariably results in α-quartz regions having oppositely-directed x_1 axes: "electrical twinning." Transitions between polymorphs require breaking of bonds; these reconstructive transformations are slow, and subject to kinematic constraints.

Hydrothermal growth of cultured quartz takes place in aqueous NaOH or Na_2CO_3 solutions at temperatures of about 350°C and pressures between 80 and 200 MPa, driven by a temperature gradient from 4 to 10 K. The process

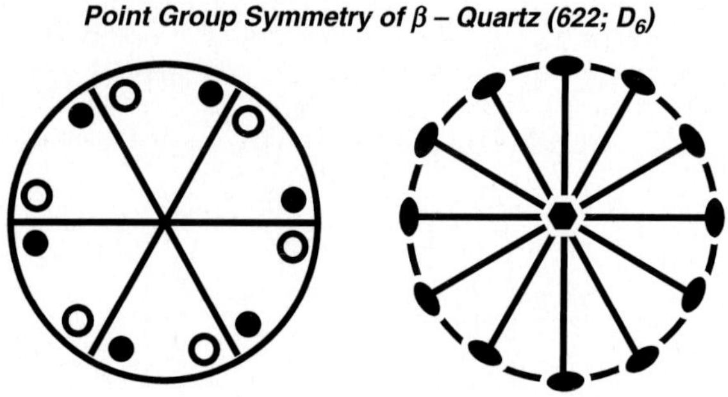

Fig. 2.4. Symmetry elements for β-quartz, point group 622

takes from one to nine months. Processing of quartz crystals includes the steps of cutting, X-ray orienting, lapping, polishing and etching (usually $NH_4F \cdot HF$ or $HF_{(aq)}$). Growth and etch rates are highly anisotropic, being most rapid along z, with the z rate between 100 and 1,000 times that along x or y axis.

2.2 Phenomenological Coefficients of Quartz

2.2.1 Matter Tensors and Symmetry

Voigt [25] considered field tensors, such as temperature, electric displacement, stress, and strain, and arranged the phenomenological coefficients (matter tensors) relating them according to their ranks. He made the simplification of converting the description of stress and strain from three-dimensional, second-rank tensors to six-dimensional vectors (first-rank tensors). This permits the description of linear elasticity, and related effects, such as piezoelectricity, in simple matrix form. Pertinent here are the intensive field tensors mechanical stress T, electric intensity E, the extensive field tensors mechanical strain S, and electric displacement D. Matter tensors appearing in the constitutive relations are, therefore, elastic stiffness c, and compliance s; dielectric permittivity ε, and impermeability β; and piezoelectric constants d, e, g, and h. A graphical portrayal of these relationships was first devised by Heckmann [26], and is now commonly used [1,27,28].

The extension of Hooke's Law to piezoelectric coupling between quasi-static mechanical and electrical variables is represented by 9×9 Van Dyke matrices. Four sets are commonly used, depending on the physical structure under consideration, as follows. A prime denotes matrix transpose.

Homogeneous sets:

(1) $S = s^{\mathrm{E}}\, T + d'\, E$ and $D = dT + \varepsilon^{\mathrm{T}} E$
(2) $T = c^{\mathrm{D}}\, S - h'\, D$ and $E = -hS + \beta^{\mathrm{S}} D$

Table 2.2. Van Dyke matrix for the mixed coefficients c^{E}, e, and ε^{S} of α-quartz

$c_{11}{}^{E}$	$c_{12}{}^{E}$	$c_{13}{}^{E}$	$c_{14}{}^{E}$	0	0	$-e_{11}$	0	0
$c_{12}{}^{E}$	$c_{11}{}^{E}$	$c_{13}{}^{E}$	$-c_{14}{}^{E}$	0	0	e_{11}	0	0
$c_{13}{}^{E}$	$c_{13}{}^{E}$	$c_{33}{}^{E}$	0	0	0	0	0	0
$c_{14}{}^{E}$	$-c_{14}{}^{E}$	0	$c_{44}{}^{E}$	0	0	$-e_{14}$	0	0
0	0	0	0	$c_{44}{}^{E}$	$c_{14}{}^{E}$	0	e_{14}	0
0	0	0	0	$c_{14}{}^{E}$	$c_{66}{}^{E}$	0	e_{11}	0
e_{11}	$-e_{11}$	0	e_{14}	0	0	$\varepsilon_{11}{}^{S}$	0	0
0	0	0	0	$-e_{14}$	$-e_{11}$	0	$\varepsilon_{11}{}^{S}$	0
0	0	0	0	0	0	0	0	$\varepsilon_{33}{}^{S}$

Table 2.3. Van Dyke matrix for the homogeneous coefficients s^{E}, d, and ε^{T} of α-quartz

$s_{11}{}^{E}$	$s_{12}{}^{E}$	$s_{13}{}^{E}$	$s_{14}{}^{E}$	0	0	d_{11}	0	0
$s_{12}{}^{E}$	$s_{11}{}^{E}$	$s_{13}{}^{E}$	$-s_{14}{}^{E}$	0	0	$-d_{11}$	0	0
$s_{13}{}^{E}$	$s_{13}{}^{E}$	$s_{33}{}^{E}$	0	0	0	0	0	0
$s_{14}{}^{E}$	$-s_{14}{}^{E}$	0	$s_{44}{}^{E}$	0	0	d_{14}	0	0
0	0	0	0	$s_{44}{}^{E}$	$2s_{14}{}^{E}$	0	$-d_{14}$	0
0	0	0	0	$2s_{14}{}^{E}$	$s_{66}{}^{E}$	0	$-2d_{11}$	0
d_{11}	$-d_{11}$	0	d_{14}	0	0	$\varepsilon_{11}{}^{T}$	0	0
0	0	0	0	$-d_{14}$	$-2d_{11}$	0	$\varepsilon_{11}{}^{T}$	0
0	0	0	0	0	0	0	0	$\varepsilon_{33}{}^{T}$

Mixed sets:

(3) $T = c^{\mathrm{E}}S - e'\,E$ and $D = eS + \varepsilon^{\mathrm{S}}E$
(4) $S = s^{\mathrm{D}}T + g'\,D$ and $E = -gT + \beta^{\mathrm{T}}D$

Constants d and g are "strain" coefficients; e and h are "stress" coefficients. The $[s^{\mathrm{E}},\ d',\ d,\ \varepsilon^{\mathrm{T}}]$ and $[c^{\mathrm{D}},\ -h',\ -h,\ \beta^{\mathrm{S}}]$ matrices are inverses, as are the $[c^{\mathrm{E}},\ -e',\ e,\ \varepsilon^{\mathrm{S}}]$ and $[s^{\mathrm{D}},\ g',\ -g,\ \beta^{\mathrm{T}}]$ matrices.

Point-group symmetry dictates the number of components appearing in the constitutive equations coupling the field tensors, and any relations among the components. Tables 2.2–2.5 display the Van Dyke matrices (or "term schemes") representing point group symmetry 32 and 622. Factors of 2 in the s, d, and g matrices arise from a convention regarding the definition of strain, and, in these tables, $\mathrm{c}_{12} = \mathrm{c}_{11} - 2\mathrm{c}_{66}$ and $\mathrm{s}_{12} = \mathrm{s}_{11} - \mathrm{s}_{66}/2$. The Van Dyke matrices of Tables 2.2 and 2.3, and of Tables 2.4 and 2.5 are inverses of each other.

2.2.2 Numerical Values for α-Quartz

2.2.2.1 Elastic Constants

Alpha-quartz possesses six independent elastic, two piezoelectric, and two dielectric constants. Two principal and competing conventions exist for

Table 2.4. Van Dyke matrix for the mixed coefficients c^{E}, e, and ε^{S} of β-quartz

$c_{11}{}^{E}$	$c_{12}{}^{E}$	$c_{13}{}^{E}$	0	0	0	0	0	0
$c_{12}{}^{E}$	$c_{11}{}^{E}$	$c_{13}{}^{E}$	0	0	0	0	0	0
$c_{13}{}^{E}$	$c_{13}{}^{E}$	$c_{33}{}^{E}$	0	0	0	0	0	0
0	0	0	$c_{44}{}^{E}$	0	0	$-e_{14}$	0	0
0	0	0	0	$c_{44}{}^{E}$	0	0	e_{14}	0
0	0	0	0	0	$c_{66}{}^{E}$	0	0	0
0	0	0	e_{14}	0	0	$\varepsilon_{11}{}^{S}$	0	0
0	0	0	0	$-e_{14}$	0	0	$\varepsilon_{11}{}^{S}$	0
0	0	0	0	0	0	0	0	$\varepsilon_{33}{}^{S}$

Table 2.5. Van Dyke matrix for the homogeneous coefficients s^{E}, d, and ε^{T} of β-quartz

$s_{11}{}^{E}$	$s_{12}{}^{E}$	$s_{13}{}^{E}$	0	0	0	0	0	0
$s_{12}{}^{E}$	$s_{11}{}^{E}$	$s_{13}{}^{E}$	0	0	0	0	0	0
$s_{13}{}^{E}$	$s_{13}{}^{E}$	$s_{33}{}^{E}$	0	0	0	0	0	0
0	0	0	$s_{44}{}^{E}$	0	0	d_{14}	0	0
0	0	0	0	$s_{44}{}^{E}$	0	0	$-d_{14}$	0
0	0	0	0	0	$s_{66}{}^{E}$	0	0	0
0	0	0	d_{14}	0	0	$\varepsilon_{11}{}^{T}$	0	0
0	0	0	0	$-d_{14}$	0	0	$\varepsilon_{11}{}^{T}$	0
0	0	0	0	0	0	0	0	$\varepsilon_{33}{}^{T}$

determining the signs of certain of the coefficients, and it is necessary to be consistent in following whichever choice is selected [16–22]. Table 2.6 contains experimental values for room-temperature elastic stiffnesses at constant electric field, c^{E} (isagric values). Recent results of James [32], indicate that measurement precision is now sufficient to distinguish among the small differences in values occurring between natural and cultured quartz, and between cultured varieties hydrothermally grown either with sodium hydroxide or with sodium carbonate mineralizer. Table 2.7 gives values for the isagric elastic compliances, s^{E} corresponding to the entries of Table 2.6.

2.2.2.2 Temperature Coefficients of Elastic Constants

It is found experimentally that a Taylor series expansion to the third order usually suffices to characterize the behavior of the elastic constants of quartz over the often-specified temperature range, −55°C to +90°C. The reference temperature, T_0, is usually taken to be room temperature. With y standing for any of the elastic stiffnesses or compliances, its value at temperature T is then given by

$$\Delta y/y_0 = T_y^{(1)}\ \Delta T + T_y^{(2)}\ (\Delta T)^2 + T_y^{(3)}\ (\Delta T)^3,$$

Table 2.6. Elastic stiffnesses, c^{E}

$c_{\lambda\mu}{}^{\mathrm{E}}$	[25]	[29][a]	[30]	[31]	[24][b]	[32]	[33][c]
c_{11}	85.48	86.75	86.74	86.80	86.80 ± 0.04	86.790	86.7997
c_{13}	14.36	11.3	11.91	11.91	11.91 ± 0.01	12.009	11.9376
c_{14}	−16.83	−17.96	−17.91	−18.04	-18.02 ± 0.07	−18.116	−18.0612
c_{33}	105.6	106.8	107.2	105.75	106.2 ± 0.8	105.79	105.7816
c_{44}	57.13	57.86	57.94	58.20	58.17 ± 0.13	58.212	58.2231
c_{66}	39.11	39.94	39.88	39.88	39.85 ± 0.03	40.000	39.8817
c_{12}	7.25	6.87	(6.98)	(7.04)	7.10 ± 0.09	6.7901	(7.0363)

unit: GPa
[a]Values of [34] corrected for piezoelectricity
[b]Averages gleaned from the literature
[c]Cultured quartz at 23°C

Table 2.7. Elastic compliances, s^{E}

$s_{\lambda\mu}{}^{\mathrm{E}}$	[25]	[29]	[30][a]	[31]	[24]	[32]	[33]
s_{11}	12.95	12.76	12.77	12.78	12.78	12.779	12.779
s_{13}	−1.53	−1.16	−1.22	−1.24	−1.23	−1.249	−1.238
s_{14}	4.31	4.52	4.50	4.52	−4.52	4.528	−4.525
s_{33}	9.88	9.61	9.60	9.74	9.69	9.736	9.733
s_{44}	20.04	20.09	20.04	19.98	19.99	19.997	19.983
s_{66}	29.28	29.10	29.12	29.17	29.18	29.102	29.172
s_{12}	−1.69	−1.79	−1.79	−1.81	−1.81	−1.772	−1.807

unit: $(\mathrm{TPa})^{-1}$
[a]Slightly different values are given in [36] and [37]

where $\Delta y = (y - y_0)$, $\Delta T = (T - T_0)$, and y_0 is the value of y at $T = T_0$. The coefficients $T_{\mathrm{y}}^{(\mathrm{n})}$ are the nth-order temperature coefficients (TCs) of y. Tables 2.8 and 2.9 contain values for the first-order TCs of the stiffnesses and compliances of quartz, respectively. In Tables 2.10 and 2.11 are given the second and third-order stiffness coefficients.

2.2.2.3 Piezoelectric Coefficients and Dielectric Permittivities

Table 2.12 contains values for the piezoelectric coefficients. Table 2.13 lists values for the dielectric permittivities and the impermeabilities, while Table 2.14 gives values for the thermoelastic (thermal expansion) coefficients.

2.2.2.4 Mass Density (ρ)

Natural quartz: $\rho = 2.6487$ at 20°C; 2.649 at 25°C [6]

Cultured quartz: $\rho = 2.6484 \pm 0.0002$ at 25°C [32]; unit: $\mathrm{Mg\ m^{-3}}$

Table 2.8. First-order temperature coefficients of stiffnesses, $Tc_{\lambda\mu}{}^{(1)}$

$\lambda\mu$	[31]	[38]	[39]	[24][a]	[40]	[32]	[33][b]
11	−48.5	−44.3	−48.5	-46.4 ± 2.1	−68.2	−43.591	−45.5
13	−383	−492	−550	-476 ± 111	−705	−592.27	−745
14	105	98	101	105 ± 8	84.2	102.59	98.7
33	−153	−188	−160	-172 ± 24	−197	−190.32	−193
44	−158	−172	−177	-170 ± 13	−186	−171.59	−160
66	169	180	178	177 ± 8	158	176.79	166
12	−2,703	−2,930	−3,000[c]	$-2,901 \pm 183$		−2,640.1	(−2,443)

unit: 10^{-6}/K
[a]Averages gleaned from the literature
[b]Cultured quartz at 23°C
[c]−2,637 for consistency

Table 2.9. First-order temperature coefficients of compliances, $Ts_{\lambda\mu}{}^{(1)}$

$\lambda\mu$	[31]	[39]	[41]
11	16.5	15.5	8.5
13	−678	−166	−168.8
14	139.5	134	140
33	134.5	140	139.7
44	201	210	211.1
66	−138	−145	−151.9
12	−1,270	−1,370[a]	−1,296.5

unit: 10^{-6}/K
[a]−1,290 for consistency

Table 2.10. Second-order temperature coefficients of stiffnesses, $Tc_{\lambda\mu}{}^{(2)}$

$\lambda\mu$	[31]	[38]	[39]	[40]	[32]
11	−75	−407	−107	−117	−110.98
13	−2,000	−596	−1,150	−1,022	−1,218.5
14	−270	−13	−48	−54.4	−30.802
33	−187	−1,412	−275	−158	−165.31
44	−212	−225	−216	−272	−254.49
66	−5	201	118	152	156.18
12	−1,500	−7,245	−3,050[a]		−3,258.7

unit: $10^{-9}/\mathrm{K}^2$
[a]−2,678 for consistency

Table 2.11. Third-order temperature coefficients of stiffnesses, $Tc_{\lambda\mu}^{(3)}$

λμ	[31]	[38]	[39]	[40]	[32]
11	−15	−371	−70	−61.9	−99.879
13	600	−5,559	−750	−43.4	686.06
14	−630	−625	−590	−816	−459.28
33	−410	−243	−250	93.7	−19.089
44	−65	−190	−216	−45.6	−244.95
66	−167	−777	21	−239	102.80
12	1,910	4,195	−1,260[a]		−2,487.8

unit: $10^{-12}/K^3$
[a]−1,110 for consistency

Table 2.12. Piezoelectric constants

	[38]	[30][a]	[33][b]	β-quartz [42]
d_{11}	2.37	2.31		
d_{14}	0.77	0.727		−1.86[c]
e_{11}	0.175	0.171	0.1719	
e_{14}	−0.0407	−0.0406	−0.0390	−0.067[c]
g_{11}		0.0578		
g_{14}		0.0182		
h_{11}		4.36		
h_{14}		−1.04		

units: d in pC/N; e in C/m^2; g in m^2/C; h in N/nC
[30] derived from cgs, rounded
[a]Slightly different values are given in [36] and [37]
[b]Cultured quartz at 23°C; $\rho = 2.64867\,Mg/m^3$
[c]At 612°C

Table 2.13. Dielectric permittivities, ε, and impermeabilities, β

	[30][a]	[32]	[33][a,b]
ε_{11}	39.21	39.16	39.136
ε_{33}	41.03	41.04	40.977
β_{11}	25.51	25.54	25.552
β_{33}	24.37	24.37	24.404

units: ε in pF/m; β in m/nF
[a]ε^S, β^S
[b]Cultured quartz at 23°C

Table 2.14. Thermoelastic constants, $\alpha^{(n)}$

	[39]	[43]	[32]	[44][a]	[33]
$\alpha_{11}{}^{(1)}$	13.71	13.92	13.77	13.65	13.74
$\alpha_{33}{}^{(1)}$	7.48	6.79	7.483	7.50	7.48
$\alpha_{11}{}^{(2)}$	6.50	15.09	13.03	11.02	
$\alpha_{33}{}^{(2)}$	2.90	8.69	9.405	8.00	
$\alpha_{11}{}^{(3)}$	−1.90	−7.86	−6.329	−19.32	
$\alpha_{33}{}^{(3)}$	−1.50	6.88	−5.440	−10.44	

units: $\alpha^{(1)}$ in 10^{-6}/K; $\alpha^{(2)}$ in $10^{-9}/\text{K}^2$; $\alpha^{(3)}$ in $10^{-12}/\text{K}^3$
[a] −50°C to +150°C

Table 2.15. Acoustic viscosities, η [46]

$\lambda\mu \rightarrow$	11	13	14	33	44	66	(12)
η	1.37	0.72	0.01	0.97	0.36	0.32	0.73

units: η in 10^{-3} Pa s (dekapoise)

- Thermoelastic constants, $\alpha^{(n)}$ See Table 2.14.
- Thermal diffusivity constants, $\alpha_{(\text{td})}$; $\alpha_{(\text{td})11} = 3.3$; $\alpha_{(\text{td})33} = 6.1$; unit: $10^{-6}\,\text{m}^2\,\text{s}^{-1}$
- Specific heat capacity, c_p; $c_\text{p} = 740$; unit: J (kg-K)$^{-1}$
- Thermal conductivity, $k_{(\text{th})}$; $k_{(\text{th})} = \rho c_\text{p} \alpha_{(\text{td})}$; $k_{(\text{th})11} = 6.47$; $k_{(\text{th})33} = 12.0$; unit: W (m-K)$^{-1}$
- Acoustic viscosity, η; unit: Pa s

Crystal quartz is often treated as lossless [16–21]; but at frequencies above 20 or 30 MHz, loss is usually dominated by internal friction. Lamb and Richter [46] determined acoustic viscosity η as the imaginary part of complex stiffness, $(c^\text{E} + j\omega\eta)$. Values for the matrix elements of η are given in Table 2.15.

The quotient η/c is the motional time constant, τ_1, and its value is related to the limiting resonator Q due to intrinsic acoustic attenuation by the relation $Q = (\omega_0\tau_1)^{-1}$.

2.2.2.5 Third-Order Stiffnesses

Quartz resonator frequency shifts caused by environmental disturbances such as acceleration, static forces, and thermal transients have their genesis primarily in elastic nonlinearities. Plane-stress and thermal transient disturbances in quartz resonators may be reduced, or eliminated, by use of certain doubly rotated cuts, notably the BAW SC and SBTC cuts [2–4, 51–54], and the SAW STC cut [55–57]. The necessary measurements of the 14 independent non-Hookean elastic stiffnesses of quartz were first made by Thurston, et al. [58]; see also [59, 60].

2.2.2.6 Recommended Values for α-Quartz

The comprehensive and readable works of Brice [24] and James [32] contain careful and insightful reviews of measurement methods, and of various data sets appearing in the literature. For the linear elastic, piezoelectric, and dielectric constants of quartz, the set obtained by Bechmann [30] using resonator methods is a good choice. The averaged values of Brice [24], and the set derived by James [32] from time-domain method measurements are good alternatives. For the temperature coefficients, the set given in [39], obtained from resonator measurements is a good choice, with the averaged values of Brice [24], and the set derived by James [32] from time-domain method measurements as good alternatives.

2.2.2.7 Beta-Quartz

Beta-quartz possesses five-independent elastic, one piezoelectric, and two dielectric constants, as well as ten independent third-order elastic constants. Table 2.16 lists the measured linear elastic stiffnesses and compliances of β-quartz. Table 2.12 contains the corresponding piezoelectric constants.

2.2.3 Piezoelectric Resonators

By far the greatest technological use of quartz is in piezoelectric resonators and transducers, including sensors. In the 1920s, and into the 1930s, extensional and flexural modes of thin bars and rods were used; in the 1930s into the 1940s, contour modes of plates and discs were used to achieve higher frequencies, since frequency is inversely proportional to the dimension determining the frequency. By the 1950s, driven by the need to reach still higher frequencies, thickness modes of thin plates were used almost exclusively [49, 110–114]. Obtaining requisite thinnesses by ordinary lapping and polishing techniques proved to be an obstacle to achieving fundamental frequencies in excess of about 50 MHz. Frequency requirements in the VHF and UHF ranges, and beyond, led to use of more modern microfabrication technologies that have enabled thickness modes to be realized in thin piezoelectric films.

Table 2.16. Elastic stiffnesses and compliances of β-quartz

$c^E_{\lambda\mu}$	[50]	$s^E_{\lambda\mu}$	[50]
c_{11}	116	s_{11}	9.41
c_{13}	33	s_{13}	−2.6
c_{33}	110	s_{33}	10.6
c_{44}	36	s_{44}	27.7
c_{66}	50	s_{66}	20.02
(c_{12})	16	s_{12}	−0.60

unit: c in GPa; s in $(\mathrm{TPa})^{-1}$
$T = 600°\mathrm{C}$

2.2.4 Plate Resonator Cuts

To obtain desirable properties such as temperature compensation, high frequency resonator plates and films must have certain orientations with respect to the crystallographic axes of the crystal. A plate or film, with lateral dimensions large compared with the thickness, can be regarded as a plane, and specified by two orientation angles. Starting from a reference state where the plane normal is along the crystallographic y axis, one rotates the plane about the crystallographic z axis by an angle φ; the x axis of the plane, originally parallel to the crystallographic x axis, becomes the x' axis. Then, a second rotation is made about the new x' axis by angle θ, creating a compound, or doubly-rotated cut, or orientation. The notation for this operation is $(YXw\ell)\varphi/\theta$ [16]. When angle $\varphi = 0$, the cut is a rotated-Y-cut, with notation: $(YX\ell)\theta$. For further details, see [1,4,16,48,105,106].

2.2.5 Equivalent Electrical Circuits

Piezoelectric resonators, filters, transducers, signal processors, sensors, and other acoustic components are invariably incorporated into using systems that are purely electrical. However, because the piezoelectric effect mediates between electrical and mechanical quantities, it is possible to describe the mechanical behavior of these components in purely electrical terms, and thereby to optimize the performance of the overall system. The simplest representation of a resonator is called the Butterworth-Van Dyke (BVD) circuit [107–109], which consists of a capacitor (C_0), representing the driving electrodes, shunted by a "motional" arm consisting of R_1 L_1 C_1 elements in series, representing acoustic loss, mechanical mass, and elastic spring, respectively, of the vibrating crystal. Circuit values depend on the material parameters, mode of motion, and geometry. For transducer applications, provision is made for the presence of mechanical ports in the appropriate equivalent circuits [35,104–106].

A resonator, represented by the BVD circuit, appears as a capacitive reactance at frequencies below the region of resonance. As resonance is approached, the reactance first becomes zero ("resonance" frequency, f_R), passing into a region of inductive reactance, and reaching a very large value before returning to zero ("antiresonance" frequency, f_A), finally becoming capacitive again at higher frequencies. The region between f_A and f_R (the "pole – zero" spacing) is the normal operating region for an oscillator or filter. This spacing is proportional to the square of the piezoelectric coupling factor (k), discussed in the next paragraph. If this value is small, the oscillator frequency will be confined to a small range, and hence will be very stable; similarly, narrowband filters require small k values, and wideband filters require large k values.

2.2.6 Piezoelectric Coupling Factors

Piezoelectric coupling factors, k, are dimensionless measures of electromechanical energy transduction. In transducers and filters, they determine bandwidth and insertion loss; in resonators and resonant MEMS devices, they establish pole-zero spacing and adjustment range; in actuators and sensors, they control electromechanical conversion efficiency. In applications, they usually appear as k^2. For a given type of resonant motion, effective values of elastic, piezoelectric, and dielectric coefficients arise. There are four basic forms expressing coupling, depending upon the constitutive equation set chosen. The generic homogeneous forms are the following:

- $k = |d|/\sqrt{(\varepsilon^{\mathrm{T}} s^{\mathrm{E}})}$ and $|h|/\sqrt{(\beta^{\mathrm{S}} c^{\mathrm{D}})}$. The generic mixed forms are:
- $k = |e|/\sqrt{(\varepsilon^{\mathrm{S}} c^{\mathrm{E}})}$ and $|g|/\sqrt{(\beta^{\mathrm{T}} s^{\mathrm{D}})}$ [28].

Explicit formulas for some α-quartz resonator plates are the following:

- X-cut (plate normal to x_1 axis): $k_{11} = |e_{11}|/\sqrt{[\varepsilon_{11}{}^{\mathrm{S}} c_{11}{}^{\mathrm{E}} + (e_{11})^2]} = 9.23\%$; pure extensional mode, driven by electric field in plate thickness direction.
- Y-cut (plate normal to x_2 axis): $k_{66} = |e_{11}|/\sqrt{[\varepsilon_{11}{}^{\mathrm{S}} c_{66}{}^{\mathrm{E}} + (e_{11})^2]} = 13.55\%$; pure shear mode, driven by electric field in plate thickness direction.
- Z-cut (plate normal to x_3 axis): $k_{44} = |e_{14}|/\sqrt{[\varepsilon_{11}{}^{\mathrm{S}} c_{66}{}^{\mathrm{E}}]} = 2.69\%$; pure shear mode, motion along direction of driving electric field in $x_2 - x_3$ plane.
- Rotated-Y-cuts (plate normal in $x_2 - x_3$ plane), pure shear mode: $k_{26}{}' = |e_{26}{}'|/\sqrt{[\varepsilon_{22}{}' c']}$; primes denote rotated values. AT cut: $k_{26}{}' = 8.80\%$ and BT cut: $k_{26}{}' = 5.62\%$. See Tables 2.17 and 2.18.

Table 2.17. Properties of AT and BT quartz cuts

Quantity	Unit	AT cut	BT cut
Angle	$(YX\ell)\theta$	$\theta = +35.25°$	$\theta = -49.20°$
Tf_{R}	$10^{-6}/\mathrm{K}$	0	0
$\partial Tf_{\mathrm{R}}/\partial\theta$	$10^{-6}/(K, \theta°)$	−5.12	+2.14
$\lvert k\rvert$	%	8.80	5.62
v	km/s	3.322	5.073
Z	Mrayl	8.800	13.438
c'_{66}	GPa	29.24	68.16
e'_{26}	$\mathrm{C/m^2}$	−0.0949	−0.0931
ε'_{22}	pF/m	39.82	40.25

Table 2.18. Properties of other rotated-Y-cut quartz cuts

Quantity	Unit	Y cut	Max Tf_{R}	Min Tf_{R}	Z cut
Angle	$(YX\ell)\theta$	$\theta = 0°$	$\theta = +4.03°$	$\theta = +69.61°$	$\theta = \pm 90°$
Tf_{R}	$10^{-6}/\mathrm{K}$	+91.11	+94.00	−96.69	−78.53
$\partial Tf_{\mathrm{R}}/\partial\theta$	$10^{-6}/(\mathrm{K}, \theta°)$	+0.583	0	0	+1.50
$\lvert k\rvert$	%	13.55	13.67	0.562	0
v	km/s	3.916	3.796	4.077	4.677

2.2.7 Properties of Selected α-Quartz Cuts

Table 2.17 contains data pertinent to the shear-mode AT and BT cuts, as these are the preponderant high frequency resonator orientations in current usage. Listed are Tf_R, resonance frequency TC; $\partial Tf_R/\partial\theta$, angle gradient of Tf_R; $|k|$, effective coupling; v, acoustic velocity; $Z = \rho v$, acoustic impedance; and the related elastic (c_{66}'), piezoelectric (e_{26}'), and dielectric (ε_{22}') values for these cuts. Table 2.18 lists $Tf_R, |k|$, and v data for four other rotated-Y-cuts.

2.3 Innovations: Past and Future

2.3.1 History of Piezoelectric Materials [61]

The phenomenon of piezoelectricity has a very distinguished history. Coulomb conjectured its existence, and later both Haüy and Becquerel conducted unsuccessful experiments. The undoubted discoverers of the phenomenon were the Curie brothers. They knew what they were looking for, had the backgrounds and facilities to bring the search to a successful conclusion, and announced their 1880 discovery as follows [62]:

> "*Those crystals having one or more axes whose ends are unlike, that is to say hemihedral crystals with oblique faces, have the special physical property of giving rise to two electrical poles of opposite signs at the extremities of these axes when they are subjected to a change in temperature: this is the phenomenon known under the name of pyroelectricity.*
> "*We have found a new method for the development of polar electricity in these same crystals, consisting in subjecting them to variations in pressure along their hemihedral axes.*"

The name "piezoelectricity" was given by Hankel; Lippmann predicted the converse effect in 1881, and it was verified by the Curies the same year. Kelvin provided an atomic model in 1893 [63], and theories were advanced by Duhem and Pockels. In 1894, Voigt introduced the term "tensor" describing the phenomenological treatment of this and other effects in crystals. Langevin used Rochelle salt for sonar in the early 1910s, and Born theoretically calculated the piezoconstant of β-ZnS. Cady invented the quartz oscillator in the early 1920s. Tartrates and other water-soluble crystals with large piezoelectric values were investigated in the 1940s and 1950s [64], as were piezoceramics such as barium titanate and similar alloys [65]. Highly piezoelectric refractory oxides such as lithium niobate were introduced in the mid-1960s [66–68], followed by piezopolymers in the late 1960s [69], and isomorphs of quartz in the late 1990s [70–73]. Table 2.19 shows some of the major applications of piezoelectricity. Figure 2.5 portrays the nominal ranges of coupling and quality factor for the various classes of piezoelectrics mentioned. The range for piezopolymers is largely coextensive with that of piezo-semiconductors.

Table 2.19. Major applications of piezoelectricity

Communications and control	Industrial	Health and consumer	Newer applications
• Cellular radio	• Transducers	• Transducers	• Smart Structures
• Television	• Sensors	• Sensors	• High Displacement Transducers
• Automotive radar	• Actuators	• Actuators	
	• Pumps	• Pumps	• Mixed-effect Devices
	• Motors	• Motors	
– Signal processing	– Ultrasonic cleaning	– Noninvasive medical diagnostics	– Microelectromechanical (MEMS) devices
– Frequency control and timing	– Sonar		
	– Nondestructive evaluation (NDE)	– Hyperthermia	– Microoptomechnaical (MOMS) devices
		– Lithotripsy	
– Correlators			– Biomimetic devices
– Convolvers	– Liquid level sensors	– Subcutaneous medication	– Composite and functionally graded devices
– Filters	– Vibration damping	– Wristwatches	– Rainbow devices
– Delay lines	– High temperature sensors	– Camera focusing/steadying/ranging	– Acousto-photonic-electronic devices
– Oscillators	– Material properties determination	– Computer timing/printing/modems	
	– Chemical/biological sensors	– Ignition of gases ("spark pump")	

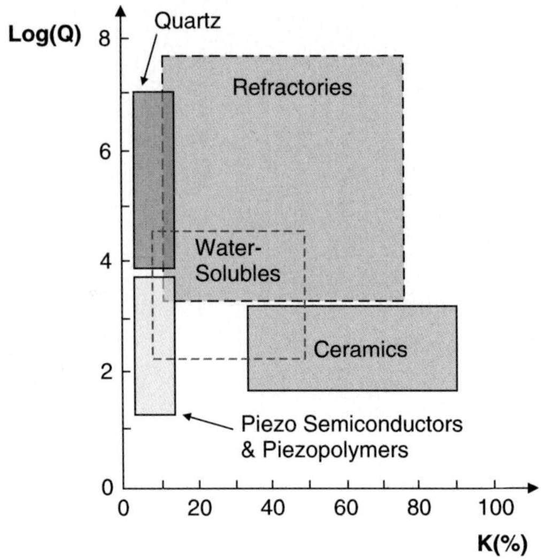

Fig. 2.5. Acoustic quality factor (Q) vs. piezoelectric coupling factor (k) for various types of piezoelectric materials

2.3.2 Current Quartz Uses

Acoustic materials are characterized primarily by elastic linearity, extremely low loss (high quality factor, Q), zero temperature coefficients (ZTCs) of frequency or delay, and particularly by the presence of piezoelectricity. Piezoelectricity provides a clean, efficient transduction mechanism mediating between mechanical motions and electric variables, using planar configurations available with conventional microelectronics fabrication technologies. Even compensation of stress and temperature-transient effects from elastic nonlinearities is possible with quartz, making available high stability bulk acoustic wave (BAW) oscillators. Surface acoustic wave (SAW) substrates provide conveniently accessible time axes for signal processing operations like convolution. The sonic/EM ratio (10^{-5}) affords severe miniaturization in both BAW and SAW devices. For many oscillator applications, it is necessary that the piezocoupling k not be large. This is because the oscillator operating point is confined to the resonator inductive region. Small coupling means a narrow inductive range, and concomitant high oscillator stability.

Acoustic device applications include wireless transceivers for voice, data, multimedia; spread-spectrum (SS) communications for wireless local area networks (WLANs) including WLAN robot, timing, and security applications; components in personal communications system (PCS) handsets, such as duplexers and voltage-controlled oscillators (VCOs); code division multiple access (CDMA) filters and timing; nonvolatile memories, and microelectromechanical (MEMS)/micro-optomechanical (MOMS) devices. Specific SAW applications are analog signal processing (convolvers, duplexers, delay lines, and filters) for the mobile telecommunications, multimedia, and industrial-scientific-medical (ISM) bands; wireless passive identification tags, sensors and transponders, and microbalances. BAW applications include resonators in precision clock oscillators; front-end GPS filters for cell phones; thin-film, solidly mounted resonators (SMRs), and stacked crystal filters (SCFs) formed as SMRs, for integration with microwave heterojunction bipolar transistor (HBT) VCOs [5, 8–11].

2.3.3 Usage Considerations

Why quartz? Quartz is the only material known that possesses the following combination of desirable attributes:

- Piezoelectric
- Zero temperature coefficient cuts
- Stress/thermal transient compensated cuts
- Low acoustic loss/high acoustic Q
- Capable of withstanding 35,000–50,000 g shocks
- Very stable
- Capable of integration with micro- and nano-electronic components
- Easy to process; hard but not brittle; under normal conditions, low solubility in everything except fluoride etchants
- Abundant in nature; easy to grow in large quantities, at low cost, and with relatively high purity and perfection. Of the man-made single crystals, quartz, at over 3,000 tonnes per year (2000), is second only to silicon in quantity grown [7].

2.3.4 History of Innovations Related to Quartz that Led to Usages

- 1880
  - Piezoelectric effect discovered: Brothers Pierre & Jacques Curie
- 1900s
  - First hydrothermal growth of quartz in laboratory
- 1910s
  - First application of piezoelectricity, in sonar (Rochelle salt)
  - First quartz crystal oscillator, filter
- 1920s
  - Introduction of quartz resonator; requires "telescope making" technology: rough X-raying; grinding, lapping, polishing
  - Quartz resonator used for frequency stabilization and control
  - First crystal control of radio station
  - BVD equivalent circuit representation
  - Electrical measurements of resonators using oscillators
  - Compound resonator for measuring constants of nonpiezoelectrics
  - First quartz crystal-controlled clock

- 1930s
  - Temperature compensated bulk acoustic wave (BAW) AT/BT cuts developed; requires X-raying to minutes of arc
  - Modal surface motions investigated using lycopodium powder patterns
  - Equivalent circuit incorporating mechanical ports for transducers
- 1940s
  - Hydrothermal growth of quartz [74]; (Czochralski growth of Si and Ge)
  - Electrical measurements using passive transmission networks
  - Electrode system: $A\ell$ evaporation directly on resonator surfaces
  - Surface finish found to affect quality of resonator
  - First contoured, high-Q, resonator designs
  - Discussion of coupled plate modes [49]
- 1950s
  - Hydrothermal growth of cultured bars commercially viable
  - Solder-sealed metal enclosures
  - Processing ambient: soft vacuum
  - Quartz crystal microbalance (QCM) developed
  - Equivalent circuit in form of acoustic transmission line
  - Temperature-compensated crystal oscillator (TCXO) described
  - Theory of vibrations of anisotropic plates [110]
- 1960s
  - "Energy trapping" theory used to reduce unwanted modes [75–80]
  - Monolithic crystal filter (MCF) developed
  - Solidly mounted resonator (SMR) described
  - Glass enclosures introduced for low aging
  - Electrolytic sweeping reduces effects of ionizing radiation and neutron flux
  - Application of microelectronics technology enabling batch-processing
  - High coupling refractories, such as lithium niobate and tantalate grown
  - Anisotropic plate theory extended [111–113]
- 1970s
  - Cultured quartz use exceeded natural quartz; (Czochralski growth of III-V binaries)
  - Electrical measurements of crystal parameters using bridges
  - Stacked-crystal filters (SCFs) described
  - UV-ozone cleaning to produce atomically clean surfaces and very low aging [81]
  - Chemical polishing reduces subsurface damage; greatly increased shock resistance [82]
  - Temperature and nonlinear effects compensation; requires X-raying to seconds of arc
  - Interdigital transducer (IDT) enables efficient transduction of surface waves [83–87]
  - Surface acoustic wave (SAW) cuts; temperature compensated ST cut
  - Stress-compensated compound cut developed (SC) [2–4]

  - Electrical measurements using microcircuit bridges
  - Cold-weld metal enclosures
  - Processing ambient: cryo-pumps
  - Miniature tuning forks (∼32 kHz) enable quartz wristwatches
- 1980s
  - Stress-compensated BAW compound cut (SBTC) [52–54]
  - Stress-compensated SAW analog of SC cut (STC) [56]
  - Acceleration reduction in SAWs [57]
  - Ceramic flatpack enclosures; high temperature bakeout
  - Microprocessor-compensated crystal oscillator (MCXO) realized
  - Dual mode self-temperature sensing feature applied to MCXO
  - Coupled-mode cuts reduce temperature effects in wristwatch tuning forks
  - Electrode system: high precision air gaps
  - Electrical measurements using network analyzers
  - Processing ambient: ultra-high vacua
- 1990s and 2000s
  - Quartz analogues investigated
  - High stability environmental sensors developed; frequency counting inherently digital
  - Thin film resonator realized (TFR) [88–97]
  - Solidly mounted resonators (SMRs), advanced fabrication makes viable [89–91, 97]
  - Theory of compound-cut, contoured resonators [114]
  - Full-scale finite element method (FEM) models available for 3D simulation of modes

2.3.5 Future Innovations

2.3.5.1 Higher Frequencies

The push to higher frequencies is often dictated by bandwidth considerations. One important criterion by which to measure an acoustic material is its upper frequency limit; the concept is somewhat analogous to the cutoff frequency used for assessing transistors. Maximum usable frequency ($f_{\max}$) for a piezoelectric resonator is determined jointly by the piezocoupling factor (k) and the acoustic time constant, τ_1. The relation is $f_{\max} = 2k^2/(\pi^3 \tau_1)$, where $\tau_1 = (2\pi f_o Q)^{-1}$, and Q is the measured acoustic quality factor at f_o, the nominal operating frequency [98]. A figure of merit (FOM) often used for piezofilters is (Qk^2) [95]. Thus $f_{\max} = [(4 f_0/\pi^2) \cdot (\mathrm{FOM})]$. Put another way, the resonant structure approaches the limit of its usefulness at a normalized frequency ($f_{\max}/f_0$) approximately 40% of the FOM. A quartz CDMA filter at the 1.9 GHz PCS band typically has an FOM of 120, indicating that such devices are intrinsically capable of operation at much higher frequencies. Values of f_{max} from measurements of attenuation and piezocoupling are much higher than frequencies encountered in practice. The discrepancy relates primarily to manufacturing technology, rather than inherent material limitations. For example, AT-cut quartz with $\tau_1 = 11.8\,\mathrm{fs}$ and $k = 8.80\%$, yields $f_{max} = 42.3\,\mathrm{GHz}$, a value exceeding present capabilities, but an achievable goal for the future [99].

2.3.5.2 High Coupling Isomorphs of Quartz [70–73]

The search for new materials is fed by practical demands for devices with improved characteristics, e.g., lower loss (higher resonator quality factor (Q), lower filter insertion loss (IL)), higher piezocoupling (k^2) for increased filter bandwidth, better temperature stability, greater miniaturization (higher acoustic velocity), etc. Materials receiving recent attention include the quartz analog langasite and its isomorphs, generically called LGX.

The langasite (LGS, $La_3Ga_5SiO_{14}$) family consists of Czochralski grown, congruently melting, class 32 materials such as langanite (LGN, $La_3Ga_{11/2}Nb_{1/2}O_{14}$) and langatate (LGT, $La_3Ga_{11/2}Ta_{1/2}O_{14}$) [70, 71]. Table 2.20 provides pertinent phenomenological data. These materials have acoustic Qs higher than quartz, but have disordered structures, due to the difficulty of satisfying both ionic size and charge compensation constraints. In the future, totally ordered $(Ca, Sr)_3(Nb, Ta)Ga_3Si_2O_{14}$ crystals, and similar members, are expected to offer higher elastic stiffnesses, as well as lower dielectric permittivity and higher piezocoupling. Candidate materials include CNGS ($Ca_3NbGa_3Si_2O_{14}$), CTGS ($Ca_3TaGa_3Si_2O_{14}$), SNGS ($Sr_3NbGa_3Si_2O_{14}$), and STGS ($Sr_3TaGa_3Si_2O_{14}$). Main market possibilities are greater bandwidth (versus quartz), high stability IF filters; high temperature sensors (no twinning); and high Q BAW and SAW resonators [72].

2.3.5.3 Epitaxial Growth of Quartz and Isomorphs

Future miniaturization goals demanded for micro and nano-electromechanical systems (MEMS, NEMS) development [100,101] will require newer methods of growth and fabrication of thin quartz films and membranes. Recent advances in low-pressure vapor-phase epitaxy [102, 103] open exciting new avenues for future development. Extension of this and other new growth methods applied to quartz and its analogues, such as the totally-ordered LGX family, may yield largely defect-free crystal lattices, and the possibility of achieving greater isotopic constituent purity. Both advances would reduce phonon scattering, realizing significantly higher acoustic Qs, and thereby also further raise the achievable maximum useable frequency [45, 47].

Table 2.20. Material constants of langasite and isomorphs

Quantity	Langasite (LGS)	Langanite (LGN)	Langatate (LGT)
Elastic stiffnesses, $c_{\lambda\mu}{}^{E}$ (GPa)			
c_{11}	188.49	192.99	188.52
c_{13}	96.88	102.25	103.36
c_{14}	14.15	14.85	13.51
c_{33}	261.68	264.65	261.80
c_{44}	53.71	49.56	51.10
c_{66}	42.21	41.16	40.32
Temperature coefficients of stiffnesses, $Tc_{\lambda\mu}{}^{E}$ (10^{-6}/K)			
Tc_{11}	−43.91	−56.34	−78.24
Tc_{13}	−61.95	−31.27	−111.40
Tc_{14}	−309.10	−478.90	−359.60
Tc_{33}	−91.90	−114.70	−102.30
Tc_{44}	−44.05	−14.14	+21.65
Tc_{66}	−22.43	+15.25	−43.63
Piezoelectric stress constants, $e_{m\lambda}$ (C/m^2)			
e_{11}	−0.402	−0.452	−0.456
e_{14}	+0.130	+0.061	+0.094
Dielectric permittivities, $\varepsilon_{ij}/\varepsilon_0$ (dimensionless)			
$\varepsilon_{11}/\varepsilon_0$	19.62	20.09	18.27
$\varepsilon_{33}/\varepsilon_0$	49.41	79.34	78.95
Mass density, ρ (Mg/m^3)			
ρ	5.739	6.029	6.150
Thermoelastic coefficients, α_{ij} (10^{-6}/K)			
α_{11}	5.63	6.67	6.09
α_{33}	4.08	5.06	3.83

References

1. W.G. Cady, *Piezoelectricity* (McGraw-Hill, New York, 1946)
2. E.P. EerNisse, in *Proc. 29th Ann. Freq. Control Symp.*, pp. 1–4, US Army Electronics Command, Ft. Monmouth, NJ, May 1975
3. J.A. Kusters, J. Leach, in *Proc. IEEE.* vol. 65, no. 2, pp. 282–284, February (1977)
4. A. Ballato, in *Physical Acoustics: Principles and Methods*, vol. 13, ed. by W.P. Mason, R.N. Thurston. (Academic Press, New York, 1977), Chap. 5, pp. 115–181, ISBN: 0-12-477913-1
5. S. R. Stein, J. R. Vig, in *The Froelich/Kent Encyclopedia of Telecommunications*, Vol. 3, ed. by F.E. Froehlich, A. Kent (Marcel Dekker, New York, 1992) pp. 445–500
6. R.B. Sosman, *The Properties of Silica* (Chemical Catalog Co., New York, 1927)
7. "Quartz crystal," Bulletin 667, Mineral Facts and Problems, US Department of the Interior, Bureau of Mines, 1975

8. A. Ballato, J. R. Vig, in *Encyclopedia of Applied Physics*, Vol. 14, ed. by G. Trigg (VCH Publishers, New York, 1996), pp. 129–145, ISBN: 3-527-28136-3
9. D.J. Jones, S.E. Prasad, J.B. Wallace, Key Eng. Mater. **122–124**, 71–144, (1996)
10. IEEE Trans. Microwave Theory Tech. **49**(4), part II, 741–847 (April 2001). Special Issue on Microwave Acoustic Wave Devices for Wireless Communications and Sensing. R. Weigel and K. Hashimoto, guest editors
11. R. Weigel, D.P. Morgan, J.M. Owens, A. Ballato, K.M. Lakin, K. Hashimoto, C. C. W. Ruppel, IEEE Trans. Microwave Theory Tech. **50**(3), 738–749 (March 2002)
12. C. Frondel, *Dana's System of Mineralogy, Volume II: Silica Minerals*, (Wiley, New York, 1962)
13. B.K. Vainshtein, *Fundamentals of Crystals: Symmetry, and Methods of Structural Crystallography*, 2nd edn. (Springer-Verlag, Berlin, 1994), ISBN 3-540-56558-2
14. B.K. Vainshtein, V.M. Fridkin, V.L. Indenbom, *Modern Crystallography II*, (Springer-Verlag, Berlin, 1982), ISBN 3–540–10517–4
15. C. Klein, C.S. Hurlbut, Jr., *Manual of Crystallography (after J.D. Dana)*, 21st edn. revised, (Wiley, New York, 1999), ISBN: 0-471-31266-5
16. "IRE Standards on piezoelectric crystals, 1949," Proc. IRE, vol. 37, no. 12, pp. 1378–1395, December 1949. (IEEE Standard no. 176)
17. "IRE standards on piezoelectric crystals: determination of the elastic, piezoelectric, and dielectric constants - the electromechanical coupling factor, 1958," Proc. IRE, vol. 46, no. 4, pp. 764–778, April 1958. (IEEE Standard no. 178)
18. "IRE standards on piezoelectric crystals – the piezoelectric vibrator: definitions and methods of measurements, 1957," Proc. IRE, vol. 45, no. 3, pp. 353–358, March 1957
19. "Standard definitions and methods of measurement for piezoelectric vibrators," IEEE Standard no. 177, (IEEE, New York, May 1966)
20. "IEEE Standard on piezoelectricity," IEEE Standard 176–1978, IEEE, New York. Reprinted in IEEE Trans. Sonics Ultrason., vol. SU-31, no. 2, Part 2, 55 pp., March 1984
21. "IEEE Standard on piezoelectricity," IEEE Standard 176–1987, IEEE, New York
22. T.R. Meeker, in *Proc. 33rd Annual Frequency Control Symp.*, pp. 176–180, (Atlantic City, NJ, May-June 1979)
23. L. Pauling, *The Nature of the Chemical Bond*, 3rd edn. (Cornell University Press, Ithaca, 1960). ISBN: 0-8014-0333-2
24. J.C. Brice, Revs. Mod. Phys. **57**(1), 105–146 (January 1985)
25. W. Voigt, *Lehrbuch der Kristallphysik* (B.G. Teubner, Leipzig, 1928)
26. G. Heckmann, Ergeb. exakt. Naturwiss. **4**, 100–153 (1925)
27. J.F. Nye, *Physical Properties of Crystals* (Oxford University Press, Oxford, 1985). ISBN: 0-19-851165-5
28. A. Ballato, IEEE Trans. Ultrason., Ferroelec., Freq. Contr., **42**(5), 916–926 (September 1995)
29. A.W. Lawson, Phys. Rev. **59**, 838–839 (1941)
30. R. Bechmann, Phys. Rev. **110**(5), 1060–1061 (June 1, 1958)
31. W.P. Mason, Bell Syst. Tech. J. **30**, 366–380 (April 1951)

32. B.J. James, "Determination of the Elastic and Dielectric Properties of Quartz," PhD Dissertation, Royal Holloway and Bedford New College, University of London, Spring 1987, 231 pp. See also B. J. James, "A new measurement of the basic elastic and dielectric constants of quartz," IEEE Intl. Frequency Control Symp. Proc. (42nd Ann.), pp. 146–154, Baltimore, MD, June 1988
33. J. Kushibiki, I. Takanaga, S. Nishiyama, IEEE Trans. Ultrason., Ferroelect., Freq. Contr. **49**(1), 125–135 (January 2002)
34. J.V. Atanasoff, P. J. Hart, Phys. Rev. **59**(1), 85–96 (1941)
35. W.P. Mason, Phys. Rev. **55**, 775–789 (April 1939)
36. R. Bechmann, Proc. Phys. Soc. (London), **B64**, 323–337 (April 1951)
37. R. Bechmann, Archiv der Elektrischen Übertragung **5**, 89–90 (1951). (present name: Archiv für Elektronik und Übertragungstechnik)
38. I. Koga, M. Aruga, Y. Yoshinaka, Phys. Rev. **109**, 1467–1473 (March 1958)
39. R. Bechmann, A. Ballato, T. J. Lukaszek, Proc. IRE **50**(8), 1812–1822 (August 1962); Proc. IRE **50**(12), 2451 (December 1962)
40. P.C.Y. Lee, Y.-K. Yong, J. Appl. Phys. **60**(7), 2327–2342 (1986)
41. J. Zelenka, P.C.Y. Lee, IEEE Trans. Sonics Ultrason. SU-**18**(2), 79–80 (1971)
42. R.K. Cook, P.G. Weissler, Phys. Rev. **80**(4), 712–716 (15 November 1950)
43. A. Ballato, M. Mizan, IEEE Trans. Sonics Ultrason. SU-**31**(1), 11–17 (January 1984)
44. J.A. Kosinski, J.G. Gualtieri, A. Ballato, IEEE Trans. Ultrason., Ferroelec., Freq. Control **39**(4), 502–507 (July 1992)
45. R. Bechmann, in *Landolt-Börnstein, Numerical Data and Functional Relationships in Science and Technology, New Series, Group III: Crystal and Solid State Physics*, ed. by K.–H. Hellwege, A.M. Hellwege (Springer Verlag, Berlin, New York, vol. III/1, pp. 40–123, 1966; and vol. III/2, pp. 40–101, 1969)
46. J. Lamb, J. Richter, Proc. Roy. Soc. (London) **A293**, 479–492 (1966)
47. W.R. Cook, Jr., H. Jaffe, in *Landolt-Börnstein, Numerical Data and Functional Relationships in Science and Technology, New Series, Group III: Crystal and Solid State Physics*, vol. III/11, ed. by K.–H. Hellwege, A.M. Hellwege (Springer Verlag, Berlin, New York, 1979) pp. 287–470
48. B. Parzen, *Design of Crystal and Other Harmonic Oscillators* (Wiley, New York, 1983). ISBN: 0-471-08819-6
49. R.A. Sykes, in *Quartz Crystals for Electrical Circuits: Their Design and Manufacture*, ed. by R.A. Heising (D. Van Nostrand, New York, 1946) Chap. 6, pp. 205–248
50. E.W. Kammer, T.E. Pardue, H.F. Frissel, J. Appl. Phys. **19**(3), 265–270 (March 1948)
51. B.K. Sinha, Ferroelectrics **41**(1), 61–73 (1982)
52. B.K. Sinha, Proc. 35th Annu. Freq. Control Symp. 213–221 (May 1981)
53. M. Valdois, B.K. Sinha, J.-J. Boy, IEEE Trans. Ultrason., Ferroelec., Frequency Contr. **36**(6), 643–651 (November 1989)
54. B.K. Sinha, IEEE Ultrason. Symp. Proc. 557–563 (December 1990)
55. B.K. Sinha, IEEE Trans. Sonics Ultrason. SU-**32**(4), 583–591 (July 1985)
56. B.K. Sinha, IEEE Trans. Ultrason., Ferroelec., Frequency Contr. **34**(1), 64–74 (January 1987)
57. S. Locke, B. K. Sinha, IEEE Trans. Ultrason., Ferroelec., Frequency Contr. UFFC-**34**(4), 478–484 (July 1987)
58. R.N. Thurston, H.J. McSkimin, P. Andreatch, Jr., J. Appl. Phys. **37**(1), 267–275 (January 1966)

59. R. Stern, R.T. Smith, J. Acoust. Soc. Am. **44**, 640–641 (1968)
60. R.F.S. Hearmon, in Landolt-Börnstein, Numerical Data and Functional Relationships in Science and Technology, New Series, Group III: Crystal and Solid State Physics, vol. III/11, ed. by K.–H. Hellwege, A.M. Hellwege (Springer Verlag, Berlin, New York, 1979) pp. 245–286
61. W.P. Mason, J. Acoust. Soc. Am. **70**(6), 1561–1566 (December 1981)
62. P. Curie et, J. Curie, Bull. Soc. Fr. Mineral. Cristallogr. **3**, 90–93 (1880); C. R. Acad. Sci. (Paris) **91**, 294, 383 (March 1880)
63. M. Trainer, Eur. J. Phys. **24**(5), 535–542 (September 2005)
64. R. Bechmann, et al., in *Piezoelectricity* (Her Majesty's Stationery Office, London, 1957), 369 pp
65. D. Berlincourt, J. Acoust. Soc. Am. **70**(6), 1586–1595 (December 1981)
66. T. Yamada, N. Niizeki, H. Toyoda, Japan. J. Appl. Phys. **6**(2), 151–155 (1967)
67. A.W. Warner, M. Onoe, G.A. Coquin, J. Acoust. Soc. Am. **42**(6), 1223–1231 (December 1967)
68. R.T. Smith, F. S. Welsh, J. Appl. Phys. **42**(6), 2219–2230 (May 1971)
69. G.M. Sessler, J. Acoust. Soc. Am. **70**(6), 1596–1608 (December 1981)
70. R.C. Smythe, IEEE Intl. Frequency Control Symp. Proc. 761–765 (May 1998)
71. D.C. Malocha, M.P. da Cunha, E. Adler, R.C. Smythe, S. Frederick, M. Chou, R. Helmbold, Y.S. Zhou, 2000 IEEE/EIA Intl. Freq. Control Symp. Proc., pp. 201–205, Kansas City, MO, June 2000
72. B.H.T. Chai, A.N.P. Bustamante, M. C. Chou, "A new class of ordered langasite structure compounds," 2000 IEEE/EIA Intl. Freq. Control Symp. Proc., pp. 163–168, Kansas City, MO, June 2000
73. P.W. Krempl, J. Phys. IV France **126**, 95–100 (June 2005)
74. F. Iwasaki, H. Iwasaki, J. Crystal Growth, **237–239**, 820–827 (2002)
75. W. Shockley, D.R. Curran, D.J. Koneval, Proc. 17th Ann. Frequency Control Symp. 88–126 (May 1963)
76. D.R. Curran, D.J. Koneval, Proc. 18th Ann. Freq. Control Symp. 93–119 (May 1964)
77. W.S. Mortley, Wireless World **57**, 399–403 (October 1951)
78. W.S. Mortley, Proc. IEE (London) **104B**, 239–249 (December 1956)
79. R.D. Mindlin, J. Acoust. Soc. Am. **43**(6), 1329–1331 (June 1968)
80. H.F. Tiersten, R.C. Smythe, J. Acoust. Soc. Am. **65**(6), 1455–1460 (June 1979)
81. J.R. Vig, J.W. Le Bus, IEEE Trans. Parts, Hybrids, Packaging **PHP-12**(4), 365–370 (December 1976)
82. J.R. Vig, J.W. LeBus, R.L. Filler, Proc. 31st Ann. Freq. Control Symp. 131–143 (June 1977)
83. R.M. White, F.W. Voltmer, Appl. Phys. Lett. **7**(12), 314–316 (1965)
84. M.B. Schulz, B.J. Matsinger, M.G. Holland, J. Appl. Phys. **41**(7), 2755–2765 (1970)
85. R.M. White, Proc. IEEE **58**(8), 1238–1276 (August 1970)
86. M.G. Holland, L.T. Claiborne, Proc. IEEE **62**(5), 582–611 (May 1974)
87. B.K. Sinha, H.F. Tiersten, Appl. Phys. Lett. **34**(12), 817–819 (15 June 1979)
88. N.F. Foster, J. Acoust. Soc. Am. **70**(6), 1609–1614 (December 1981)
89. W.E. Newell, Proc. IEEE **52**(12), 1603–1607 (December 1964)
90. W.E. Newell, Proc. IEEE **53**(6), 575–581 (June 1965)
91. W.E. Newell, Proc. IEEE **53**(10), 1305–1308 (October 1965)
92. A. Ballato, T. Lukaszek, Proc. IEEE **51**(10), 1495–1496 (October 1973)

93. A. Ballato, H.L. Bertoni, T. Tamir, IEEE Trans. Microwave Theory Tech. **MTT-22**(1), 14–25 (January 1974)
94. K.M. Lakin, G.R. Kline, R.S. Ketcham, J.T. Martin, K.T. McCarron, IEEE Intl. Freq. Control Symp. Proc. 536–543 (May-June 1989)
95. K.M. Lakin, IEEE Intl. Freq. Control Symp. Proc. 201–206 (May 1991)
96. K.M. Lakin, G.R. Kline, K.T. McCarron, IEEE MTT-S Intl. Microwave Symp. Digest **3**, 1517–1520 (June 1993)
97. K.M. Lakin, IEEE Microwave Magazine **4**(4), 61–67 (December 2003)
98. A. Ballato, J.G. Gualtieri, IEEE Trans. Ultrason., Ferroelect., Freq. Cont. **41**(6), 834–844 (November 1994)
99. H. Iwata, IEICE Electronics Express **1**(12), 346–351 (September 2004)
100. K.L. Ekinci, M.L. Roukes, Rev. Sci. Instrum. **76** art. 061101, 12pp. (2005)
101. MEMS: A Practical Guide to Design, Analysis, and Applications, ed. by J.G. Korvink, O. Paul (Springer-Verlag, Heidelberg 2006), ISBN: 3-540-21117-9
102. N. Takahashi, T. Nakumura, S. Nonaka, H. Yagi, Y. Sinriki, K. Tamanuki, Electrochemical and Solid State Lett. **6**(5), C77–C78 (2003)
103. N. Takahashi, T. Nakumura, Electrochemical and Solid State Lett. **6**(11), H25–H26 (2003)
104. A. Ballato, IEEE Trans. Ultrason., Ferroelect., Freq. Contr., **48**(5), 1189–1240 (September 2001)
105. R. A. Heising (ed.), *Quartz Crystals for Electrical Circuits: Their Design and Manufacture* (D. Van Nostrand, New York, 1946)
106. W.P. Mason, *Piezoelectric Crystals and their Application to Ultrasonics* (Van Nostrand, New York, 1950)
107. A. Ballato, Proc. IEEE **58**(1), 149–151 (January 1970)
108. B.A. Auld, *Acoustic Fields and Waves in Solids*, Vol. I and II, (Robert E. Krieger Publishing, Malabar, FL, 1990). ISBNs: 0-89874-783-X; 0-89874-782-1
109. W.P. Mason, Proc. IEEE **57**(10), 1723–1734 (October 1969)
110. R.D. Mindlin, J. Appl. Phys., **23**(1), 83–88 (January 1952)
111. R.D. Mindlin, D.C. Gazis, Proc. Fourth U. S. Natl. Congr. Appl. Math. 305–310 (1962)
112. R.D. Mindlin, P.C.Y. Lee, Intl. J. Solids and Struct. **2**(1), 125–139 (January 1966)
113. R.D. Mindlin, W.J. Spencer, J. Acoust. Soc. Am. **42**(6), 1268–1277 (December 1967)
114. B.K. Sinha, IEEE Trans. Ultrason., Ferroelec., Frequency Contr. **48**(5), 1162–1180 (September 2001)

3

The Role of Ferroelectricity for Piezoelectric Materials

W. Wersing, W. Heywang, H. Beige, and H. Thomann

Most of the technological applications of piezoelectricity used nowadays are based on ferroelectric materials. This is due to the following reasons:

1. Because of the high piezoelectric effect that can be found in these materials, a high and efficient electromechanical transformation of energy and signals can be achieved.
2. In general, the remnant electrical polarization that occurs in these materials can be oriented into a desired direction by applying an external electrical field: this means by poling, if needed, at elevated temperature. Therefore it becomes possible to imprint a unipolar direction of macroscopic preference or anisotropy to a device even after processing it. Obviously, this unipolarity is mandatory for every piezoelectric action.
3. Therefore, materials that are macroscopically isotropic by nature after the production process can be also used. This concerns, e.g., ceramics, plastics, or composites, which can be processed using well-known methods to adapt and shape the material according to the requirements of the final use.
4. For piezoelectric applications, the group of ceramics based on the so-called perovskite structure is especially important. These materials are already being used over a broad field of technological applications such as in capacitors and PTC resistors[1]. Therefore, experience in the mass production, reliability, etc. of these materials is already available.
5. Therefore, also from the point of view of production cost, ferroelectric materials offer good advantages.

Consequently, it makes sense to offer here a short introduction into this class of materials.

---

[1] PTC: Giant Positive Temperature Coefficient of resistivity within the ferroelectric Curie-Weiss region.

3.1 Fundamental Principles of Ferroelectricity

The name ferroelectricity was chosen because of some principal analogies to ferromagnetism[2]:

- Below a certain phase transition temperature, the so-called Curie temperature T_C, a spontaneous polarization P_s occurs in ferroelectric materials similar to the spontaneous magnetization in ferromagnetic materials.
- Above T_C, the temperature dependence of the dielectric permittivity ε follows a Curie–Weiss law (Eq. 3.1), just as the magnetic permeability of ferromagnetic materials does.

$$\frac{\varepsilon}{\varepsilon_0} = \frac{C}{T - T_0}. \qquad (3.1)^3$$

- To minimize depolarizing fields, different regions of the crystal are polarized in different directions of a possible P_s, each volume of uniform polarization being called a *ferroelectric domain* similar to ferromagnetic domains.
- In the ferroelectric state (below T_C), the dielectric behavior, i.e., the dependence of dielectric displacement **D** on electric field strength **E**, is highly nonlinear and shows a characteristic hysteresis loop similar to the dependence of the induction **B** on magnetic field strength **H** in ferromagnetic materials.

However, besides these analogies decisive differences also exist, differences that are important especially in the context of piezoelectricity. Basically, these differences are due to the fact that there does not exist an electrical counterpart to the atomic spin. Thus every electrical dipole requires particle displacements, which influence the structure of the lattice. Consequently, permanent polarization itself causes lattice distortions, changing the lattice symmetry remarkably. As long as we consider the polarization being the major reason for these distortions, we may speak about ferroelectric phases or phase systems.

In Sect. 3.1 we try to understand such systems and to correlate them with the atomic lattice structure conceptually. Using the model of a hypothetical single-domain crystal, the thermodynamic correlations with piezoelectricity will be deduced.

In Sect. 3.2 we will deal with the fact that single crystals, and even more polycrystalline ceramics, cannot be described by single-domain models. Apparent charges have to be compensated either by domain splitting or by ions;

---

[2] In spite of ethymologically meaningless, the name "ferroelectricity" has been accepted generally. Ferroelectric properties had been found in Seignette-salt before. Therefore in past times the name Seignette-electricity had been used, particularly in the Russian literature.

[3] ε_0: vacuum permittivity, T_0: Curie-Weiss temperature; $T_0 < T_C$ in case of a first order phase transition (see also Eq. 3.16), $T_0 = T_C$ in case of a second order (continuous) phase transition.

stresses have to be minimized. The properties of such samples are called *intrinsic*, as long as the domain configuration is not changed by the measuring signal. A conceptual overview will try to explain the interrelation between these intrinsic properties on one hand, and their dependence on composition as well as on preparation and poling procedure on the other.

Section 3.3 finally deals with the same properties in the case that a reorientation of the domains becomes possible under the influence of the measuring signals themselves. We call these properties *extrinsic*.

It should be emphasized again that the aim of this chapter is only to provide a conceptual understanding of possibilities and limitations of piezoelectric material engineering. It should be sufficient for the understanding of the different applications. A more profound description of the state of the art of materials and device design is given in Chap. 12.

3.1.1 Basic Properties of Ferroelectrics

In contrast to quartz, in which we find a preformed asymmetry due to the atomic structure of the lattice, in ferroelectric materials this asymmetry does not exist above the Curie temperature as in ferromagnetic materials. But at the Curie temperature it develops because of the so-called internal (or local) electrical field $\mathbf{E}_i$, which corresponds to the sum of the applied external field $\mathbf{E}$ and the feedback of the internal dipoles by means of the existing polarization $\mathbf{P}$:

$$\vec{E}_i = \vec{E} + \frac{\beta}{\varepsilon_0}\vec{P} \tag{3.2}$$

with β being the Lorentz factor.

So far, we have dealt with an exact analogy with magnetism. However, this analogy does not include the nature of the dipoles themselves. In the ferromagnetic case, these dipoles are mainly based on the spin of the single particles. In the electrical case, dipoles cannot be formed without the cooperation of several particles that virtually occupy the lattice. This can be the cloud of electrons surrounding an atomic nucleus or molecule-like structures exhibiting a bond angle, or ions with a displacement relative to the surrounding lattice. Sometimes, such local dipoles already exist above the Curie temperature and are ordered below it under the influence of the internal field; some times they do not form except under the influence of the internal field itself. Depending on the prevalence of the first or the second mechanism, two basically different types of ferroelectric materials are distinguished nowadays:

1. Ferroelectric materials exhibiting dipoles or dipole-like structures that are randomly oriented above the Curie temperature and that will be ordered below, where they give rise to a spontaneous polarization (order–disorder type); and

2. Ferroelectric materials where the lattice cells are not polarized except under the influence of the internal field below the Curie temperature. In this case, the spontaneous polarization is based on the relative displacement of sublattices formed by ions and their electron clouds (displacive type).

But we should state already here that in every ferroelectric material both these mechanisms coexist, even though to different extents. On one hand, dipoles that are preformed in the paraelectric state grow in size by the influence of the internal field below the Curie temperature; on the other, purely ionic bonded crystals cannot undergo a transition to a ferroelectric phase, but become antiferroelectric owing to reasons of electrostatics. Angular bonds, which are mandatory for the development of ferroelectricity, cannot be formed except under the influence of covalent contributions.

From the explanation of ferromagnetism, the thermodynamic statistics of an order–disorder transition is well known. It can be understood as a consequence of dipole orientation within the internal field.

The case of a displacive ferroelectric transition is more complex, especially since there does not exist a well-defined borderline between dipoles and the surrounding lattice. Therefore at this stage a simple introduction to features of such transitions could be helpful. A more thorough explanation and quantitative results cannot be given here, extended outlines are presented in Chap. 12.

The most representative materials among the displacive ferroelectrics are the perovskites, of which a prototype is $BaTiO_3$ (BT), the lattice structure of which is depicted in Fig. 3.1. The core of the essentially cubic lattice cell is formed by an oxygen octahedron with a 4-valent titanium ion in the center. Neutrality and steric stability of the lattice is ensured by the 2-valent barium ions at the corners of the unit cell. The essential part of the cell that concerns the ferroelectric and piezoelectric properties is represented by the TiO_6 octahedron. This can be expected not only because of the high permittivity

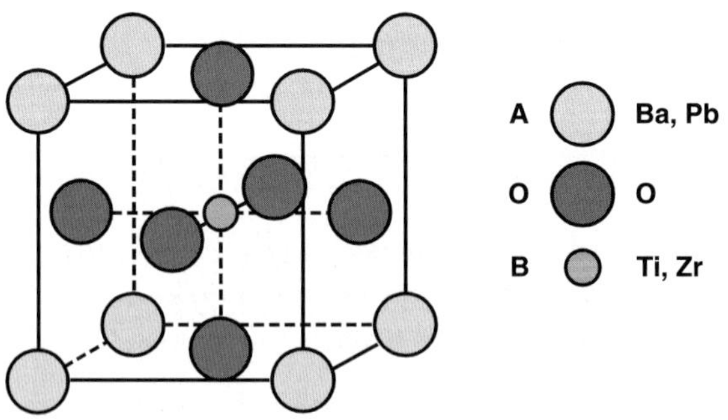

Fig. 3.1. The perovskite structure ABO_3. $BaTiO_3$ is the prototype ferroelectric that crystallizes in this structure. Other important examples are $PbTiO_3$ and $Pb(Zr_xTi_{1-x})O_3$

found in all titanium oxides themselves, but also because of the fact that the temperature-independent quotient

$$\frac{TC\varepsilon}{\varepsilon/\varepsilon_0} = \frac{1}{\varepsilon}\frac{\mathrm{d}\varepsilon}{\mathrm{d}T} \cdot \frac{\varepsilon_0}{\varepsilon} = -\frac{1}{C} \cong -10^{-5}\mathrm{K}^{-1} \tag{3.3}$$

as represented by the Curie–Weiss law of BT can be found in equal magnitude within all TiO_2 compounds independent of their specific lattice structure [1].

Additionally, the reason for the development of a ferroelectric phase instead of an antiferroelectric one below the Curie temperature can be found within the TiO_6 octahedra themselves and their interaction. For, the electrostatic preference of the mutual saturation of the apparent charges at the ends of dipole chains is overcompensated by contributions of the homopolar parts of the Ti–O bonds. It has been shown by X-ray fine structure investigations of BT [2] that already above the Curie temperature a closer binding between the Ti ion and three of its closest oxygen neighbors exists, causing a minor deviation from the central position and the formation of a tetrahedron-like TiO_3 configuration. But owing to the minor energy contribution of this deviation the Ti ion changes the possible partners, continuously oscillating from one position to the other. The average position is found in the center of the unit cell.

The preference for a smaller atomic distance between the Ti ion and its oxygen partners is due to the homopolar part of the Ti–O bond. Furthermore, this homopolar contribution involves also the tendency of oxygen towards angled valencies. This tendency is well known, e.g., from the water molecule, and plays an important role in ferroelectricity. For, it favors the parallel orientation of the polarization in neighboring lattice cells as illustrated in Fig. 3.2. In other words, it causes the transition to ferroelectricity instead of antiferroelectricity below the Curie temperature.

The transition from the paraelectric to the ferroelectric state, which we have concentrated on up to now, is not the only ferroelectric phase transition occurring in BT. This can be seen, e.g., by the results shown in Fig. 3.3. Here Fig. 3.3a shows the temperature dependence of permittivity of

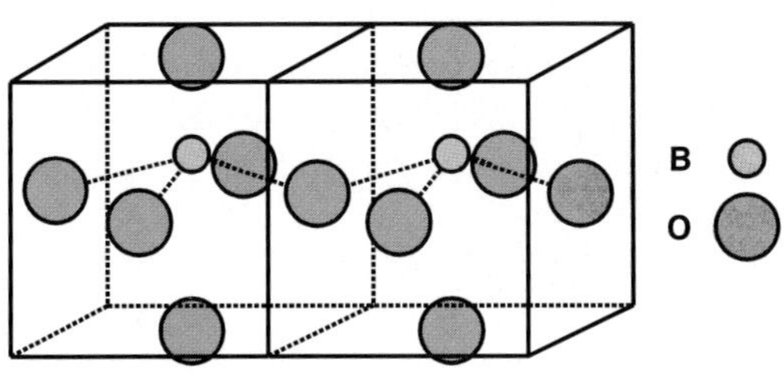

Fig. 3.2. The homopolar contribution of the Ti–O bond involving the tendency of oxygen towards angled valencies, which favors the parallel orientation of the polarization in neighbouring lattice cells

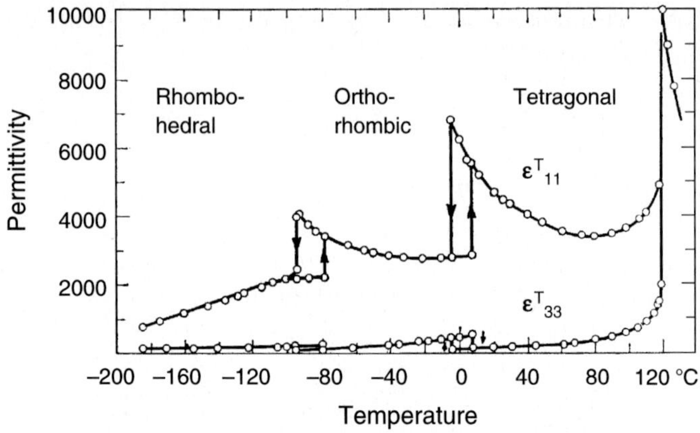

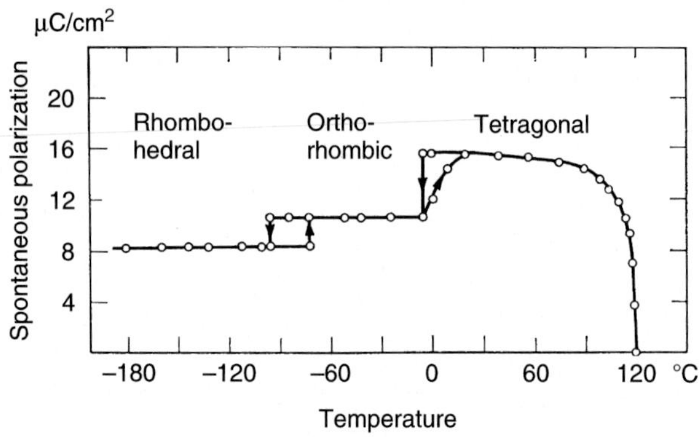

Fig. 3.3. (**a**) Dielectric constants of $BaTiO_3$ *vs.* temperature. The values of $\varepsilon^T{}_{11}$ and $\varepsilon^T{}_{33}$ in the tetragonal phase refer to single-domain crystals (according to Merz [3]). (**b**) Spontaneous polarization of $BaTiO_3$ *vs.* temperature, measured along the pseudo-cubic (001) edge, i.e., parallel to the 3-axis in the tetragonal phase (according to Merz [3]). In the orthorhombic phase or in the rhombohedral phase, the magnitude of P_s is obtained by multiplying the value in Fig. 3.3b with $\sqrt{2}$ or $\sqrt{3}$, respectively. The onset of P_s at T_C appears to be rather continuous. Although this was found to be due to crystal imperfections, it shows that the transition is only weakly first order

(100)-oriented BT single crystals as measured by Merz [3]. It reveals two more phase transitions below the Curie temperature, which occur when lowering the temperature by about 100 and 200°C, respectively, below T_C. Figure 3.3b explains more explicitly what is going on. Below the Curie temperature, first a spontaneous polarization develops in the (100) direction; at about 10°C this polarization turns into a (110) direction. At −70°C it turns again and finds its final orientation in a (111) direction. Furthermore, at every transition the absolute value of polarization increases remarkably. Corresponding to the polarization the lattice distortion changes from the undistorted cubic phase above the Curie temperature to a tetragonal phase below, then to an orthorhombic one and finally to a rhombohedral one.

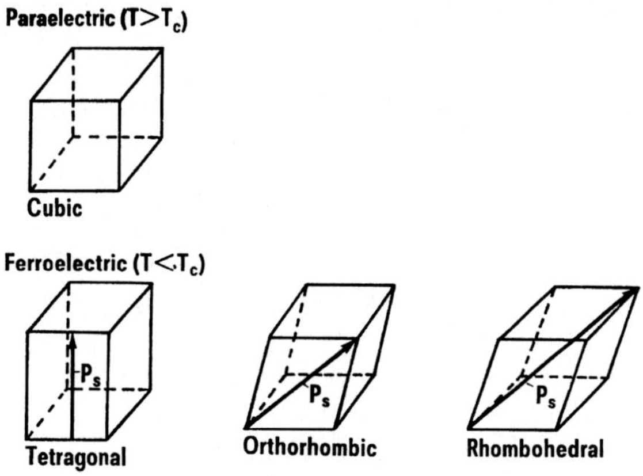

Fig. 3.4. The four phases of $BaTiO_3$. The arrows indicate the direction of the spontaneous polarization related to spontaneous cell deformation

At every transition the Ti ion adopts its privileged partnership with O ions or, in terms of chemistry or covalent bonds, its coordination number. In the cubic phase it moves statistically among all its six oxygen neighbors, in the tetragonal phase among five neighbors, in the orthorhombic phase among four neighbors, and in the rhombohedral phase, finally, it is coordinated to three oxygen ions. Owing to the lattice distortion, these coordinations can be exactly equivalent only in the cubic and in the rhombohedral case, and it should be emphasized once more that they describe the homopolar picture only and do not include the heteropolar part of the binding. The described phase transitions are summarized and visualized once more in Fig. 3.4.

In order to make the further discussions of this chapter as simple as possible, it is important that all features discussed up to now are explained also quantitatively by the phenomenological Landau–Ginsburg–Devonshire thermodynamic theory [4], which is based on relatively simple assumptions. Devonshire used the cubic state as the basic or prototype phase and treated the ferroelectric distortions as minor deviations. For this reason, he has expanded the free energy A around the cubic state ($A = A_0$) in ascending power of the polarization, which he used as measure for the distortions. Because of the symmetry of the cubic perovskite lattice, there are no terms exhibiting odd powers of $P_{1,2,3}$ and one gets an expression

$$A = A_0 + \frac{\chi}{2}\left(P_1^2 + P_2^2 + P_3^2\right) + \frac{1}{4}\xi_{11}(P_1^4 + P_2^4 + P_3^4) + \text{terms of higher order.} \tag{3.4}$$

Because of the general relation

$$\partial A \big/ \partial \vec{P} = \vec{E}, \tag{3.5}$$

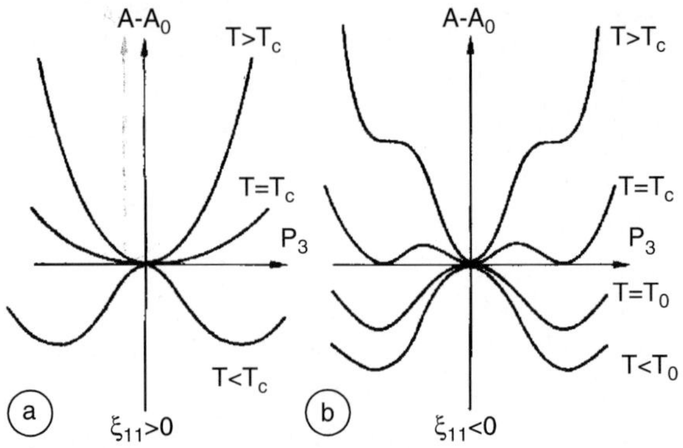

Fig. 3.5. Free energy of $BaTiO_3$ as plotted vs lattice polarization P_3 (schematically)

one gets in the case of high permittivity ($\varepsilon >> \varepsilon_0$ or $\mathbf{D} \approx \mathbf{P}$) from the Curie–Weiss law (Eq. 3.1) as valid in the cubic state

$$\chi \cong \frac{T - T_0}{\varepsilon_0 C}. \tag{3.6}$$

Now, in the Devonshire theory the validity of (3.4) is extended to the region below the Curie temperature. Furthermore, it is assumed that all terms of higher power in (3.4) do not depend on temperature significantly. Under these assumptions, (3.4) represents a family of curves as shown in Fig. 3.5, where $A - A_0$ is plotted versus the polarization for different temperatures. It should be noted that the Devonshire equation needs only the correct adaptation of three constants for a roughly correct description of the permittivity, polarization, and phase transitions observed in BT. In this description, the observed lattice distortions are nothing but the electrostrictive consequence of the spontaneous polarization (cf. Fig. 3.4).

Let us go back to Fig. 3.5, which describes the neighborhood of the Curie temperature. In the case of Fig. 3.5a, no secondary minimum exists above the Curie temperature and the curve of free energy is nearly parabolic with a curvature corresponding to the Curie–Weiss law. Below T_C, the central minimum becomes unstable and lateral minima develop in the (100) directions. This represents a thermodynamic transition to a ferroelectric state of second order. On the contrary, in case the side minima pre-exist above the Curie temperature, owing to the already mentioned influence of covalency, the diagram of $(A - A_0)$ vs. $\mathbf{P}$ transforms to the shape in Fig. 3.5b. The central minimum and the (100) minima coexist within a limited temperature region, whether they are stable or metastable. The transition from the non-ferroelectric cubic state to the tetragonal ferroelectric state is of first order.

As mentioned already, on cooling down further two more ferroelectric phase transitions occur. Both are of first order as shown in the complete

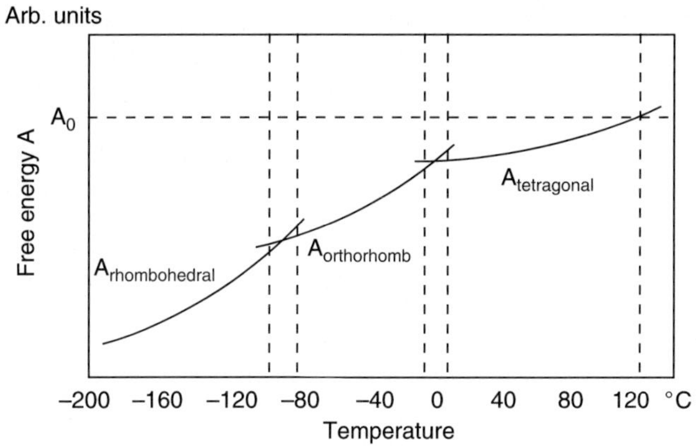

Fig. 3.6. The free energy of the different ferroelectric phases of $BaTiO_3$ as a function of temperature (schematic)

energy phase diagram of BT (Fig. 3.6), where we have plotted the free energy of the possible phases vs. temperature. Of course, the stable phase is represented by the lowest curve. But owing to the fact that we are dealing with very small energy differences, especially near the transitions themselves, the apparent properties often are co-determined by the neighboring metastable phases.

Generalizing, we may call a phase ferroelectric when it has two or more polar states (in the absence of an electric field) that can be shifted or switched from one to another by an electric field. The polar character of these orientation states must represent a completely stable configuration in zero field. Ferroelectrics are therefore often considered to belong to the subgroup of the 10 polar crystal classes[4] that are capable of being switched in some manner. The possibility of switching usually implies a polar structure characterized by small distortions from a higher-symmetry non-ferroelectric phase, which is termed the *prototype phase.* Although the prototype phase is not necessarily of nonpolar character, it proves to be so for the vast majority of known ferroelectrics, and in the following we shall make this assumption generally.

As a result of the small structural displacements from the prototype phase, a typical ferroelectric possesses a spontaneous polarization P_s, which decreases with increasing temperature T, to disappear continuously, or more often discontinuously, at a certain temperature, the so-called Curie temperature T_c. The phase above the Curie point is usually termed *paraelectric.* It is not, of course, necessary that the paraelectric phase be the prototype phase, although this is quite often the case.

A ferroelectric transition is usually connected with the condensation of a soft polar or optically active mode of lattice motion at the Brillouin-zone

[4] These classes are called polar because they possess a spontaneous polarization or electric moment per unit volume.

center, which causes the appearance of a long-range polar order. Because of the existence of coupling between modes, it is not a necessary condition for ferroelectricity that a zone-center polar mode should be the driving instability. A driving zone-boundary mode can couple to a zone-center polar mode and, upon condensation, induce a small spontaneous polarization in an indirect fashion. Such ferroelectrics are often termed *improper ferroelectrics* [5].

3.1.2 Ferroelectrics Important for Piezoelectric Devices

We consider here the most important technological ferroelectrics used today as single crystals or ceramics as well as thin films. All these materials are based on corner-linked oxygen octahedral structures and exhibit strong ferroelectric properties. The simplest configuration is the very well known perovskite structure.

In the perovskite structure (Fig. 3.1) with the general formula ABO_3, the octahedra are linked in a regular cubic array forming the high-symmetry cubic $m3m$ prototype structure. The small, six-fold-coordinated B-site in the center of the oxygen octahedron is filled by a small highly charged (3-, 4-, 5-, or 6-valent) cation, and the larger 12-fold coordinated A-site between octahedra carries a larger (1-, 2-, or 3-valent) cation.

The first ferroelectric perovskite to be discovered was BT. This was a particularly important event since the extreme simplicity of the structure gave for the first time an opportunity to study the onset of ferroelectricity from a simple, highly symmetric prototype phase. It is interesting to note that BT, which was the highlight at that time, is still the base for the composition of most of the world's ceramic capacitors and also plays a very important role in piezoelectric devices.

The core of the essentially cubic BT lattice cell is formed by an oxygen octahedron with a 4-valent titanium ion in the center. Neutrality and steric stability of the lattice is ensured by the 2-valent barium ions at the corners of the unit cell. The essential part of the cell concerning ferroelectric and piezoelectric properties is represented by the TiO_6 octahedron. BT is cubic above 120°C[5]. Below 120°C, it transforms successively to three ferroelectric phases: first to $4\,mm$ tetragonal, then to mm orthorhombic at about 5°C, and finally to a $3\,m$ trigonal phase below −90°C (Fig. 3.3b). The polar axis in the three ferroelectric phases is in the (001), (011), and (111) directions, respectively. All three transitions are of first order ($T_0 < T_C$), and the dielectric constant shows peak values at the transitions (Fig. 3.3a). Above T_C the dielectric constant follows a Curie–Weiss law. The permittivity maxima in Fig. 3.3a would be even more pronounced if instead of the ε_{11}^{T}[6] component, which refers to the axis system of the cubic prototype phase, the principal axis component ε'^{T}_{11} were shown.

[5] Up to 130°C, depending on crystal quality.

[6] Superscript T denotes "at constant stress", see also Eqs.: (3.9)–(3.11).

As will be seen in Sect. 3.2, together with ε'^{T}_{11}, also the piezoelectric shear constants d'_{15} reach a maximum at the ferroelectric phase transitions and, as a consequence, in a polycrystalline material the application-relevant piezoelectric constants achieve maximum values close to ferroelectric phase transitions. Unfortunately, this effect cannot be exploited in BT because the material properties are strongly temperature dependent. Therefore, modifications of BT and solid solutions such as $(Ba_{1-u}Sr_u)TiO_3$ (BST, Fig. 3.7) and $Ba(Zr_xTi_{1-x})O_3$ (Fig. 3.8a) have been investigated to overcome this problem, unfortunately without success.

However, the substitution of Ba by Pb changes the phase diagram in Fig. 3.8a (left) to that of Fig. 3.8b (right) [6]. The orthorhombic phase disappears and the phase boundary tetragonal–rhombohedral at $x \approx 0.5$ becomes

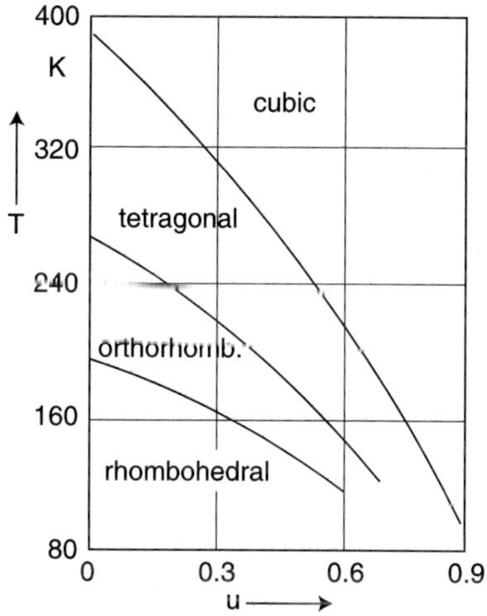

Fig. 3.7. Temperature–composition phase diagram of $(Ba_{1-u}Sr_u)TiO_3$

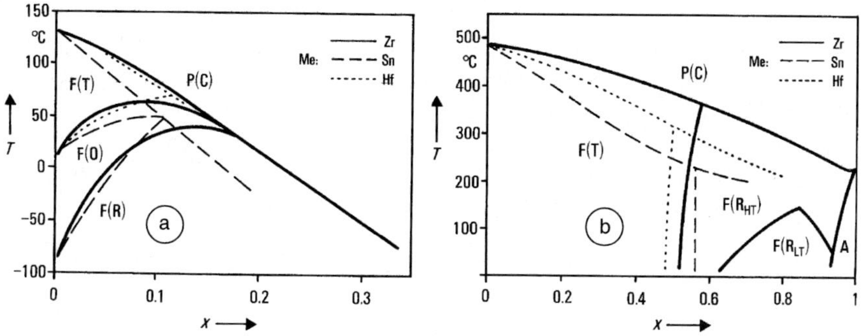

Fig. 3.8. Phase diagrams of $Ba(Me_xTi_{1-x})O_3$ and $Pb(Me_xTi_{1-x})O_3$

almost vertical, i.e., it is not dependent on temperature. The transition from tetragonal to rhombohedral phase is not abrupt but continuous within a certain region in which the coexistence of both phases is observed (morphotropic phase boundary). The physical origin of the phase coexistence is, on one hand, due to the first-order phase transition between the tetragonal and rhombohedral phases, and the small enthalpy difference allows the metastable tetragonal phase to coexist on the rhombehedral side of the phase boundary and vice versa, and on the other, due to a fluctuation of the Zr concentration in (lead–zirconate–titanate) PZT ceramics [7].

The fact that the morphotropic phase boundary is almost independent of temperature makes this material very favorable, because the maximum dielectric and piezoelectric constants as well as other anomalous properties in the vicinity of the phase boundary can be widely used for technological purposes. Therefore, extensive investigations and developments in the area of $Pb(Zr_xTi_{1-x})O_3$ (PZT) ceramics have been undertaken to control this feature (*see* Chaps. 4 and 6). Nevertheless, the results are important for these kinds of transitions also from a scientific point of view.

$KNbO_3$ (KN) is qualitatively similar to BT in many ways. It undergoes the same series of ferroelectric transitions in the same sequence as the temperature is lowered. The cubic–tetragonal transition is at $T = 435°C$, the tetragonal–orthorhombic one is at 225°C, and the orthorhombic–rhombohedral one is at −10°C, and the transitions are all of first order. The solid solutions $K(Ta_xNb_{1-x})O_3$ (KTN) are ferroelectric for $x < 0.95$ and exhibit the same ferroelectric phases as BT and KN. The transition temperature varies fairly linearly with composition, with a cubic–tetragonal transition near room temperature when $x \approx 0.63$.

KN and KTN have very high electro-optic coefficients (a factor of 10 higher than those of $LiNbO_3$), which make them very interesting for high-speed electro-optic modulation applications. Very recently, these materials have become increasingly important as base materials for the development of novel lead-free piezoelectrics [8].

Although $LiNbO_3$ and $LiTaO_3$ do not have perovskite structures, these ABO_3 lattices are composed of oxygen octahedra involving similar instabilities as in BT. Figure 3.9 shows the basic room-temperature structure consisting of a sequence of distorted oxygen octahedra joined by their faces along a trigonal, polar *c*-axis. Within the oxygen cages, as one progresses along the polar axis, the cations appear in the sequence Nb (Ta), vacancy, Li, Nb (Ta), vacancy, Li,..., and unlike the perovskite ferroelectrics the relative offset of cations from possible nonpolar positions is quite large. This implies a large spontaneous polarization ($\approx 70\,\mu C\ cm^{-2}$ in $LiNbO_3$ and $\approx 50\,\mu C\ cm^{-2}$ in $LiTaO_3$) and a high Curie temperature (≈1200°C for $LiNbO_3$ and 620°C for $LiTaO_3$). The observed Curie constants for $LiNbO_3$ and $LiTaO_3$ are of the order 10^5 K, which is typical for systems of preferentially displacive

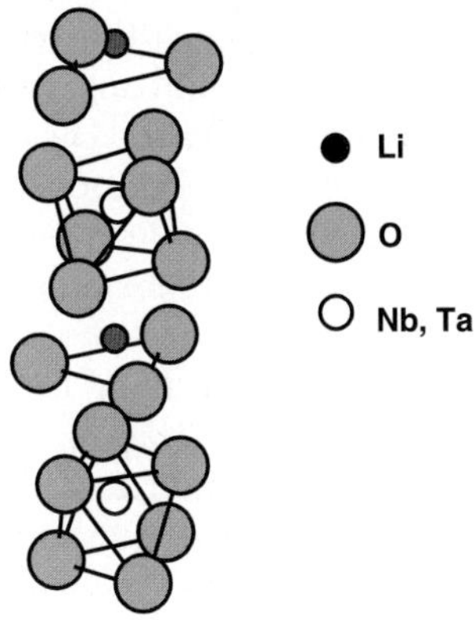

Fig. 3.9. The ferroelectric structure of $LiNbO_3$ and $LiTaO_3$. The structure consists of a sequence of distorted oxygen octahedra joined by their faces along the trigonal polar axis

character[7]. Both compounds are uniaxial at all temperatures, with only a single second-order transition from the paraelectric phase (point group $R3m$) to the ferroelectric ($R\bar{3}m$) phase.

Large optical-quality single crystals of $LiNbO_3$ and $LiTaO_3$ can be grown by the Czochralski method and are commercially available. These materials are generally used in the single-crystal form. The robustness, good performance, and wide availability of these materials make them a popular choice for commercial nonlinear optical and electro-optic devices, such as modulators and harmonic generators as well as pyroelectric detectors. Particularly, they are important for surface acoustic wave (SAW) devices (see Chap. 14).

3.1.3 Origin of Ferroelectricity in Perovskite Oxides

The soft-mode picture introduced by Cochran [9] during the 1960s is considered as a key concept to explain the ferroelectric phase transitions in perovskite (ABO_3) compounds, since soft modes are strong indicators of lattice instabilities and often precursors of such phase transitions. For a microscopic understanding, we have to take into account furthermore the interaction between binding forces (attraction and repulsion) on one hand, and long-range Coulomb forces on the other. It is now generally agreed that this delicate balance between short-range and long-range forces is connected to the homopolar part of the Ti–O bond, i.e., the orbital hybridization between O 2p and Ti 3d states (for a detailed discussion see Chap. 21). This is also the reason for the preference of a smaller atomic distance between the Ti ion and its oxygen partners, which has been shown by X-ray fine structure investigations on BT.

[7] Order-disorder ferroelectrics such as TGS have Curie constants up to two orders of magnitude smaller.

The homopolar contribution involves also the tendency of oxygen towards angled valencies, which plays an important role in ferroelectricity because it favors the parallel orientation of the polarization in neighboring lattice cells. In other words, it causes the transition to ferroelectricity instead of antiferroelectricity below the Curie temperature.

Both the substitution of Ba by smaller ions such as Sr, and of Ti by larger ions such as Zr, Sn, and Hf lead to a depression of the Curie temperature, as can be seen from the phase diagrams in Figs 3.7 and 3.8. In the first case, both Sr and Ca are smaller than Ba, the lattice constant decreases, and the phase transitions are shifted towards lower temperatures like by an equivalent hydrostatic compression of the whole lattice. In the second case, Zr, Sn, and Hf are larger than Ti, and therefore the average free space inside the O_6 octahedron decreases in the same way as in the case of a hydrostatic compression. These results clearly demonstrate that the major player concerning the ferroelectric polarization in BT is the O_6 octahedron itself, and that in BT the A-site ions do not actively participate in the ferroelectric polarization mechanism.

Contrary to the substitutions discussed up to now, the Curie temperature increases remarkably when replacing barium by lead. However, this cannot be understood by a similar explanation as given above because the ionic radii of Pb and Sr are nearly equal and the lattice contraction due to the substitution is also the same. Therefore, the much higher Curie temperature and the higher spontaneous polarization and strain in $PbTiO_3$ compared to BT can be explained only by a strong orbital hybridization between Pb 6s and O 2p states, whereas, on the contrary, Ba 5p does not hybridize with the valence band (see Chap. 21). We can get hints as to the relevant mechanism [10] when looking at the stereochemistry of compounds with divalent lead ions, especially of PbO itself. There, concerning the nearest neighbors, configurations are found in which the lead ions form the apices of pyramids with quadratic or also triangular bases (cf. Fig. 3.10). The average distance between the lead and the oxygen ions there is 230 pm. This value is smaller than the Pb–O

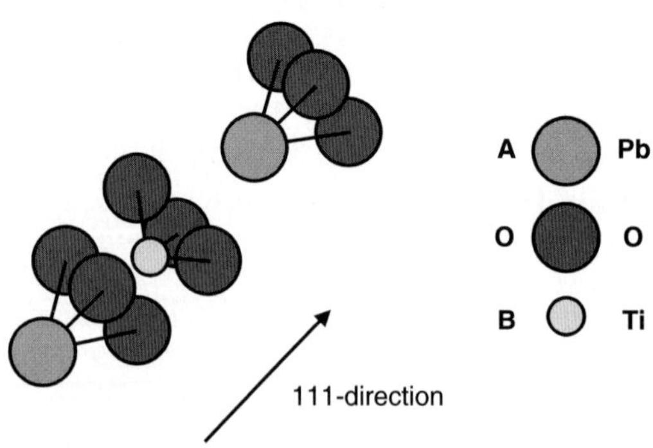

Fig. 3.10. Local dipole moments in (111) direction induced by covalent Pb–O and Ti–O bonds in rhombohedral $Pb(Zr_xTi_{1-x})O_3$

distance in cubic $PbTiO_3$, which is 252 pm. Therefore, a reduction of this distance is required for getting a contribution from the homopolar s-electron pair to the binding energy. Since such a big compression of the lattice cannot be expected, only a displacement of the lead ion towards selected partners out of its 12 oxygen neighbors makes sense. Two possibilities result in configurations that have been found in PbO similarly. In the first case, the lead sublattice is displaced in the (100) direction so that quadratic lead oxygen pyramids are formed. The whole lattice is deformed tetragonally, and the lead oxide pyramids contribute to the overall polarization of the whole lattice remarkably. In the second case, the lead sublattice is displaced in the (111) direction. Three neighboring oxygen ions are approached by lead, and triangular pyramids with Pb apices are formed. These pyramids represent local dipoles in the (111) direction. In each case, the interaction between TiO_6- and Pb-controlled ferroelectric polarization results in an elevated Curie temperature. Furthermore, in lead-containing perovskites, the strong orbital hybridization between Pb 6s and O 2p states favors the tetragonal or rhombohedral ferroelectric phase and this explains the suppression of the orthorhombic phase in material systems such as $Pb(Zr_xTi_{1-x})O_3$ (see Chap. 21).

3.1.4 Thermodynamics of Piezoelectrics and Ferroelectrics

To form a bridge between ferroelectricity and piezoelectricity, it is not sufficient to use the appearance of a spontaneous polarization as a measure of piezoelectricity, like we have done so far for studying ferroelectric phase systems of BT. Obviously, we have to include mechanical and electrical boundary conditions. For this reason, we assume that the properties of such a material can be described by six parameters:

- temperature T and entropy S
- stress $\mathbf{T}$[8] and strain $\mathbf{S}$ (components T_α and S_β)
- electric field $\mathbf{E}$, and displacement $\mathbf{D}$ (components E_i and D_k)

To generate the equations of state, thermodynamic potentials are defined. Since three independent variables can be chosen in eight different ways from among the conjugate pairs $(T,\ S)$, $(T_\alpha,\ S_\beta)$, and $(E_i,\ D_k)$, this implies eight thermodynamic potentials, each containing a complete description of the material of interest.

It is convenient to use the following thermodynamic potentials in addition to the internal energy U (per unit volume):

$$\text{Helmholtz free energy } A = U - TS \tag{3.7a}$$

[8] Though the same letters are used to denote temperature and stress and entropy and strain, respectively they can be well distinguished because stress and strain are typed as tensors (bold type) or as tensor components (with indices).

Elastic Gibbs free energy $G_1 = U - TS - T_\alpha S_\alpha$ (3.7b)[9]
Electric Gibbs free energy $G_2 = U - TS - E_i D_i$ (3.7c)
Gibbs free energy $G = U - TS - T_\alpha S_\alpha - E_i D_i$ (3.7d)

The differential forms describing infinitesimal changes in these thermodynamic potentials are

$$\begin{aligned} \mathrm{d}U &= T\mathrm{d}S + T_\alpha \mathrm{d}S_\alpha + E_i \mathrm{d}D_i \\ \mathrm{d}A &= -S\mathrm{d}T + T_\alpha \mathrm{d}S_\alpha + E_i \mathrm{d}D_i \\ \mathrm{d}G_1 &= -S\mathrm{d}T - S_\alpha \mathrm{d}T_\alpha + E_i \mathrm{d}D_i \\ \mathrm{d}G_2 &= -S\mathrm{d}T + T_\alpha \mathrm{d}S_\alpha - D_i \mathrm{d}E_i \\ \mathrm{d}G &= -S\mathrm{d}T - S_\alpha \mathrm{d}T_\alpha - D_i \mathrm{d}E_i \end{aligned} \tag{3.8}$$

and explicitly indicate the three principal variables for each potential.

When, according to the rule given, a particular set of independent variables is held constant, a system will come to thermodynamic equilibrium in such a way that the free energy – for which the constrained variables are the principal ones – is minimized. Thus, for example, in the common experimental situation for static measurements with T, T_α, and E_i fixed, irreversible processes bring about an approach to equilibrium, which minimizes the Gibbs free energy G. If the system undergoes a phase transition from phase A to phase B, the transition occurs when $G_\mathrm{A}(T,\ T_\alpha,\ E_k) = G_\mathrm{B}(T,\ T_\alpha,\ E_k)$. Therefore, this equation determines the transition temperature, i. e., $T_\mathrm{t} = T_\mathrm{t}(T_\alpha,\ E_k)$.

If we take temperature, stress, and electric field as independent variables, we get from the Gibbs potential the following relations:

$$-S = \left(\frac{\partial G}{\partial T}\right)_{T_\alpha, E_i}, \quad -S_\alpha = \left(\frac{\partial G}{\partial T_\alpha}\right)_{T, E_i}, \quad -D_i = \left(\frac{\partial G}{\partial E_i}\right)_{T, T_\alpha}. \tag{3.9}$$

The linear differential form of these equations, e.g.,

$$\Delta S_\alpha = \left(\frac{\partial S_\alpha}{\partial T}\right)_{T_\alpha, E_i} \Delta T + \left(\frac{\partial S_\alpha}{\partial T_\beta}\right)_{T, E_i} \Delta T_\beta + \left(\frac{\partial S_\alpha}{\partial E_i}\right)_{T, T_\alpha} \Delta E_i \tag{3.10}$$

leads to the equations

$$\begin{aligned} \Delta S_\alpha &= \alpha_\alpha^E \Delta T + s_{\alpha\beta}^E \Delta T_\beta + \mathrm{d}_{i\alpha} \Delta E_i \\ \Delta D_i &= p_i^T \Delta T + \mathrm{d}_{i\alpha} \Delta T_\alpha + \varepsilon_{ik}^T \Delta E_k \end{aligned} \tag{3.11}$$

which, in their isothermal form ($\Delta T = 0$), are called *basic piezoelectric equations.* In addition, six other linear equations, i.e., three further sets of basic

[9] i = 1, 2, 3 and $\alpha = 1, \ldots, 6$ (Voigt notation); $T_\alpha S_\alpha \equiv \sum_\alpha T_\alpha S_\alpha$ (summation convention).

piezoelectric equations, can be derived (either isothermally or adiabatically) by use of the other thermodynamic potentials (see Chap. 18).

The coefficients in (3.11) are thermal expansion coefficients $\left(\alpha_{\alpha}^{E}\right)$, elastic compliances at constant electric field $(s_{\alpha\beta}^{E})$, piezoelectric constants $(d_{i\alpha})$, pyroelectric coefficients $\left(p_{i}^{T}\right)$, and permittivities at constant stress $\left(\varepsilon_{ik}^{T}\right)$. The crystallographic symmetry of the material imposes conditions on the components of these tensors or matrices. In particular, some components may be required to be zero and others to be interrelated.

This thermodynamic derivation of the basic piezoelectric equations offers the advantage of not depending on a specific model. It clearly shows that the material constants must depend on the amplitude of the applied stress and the electric field. In spite of the lack of a model-related interpretation, it offers an insight into the relevant mechanisms of the nonlinearities observed and the interrelation between dielectric, elastic, and piezoelectric coefficients. These interrelations have become the basis for every interpreting model and even for first-principles calculations (cf. Chap. 21).

Twenty of the 32 point groups are piezoelectric groups, which allow a crystal to exhibit electric polarity when subjected to stress. In a similar fashion, a strain is produced by the application of an electric field, which, in lowest order, is linearly proportional to that field. Ten of these piezoelectric groups have a unique polar axis (spontaneous polarization) and are termed *pyroelectric* or *polar*.

Among the 12 non-piezoelectric groups, 11 are centrosymmetric. These are characterized by a center of symmetry and can therefore possess no polar properties, i. e., they are neither pyroelectric nor piezoelectric. In such crystals, an applied electric field produces a strain that is not changed when the field is reversed, i. e., the effect is quadratic. This property is termed *electrostriction* and occurs naturally in all substances. This characteristic is fundamentally nonlinear (like the most important characteristics of ferroelectrics) and hence requires an extension of the differential equation of state to higher orders. For example, the elastic nonlinear equation of state for centrosymmetric crystals resulting from (3.9) is:

$$S_{\alpha} = s_{\alpha\beta}^{E} T_{\beta} + M_{ik\alpha} E_i E_k . \tag{3.12}$$

From the elastic Gibbs potential G_1 one can get a similar equation of state for centrosymmetric crystals:

$$S_{\alpha} = s_{\alpha\beta}^{D} T_{\beta} + Q_{ik\alpha} D_i D_k . \tag{3.13}$$

The coefficients $M_{ik\alpha}$ and $Q_{ik\alpha}$ are termed electrostrictive constants.

As mentioned above, we consider here only ferroelectrics with a centrosymmetric prototype phase. When such a crystal transforms from the paraelectric phase to a polar phase, the crystal will deform spontaneously, thereby changing its symmetry. For constant T and for $T_\beta = 0$ and $E_i = 0$ ($D_i = P^s{}_i$), the spontaneous deformation $S^s{}_\alpha$[10] resulting from (3.13) is:

$$S^s_\alpha = Q_{ik\alpha} P^s_i P^s_k. \tag{3.14}$$

It follows from (3.14) that the spontaneous strain is of similarly low order as P^{s2}_i. This justifies the expansion of the thermodynamic potential in terms of deviations from the centrosymmetric prototype state $D_i = 0,\ S_\alpha = 0$. Normally, the most convenient potential to be expanded is the elastic Gibbs function G_1.

$$G_1(D_i, T_\alpha) = G_0 + \frac{1}{2}\chi^T_{ij} D_i D_j + \frac{1}{4}\xi^T_{ijkl} D_i D_j D_k D_l + \frac{1}{6}\zeta^T_{ijklmn} D_i D_j \ldots D_n$$
$$-Q_{ik\alpha} T_\alpha D_i D_k - \frac{1}{2} s^D_{\alpha\beta} T_\alpha T_\beta \tag{3.15}$$

G_0 is the potential of the nonpolar phase. The Devonshire[11] coefficients χ, ξ, and ζ (at constant T_α) in general depend on temperature and on material composition, but many examples of ferroelectric transitions can be described assuming the coefficients ξ and ζ to be independent of temperature.

Whether the transition from the nonpolar phase to a ferroelectric phase is of first or second order depends on the contribution of the third term on the right-hand side of (3.15). If this is negative, then the transition is of first order. In this case, higher-order terms with a positive contribution to the thermodynamic potential are necessary for stability reasons (fourth term in (3.15)). Therefore, such higher-order terms are normally included in the potential.

According to the crystallographic symmetry, several components of χ, ξ, ζ, Q, and s in (3.15) are required to be zero or to be interrelated. In the important case of a cubic prototype phase, there is only one χ component ($\chi_{ij} = \chi\delta_{ij}$), two independent ξ components (termed $\xi^T{}_{11}$ and $\xi^T{}_{12}$), three independent ζ components (ζ_{111}, ζ_{112}, ζ_{123}), three independent Q components (Q_{11}, Q_{12}, Q_{44}[12]), and three independent compliance components $s^D{}_{11}$, $s^D{}_{12}$, $s^D{}_{44}$ at constant displacement.

In this case, (3.15) is simplified to

$$G_1(D_i, T_\alpha) = G_0 + \frac{1}{2}\chi^T D_i D_i + \frac{1}{4}\xi^T_{11} \cdots \tag{3.15a}$$

which is nothing but an expansion like that of (3.4). We have to take into account only that in ferroelectrics $D_i \approx P_i$ and that owing to the inclusion of elasticity, A has been replaced by G.

---

[10] In order to avoid a collission between tensor subscripts and the label s (for spontaneous) a superscript of s is used here instead of a subscript.

[11] The phenomenological theory based on polynomial thermodynamic potential is often termed Devonshire theory (Devonshire 1949, 1951, 1954) and the coefficients χ, ξ, ζ are therefore termed Devonshire coefficients.

[12] $Q_{111} = Q_{222} = Q_{333} = Q_{11}$, $Q_{221} = Q_{331} = Q_{112} = Q_{332} = Q_{113} = Q_{223} = Q_{12}$, $Q_{234} = Q_{135} = Q_{126} = Q_{44}/2$, zero otherwise.

This phenomenological theory provides a convenient and simple framework within which the phase transitions and the dielectric, elastic, and piezoelectric properties of ferroelectrics as well as their dependence on temperature, composition, applied mechanical stress, and electric field can be interpreted. In order to do this, 10 independent parameters (as shown above) and their dependence on composition must be determined. Because these parameters are directly related to macroscopically observable quantities, they can be determined in a self-consistent way from a number of experimental measurements [11].

This interpretation is generally used for the description of ceramic materials in technological use. But here many additional parameters have to be taken into account, such as grain size, density, inter- and hetero-faces, dopants, homogeneity, etc.

Thus we will first deduce the relevant formulas for dielectric, piezoelectric, and elastic properties of ferroelectric crystals, and after this in Sects. 3.2 and 3.3 proceed to the macroscopic properties in the intrinsic and extrinsic case.

3.1.4.1 Dielectric Properties

From (3.15), the permittivity of the cubic phase can be easily calculated (as shown below) to be $\varepsilon = 1/\chi$. Therefore, the temperature dependence of χ can be written in the Curie–Weiss form

$$\chi^T = \frac{1}{\varepsilon_0 C}(T - T_0) = \chi_0 (T - T_0)\,. \tag{3.16}$$

Minimizing (3.15a) with respect to D_i at constant temperature generates the description of the different possible ferroelectric phases, expressed in terms of their spontaneous polarization. For ferroelectrics with a cubic prototype phase, such as BT, beside the cubic phase, three stable phases can exist principally: a tetragonal, an orthorhombic, and a trigonal (rhombohedral) phase. The polar axis in these three phases is (001), (011), and (111) respectively, and the spontaneous polarization P^s and Gibbs potential G_1 of each phase follows as:

$$P^{s^2} = \frac{-a}{2b}\left(1 \pm \sqrt{1 - \frac{4b\chi_0(T-T_0)}{a^2}}\right), \qquad (3.17)^{13}$$

$$G = G_0 - \frac{1}{4}P^{s4}\left(a + \frac{4}{3}bP^{s2}\right) \tag{3.18}$$

where the coefficients a and b can be expressed for the different phases as follows:

---

[13] This eqn for P^s and the following eqns for material properties hold only for a single domain or for a single crystal with completely aligned spontaneous polarization.

$$\text{tetragonal phase:} \quad \begin{aligned} a_t &= \xi_{11}, \\ b_t &= \varsigma_{111}, \end{aligned} \tag{3.19}$$

$$\text{orthorhombic phase:} \quad \begin{aligned} a_o &= \tfrac{1}{2}\left(\xi_{11} + \xi_{12}\right), \\ b_o &= \tfrac{1}{4}\left(\varsigma_{111} + 3\varsigma_{112}\right), \end{aligned} \tag{3.20}$$

$$\text{rhombohedral phase:} \quad \begin{aligned} a_r &= \tfrac{1}{3}\left(\xi_{11} + 2\xi_{12}\right), \\ b_r &= \tfrac{1}{9}\left(\varsigma_{111} + 6\varsigma_{112} + 2\varsigma_{123}\right). \end{aligned} \tag{3.21}$$

The sign of the a coefficient determines the order of the corresponding transition. It is, e.g., of first order if the corresponding a is negative. Then, the upper sign in (3.17) holds.

In the case of a second-order transition, the terms of sixth order in (3.15) are often neglected. This approximation yields a simplified expression for the spontaneous polarization

$$P^{s2} = \frac{\chi_0\left(T_0 - T\right)}{a}. \tag{3.22}$$

The transition temperature T_C from the nonpolar phase to a ferroelectric phase in the case of a second-order transition is identical with T_0. In case of a first-order transition, T_C can be calculated by imposing that the Gibbs free energy (for zero stress and field) of the polar and nonpolar phase are equal, which leads to

$$T_C = T_0 + \tfrac{3}{16}\tfrac{a^2}{\chi_0 b}. \tag{3.23}$$

Differentiating the elastic Gibbs function (3.15) with respect to D_i, one derives the components of the electric field E_i, which are nonlinear functions of D_k. With the substitution

$$D_k = P_k^s + \Delta D_k, \tag{3.24}$$

where ΔD_k is the displacement induced by an applied electric field, and an elastic stress at constant temperature, it is possible to linearize the field components for each individual ferroelectric phase specified by P_k^s. The resulting linear equations between E_i and ΔD_k (for zero stress) in general deliver the reciprocal permittivity matrix, which can be inverted to give ${\varepsilon^T}_{ik}$. The permittivity matrices of the ferroelectric phases are related to the axis of the nonpolar prototype phase and therefore are not normally diagonal. They can be diagonalized by transforming them to the principal axis system (3-axis parallel to P_s)[14].

For ferroelectrics with a cubic nonpolar phase, the principal axis permittivities can be written in the following way:

$$\text{Components parallel to } P^s\text{: } \varepsilon_{33}^{\prime\, T} = \frac{1}{2P^{s2}} \cdot \frac{1}{a + 2bP^{s2}}, \tag{3.25}$$

[14] Here the principal axis components are marked with a prime.

where a and b for the different phases are given in (3.19–3.21). The permittivity components normal to P^s are

$$\text{for the tetragonal phase: } \varepsilon'^T_{11} = \varepsilon'^T_{22} = \frac{1}{P^{s2}} \cdot \frac{1}{(\xi_{12} - \xi_{11}) + (\varsigma_{112} - \varsigma_{111}) P^{s2}}, \tag{3.26}$$

$$\text{for the orthorhombic phase: } \begin{aligned} \varepsilon'^T_{11} &= \frac{1}{P^{s2}} \cdot \frac{1}{\frac{1}{2}(\xi_{12} - \xi_{11}) + \frac{1}{4}(2\varsigma_{123} - \varsigma_{112} - \varsigma_{111}) P^{s2}}, \\ \varepsilon'^T_{22} &= \frac{1}{P^{s2}} \cdot \frac{1}{(\xi_{11} - \xi_{12}) + (\varsigma_{111} - \varsigma_{112}) P^{s2}}, \end{aligned} \tag{3.27}$$

$$\text{for the rhombohedral phase: } \varepsilon'^T_{11} = \varepsilon'^T_{22} = \frac{1}{P^{s2}} \frac{1}{\frac{2}{3}(\xi_{11} - \xi_{12}) + \frac{4}{9}(\varsigma_{111} - \varsigma_{123}) P^{s2}}. \tag{3.28}$$

Figure 3.13 shows the permittivities calculated with the aid of (3.25–3.28) in the vicinity of the technologically most interesting phase transition between the tetragonal and the rhombohedral ferroelectric phase of PZT known as the *morphotropic phase boundary* (MPB). The strong peak of the permittivity components normal to the spontaneous polarization is characteristic of this boundary (see below).

3.1.4.2 Piezoelectric and Elastic Properties

Applied field E_k and stress T_α are assumed to be small enough to make the induced displacement ΔD_i in (3.11) considerably smaller than spontaneous polarization. Therefore, (3.24) can be inserted into (3.13) with ΔD_i taken from (3.11) for constant T ($\Delta T_\alpha = T_\alpha$ and $\Delta E_k = E_k$) to give:

$$S_\alpha = S^s_\alpha + s^D_{\alpha\beta} T_\beta + 2Q_{ik\alpha} P^s_k d_{i\beta} T_\beta + 2Q_{jk\alpha} P^s_k \varepsilon^T_{ij} E_i. \tag{3.29}$$

Comparing (3.29) with the first equation of (3.11) for constant T and zero stress and bearing in mind that the field- and stress-induced strain in the polar phase $\Delta S_\alpha = S_\alpha - S_\alpha{}^s$, one finds:

$$\mathrm{d}_{i\alpha} = 2Q_{jk\alpha} P^s_k \varepsilon^T_{ij}. \tag{3.30}$$

Inserting (3.30) into (3.29), one finds similarly for $T_\beta \neq 0$:

$$s^E_{\alpha\beta} = s^D_{\alpha\beta} + 4Q_{ik\alpha} Q_{jl\beta} P^s_k P^s_l \varepsilon^T_{ij}. \tag{3.31}$$

From (3.30) we conclude that the piezoelectric effect of ferroelectrics with a centrosymmetric prototype phase is the result of electrostriction linearized by spontaneous polarization. However, the most important result of (3.30) and (3.31) is the fact that the dielectric anomalies occurring near ferroelectric phase transitions are directly reflected in the piezoelectric and elastic behavior, too.

Today's most important piezoelectric materials are based on the solid solution PZT, which transforms only to two ferroelectric phases; it is tetragonal at low Zr concentrations x and rhombohedral at high x. The main nonzero principal axis components of the piezoelectric matrices for both of these phases are given below:

$$\text{tetragonal phase:}\quad \begin{aligned} d'_{31} &= d'_{32} = 2Q_{12}P^s\varepsilon'^T_{33} \\ d'_{33} &= 2Q_{11}P^s\varepsilon'^T_{33}, \\ d'_{15} &= d'_{24} = Q_{44}P^s\varepsilon'^T_{11}, \end{aligned} \tag{3.32}$$

$$\text{rhombohedral phase:}\quad \begin{aligned} d'_{31} &= d'_{32} = \tfrac{1}{3}\left(2Q_{11} + 4Q_{12} - Q_{44}\right)P^s\varepsilon'^T_{33}, \\ d'_{33} &= \tfrac{2}{3}\left(Q_{11} + 2Q_{12} + Q_{44}\right)P^s\varepsilon'^T_{33}, \\ d'_{15} &= d'_{24} = \tfrac{1}{3}\left(4Q_{11} - 4Q_{12} + Q_{44}\right)P^s\varepsilon'^T_{11}. \end{aligned} \tag{3.33}$$

3.2 Intrinsic Piezoelectric Effect

In non-ferroelectric piezoelectric materials, we find a predetermined asymmetry necessary for the appearance of piezoelectricity due to the atomic structure of the lattice; in ferroelectric materials this asymmetry usually does not exist except below the Curie temperature. However, to minimize depolarizing fields and mechanical stresses, ferroelectric crystals or crystallites also split into ferroelectric domains, i.e., into regions with different directions of P_s. In case of multiaxial ferroelectrics, 180° domains and non-180° domains (e.g., 90° domains in tetragonal BT and PZT) are formed. Thus, in most cases a poling procedure is required to imprint the macroscopic asymmetry by proper orientation of these domains (*see* Fig. 3.11). In non-ferroelectric piezoelectric materials, the observed piezoelectric response originates from atomic displacements within the individual unit cells of the material. A similar piezoelectric response occurs in ferroelectric piezoelectrics as well. It dominates as long as the domain configuration in the material remains unaffected by the applied electric field or mechanical stress and is called the *intrinsic response*. Unlike in non-ferroelectric piezoelectric materials, in ferroelectric piezoelectrics there exists an additional contribution to the piezoelectric response called *extrinsic piezoelectric effect*. This extrinsic response mechanism results principally from non-180° (e.g., 90°, Fig. 3.11) domain wall movements driven by the application of an external electric field or a high enough mechanical stress.[15]

It should be emphasized that the intrinsic piezoelectric response is not identical to the piezoelectric response at low driving fields. In the low-field region ($E_{\text{drive}} \ll E_{\text{c}}$), the magnitude of the driving field is insufficient to promote significant domain wall motion and the behavior is dictated (primarily) by the intrinsic ionic response and reversible domain wall vibration (see Sect. 3.3). In the high-field region ($E_{\text{drive}} \approx E_{\text{c}}$), the piezoelectric response strongly increases owing to non-180° domain wall switching processes. Furthermore, the response becomes increasingly nonlinear associated with a considerable hysteretic behavior.

[15] Principially, also the movement of a phase boundary (e.g. the boundary between the tetragonal and rhombohedral phase in PZT) driven by an external field amplitude can contribute to the extrinsic response.

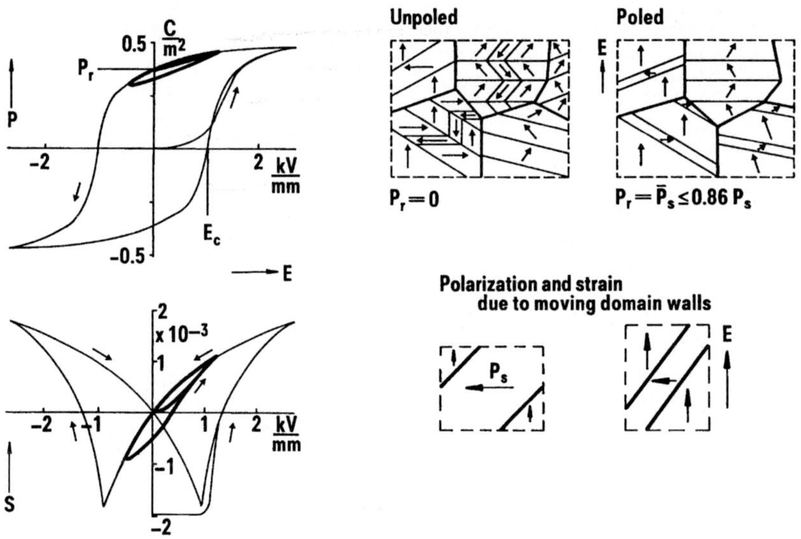

Fig. 3.11. Polarization and strain hysteresis loops of tetragonal $Pb(Zr_xTi_{1-x})O_3$ ceramics (left-hand side). The bold curves are obtained when a field of magnitude smaller than the coercive field strength E_c is applied. The upper right figure shows schematically the microstructure (grains and domains) before and after poling. The highest possible degree of poling results in a remnant polarization $P_r = \langle P_s \rangle \leqslant 0.86\, P_s$. The lower right figure shows that any shift of 90° (non-180°) domain walls due to an applied electric field results in a change of polarization and deformation (extrinsic contribution to the piezoelectric effect)

3.2.1 Anomalous Material Properties of PZT in the Vicinity of the Ferroelectric Tetragonal–Rhombohedral Phase Transition

In this section, we deal with the intrinsic piezoelectric effect mainly of lead-containing perovskites; the extrinsic effect is discussed in the following section.

When comparing different piezoelectric material classes (Table 3.1), one finds that all non-ferroelectric piezoelectric materials exhibit only very small piezoelectric constants $d_{i\alpha}$ and small electromechanical coupling constants $k_{i\alpha}$. Furthermore, these materials are only piezoelectric in the form of single crystals or textured thin films. On the other hand, these materials show extremely good long-term stability, and in many cases it is possible to find crystal cuts that also give high temperature stability. Both these properties make them important for frequency and time standards.

In applications where higher piezoelectric constants or higher coupling coefficients are required together with suitable high stability, as for example for broadband frequency filters (see Chap. 14), ferroelectric materials can be used with high Curie temperatures as compared to the operating temperature and with a material composition far away from any possible phase transition

Table 3.1. Comparison of the most important piezoelectric material classes by means of typical examples

Material	Example	$\varepsilon/\varepsilon_r$	$s\ (10^{-12}m^2/N)$	k	$d\ (pm/V)$
Non ferroelectric single crystals	Quartz	5	10	0.1	2.3
Ferroelectric single crystals (far away from phase transitions)	$LiNbO_3$	30–80	6	...0.3	...20
Ferroelectric ceramics/single crystals (in the vicinity of phase transitions)	PZT	200–5,000	10–20	0.3–0.8	100–1,000
Piezoelectric (ferroelctric) polymers	PVDF	10	400	0.15	20

in the material system. Important examples are the uniaxial ferroelectrics $LiNbO_3$ and $LiTaO_3$ and the multiaxial material $PbTiO_3$. These materials show medium permittivities and piezoelectric constants. Owing to the fact that in uniaxial materials poling is only possible via 180° domain switching, it is clear that $LiNbO_3$ and $LiTaO_3$ can be used only in form of single crystals or textured thin films; and even with single crystals, special poling procedures are required (see Chap. 14) to achieve single-domain crystals. They combine good piezoelectric performance with high spatial homogeneity.

However, the development of the modern piezo technology was not possible until BT was discovered to be ferroelectric by von Hippel and coworkers [12] and until Gray recognized that a poling process is necessary and suitable to make BT ceramics piezoelectric [13]. A breakthrough for technological applications was the discovery of very strong piezoelectric effects in PZT solid solutions by Jaffe and coworkers in 1954 [14]. It is well known that the piezoelectric constants of PZT ceramics exhibit peak values in the vicinity of the morphotropic phase boundary between the tetragonal and rhombohedral ferroelectric phases. As can be seen from Fig. 3.12, further material anomalies occur in the vicinity of this phase transition. The crucial advantage for many technological application is the fact that the phase boundary tetragonal–rhombohedral of $Pb(Zr_xTi_{1-x})O_3$ is very steep, i.e., nearly independent of temperature (Fig. 3.8b). As a consequence, the material anomalies such as maxima of certain dielectric, elastic, and piezoelectric constants that appear at this phase boundary are almost independent of temperature in a wide temperature range and, therefore, very favorable for technological use.

There exist different competing explanations for these anomalies. The usual one – which is still widely accepted – of the origin of the peak in the piezoelectric and dielectric properties at the morphotropic phase boundary

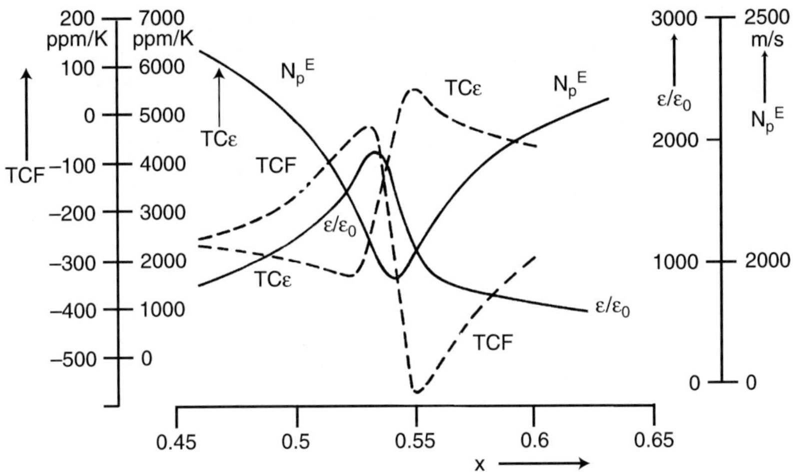

Fig. 3.12. Permittivity $\varepsilon/\varepsilon_0$, planar frequency constant N_p^E, and their temperature coefficients of $Pb(Zr_xTi_{1-x})O_3$ + 2% Nd ceramics

is mainly based on two arguments. First, a facilitated switching of domains during poling among six tetragonal and eight rhombohedral directions, which is possible because the free energies of tetragonal and rhombohedral phases cross in this phase transition region (similar to the case shown in Fig. 3.6), makes polarization switching easy even with rather small electric fields. And, second is a facilitated movement of domain walls in the morphotropic phase boundary region [15]. Since the discovery of a monoclinic phase in PZT between the tetragonal and rhombohedral phase [16], the first argument has been reinforced by the fact that 24 new monoclinic possible domain orientations are coming into play.

However, there are several arguments against this explanation of the origin of the anomalous material properties in the vicinity of the morphotropic phase boundary, and these arguments indicate the intrinsic nature of the anomalous properties at the morphotropic phase boundary. The important arguments are as follows:

- The development of the peaks in dielectric and piezoelectric properties starts already quite far away from the morphotropic phase boundary in the tetragonal or rhombohedral phase. For example, in Fig. 3.12 the phase transition tetragonal–rhombohedral occurs at $x = 0.54$, whereas the increase in the permittivity, compliance (decrease in N_p^E), and electromechanical coupling factor (not shown in Fig. 3.12) can already be seen at $x < 0.5$ and $x > 0.6$ in the pure tetragonal phase and in the pure rhombohedral phase, respectively. As shown below, this behavior can be interpreted as an intrinsic effect [17].

- The characteristic behavior of the temperature coefficients of permittivity $TC\varepsilon$ and of frequency constant TCF can also be interpreted as an intrinsic effect [18].
- The peak in permittivity and the characteristic behavior of the temperature coefficient of unpoled ceramics in the vicinity of the morphotropic phase boundary are similar to those shown for poled ceramics (Fig. 3.12) [18]. Obviously, facilitated switching of domains during poling cannot play any role. Furthermore, these anomalies are also observed in heavily acceptor-doped ceramics (see Chap. 4), where domain walls are strongly pinned and, therefore, hardly influence the material properties.

In any case, both arguments, higher mobility of domain walls on one side or increase of intrinsic parameters on the other, refer to the energetic neighbourhood of the ferroelectric phases involved. These arguments are not mutually exclusive, but should be regarded as complementary. Anyway, a better insight into the prevalent mechanisms is crucial, as the interaction of these phases is used for achieving the highest energy conversion.

In search for optimized piezoelectric materials, Yamashita and coworkers have found a general tendency towards higher coupling with decreasing B-site ionic mass (cf. Chap. 9). Such a change favors larger elongations in the corresponding soft mode, provided the restoring forces do not change. Thus for generalizing this rule, the energy levels of the involved phases have to be taken into account, which include every substitution that changes the distance to the Curie temperature.

Finally, we must have an even closer look at the influence of the polarization process itself on the properties near the MPB. In the following subsections, we shall show that the anomalous material properties in the vicinity of the MPB are mainly due to the large dielectric anisotropy ($\varepsilon_{11}{}^{T} >> \varepsilon_{33}{}^{T}$)[16] within the crystallites, which itself is a consequence of the fact that the phase transition ferroelectric–paraelectric is only weakly of first order, i.e., it is nearly continuous. Moreover, it is important that the large dielectric anisotropy primarily causes very large shear constants d_{15} and $s_{44}{}^{E}$ within the crystallites. Thus, the high effective dielectric and piezoelectric properties observed at the MPB of ceramics are due to the contribution of those crystallites with a P_s direction that considerably deviates from the 3-axis (poling direction) of the ceramic sample; crystallites with a P_s direction of 57° to the 3-axis of the sample contribute optimally.

3.2.2 Single-Crystal Properties of PZT

We have already seen in Fig. 3.2 that in BT single crystals the permittivity components perpendicular to the spontaneous polarization approach a strong maximum when going from the tetragonal to the orthorhombic phase and a

[16] In this section we generally consider principal axis components i.e. the primes to characterize principal axis components have been omitted for simplicity.

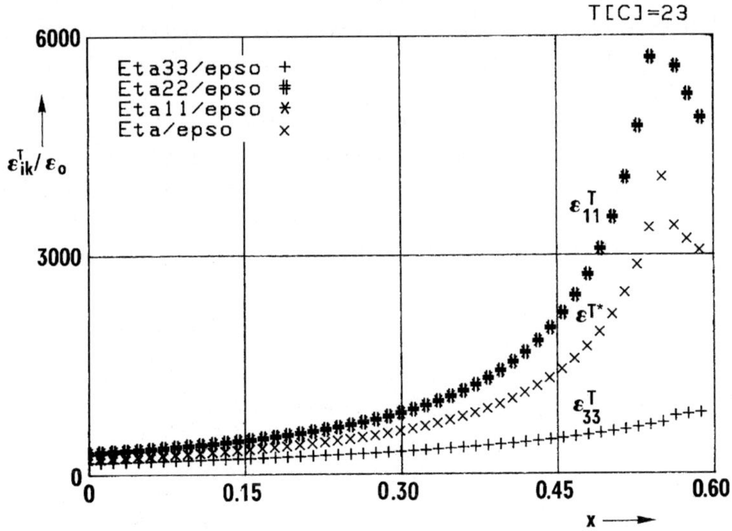

Fig. 3.13. Single-crystal permittivities $\varepsilon_{11}{}^T = \varepsilon_{22}{}^T$ and $\varepsilon_{33}{}^T$ and effective permittivity ε^{T*} of unpoled ceramics of $Pb(Zr_xTi_{1-x})O_3$

similar peak seems to occur when going from the orthorhombic to the rhombohedral phase. On the other hand, the permittivity component $\varepsilon^T{}_{33}$ parallel to the spontaneous polarization shows almost no change at the phase transitions and its value is very much smaller than $\varepsilon^T{}_{11} = \varepsilon^T{}_{22}$. The permittivity maxima in Fig. 3.3a would be even more pronounced if instead of the permittivity components, which refer to the axis system of the cubic prototype phase, the principal axis components were shown.

A similar strong dielectric anisotropy is observed in the vicinity of the MPB of PZT, i.e., a pronounced permittivity peak for $\varepsilon^T{}_{11} = \varepsilon^T{}_{22}$ and only a very flat curve for $\varepsilon^T{}_{33}$ (Fig. 3.13). As a result of the strong maximum of $\varepsilon^T{}_{11} = \varepsilon^T{}_{22}$, one obtains together with this maximum also a maximum in the effective permittivity ε^{T*} of unpoled ceramics and similarly (not shown in Fig. 3.13) also in the effective permittivities of poled ceramics.

This important property of the MPB has been well known for a long time [19]; however, in recent years it has been rediscovered and has initiated intensive discussions [20]. The large dielectric permittivity perpendicular to the spontaneous polarization facilitates polarization rotation by electric fields applied along nonpolar axes [21], i.e., the effect of strong polarization rotation often discussed in relaxor ferroelectrics such as $PbTiO_3$–$Pb(Mg_{1/3}Nb_{2/3})$ is nothing but the dielectric anisotropy discussed here.

The strong dielectric anisotropy, as well as the permittivity maxima at the morphotropic phase boundary, is not only of theoretical interest but are also very important for the tailoring of application-adapted piezoelectric ceramics. The strong dielectric anisotropy at the MPB can be explained with the aid of the results already obtained for BT in Sect. 3.1.

All ferroelectric phase transitions are induced by an instability of the crystal lattice that results in a soft lattice mode, and, analogous to Curie–Weiss law, a strong increase of permittivity. During the phase transition, the crystal lattice stabilizes because of the development of the spontaneous polarization P_s and therefore the permittivity drops down, particularly in the direction of P_s (*cf.* Fig. 3.3a). A new instability can occur perpendicular to P_s, which results in a strong increase of permittivity perpendicular to P_s. This instability induces a transition leading away from the first ferroelectric phase towards a second ferroelectric phase with an additional polarization component perpendicular to the spontaneous polarization of the first ferroelectric phase. The strong dielectric anisotropy in the vicinity of a MPB can be similarly understood. Here, the phase transitions are not driven by temperature but by composition. Both these parameters characterize equivalently the structural state of the ferroelectric mixed crystal. Already, far away from the MPB the transition from the tetragonal to the rhombohedral phase or vice versa can be identified by a lattice instability accompanied by an increase of the permittivity component perpendicular to P_s or an increased sensitivity to polarization rotation.

An explanation of the strong dielectric anisotropy at the MPB can be obtained more quantitatively with the aid of thermodynamic potentials [20], e.g., the elastic Gibbs potential G_1 (3.15a). We directly use (3.17–3.28), which have been derived from G_1 within Devonshire's theory.

First we consider permittivity $\varepsilon^T{}_{33}$ parallel to P_s. At the Curie temperature, P_s jumps from 0 to $P^2{}_s(T_\mathrm{C}) = -3a/4b$[17] and then increases only slightly with decreasing temperature, and in good approximation $P^2{}_s(T_0) = a/b \approx P^2{}_s(T_\mathrm{room})$. Thus from (3.25) we have $\varepsilon^T{}_{33} \approx 1/2bP_s^4$. Because b is not (or only weakly) dependent on the zirconium concentration x, and because P_s only slightly decreases from $x = 0$ to the phase boundary by a factor of 0.8[18] [22,32] to 0.7 [17], it follows that $\varepsilon^T{}_{33}$ increases within this interval only by a factor of 2.4 to 4, as can also be seen from Fig. 3.13.

Let us now consider the permittivities $\varepsilon^T{}_{11} = \varepsilon^T{}_{22}$ perpendicular to P_s. Here our arguments are based on the experimental fact that the reciprocal permittivities of unpoled ceramic samples decrease linearly on both sides of the MPB when x approaches the phase boundary value $x = x_\mathrm{MPB}$ from $x < x_\mathrm{MPB}$ or $x > x_\mathrm{MPB}$ [7]. This behavior is exactly described by (3.26) and (3.28) if we put

$$\xi_{11} - \xi_{12} = \xi_0 \, (x - x_0) \, , \tag{3.34}$$

where ξ_0 is a positive constant, and the concentration parameter x_0 has to be chosen in the vicinity of the MPB, x_MPB. We see from (3.26) and (3.28) that for $x = x_0$ the rhombohedral phase does exist, whereas the tetragonal

---

[17] In case of a first order phase transition (cf. (3.17) and (3.22)).

[18] Calculated from the lattice parameter ratio [22] and the electrostrictive coefficients [32].

phase does not[19,20]. For $x > x_0$ the permittivity $\varepsilon^T{}_{11} = \varepsilon^T{}_{22}$ in the rhombohedral phase decreases. For $x < x_0$, it increases strongly until the phase boundary ($x = x_{\mathrm{MPB}}$) is reached (cf. Fig. 3.13), just before approaching infinity. If x is further decreased ($x < \mathrm{x}_{\mathrm{MPB}}$), we are on the tetragonal side of the phase boundary, i.e. $(\xi_{12} - \xi_{11}) \geqslant |(\zeta_{112} - \zeta_{111})P^{s2}|$, $\varepsilon^T{}_{11} = \varepsilon^T{}_{22}$ is just below infinity (cf. Fig. 3.13.) and decreases with a further decrease of x. The composition dependence of the coefficients ξ_{11} and ξ_{12} can easily be chosen in such a way that (3.34) is fulfilled, and in addition, in the vicinity of the MPB the coefficients $a_t = \xi_{11}$ or $a_r = (\xi_{11} + 2\xi_{12})/3$ (cf. (3.19) and (3.21)) are only slightly negative, i.e., the phase transition to the cubic phase in the vicinity of the MPB is only weak and of first order, i.e., $T_0 \approx T_{\mathrm{C}}$ (cf. (3.23)). The simplest choice of ξ_{11} and ξ_{12} that fulfils these requirements is $\xi_{11} = -\xi_{12} = \xi_0(x - x_0)/2$. The tendency that the phase transition to the cubic phase of PZT becomes of second order in the vicinity of the MPB has been found experimentally from the temperature dependence of lattice parameters [23] and from Raman spectroscopy [24]. Obviously, the fact that the phase transition to the cubic phase is only weak of first order is the true physical origin for the strong dielectric anisotropy in the vicinity of the MPB. This similar to the dielectric anisotropy in BT at the transition from the tetragonal to the orthorhombic phase (cf. Fig. 3.3b).

The dielectric anisotropy in the vicinity of the morphotropic phase boundary can be determined also experimentally by dielectric measurements [17,25]. This is due to the fact that the single-crystal permittivity ε_{33}^T can be determined from the Curie–Weiss constant C, the Curie–Weiss temperature T_0, and the Curie temperature T_{C}. All these parameters can be obtained merely by dielectric measurements in the cubic phase of the material [25]. Therefore, no domain wall contributions are included in these results. If ε_{33}^T vs. composition has been determined, one can calculate ε_{11}^T from the effective permittivities $\varepsilon^{T*}(x)$ using (3.35).

$$\varepsilon^T{*} = \tfrac{1}{4}\left(\varepsilon_{11}^T + \sqrt{\varepsilon_{11}^{T2} + 8\varepsilon_{11}^T\varepsilon_{33}^T}\right). \tag{3.35}$$

The effective permittivities $\varepsilon^{T*}(x)$ are obtained from the permittivities (vs. x) measured on unpoled ceramics after the measured values have been corrected for the influence of domain walls and porosity as described in [27, 28, and 29].

Results obtained in this way are shown in Fig. 3.14 [17][21]. The dielectric anisotropy is very pronounced, similar as in Fig. 3.13. In addition, it can be

[19] The existence range of the different phases is identical to the range where the principal axis permittivities $\varepsilon^T{}_{11} = \varepsilon^T{}_{22} > 0$.

[20] For PZT holds $\zeta_{123} < \zeta_{112} < \zeta_{111}$, with $\zeta_{111} > 0$. A good description of $BaTiO_3$ is already obtained in the case that only $\zeta_{111} \neq 0$.

[21] In order to avoid an influence of conductivity effects on the dielectric measurements in the cubic phase of the investigated ceramics, their Curie temperatures have been sufficient lowered by modifying them with Sr and Ba.

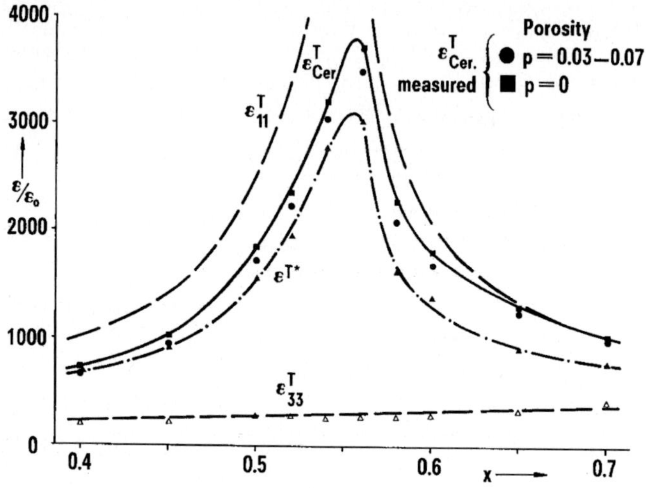

Fig. 3.14. Permittivities of $(Pb_{0.83}Ba_{0.05}Sr_{0.12})(Zr_xTi_{0.98-x}Nb_{0.02})O_3$. Single crystal permittivities $\varepsilon_{11}{}^T = \varepsilon_{22}{}^T$ and $\varepsilon_{33}{}^T$, effective permttivity ε^{T*} of unpoled ceramics, compared to measured permittivity $\varepsilon^T{}_{\mathrm{Cer}}$ of unpoled ceramics

seen that in the investigated ceramic system the domain wall contribution to permittivity is higher on the rhombehedral side of the MPB than on the tetragonal side (see Chap. 3.3). This agrees with the experimental result that the dielectric losses on the rhombohedral side of the MPB are about twice as large as those on the tetragonal side.

The strong anisotropy of single-crystal permittivities determines also the piezoelectric and elastic properties of single crystals and hence those of ceramics, thin films, and composites.

From the result obtained for the single-crystal permittivities, the principal axis components of the piezoelectric constants can be calculated using (3.32) and (3.33) with the aid of proper electrostrictive coefficients. Usually, measurement techniques to determine electrostrictive coefficients require oriented single crystals. However, a new, highly sophisticated method has been developed to measure electrostrictive tensor components of ferroelectric ceramics above their Curie temperature, based on X-ray diffraction using a position-sensitive detector in combination with a time-resolved registration technique [30].

Apparently, the maximum of $\varepsilon_{11}^T = \varepsilon_{22}^T$ in the vicinity of the MPB causes the piezoelectric shear modulus $d_{15} = d_{24}$ to reach a pronounced maximum at the same time. On the other hand, the components $d_{31} = d_{32}$ and d_{33} show only a very flat behavior at the phase boundary. For example, Fig. 3.15 shows the piezoelectric constants of single-domain and polycrystalline PZTs, which have been calculated with the aid of Devonshire's theory [17] using electrostrictive coefficients from measurements on PZT ceramics [30].

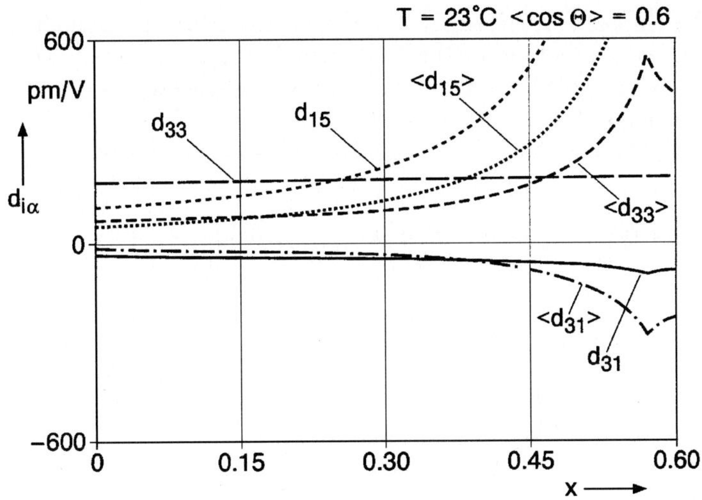

Fig. 3.15. Piezoelectric constants of $Pb(Zr_xTi_{1-x})O_3$. Single-crystal constants d_{31}, d_{33}, d_{15} and mean values $<d_{31}>$, $<d_{33}>$, $<d_{15}>$

Quite similarly, one finds that the maximum of $\varepsilon_{11}^T = \varepsilon_{22}^T$ in the vicinity of the MPB causes the shear compliance modulus $s_{44}^E = s_{55}^E$ to reach a pronounced maximum, whereas the nonshear compliance moduli again stay quite flat.

The calculation of the temperature coefficients of the dielectric, elastic, and piezoelectric single-crystal properties such as $TC\varepsilon^T{}_{11} = (1/\varepsilon^T{}_{11})(d\varepsilon^T{}_{11}/dT)$ is also straightforward now with the aid of Devonshire's theory, i.e., with the equations given in Sects. 3.1.4.1 and 3.1.4.2 (see also Chap. 7).

3.2.3 Properties of PZT Ceramics

The properties of ferroelectric ceramics, the so-called effective material properties can be calculated quite accurately from the properties of single grains or crystallites, with the aid of the "effective medium approximation" (see Chap. 23). This method, however, involves quite complex mathematical expressions. Therefore, we are limiting this discussion to the first and most essential term of this approximation, i.e., the averaging over the various orientations of the domains in the crystallites. In the case of piezoelectric constants, we have, for example:

$$d*_{i\alpha} = \langle d_{i\alpha} \rangle + \cdots . \tag{3.36}$$

Before using this approximation to calculate the intrinsic piezoelectric constants of ceramics, we use it to calculate the remnant polarization and remnant deformation of ceramics. The remnant polarization $P_3{}^r$ in the ceramic is calculated from the spontaneous polarization of the domains as:

$$P_3^r \approx \langle P^s \rangle = P^s \langle \cos\theta \rangle , \tag{3.37}$$

where θ is the angle between the 3-axis of a crystallite or a domain (direction of P_s), and the direction of the field that was applied to pole the ceramic, usually the 3-axis of the ceramic sample or the piezoelectric transducer under consideration.

The remnant strain is found from the spontaneous strain within the domains:

$$S_1^r = S_2^r \approx \langle S_1^s \rangle = S_1^s + \tfrac{1}{2}\left(S_3^s - S_1^s\right)\left(1 - \langle \cos^2\theta \rangle\right), \tag{3.38a}$$

$$S_3^r \approx \langle S_3^s \rangle = S_1^s + \left(S_3^s - S_1^s\right)\langle \cos^2\theta \rangle . \tag{3.38b}$$

For unpoled ceramics we have $\langle \cos^2\theta \rangle = 1/3$ and therefore:

$$S_1^r = S_2^r = S_3^r = \tfrac{1}{3}\left(S_3^s + 2S_1^s\right). \qquad (3.39)^{22}$$

The change in the remnant strain caused by poling is:

$$\Delta S_1^r = \Delta S_2^r = S_{1poled}^r - S_{1unpoled}^r = -\tfrac{1}{2}\left(S_3^s - S_1^s\right)\left(\langle \cos^2\theta \rangle - \tfrac{1}{3}\right), \tag{3.40a}$$

$$\Delta S_3^r = S_{3poled}^r - S_{3unpoled}^r = \left(S_3^s - S_1^s\right)\left(\langle \cos^2\theta \rangle - \tfrac{1}{3}\right). \tag{3.40b}$$

The difference in the components of spontaneous strain can be directly expressed by the tetragonal and the rhombohedral distortion of the crystal lattice, respectively:

$$\text{tetragonal phase: } S_3^s - S_1^s = \frac{c}{a} - 1 \tag{3.41a}$$

$$\text{rhombohedral phase: } S_3^s - S_1^s = \tfrac{3}{2}\left(\pi/2 - \alpha\right). \qquad (3.41\text{b})^{23}$$

Both the tetragonal crystal distortion c/a and the rhombohedral distortion expressed by the angle α of the rhombohedral cell can be measured directly by means of X-ray diffraction. Often, the rhombohedral distortion is expressed by the deviation of the angle α from $\pi/2$:

$$\delta = \left(\pi/2 - \alpha\right). \tag{3.42}$$

From (3.40b) and (3.41b) one can see that in the hypothetical case of infinite poling ($\theta = 0$) the remnant strain in a rhombohedral ceramic would be equivalent to δ. Thus sometimes δ is called the *homogeneity parameter* (cf. Chap. 4) and is generally used to characterize the spontaneous cell deformation along the polar axis. By comparing (3.41a) and (3.41b) one can see that in the tetragonal phase δ is then given by:

$$\delta = \tfrac{2}{3}\left(c/a - 1\right). \tag{3.43}$$

[22] This deformation is zero in case of a phase transition without change of volume.
[23] If the strain components are not referred to principal axis but to the cubic prototype phase's axis we have instead $S_4^s = \pi/2 - \alpha$.

3.2.4 Intrinsic Piezoelectric Constants of PZT Ceramics

Now we use the approximation of (3.36) to calculate the intrinsic piezoelectric constants of ceramics. By averaging the orientations, we have found in accordance with [31] the following:

$$\begin{aligned}
2d_{31}^* &\approx 2\langle d_{31}\rangle = (d_{31}+d_{33}-d_{15})\langle\cos\theta\rangle + (d_{31}-d_{33}+d_{15})\left\langle\cos^3\theta\right\rangle \\
d_{33}^* &\approx \langle d_{33}\rangle = (d_{31}+d_{15})\langle\cos\theta\rangle + (d_{33}-d_{31}-d_{15})\left\langle\cos^3\theta\right\rangle \\
d_{15}^* &\approx \langle d_{15}\rangle = (d_{33}-d_{31})\langle\cos\theta\rangle + (d_{15}-d_{33}+d_{31})\left\langle\cos^3\theta\right\rangle
\end{aligned} \tag{3.44}$$

where θ is again the angle between the 3-axis of a crystallite or a domain (direction of P_s), and the 3-axis of the ceramic sample or the piezoelectric transducer under consideration.

The effective piezoelectric constants of ceramics can be easily discussed by rewriting Eqs. (3.44) in the following form:

$$\begin{aligned}
4d_{31}^* &\approx 4\langle d_{31}\rangle = \langle\cos\theta\rangle\left[(3d_{31}+d_{33}-d_{15})+\kappa\,(d_{31}-d_{33}+d_{15})\right], \\
2d_{33}^* &\approx 2\langle d_{33}\rangle = \langle\cos\theta\rangle\left[(d_{33}+d_{31}+d_{15})+\kappa\,(d_{33}-d_{31}-d_{15})\right], \\
2d_{15}^* &\approx 2\langle d_{15}\rangle = \langle\cos\theta\rangle\left[(d_{15}+d_{33}-d_{31})+\kappa\,(d_{15}-d_{33}+d_{31})\right],
\end{aligned} \tag{3.45}$$

with $\kappa = 2\frac{\langle\cos^3\theta\rangle}{\langle\cos\theta\rangle} - 1$.

Figure 3.16 shows the parameter κ as a function of the mean degree of orientation of the spontaneous polarization of all domains relative to the external poling field. In general, κ decreases with increasing degree of orientation from 1 to 0 (at $<\cos\theta> = 0.5$), and then increases again. The exact behavior of κ depends on which distribution function for the orientation of polarization was used to calculate the average, or which domain processes were assumed to be applicable. In the simplest case, the averaging can be performed if one assumes that, starting at a statistical orientation of polarization directions in the unpoled ceramic, with increasing degree of poling the polarization directions continue to be statistically distributed, but in a confined angle range θ_{p} (Fig. 3.17(b)) where θ_{p} is a function of poling degree. This leads to a behavior of κ as described by the solid line in Fig. 3.16. It is more realistic, however, to assume that with increasing poling field the distribution of polarization directions transforms from the behavior shown in Fig. 3.17(a) to that shown in Fig. 3.17(c) because 180° and 90°[24] domain walls (Fig. 3.17(c′)) with fitting orientations are shifted. In this case, $\theta_{2\mathrm{D}}$ increases to 90° and a further increase of poling field leads to Fig. 3.17(d) after 90° domain walls (Fig. 3.17(d′)) have been shifted. This behavior of κ is described by the dashed line in Fig. 3.16. Since the difference between the two described distributions of polarization

[24] We use here 90° walls synonymously for 71° and 109° walls in rhombohedral materials.

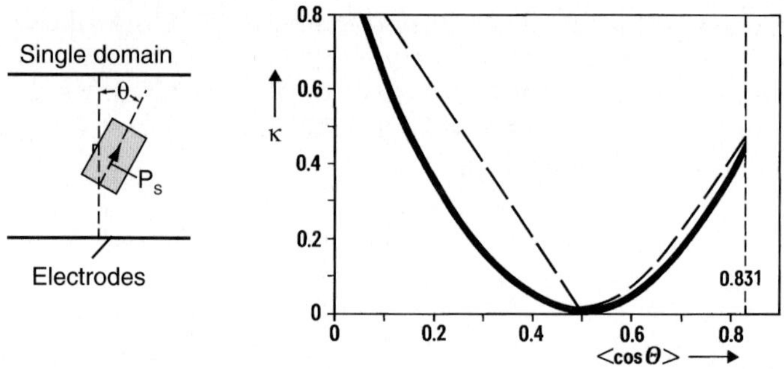

Fig. 3.16. $\kappa = 2(<\cos^3\theta>/<\cos\theta>) - 1$ as function of mean degree of orientation of spontaneous polarization $<\cos\theta>$ of all domains in relation to an external poling field. The two curves correspond to two different distributions of polarization orientations

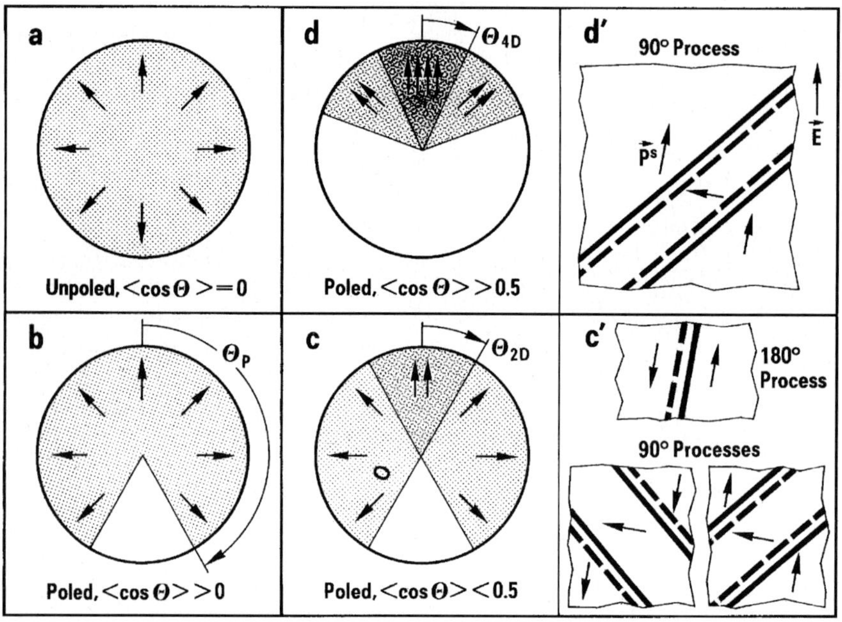

Fig. 3.17. Distributions of polarization orientations used for calculation of mean values: **(a)** Unpoled ceramics. **(b)** Simplest distribution for poled ceramics that leads to the solid curve shown in Fig. 3.16. **(c)** More realistic distribution that develops from the one shown in **(a)** by 180° and 90° domain wall shifts as shown in **(c′)**. **(d)** Distribution that is developed from the one shown in **(c)** by further 90° domain wall shifts as shown in **(d′)**. The distributions in **(c)** and **(d)** yield the dashed curve in Fig. 3.16

directions does not essentially influence the behavior of κ, the simpler first distribution has been used for the calculation of piezoelectric mean values in Fig. 3.15.

A review of (3.44) and (3.45) enables the understanding of the piezoelectric properties of PZT ceramics not only in the vicinity of the morphotropic phase boundary but also far away from it, e.g., that of $PbTiO_3$ ceramics. Furthermore, the dependence of these properties on the degree of domain orientation, i.e., on the degree of poling, can be seen.

Let us first consider the piezoelectric properties far away from the morphotropic phase boundary, e.g., those of $PbTiO_3$. Typical properties of modified $PbTiO_3$ are around $d_{33}/d_{31} \approx -5$ and $d_{33}/d_{15} \approx 2$. Therefore, with certain material modifications it can be achieved that the term $(3d_{31} + d_{33} - d_{15})$ in the first line of (3.45) tends to zero. As a result, at a certain degree of poling, such that $<\cos\theta> \approx 0.5$, $<d_{31}>$ will assume very small values. Because at the same time $<d_{33}>$ can have quite normal values, the ratio $|d_{33}/d_{31}|$ can achieve very high values ($\geqslant$20). This effect is known as *anisotropic piezoelectric effect* [32] and is found, e.g., in lead titanate ceramics modified with Ca [33] and in those modified with Sm [34, 35]. The large anisotropy of the piezoelectric effect implies that these ceramics possess an especially high hydrostatic piezoelectric effect $d_\mathrm{h} = d_{33} + 2d_{31}$. They are therefore very interesting for underwater transducers, e.g., for hydrophones [36].

Now we consider the piezoelectric properties in the vicinity of the MPB. As mentioned above, in the vicinity of the MPB the maximum of $\varepsilon_{11}^T = \varepsilon_{22}^T$ causes the piezoelectric shear modulus $d_{15} = d_{24}$ to reach a pronounced maximum as well. On the other hand, the components $d_{31} = d_{32}$ and d_{33} show only a flat behavior at the phase boundary (Fig. 3.15). From (3.45) it can be seen that for an adequate degree of poling, i.e., $\kappa \approx 0$ (Fig. 3.16), the large value of d_{15} drives the effective piezoelectric constants $d^*{}_{31}$ and $d^*{}_{33}$ of a polycrystalline material into a pronounced maximum, too. On the other hand, in those crystallites that are completely oriented parallel to the external field ($\theta = 0$, $\kappa = 1$), e.g., (001)-oriented tetragonal crystallites or (111)-oriented rhombohedral crystallites (Fig. 3.18), the large d_{15} values cancel in (3.44) and (3.45) and do not contribute to the effective constants $d^*{}_{31}$ and $d^*{}_{33}$. Obviously, the contribution of those crystallites with a P_s direction that considerably deviates from the 3-axis (poling direction) of the ceramic sample contribute strongly; particularly crystallites with a P_s direction of 57° to the 3-axis of the sample contribute optimally. Furthermore, it is quite interesting (Fig. 3.18) that just those crystallites that do not contribute much to the intrinsic piezoelectric effect, i.e., (001)-oriented tetragonal and (111)-oriented rhombohedral crystallites, optimally contribute to the extrinsic piezoelectric effect via 90° domain wall processes. On the other hand, those crystallites that optimally contribute to the intrinsic piezoelectric effect, i.e., (111)-oriented tetragonal and (001)-oriented rhombohedral crystallites, cannot contribute to the extrinsic effect because the applied driving field will not favor any of the other possible domain states, and the domain structure remains stable.

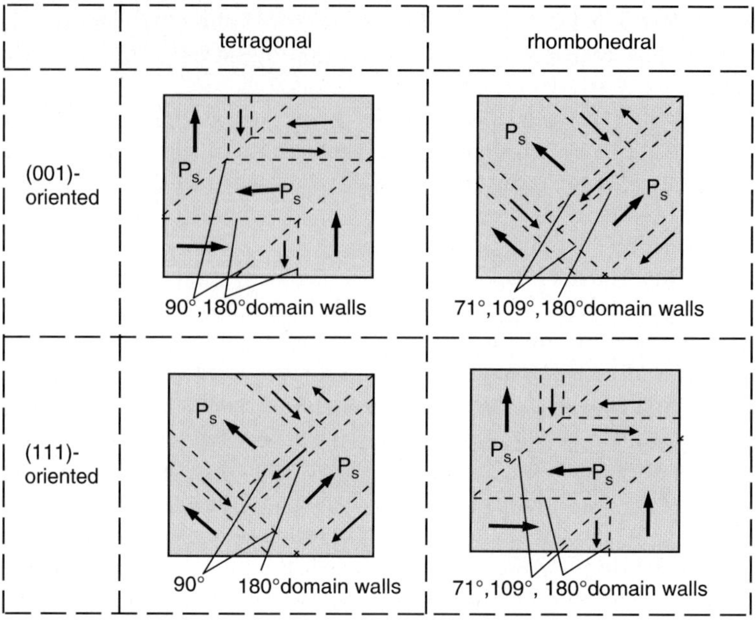

Fig. 3.18. Domains in tetragonal and rhombohedral crystallites with selected orientations

Finally, we would like to stress here that with a similar consideration the elastic behavior of PZT ceramics in the vicinity of the MPB can be understood. It can be shown that in polycrystalline PZT non-shear compliance components such as ${s_{11}^{E}}^{*}$ or ${s_{33}^{E}}^{*}$ go through a considerable maximum at the MPB driven by the large single-crystal shear value s_{44}^{E}: a fact which is well known and widely used for many piezoelectric applications. Furthermore, the temperature dependence of dielectric and elastic properties of PZT ceramics in the vicinity of the MPB can be understood similarly, a fact that is very important for the application of piezoelectric ceramics in mechanical frequency filters [37] (see also Chap. 7).

3.2.5 Intrinsic Piezoelectric Constants of Single Crystals and Textured Thin Films

Returning to the considerations of the last section, the exceptional dielectric, elastic, and piezoelectric properties of single crystals such as $Pb(Zn_{1/3}Nb_{2/3})O_3$ (PZN) and $Pb[(Zn_{1/3}Nb_{2/3})_xTi_{1-x}]O_3$ (PZN-PT) and ferroelectric thin films can be understood as well.

In 1997, Park und Shrout [38] found exceptionally large longitudinal piezoelectric coefficients (${d_{33}}^{*} > 2000$) and electromechanical coupling constants (${k_{33}}^{*} > 0.9$) in rhombohedral $Pb[(Zn_{1/3}Nb_{2/3})_xTi_{1-x}]O_3$ (PZN-PT) and $Pb[(Mg_{1/3}Nb_{2/3})_xTi_{1-x}]O_3$ (PMN-PT) in (001)-oriented single crystals

Table 3.2. Properties of complex perovskite single crystals [ ]

Compound	Structure	Orientation	$\varepsilon_{33}{}^{T}/\varepsilon_0$	$s_{33}{}^{E}$ $(10^{-12}m^2/N)$	d_{33} (pm/V)
$Pb(Zn_{1/3}Nb_{2/3})O_3$	Rhombohedral	111	900	7.4	83
		001	3,600	48	1,100
$Pb[Ti_{0.08}(Zn_{1/3}Nb_{2/3})]O_3$	Rhombohedral	111	2,150	7.4	82
		001	4,200	130	2,200
$Pb[Ti_{0.095}(Zn_{1/3}Nb_{2/3})]O_3$	MPB	111	4,300	10.4	600
		001	1,600	77	1,600
$Pb[Ti_{0.12}(Zn_{1/3}Nb_{2/3})]O_3$	Tetragonal	001	900		500

which have been poled in (001) direction. On the other hand, the coefficients measured along the direction of the polar crystal axis were found to be up to one order of magnitude smaller [38]. These results are listed in Table 3.2. As can be seen from Fig. 3.18, 180° domain wall switching in a (001)-oriented crystal leads to a domain wall distribution where the spontaneous polarization of each domain points in the direction of one of the four possible cell diagonals ($\theta = 54.7°$), whereby all these four directions are energetically equivalent. From Fig. 3.16, we see that in this case, i.e. for $<\cos 54.7°> = \cos 54.7° = 1/\sqrt{3}$, the parameter $\kappa \approx 0$ holds and, thus, the huge piezomodulus d_{15} optimally contributes to $d_{33}{}^{*}$. On the other hand, in case of an applied field parallel to the spontaneous polarization ($\theta = 0°$), the parameter $\kappa \approx 1$ holds and d_{15} does not contribute to $d_{33}{}^{*}$.

Similarly, the piezoelectric properties of ferroelectric thin films can be understood. It has been shown that the intrinsic piezoelectric effect of films oriented along their polar axis must be smaller than that of randomly oriented films [39].

David et al. investigated rhombohedral PZT films with (001), (111), and randomly oriented crystallites [40]. In case of an applied field amplitude of $0.1\,\mathrm{V\mu m^{-1}}$, these films showed effective d_{33} values of 100, 77, and 63 pm V^{-1}, respectively. It should be mentioned that these values have not been corrected for the clamping caused by the substrate. With increasing field amplitude, the effective d_{33} value of the (111)-oriented film increased especially strongly, obviously because of domain wall motions which are possible in this orientation, as has been discussed in the last section.

Kügeler et al. [41] studied (111)-oriented $Pb(Zr_xTi_{1-x})O_3$ films for microelectromechanical system (MEMS) applications. A maximum of the effective piezoelectric coefficients d_{33} ($d_{33} \approx 100\,\mathrm{pm\ V^{-1}}$) has been found for the film composition $x = 0.45$ slightly on the tetragonal side of the MPB. For more tetragonal compositions, $x < 0.45$, d_{33} decreased very rapidly. On the other hand, on the rhombohedral side of the MPB, the effective piezoelectric module decreased only very slightly; even at $x = 0.6$, the d_{33} value was only less than 10% smaller than the maximum value at $x = 0.45$. This result fully agrees with the discussion of the last section. Near the MPB, the intrinsic effect of tetragonal films reached a maximum owing to the strong contribution of the shear effect to the effective d_{33}. Although the intrinsic effect was low at the rhombohedral side of the MPB, a strong piezoelectric response was found as a result of 90° (72°/108°) domain wall movements that could take place over a wide composition range because of the relatively high amplitude (700 kV m^{-1}) of the measuring field.

3.3 Extrinsic Piezoelectric Effect

As already emphasized in the last section, in ferroelectric piezoelectrics there exists an additional contribution to the piezoelectric response called the extrinsic piezoelectric effect. This extrinsic response mechanism results principally from non-180° (e.g., 90°, Fig. 3.11) domain wall movements driven by the application of an external electric field or a high enough mechanical stress. Thus investigation of the relationship between microstructure, domain configuration, and macroscopic properties must be a main topic in the research field of ferroelectric ceramics.

3.3.1 Ferroelectric Domains

One of the most evident aspects of the microstructure in ferroelectrics is the existence of domains. The domain boundary in a ferroelectric crystal is the layer separating domains polarized in different directions, as shown in Fig. 3.19. After a phase transition to a ferroelectric state takes place (in the absence of an external force), the crystal breaks up into separate regions called *domains*, in which the polarizations are uniform but the directions of the polarizations are different. The formation of domains is caused by a decrease of the free energy of the crystal; i.e., the free energy of a polydomain state is lower than that of a single-domain state. The boundaries separating domains are referred to as *domain walls.*

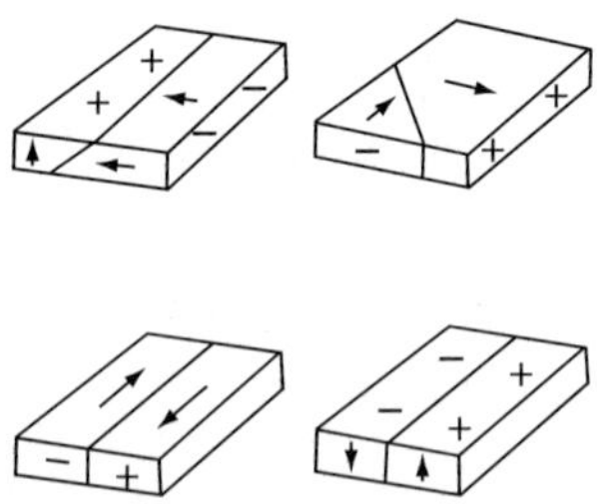

Fig. 3.19. Basic structures of domain configurations of $BaTiO_3$ in the tetragonal phase

Ferroelectric domains are one of the fundamental properties of ferroelectrics and related materials. Macroscopically, the structures and dynamics of ferroelectric domains determine, or at least influence, the electrical, optical, acoustic, pyroelectric, and mechanical properties of ferroelectric materials. Thus the study of domain structures is important, both from the technological and the scientific points of view. For example, the dielectric properties of high-permittivity multilayer ceramic capacitors are remarkably influenced by domain wall contributions; the performance of many actuators depends strongly on the domain structure variation under an electric field, etc.

3.3.2 Domain Structure in Ferroelectric Ceramics

As mentioned before, the specific domain structure in multiaxial ferroelectric crystals is a consequence of the mechanical, electrical, and thermal conditions at the phase transition. When multiaxial ferroelectric crystals undergo phase transitions, 180° domains (the electrical twins) and non-180° domains (the mechanical twins) are formed (see Fig. 3.19). The real structure of the domains in a single crystal can be understood on the basis of the domain geometry relations by consideration of the symmetry changes that occur at the transition from the paraelectric to the ferroelectric phase [42].

However, the formation of domains in ceramics is different from that in single crystals because the domain structure in a ceramic grain forms under clamped conditions, whereas in a single crystal it forms free of stress. The domain twinning in ferroelectric ceramics depends on the grain size. A grain in a ceramic is clamped by its neighboring grains in all three dimensions. It can deform only by a cooperative motion of the adjacent grains. Maintaining the shape requires either high internal stresses or mechanical twinning, which at least preserves the gross shape of the grains.

Arlt et al. [43] have studied in detail the domain structure in ferroelectric ceramics as well as the influence of the domain structure on the properties of ferroelectric ceramics. It appears that a simple lamellar structure is favored in fine-grained ceramics, which allows a two dimensional stress relief. In coarse-grained ceramics, a banded lamellar structure will prevail in order to take away homogeneous stress in three dimensions and to allow grain adjustment in the third dimension as well, as shown in Fig. 3.20. These different domain structures obviously affect the domain motion processes.

3.3.3 The Dynamics of Domain Wall Motions in Ceramics

If the domain walls are fixed and no other defects contribute to piezoelectricity in ceramics, then an averaging calculation would predict all the main intrinsic electromechanical properties when using appropriate values of single-domain crystals and a simple orientation distribution. This calculation [45] has to take into account the dielectric, piezoelectric, and elastic interactions

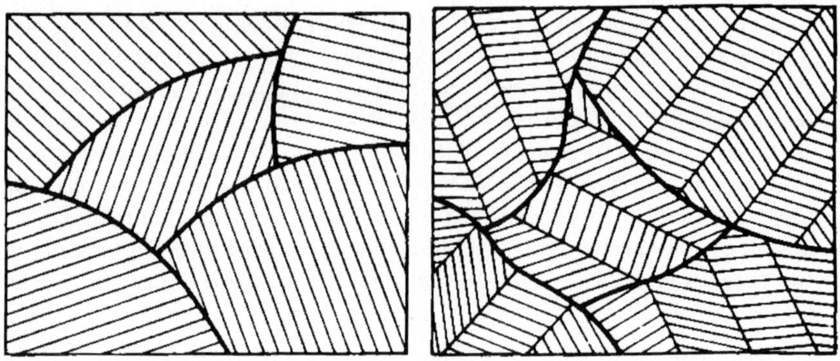

Fig. 3.20. Domain structures in polycrystalline materials: **(a)** simple lamellar twinning. **(b)** banded twin structure in ceramics [44]

between the domains, their shape and configuration, and their degree of orientation. Muratake [46] first introduced a method that takes into account some interaction effects by considering electric and elastic boundary conditions of the grain. Turik [47] improved the Muratake model by using a laminar grain consisting of a stack of 90° or 180° domains. For the laminar grain of BT, the effective piezoelectric, elastic, and dielectric coefficients were calculated considering the boundary conditions between the domains. The experimentally determined piezoelectric constants d_{31} and d_{33} of BT ceramics are, however, much larger than the calculated ones. This deviation between calculated and measured piezoelectric constants had led Turik to suppose additional "orientational" contributions that may be caused by domain wall motions.

Since the 1980s, Arlt et al. (e.g., [43] and [44]) have systematically studied the microstructure and dynamics of 90° domain wall motion in BT and PZT ceramics. They could explain the strong increase in the piezoelectric and dielectric coefficients by shifts and oscillations of 90°-type domain walls (90° in the tetragonal, 71° and 109° in the rhombohedral phase). They used Fousek's model [48] to evaluate the dielectric, piezoelectric, and elastic coefficients referring to a special distribution function and by averaging the domain contributions in every grain of the ceramic samples. A similar work was done by some Russian scientists (e.g., [49]). They considered some special distributions of domain structures in ceramics (the laminar twin structure) to assess the physical properties of ferroelectric ceramics.

All these models describe the average dynamic behaviour of the domain wall motion in the single crystal with respect to the different orientations of polarizations. All these averaging methods have drawbacks and are of limited value for the calculation of the volume part of contributions to piezoelectric constants of ceramics. The experimentally determined piezoelectric constants d_{31} and d_{33} are effectively larger than those calculated by the averaging methods. A similar disagreement was found for the dielectric constants, which in ceramics are much larger than expected from theoretical calculations.

Generally speaking, the electromechanical properties of ferroelectric ceramics are caused by two mechanisms: (1) the intrinsic piezoelectric effect connected with the deformation of each unit cell of the ferroelectric material by an electric field, and (2) the extrinsic piezoelectric effect based on domain wall motion. In PZT this motion refers to non-180° domain walls and the movement of the interphase boundaries between the tetragonal and rhombohedral phase regions. It is believed that the piezoelectric effect in polycrystalline ferroelectrics is caused not only by the ionic displacement in connection with the change of the polarization magnitude but also by the movements of domain walls and interphase boundaries.

Studies on materials such as BT and PZT [50] have shown that 60–70% of the piezoelectric coefficient values may originate from extrinsic contributions. In fact, the performance of many capacitors, transducers, and actuators is based on the control of the domain structure under the applied electric field.

Microstructural investigations (e.g. [51] and [52]) have also shown that poled ferroelectric ceramics contain a large number of non-180° domain boundaries which strongly affect their electromechanical behavior. In general, these domain structures give rise to complex linear and nonlinear behavior, which is very sensitive to the quality of the sample and its defect concentration as well as to the external boundary conditions.

3.3.4 A Physical Picture of Domain Wall Motions

Now let us consider a physical model in order to obtain an insight into the relationship between electromechanical properties and domain wall motions as well as into the movement of interphase boundaries in ferroelectric ceramics.

The movements of the domain walls are not independent, but are strongly interacting. The microscopic process for an individual domain wall motion is rather complicated and not yet well understood. However, in order to describe the macroscopic effects resulting from the collective motion of domain walls, the microscopic details of each individual domain may be neglected.

Let us assume that the average displacement of a domain wall in a certain direction is quasi one-dimensional, i.e., the wall moves as a plane without bending. Figure 3.21 shows the variation of the wall energy as a function of the position generated by the interaction of the domain wall with the lattice, point defects, dislocations, and neighboring walls in a ceramic sample. The height of the potential barrier $(\phi_1 + \phi_2)$ is the minimum energy required for the wall to relocate from a depoling state to a poled state. ϕ_w is the height of the potential barrier for a non-180° domain wall to cross one lattice distance.

In a depolarized ceramic sample, all domain walls are randomly oriented and the distribution of the domain walls possesses the symmetry $\infty\infty m$. Obviously, the domain walls are located at the minima of the potential energy $(z = 0)$. When an external a.c. field is applied, the domain walls oscillate.

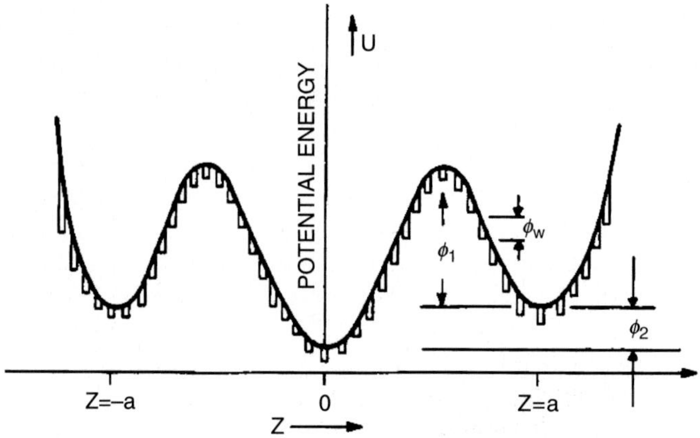

Fig. 3.21. Energy levels in the domain wall potential of ferroelectric ceramics

However, as long as the magnitude of the domain wall vibration is not very large, the macroscopic potential energy for the domain wall motions is symmetric with respect to the z-axis, i.e., $U(\Delta z) = U(-\Delta z)$.

The situation is different for poled ceramics. During the poling process, the domain walls overcome neighboring energy barriers and move into new equilibrium positions. As shown in Fig. 3.21, the new positions are metastable states ($z = a$ or $z = -a$). After the field is removed, the orientation distribution of the domain walls becomes conical in the poled samples. The potential for the domain wall motion is no longer symmetric in the poled state, i.e., $U(a - \Delta z) \neq U(a + \Delta z)$. It also has been shown [53] experimentally that the domain wall motion in poled ceramics is highly nonlinear. Therefore, the potential energy for the domain wall motion may be expanded as a polynomial function of the domain wall displacement Δl [54]:

$$U = U_0 + \frac{c_1}{2}\Delta l^2 + \frac{c_2}{3}\Delta l^3 + \frac{c_3}{4}\Delta l^4 + \text{higher-order terms}, \tag{3.46}$$

where U_0 is the residual energy of a domain boundary, which is assumed to be independent of the domain wall motion. The presence of the third-power term describes the asymmetric feature of the domain wall motion in ferroelectric ceramics. The odd terms will be zero if the domain wall vibrations take place around a potential energy minimum of the domain wall. Therefore, the differential equation of the forced vibration of a non-180° domain wall in ferroelectric ceramics may be expressed as follows (e.g., [48] and [54]):

$$Am\Delta\ddot{l} + Ab\Delta\dot{l} + \frac{\partial U}{\partial \Delta l} = -\left(\frac{\partial W_E}{\partial \Delta l} + \frac{\partial W_M}{\partial \Delta l}\right), \tag{3.47}$$

where m represents the effective mass of the domain wall. A is the area of the domain wall, b is the damping constant, and the dots on top of Δl represent the derivates with respect to time. The third term is the restoring force.

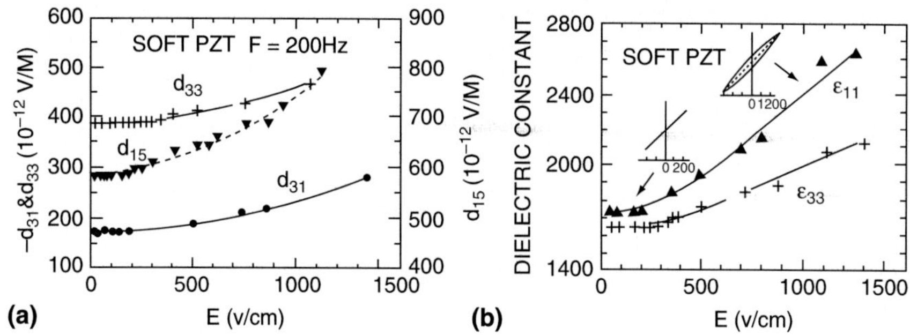

Fig. 3.22. (a) Depencence of the piezoelectric coefficients d_{33}, d_{15}, and d_{31} on the a.c. electric field amplitude. **(b)** dependence of the permittivities $\varepsilon_{11}{}^T$ and $\varepsilon_{33}{}^T$ on the a.c. electric field amplitude. The two insertions show the P–E loops at low and medium-strong electric fields, respectively

W_E and W_M are the energies of the average electric and elastic dipoles induced by the electric field and the mechanical stress, respectively. In general, the electric and elastic driving frequencies are different.

Equations (3.46) and (3.47) show that the nonlinearities in ferroelectric ceramics can be evaluated from the non-180° domain wall motion. Furthermore, the domain wall motion can be influenced by pinning effects.

Figure 3.22 shows the a.c. electric field dependence of the piezoelectric and dielectric coefficients of soft PZT ceramics [54].

3.3.5 The Rayleigh Law, a Phenomenological Model

The Rayleigh law is a very useful tool to describe the displacement of domain walls (or other interfaces) in a medium with randomly distributed pinning centers (e.g., [55] and [20]). The nonlinearity of the piezoelectric response (field dependence of the piezoelectric coefficient) and accompanying hysteresis can be described by the following set of equations:

$$d\left(E_0\right) = d_i + \alpha E_0, \tag{3.48}$$

$$S\left(E\right) = \left(d_i + \alpha E_0\right) E \pm \frac{\alpha}{2}\left({E_0}^2 - E^2\right), \tag{3.49}$$

where d_i indicates the reversible, field-independent component of the piezoelectric coefficient d, α is the nonlinear Rayleigh coefficient, $E = E_0 \sin \omega t$ is the applied alternating field strength with frequency ω, and S is the mechanical deformation. The field-dependent component of the material response (terms containing α) is considered to be a consequence of the pinning of domain walls on randomly distributed pinning centers and is responsible for the medium-field hysteresis described by (3.49). Equations (3.48) and (3.49) describe the simplest case of nonlinearity and the associated hysteresis; they can be extended easily to include higher-order terms [57]. The most important aspect of the Rayleigh-like behavior is the fact that hysteresis and nonlinearity

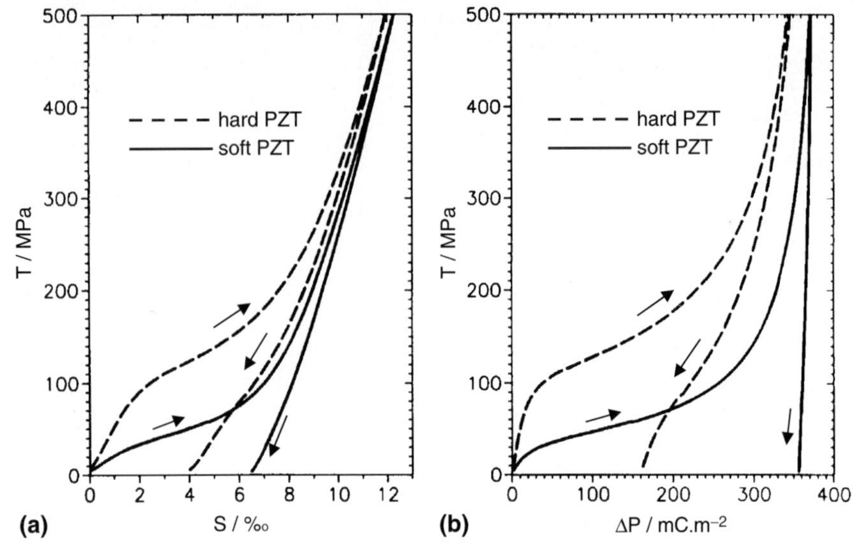

Fig. 3.23. **(a)** strain and **(b)** depolarization of hard and soft PZT ceramics under uniaxial stress [57]

are essentially linked, as can be seen from (3.48) and (3.49) and in Fig. 19.7 of Chap. 19; i.e., both hysteresis and nonlinearity are results of the same domain wall pinning processes. This link between hysteresis and nonlinearity is characteristic for soft PZT ceramics in contrast to the behaviour of hard PZT ceramics in which the response hysteresis cannot be predicted from the nonlinearity of the piezoelectric coefficient [20].

Also, a large mechanical stress applied parallel to the poling direction induces a significant domain wall motion. With increasing load, the ferroelastic domain switching causes an additional strain and depolarization ΔP [56]. Figure 3.23 shows the strain and depolarization of hard and soft PZTs under uniaxial stress.

3.3.6 Fatigue and Microstructure

Fatigue in ferroelectric ceramics is the gradual change of the material properties during cyclic loading. It is caused by microscopic material modifications of mechanical or electrical origin. Owing to the electromechanical coupling, both mechanisms can yield similar or even identical changes in the properties.

The two major effects arising during fatigue are the reduction of the domain switchability and the development of internal fields and offsets. Figure 3.24 displays the reduction in the height of ferroelectric hysteresis loops and the asymmetric strain response due to bipolar fatigue [57].

The defect size becomes an important element of the fatigue effect. Isolated point defects interact with the domain wall, but they usually represent only a

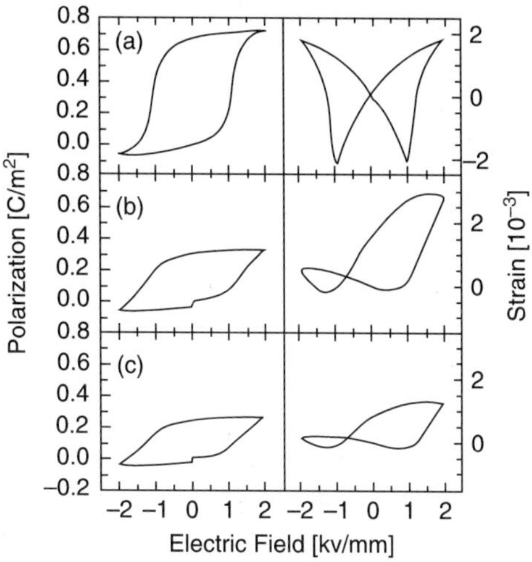

Fig. 3.24. Macroscopic polarization and strain in soft PZT measured at 10^{-3}HZ, after **(a)** 0, **(b)** $3 \cdot 10^6$, and **(c)** 10^8 bipolar electric fatigue cycles at 50 Hz, 2 kV mm^{-1} [57]

small barrier to domain wall motion. Their average effect is of a frictional type. Thus a viscosity for domain wall motion should be defined. In this regime, the domain wall motion is essentially smooth at varying velocities yielding slightly modified dielectric constants. Further increase of defect size alters this smooth motion into discrete jumps of the domain wall. Large changes in the dielectric constant as well as offsets arise. Finally, a further increase in defect size may be sufficient to block the polarization within a certain volume. No further switching takes place in that volume, the size of which has not yet been determined [58].

It is generally agreed that a certain degree of clustering of defects develops during fatigue. Whether such clusters consist of oxygen vacancies, or defect states at grain boundaries, at domain walls, at microcracks, or at electrode contacts, is still an open question [59].

3.3.7 Fracture Behaviour

PZT is one of the most commonly used materials for electromechanical transducers and actuators. In these applications, the material is subject to high electric fields and mechanical stress under which it may fracture. The limited performance of the components due to cracking still impedes their large-scale use in highest power applications. Therefore, the fracture behavior of the material needs to be well understood in order to improve their lifetime and reliability.

In a cracked PZT component, the high stresses in the vicinity of the crack tip cause ferroelastic domain switching within a finite zone ahead of the crack front. Once the crack penetrates the zone, toughening takes place in the region behind the crack tip, where compressive stresses close the crack and shield the crack tip from the external load.

A few attempts have been made to experimentally quantify the fracture behaviour of PZTs. Some of them have been devoted to the distribution of the shielding stresses in the wake of a growing crack, while others were focused on the measurement of the domain switching zone using experimental methods such as atomic force microscopy, interference contrast, and X-ray diffraction (e.g., [60]).

In polarized ferroelectric ferroelastic materials, the switching mechanism is accompanied by a considerable change of the electric potential on the specimen surface. This electric potential can be imaged with the help of liquid crystals. Ferroelectrics exhibiting fairly large domains have been characterized on a local scale using this approach. This technique can be extended to visualize the ferroelastic domain switching zone in front of a crack tip in PZT [60]. Using this technique, it is further possible to track the evolution of the switching zone during crack propagation. For a better understanding of the mechanical behavior of ferroelectric ferroelastic materials, an analysis of the crack tip field within the switching zone is desirable. As soon as the domain switching takes place, the local poling condition and thereby the macroscopic piezoelectric coefficients are altered.

The relation between switching behavior and fracture toughness under applied electric and elastic fields in ferroelectric materials has not been fully explored.

3.3.8 Conclusions

The electromechanical properties in ferroelectric ceramics are caused by the intrinsic and extrinsic piezoelectric effect.

In the past years, considerable efforts were made to understand details of extrinsic (non-lattice) contributions to the piezoelectric effect in a given material. The extrinsic contributions are believed to originate mostly from the displacement of domain wall boundaries and interphases.

The understanding of extrinsic contributions to the response in ferroelectric ceramics is far from complete. Therefore, both systematic theoretical and experimental studies in this area are still desirable and of great significance, and of utmost importance for practical applications.

3.4 Conclusions for Materials Engineering

In contrast to the dominance of silicon in semiconductor microelectronics, no unique material exists in the field of piezoelectricity. Here we can discern at least four different classes of materials as listed in Table 3.1 of Sect. 3.2. From the outlines as given in Chaps. 2 and 3, up to now, the following rules can be deduced for the development and application of these different classes of piezoelectric materials:

1. Non-ferroelectric materials cannot be piezoelectric, unless they are single crystalline or at least produced by oriented, e.g., epitaxial, growth. They exhibit only small piezo-moduli and small permittivities, a typical example being quartz. The elastic properties resemble those of other inorganic materials. In single crystals, all material properties are stable over time and depend on crystal quality only. Owing to the existing anisotropy, normally orientations of minor temperature dependence can be found. Thus specific cuts are widely used as frequency- or time standards (cf. Chap. 2). In addition, we would like to mention langasite as a material with the same crystal structure as α-quartz, but applicable up to extreme temperatures, e.g., for sensors working up to 900°C. Epitaxial layers are by far cheaper than specific cuts of single crystals. Only polar substances should be used because normally they cannot be deposited as single crystals, but as oriented, texturized layers. Important examples are hexagonal ZnO and AlN. Both are used for integrated microwave resonators and filters in telecommunications as well as increasingly in biosensor arrays.
2. Higher piezoelectric coupling than in these polar substances cannot be found except in ferroelectric materials. In case the highest stability is required, normally single-crystalline slices or at least textured epitaxial layers are used. They must exhibit a Curie temperature that is high enough, or more precisely, no phase transition should occur in the neighbourhood, both with respect to a change of temperature and to a variation of composition. Important examples in the case of uniaxial ferroelectrics are $LiNbO_3$ and $LiTaO_3$, and in the case of polyaxial ferroelectrics $PbTiO_3$. Such materials exhibit a piezoelectric coupling that is about 10 times higher than that of quartz. They have medium permittivity and no anomalies of the elastic properties. In uniaxial ferroelectrics, 90° domain rotation cannot occur, but only single crystals or texturized layers can be used. Extrinsic effects, which could affect long-term stability, are excluded. By specific polarization means, e.g., by crystal growth within an electric field, even single-domain substrates can be produced. Thus, from the point of view of piezoelectricity, the highest long-term stability can be warranted, which is mandatory for frequency control and band filtering as well as for signal transmission or identification (cf. Chap. 14).
3. In case an even higher energy coupling is desired, one has not only to use ferroelectric substances; additionally we cannot avoid any more the neighbourhood of ferroelectric phase transitions; such a neighbourhood can enhance the electromechanical coupling by almost two orders of magnitude. Stability is no more a decisive criterion for technological applicability, it is only a part of the decisive figure of merit.

These concessions to highest stability bring about not only applications where highest coupling is desired; even more they open the way to the use of polycrystalline substances, which means to ceramic production procedures. These procedures are cheaper than growing single crystals or textured layers by orders of magnitude. Their technology has been well elaborated for the mass production of electrical devices such as resistors, capacitors, or magnetic cores. Their shape can be adapted exactly to the requirements of the user. The well-known ceramic thin-layer and thick-film technologies can be applied. Double layers that bend under applied voltage have been produced, stapling technologies can be used to achieve higher elongation, and so on.

By nature ceramics are isotropic and do not become piezoelectric without a specific polarization process, which means applying a high voltage at elevated temperature. On the one hand this causes some restrictions, but on the other offers also novel opportunities. The final polarization need not necessarily be homogeneous throughout the whole device. A polarization process in situ can be adapted to all final requirements.

We recognize immediately that piezoelectric materials engineering is not restricted to the material itself; it must comprise the whole device or even the system in which it is used. Thus it is advisable that the designer and the end-user also have a certain insight into the basic principles of this engineering. For this reason, the conclusions of this paragraph do not intend to discuss the materials engineering of piezoelectric devices in detail for any specific application. This will be done in the chapters to follow. They intend to offer general rules that are helpful not only for the producer but also for the user, giving an insight into possibilities, interdependencies, and limitations. They should promote a fruitful cooperation.

The example of a specific application could be helpful to elucidate the problems to be solved within such a cooperative engineering procedure. As already mentioned, concessions have to be made, e.g., as to the temperature dependence. Nevertheless, it has been possible to take advantage of these highly piezoelectric materials, e.g., for channel filters in telecommunication. The characteristic of these high-quality filters is determined by specifically shaped metal resonators; nevertheless, the piezoelectric transducers are not allowed to affect this high quality remarkably by the temperature dependence of their piezoelectric, dielectric, or elastic properties or even by aging. Thus extrinsic piezoelectric effects must not occur. Any neighbourhood of the Curie temperature by warming cannot be tolerated, but for high enough energy conversion the neighbourhood of an MPE is mandatory. Any deviation from this boundary due to temperature change must be compensated by other parts of the filter device, but on the contrary, temperature coefficients in the metallic parts or in the address circuitry can be compensated by the material choice of the piezoelectric transducer.

For more detailed information we would like to refer to Chap. 7, which is devoted to this topic. We have only used this example for demonstrating that in the piezoelectric case materials engineering depends on the design of the specific application by far more than in many other fields. It is a must that the materials engineer understands the requirements of the user. On the other hand, these remarks should motivate the system designer to understand more about the possibilities, interdependencies, and also limitations of piezoelectricity. Only this way can he become an active partner of the materials engineer.

Concerning such a cooperation, we should not forget that production costs are a major factor of the figure of merit for applicability. Here the simple structure and compactness of ceramic piezoelectric transducers offer many interesting advantages as compared to conventional electromechanical devices.

Returning to materials engineering itself, we have seen that above all, surroundings of ferroelectric phase boundaries are of interest; but there not only large piezoelectric effects can be found, but external parameters like stress or temperature have an increasing influence as well. Therefore, it is of major technical interest that the phase boundary in question is as vertical as possible referring to the composition–temperature phase diagram. Only then the critical distance to the boundary does not change remarkably within an acceptable temperature range. This is the case in the PZT phase diagram and explains the extraordinary technical relevance of PZT materials.

Precursors of every ferroelectric phase transition are lattice instabilities which can be recognized by a corresponding soft mode or as a Curie–Weiss-like increase of permittivity. Consequently, the transition from one ferroelectric phase to another is accompanied by an increase of the permittivity perpendicular to the existing spontaneous polarization, whereas only minor changes are found parallel to it. The resulting pronounced dielectric anisotropy explains the specific intrinsic anomalies near the phase boundary. It is directly related to the maxima of the piezoelectric and elastic shear moduli, by which especially inclined domains can contribute to the intrinsic longitudinal and transverse piezoelectric effect. This underlines the importance of domain engineering, not only in ceramics.

Obviously, the extrinsic contributions to piezoelectricity are also influenced by phase transitions. Owing to the small lattice distortions in the neighbourhood of phase transitions, 90° walls in the tetragonal phase (correspondingly 72° or 108° walls in the rhombohedral phase) can be moved easily by an external field. Only in such a class of materials can major extrinsic contributions be expected. Interestingly, these contributions become higher the less the crystallite orientation favors the intrinsic contributions and vice versa. Domain wall movements in a small field are reversible. They become more irreversible with higher external signals. The nonlinear behavior and hysteresis are connected with it.

It is possible to pin the domain walls by specific modifications of the material and to suppress the movement of 90° walls (cf. Chap. 4) omitting mostly extrinsic contributions and reducing dielectric and elastic losses remarkably. This is important not only in the already mentioned case of filters but also for every power application.

It should be mentioned that pinning is an important means for long-term stabilization, too. The domain configuration that has been produced by the polarization procedure can relax only very slowly; thus aging can be avoided or at least drastically diminished.

4. The last class of piezoelectric materials as mentioned in Table 3.1 of Sect. 3.2 are organic plastics. They are normally used as foils and are very useful for cheap mass products or sensor arrays. Their technology is quite different from the inorganic piezoelectric materials. They are treated extensively in Chap. 6.

References

1. W. Wersing, in *Electronic Ceramics*, ed. by B.C.H. Steele (Elsevier, London, New York, 1991)
2. R. Comes, M. Lambert, A. Guinier, Solid State Commun. **6**, 715 (1968)
3. W.J. Merz, Phys. Rev. **76**, 1221 (1949)
4. A.F. Devonshire, Phil. Mag. **40**, 1040 (1949); Phil. Mag. **42**, 1065 (1951)
5. V. Dvorak, Phys. Stat. Sol. **45**, 147 (1971)
6. W. Wersing, Ferroelectrics **54**, 207 (1984)
7. W. Wersing, Ferroelectrics **7**, 163 (1973)
8. Y. Saito, H. Takao, T. Tani, T. Nonoyama, K. Takatori, T. Homma, T. Nagaya, M. Nakamura, Nature **432**, 84 (2004)
9. W. Cochran, Phys. Rev. Lett. **3**, 412 (1959)
10. H. Thomann, Ferroelectrics **73**, 183 (1987)
11. M.J. Haun, Z.Q. Zhuang, E. Furman, S.J. Jang, L.E. Cross, Ferroelectrics **99**, 45 (1989)
12. A. von Hippel, R.G. Breckenridge, F.G. Chesley, L. Tisza, Ind. Eng. Chem. **38**, 1097 (1946)
13. R.B. Gray, U.S. Patent **2**, 486, 560, (Nov. 1, 1949); filed (Sept. 20, 1946)
14. B. Jaffe, R.S. Roth, S. Marzullo, J. Appl. Phys. **25**, 809 (1954)
15. D. Damjanovic, M. Demartin, J. Phys. Condens. Matter **9**, 4943 (1997)
16. B. Noheda, D.E. Cox, G. Shirane, J.A. Gonzalo, L.E. Cross, S.-E. Park, Appl. Phys. Lett. **74**, 2059 (1999)
17. W. Wersing, G. Zorn, K. Lubitz, J. Mohaupt, Jpn. J. Appl. Phys. **24**(Suppl. 24–2), 724 (1985)
18. W. Wersing, Ferroelectrics **37**, 611 (1981)
19. K. Carl, K.H. Härdtl, Phys. Stat. Sol. (a) **8**, 87 (1971)
20. D. Damjanovic, J. Am. Ceram. Soc. **88**, 2663 (2005)
21. L. Bellaiche, A. Garcia, D. Vanderbilt, Phys. Rev. B **64**, 060103 (2001)
22. B. Jaffe, W.R. Cook, H. Jaffe, *Piezoelectric Ceramics* (Academic, London, New York, 1971)
23. Y. Fetiveau, Thesis, University Lyon, 1969
24. G. Burns, B.A. Scott, Phys. Rev. Letters **25**, 1191 (1970)
25. W. Wersing, Proc. 4th Int. Meeting Electro- and Magnetoceramics, Celle, 162–182 (1981)
26. D.A.G. Bruggemann, Ann. Phys. (Germany) **24**, 636 (1935)

27. G. Arlt, H. Dederichs, Ferroelectrics **29**, 47 (1980)
28. G. Arlt, Ferroelectrics **40**, 149 (1982)
29. W. Wersing, K. Lubitz, J. Mohaupt, Ferroelectrics **68**, 77 (1986)
30. G. Zorn, W. Wersing, H. Göbel, Jpn. J. Appl. Phys. **24**(Suppl. 24–2), 721 (1985)
31. A.G. Luchaninov, A.V. Shil'nikov, L.A. Shuvalow, Ferroelectrics **41**, 181 (1982)
32. W. Wersing, K. Lubitz, J. Mohaupt, IEEE Trans. Ultrason. Ferroelectr. Frequ. Control **36**, 424 (1989)
33. Y. Yamashita, K. Yokoyama, H. Honda, T. Takahashi, Jpn. J. Appl. Phys. **20**(suppl. 20–4), 183 (1981)
34. H. Takeuchi, S. Jyomura, E. Yamamoto, Y. Ito, J. Acoust. Soc. Am. **72**, 1114 (1982)
35. H. Takeuchi, S. Jyomura, C. Nakaya, Proc. 1985 IEEE Ultrason. Symp. 605–613 (1986)
36. K.M. Rittenmyer, R.Y. Ring, T.A. Henriques, J. Acoust. Soc. Am. **79**, 2073 (1986)
37. H. Thomann, W. Wersing, Ferroelectrics **40**, 189 (1982)
38. S.E. Park, T.R. Shrout, J. Appl. Phys. **82**, 1804 (1997)
39. W. Wersing, R. Bruchhaus, SPIE **2364**, 12 (1994)
40. D.V. Taylor, D. Damjanovic, Appl. Phys. Lett. **76**, 1615 (2000)
41. C. Kügeler, M. Hoffmann, U. Böttger, R. Waser, SPIE **4699**, 114 (2002)
42. L.A. Shuvalov, J. Phys. Sci. Japan **28**(Suppl. 28), 38 (1970)
43. G. Arlt, Ferroelectrics **104**, 217 (1990)
44. G. Arlt, IEEE Ultrasonics Symposium 733–741 (1990)
45. G. Simpson, Ferroelectrics **6**, 283 (1974)
46. M. Muratake, J. Phys. Soc. Japan **11**, 807 (1956)
47. A.V. Turik, G.I. Bondarenko, Ferroelectrics **7**, 303 (1973)
48. J. Fousek, B. Brezina, J. Phys. Soc. Jpn. **19**, 830 (1964)
49. V.I. Aleshin, Sov. Phys. Tech. Phys. **35**, 107 (1990)
50. A.G. Luchaninov, A.V. Shil'Nikov, L.A. Shuvalov, I.J.U. Shipkova, Ferroelectrics **98**, 123 (1989)
51. E.K.W. Goo, R.K. Mishra, G. Thomas, J. Appl. Phys. **52**, 2939 (1981)
52. P.G. Lucata, V. Teodoresu, F. Vasiliu, Appl. Phys. **A37**, 237 (1985)
53. G. Arlt, N.A. Pertsev, J. Appl. Phys. **70**, 2283 (1991)
54. S. Li, W. Cao, R.E. Newnham, L.E. Cross, Ferroelectrics **139**, 25 (1993)
55. H. Kronmüller, Z. Angew. Phys. **30**, 9 (1970)
56. M. Davis, D. Damjanovic, N. Setter, J. Appl. Phys. **95**, 5679 (2004)
57. A.B. Schäufele, K.H. Härdtl, J. Am. Ceram. Soc. **79**(10), 2637 (1996)
58. D.C. Lupascu, J. Rödel, Adv. Eng. mater. **7**, 882 (2005)
59. D.C. Lupascu, *Fatigue in Ferroelectric Ceramics and Related Issues* (Springer, Heidelberg, 2004)
60. A.B.K. Njiwa, D.C. Lupascu, J. Rödel, Acta Materialia **52**, 4919 (2004)

4

Piezoelectric PZT Ceramics

G. Helke and K. Lubitz

4.1 Physical Characteristics of Ferroelectric Ceramics

Today, the most important piezoelectric materials that are technologically important are ferroelectric ceramics based on Pb-containing perovskites. They offer the advantages of most ceramic materials such as ease of fabrication, possibility of variable and application-adapted shaping, as well as low-cost manufacturing. In addition, diverse chemical modifications are available in order to tailor the piezoelectric properties to different applications. The precondition, however, is the possibility to impose a unipolar anisotropy into the otherwise macroscopically isotropic ceramic. This is possible only if the spontaneous polarization can be oriented by a poling process using an external electric field for generating a remanent polarization. In addition, the special behaviour in the vicinity of ferroelectric phase transitions promotes extremely high piezoelectric effects, where the intrinsic ones are connected with the dielectric anisotropy and the extrinsic ones with increased domain wall mobility. Especially, in the periphery of the morphotropic phase boundary (MPB) present in a series of Pb-containing perovskites, such as lead–zirconate–titanate (PZT), this effect technically can be used extensively because the MPB in the temperature–composition phase diagram is nearly vertical, what implies nearly temperature independence.

In this chapter, the properties of ferroelectric ceramics with very high piezoelectric effects are discussed. It will be shown how optimal properties for specific applications can be tailored by sophisticated materials and process engineering.

In Sect. 4.1, the most important properties that characterize piezoelectric ceramics in general are summarized. In Sect. 4.2 all application-relevant material properties of PZT are described and discussed, which occur in the vicinity of the MPB between the tetragonal and the rhombohedral phase. Obviously, the MPB can be shifted by chemical modifications. However, this causes not only a simple shift of the MPB-related properties; the properties are changed principally as outlined in Sect. 4.3. There, these modifications are classified first according to whether we are dealing with isovalent or heterovalent ions (Sects. 4.3.1 and 4.3.2, respectively). Again, especially the group of heterovalent substitutions splits into subgroups. Here we find basically different characteristic properties within the so-called soft or hard piezoceramics, applicable for entirely different fields of usage. Of course, many attempts have been made to combine various types of substitutions with additives. Therefore, in Sect. 4.3.3 an attempt is made to show and compare the technical benefits and drawbacks of different material systems. Actually this is difficult, for till now most data available in the literature are not fully comparable because of unknown parameters such as grain size and distribution, relative density, final stoichiometry, chemical homogeneity, and the exact position relative to the MPB. This is the more problematic, since all these parameters depend not only on the overall composition but also on the specific fabrication procedures.

In Sect. 4.4, preparation aspects of PZT material are discussed, as well as the influence of material modifications by increasing the hardness or softness of the PZT and by changing the sintering behavior and the microstructure of the ceramics. Finally in Sect. 4.5, the processing techniques for different piezoelectric devices are summarized.

4.1.1 Perovskite Structure, Spontaneous Polarization, and Ferroelectric Domains

The best-known group of piezoelectric ceramic materials are the polycrystalline perovskites with a purely cubic structure in the paraelectric state. Owing to the spontaneous polarization of the crystal lattice below the Curie temperature, T_C, dipole formation and a structural phase transition into the ferroelectric state by spontaneous distortion of the unit cell take place [1–3].

As a result of the sintering process, all crystalline grains ("crystallites") are connected with differently oriented grains. Generally, these grains are separated by well-defined grain boundaries. When cooling down under the transition into the ferroelectric state, the spontaneous unit-cell distortion in a grain is hindered by the differently oriented neighboring grains, so that internal stresses cannot be minimized except by fragmentation of the grains into domains [4]. A complicated domain structure is created within each grain according to the allowed directions of the spontaneous polarization. Thus, in the tetragonal phase the so-called separated 180° and 90° domain walls are formed, and in the rhombohedral phase 180°, 109°, and 71° walls.

In the ferroelectric domains themselves, a uniform orientation of the dipoles takes place. Since the crystallites and the domains contained in them are randomly oriented, the characteristics of ferroelectric ceramics are isotropic both after the synthesis and after cooling below the Curie temperature T_C in the first stage. The dipole moments of the domains compensate each other because of the statistical distribution of the polarization directions. But ferroelectricity itself offers the mechanism by which the domains can be oriented in a d.c. electric field, so that ferroelectric ceramics finally behave

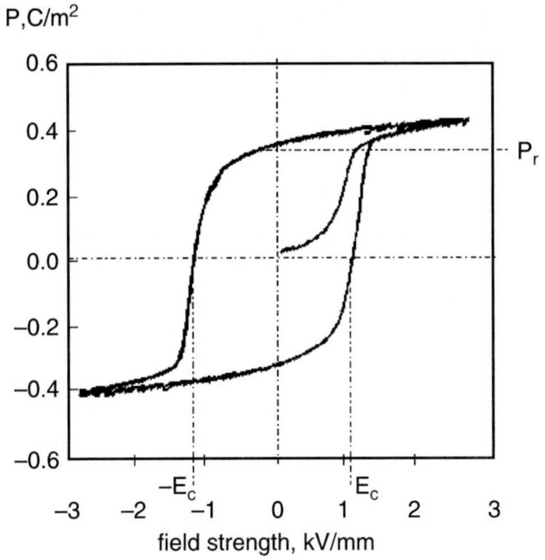

Fig. 4.1. Polarization–field strength characteristics: the ferroelectric hysteresis loop

as polar. When switching off the electric field after this poling procedure, a metastable state of domain configurations remains, resulting in a finite remanent polarization P_r.

The dependence of the polarization on the magnitude and direction of the electric field is described by the ferroelectric hysteresis loop. The specific processes that determine the shape of the hysteresis loop and the characteristics such as P_r or the electrical coercive field strength E_c are complicated, and their nature depends on the material parameters. The shape of a typical hysteresis loop of a ceramic is shown in Fig. 4.1.

In the remanent state, the polarization P_r is the average dipole moment of the domains. Owing to the shift and decrease of the number of the domain walls and the statistical orientation of the domains, a maximum $P_r = 0.831\ P_s$ for the tetragonal phase and $P_r = 0.861\ P_s$ for rhombohedral phase (P_s is the spontaneous polarization of the domains) can be achieved [5]. The polarization direction of the individual domains is limited to the minor angles Θ around the direction of the polarization field. The stability of the remanent polarization depends on the domain configuration. The anisotropy of the physical constants will also be influenced by the mobility of the non-180° domains.

For domain reorientation, an external poling field must be applied on the sample for a certain duration. Theoretically, most of the domains should be reoriented when the electric poling field is larger than the coercive field of the ferroelectric sample. However, a practical poling field must exceed 3 or 4 times the coercive field before maximum piezoelectricity can be obtained. For a given poling field and poling time, better domain rearrangement results at higher poling temperatures. It must be pointed out that both the coercive field and the saturation field decrease when the temperature increases. This is due to the higher mobility of ferroelectric domains at higher temperatures. It can be explained mainly by the decrease of crystalline anisotropy. Furthermore, the strain caused by reorientation of the non-180° domain is smaller and thus the poling becomes easier [1].

4.1.2 The Field-Induced Strain and the Piezoelectricity of Ferroelectric Ceramics

4.1.2.1 Remanent Polarization, Electrostriction, and Piezoelectricity of Ferroelectric Ceramics

One of the most valuable features of ferroelectricity is the fact that ferroelectric ceramics can be transformed into polar materials by applying a d.c. electric field. The reorientation of most domains persists also after switching off the d.c. electric field, and the existence of the remanent polarized state is a necessary condition for the piezoelectricity of ferroelectric ceramics.

Ab initio computations of the effective properties of the piezoceramics are very much limited because of the complicated domain structure.

A basic phenomenological approach (*see* also Sect. 3.1.4) leads to the following relationship using the tensor form, and adding subscripts to distinguish induced polarization P_3 from remanent polarization P_r

$$\mathrm{S}_3 = Q_{33}{}^*(P_r + P_3)^2 = Q_{33}{}^* P_r^2 + 2Q_{33}{}^* P_r P_3 + Q_{33}{}^* \mathrm{P}_3^2 \tag{4.1}$$

where $Q_{33}{}^*$ is the electrostriction constant for poled ceramics. Here small-signal parameters and quasistatic conditions without hysteresis and domain effects are used.

The term $Q_{33}{}^*\ P_r{}^2$ represents the remanent strain $S_3{}^{\mathrm{pr}}$ of a polarized ceramic. The term $2\ Q_{33}{}^*\ P_r\ P_3$ corresponds to the piezoelectric effect as linearized electrostriction. $Q_{33}{}^* P_3{}^2$ describes the electrostriction caused by the induced polarization, which can be neglected here.

$$\mathrm{S}_3{}^{\mathrm{pr}} = 2\ Q_{33}{}^* P_r P_3 = g_{33} P_3, \tag{4.2}$$

$$\mathrm{g}_{33} = 2\ Q_{33}{}^* P_r, \tag{4.3}$$

$$\mathrm{S}_3{}^{\mathrm{pr}} = \text{strain at constant remanent polarisation } P_r.$$

The term $Q_{33}{}^* P_r{}^2$ is the strain resulting from poling. The second term $2Q_{33}{}^*\ P_r P_3$ is the piezoelectric response. Following the conventional practice, the piezoelectric voltage coefficient is defined via $g_{33} = 2Q_{33}{}^* P_r$, neglecting higher-order terms. Combining the above equations shows that the piezoelectric charge coefficient d_{33} is directly related to the remanent polarization P_r and the dielectric coefficient $\varepsilon^T{}_{33}$ by

$$\mathrm{d}_{33} = \varepsilon^{\mathrm{T}}{}_{33}\ \mathrm{g}_{33}, \tag{4.4}$$

$$\mathrm{d}_{33} = 2\ Q_{33}{}^* \varepsilon^{\mathrm{T}}{}_{33} P_r. \tag{4.5}$$

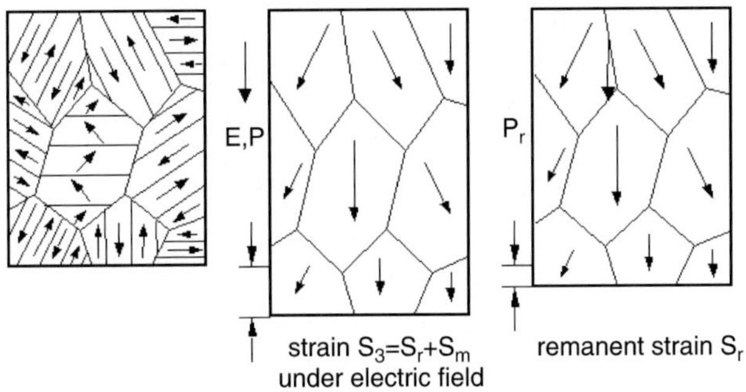

Fig. 4.2. Strain of ferroelectric ceramics (piezoceramics) caused by the field and remanent strain

4.1.2.2 Extrinsic Contributions to the Strain

The term "hysteresis loop" of ferroelectrics usually means the $P(E)$ dependence. However, the large-signal physical quantities vary with electric field strength, polarization, and domain reorientation. From the practical point of view, the most important features are the field-dependent piezoelectric coefficients and, especially, the strain. The extrinsic contributions are described by effective piezoelectric properties, which consider domain processes and reorientations additionally [6, 7].

A simple illustration of the poling process is shown in Fig. 4.2 [1]. The unpoled sample (left) is elongated by the electric field and shows remanent strain after field removal.

Non-180° wall motion, that is motion of 90° walls in the tetragonal phase and motion of 71° and 109° walls in the rhombohedral phases, will give rise to shape changes [6, 7].

In Fig. 4.3, the strain is described both for poling and the repeated electric field change depicted by the so-called butterfly loop. During repeated field change without change of direction, the strain shows the behavior represented in the section S_{m}.

The coercive field strength E_{c} is defined by the point $P = 0$ (see Fig. 4.1). It should be mentioned that the minimum of the strain does not coincide with $P = 0$.

4.2 The Morphotropic Phase Transition of Piezoelectric Ceramics Based on Lead–Zirconate–Titanate

4.2.1 Structural Properties

The importance of perovskites with ferroelectric characteristics is mainly due to the nearly unlimited isomorphism of the perovskite structure. This has

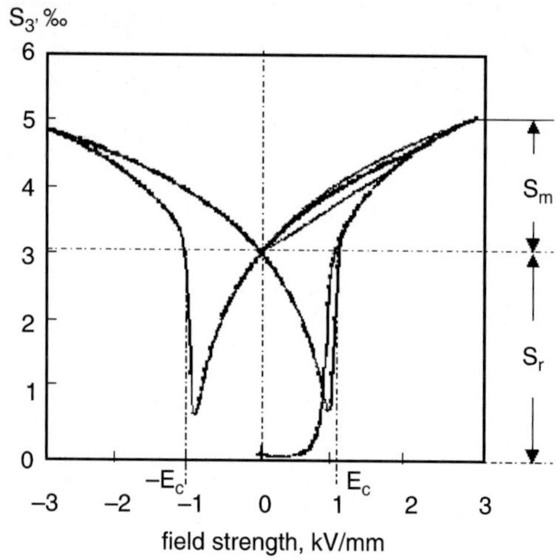

Fig. 4.3. Strain field strength characteristic of ferroelectric ceramics with perovskite structure

made possible a continuous change of composition by substitution of different elements at equivalent positions. The isomorphism leads finally to the formation of solid solutions, the characteristics of which can be varied within a wide range.

Particularly interesting are such compositions with perovskite structure in which almost equal parts of lead zirconate ($PbZrO_3$) and lead titanate ($PbTiO_3$) are present as a binary solid solution, generally called PZT [8–10]. Corresponding to the Curie temperatures $T_c = 490°C$ for $PbTiO_3$ and $T_c = 235°C$ for $PbZrO_3$, the Curie temperatures of binary solid solutions of the system PZT lie in the range between these limiting temperatures. For the most interesting compositions, therefore, the Curie temperature T_c is >300°C. This permits the high operating temperatures of piezoelectric PZT ceramics.

Crucial for the system lead-zirconate-titanate is the occurrence of a morphotropic phase transition, i.e., a composition-dependent phase transition from a ferroelectric tetragonal structure in the Ti-rich region to a ferroelectric rhombohedral structure in the Zr-rich region (Fig. 4.4).

The morphotropic phase transition can be described as phase transition of the first order within a finite two-phases range [7] where both tetragonal and rhombohedral ferroelectric phases (Fig. 4.5) coexist. In general, this coexistence range extends over 2–5 mol% $PbZrO_3$. Formally, the MPB is defined as a temperature-dependent line where tetragonal and rhombohedral phase percentages are equal.

The coexisting ferroelectric phases near the MPB are sensitive to external effects (temperature, electric field, pressure) and can change into one another [11–14].

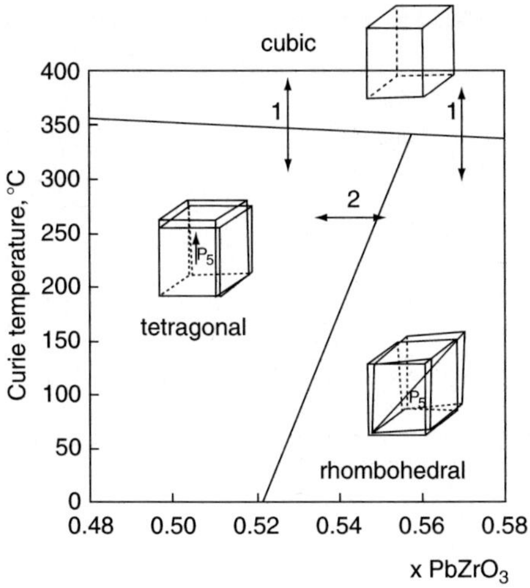

Fig. 4.4. Phase transitions of lead zirconate titanate ceramics at the polymorphic (1) and at the morphotropic (2) phase boundary. Note the extension of the abscissa

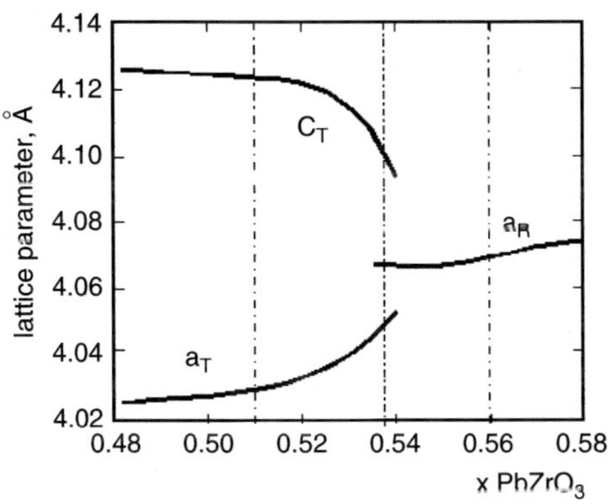

Fig. 4.5. Lattice constants of a continuous series of solid solutions of the system PZT in the neighborhood of the morphotropic phase transition

Process parameters during the fabrication of the PZT ceramics determine basically the width of the morphotropic phase transition region. This range may also be shifted to different Zr/Ti ratios by modification of the composition of the solid solution.

4.2.1.1 The Homogeneous Strain Parameter

The morphotropic phase transition in a relatively narrow range of the Ti/Zr ratio close to the ratio 0.5/0.5 of the molecular portions is connected with the coexistence of two phases of very different spontaneous unit-cell distortions within this range.

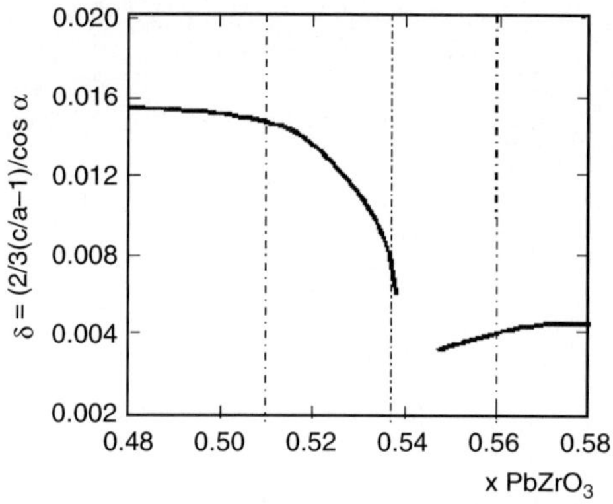

Fig. 4.6. Homogeneous strain parameter δ within the range of morphotropic phase transition

The deformation of the cubic crystal lattice below T_C is usually described as a tetragonal distortion by $c/a - 1$ ("tetragonality") and as a rhombohedral distortion – by the angle α or better by the deviation of the angle α from 90°. The different character of these parameters does not permit a direct comparison of the deformations within the adjacent tetragonal and rhombohedral compositions in lead-zirconate-titanate.

As the so-called homogeneous strain parameter, which characterizes the spontaneous unit-cell distortion, the relative extension δ of the perovskite unit cell along the polar axis can be used (Fig. 4.6) [15,16]. For the deduction, see also Sect. 3.2 (3.42) and (3.43).

In the tetragonal lattice, the c-axis is the polar axis, and one gets the homogeneous strain

$$\delta = 2/3(\mathrm{c}/\mathrm{a} - 1). \tag{4.6}$$

For the rhombohedral lattice with the diagonal of the unit cell as polar axis

$$\delta \approx \pi/2 - \alpha. \tag{4.7}$$

Furthermore, a characteristic decrease of the homogeneous strain parameter δ should be noticed when approaching the MPB from the tetragonal as well as from the rhombohedral side (Fig. 4.6, for $x = 0.538$, $PbZrO_3$) [17]. This has been observed also for other systems [18] similarly.

The crystal tetragonality or the homogeneous strain parameter $\delta = 2/3(c/a - 1)$ increases with increasing grain size. These dimensional changes of the unit cell are the result of the decrease of the internal stresses in the ceramic that occur during the transition from the paraelectric to the ferroelectric state [19–21]. The crystal tetragonality also changes with the fraction F_T of the tetragonal phase in the region of the morphotropic phase transition [22]. The spontaneous unit cell distortion is mutually clamped in fine-grained materials, which reduces the intrinsic polarization and increases the internal stress in the ceramic furthermore [19].

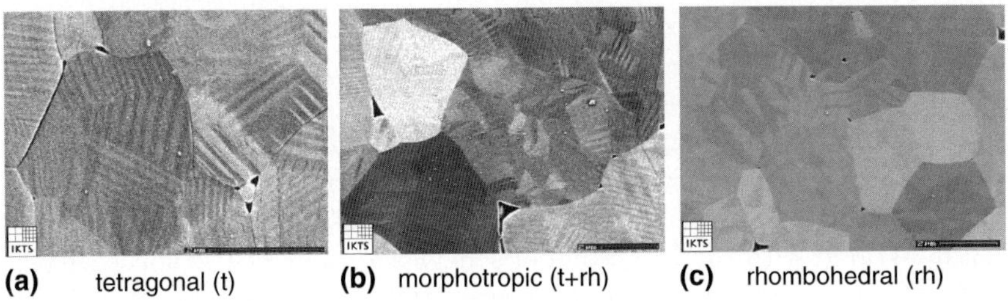

Fig. 4.7. Domain configuration within the range of the morphotropic phase transition

Thus the homogeneous strain parameter is a good additional means for understanding the properties of ferroelectric ceramics, which depend in a complex manner on composition, fabrication, grain size, and poling conditions. There are many causes for these grain size effects. One of them is the increase of the surface-to-volume ratio so that the grain boundaries play a substantial role when determining the material properties.

4.2.1.2 Domains of the Ferroelectric Phases

The morphotropic phase transition from tetragonal to rhombohedral structure can be illustrated and traced by changes of the ferroelectric domain structure, as illustrated in Fig. 4.7. A more complicated domain structure with a large number of domain walls corresponds to the tetragonal phase (Fig. 4.7a). In the transient region to the rhombohedral phase, the domain structure changes its appearance considerably: tetragonal and rhombohedral phases (Fig. 4.7b) coexist. In the rhombohedral phase, however, one observes a much smaller number of domain walls (Fig. 4.7c) [23].

Because of the much higher lattice distortion during formation of the spontaneous polarization of tetragonal compositions, the internal stresses are higher. This leads to an increased number of domains per unit volume as compared to rhombohedral compositions.

Anyway, it should be kept in mind that the configuration and number of domains change considerably by the poling process.

4.2.2 Ferroelectric Properties of Poled PZT Ceramics

The correlations of structural and electromechanical properties will be presented exemplarily for the following system of solid solutions: $x[PbTiO_3 - PbZrO_3] - (1-x)Sr(K_{1/4}Nb_{3/4})O_3$ (PZT-SKN). This system has been chosen because of its excellent piezoelectric properties (ceramics sintered at $1,150°C$, with a mean grain size of about $3\,\mu m$ [21]).

The considered anomalies of structural and ferroelectric/piezoelectric characteristics within the range of the morphotropic phase transition arise similarly within all possible modifications of PZT ceramics (see below: binary, ternary, and quaternary solid solutions as well as modifications with acceptor and donor dopings) and have a profound impact in practical applications.

The evaluated system PZT-SKN belongs to the category "soft piezoceramics". Materials of this compositional series are characterized by a relatively low coercive field and high piezoelectric activity. The interaction of electric field and ferroelectric domains at different spontaneous lattice distortions can be well illustrated with such a system of soft piezoceramics. The specification of piezoceramics modified in different ways results basically from intrinsic crystal properties of the perovskite structure and extrinsic effects due to domain wall motions.

4.2.2.1 Dielectric Properties, Remanent Polarization, and Unidirectional Strain

Above all, piezoelectric ceramics can be characterized by the piezoelectric, elastic, and dielectric coefficients and by the coefficients of the electromechanical coupling (see Sect. 13.3). Because of the anisotropy of polarization, these coefficients depend on the direction, leading to a piezoelectric-elastic-dielectric matrix, which in ceramics consists basically of three piezoelectric, five elastic, and two dielectric independent coefficients [9, 24–26].

For characterization and comparison of different piezoelectric ceramics, commonly the dielectric permittivity ${\varepsilon_{33}}^T/\varepsilon_0$ and the piezoelectric coefficients d_{33} and g_{33} are used. They can be combined with the electromechanical coupling coefficients, and in particular with k_p. They describe the low-frequency energy conversion in piezoelectric ceramics and are the most important characteristics measuring the effectiveness of piezoelectric ceramics.

It should be noted that not only piezoelectric properties but also other intrinsic parameters, such as dielectric or elastic properties, contribute to the small-signal parameters of a material remarkably. They can be measured at fields at least 100 times smaller than the coercive field. In the mostly used ferroelectric materials such as lead zirconate titanate, however, the extrinsic properties due to large-signal contributions can be up to a factor 2 larger than the small-signal properties.

4.2.2.2 Dielectric Permittivities Before and After Poling

According to Fig. 4.8, the dielectric permittivity $\varepsilon/\varepsilon_0$ before the poling (1) passes through a smooth maximum within the range of the morphotropic phase transition. After poling, ${\varepsilon_{33}}^T/\varepsilon_0$ (2) reaches a more distinct maximum, which appears in the more tetragonal section of the morphotropic phase transition region, where its value increases. This somewhat surprising result can be explained using the above-mentioned features: owing to the minor energy difference, the morphotropic phase transition in PZT is very soft resulting in a broad region of coexistence of both phases changing from grain to grain

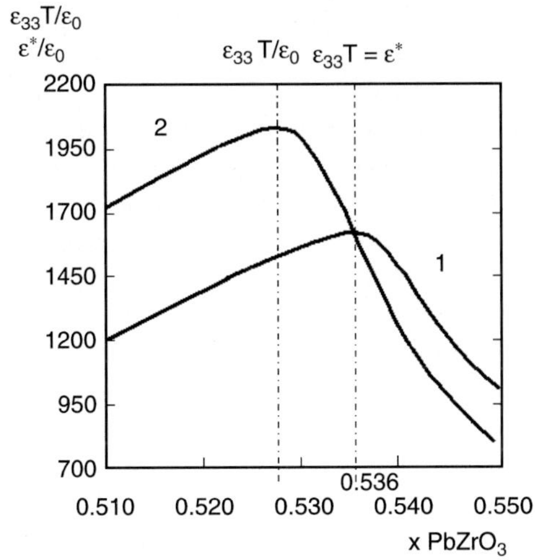

Fig. 4.8. Dielectric permittivity (1) before and (2) after the polarization

according to the fluctuations of the Ti/Zr concentration. This can be proved by the configuration of domains (Fig. 4.7) and by X-ray diffraction [17].

In ceramics, the measured permittivity depends on the different mechanisms within both these phases and on their relative concentrations: within the tetragonal range, dielectric mechanisms and the dielectric anisotropy prevail, 90° domain wall motion is low, and rotations of 180° domains predominate (see Sect 3.2.2). This explains both the increase of $\varepsilon_{33}{}^{T}/\varepsilon_0$ when approaching the phase boundary, due to the removal of the "clamping effect", and the increased permittivity maximum after poling. Within the rhombohedral range, the high mobility of the 71° and 109° domain walls prevails. The mobility reduction by clamping explains the decay of $\varepsilon_{33}{}^{T}/\varepsilon_0$ with increasing Zr content in the unpoled state. After poling, the concentration of these mobile walls is diminished, explaining the permittivity decay in the rhombohedral region.

The change of the dielectric permittivities by poling goes through zero at a critical composition $x = 0.536$ $PbZrO_3$ [27]. This composition is almost identical with that whose tetragonal phase fraction is 50% according to X-ray diffraction [17].

4.2.2.3 Remanent Polarization P_r

In the tetragonal (titanium-rich) phase of PZT ceramics, both 180° and 90° domains exist. The 180° domains are switched nearly completely by an external field and are sufficiently stable after poling. In the rhombohedral phase, 180°, 71°, and 109° domain walls prevail, among which only the 180° domains remain stable, while reorientations of the 71° and 109° domains are unstable because of internal stress.

Important parameters of PZT ceramics are the homogeneous strain parameter δ and the associated degree of reorientation of domains during the poling process.

The quality of the remanent polarization P_r can be attributed to the strain parameter δ (intrinsic contribution) and to the mobility η of the non-180° domains (extrinsic contribution). In the tetragonal phase, the strain parameter δ plays the crucial role owing to a decrease of the wall mobility η of the non-180° domains within an external electric field. In the rhombohedral phase, the decrease in the number of the domain walls and an increase of the mobility η prevail as long as δ remains small [15, 16]. One gets

$$P_r \cong 6(1/3 + 2/3\eta)\sqrt{\delta}\ \mathrm{C/m^2}, \tag{4.8}$$

for the tetragonal phase (e.g., $P_r \cong 37.5\,\mu\mathrm{C\ cm^{-2}}$ for $\eta = 0.2$ and $\delta = 0.018$) and

$$P_r \cong 8(1/4 + 3/4\eta)\sqrt{\delta}\ \mathrm{C/m^2}. \tag{4.9}$$

for the rhombohedral phase (e.g., $P_r \cong 43.3\,\mu\mathrm{C\ cm^{-2}}$ for $\eta = 0.6$ and $\delta = 0.006$).

It must be noted that the ratio of the mean number of domain reorientations in the rhombohedral and tetragonal phases η_R and η_T is equal to the ratio of the reciprocal values of the strain of the appropriate phases: $\eta_R/\eta_T = 0.6/0.2 = 3 \approx \delta_T/\delta_R$ [16, 28]. This ratio is consistent with the measured mobility of non-180° domains for a rhombohedral and tetragonal composition of the considered "PZT-SKN" containing x% $PbZrO_3$: in the polarized state of a rhombohedral composition ($x = 56\%$) 22–25% domains were switchable, in the tetragonal composition ($x = 51\%$) only 7–8% [18, 22].

In the rhombohedral phase, both the number of the reorientations of non-180° domains and the remanent polarization reach maximum values wherever δ is minimum. The coercive field E_c, however, decreases continuously with the transition from a more tetragonal to a more rhombohedral structure. It reaches a slight minimum on the rhombohedral side (Fig. 4.9).

From Figs 4.8 and 4.9 it follows that the position of the maxima of the dielectric permittivity $\varepsilon_{33}{}^T/\varepsilon_0$ and of the remanent polarization P_r does not correspond within a composition series; the maxima rather lie near the tetragonal and/or rhombohedral edge of the morphotropic transition range.

4.2.2.4 The Unidirectional Strain S_m

Coming from the tetragonal side, the lattice distortion (strain δ) decreases with increasing $PbZrO_3$ content in the MPB region and further increases slightly in the rhombohedral $PbZrO_3$-rich region (Fig. 4.6). This suggests that the domains near the MPB are aligned most easily by poling owing to the small spontaneous strain (Fig. 4.6).

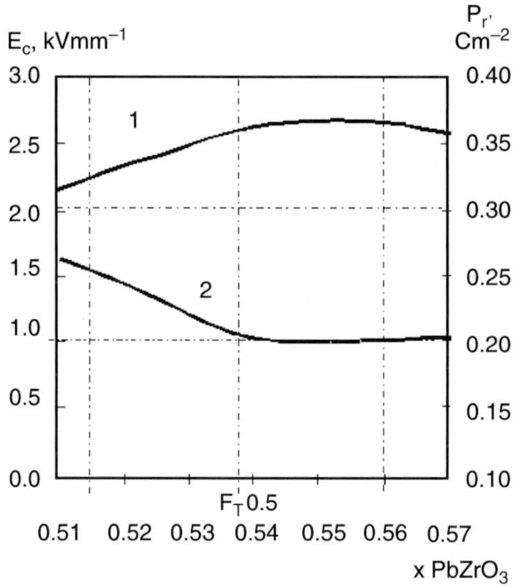

Fig. 4.9. Remanent polarization P_r (1) and coercive field E_c (2) in the range of morphotropic phase transition

Since the phase composition ratio can also be changed by an electric field, furthermore, a sample within the range of the morphotropic phase transition is poled more easily because there are more states available to achieve a higher degree of polarization alignment. Also, the unidirectional strain S_m of poled lead zirconate titanate ceramics attains a maximum in samples within the morphotropic phase transition region (Fig. 4.10).

The largest values of the unidirectional strain S_m are observed near the center of the coexistence range of the phases resulting from the increasing number of domain reorientations within an electric field.

In summary, the number of reorientations of non-180° domains in the coexistence range is not only connected with the spontaneous strain but also with induced transformations of the coexisting phases by an electric field [11, 12, 29, 30], especially when strong d.c. electric fields and/or pressure are applied.

Figure 4.10 shows the unidirectional strain S_m as function of the $PbZrO_3$ content (Zr/Ti ratio) of lead zirconate titanate solid solutions in the range of the morphotropic phase transition in comparison to the maximum of the dielectric permittivity $\varepsilon_{33}{}^T/\varepsilon_0$ of the remanent polarized ceramics. The maximum of the unidirectional strain S_m is not obtained at the MPB itself (composition $x = 0.538$ $PbZr0_3$) but with samples in which the effective volume fraction of tetragonal phase is about 60% (F_T 50%, Fig. 4.11) [17]. Corresponding results were obtained also for other PZT solid solution systems [18].

The maximum of the unidirectional strain S_m is found already on the tetragonal side of the coexistence range and corresponds to spontaneous lattice

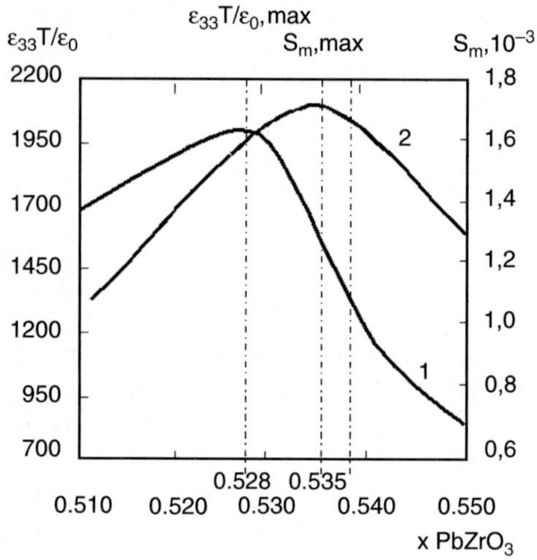

Fig. 4.10. Dielectric permittivity $\varepsilon_{33}{}^{T}/\varepsilon_0$ (1) and unidirectional strain S_{m} (2) at a maximum field strength of 2 kV/mm

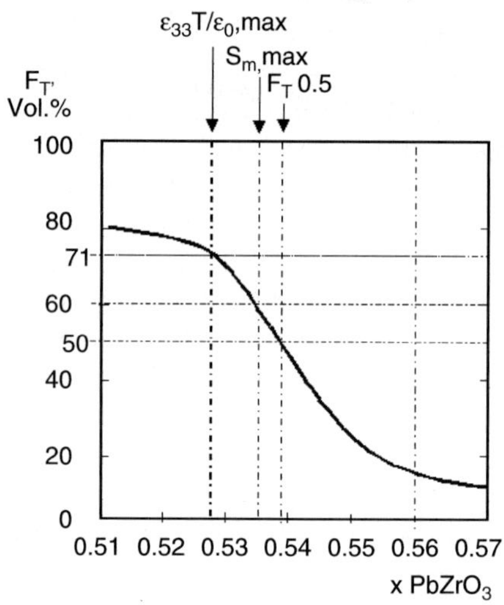

Fig. 4.11. Concentration of the tetragonal phase within the range of the morphotropic phase transition

distortion of about 1.5% (homogeneous strain parameter $\delta = 0.010$). An increased volume fraction of the 90° domains and a high activity of the non-180° domain reorientations in the tetragonal phase result merely in a high strain. With further increasing tetragonality, domain switching is more difficult, so that a volume fraction of 60–70% of tetragonal phase in the coexistence field seems to be the optimum for macroscopic deformation [18, 22].

The maximum of $\varepsilon_{33}{}^T/\varepsilon_0$ is shifted into the more tetragonal section by about 1 mol% corresponding to $x = 0.528$ $PbZrO_3$. According to Fig. 4.11, this corresponds to $F_T \approx 71\%$.

The dependence in Fig. 4.11 was measured by X-ray structure analysis for defining the fraction of tetragonal phase in the coexistence field [18, 31, 32]. This typical behavior was found also by X-ray measurements of other similar lead zirconate titanate compositions [12, 22, 31].

4.2.2.5 Piezoelectric Properties of Poled PZT Ceramics

When comparing the two most important ferroelectric characteristics – $\varepsilon_{33}{}^T/\varepsilon_0$ and P_r – on one hand with the substantial structure parameter δ and on the other with the configuration and mobility η of the domains, we can get an overall insight into the performance of the piezoelectric d and g coefficients and of the coupling coefficient k_p:

- Simple 180° domain wall motions do not contribute to the piezoelectric response d [6]. The behavior of the $\varepsilon_{33}{}^T/\varepsilon_0$ (Fig. 4.8) and P_r (Fig. 4.9) can be used to describe further piezoelectric characteristics (Fig. 4.12).
- The piezoelectric characteristics k and d correlate with the remanent polarization and the dielectric permittivities using the (4.5) and (13.22) by the following relationships [15, 16]:

$$\mathrm{d} \sim \varepsilon^{\mathrm{T}} P_r, \tag{4.10}$$

$$\mathrm{k} \sim (\varepsilon^{\mathrm{T}})^{1/2} P_r. \tag{4.11}$$

  From these relationships, it follows that the maxima of the piezoelectric coefficients d and of the electromechanical coupling factors are determined by the proximity of maximum permittivity and remanent polarization in the neighborhood of the MBP [15, 16].
- As already mentioned, usually the maximum $\varepsilon_{33}{}^T/\varepsilon_0$ belongs to the tetragonal side of the morphotropic phase transition (Figs. 4.8, 4.10, 4.12). The maximum values of d_{33} und k_p (Fig. 4.12) occupy intermediate positions between the maxima of $\varepsilon_{33}{}^T/\varepsilon_0$ and P_r. The maximum values of the electromechanical coupling coefficient k_p and the unidirectional strain S_m coincide at the Zr concentration of $x = 0.535$ still in the more tetragonal section of the range of the morphotropic phase transition (about 60% F_T) [18, 22].

The maximum of the piezoelectric coefficient g_{33} coincides with the maximum of P_r within the rhombohedral range (Fig. 4.13).

This results from the apparent direct proportionality of the g coefficient to P_r:

$$\mathrm{g} = \mathrm{d}/\varepsilon^{\mathrm{T}} \sim P_r. \tag{4.12}$$

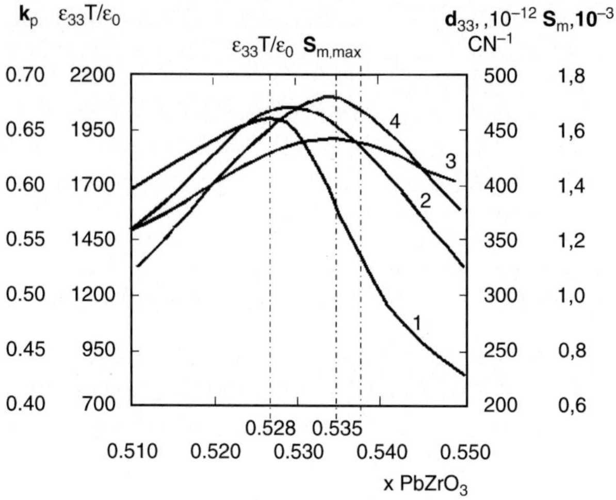

Fig. 4.12. Comparison of the dielectric permittivity $\varepsilon_{33}{}^{T}/\varepsilon_0$ (1), the piezoelectric coefficient d_{33} (2), the electromechanical coupling coefficient k_p (3), and the large-signal unidirectional strain S_m (4) in the range of the morphotropic phase transition

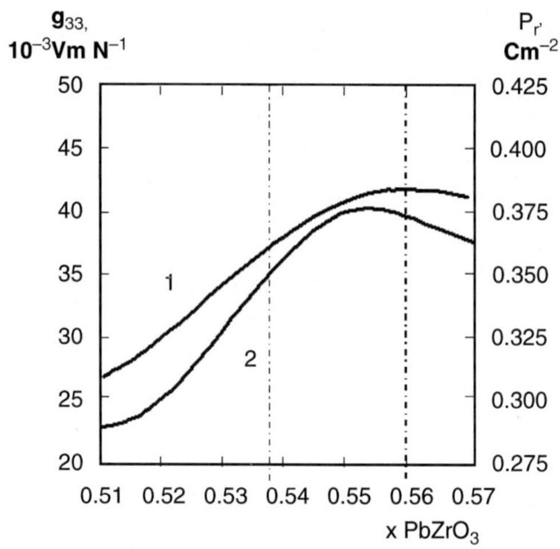

Fig. 4.13. Performance of remanent polarization P_r (1) and piezoelectric coefficient g_{33} (2) the range of the morphotropic phase transition

4.2.2.6 Stability of Dielectric and Electromechanical Characteristics: Correlation of Stability with Structural Characteristics

Both intrinsic and extrinsic factors determine the stability of piezoelectric properties against aging and different types of external stress. This will be discussed in the following paragraphs.

4.2.2.7 Curie Temperature

The thermal stability of the characteristics of piezoceramics correlates with three main factors: the value of the Curie temperature, the strain parameter δ, and the character of the domain structure, especially the number and mobility η of the domain walls [83].

With increasing Curie temperature T_C, the stability of the characteristics at the operating temperature must increase, too. In the vicinity of T_C, the spontaneous polarization decreases and the dielectric permittivities increase, accompanied by a change of all piezoelectric characteristics depending on them.

When comparing tetragonal and rhombohedral compositions with approximately the same T_C, one must consider the character of the domain structure, i.e., the mobility and number of domain walls and, in addition, the role of the spontaneous strain. Within a given doping system, the Curie temperature T_C correlates with the homogeneous strain parameter δ (e.g., PZT-SKN of Table 4.1) and increasing Ti content.

4.2.2.8 Temperature Coefficients in the Range of the Morphotropic Phase Transition

The ferroelectric tetragonal compositions show lower temperature coefficients of the dielectric and piezoelectric coefficients (Fig. 4.14.). For many applications, materials with high thermal stability or low temperature coefficient of the resonant frequency (2) are required. The stability of the resonant frequency f_r is related to the smaller number of domain walls and their reduced mobility.

The temperature coefficient of the piezoelectric coefficient d_{33} (3) changes similar to α_e (1). A first approximation gives $\alpha_d \gg 1/2\alpha_e$ [34].

Table 4.1. Homogeneous strain parameter and Curie temperatures [33]

Material	$\delta = 2/3(c/a - 1)$	$T_C, ^\circ C$
PZT–NiSb 44	0.0102	263
PZT–NiSb 42	0.0133	268
PZT–SKN 530	0.0149	348
PZT–SKN 525	0.0152	351
PZT–SKN 520	0.0154	352
PZT–SKN 515	0.0160	353
PZT–SKN 510	0.0161	354
PZT 52	0.0187	386

PZT–NiSb 44: $Pb[(Zr_{0.44}Ti_{0.56})_{0.92}(Ni_{1/3}Sb_{2/3})_{0.08}]O_3$,
PZT–SKN 530: $[Pb_{0.975}Sr_{0.02}K_{0.005}][(Zr_{0.53}Ti_{0.47})_{0.985}Nb_{0.015}]O_3$.
PZT 52: $Pb(Zr_{0.52}Ti_{0.48})O_3$.

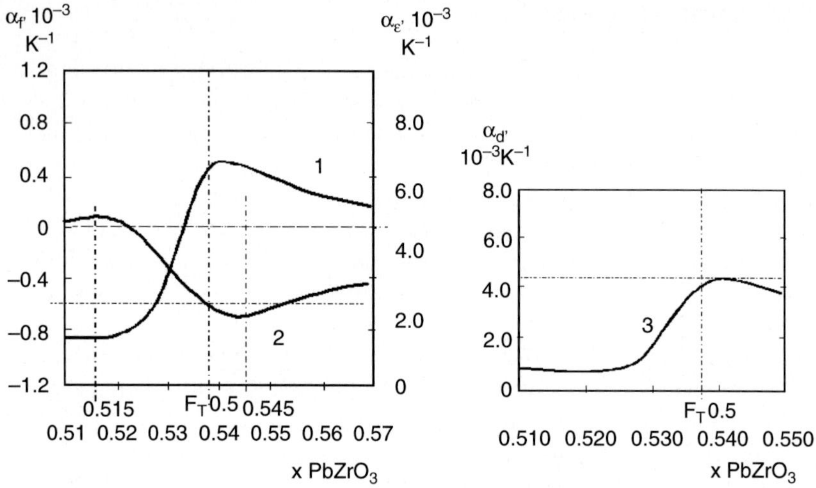

Fig. 4.14. Temperature coefficients α_e of the dielectric permittivity (1), α_f of the resonant frequency (2), and α_d of the electromechanical characteristics (3) within the temperature range 0–85°C

An increased mobility of the domain walls with the simultaneous decrease of δ within the rhombohedral range is coupled with higher temperature coefficients. This means, in particular, that rhombohedral compositions with maximum remanent polarization P_r and accordingly maximum voltage sensitivity g_{33} do not exhibit the highest thermal stability (α_g).

In the range of the morphotropic phase transition itself, the dielectric anomalies induce also elastic anomalies via electrostrictive interaction [35,36]. The temperature dependences in Fig. 4.14 show that both the temperature coefficients of the resonant frequency and of the dielectric permittivity as well as the temperature coefficient of the piezoelectric coefficient reach their largest values within the near-rhombohedral range.

An anomalous feature of the temperature coefficient of the resonant frequency, however, is the "zero passage" and change of sign. However, the reproduction of minimum or zero temperature coefficients in the coexistence range is difficult, because the phase fractions are also influenced by external effects such as electric fields [37], e.g., during poling, due to field-induced phase transitions.

4.2.2.9 Aging Rates in the Range of the Morphotropic Phase Transition

A further unusual feature of ferroelectric ceramics concerns the aging even in the absence of either external electrical or mechanical stresses or temperature changes [3,38]. Aging is a spontaneous process. The stability of the metastable

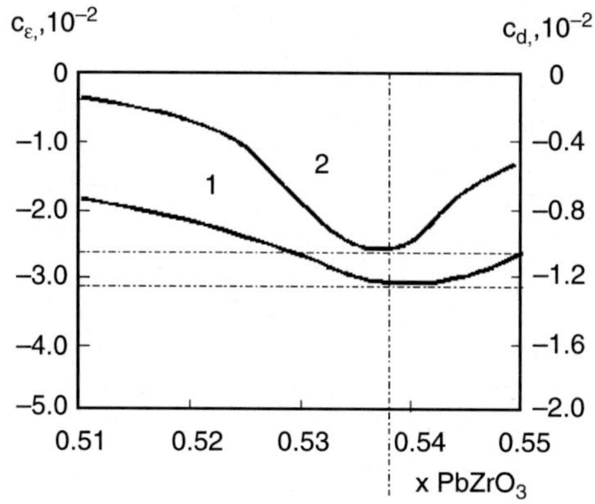

Fig. 4.15. Aging behaviour of the dielectric permittivity $\varepsilon^{\mathrm{T}}/\varepsilon_{\mathrm{o}}$ (1) and the piezoelectric coefficients d_{33} (2) within the range of the morphotropic phase transition. (1000 h after poling)

domain structures and of the remanent polarization after poling determines the aging rate of the respective piezoelectric, elastic, or dielectric coefficients. The aging process restarts after every significant perturbation of the material, including exposing the material to electric fields, stresses, or temperatures high enough to disturb the domain distribution.

The time stability of the compositions can be characterized by an aging rate c_x in percentage per decade of its characteristics, e.g., c_ε or c_d.

For an explanation, one should consider the correlation with the parameter δ and the mobility η of the domain walls. The tetragonal compositions show an increased time stability of the characteristics (Fig. 4.15) with increased δ and reduced mobility η of the domain walls.

4.2.2.10 Nonlinear Behaviour of PZT

Most mechanical, electrical, and piezoelectric coefficients exhibit a nonlinear behavior when the piezoelectric material is subjected to large electric fields and/or mechanical stresses.

Nonlinear behavior in piezoelectric materials arises from the influence of mechanical stresses and electrical fields on domains. The limits of linear behavior vary for the different ceramic compositions and are related to the coercive force. Under dynamic conditions nonlinearities cause losses, lower the efficiency, and generate heat.

However, the difference in nonlinear behavior of the various ceramic compositions is not the only criterion for a given power application. In addition, there are further limiting factors such as dynamic strength of the ceramic, reduction in efficiency due to internal dielectric and mechanical losses, or even depolarization.

4.3 Compositional Modifications

Within the field of piezoceramics, the by far most used and widespread group is that derived from PZT. Not the least, this is due to the fact that the properties of PZT can be adapted to a broad spectrum of different requirements by minor amounts of oxides. In this context, the competition between the two phases (tetragonal and rhombohedral) plays a major role.

Many results on the impact of these additives have been found empirically. Thus, it is not easy to find the relevant correlation of the different factors such as grain size, lattice defects, domain structure, or actual incorporation of the additives. Nevertheless, we try to offer a systematic survey on the data available referring to the basic considerations of Chap. 3.

Research and development of such additives started by the use of isovalent substitutions of Pb^{2+} or Zr^{4+} respectively by Ti^{4+} on A- and B-sites. Then heterovalent substitutions were included as well. Nowadays very complex ternary and quaternary systems are used including specific dopants.

When discussing these systems, we will follow mainly their historical development.

4.3.1 Isovalent Substitutions

The first additives used for the modification of PZT were the isovalent ones [10]. Possible modifications of lead zirconate titanate ceramics are compositions with isovalent substitutions of the cations Pb^{2+} by Ba^{2+}, Ca^{2+}, and/or Sr^{2+} on A-sites and with substitutions of Ti^{4+} and Zr^{4+} by Sn^{4+} on B-sites. This leads to solid solutions of

$$({A_1}^{2+}{A_2}^{2+})({B_1}^{4+}{B_2}^{4+})O_3.$$

The modifying elements are added as oxides or carbonates into the composition during mixing the components. Usually, the concentration of the modifiers is varied from 1 to 5 at. %. The effects of isovalent modifiers are classified according to the individual characteristics of the modifiers: radii of the atoms, electron configuration, and nature of chemical bonding.

The first examples for modification of PZT by substitution of isovalent ions have been the composition series $Pb_{1-x}Sr_x(Zr_{0.53}Ti_{0.47})O_3$ [39]. The substitution of strontium ions at lead positions results in an increase of the room-temperature dielectric permittivities ${\varepsilon_{33}}^T/\varepsilon_0$, a decrease of the Curie temperature T_C, and a maximum electromechanical coupling coefficient k_p for instance with $x = 0.10$ (Table 4.2). That maximum correlates to the MPB shifted by the Sr doping. Nevertheless, the piezoelectric parameters are not high enough for any application.

Table 4.2. Modification of PZT by substitution of Sr^{2+} ions in $Pb_{1-x}Sr_x Zr_{0.53}Ti_{0.47})O_3$

x	T_C, °C	$\varepsilon_{33}{}^T/\varepsilon_0$	k_p
0	385	544	0.48
0.01	—	584	0.49
0.05	360	1,002	0.50
0.10	328	1,129	0.59
0.125	290	1,237	0.47
0.15	265	1,260	0.43
0.20	242	1,257	0.34

Table 4.3. Substitution of heterovalent ions in the perovskite lattice

Designation	Small ions: B^{4+}-positions		Large ions: A^{2+}-positions	
	Ion	Ion radius[a]	Ion	Ion radius*
PZT	Ti^{4+}	0.68	Pb^{2+}	1.20
	Zr^{4+}	0.80	Sr^{2+}	1.13
Donors, "softener"	Nb^{5+}	0.70	La^{3+}	1.22
	Sb^{5+}	0.62	Bi^{3+}	0.96
	W^{6+}	0.68	Nd^{3+}	1.04
Acceptors, "hardener"	Mn^{2+}	0.80	K^{+}	1.33
	Fe^{3+}	0.64	Ag^{+}	1.26
	Al^{3+}	0.50		

[a]Ion radius in 10^{-10} m

4.3.2 Modification by Heterovalent Ions

4.3.2.1 Donor and Acceptor Dopings: "Soft" and "Hard" Piezoceramics

The application of piezoceramics outside of the PZT binary system and doped with heterovalent ions has been favored for a long time because of their excellent physical characteristics [1].

Dopants are usually added in concentrations $\leqslant 3$ at. %. Actually, two different groups of ions have been identified for doping: ions with higher valencies as donors, and ions with lower valencies as acceptors (Table 4.3). Donors and acceptors influence the PZT ceramics antagonistically and are called *softeners* or *hardeners*, respectively. The antagonism can be understood considering the fact that vacancies in general are the mediators.

The violation of neutrality by the substitution of ions with deviating valencies is compensated in PZT by creation of charged lead or oxygen vacancies.

Dopings by 3-valent ions at A^{2+}-positions replacing Pb^{2+} resp. 5-valent or even 6-valent ions at B^{4+}-positions replacing $(Zr_x^{4+} + Ti_{1-x}^{4+})$ act as donors and form double negatively charged (V_A^{2-})-lead vacancies for compensation [40, 41].

Table 4.4. Modification of PZT with donor and acceptor material

	Dopant	T_C, °C	$\varepsilon_{33}{}^T/\varepsilon_0$	Tan δ, 10^{-3}	k_p	d_{33}, pC/m	Q_m
	soft						
(1)	Nb^{5+}	365	1700	15	0.60	374	85
(2)	Sb^{5+}	>350	1510	15	0.46	410	95
(3)	Nd^{3+}	330	1600	20	0.60	355	100
	hard						
(4)	Fe^{3+}	300	820	4	0.59	240	500
(5)	Ni^{3+}	330	1000	8	0.50	200	350

(1) $Pb_{0.98}(Zr_{0.52}Ti_{0.48}Nb_{0.024})O_3$, (2) $Pb_{0.96}Sr_{0.05}(Zr_{0.52}Ti_{0.46}Sb_{0.02})O_3$ [42], (3) $Pb_{0.97}Nd_{0.02}(Zr_{0.54}Ti_{0.46})O_3$ [43], (4) $Pb(Zr_{0.525}Ti_{0.472}Fe_{0.003})]O_3$ [44], (5) $Pb_{0.95}Sr_{0.05}[(Zr_{0.52}Ti_{0.44})Ni_{0.04})]O_3$ [42]

With the formation of lead vacancies the homogeneous strain δ is reduced, as derived from X-ray diffraction analysis. This results in a decrease of the coercive field E_c due to the increase of the domain wall mobility, facilitating domain reorientation [10], as well as in an increase of the remanent polarization and remanent strain. A typical set of dielectric and piezoelectric data is given in Table 4.4.

Compared to soft materials, in hard materials $\varepsilon_{33}{}^T$, tan δ, and d_{33} are smaller, whereas the mechanical quality factor Q_m is higher.

On the other hand, acceptor doping such as Ag or K at A^{2+}-sites or Fe, Ni, or Mn at B^{4+}-sites, respectively, forms double positively charged ($V_O{}^{2+}$) oxygen vacancies [7, 10].

Owing to the interaction with oxygen vacancies ($V_A{}^{2-}$), the mobility of the domain walls is reduced and the spontaneous strain δ and the coercive field E_C are increased. This explains the hardening. A typical set of dielectric and piezoelectric data is also given in Table 4.4.

In acceptor-modified compositions, domains and domain walls are stabilized. The mechanism is very complex and not always clear [6, 45]. Since the charged vacancies and the spontaneous polarization interact, their distribution may also determine the characteristics of the domain structure. It has to be emphasized that the explanation of the doping effects given here cannot be regarded as being entirely complete.

4.3.2.2 Coupled Substitution of Heterovalent Acceptor and Donor Ions

Primarily donors and acceptors are antagonists and therefore can cancel this influence mutually. When the charge concentrations of donors and acceptors in the perovskite structure are equal, the effects can cancel out [46,47]. A coupled substitution of heterovalent ions can be used for controlling the properties. Thus, for example, it will be possible to compensate the charge deficits caused

Table 4.5. Modification PZT by coupled substitution of donors and acceptors

	Doping	T_C, °C	$\varepsilon_{33}{}^T/\varepsilon_0$	k_p	d_{33}, pC/m	T_S, °C
(1)	Al + Nb	350	800	0.43	289	1250
(2)	Ag + Bi	-	1580	0.61	-	-
(3)	Al + Nb	343	1980	0.66	494	1180
(4)	Fe + Nb	343	1960	0.66	495	1200

(1) $Pb_{0.98}Sr_{0.02}[(Zr_{0.53}Ti_{0.47})_{0.98}Al_{0.01}Nb_{0.01}]O_3$,
(2) $Pb_{0.95}(Ag_{0.025}Bi_{0.025})(Zr_{0.52}Ti_{0.48})O_3$ [48],
(3) $[Pb_{0.995}Sr_{0.02}][(Zr_xTi_{1-x})_{0.975}Al_{0.005}Nb_{0.02}]O_{3.06}$ [49],
(4) $[Pb_{0.995}Sr_{0.02}][(Zr_xTi_{1-x})_{0.975}Fe_{0.005}Nb_{0.02}]O_{3.06}$ [49]

by acceptor doping by means of charge surplus resulting from donor doping completely (see compositions 1 and 2 in Table 4.5)

By coupled substitution of acceptor and donor ions at Zr/Ti positions, a high level of piezoelectric activity is achieved within the system PZT (see compositions 3, 4 in Table 4.5). In addition, it is possible to increase the stability of piezoceramics when maintaining the piezoelectric activity and the high dielectric permittivity. The materials of such a system are also characterized by a high Curie temperature and, in particular, by an improved sintering activity, a decreased sintering temperature T_S (Table 4.5), as outlined in Sect. 4.4.

The dielectric and electromechanical characteristics of compositions with partial valency compensation (examples 3, 4) are much better and are more often used than those with valency-compensated coupled substitution (examples 1, 2) where Pb and oxygen vacancics are minimized.

A coupled substitution (double substitution) with unequal amounts of heterovalent donor and acceptor ions can stabilize ferroelectric "soft" ceramics when donor doping is dominant. On the other hand, the influence of acceptor doping predominates if the acceptor concentration is higher than the donor concentration. Therefore, it is also possible to synthesize piezoceramic materials with "middle hardness", i.e., with higher coercive field E_c than "soft" materials, but with a high Curie temperature T_C and high piezoelectric activity.

4.3.2.3 Thermal Stability and Aging of the Resonant Frequency

The addition of small concentrations of dopants permits modification of the ferroelectric properties and also thermal stability. Depending on the type of cations substituted into the solid solutions, the degree of ferroelectric hardness and stability of the domain structure against external influences can be controlled as well.

Low temperature coefficients of the resonant frequency can be obtained by substituting small amounts of alkaline-earth cations in acceptor-doped material and by adjusting simultaneously the ratio of zirconium to titanium, which governs the ratio of tetragonal to rhombohedral phases in the ceramic.

The addition of Mn or Cr [27,50,51] as modifiers results in a strong decrease of $\delta f/f_r$. The modifiers Mn and Cr have variable valencies. Thus, for an interpretation of the characteristics of the modified solid solutions one must consider the possible change of these oxidation states of the elements during the synthesis of the ceramic. For example, in the case of Mn doping, one uses Mn^{2+} compositions as raw materials, which may oxidize already in air to Mn^{3+} or change to the higher valence state Mn^{4+} in the presence of lead oxide. The outstanding features of such modified compositions are very low aging and low temperature coefficient of the resonant frequencies [10].

Same effects can be obtained by substituting <1 wt.% Cr^{3+} at the B-site, although the mechanism has not been elucidated. The ions are probably present as Cr^{3+} and since they slightly reduce the permittivity and the piezoelectric activity, it can be assumed that they promote an orientation of the acceptor ion–oxygen vacancy dipoles during and after poling [3].

By the addition of the oxides of Mn^{4+} or Cr^{3+} additional oxygen vacancies develop, as well. It is assumed that this is the source of a space charge field E_{sp} that stabilizes the remanent field. This so-called stabilization field can be detected within the hysteresis and derived from the imbalance of the full hysteresis loop of a poled ceramic. Free charges at the grain boundaries or defects oriented relative to the spontaneous polarization are responsible for E_{sp}.

The existence of a space charge field influences thermal stability, too. The increase of E_{sp} is coupled with a decrease of the wall mobility and an increase of E_C. Since the values of the space charge field correlate with the improvement or deterioration of thermal stability, it is obvious that the thermal stability of the resonant frequency is determined not only by the mobility of the domain walls but also by their number. Modifiers that reduce the mobility and also the number of the domain walls increase the stability.

4.3.3 Multicomponent Systems: Ternary and Quaternary Solid Solutions

4.3.3.1 Existence Criteria of Complex Perovskites

An even larger variety of modifications is possible if combinations of ions are substituted. Multicomponent doping follows well-known rules. Already, perovskites (structure family) without atomic defects [52]

$$(A_1, A_2, \ldots, A_m)(B_1, B_2, \ldots, B_n)O_3$$

offer a large variety of modifications.

Such perovskites with cubic structure in the paraelectric phase can be composed by isovalent substitution: $A^{2+} = Pb^{2+}$, Ba^{2+}, Ca^{2+}, or Sr^{2+}; $B^{4+} = Ti^{4+}$, Zr^{4+}, or Sn^{4+} (see isovalent substitutions, Sect. 4.3.1).

The electrical neutrality condition leads to

$$\sum_{i=1}^{m} x_{Ai} n_{A_i} + \sum_{k=1}^{n} x_{Bk} n_{Bk} = 6, \tag{4.13}$$

with x_{Ai}, x_{Bi} the fractions of the ions and n_{Ai}, n_{Bi} the valencies of the ions.

The stoichiometry condition gives

$$\sum_{i=1}^{m} x_{Ai} = 1, \quad \sum_{i=1}^{n} x_{Bi} = 1. \tag{4.14}$$

Perovskite structures cannot exist except the under geometrical condition fulfilled by the radii of oxygen and of the ions at the A- and B-sites (compare Fig. 4.1 of Sect. 3.1):

$$\frac{R_A + R_O}{\sqrt{2}(R_B + R_O)} - t \tag{4.15}$$

For the tolerance factor t, the condition $0.8 < t < 1.05$ should be fulfilled [53]. The deviation of t from the simple geometrical condition $t = 1$ shows that not only a pure ionic binding must be considered.

The condition for the radii of the cations A^{2+} and B^{4+} applies also to more complicated compositions with statistically distributed cations. In this case, we have to replace R_A and R_B by their average values

$$\sum_{i=1}^{m} x_{Ai} R_{Ai} = \overline{R_A} \tag{4.16}$$

$$\sum_{k=1}^{n} x_{Bk} R_{Bk} = \overline{R_B}. \tag{4.17}$$

The above conditions can be fulfilled in the case of cation valencies 1, 2, and 3 on the A-site and 1, 2, 3, 5, 6 on the B-site. Since R_O is unchanged, the following conditions have been found experimentally in practice [52]:

$$\overline{R_A} \geqslant 0,90 \times 10^{-10}\text{m}$$
$$0,51 \times 10^{-10}\text{m} < \overline{R_B} < 1,1 \times 10^{-10}\text{m}$$
$$\overline{R_A} \geqslant \overline{R_B}.$$

To date, more than 1,000 compositions of complex compounds have been identified [16, 45, 54] that fulfill the above criteria.

However, the role of vacancies remains an unsolved problem. Also, no unanimous opinion exists concerning the resulting stoichiometry in case of a simultaneous acceptor and donor doping. Here it should be noted that the concentration and impact of vacancies cannot be determined by the initial chemical composition alone. Evaporation of Pb and valency changes during sintering are mostly not known. Of course, the neutrality condition must be fulfilled and is helpful for this problem. A consistent terminology is still lacking concerning definitions such as structure with vacancies, cation–anion complexes, defect structure, non-stoichiometric composition, substructure, and second phases. Clear definitions would be helpful.

4.3.3.2 Multicomponent Systems: Substitution of Complex Compounds

Multicomponent systems consist of PZT with additions of complex perovskitic compounds. In Table 4.6, some possible ion combinations that are applicable for such complex compounds are summarized. Often these complex compounds are synthesized as precursors for the preparation of the ceramics.

The combinations of ions of different valency states, stated in the first column of the Table 4.6, according to the general existence conditions for perovskites are introduced into the complex compounds. These complex compounds are perovskites, too. They are isomorphic with the lead titanate and lead zirconate and, therefore, also go into solution.

With the modification of the binary PZT by complex compounds one succeeds in synthesizing multiple component systems exhibiting unusual characteristics. An increase of the electromechanical effects, piezoelectric coefficients, and unidirectional strain can be reached, as well as a decrease of the temperature coefficients and aging. Ternary and also quaternary solid solutions offer new degrees of freedom for tailoring piezoceramics. Furthermore, it is possible to achieve lower sintering temperatures, low loss of PbO, lower porosity, higher density, and finer and more uniform grains [55].

This is probably due to this fact that heterovalent ion combinations favor the creation of vacancies at both A- and B-sites. This promotes the diffusion processes during sintering and consequently an optimum ceramic microstructure and optimum physical characteristics.

Ternary Solid Solutions

Of increasing interest are three-component systems or ternary solid solutions for piezoceramic materials with maximum piezoelectric activity, such as

$$PbTi0_3 - PbZr0_3 - (PbB'_{1-\alpha}B''{}_{\alpha}0_3)$$

with cation complexes $B'_{1-\alpha}B''_{\alpha}O_3$ ($\alpha = 1/4$, $1/3$ or $1/2$ as a function of the valencies of the cations B' and B'').

Table 4.6. Composition of complex compounds [1, 10, 54, 55, 62]

Complex compounds	Ions	
$(A^{1+}{}_{1/2}A^{3+}{}_{1/2})TiO_3$	A^{1+}	Li, Na, K, Ag
	A^{3+}	Bi, La, Ce, Nd
$Bi(B^{2+}{}_{1/2}Mn_{1/2})$	B^{2+}	Sr, Zn, Ni
$A^{2+}(B^{1+}{}_{1/4}B^{5+}{}_{3/4})TiO_3$	A^{2+}	Pb, Sr
	B^{1+}	Li, K, Ag
$Pb(B^{2+}{}_{1/3}B^{5+}{}_{2/3})TiO_3$	B^{2+}	Mg, Ni, Zn, Mn, Co, Fe, Cu
$Pb(B^{3+}{}_{1/2}B^{5+}{}_{1/2})TiO_3$	B^{3+}	Mn, Sb, Al, Fe, Co, Yb, In
	B^{5+}	Nb, Sb, Ta
$Pb(B^{3+}{}_{2/3}\,B^{6+}{}_{1/3})\,TiO_3$	B^{6+}	W, Te, Re
$Pb(B^{2+}{}_{1/2}\,B^{6+}{}_{1/2})\,TiO_3$		

Most of the complex compounds are so-called *relaxor ferroelectrics* with diffuse paraelectric–ferroelectric phase transition because of fluctuations in the ratio of B′ and B″ ion concentrations [6, 56]. Such relaxor ferroelectrics offer a great variety of improved soft piezoceramics that combine high relative permittivity, high piezoelectric activity, and reduced Curie temperatures. In general, the used compositions are adjusted to the range of the MPB.

Often ternary systems are doped with a stabilizer such as Mn or Cr additionally [57]. They show a wider variety of characteristics, especially zero temperature coefficients and lower aging. Furthermore, it is advisable to use Mn directly as a constituent of the complex oxide precursors.

Owing to the large number of possible ternary piezoelectric ceramic systems, it is impossible to discuss all of them in detail. In Table 4.7 some complex ionic combinations forming lead-based piezoelectric ceramics are listed. The ternary system $Pb[(Zr_xTi_{1-x})_{1-z}(Mg_{1/3}Nb_{2/3})_z]O_3$ was one of the first attractive systems offering excellent piezoelectric properties [64]. Compositions near the morphotropic phase transition show large values of the dielectric permittivity ${\varepsilon_{33}}^T/\varepsilon_0$ and the electromechanical coupling coefficient k_p, (see examples 2 and 3). Solid solutions of this system offer the advantage of a broader sintering interval and smaller evaporation of lead oxide.

In order to improve the properties of PZT ceramic, other non-lead compounds, such as $({Ag^{1+}}_{1/2}{Bi^{3+}}_{1/2})TiO_3$, $({Ag^{1+}}_{1/2}{La^{3+}}_{1/2})TiO_3$, $Sr({K^{1+}}_{1/4}{Nb^{5+}}_{3/4})TiO_3$, and $Sr({Ag^{1+}}_{1/4}{Nb^{5+}}_{3/4})TiO_3$ or $Bi(Zn_{1/2}Mn_{1/2})$, perovskite $A^{2+}({Zn^{2+}}_{1/2}\ {Mn^{4+}}_{2/3})O_3$ can also be added to PZT as a third element in order to form ternary piezoelectric ceramics and to improve PZT ceramics.

Quaternary Solid Solutions

Four-component systems (quaternary solid solutions) of the type

$$PbTiO_3 - PbZrO_3 - (PbB'_{1-\alpha}B''{}_{\alpha}O_3) - (PbB'''{}_{1-\beta}B''''{}_{\beta}O_3)$$

as well as their essential characteristics and the homogeneous strain parameter δ are shown in Table 4.8. Soft and hard properties can be distinguished.

Cation complexes with niobium and/or tungsten as a fourth component lead to very soft materials. This is a result of more complicated structural changes with heterovalent substitutions in the PZT and with a possible increase in the concentration of vacancies.

In the second group, hard materials with Mn addition are important. They offer a combination of small dielectric losses, tan δ, high mechanical quality factor Q_m, and a high coupling constant k_p.

The homogeneous strain parameter δ of the two indicated material categories is quite similar. The modification of the composition with manganese-containing complexes becomes obviously more effective because of the strong reduction of the domain mobility.

Table 4.7. Dielectric and electromechanical characteristics of ternary solid solutions based on PZT

	Substitutions to PZT	δ, 10^3	E_c, kV/mm	T_C, °C	$\varepsilon_{33}{}^T/\varepsilon_0$	k_p	d_{33}, pc/m	Q_m	T_S, °C
	Soft								
(1)	$(Ni_{1/2}W_{1/2})$	—	—		1600	0.55	—	75	—
(2)	$(Mg_{1/3}Nb_{2/3})$	—	—	250	2406	0.48	—	96	1280
(3)	$(Mg_{1/3}Nb_{2/3})$	—	—	196	3856	0.58	(570)	61	—
(4)	$(Ni_{1/3}Sb_{2/3})$	13.3	0.85	273	2050	0.61	460		1260
(5)	$(Li_{1/4}Nb_{3/4})$	—	—	380	1800	0.78	—		—
(6)[a]	$(K_{1/4}Nb_{3/4})$	15.3	1.12	349	1950	0.63	470	90	1120
(7)[a]	$(Ag_{1/4}Nb_{3/4})$	15.7	1.08	349	2080	0.66	498		1060
	Hard								
(8)	$(Zn_{1/3}Nb_{2/3})$	—	—	.	1370	0.55	—	1410	—
(9)[b]	$Bi(Zn_{1/2}Mn_{1/2})$	—	—	357	1000	0.55	240	1000	1020
(10)	$(Ni_{1/3}Sb_{2/3})$	13.0		281	1250	0.38	200	720	1260

T_S sintering temperature
(1) $Pb[(Zr_{0.40}Ti_{0.40})(Ni_{1/2}W_{1/2})_{0.2}]O_3$,
(2) $Pb[(Zr_{0.245}Ti_{0.3850})(Mg_{0.125}Nb_{0.250})]O_3$ [58],
(3) $Pb_{0.95}Sr_{0.05}[(Zr_{0.25}Ti_{0.375})(Mg_{1/3}Nb_{2/3})_{0.375}]O_3$ [59],
(4) $Pb[(Zr_{0.45}Ti_{0.47})Ni_{0.027}Sb_{0.054}]O_3$, (5) $Pb[(Zr_{0.51}Ti_{0.47})(Li_{1/4}Nb_{3/4})_{0.02}]O_3$ [60],
(6) $[Pb_{0.975}Sr_{0.02}K_{0.005}][(Zr_{0.53}Ti_{0.47})_{0.985}Nb_{0.015}\ O_3$,
(7) $[Pb_{0.975}Sr_{0.02}Ag_{0.005}][(Zr_{0.53}Ti_{0.47})_{0.985}Nb_{0.015}\ O_3$,
(8) $Pb[(Zr_{0.0.440}Ti_{0.35})(Zn_{1/3}Nb_{2/3})_{0.125}]O_3 + 0.5$ wt.% $Mn0_2$ [61],
(9) $0.98Pb(Zr_{0.50}Ti_{0.50})O_3 - 0.02\ Bi(Zn_{1/2}Mn_{1/2})O_3$ [62,63],
(10) $Pb[(Zr_{0.41}Ti_{0.51})Ni_{0.027}Sb_{0.054}]O_3 + 0.5$ wt.% $Mn0_2$ [27].
[a]The coupled substitution of the acceptor ion K^+ and/or Ag^+ with large ion radius at A-positions and the donor ion Nb^{5+} with a small ion radius at B-sites represents a partial compensation with reduced PbO loss and high dielectric and electromechanical values. With the synthesis of the composition $[Pb_{0.98}Sr_{0,02}Ag_{0.005}][(Zr_{0.53}Ti_{0.47})_{0.985}Nb_{0.015}]O_3$, a further decrease of the sintering temperature to $T_S \cong 1060$°C is observed.
[b]With addition of bismuth and manganese in connection with two-valent ions in the complex compound $Bi(Zn_{1/2}Mn_{1/2})$ the sintering temperature T_S of ceramics can be reduced remarkably. This permits the fabrication of lead zirconate titanate ceramics without larger losses of lead oxide. These piezoceramics sintering at low temperatures show a high mechanical quality factor, and some of them a high thermal stability of the resonant frequency [62].

As can be seen from both Tables 4.7 and 4.8, specific materials are generally available and partially already standardized. Nevertheless, specific adaptations to special requirements are often necessary. They should remain within the given framework and obey the general rules given above.

Table 4.8. Dielectric and electromechanical characteristics of quaternary solid solutions based on PZT

Material	δ, 10^3	T_C, °C	$\varepsilon_{33}^T/\varepsilon_0$	Tanδ, 10^3	k_p	d_{33}, pC/m	Q_m	T_S, °C
Soft								
(1) $Nb_{2/3}Mg_{1/3}$-$W_{1/3}Ni_{2/3}O_3$	8.0	212	3800	16	0.65	690	85	1150
(2) $PbW_{1/2}Ni_{1/2}O_3$-$PbNb_{2/3}Ni_{1/3}O_3$	—	238	3010	15	0.637	—	89	—
(3) $PbW_{1/2}Mg_{1/2}O_3$-$PbW_{1/2}Zn_{1/2}O_3$	12	250	2635	15	0.675	—	88	1100
Hard								
(4) $Nb_{2/3}Zn_{1/3}O_3$-$PbNb_{2/3}Mn_{1/3}O_3$	13.3	324	1310	2.4	0.577	—	1110	1240
(5) $Nb_{2/3}Zn_{1/3}O_3$-$PbNb_{2/3}Mn_{1/3}0_3$	11	316	1000	6	0.562	—	1500	1240
(6) $Nb_{2/3}Zn_{1/3}O_3$-$PbNb_{2/3}Sn_{1/3}O_3$	9.8	320	1165	4	0.60	295	1570	1240

T_S sintering temperature
(1) $Pb[(Zr_{0.240}Ti_{0.385})(Nb_{0.250}Mg_{0.125})]_{0.97}(W_{0.01}Ni_{0.02})]O_3$ [33],
(2) $Pb[(Zr_{0.3664}Ti_{0.400})(Nb_{2/3}Ni_{1/3})_{0,1146}(W_{1/2}Ni_{1/2})_{0.1194}]O_3$ [16,65],
(3) $Pb[(Zr_{0.378}Ti_{0.400})(W_{1/2}Mg_{1/2})_{0.066}(W_{1/2}Zn_{1/2})_{0.154}]O_3$ [16,65],
(4) $Pb[(Zr_{0.3973}Ti_{0.425})(Nb_{2/3}Zn_{1/3})_{0.133}(Nb_{2/3}Mn_{1/3})_{0.0448}]O_3$ [16,65],
(5) $Pb[(Zr_{0.3954}Ti_{0.450})(Nb_{2/3}Zn_{1/3})_{0.0544}(Nb_{2/3}Mn_{1/3})_{0.1002}]O_3$ [16,65],
(6) $Pb[(Zr_{0.44}Ti_{0.44})(Zn_{1/3}Nb_{2/3})_{0.06}(Sn_{1/3}Nb_{2/3})_{0.06}O_3] + 0.5$ wt.% MnO_2 [66].

4.4 Specific Aspects of Pb-Containing Piezoelectric Ceramics

To date, PZT-based ceramic is the most used technological piezoelectric material. Owing to its high piezoelectric effects and to its simple fabrication, its triumphant progress started about fifty years ago encompassing a large number of applications. Since that time, many thousands of publications have dealt with their different fabrication methods, doping systems, physical properties, and specific application profiles. Their geometry ranges from thin film elements to heavy ceramic bodies [67].

Even today a vibrant R&D activity can be observed worldwide. This is due to the numerous new applications that require specific adaptation. It is very favorable that within this synthetic material the method of fabrication, selection of dopants, the design of components, and the application modalities are closely related and interlinked. Therefore, besides soft standard PZT and hard power PZT, most manufacturers offer a series of custom-tailored ceramics for specific applications. In order to understand the close connection, the essential material aspects are summarized in the following section.

Mostly PZT ceramics are manufactured using the mixed-oxide method. PbO, ZrO_2, TiO_2, and doping oxides are wet-milled. After drying, the samples are calcined at peak temperatures of between 600 and 1,000°C. During heating, at first $PbTiO_3$ is formed, followed by $PbZrO_3$ with simultaneous interdiffusion of Ti and Zr, and finally by the formation of the homogeneous PZT [8]. Because the transport of PbO takes place over the gas phase, the calcination cake is blown up due to the Kirkendahl effect. The doping cations are incorporated at the A- or B-site of the perovskite depending on atomicity and ion size. They also may be present as second phases, such as Nb in a pyrochlore phase. The quantitative reaction during calcination can be controlled by X-ray methods and promoted by a small excess of PbO.

In order to achieve homogeneous calcinations, the starting ZrO_2 powder must be as fine as possible, because the Zr ion has the lowest diffusion coefficient, lower by orders of magnitude, within the perovskite. A practical limit for the fineness of the ZrO_2 corresponds to a specific surface area of about $30\,m^2\,g^{-1}$ for avoiding segregation problems during milling and drying. Owing to the statistics of the particle mixing of TiO_2 and ZrO_2, after calcinations PZT particles with a widely varying Zr/Ti ratio are present. In general, good piezoelectric properties of sintered PZT can be achieved only if the ratio Zr/Ti within the grains and from grain to grain near the MPB is nearly constant. Therefore, a second wet-milling is necessary which improves the chemical homogenization and the sinterability. A typical mean primary particle size of 0.5 µm after second milling requires the incorporation of about 500 primary particles within one single growing grain of 4 µm diameter during sintering. This leads to a homogeneous Ti and Zr distribution.

Even now the question of the mechanisms of densification and grain growth during sintering is a matter of discussions and investigations because of the complex relation between the preparation methods and chemical composition. In most cases, a combination of nucleation and Ostwald ripening determines the microstructure. The nucleation of the finally growing PZT grains is promoted by a local melt of PbO and PT [68]. Experimental hints supporting that assumption are, e.g., etch pits that can be seen after thermal etching in the middle of the grains. Similarly, microanalyses of single grains show a compositional gradient from inside to outside [69]. Further growth of the enlarged nuclei is assumed to be due to normal Ostwald ripening.

By doping, in most cases, a grain growth inhibition is superimposed on both these mechanisms. Because the defect equilibria of the doped PZT depend on temperature, disequilibria can be formed during heating and sintering at the surface of the growing grains so that not all present cations can be incorporated directly and stoichiometrically within the growing grain.

This behaviour will be illustrated by three examples:

In Fig. 4.16 [70], the grain size of a PZT sample prepared by the mixed-oxide method is shown as a function of the Nd content. At small Nd concentration, good densification at low sintering temperature and maximum grain size are observed. It diminishes drastically above 0.8 mol%. Probably, this can

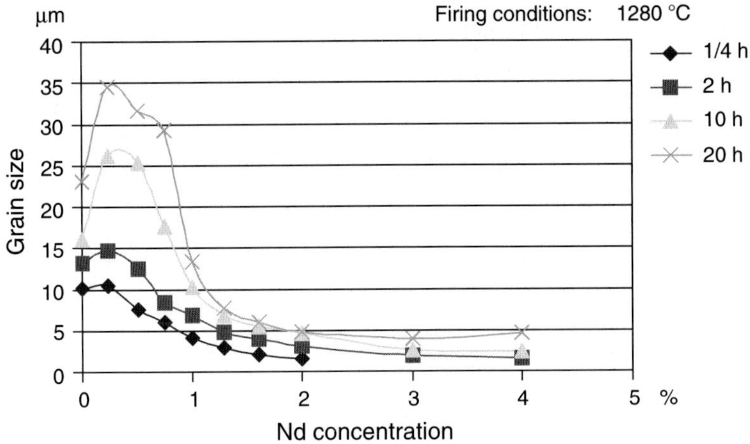

Fig. 4.16. Grain size as a function of Nd_2O_3 concentration for $Pb_{1.00}Nd_xZr_{0.53}Ti_{0.47}O_3$ at 1280°C sintering temperature and different sintering times

be explained by the assumption that only below this concentration the existing natural impurities of the raw materials are homogeneously incorporated on A- and B-sites together with the Nd^{3+} on the A-sites. Thus they ensure electro-neutrality by their different valencies without formation of additional vacancies.

A further example is undoped lead titanate (PT). This material sinters at 700°C already forming large grains, but cracks into a powder of single grains during cooling down because of the large tetragonal anisotropy.

A third example is shown in Fig. 4.17. Extreme grain growth at reduced sintering temperature can be obtained even in complex doping systems if the doping concentration is varied empirically until defect equilibrium and stoichiometric incorporation of all offered cations at sintering temperature is achieved [71, 72]. In general, this doping concentration is different from the so-called stoichiometric hypothesis, which in the present case would be 1.0 at.% Mn^{2+} on B-sites and 2 at.% Nd^{3+} on A-sites. It must be assumed that temperature-dependent defect equilibria exist where valency changes of the cations (for example Mn^{3+} or Mn^{4+}), incorporation of even smaller ions on A- and B-sites, and the formation of Pb and O vacancies have to be included. Such a temperature shift of equilibrium may be the reason for the minor temperature shift of the maxima as found with sintering temperature (cf. the example of Fig. 4.17).

In Fig. 4.18, the grain size shows a clear maximum for an Fe addition of 1.5 at.% independent of the sintering temperature. For a compensation of 2 at.% Nd^{3+} on A-site using Fe_2O_3 as a dopant, the maximum expected is 2 at.% Fe. The fact that the point of stoichiometric incorporation and maximum grain growth lies at 1.5 at.% Fe shows that Fe is incorporated with

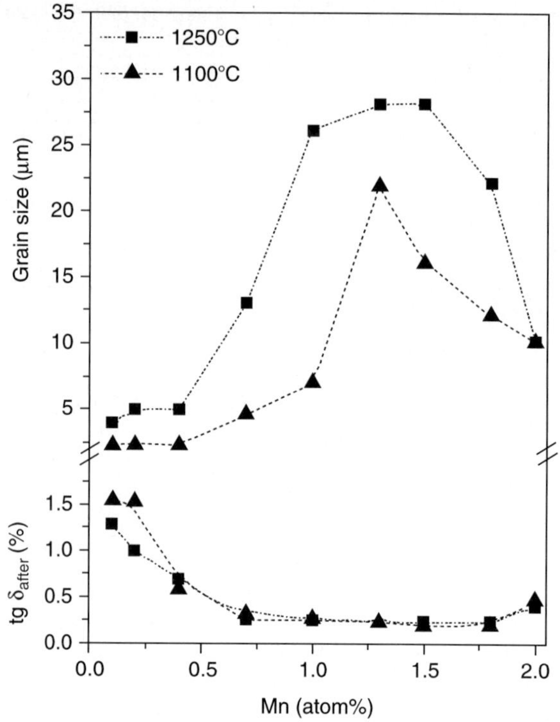

Fig. 4.17. Grain size and loss angle as a function of Mn concentration for the composition $Pb_{1.00}Nd_{0.02}(Zr_{0.545}Ti_{0.455})_{1-y}Mn_yO_3$ for two sintering temperatures [71]

valencies 2+ and 3+. It also suggests that depending on doping the grain growth inhibition in PZT is caused by the formation of A-site or oxygen vacancies and by the diffusion processes by which Pb and/or oxygen leave the sample.

Further results of the doping variations with Mn and Fe in Figs. 4.17 and 18 show another important behavior. Both dopings are used for the fabrication of hard PZT (see Chaps. 7, 10 and 11). It is generally agreed that the effect of these dopings is based on the fixing of the domain structure and on the reduction of the domain wall mobility. Till now, two different reasons have been discussed [73, 74]. As a bulk effect, complexes of dopants or vacancies within the domains or within the domain walls may be oriented and stabilize the polarization. Charged and polarized layers at the grain boundaries may have the same effect.

In Fig. 4.19 the mechanical quality factor Q is shown as a function of the Mn and Fe doping. With Fe doping up to 1.5 at.%, the ceramic shows all characteristics of a soft PZT. For higher doping concentrations, Fe enrichment at the grain boundaries and a steep Q increase can be found, indicating a stabilization by internal electric fields. In case of Mn, the quality factor

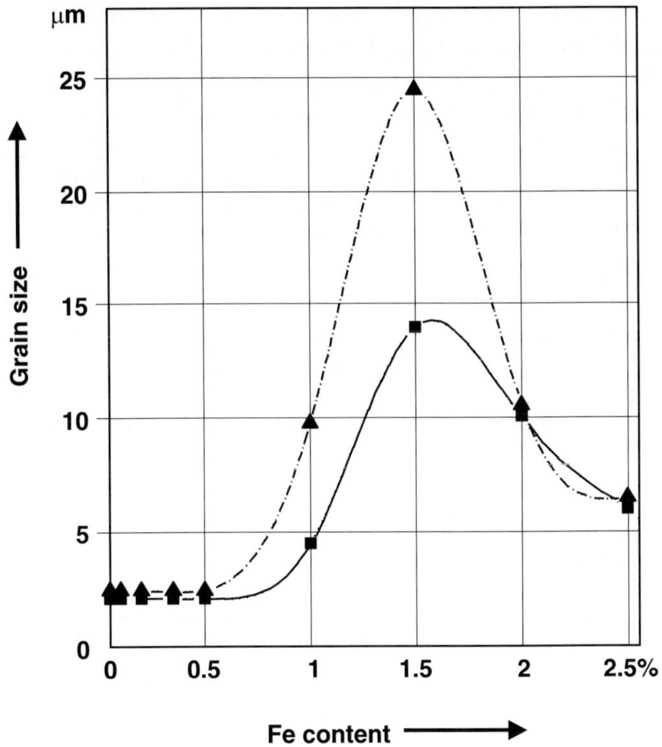

Fig. 4.18. Grain size as a function of Fe concentration for the composition $Pb_{1.00}Nd_{0.02}(Zr_{0.545}Ti_{0.455})_{1-y}Fe_yO_3$ for two sintering temperatures [71]

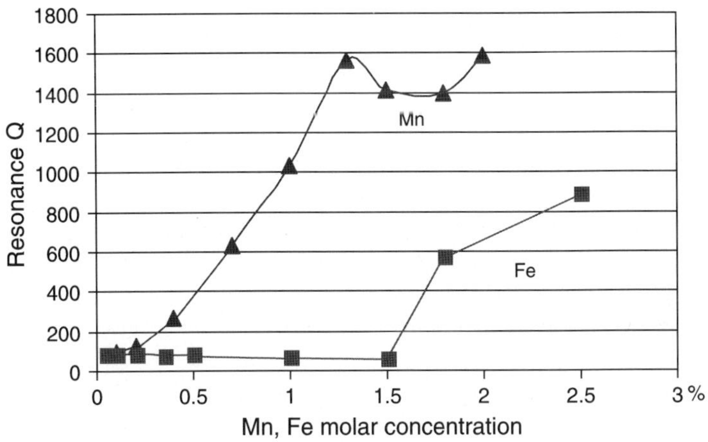

Fig. 4.19. Mechanical quality factor versus Mn or Fe doping

continuously increases from the smallest doping concentrations. This can be ascribed to bulk effects. The details of the defect structure inside the samples, for example whether doping complexes or pairs of vacancies are formed, are being investigated presently [75], but future research is still necessary. Suitable methods are paramagnetic measurements, electron spin resonance, and thermogravimetry [74]. Even after-effects can bring new information: over a wide temperature range down to 4 K, the time dependence of the small-signal permittivity is measured immediately after a new distribution of the domain structure. Hence activation energies and possible structures of defect complexes can be derived.

Another important aspect must be mentioned here. At the sintering temperature, PZT is already metastable and decomposes into volatile PbO as well as lead titanate and ZrO_2. Therefore, the PbO partial pressure is a further important parameter for the formation of the defect equilibrium and for the homogeneity of the sample [76]. Therefore, since the beginning of the PZT development, the PbO partial pressure during sintering is defined by using saggers or PbO donors such as $PbZrO_3$.

Finally, the role of oxygen must be mentioned. Oxygen ions diffuse through the perovskite lattice very easily even down to room temperature. Nitrogen, however, cannot diffuse. Therefore sintering in air gives closed porosity in general, and a final relative density of 96% can be reached. Densities near 100% can result when sintering in pure oxygen, as it is used, e.g., for the fabrication of transparent PLZT ceramics [77]. A lowering of the oxygen partial pressure during sintering, sometimes caused by the remaining binder due to insufficient burning, may reduce Ti^{4+} to Ti^{3+}, thereby creating oxygen vacancies and resulting in an undesirable electrical conductivity of the ceramic.

Owing to the complex influences of many process parameters and of the doping, most published experimental data cannot be used except as qualitative hints (this includes also data reported in Sect. 4.3). In any case, from the data available, it is not fully clear whether the defect structure, the doping distribution, and the concentration of Ti and Zr are completely homogeneous both inside each grain and from grain to grain of the sintered ceramic. Especially in the range of the MPB, where all piezoelectric data change dramatically and partially have maximum within a range of only a few mol% of Ti or Zr, the homogeneity of the Ti/Zr distribution is the most important parameter to be considered. In Fig. 4.20 some piezoelectric data are compared in the range of the MPB for a normal (MO) and a very homogeneous (HTS) preparation. The sharpness of the changes and the maximum of the values are a direct measure of the more or less homogeneous preparation. If grains with different Ti/Zr concentrations are present within the sample, the principally present coexistence region of rhombohedral and tetragonal phases is broadened additionally, and the maximum piezoelectric values are lowered according to the mixing rules. An important homogeneity criterion is the height and sharpness of the permittivity curve near the Curie temperature.

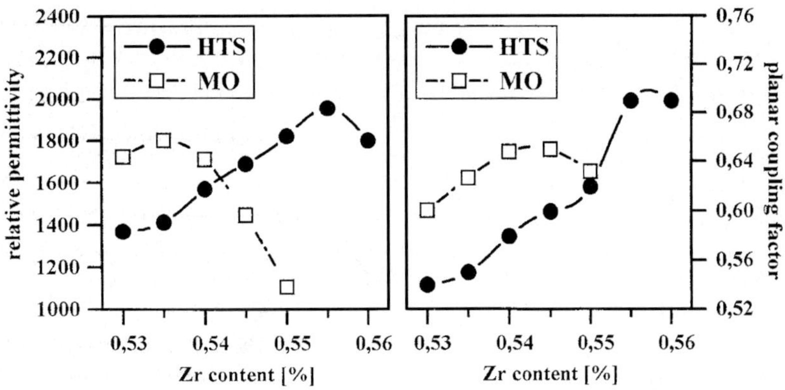

Fig. 4.20. Relative permittivity and coupling factor as a function of Zr content [79]

4.5 Powder Preparation Methods and Technologies for Different Components

For getting optimum piezoelectric values in the range of the MPB, in the past a large variety of fabrication methods were examined in order to reach high homogeneity and reproducibility within tolerable costs. The following methods should be mentioned especially: the chemical precipitation of all or some special cations, hydrothermal methods, spray-drying with simultaneous precipitation from a solution, and the sol gel method. Seldom, however, within this investigation was the improvement of the homogeneity verified by the narrowing the MPB range or by microanalytical methods (see Sect. 13.1). Besides poor reproducibility and, particularly, drastically increased production costs, the typical PZT problems of second phase formation, incomplete incorporation of cations during grain growth, and difficult grain size control have not yet been solved satisfactorily.

Therefore, for the most applications up to now the above-described mixed-oxide method is used. It is a compromise between low costs, high reproducibility, and good, but not best, piezoelectric data. Consciously the doping combinations are adjusted to compromise between moderate grain growth inhibition and a broader range of MPB. At sintering temperatures between 1,200–1,300°C, relative densities >96% and typical mean grain size of 4 μm are obtained. At larger grain sizes, the mechanical toughness is reduced. This causes drawbacks for the polarization process because of internal mechanical stress between adjacent grains as well as for machining and for long-term stability under large-signal operation. At a mean grain size <1 μm, the piezoelectric activity decreases drastically. Different reasons are given: compositional fluctuations, mechanical clamping of the grains and change of domain structure, and domain wall mobility [19].

As dopants, preferentially oxides of such cations are used that are not volatile during sintering, have a well-defined valence state, and are incorporated at a well-defined lattice site. As to standard PZT, the large rare earth ions La^{3+} and Nd^{3+} on A-site and Nb^{5+} on B-site are preferred, while for hard or power PZT, Fe or Mn on B-sites are preferred additionally as outlined above.

Often the mixed-oxide method is modified by using precursors instead of oxides. Such precursors are, e.g., $NiNbO_2$ or $MnNbO_2$. They are used for avoiding second phases or the formation of pyrochlore. An especially positive influence on the homogeneous Zr/Ti distribution is found when using a $Zr_xTi_{1-x}O_2$ as precursor. It can be produced by separate calcination of TiO_2 and ZrO_2 or, for high homogeneity and small particle size, by hydrothermal precipitation [78,79]. A narrower MPB range and best piezoelectric values are the result (Fig. 4.20).

The pre-reacted or calcined and milled PZT powder is the starting material for the shaping of the different components. In this case, most traditional ceramic technologies can be used.

For all larger components such as disks, blocks, plates or calottes, the PZT powder is granulated under addition of a suitable binder. For small quantities this is done in a mortar, but for industrial series by spray drying. For shaping, uniaxial or isostatic pressure is applied. Typical relative green densities of 60% are achievable. The next steps are debindering and sintering, often under controlled O_2 and PbO partial pressure. The machining to final dimensions is done by dicing, grinding, lapping, and polishing. Owing to the considerable softness of the PZT ceramic, in general diamond tools are used. For metallization, in the simplest case metal pastes (especially Ag, but also Cu, Ni) are screen printed and burned in at 400–600°C. Higher-quality electrodes are sputtered and may consist of many different layers depending on the type of further contacting. An example of a solderable metallization is the triple sputtering of Cr for adhesion, Pt as diffusion barrier when soldering, and finally Au because of its excellent wetting during soldering.

For optimum adhesion of the metallization, it is recommended to anneal the components after machining at 950–1,000°C. The removal of contaminants, the annealing of mechanical damages, and a certain degree of recrystallization of the surface can be achieved.

For the final polarization, there is no uniform choice of the parameters such as time, temperature, and electric field strength. It depends strongly on the type of PZT and on the design of the component. Thin ceramic foils of soft PZT can be polarized completely in air at $2\,\mathrm{kV\ mm^{-1}}$ within 1 s. For thicker components, the necessary voltage comes up to the kilovolt range, where poling is performed in oil. If instead of oil SF6 gas is used, no further cleaning is necessary. Because the switching of 90° domains is connected with internal mechanical stress, for bigger components in general the poling temperature is raised. This way the coercive force is decreased (see Chap. 3) and thermally activated processes for compensation are promoted. Especially for hard PZT,

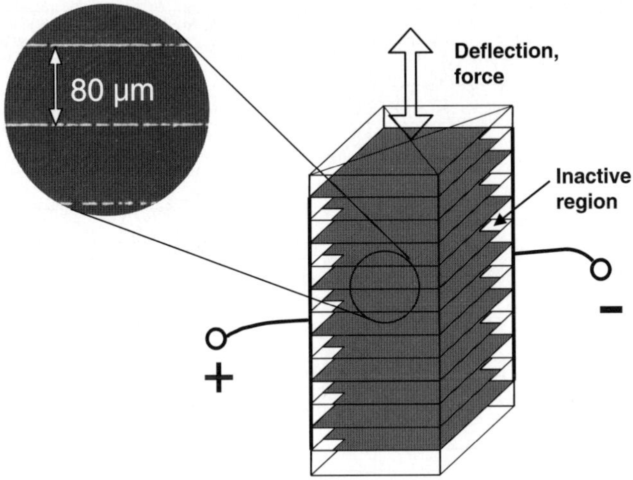

Fig. 4.21. Principle of design of a multilayer stack. The inner AgPd electrodes are gray, the electrical connections are black, and the cross-sections of inactive regions are white

increased poling temperature and time are necessary. Also, poling by a reduced electric field is possible when heating above the Curie temperature and cooling down afterwards.

For the fabrication of thin piezoelectric components, wet-shaping is used. The PZT powder is mixed with the binder and an aqueous or organic solvent. With that slurry, green tapes are manufactured by doctor-blading or tape casting with thicknesses between 20 and 200 µm [80]. Out of these green tapes, round parts, for example, are punched out, sprayed with ZrO_2, stacked, sintered, separated, metallized, and poled. Several millions of low-cost buzzers are produced annually this way worldwide.

For the fabrication of stack actuators and bimorphs, a multilayer technology has been adapted to PZT [81]. It was developed initially for multilayer capacitors of $BaTiO_3$. Figure 4.21 shows the design of a multilayer stack actuator with thin ceramic sheets and internal electrodes. The key advantage of this design is the reduction of the operational voltage. With values <200 V, an electrical field strength of 2–3 kV mm^{-1} can be realized, which is necessary for the large-signal driving of such stacks. The fabrication of PZT multilayer components comprises the following steps: tape casting, punching of square, thin plates, screen printing of two different designs with metal pastes (in general AgPd 70/30), stacking of unprinted and printed (alternately design 1 or 2) plates, lamination under pressure at elevated temperature, dicing of single components, debindering, sintering, machining, and finally metallization in order to connect all internal electrodes of same polarity.

The application of the multilayer technology on PZT requires the solution of three specific material problems [82]. The sintering temperature must be

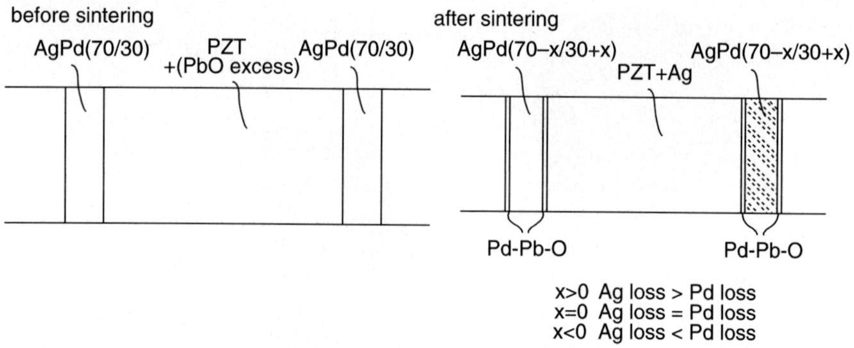

Fig. 4.22. Schematic reaction of AgPd inner electrodes and PZT ceramic during sintering of a multilayer actuator

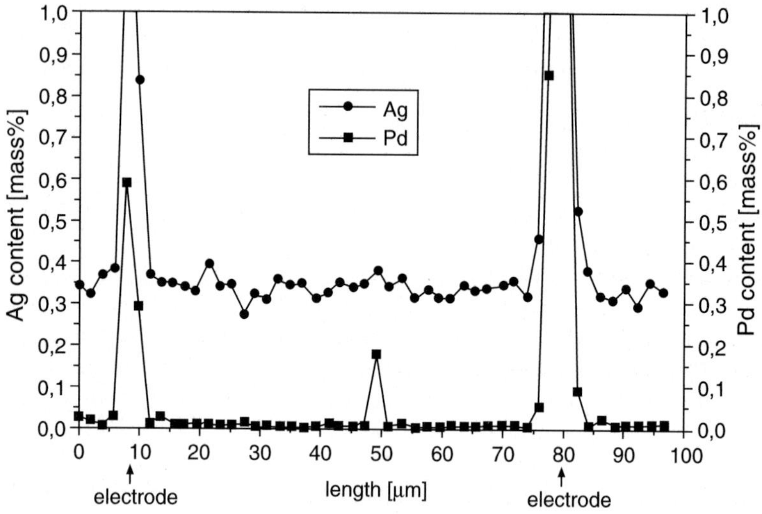

Fig. 4.23. Microprobe analysis of Ag and Pd inside a sintered multilayer PZT

reduced by about 200°C when using AgPd 70/30 as the electrode metal. A PbO excess normally leaving the sample during sintering may form a Pd–Pb compound at the ceramic–electrode interface (see Fig. 4.22). In addition, Ag may diffuse from the electrode inside the ceramic (see Fig. 4.23), where it is incorporated on A-sites of the perovskite with atomicity +1, changing the stoichiometry, the grain size, and the position of the MPB with all positive consequences for the piezoelectric data [84]. As mentioned in Sect. 4.3, during the past years specific recipes have been developed that fulfil all the requirements. They allow, e.g., the mass production of multilayer stack actuators for diesel injection valves, described in detail in Chap. 13.

4.6 Conclusion

Dielectric, elastic, and piezoelectric coefficients are the fundamental characteristics of the piezoelectric phenomenon. But only the knowledge of these values is insufficient. For a complete understanding of material properties and optimum application, additionally all boundary conditions such as temperature, time, and external mechanical and electrical loads have to be taken into account. Only all these factors together offer the basis for optimum choice of materials and practical application. Owing to the variety of modifications, the behavior of these characteristics described above can be transferred to any system with variable cations as modifiers when considering the range of the morphotropic phase transition.

Presently, standard piezoceramic materials exist that are tailored for certain ranges of application and fulfil user-specific requirements. This particularly applies to materials for sensors and actuators, frequency filters, or high-frequency ultrasonic transducers.

As for practical applications, the most relevant characteristics, T_C and ${\varepsilon_{33}}^T/\varepsilon_o$, k_p, d_{33}, g_{33}, temperature coefficients, and aging rates, must be considered for small- and large-signal load.

Within the range of MPB, the area closest to the rhombohedral side is characterized by reduced thermal and aging stability, but by maximum voltage sensitivity g_{33}. Such materials are important for accelerometers, defectoscopes, and devices of the medical diagnostics.

The area close to the tetragonal side of the MPB is characterized by high values of dielectric permittivities and the piezoelectric characteristics k_p and d_{33}. Soft materials in this range can be effectively used for low-frequency transducers, for example, hydrophones, sound receivers, and microphones as well as actuators.

Materials of middle hardness with tetragonal structure are characterised by high thermal and aging stability of the piezoelectric characteristics and are preferably used for sensors at higher operating temperature. Materials of middle hardness with maximum value of the deformation S_3 within the range of the morphotropic phase transition are also of exceptional interest for actuators.

Ferroelectric hard materials with preferably tetragonal structure are preferred for ultrasonic transducers with high efficiency.

Depending on the application and on the given specification, a wide variety of fabrication methods and compositions are used. Compared with other electroceramics, PZT components must be developed in close cooperation of application engineers, material scientists and production engineers, especially. In Sect. 4.3 a survey is given of the different PZT classes in use today. In the future, further improvements of the piezoelectric properties and a widening of the application limits, especially in the case of large-signal applications, can be expected. But it is still necessary to improve the homogeneity of the materials with respect to all specific characteristics. Simultaneously, the grain size distribution, the homogeneity of the Ti/Zr incorporation within single grains and from grain to grain, the doping distribution, the composition and structure of the grain boundaries, the position and width of the MPB, and the presence of doping or vacancy complexes must be controlled and optimized. These improvements cannot be achieved without a combination of technology and modern analytical techniques taking into account the future requirement of applications and market.

References

1. Y. Xu, *Ferroelectric Materials and Their Applications* (North Holland, New York, 1991)
2. I.S. Jeludev, *Ferroelectricity and Symmetry, Solid State Physics*, vol. 26 (Academic Press, New York, 1971)
3. A.J. Moulsen, J.M. Herbert, *Materials, Properties, Applications* (Chapman & Hall, London, 1996 reprinted)
4. G. Arlt, Bol. Soc. Esp. Ceram. Vidrio. **34**(5/6), 267–271 (1995)
5. N. Uchida, T. Ikeda, Jpn. J. Appl. Phys. **6**(9), 1079–1088 (1967)
6. L.E. Cross, In *Ferroelectric Ceramics* (Birkhäuser Verlag, Basel, 1993)
7. L.E. Cross, Jpn. J. Appl. Phys. **34**, 2525–2532 (1995)
8. B. Jaffe, W.R. Cook, H. Jaffe, *Piezoelectric Ceramics* (Academic Press, London, 1971)
9. T. Ikeda, *Fundamentals of Piezoelectricity.* (Oxford University Press, Oxford, 1990)
10. Don Berlincourt, J. Acoust. Soc. Am. **91**(5), 3034–3040 (1992)
11. V.A. Isupov, Y.u.E. Stolypin, J. Phys. Soc. Jpn. **28**(Suppl.), 312 (1970)
12. V.A. Isupov, Ferroelectrics **46**, 217–225 (1983)
13. G.M. Konstantinov, M.F. Kupriyanov, V.A. Serduli, A.A. Lebedinskij, S.S. Zaycev, E.G. Fesenko, J. Tech. Phys. **59**(7), 80–84
14. M. Kupriyanov, G. Konstantinov, A. Panich, Ferroelectrics **127**, 77–82 (1992)
15. E.G. Fesenko, A.Y.A. Dantsiger, L.A. Reznitchenko, M.F. Kupriyanov, Ferroelectrics **41**, 137–142 (1982)
16. E.G. Fesenko, A.Y.A. Dantsiger, O.N. Razumovskaya, New piezoceramic materials, in Russian: Novye p'ezoelektricheskie materialy. *Univ. Rostov on Don* **1983**
17. U. Lange, Internal Report Fraunhofer ISC, 07/2000 (2000)
18. M.J. Hoffmann, H. Kungl, Curr. Opin. Solid State Mater. Sci. **8**, 52–17 (2004)
19. C.A. Randall, N. Kim, J.P. Kucera, W. Cao, T.R. Shrout, J. Am. Ceram. Soc. **81**(5), 677–688 (1998)
20. G. Arlt, Ferroelectrics **104**, 217–227 (1990)
21. G. Helke, S. Seifert, S.J. Cho, J. Eur. Ceram. Soc. **19**, 1265–1268 (1999)
22. J.T.H. Reszat, *Dissertation*, (University of Karlsruhe, Germany, 2003)
23. G. Helke, A. Schönecker, P. Obenaus, U. Keitel, L. Seffner, T. Scholehwar, U. Lange, *Paper ISAF* (2000)
24. H. Jaffe, Proc. IEEE. **5310**, 1372–1386 (1965)
25. D.A. Berlincourt, D.R. Curran, H. Jaffe, In *Physical Acoustics*, vol. IA, ed. by W.P. Mason, (Academic Press, New York, 1964)

26. ANSI/IEEE Std 176–1987 IEEE Standard on Piezoelectricity – Description
27. G. Helke, *Proc. EPCC* 83, Liberec, CSSR
28. Y.A. Kvapulinskii, Z. Surovjak, M.F. Kupriyanov, S.M. Zayzev, A.Y.A. Dantsiger, E.G. Fesenko, J. Techn. Fizikoj **49**, 1049–1052 (1979)
29. V.A. Isupov, Solid State Phys. **22**, 172–177 (1980)
30. M. Fukuhara, A.S. Bhalla, R.E. Newnham, Phys. Stat. Sol. (a) **122**, 677–682 (1990)
31. H. Kungl, *Dissertation*, (University Karlsruhe, Germany, 2005
32. M.J. Hoffmann, H. Kungl, J.T.H. Reszat, S. Wagner, In *Polar Oxides- Properties, Characterization, and Imaging* (Wiley, Weinheim, 2005)
33. G. Helke, Teil I/II. Keram. Z **54**(11), 936–942/12, 1034–1036 (2000)
34. European Standard EN 50324–2. Piezoelectric properties of ceramic materials and components. Part 2: Methods of measurement and properties-Low power
35. H. Thomann, W. Wersing, Ferroelectrics **40**, 189–202 (1982)
36. W. Wersing, Proc.4th Intern. Meeting Electro- and Magnetoceramics, 162–182 1981
37. H. Banno, T. Tsunooka, Jpn. J. Appl. Phys. **6**(8), 954–962 (1967)
38. IEEE Standard Definitions of Terms Associated with Ferroelectric and Related Materials. IEEE (Draft). Trans. on Ultrasonics, Ferroelectrics, and Freq. Control (UFFC), **50**(12), (December 2003)
39. F. Kulcsar, J. Am. Ceram. Soc. **42**(1), 49–51 (1959)
40. R.B. Atkin, R.L. Holman, R.M. Fulrath, J. Am. Ceram. Soc. **54**(2), 113–115 (1971)
41. S. Takahashi, Ferroelectrics **41**, 143–156 (1982)
42. G. Helke, (Review, unpublished). DKG- Jahrestagung, (Annual Meeting) 08.-10.10.2001, Bayreuth
43. W. Rossner, K. Lubitz, G. Tomandl, Silicates Industriels **3–4**, 31–34 (1985)
44. T.B. Weston, A.H. Webster, V.M. McNamara, J. Am. Ceram. Soc. **52**(5), 252–257 (1969)
45. A.V. Gorish, *Physics of Ferroelectric Ceramics* (Gordon & Breach, Moscow, 1999)
46. S. Takahashi, Ferroelectrics **27**, 109–112 (1980)
47. R.B. Atkin, R.M. Fulrath, J. Am. Ceram. Soc. **54**(5), 265–270 (1971)
48. W.R. Cook, H. Jaffe, *Crystal and Solid State Physics*, vol. 11 (Springer, New York, 1979)
49. DE 102 29 086 A1 $Pb(Al_{0.5}\,Nb_{0.5})O_3$-$PbTiO_3$-$PbZrO_3$
50. H. Thomann, Ferroelectrics **4**, 141–146 (1972)
51. N. Uchida, T. Ikeda, J. Appl. Phys. **6**(11), 1292–1299 (1967)
52. E.G. Fesenko, *Perovskite Family and Ferroelectricity* (Atomizdat, Moscow, 1972)
53. V.M. Goldschmidt, Geochem. Vert. Elem. VII–VIII Norsk. Vid. Akad., Math.-Naturvid. Kl., **3**, 1–17 (1923)
54. F.S. Galasso, *Structure, Properties and Preparation of Perovskite-Type Compounds* (Pergamon Press, Oxford, (1969))
55. T. Tanaka, Ferroelectrics **40**, 167–187 (1982)
56. G.A. Smolenskii, V.A. Isupov, A.I. Agranovskaya, N.N. Krainik, Sov. Phys.-Solid State 2, 2651 (1961)
57. H. Ouchi, M. Nishida, S. Hayakawa, J. Am. Ceram. Soc. **49**(11), 577–582 (1966)
58. DAS 1 646 690 Verbesserte piezoelektrische Keramik und Verfahren ihrer Herstellung. $Pb(Mg_{1/3}Nb_{2/3})O_3$-$PbTiO_3$-$PbZrO_3$

59. H. Ouchi, J. Am. Ceram. Soc. **49**(3), 169–176 (1968)
60. DAS 1 646 817 Piezoelektrischer Keramikwerkstoff $Pb(Li_{1/4}Nb_{3/4})O_3$-$PbTiO_3$-$PbZrO_3$
61. M. Nishida, S. Kawashima, I. Ueda, H. Ouchi, S. Hayakawa, Piezoelectric Properties of $Pb((Zn_{1/3}Nb_{2/3})O_3$-$PbTiO_3$-$PbZrO_3$ Ceramics Modified with MnO_2 and Al_2O_3. Proc. 1st Meeting on Ferroelectric Materials and Their Applications F-13, 333–338. Kyoto (May 1996)
62. G.E. Savenkova, O.S. Didkovskaya, V.V. Klimov, Y.U.I. Venevtsev, J. N. Neorg. Mat. **7**(6), 996–1000 (1971)
63. Brit. Patent 1 401 389 Piezoceramic Materials. [56]
64. H. Ouchi, K. Nagano, S. Hayakawa, J. Am. Ceram. Soc. **48**(12), 630–635 (1965)
65. Pat. US 4,313,839 Piezoceramic Material
66. Pat. US 3,890,241 Piezoelectric Ceramic compositions $Pb[(Zn_{1/3}Nb_{2/3})O_3 - (Sn_{1/3}Nb_{2/3})O_3 - PbTiO_3 - PbZrO_3 + 0.05$ to 5 weight % MnO_2
67. W. Wersing, *Piezoelectric Materials in Devices*, ed. by N. Setter (EPFL Swiss Federal Institute of Technology, Lausanne 2002), pp. 29–66
68. A.I. Kingon, J.B. Clark, J. Am. Ceram. Soc. **66**, 253–256 (1983)
69. M. Hammer, M.J. Hoffmann, J. Am. Ceram. Soc. **81**, 3277–3284 (1998)
70. W. Rossner, PhD Thesis, Erlangen, 1985
71. E. Brzozowski, Private Communication
72. R.B. Atkin, M. Fulrath, J. Am. Ceram. Soc. **54**, 265–270 (1971)
73. W. Heywang, H. Thomann, Ann. Rev. Mater. Sci. 27–47 (1984)
74. D. Hennings, H. Pomplun, J. Am. Ceram. Soc. **57**, 527–530 (1974)
75. C.L. Huang, B.H. Chen, L. Wu, Solid State Commun. **130**, 19–23 (2004)
76. K.H. Haerdtl, H. Rau, Solid State Commun. **7**, 41–45 (1969)
77. G.H. Haertling, J. Am. Ceram. Soc. **82**, 797–818 (1999)
78. S. Kim, G.F. Lee, T.R. Shrout, S. Venkataramani, J. Mater. Sci. **26**, 4411–4415 (1991)
79. M. Kulig, G. Preu, D. Cramer, K. Lubitz, *Ceramics Charting the Future* (P. Vincenzini, Techna, Firenze 1995), pp. 2493–2498
80. H. Hellebrand, *Material Science and Technology*, ed. by R.W. Cahn, P. Haasen, E.J. Kramer (VCH Weinheim, New York, 1996), pp. 191–265
81. S. Takahashi, A. Ochi, M. Yonezawa, T. Yano, Ferroelectrics **50**, 181–190 (1983)
82. K. Lubitz, *Piezoelectric Materials in Devices*, ed. by N. Setter, (EPFL Swiss Federal Institute of Technology, Lausanne, 2002), pp. 183–194
83. G. Helke, Piezoelectricity of Ferroelectric Ceramics. *Adaptronic Congress* 23–24 (April 2002)
84. M. Laurent, PhD Thesis, (Karlsruhe, 2002)

5

Relaxor Ferroelectrics

L.E. Cross

5.1 Introduction

It is interesting to note that as with almost all initial advances in the study of ferroelectric oxides the impetus for studies of relaxor ferroelectrics stemmed from early work on polycrystalline ceramic systems. In the case of the relaxor ferroelectrics, the advance originated from studies in Professor Smolensky's group at the Ioffe Institute in Leningrad (now St. Petersburg) on perovskite structure electroceramics of complex composition. Originally classified as ferroelectrics with diffuse phase transitions, it slowly became clear that the very high dielectric maximum being highly dispersive could not mark a classical ferroelectric phase transition. Now, following studies at Penn State University, the suggested designation as relaxor ferroelectrics has become internationally adopted. The name is nicely compact and does highlight two key features, the combination of massive dielectric relaxation with almost classical lower-temperature ferroelectric response.

Over the last 20 years there has been an explosive growth of interest in the topic. Not surprising perhaps, as relaxor studies combine a focus on materials with properties of very strong practical importance, but underpinned also with challenging fundamental considerations, a nearly ideal mix for current applied research funding. The current, very high level of interest is nicely reflected in a recent excellent review by G. Samara "Ferroelectrics Revisited: Advances in Materials and Physics" [1], which devotes some 30% of the text to the topic of relaxor ferroelectrics. In this short chapter, the focus will be upon the relevance of relaxor studies to the basic understanding of elasto-dielectric transducing functions in oxide ferroelectrics.

Early realization of the importance of relaxor ferroelectrics for transducing structures came with the exploration of the large, stable electrostrictive strains that could be controlled in lead magnesium niobate ($PbMg_{1/3}Nb_{2/3}O_3$) (PMN): a Smolensky group discovery [2] that has became the prototype for perovskite-structure relaxors. The utility of this ceramic was highlighted by the incorporation of PMN-based multilayer actuators to drive remote-controlled tilt mirrors in the replacement wide-field and planetary camera II of the Hubble Space Telescope. These PMN tilt mirrors permitted ground control to make final, precise alignment of the corrector optics required to compensate for the original primary mirror problem and bring the Hubble to its present fantastic performance. Relevance of the relaxor ferroelectric studies to the deeper understanding of the strong piezoceramics in the lead zirconate titanate (PZT) family, which still underpin all most important electro-elastic transducer applications, is further clarified by discussion of the relaxor ferroelectric lead titanate solid solutions which incorporate morphotropic phase boundaries (MPBs), where, unlike the PZTs, respectable single-crystals can be grown and the orientational behavior of the intrinsic elasto-dielectric properties fully studied.

5.2 Relaxor Ferroelectric Compositions

Relaxor ferroelectric properties occur in a number of perovskite and tungsten-bronze structure compounds and solid solutions. In perovskites, the properties to be described are most often associated with mixed cations of different valencies occupying one of the lattice sites in this ABO_3 structure. Typical examples of A-site substitutions are bismuth strontium titanate, $(Bi_{2/3x}Sr_{1-x})TiO_3$, and lead lanthanum zirconate titanate, $(Pb_{1-x}La_{2/3x})$ $(Zr_yTi_{1-y})O_3$ (PLZT), and for the B-site substitution the compounds lead magnesium niobate, $(PbMg_{1/3}Nb_{2/3})O_3$(PMN), and lead scandium tantalite, $Pb(Sc_{1/2}Ta_{1/2})O_3$ (PST). It should be noted that in the A-site system x must be above a critical value for the relaxor properties to be manifest.

In the tungsten bronze structure, which is much more open than the perovskite, aliovalent cations are not necessary, and typical examples are the solid solution barium strontium niobate, $(Ba_xSr_{1-x})Nb_2O_6$, and lead barium niobate. Neither barium nor strontium forms a stable bronze, but in solid solution in the range $0.25 < x < 0.6$ relaxor properties occur, while in lead barium niobate, $Pb_{1-x}Ba_xNb_2O_6$, relaxation is confined to solid solution near the $x = 0.4$ composition. From its early discovery and because of outstanding properties and ease of preparation in both ceramic and single-crystal forms, lead magnesium niobate has become the prototype for perovskites and, similarly, barium strontium niobate for the bronzes.

5.2.1 *B*-Site-Substituted Perovskite Relaxor Ferroelectrics

5.2.1.1 Lead Magnesium Niobate (PMN)

Three of the key features of this typical relaxor are illustrated in Fig. 5.1. In the weak-field dielectric response (Fig. 5.1a), the permittivity reaches a high peak value ~20,000 typical for a ferroelectric perovskite near T_c, but

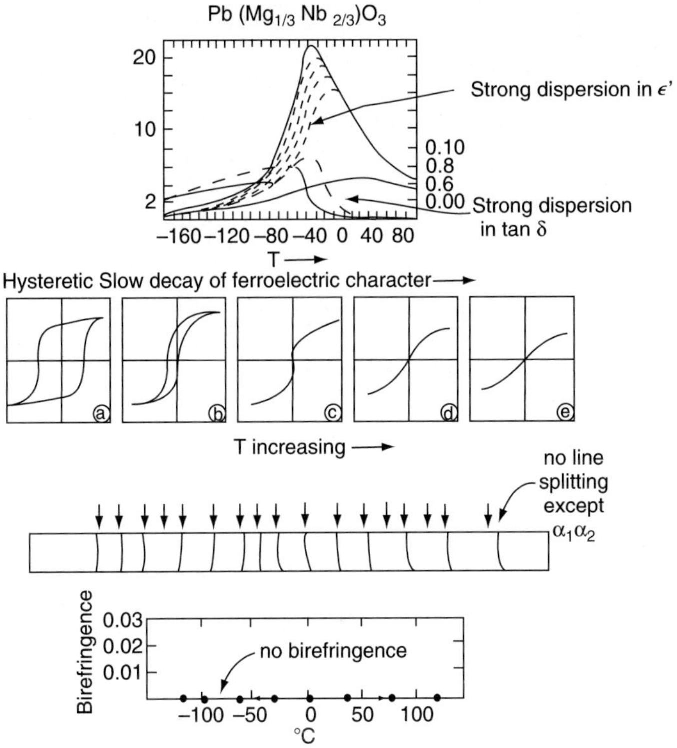

Fig. 5.1. Characteristic features of the ferroelectric relaxor, lead magnesium niobate, $Pb(Mg_{1/3}Nb_{2/3})O_3$ (PMN). (**a**) Dielectric dispersion in PMN as a function of temperature and frequency. (**b**) Dielectric hysteresis in PMN as a function of temperature. (**c**) Optical and X-ray evidence for the absence of a macroscopic phase change below the dielectric maximum at T_m

the maximum clearly does not mark a phase change to a ferroelectric form as the maximum temperature T_m moves to higher temperature with increasing measuring frequency in a manner typical for a relaxation dielectric. There is also an associated maximum in tan δ, again typical of relaxor response (not shown). Behavior at lower temperatures under high alternating fields does, however, exhibit bonafide hysteretic response, a necessary and sufficient condition for ferroelectricity. However, unlike a normal ferroelectric, the hysteretic response is not lost suddenly at T_m but decays more slowly into just nonlinear response (Fig. 5.1b).

The third surprising feature is that to longer-coherence-length probing radiation, samples even at very low temperature show no evidence of optical anisotropy or X-ray line splitting, which would be characteristic of the domains in a normal ferroelectric (Fig. 5.1c). The early explanation that this could be due to a nanoscale structure of polar regions has proven correct, but only recently has its origin been correctly diagnosed.

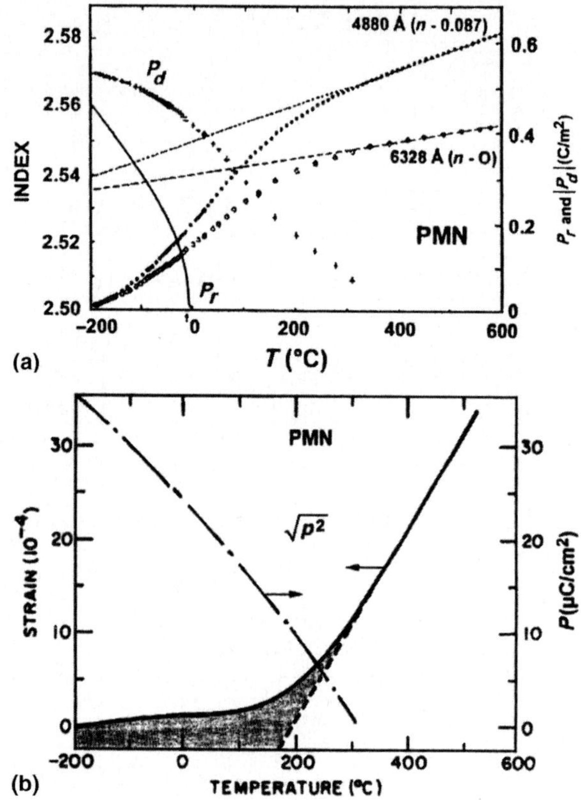

Fig. 5.2. (**a**) Refractive index of PMN at optical wave length showing the departure from linear behavior due to the onset of polar nano regions. (**b**) Thermal contraction in PMN showing the change from linearity induced by electrostriction in the polar nano region

Key forward progress came from the realization that even though in such a structure of compensating nanoscale polarizations, macro $\overline{P} = 0$, RMS polarization $\sqrt{P^2} \neq 0$ so that examination of properties that depend on P^2 such as quadratic electro-optic effects or electrostriction could indicate the onset of nanoscale polarization fluctuations.

The first clear evidence for the presence of nanoscale polar regions came from refractive index measurement on PMN by Burns and Dacol [3] (Fig. 5.2a). Normal, linear decrease of index n with temperature is enhanced beginning at a temperature T_B much above T_m through contribution from the electro-optic constants g_{33} and g_{31} of the form

$$\Delta n = -\frac{n_o^3}{2}\left(\frac{g_{33} + 2g_{31}}{3}\right) P_d^2$$

showing the onset of nanoscale polarization that did not perturb optical isotropy.

This diagnosis was supported by thermal contraction measurements by Bhalla et al. [4] (Fig. 5.2b), which confirmed a departure from linear contraction again at ~345°C so that the total strain becomes

$$S_{11}(T) = \alpha(T - T_o) + (Q_{11} + 2Q_{12})P_d^2$$

from which

$$P_d^2 = \frac{S_{11}(T) - \alpha(T_o - T)}{(Q_{11} + 2Q_{12})}$$

A microscopic picture of the polarization situation is now also emerging from X-ray [5] and neutron diffraction [6], high-resolution transmission electron microscopy [7], and neutron pair distribution function analysis [8]. The consensus appears to be that there are nanopolar regions that begin to appear at ~618 K, which grow from 2–3 nm in size near T_B to ~10 nm at 160 K, and that the volume fraction increases to near 30% at low temperature. It does appear generally agreed that above T_m the polar nanoregions are dynamic and that they do probably carry a major fraction of the system polarizability so that the dynamics of the nanopolar regions must tie in closely with the massive dielectric relaxation near T_m. Analysis of the major dispersion reveals departure from Arrhenius behavior, which can be very satisfactorily described by the empirical Vogel–Fulcher relation [9] that takes the form:

$$\tau^{-1} = \omega = \omega_o \exp(E/k(T - T_f))$$

where ω_o is the attempt frequency, E the energy barrier between equivalent polar orientations, and T_f can be viewed as a dipolar freezing temperature for the relaxation process. In PMN the fit is remarkably good [10] (Fig. 5.3) with $E = 0.079$ eV, $T_f = 217.7$ K, and k and $\omega_o = 10^{12}$ Hz. It should be noted, however, that in fact there is a very broad distribution of relaxation times in PMN that changes at T_f, the freezing temperature, so it is the largest

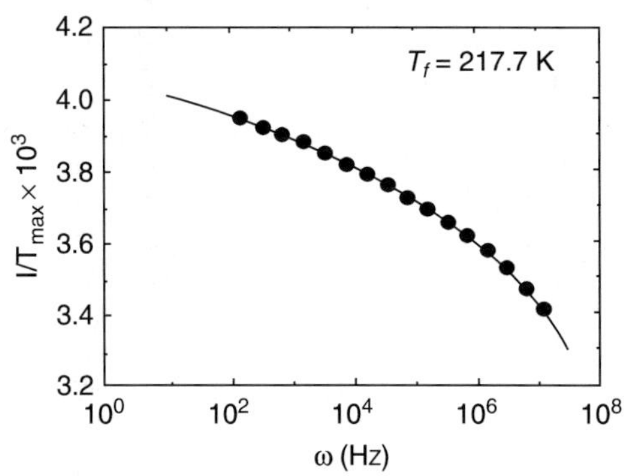

Fig. 5.3. The non-Arrhenius character of the dielectric relaxation in PMN. Solid line is a Vogel–Fulcher fit to the data. $E = 0.079$ eV, $T_f = 217.7$ K, $\omega_o = 10^{12}$ Hz

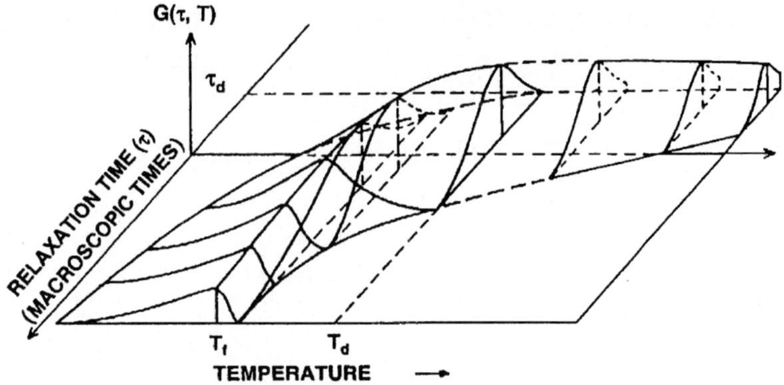

Fig. 5.4. Suggested temperature-dependent relaxation time spectrum for PMN. $G(\tau, T)$ is the number of polar micro regions having a relaxation time τ, T_f is the freezing temperature, and τ_d is the Debye relaxation time

relaxation times that became macroscopic. A suggested response profile is given in Fig. 5.4.

Thus the Vogel–Fulcher relation suggests that the zero-field, cooled PMN freezes into a random array of micropolar regions in a spin-glass-like state [11]. The micropolar regions are dominantly polarized along the (111) orientation, but because of the random arrangement there is no macroscopic anisotropy. Under high electric field, however, a static change can be induced, and the nanopolar region can be reoriented to form macro-domains, which then lead to the observed ferroelectricity at lower temperatures.

The importance of the local *B*-site cation arrangement for the onset of the relaxor nanopolar structures is very nicely demonstrated in lead scandium tantalate. Material quenched from high temperature shows a disordered Sc:Ta arrangement and initial relaxor ferroelectric response [12]; however, under high-temperature annealing, the Sc:Ta order 1:1 in NaCl-like ordering and the ordered structure exhibits a sharp first-order phase change at T_m into a conventional ferroelectric [13, 14]. In fact, it does appear in PST that even for the disordered structure with careful preparation to avoid oxygen vacancy effects, the more perfect disordered structure goes through the major relaxor dispersion, but at lower temperature crosses over into the ferroelectric form as evidenced by the loss of dispersion and the onset of a macrodomain structure [15].

In PMN, because of the 1:2 (Mg:Nb) composition on the *B*-site and the fact that the structure chooses to order 1:1 in (111)-oriented (Mg_2 Nb) and pure Nb composition sheets in which the Mg-containing sheet is still disordered, it is not possible to induce a highly ordered state and even 1:1 order is strongly kinetically limited. None the less, it does appear that limited ordering is an aid to relaxor behavior [16] and there is gathering evidence that the first onset of micropolar regions at T_B occurs in the most highly ordered regions [17] and that it is in these regions that the nanopolarization is most mobile.

5.2.2 A-Site-Substituted Perovskite Relaxor Ferroelectrics

5.2.2.1 Lead Lanthanum Zirconate Titanate (PLZT)

Probably the most widely studied mixed *A*-site perovskite system is lanthanum-modified lead zirconate titanate (PLZT). One of the first electronic ceramics to be hot-pressed to optical transparency, much original work was focused on the optical properties, which hinge upon relaxor ferroelectric behavior closely similar to PMN. As in pure PZT, there is a pseudo-morphotropic phase boundary separating rhombohedral and tetragonal phases at lower La^{3+} concentration. Most interesting are compositions in the $Pb_{1-3/2x}La_x$ ($Zr_{0.65}$ $Ti_{0.35}$)O_3 system, with x between 0.06 and 0.12. $PbZr_{0.65}Ti_{0.35}O_3$ is in the rhombohedral phase field of the solid solution and completely free of radio frequency dispersion in dielectric properties. The 0.06 mole fraction La^{3+} induces weak relaxor properties (Fig. 5.5) and, as in disordered lead scandium tantalite, this composition is close to cross-over into ferroelectricity. The relaxor phase can be stabilized by additional La^{3+} or by increasing hydrostatic pressure (Fig. 5.6). A most interesting feature of the whole range of La-substituted 65/35 PZTs is the manner in which the phase diagram changes with La concentration (Fig. 5.7). Again, thermal expansion and optical index show that

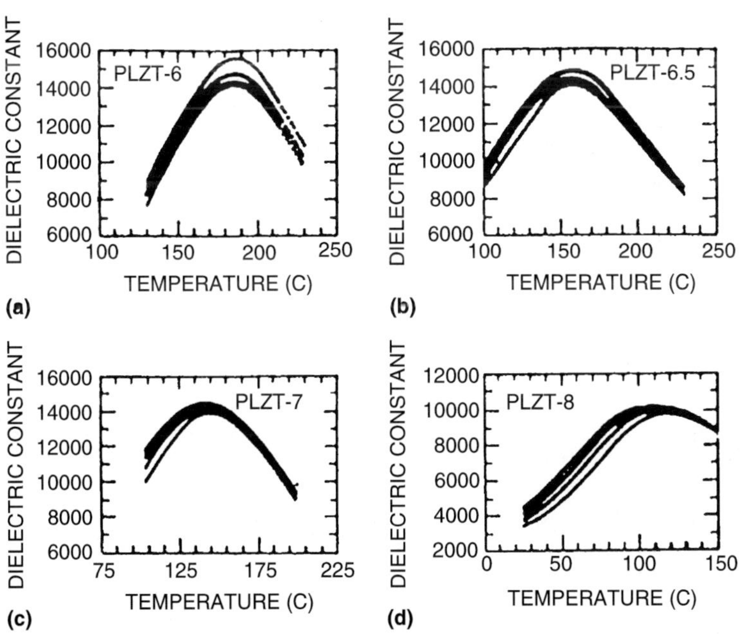

Fig. 5.5. Permittivity as a function of temperature at measurement frequencies of 0.1, 1.0, 10, and 100 kHz for compositions of PLZT with (**a**) 6%, (**b**) 6.5% (**c**) 7% and (**d**) 8%. It shows the onset of dispersive response in the 6% composition, and how the dispersion is enhanced with increasing lanthanum concentration

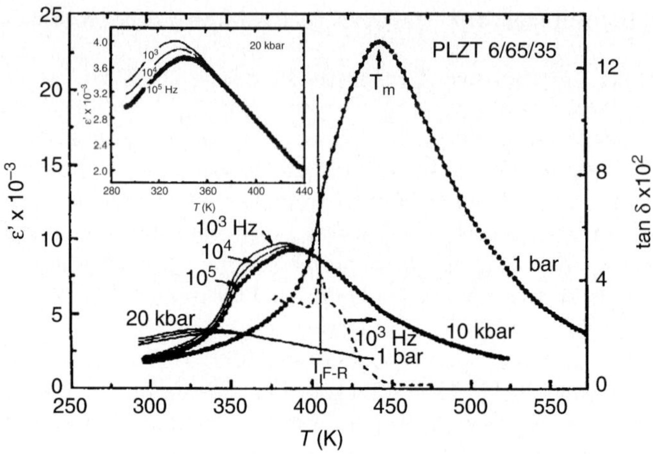

Fig. 5.6. Dielectric response as a function of temperature in PLZT 6/65/35 showing how dispersion is enhanced under increasing hydrostatic pressure

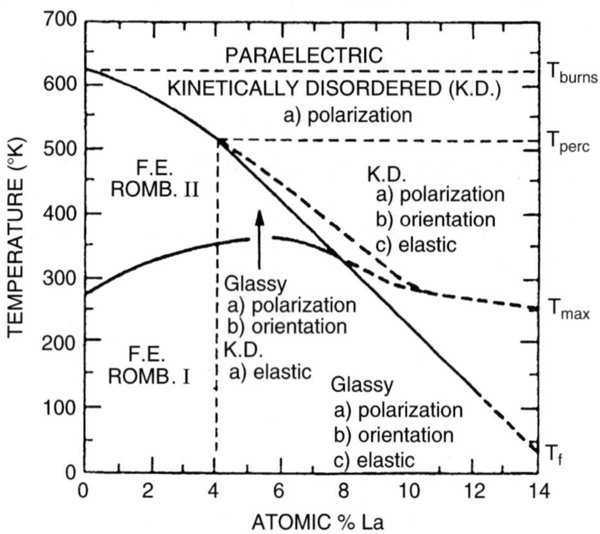

Fig. 5.7. Phase diagram for $x/65/35$ PLZT over the range of x from 0 to 14%. Note that the Burns temperature does not change over the whole composition range and the divergence of $T_{\max}$ and T_{f} at 4% *L*anthanum with the onset of relaxor ferroelectric response

nanopolar regions onset at a Burns temperature T_B, which is identical to the ferroelectric Curie temperature of the unsubstituted 65/35 PZT. Up to 4 mol% La, T_m marks a dispersion-free dielectric maximum, marking the onset of normal ferroelectricity. Beyond 4 mol%, La begins to become dispersive and diverges from the freezing temperature T_f, which continues to plunge with

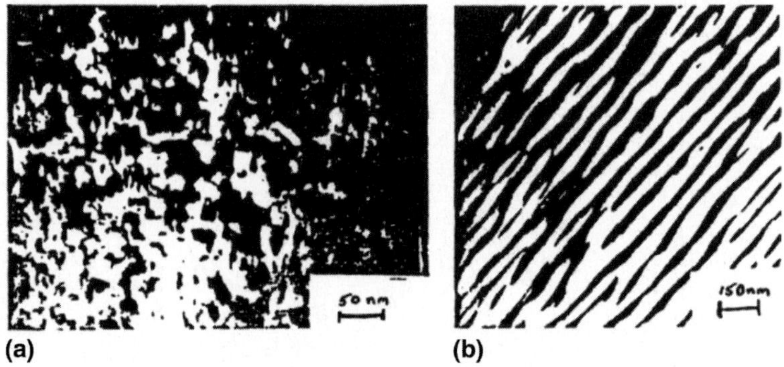

Fig. 5.8. (**a**) Structure of nanoscale domains revealed in PLZT 8.2/65/35 composition by dark-field TEM. (**b**) Modification of the domain pattern by beam charging showing the induction of larger macrodomain stripes

added La. One of the best demonstrations of NPR in PLZT is the TEM study by Randall [18] showing for an 8.2:65:35 PLZT composition dark field contrast from NPR (Fig. 5.8a) and by charging from the beam current the evolution towards more normal continuous ferroelectric macro-domains (Fig. 5.8b).

5.2.2.2 Bismuth Strontium Titanate

A second *A*-site-substituted system that gives additional insight into relaxor ferroelectric behavior is bismuth-substituted strontium titanate, $(Bi_{2/3x}Sr_{1-x})$ TiO_3, originally classified as Skanavi dielectrics in honor of its Russian discoverer. More recent studies have again shown clear relaxor ferroelectric behavior for higher bismuth concentrations.

In pure strontium titanate, $SrTiO_3$, the weak-field dielectric permittivity follows the Curie–Weiss law from high temperature down to 50 K with a Curie–Weiss temperature $\theta = 35.5$ K; however, no permittivity peak appears below 50 K and the permittivity levels off at a very high value that persists to a very low temperature. The suppression of ferroelectricity is attributed to quantum fluctuations and often described by a model due to Barrett [19, 20]. Interestingly, at the first introduction of Bi^{3+} at low levels, the quantum paraelectric behavior is modified by two weak dispersions (A, B) exhibiting nicely separated $\tan\delta$ maxima (Fig. 5.9). At higher Bi^{3+} concentrations, a more massive relaxation (C) appears, and with further increase in Bi^{3+} concentration digests both A and B.

Dispersions A and B follow the simple Arrhenius law over the frequency range 10–10^9 Hz : $\tau = \tau_o \exp(E/kT)$, and display nicely symmetric Cole-Cole arcs, typical for dielectric dispersion. Relaxation C, however, is not of the Debye type and follows the Vogel–Fulcher law:

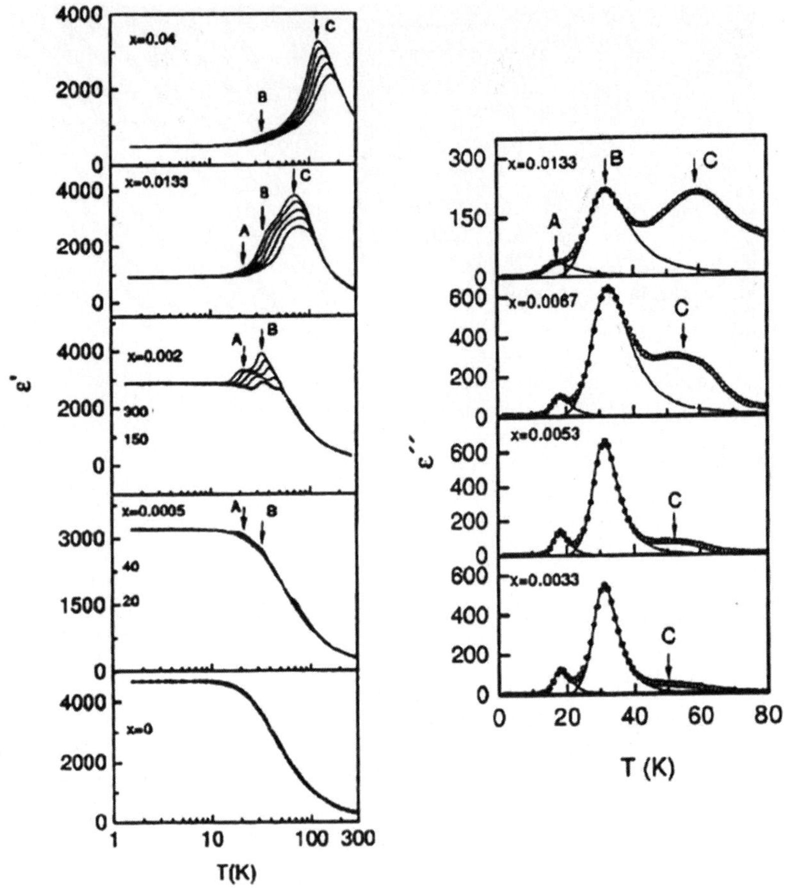

Fig. 5.9. (**a**) Temperature dependence of ε' for $Sr_{1-1.5x}Bi_xTiO_3$ with x ranging from 0 to 4% at 0.1, 1, 10, 100 and 1,000 kHz. Note the log T scale. (**b**) Temperature dependence of ε'' for similar compositions and temperature range

$$\tau = \tau_o \exp\left(\frac{E}{k(T - T_f)}\right) \text{ of a relaxor ferroelectric.}$$

Dispersions A and B increase in amplitude with Bi^{3+} concentration but do not move in frequency with Bi^{3+} concentration or applied bias field. Dispersion ε also increases markedly with Bi^{3+} concentration, but moves to lower temperatures with concentration and with bias field.

For a simple defect-dipole-induced relaxation as in A and B, it is expected that the dielectric step height follows a simple Debye–Langevin relation:

$$\Delta\varepsilon = (n(q\delta)^2/3kT)\varepsilon_o$$

For both A and B relaxations the observed $\Delta\varepsilon$ would require a totally unrealistic displacement δ_s, suggesting that the effective electric movement is massively augmented by high lattice polarizability, i.e.,

$$\Delta\varepsilon = (n(Nq\delta)^2/3KT)\varepsilon_o$$

where for realistic δ, $N \sim 100$ is required.

It is suggested that A and B dispersions are associated with local displacements of the Bi^{3+} ion and are distinguished by the absence or near-presence of the associated strontium vacancy, which appears reasonable. For the C dispersion, however, the very different character might suggest that, as in the original Skanavi model, Ti^{4+} is participating. Again, polar clusters form at the higher Bi^{3+} concentration, but now the clusters communicate much more effectively through the oxygen framework to give positive feedback and the slowing down evidenced in the Vogel–Fulcher model.

Clearly, the system offers important clues as to the origin of relaxor response, but much more work is needed to pin down the possibilities.

5.3 Polycrystalline Ceramics Relaxor Based Actuators

Realization of the absence of stable domain states in PMN in the temperature range from T_m to T_B suggested that the carefully prepared ceramic should be free from aging effects. In fact, the ceramic is not easy to prepare from mixed oxide powders because of the very refractory nature of magnesium oxide, and often early ceramics were heavily contaminated by a pyrochlore phase or by the excess MgO that was used to alleviate the problem. This difficulty was nicely solved by the columbite method of fabrication [21], in which the necessary MgO was pre-reacted with niobium pentoxide (Nb_2O_5) to form the columbite phase, which could then be reacted with PbO to form PMN. In the pure stoichiometric PMN, absence of aging is evidenced over a wide temperature range (Fig. 5.10), and field-induced polarization leads to useful electrostrictive strain (Fig. 5.11). As a function of electric field, the strain is not quadratic because of dielectric nonlinearity; none the less, it leads to impressive induced piezoelectricity at modest bias field levels and is age-free over a suitable range of temperature. Fortunately, the age-free range of T can be manipulated by limited solid solution with lead titanate ($PbTiO_3$) to give performance superior to the best hard PZT at room temperature. It was this longer-term stability of the bias-induced displaced states that led to PMN being chosen for Hubble corrector tilt mirrors, a decision that has been amply justified by its subsequent performance. Interesting to note also is that many of the actuator controlled atmospheric density fluctuation corrector mirrors now incorporated into most very large ground-based telescopes also use ceramic PMN-based compositions.

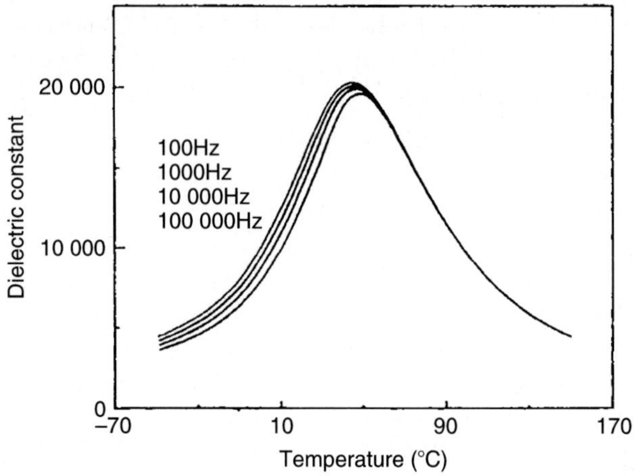

Fig. 5.10. Absence of aging in 0.9 PMN:0.1PT over the temperature range −60°C to +160°C after 1,000 h of aging at 20°C

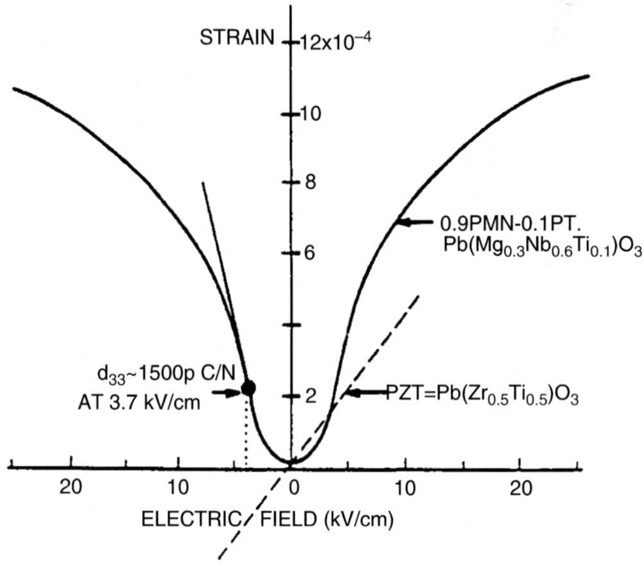

Fig. 5.11. Electrostrictive strain as a function of applied E field in 0.9 PMN:0.1PT. Slope gives the effective d_{33}, which reaches ∼1, 500 pC/N at 3.7 kV cm^{-1} bias field

Aging can be re-introduced in PMN in a controlled manner by small manganese doping [22], which gives most interesting insight into micropolar behavior and suggests a very strong caution in the interpretation of micropolar dynamics from samples in which aging may not have been investigated and controlled. Unlike conventional ferroelectrics, the weak field permittivity does not exhibit log–linear aging (Fig. 5.12), which is most evident at the

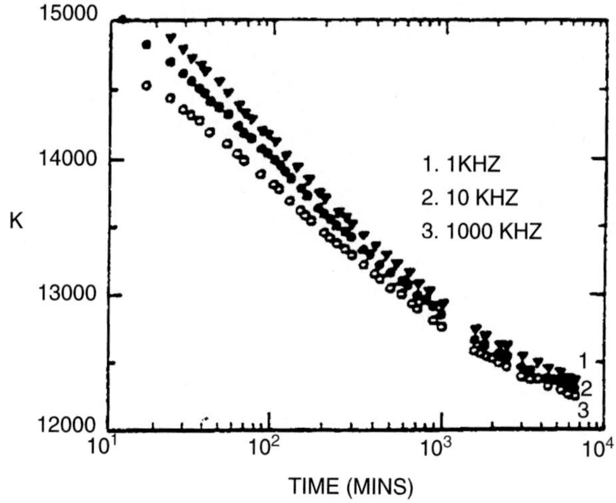

Fig. 5.12. Absence of log linear aging in 0.9 PMN:0.1PT ceramic modified by a small addition of MnO to introduce the aging

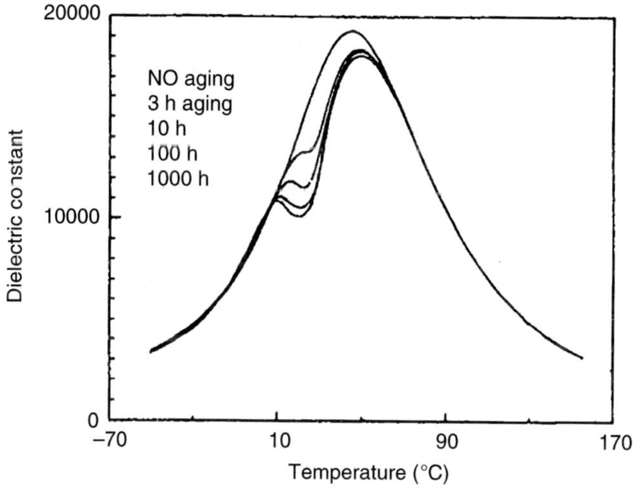

Fig. 5.13. Influence of aging at 20°C on the dielectric response from −60 to +160°C in the MnO-modified PMN:0.1PT measured at 100 Hz

aging temperature (Fig. 5.13), and markedly reduces the dispersive component (Fig. 5.14). Most intriguing is the asymmetry about the aging temperature, where dispersion is lost on heating up to the temperature of the dielectric maximum but is recovered rapidly to the original virgin condition on cooling.

Since the micropolar structure is disorder related, it is unlikely that alternative orientations of the NPR will be energetically identical, so it is probable that the micropolar vector will spend longer time in deeper minima. In this case, mobile defects will probably rearrange to accommodate

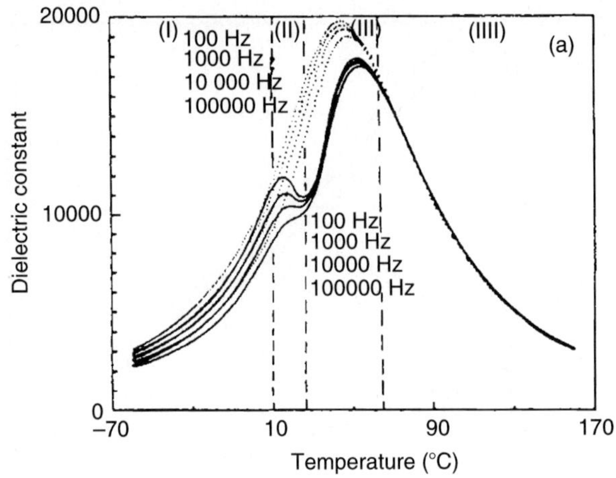

Fig. 5.14. Dielectric response as a function of temperature in MnO-modified PMN:0.1PT after aging for 1,000 h at 23°C. *Dotted curves* show data for a sample freshly de-aged at 160°C for 30 min

this preference, further stabilizing the orientation and reducing the associated "hopping" dispersion. In this model, the asymmetry between heating and cooling would appear logical, as on heating the whole family of accessible mobile NPRs is already stabilized and of reduced dispersion, while on cooling new unstabilized NPRs will appear to rebuild the dispersion. The model that suggests a "breathing mode" of the NPRs as contributing strongly to dispersion would depend on the stabilization of an annulus around the NPR by defect migration, in which case either heating or cooling would move the boundary region and thus restore dispersion.

Although the suggestions are not definitive, aging does suggest the need for real caution in studies aimed at exploring the dynamics of the NPR. Surface studies will always favor static models owing to the huge anisotropy induced by any real surface. Volume studies must clearly be carried forward on age-free materials or on freshly quenched samples in which the dispersion has not been attenuated.

5.4 Single-Crystal Systems

5.4.1 Perovskite Relaxor Ferroelectric Single Crystals

Useful properties of relaxor ferroelectric single crystals were first reported by Nomura and coworkers in early 1970 for the lead zinc niobate: lead titanate ($PbZn_{1/3}Nb_{2/3}O_3$:$PbTiO_3$); PZN:PT) solid solutions system, where single crystals could be readily grown by flux methods. Crystals that could be poled into a ferroelectric state exhibited a morphotropic phase boundary

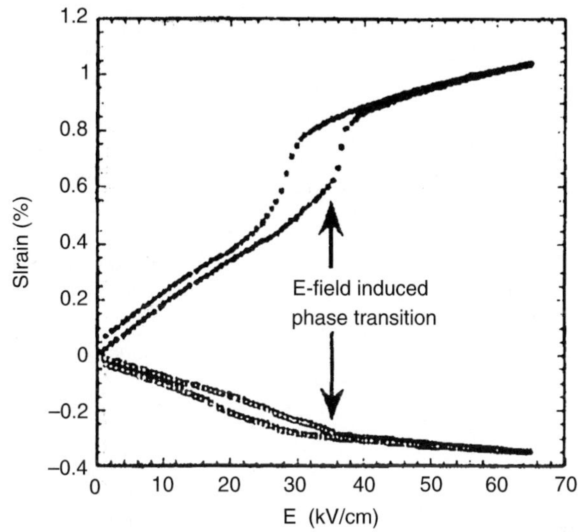

Fig. 5.15. Longitudinal and transverse electric-field-induced elastic strains in (001)-oriented PZN-4.5 PT single crystal

(MPB) between rhombohedral and tetragonal from around 9% $PbTiO_3$ with very strong piezoelectric properties for some orientations [23–25].

The subject, however, really "took off" with the systematic measurements by Park and Shrout [26] on the piezoelectric properties of compositions of single crystals in both the PMN:PT and the PZN:PT systems. For composition in the ferroelectric rhombohedral phase field close to the MPB, $d_{33} >$ 2,500 pm V^{-1} and electromechanical coupling $K_{33} > 0.94$ were demonstrated for (001) field poled crystals. Only modest properties were, however, induced in (111) poling, which permits full orientation to the single domain state.

A second very unusual feature is the almost anhysteric quasi-linear strain ∼0.6% for electric fields (E) up to 25 kV cm^{-1} (Fig. 5.15) suggesting the absence of extrinsic domain-related contributions to strain in the (001)-poled crystal (Fig. 5.16a). For the fully (001)-poled rhombohedral state (Fig. 5.16b), the resultant 111, $1\bar{1}1$, $\bar{1}11$, and $\bar{1}\bar{1}1$-oriented polar domains are equally changed in energy by a (001)-oriented E-field, so there is no driving force to move the domain walls. Since the single-domain permittivity along the (111) domain polar axis is modest, it is natural to suggest that field-induced rotation of the domain vectors (as in Fig. 5.16c) is the intrinsic strain mechanism, with a field-induced switch over to a tetragonal form at a very high field as observed (Fig. 5.16d).

Now, it is natural also to question whether, on the basis of the observed polarization changes, the tilt would be adequate to generate the observed strain levels and if so, what would be the required tilt of P_s from (111) to produce the strain level observed before the switchover. Early phenomenological analysis [27] using strain measured in the tetragonal state to characterize

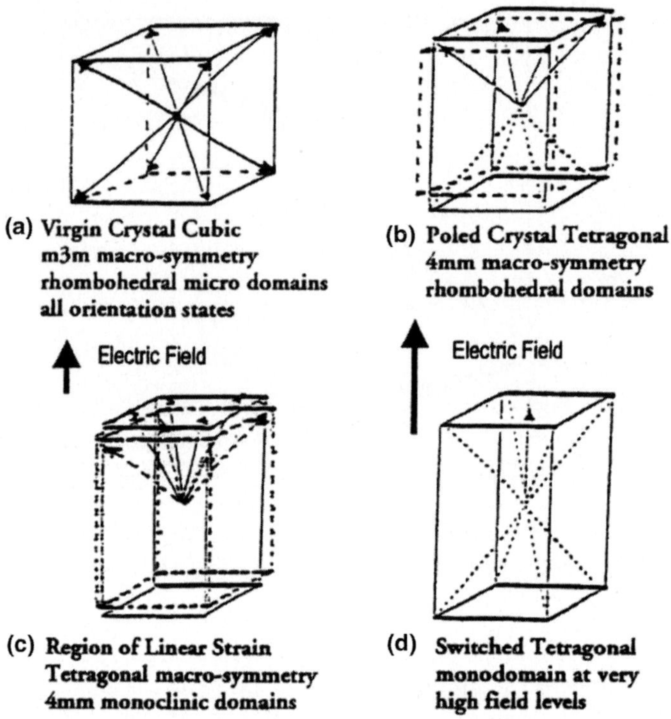

Fig. 5.16. Sequence of states in an (001)-oriented PZN:4.5 PT single crystal. In virgin, poled, under modest (001) E field, and after high switching field

Q_{11} show that the induced strain is more than adequate. More recent calculation [28] using total strain measured for the tetragonal state after switching to determine Q_{11} agrees closely with the observed S_3 and S_1 values, and the calculated tilt of P_s from (111) before switchover is similar to that measured in the tilt of the optical indicatrix under (001)-oriented E-field in a rhombohedral single domain [29]. With support also from ab initio calculation [30], there is now little doubt that the driving mechanism for high d_{33} and large S_3 under (001)-oriented E-field is rotation of the polarization vector away from the (111) direction in the rhombohedral domain towards the orientation of the driving E-field.

The basic explanation is quite clear; however, there is much complexity in the detail. In a relaxor ferroelectric, a poling operation is required to put the low-temperature form into the ferroelectric phase with long-range order. The resulting phase may depend on the magnitude, orientation, and temperature of the poling operation. The original phase diagram is oversimplified, and in some compositions close to the MPB there is evidence of stable monoclinic and orthorhombic forms. Unfortunately, much confusion has also been generated by overemphasis on the role of the monoclinic phase in these crystals.

It is absolutely true that under high (001) E-field the P vector in the domain is tilted more than 20° away from the rhombohedral (111) and the symmetry is unquestionably monoclinic, but of critical importance is the fact that the monoclinic form is metastable so that when the field is removed the domain reverts back to its equilibrium rhombohedral form giving the recoverable anhysteretic quasi-linear actuation strain under the E-field. It is perfectly true that in the monoclinic state P_s may lie anywhere in the mirror plane between rhombohedral and tetragonal orientation, but it is absolutely untrue that it is free to rotate in this plane in that for any given composition, temperature, and pressure, the vector will always occur at the same orientation in one of the family of 24 permitted orientation states. What is unusual is that the P_s vector will change its orientation in the mirror as a function of temperature, giving an unusual component of pyroelectricity orthogonal to the direction of P_s in the single domain. Clearly, for the type of actuation achieved in the (001)-poled rhombohedral form of both PMN:PT and PZN:PT, in rhombohedral compositions close to the MPB the last thing one would want is a stable intervening monoclinic form between tetragonal and rhombohedral, as this would necessarily introduce massive hysteresis in the strain because the P_s vector drops into and has then to be pulled out from the stable tilted orientation. A key basic unresolved problem in the PMN:PT and PZN:PT systems is why the tilting mechanism is so astoundingly successful in raising piezoelectric response. Other simpler perovskites such as $BaTiO_3$ and $KNbO_3$ have ferroelectric/ferroelectric phase boundaries, and at temperature close to the boundary exhibit appropriate high dielectric anisotropy [31, 32]. For simple domain-average-engineered structures, however, both measured and calculated properties, although significantly enhanced, are almost an order of magnitude less than in the relaxor ferroelectrics.

For future development based on the relaxor single crystals, it would appear that a very important question is how the residual disorder and random fields soften both the dielectric and elastic properties without destroying the high-temperature reach and the strong remanent polarization in the poled ferroelectric forms, as discussed further in Sect. 5.

5.4.2 Tungsten-Bronze Structure Relaxor Ferroelectrics

A substantial number of corner-linked-octahedron-type complex ferroelectric oxides that crystallize in the tetragonal tungsten-bronze crystal structure exhibit the relaxor ferroelectric character. A cross-section of the structure normal to the fourfold axis (Fig. 5.17) illustrates the corner linking of octahedra, which is much less regular than in the perovskite, leading to five-, four-, and threefold tunnels for the A-sites running parallel to the fourfold symmetry axis. The general chemical formula is of the form $(A_1)_4(A_2)_2\,C_4B_{10}O_{30}$, where A_1 are 15-fold coordinated sites in the fivefold tunnels, A_2 are 12-fold coordinated in the fourfold tunnels, C are ninefold coordinated in the threefold tunnels, and B are the sixfold coordinated B-site cation as in the perovskite.

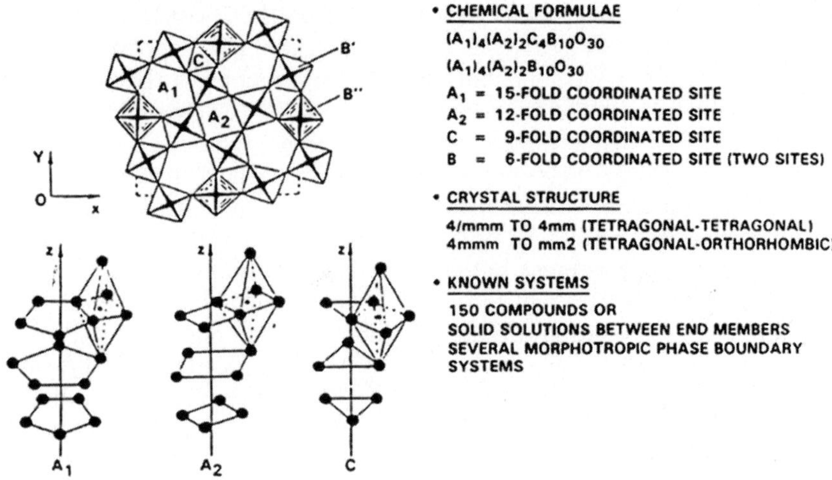

Fig. 5.17. Cross-section of a ferroelectric tetragonal tungsten-bronze taken normal to the fourfold axis. Identifying fivefold (A_1), fourfold (A_2), and threefold tunnels and the 15-, 12-, and 9-fold coordinated sites in these tunnels

Table 5.1. Ferroelectric species that could occur from the 4/mmm prototypic point symmetry in Shuvalov notation. Non-zero components of P_s delineated for each species

1	4/mmm (1) D 4 F 4 mm	$p_Z^2 \neq 0,\ p_X^2 = p_Y^2 = 0$
2	4/mmm (2) D 2 F mm2	$p_X^2 = p_Y^2 \neq 0,\ p_Z^2 = 0$
3	4/mmm (4) A 2 F m	$p_X^2 = p_Y^2 \neq 0,\ p_Z^2 = 0$
4	4/mmm (4) A.4 F m	$p_X^2 \neq p_Y^2 \neq 0,\ p_Z^2 = 0$
5	4/mmm (8) A 1 F 1	$p_X^2 \neq p_Y^2 \neq p_Z^2 \neq 0$

An example of a ferroelectric (non-relaxor) bronze in which all sites are occupied is $(K_4)_{A1}(K_2)_{A2}Li_4(Nb)_{10}O_{30}$. Very frequently, however, the structure is quite stable when not all cites are filled as, for example, in the solid solution $(Sr_xBa_{1-x})_5Nb_{10}O_{30}$, in which the A-sites are only 5/6 filled, there is charge unbalance, disorder, and strong relaxor behavior.

All bronze structured ferroelectrics and relaxor ferroelectrics have high-temperature paraelectric point symmetry $4/mmm$, which by Aizu/Shuvalov symmetry considerations limits possible ferroelectric species and polar direction to those given in Table 5.1. That the setting is with the fourfold axis along z, $P_i{}^2 \neq 0$, means that for the ith direction $\pm Ps_i$ occurs with equal probability in the domain structure. To date, only ferroelectrics in $4mm$ and $mm2$ point symmetries have been found; however, there is mounting evidence that in several relaxor compositions nanopolar regions may have lower symmetry involving a small tilting of the P vector away from the four- or twofold axis.

5.4.2.1 Barium Strontium Niobate $(Sr_{1-x}Ba_x)Nb_2O_6$

Neither pure barium niobate nor pure strontium niobate will form in the bronze structure; however, in the solid solution they form 0.25Ba/0.75Sr to 0.75Ba/0.25Sr stable tungsten-bronze structure forms. The most widely studied is $(Sr_{0.61}Ba_{0.39})Nb_2O_6$, which is pseudo-congruently melting so that excellent-quality, large, and striation-free single crystals can be grown. It has been shown from X-ray analysis by Abrahams [33] that the large Ba^{2+} ions are exclusively in the 15-fold A_2-sites, while the strontium ions are mixed between A_1 and A_2 sites and the smaller C site is completely empty.

Massive dispersion in the weak-field dielectric permittivity (Fig. 5.18) immediately suggests relaxor ferroelectric behavior, which is supported by the Vogel–Fulcher analysis (Fig. 5.19), and there is clear evidence from both optical birefringence and thermal contraction studies of the presence of c-axis micropolar regions up to a Burns temperature some 200°C above the dielectric maximum at T_m. The uniaxial character of the polar behavior permits a very simple macroscopic demonstration that the polar nanoregions between T_m

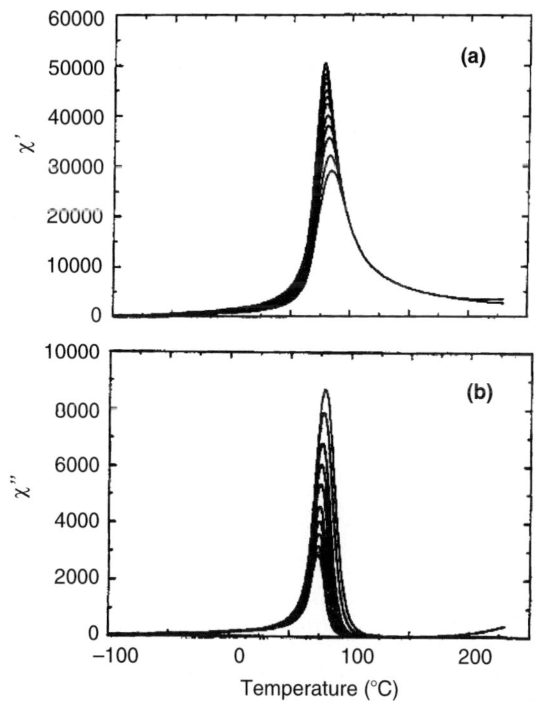

Fig. 5.18. Massive dielectric dispersion in real and imaginary components of the weak-field permittivity C_c as a function of temperature and frequency. Measured frequencies are $2\times10^2, 4\times10^2, 10^3, 2\times10^3, 4\times10^3, 10^4, 2\times10^4, 4\times10^4, 10^5, 2\times10^5$, 4×10^5, and 10^6 Hz. *Top curve* is at the lowest frequency, the *bottom* at the highest

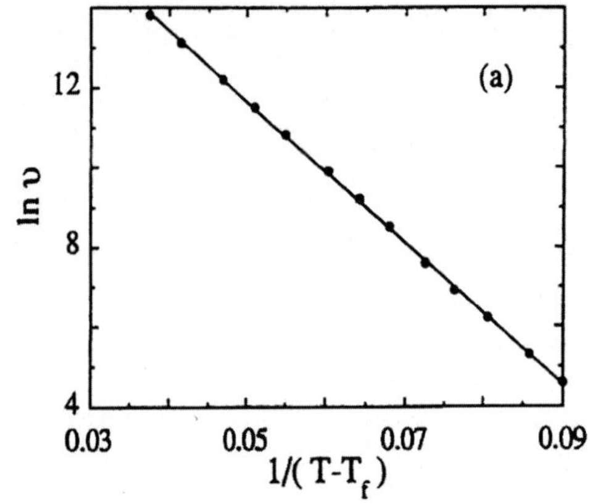

Fig. 5.19. Vogel–Fulcher fitting of dielectric data for the more strongly dispersive $(Sr_{0.75}Ba_{0.25})Nb_2O_6$

and T_B, the Burns temperature, are in fact dynamic. As the enhanced thermal contraction above T_m shows, the onset of nano-polarity contributes an electrostrictive component to the elastic strain of the form $S_3 = Q_{33}\,P^2_{s(3)}$; one notes also that the local strain S_3 does not change with the sign of $P_{s(3)}$. Now, if we measure an effective $Q_{33(e)}$ by the induced strain $S_{3(e)}$ vs. total polarization charge, i.e., $S_{3(e)} = Q_{33e}\,{P_T}^2$, and an increasing part of P_T comes from the switching of the already polarized and therefore non-strain-generating regions, then $Q_{33e} < Q_{33}$ and will decrease as the fraction of polarization carried by the NPR switching increases.

The strain S_{11} induced by P_3 through the constant Q_{31} will be similarly affected; however, the polarization P_1 cannot be influenced by the nanopolar P_3, and Q_{11} should not be effected, and this is indeed the case in Fig. 5.20. Thus, as expected, the kinetics of nanopolar switching may be responsible for the large dispersion and interaction between the NPR responsible for the Vogel–Fulcher slowing down evidenced in relaxor behavior.

5.5 Single-Crystal Insights for Polycrystalline Ceramic Piezoelectrics

5.5.1 Monoclinicity

It has long been known from the admirable symmetry studies of Aizu [34] and Shuvalov [35] that monoclinic and triclinic phases were feasible ferroelectric species from the perovskite $m\bar{3}m$ prototypic point symmetry. Ceramist unfortunately are not avid readers of symmetry literature, but the outstanding properties and clearly evident complexity in the MPB phase region in the PMN:PT and the PZN:PT solid solution systems helped to promote a closer

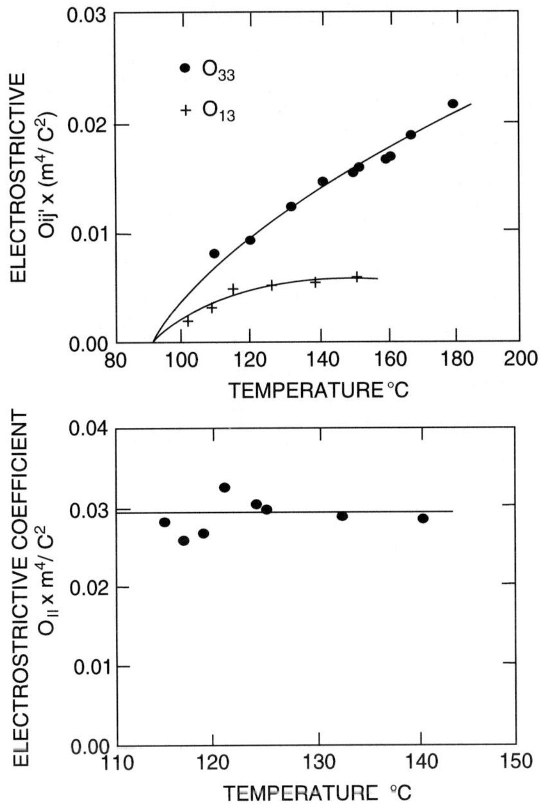

Fig. 5.20. Apparent electrostriction constants Q_{33}, Q_{13}, and Q_{11} measured as the ratio of total induced strain to the square of total electrical polarization

look at the PZT solid solution which underpins all polycrystalline piezoceramics. Much earlier, it had been suspected by the pioneer development group under Hans Jaffe at Clevite Corporation that there was a low-temperature phase transition for the MPB composition [36]. A possible candidate was the rhombohedral I – rhombohedral II ($R3m - R3c$) phase transition in the rhombohedral region which might cross the MPB at lower temperature. Structure studies by Amin [37] initiated to check this hypothesis soon disproved the possibility, but unfortunately at that time neither samples nor equipment were adequate for more detailed symmetry studies.

It needed Dr. Noheda from Professor Gonzalo's ferroelectric group in Madrid working with Dr. Shirane at Brookhaven to uncover the narrow insert of monoclinic phase that interleaves the tetragonal and rhombohedral phases at lower temperature at the PZT MPB composition [38] (Fig. 5.21). Truly, however, how this interleaving monoclinicity in the pure PZT relates to the exciting family of properties available in the modified commercial PZT composition will need more careful study. In commercial compositions, the

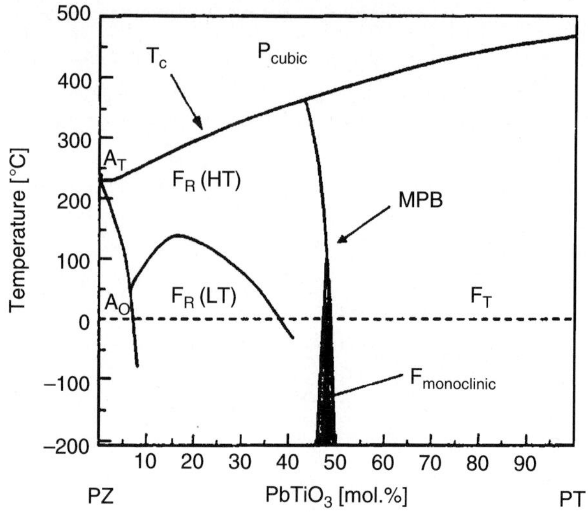

Fig. 5.21. Revised phase diagram for the PZT composition field showing the narrow insert of monoclinic phase between rhombohedral and tetragonal phases at the morphotropic phase boundary

useful properties are often a complex ensemble of intrinsic domain strains, extrinsic contribution from ferroelastic-ferroelectric domain wall motions, and possible phase switching. Recent high-precision X-ray studies [39] suggest that as in the PMN:PT and PZT:PT systems the major intrinsic contribution from both tetragonal and rhombohedral grains comes largely through tilting of the P_s vectors. Since through its 24 permitted orientation states the monoclinic form may provide bridging polarization continuity, one suspects that this will permit switching of a wider cone of orientation states as evidenced by the large P_r values. This flatter cone will then expose more suitably tilted polar vectors to measuring fields applied in the original poling directions.

5.5.2 Intrinsic Polarization Mechanisms

Since for the single-crystal systems the model of enhanced piezo activity by off-axis fields tilting the P_s vector is now very widely accepted and well authenticated in the relaxor–ferroelectric MPB compositions, a natural question is how general this model is and how it will apply to simple classic proper ferroelectrics.

Extensive studies of $BaTiO_3$ [40] and $KNbO_3$ [41], for which excellent single crystals have been grown, certainly confirm the enhancement of d^* constants by off-axis fields, but in both these systems the level of d values and coupling constants that can be achieved falls far short of those in the best relaxor ferroelectric solid solutions.

The studies in $BaTiO_3$ have, however, been very useful in highlighting a heretofore unexpected extrinsic contribution to response in that in a crystal in the tetragonal phase by poling with (111)-oriented E-fields it was possible to develop domain-engineered poled states with very different densities of 90° domain walls [42]. Also the (111) E-field changes the energy of (100), (010), and (001) twin components in an equivalent manner so that once the crystal is poled, application of the field in the same orientation can produce no energy difference between domain pairs to drive wall motion. In spite of this, it has been shown [43] that these static walls enhance piezoelectric response in proportion to the density of walls present.

Since in the proper ferroelectric crystals even at compositions close to ferroelectric–ferroelectric phase transitions the response is so much lower, a natural question is why are the relaxor ferroelectric solid solutions so effective. One possible explanation is the adaptive phase model proposed by Yin, Wang, and Khachatuyan [44]. Emphasizing the similarities between martensites and relaxor ferroelectrics, these authors prepare a model based on very fine scale twining of tetragonal or rhombohedral domains. If the twin wall energy is very low and the twining is fine enough to be invisible in optics, the observed monoclinic phase derived from X-ray diffraction with a finite beam size can actually be the volume-averaged parameters of the finely twinned structure. Now, under an electric field the progressive switching of domains leads to a rotation of the volume-averaged polarization rather than of an individual polar vector and, therefore, the observation of the strongly increased monoclinic distortion in the diffraction experiment. The model is indeed in agreement with the observed evolution of the lattice parameters as a function of the E-field and also satisfies the required martensite invariance conditions [45].

In view of the micropolar structures that are endemic to relaxor ferroelectric systems, the model does draw in an interesting manner the likely difference between polarized relaxors and normal ferroelectrics, but much more work is needed to reconcile with the extensive TEM and scanning force microscopy (SFM) as well as X-ray and neutron scattering data for these crystals and to explain why the very low energy walls can move in a completely anhysteretic manner.

A second new and very exciting insight into the major importance of polarization fluctuations in the relaxor–ferroelectric MPB systems [46] is evidence of destabilization of the cubic–tetragonal phase transitions in the PMN:PT by very modest (111)-oriented E-fields applied to the PMN 0.705:PT 0.295 composition adjacent to the MPB. Only 1.3 kV cm^{-1} is needed to destabilize the first-order cubic–tetragonal phase change leading to a critical end point of the liquid–vapor type.

In the PZ:PT composition phase diagram, a line of critical end points is delineated where the fields needed for polarization rotation become exceedingly small. It is suggested that this criticality is even more important than the phase boundary in raising the intrinsic piezoelectric response from vector rotation.

Certainly, the "softness" exhibited critically at the tetragonal–cubic transition under 1.33 kV cm^{-1}, showing ε' approaching ~100,000 at the transition, is most impressive. This origin for the observed softening in the relaxor solid solution gibes with the much stiffer response evidence in the simpler ferroelectric–ferroelectric phase behavior.

In both these proposed explanations for the enhanced "softness" in the relaxor ferroelectric–ferroelectric solid solutions, the charge and structural disorder that brings about the obvious effects in the original paraelectric shows its importance in the lower-temperature ferroelectric phases. Such considerations must clearly be of importance in the intensified search for environmentally more benign lead-free piezoceramics that can be fabricated in the most useful and inexpensive random-axis polycrystalline ceramic form.

References

1. G.A. Samara, in *Solid State Physics*, vol. 56, ed. by H. Ehrenreich (Academic, New York
2. G.A. Smolensky, A.I. Agranovskaya, *Sov. Phys. Solid State* **1429**, 1959
3. G. Burns, F.H. Dacol, *Solid State Comm.* **48**, 853, 1993; *Phys. Rev. B* **28**, 2527, 1983
4. L.E. Cross, *Ferroelectrics* **76**, 241, 1987
5. N.D. Mathan, E. Husson, G. Calvarin, J.R. Gavarri, A.W. Hewat, A. Morell, *J. Phys. Cond. Matter* **3**, 8159, 1991
6. A. Naberezhnov, S. Vakhrushev, B. Dorner, D. Stauch, H. Moudden, *Eur. Phys. J. B* **11**, 13, 1999
7. D. Viehland, M.C. Kien, Z. Xu, J.F. Li, *Appl. Phys. Letters* **67**, 2471, 1995
8. I.K. Jeong, T.W. Darling, J.K. Lee, Th. Proffen, H. Heffner, J.S. Park, K.S. Hong, W. Dmowski, T. Egami, *Phys. Rev. Lett.* **94**, 147602, 2005
9. D. Viehland, S.J. Jeong, L.E. Cross, *J. Appl. Phys.* **68**, 2916, 1990
10. D. Viehland, J.F. Li, S.J. Jong, L.E. Cross, *Phys. Rev. B* **43**, 8316, 1991
11. E.V. Colla, N.K. Yuskin, D. Viehland, *J. Appl. Phys.* **83**, 3298, 1998
12. N. Setter, L.E. Cross, *J. Appl. Phys.* **51**, 4356, 1980
13. C.G.F. Stenger, F.L. Scholten, A.J. Burggraaf, *Solid State Comm.* **32**, 989, 1979
14. C.G.F. Stenger, A.J. Burggraaf, *Phys. Stat. Solidi* **61**, 653, 1980
15. F. Cho, J.M. Reaney, N. Setter, *J. Am. Ceram. Soc.* **78**(7), 1947, 1995
16. C.A. Randall, A.S. Bhalla, T.R. Shrout, L.E. Cross, *J. Mat. Res.* **5**(4), 829, 1990
17. T. Egami, *Ferroelectrics*, **267**, 101, 2002
18. C.A. Randall, D.J. Barber, R.W. Whatmore, *J. Microscopy* **145**, 275, 1987
19. J.H. Barrett, *Phy. Rev.* **86**, 118, 1952
20. E. Pytte, *Phys. Rev. B* **5**, 3758
21. S. Swartz, T.R. Shrout, *Material Res. Bull.* **17**, 1245, 1980
22. W. Pan, E. Furman, G.O. Dayton, L.E. Cross, *J. Mat. Sci. Lett.* **5**, 647, 1986
23. S. Nomura, H. Arima, F. Kojima, *Jpn. J. Appl. Phys.* **12**, 531, 1973
24. J. Kuwata, K. Uchino, S. Nomura, *Ferroelectrics* **37**, 579, 1981
25. J. Kuwata, K. Uchino, S. Nomura, *Jpn. J. Appl. Phys.* **21**, 1298, 1982
26. S.E. Park, T.R. Shrout, *J. Appl. Phys.* **82**, 1804, 1997

27. L.E. Cross, P. Hana, Ninth US Japan Seminar on Dielectric and Piezoelectric Ceramics, Okinawa, November 2, 1.1.15, 1999
28. L.E. Cross, Fundamental Physics of Ferroelectrics, Aspen Colorado, AIP Proc 535, 1, 2000
29. A.A. Bokov, Z.-G. Ye, *Ceram. Trans.* **136**, 37, 2003
30. H. Fu, R.E. Cohen, *Nature* **403**, 281, 2000
31. S. Wada, S. Suzuki, T. Noma, T. Suzuki, M. Osada, M. Kakihana, S.E. Park, L.E. Cross, T.R. Shrout, *Jpn. J. Appl. Phys. Pt 1* **38**, 5505, 1999
32. N. Nakamura, T. Tokiwa, Y. Kawamura, *J. Appl. Phys.* **91**, 9272, 2002
33. P.B. Jamiesan, S.C. Abarahams, J.L. Bernstein, *J. Chem. Phys.* **48**, 5048, 1968
34. K. Aziu, *Phys. Rev.* **146**, 423, 1966
35. L.A. Shuvalov, *J. Phys. Soc. Japan* **28**, 38, 1970
36. H. Jaffe (private communication)
37. A. Amin, Thesis in Solid State Science, Penn State, November 1979
38. J. Noheda, A. Gonzalo, L.E. Cross, R. Guo, S.E. Park, D.E. Cox, G. Shirane, *Phys. Rev. B* **61**, 8687, 2000
39. M.J. Hoffman, H. Kungl, J-Th. Reszat, S. Wagner, in *Polar Oxides Properties, Characterization, and Imaging*, ch. **7**, p. 137, ed. by R. Waser, U. Böttger, S. Tiedke (Wiley-VCH)
40. D. Damjanovic, F. Brem, N. Setter, *Appl. Phys. Lett.* **80**(4) 652, 2002
41. S. Wada, K. Muraoka, H. Kakemoto, T. Tsurumi, H. Kwmagai, *Jpn. J. Appl. Phys.* **40**(98), 5690, 2000
42. S. Wada, T. Tsurumi, *Br. Ceram. Trans.* **103**, 93, 2004
43. S. Wada, K. Yaho, K. Yakoo, T. Tsurumi, in *Proceedings of 12th US:Japan Seminar on Dielectric and Piezoelectric Ceramics*, Annapolis, MD, November 2005, 177
44. Y.M. Jin, Y.U. Wang, A.G. Khachaturyan, *J. Appl. Phys.* **94**, 3629, 2003
45. D. Viehland, *J. Appl. Phys.* **88**, 4794, 2000
46. Z. Kutnjak, J. Petzelt, R. Blinc, *Nature* **441**, 956, 2006

6

Piezoelectric Polymers and Their Applications

S. Bauer and F. Bauer

6.1 Introduction

The science and technology of piezoelectric polymers has long been dominated by ferroelectric polymers from the polyvinylidene fluoride (PVDF) family. The piezoelectricity in this polymer class arises from the strong molecular dipoles within the polymer chain and from the resulting change of the dipole density upon application of a mechanical stimulus. Ferroelectric polymers show moderate piezoelectric coefficients (d_{33} around 20–30 pC/N) in comparison to ceramic piezoelectrics, with an acoustic impedance comparable to that of water and other liquids. The thermal stability of the piezoelectric effect is limited to below 100°C [1]. Applications of ferroelectric polymers emerged in many niches [2], examples are hydrophones [3] and clamp-on transducers [4] used as pressure sensor for diesel injection lines, with selling numbers over 50 million pieces per year, as well as piezoelectric ignition systems [5] for measuring the mechanical and physical state of matter under shock loading. Printable polymer ceramic composites [6] and nanocomposites [7] further enlarge the material basis available to application engineers.

A relatively new development are relaxor ferroelectric polymers, based on electron-irradiated poly(vinylidene fluoride) trifluoroethylene copolymers [8] or on terpolymers of vinylidene fluoride, trifluoroethylene, and chlorofluoroethylene [9]. Relaxor ferroelectric copolymers exhibit strong electrostriction and thus large piezoelectric coefficients, when used under electric dc-bias fields.

Around 20 years ago, internally charged cellular polymer electrets were shown to display large piezoelectric d_{33} coefficients [10]. Recent work demonstrated close similarities between cellular polymers and ferroelectric materials, hence they are now commonly referred to as ferroelectrets [11]. Ferroelectrets display large intrinsic piezoelectric d_{33}-coefficients well above 100 pC N^{-1}, very small d_{31} and d_{32} coefficients and an acoustic impedance well suited for coupling to air and other gases. The best characterized ferroelectret, based on cellular polypropylene films, shows a limited thermal stability of the piezoelectric response up to 50°C. The materials have been pioneered in Finland [12] and have already entered the market in several niches, for example in flexible keyboards and musical pick-ups [13–15]. They promise large area applications, for example in surveillance and intruder systems. Cellular polymers based on cellular polytetrafluoroethylene and its copolymers with a thermal stability of the piezoelectric coefficient up to 100°C are currently available on the laboratory scale [16–18]. The field of piezoelectric polymers, therefore, received many new stimuli recently, and the material class of piezoelectric polymers has been significantly broadened.

This book chapter briefly describes the origin of the electromechanical effects in ferroelectric, relaxor ferroelectric, and ferroelectret polymers in Sect. 6.2. A few selected applications, some successfully implemented on the market will be used to illustrate the evolution of piezoelectric polymer technology to the current state of maturity in Sect. 6.3. Recent developments available in form of laboratory demonstrators are used to outline future directions of research and applications in Sect. 6.4. In the final Sect. 6.5, the history and evolution of PVDF gauges illustrates that piezoelectric polymers exhibit unique material properties that make them the material of choice for piezoelectric shock wave sensors.

6.2 Piezoelectric Polymers

A brief description of the approximately 40-years history of piezoelectricity in polymers introduces the wide range of piezoelectric polymers currently available to application engineers. The origin of the piezoelectric response in polymers ranges from dipole density piezoelectricity in ferroelectric polymers to intrinsic like piezoelectricity in ferroelectrets. Printable piezoelectric lacquers and multifunctional polymer ceramic nanocomposites further enlarge the material basis by combining to advantage the properties of polymeric and ceramic piezoelectric materials.

6.2.1 History of Piezoelectric Polymers

Direct piezoelectricity was first described by Jaques and Pierre Curie in 1880, at the time of their discovery being 25 and 21-years old at the beginning of their scientific careers. The discovery was not made by chance, the Curie brothers have looked for the piezoelectric phenomenon, possibly inspired by their familiarity with the phenomenon of pyroelectricity. The early experiments of the Curie brothers dealt only with direct piezoelectricity, so the introduction of the converse piezoelectric effect is credited to the theoretical work of Lippmann on the basis of general thermodynamic arguments. However, after becoming aware of the theoretical predictions, converse piezoelectricity was verified experimentally also by the Curie brothers. The early studies of piezoelectricity were simply driven by scientific curiosity, and not at all by expectations of technological applications to come. Since then, the situation changed drastically, now a few thousands of papers are produced per year, which deals with piezoelectricity and with applications of this phenomenon. The fascinating discovery of the piezoelectric effect has been recently outlined in an excellent contribution by the historian Katzir [19].

The discovery of a strong piezoelectric effect in polymers, significantly large to be of interest for technical applications, dates back to the pioneering work of Kawai on poly(vinylidene fluoride) in 1969 [20], followed by the finding of pyroelectricity in the same polymer two years later by Bergman et al. [21] and by Nakamura and Wada [22]. Although piezoelectricity was found much later than pyroelectricity in crystals, the converse happened in polymers. Pyroelectricity belongs to one of the earliest phenomenon known to mankind, with first written reports dating back to ancient Greek philosophers [23]. Later on, even ferroelectricity could be unambiguously confirmed in the PVDF polymer family, as discussed in depth in an early review of Lovinger from 1982 [24].

Since then, the science and technology of piezoelectric polymers has been dominated by polymers from the PVDF family, especially the copolymers of vinylidene fluoride and trifluoroethylene P(VDF-TrFE). PVDF polymers with a thermal stability of the piezoelectric effect up to 125°C were announced in the 2005 fall meeting of the Materials Research Society by the company Measurement Specialties [2]. Relaxor ferroelectric polymers entered the scene in 1998, in a seminal contribution by Zhang et al. [8]. Most recently, terpolymers of vinylidene fluoride, trifluoroethylene, and chlorofluoroethylene were shown to exhibit relaxor ferroelectric behavior, in a collaborative effort of the groups of Francois Bauer in France and Qiming Zhang in US [9]. Relaxor ferroelectric polymers display large strains on the order of several percent when stressed with electric fields on the order of $100\,\mathrm{V\mu m^{-1}}$. They, therefore, reveal large electric-field induced piezoelectric coefficients under applied electric dc fields. In addition, these polymers also display a large energy density, making them ideal as dielectric medium in capacitors [25].

Applications of ferroelectric polymers appeared in niches, some of the lesser known use of ferroelectric polymers in tactile devices, energy conversion, etc. are discussed by Lang and Muensit in a recent review [26]. There are also some successful industrial applications, examples are hydrophones [3] and clamp on piezoelectric pressure sensors based on PVDF [4], besides shock wave sensors discussed in detail later [5]. A large number of applications, including piezoelectric cables for traffic counting are listed in [2]. Many more applications may be found on the market, where the use of piezoelectric polymers is not addressed by the companies.

Printable piezoelectric materials, based on composites, are developed by the Swiss company Algra with sensitivities up to $1{,}000\,\mathrm{pC\,N^{-1}}$ in device configurations [6]. Details of the material are not available in open literature, but the materials are currently employed by several electronic companies in products that are found on the market, for instance in piezoelectric switches and keyboards. Nanocomposites based on ferroelectric polymers and piezoelectric ceramic nanocrystals have been shown to be truly multifunctional by B. Ploss in 2000 [7]. By electric field poling such nanocomposites can be shown to be piezoelectric but not pyroelectric and vice versa, applications of these multifunctional composites still await demonstration.

The work on cellular polymers dates back to the pioneering contribution of Savolainen and Kirjavainen in 1989 [10]. Cellular polymers are promoted in Finland by research done at the national Finnish research center VTT [12] and by several Finnish companies [14–16]. So far, the thermal stability of the piezoelectric effect is guaranteed up to 60°C, a somewhat lower stability in comparison to ferroelectric polymers like PVDF. Recently, cellular polymers were shown to display close similarities to ferroelectrics, hence they have been called ferroelectrets [11]. A couple of reviews are available, discussing the state of the art and the development of ferroelectrets [12, 27–29]. Progress in the field is rapid, and materials with a stable piezoelectric response up to 100°C are currently available on the laboratory scale. Successful implementations into the market comprise example keyboards and musical pick-ups [13–15].

6.2.2 Origin of the Piezoelectric Response in Ferroelectric and Ferroelectret Polymers

The direct and converse piezoelectric d-tensor is defined by $d_{ip} = \partial D_i/\partial T_p = \partial S_p/\partial E_i$, where D is the displacement, E the electric field, S the mechanical strain, and T the mechanical stress, respectively [30]. For S and T the contracted notion is employed, with values of p ranging from 0 to 6. This is a commonly adopted practice to represent the third rank piezoelectric tensor in matrix form. There are in total four possibilities to define piezoelectric tensors, d, e, g, and h each of them being used for specific purposes, d is the charge coefficient and $g = d/\varepsilon$ the voltage coefficient, where ε is the dielectric "constant" of the material. For hydrophones, the dg product is often used as a figure of merit (FOM). Further figures of merit are the acoustic impedance Z, describing the efficiency of coupling sound waves from the piezoelectric material to adjacent media and vice versa. For air-borne ultrasound, another application niche of piezoelectric polymers, a FOM based on energy considerations has been introduced recently [31]. In this FOM, the proportion of the acoustical output and the total input energy delivered to the material is considered. In the case of a negligible acoustic impedance of the medium (air or any other gas) in comparison to the acoustic impedance of the piezoelectric material, the FOM is $f = 10^4 k_{33}^4/Z_P^2$. Here k_{33} is the piezoelectric thickness extension coupling factor and Z_{P} is the acoustical impedance of the piezoelectric material. The factor 10^4 is only used to make numbers for different materials reasonable in size.

Polymers belong to the orthorhombic mm2 symmetry group. Therefore, the piezoelectric tensor of polymers is described by five piezoelectric

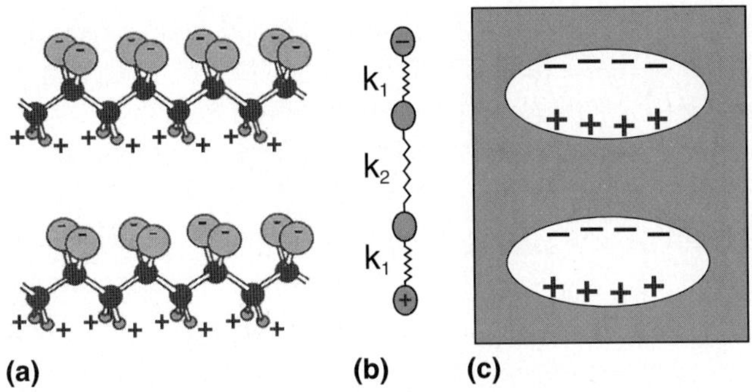

Fig. 6.1. Scheme of the chain structure in ferroelectric polymers (**a**), of the most simple model for piezoelectricity (**b**) and of the voided structure of ferroelectrets (**c**) (after [32])

coefficients: longitudinal piezoelectric coefficients d_{31} and d_{32}, the transverse piezoelectric coefficient d_{33} and the shear coefficients d_{15} and d_{24}.:

$$d = \begin{pmatrix} 0 & 0 & 0 & 0 & d_{15} & 0 \\ 0 & 0 & 0 & d_{24} & 0 & 0 \\ d_{31} & d_{32} & d_{33} & 0 & 0 & 0 \end{pmatrix}$$

Figure 6.1 schematically depicts the chain structure of the ferroelectric polymer PVDF (a), a simple model for piezoelectricity in polymers (b) and the voided structure of cellular ferroelectrets (c) [32]. According to the model shown in Fig. 6.1b, piezoelectricity arises, when the spring constants k_1 and k_2 are different. It is obvious from Fig. 6.1 that the origin of piezoelectricity is quite different in ferroelectric polymers and in ferroelectrets. Although ferroelectric polymers display dipole density piezoelectricity ($k_1 > k_2$), ferroelectrets show intrinsic piezoelectricity ($k_1 < k_2$), due to the softness of the voids in comparison to the much more stiffer surrounding polymer matrix. Because of the difference in the relative size of the force constants in ferroelectric and ferroelectret polymers, the dipole density piezoelectricity of ferroelectric polymers is described by a negative d_{33} coefficient, whereas the intrinsic like piezoelectric effect of cellular polymers results in positive d_{33} coefficients.

Figure 6.1 provides a simplified view on ferroelectric polymers, and the microscopic structure of ferroelectric polymers is much more complex, as depicted in Fig. 6.2 for a polarized sample. PVDF and its copolymers are semicrystalline, so ferroelectric crystallites with oriented dipoles in the chains are surrounded by an amorphous matrix. The polarization is compensated by space charges at the end of the crystallites, so ferroelectric polymers also show electret effects [33]. Typical values of the piezoelectric coefficients are $d_{33} = -20$–$23\,\mathrm{pC\,N^{-1}}$, $d_{31} = 16\,\mathrm{pC\,N^{-1}}$ and $d_{32} = 3\,\mathrm{pC\,N^{-1}}$ for uniaxially stretched, and $d_{33} = -20$–$25\,\mathrm{pC\,N^{-1}}$, $d_{31} = d_{32} = 5\,\mathrm{pC\,N^{-1}}$ for biaxially stretched PVDF, with a maximum temperature of use around

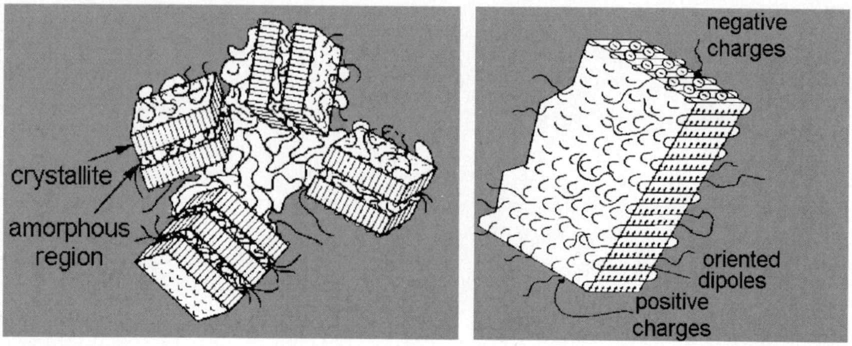

Fig. 6.2. Stacks of amorphous and crystal lamellae in polyvinylidene fluoride, and one poled stack with compensation charges at the interface between crystallites and amorphous regions

75–80°C. Copolymers with trifluoroethylene depict slightly larger values of $d_{33} = -24$–$30\,\mathrm{pC\,N^{-1}}$ and $d_{31} = d_{32} = 7\,\mathrm{pC\,N^{-1}}$ and an extended temperature range of use up to 110°C [34]. Slightly larger values are obtained on ferroelectric copolymers of VDF and TrFE [35]. In comparison to ceramic materials, these values are much smaller, and the advantages of piezoelectric polymers are their smaller acoustical impedance and flexibility. On the shear coefficients, there is practically no information available from manufacturers, only sparse reports are available in the scientific literature.

The cross-section of a typical ferroelectret polymer, based on biaxially stretched and inflated cellular polypropylene, is depicted in Fig. 6.3, together with a schematic of the charged material [11]. Figure 6.3 reveals the large difference in morphology and structure of ferroelectric and ferroelectret polymers. Typical values for d_{33} in cellular polypropylene are around $200\,\mathrm{pC\,N^{-1}}$, whereas d_{31} and d_{32} are very small on the order of $1\,\mathrm{pC\,N^{-1}}$ only [36]. Shear coefficients have not been determined yet, but are most probably extremely small like the d_{31} and d_{32} coefficients.

Table 6.1 shows a comparison of different piezoelectric materials in descending order of stiffness (material data from [11] and [31]). Quartz is the stiffest piezoelectric material, and shows only small values for d_{11} and g_{11}. Albeit the small piezoelectric coefficients, quartz is used in many technological applications, because of its excellent mechanical properties, yielding for example resonators with a very high quality factor. The chosen ceramic, one of the many forms of lead zirconate titanate PZT is softer in comparison to quartz, and shows large piezoelectric d_{33} (and d_{31}) charge coefficients, and correspondingly small g voltage coefficients due to the large dielectric constant of ferroelectric materials. The ferroelectric polymer PVDF is much softer than the ceramic materials, and it also displays only moderate values for the d and g coefficients. Cellular polymers are extremely soft in comparison to all the other materials listed in the table, but they show both large d_{33} and g_{33}

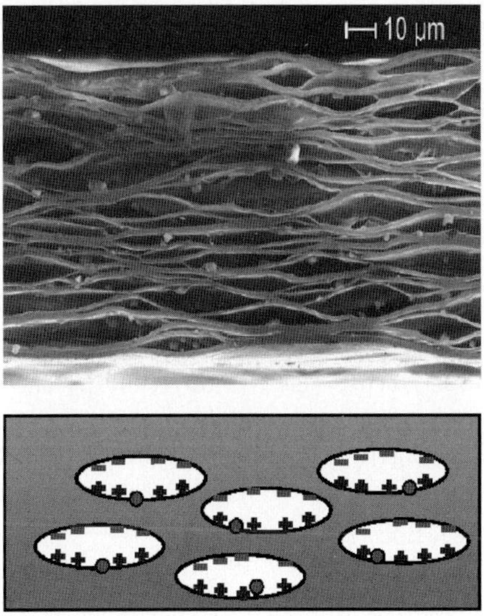

Fig. 6.3. Cellular polypropylene ferroelectret and internally charged voided polymer (after [11])

Table 6.1. Comparison of piezoelectric polymers with quartz and the ceramic PZT

Material	E (GPa)	d_{33} [pC N^{-1}]	g_{33} [Vm N^{-1}]
Quartz	72	2 (d_{11})	0.05 (g_{11})
PZT	50	171	0.01
PVDF	2	20	0.2
Cellular polypropylene	0.002	200	30

Material data taken from [11, 31]

coefficients, due to the softness of the material and the low dielectric constant of foams. However, in comparison to PVDF, both the d_{31} and d_{32} coefficients, as well as the corresponding g values are negligible in cellular polymers.

Most remarkable is the very large value for the airborne ultrasound FOM obtained for cellular polymers (160 with an acoustic impedance of 0.028 MRayl), whereas the values for piezoelectric polymers like PVDF are much smaller (0.2 with an acoustic impedance of 3.9 MRayl) [31]. Only porous piezoelectric ceramics may compete with cellular polymers, with a FOM for airborne ultrasound of 40 and an acoustic impedance of only 6.5 MRayl. Bulk ceramics have values comparable to that of PVDF (0.4) with a very large acoustic impedance of 25–37 MRayl [31].

6.2.3 Relaxor Ferroelectric Polymers

The low responses in traditional ferroelectric materials, such as the low strain level (~0.1% strain) and low sensitivity to external mechanical stimuli that can be achieved limit the performance of devices and systems made of ferroelectric polymers. As has been observed in many ferroelectric materials, by operating near the instability regions such as ferroelectric-paraelectric (F-P) transitions, these responses can be significantly enhanced [37]. Unfortunately, phase transitions occur over a narrow temperature range and, in most cases, involve large hysteresis, preventing the enhanced responses to be used in applications. It has been found that by introducing defects into P(VDF-TrFE) copolymers, it is possible to convert the polymer from a normal ferroelectric to a relaxor ferroelectric [8]. A new class of ferroelectric polymers, i.e., the terpolymers of P(VDF-TrFE-CFE), were developed from the normal ferroelectric PVDF polymer by employing proper defect modifications, which eliminate detrimental effects associated with a normal first order F-P transition while maintaining high material responses [9, 38]. The introduction of the third monomer into the polymer chain serves to interrupt the ferroelectric domains, thereby reducing their size [39]. Random defect introduction, as in the irradiated copolymer samples [8], broadens the ferroelectric transition and reduces the ferroelectric-paraelectric transition temperature. The random incorporation of the bulky third monomer into the polymer chains forces a conformation change from the all-trans (Tm $\geqslant$ 4) or conformation to the transgauche (TG) and T_3G conformations [40]. Addition of defects in the form of chemical monomers, in this case 1,1-chlorofluoroethylene (CFE), which are copolymerised with VDF-TrFE, eventually favor the TG conformation and eliminate the normal ferroelectric phase, leading to a relaxor ferroelectric with electromechanical strain greater than 7% and an elastic energy density of 0.7 J cm^{-3} at 150 MV m^{-1} [41]. It is the change in conformation from the disordered TG and T_3G to the all *trans* conformation that leads to the observed large electrostrictive strain, shown in Fig. 6.4. This high electrostrictive strain coupled with a relatively high elastic modulus (>0.3 GPa) makes the polymers suitable as a structural component in addition to the electroactive functions (actuators and sensors). The relaxor-ferroelectric terpolymer P(VDF-TrFE-CFE) exhibits a room temperature dielectric constant greater than 50 [38, 39]. This value is among the highest reported in literature for polymeric materials [42]. Relaxor ferroelectric polymers show practically no piezoelectric activity, but they can display very large field-induced piezoelectricity when biased with a suitable large electric dc-field. From Fig. 6.4, a large field-induced piezoelectric d_{33} coefficient around 400 pC N^{-1} is estimated at a dc-bias field of 100 Vμm^{-1}. Relaxor ferroelectric polymers, therefore, may find applications in certain niches, where large piezoelectric coefficients are desired and where dc-bias operation is not detrimental.

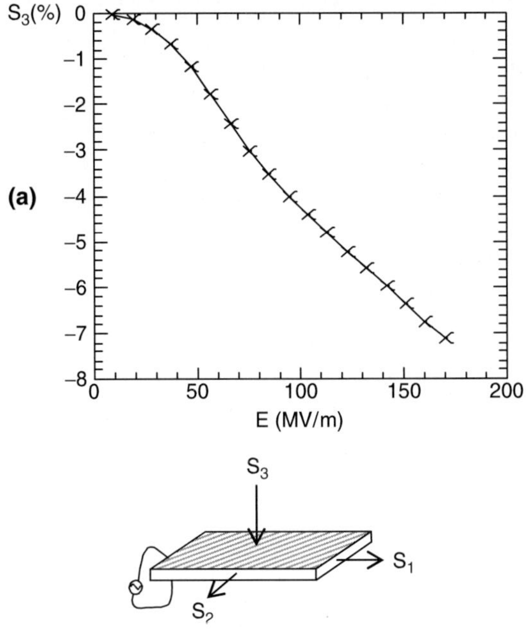

Fig. 6.4. Strain vs. electric field for a relaxor ferroelectric P(VDF-TrFE-CFE) terpolymer. The large strain that can be achieved with relaxor ferroelectric polymers make these materials interesting for inducing large piezoelectric effects by field biasing (after [41])

6.2.4 Printable Piezoelectrics and Multifunctional Composites

Piezoelectric laquers have been developed by the swiss company ALGRA [6], for example. Their piezoelectric laquers are based on the incorporation of piezoelectric ceramics into laquer matrices. The piezoelectric laquers can be printed on many different substrates. After printing, the piezoelectric elements are poled with a not revealed poling process. No details on the fabrication process are available in open literature, but the technical data information provided sound impressive with sensitivities reaching $1\,\mathrm{nC\,N^{-1}}$ in a typical element. Piezoelectric laquers are certainly an interesting class of materials, since they combine the advantageous features of polymers (flexibility and low brittleness) with the advantages of piezoelectric ceramics (large piezoelectric coefficients).

Piezoelectric laquers belong to the family of composites. Although this chapter primarily deals with polymers alone, it is interesting to mention that by combining piezoelectric ceramics with ferroelectric copolymers, true multifunctionality can be easily achieved [7]. 0–3 composites of lead titanate nanoparticles in PVDF-TrFE copolymers have been poled in a sequential manner to make the material piezoelectric but not pyroelectric, or vice versa.

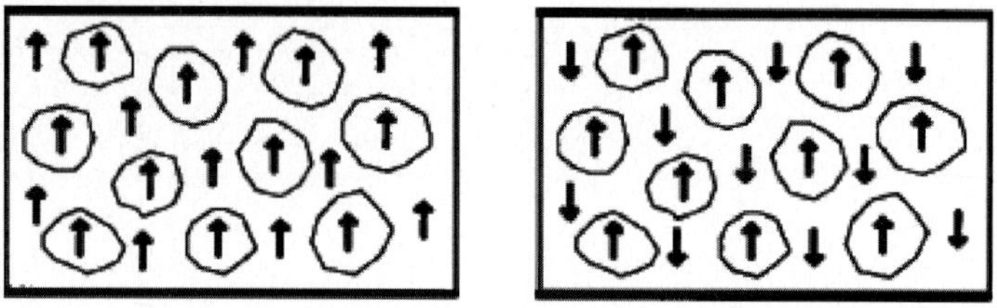

Fig. 6.5. Multifunctional composite of a ferroelectric polymer and nanosized piezoelectric ceramic inclusions. Indicated are the directions of the polarization vector in the polymer and in the ceramic, for a piezoelectric but not pyroelectric element (**a**) and vice versa (**b**) (after [7])

A schematic view of the composites with the different orientation of the polarization vectors is shown in Fig. 6.5. Compensation of effects is possible, since the piezoelectric coefficients are different in sign in piezoelectric ceramics (intrinsic piezoelectricity) and ferroelectric polymers (dipole density piezoelectricity), while the pyroelectric coefficients are both negative in polymers and ceramics. Applications may include piezoelectric transducers insensitive to temperature changes and drift, and pyroelectric sensors insensitive to microphony.

6.3 Applications of Piezoelectric Polymers

Table 6.1 is also useful in assessing the possible application niches for the different materials. PVDF and its copolymers are excellent materials for use in hydrophones, because of its nearly perfect match with water and other liquids. Commercially available are hydrophones made of PVDF used for the calibration of medical ultrasound equipment [3], examples of such hydrophones are depicted in Fig. 6.6. PVDF-based hydrophones are sold by many companies worldwide [43], and provide an excellent niche application of this piezoelectric polymer. Another success story for the use of PVDF in a large scale commercial application, with production rates of several million pieces per year is a pressure sensor, developed by the Austrian company AVL in Graz for Diesel injection lines [4], as illustrated in Fig. 6.7. Also noteworthy are the applications of PVDF in shock wave sensors, work pioneered by Francois Bauer of ISL, France, R.A. Graham of Sandia National Laboratories and L.M. Lee of Ktech Corporation, Albuquerque [5,34] (see also Chap. 6.5 for further details). The flexibility and the ease of production in any kind of form also allow for unusually shaped piezoelectric elements, the piezoelectric cable described in [2] and shown in Fig. 6.8 being an example. Such cables are extremely rugged and can stand heavy loads, for example in truck axle counting. Application areas include physical security, proximity sensing, traffic counting, and transducers for security fences.

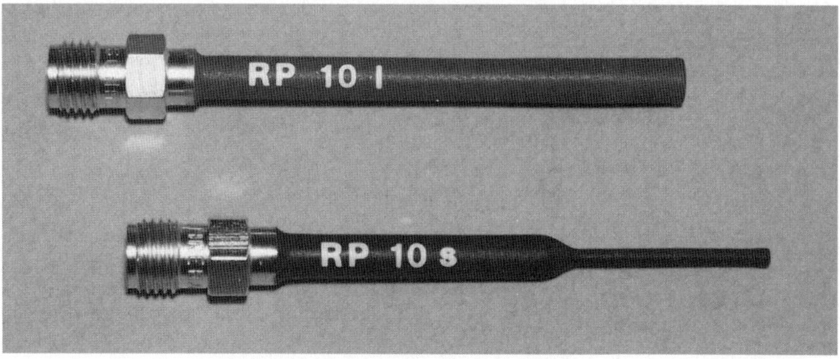

Fig. 6.6. Commercially available PVDF hydrophones (taken from [3])

Fig. 6.7. Pressure sensor for Diesel injection lines. Such pressure sensors have a multimillion pieces per year market (taken from [4])

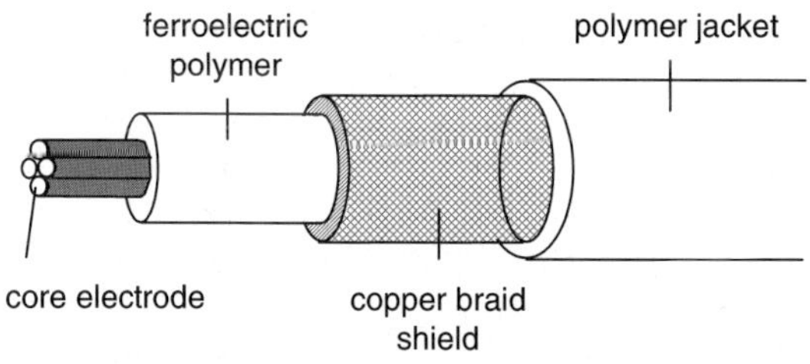

Fig. 6.8. Piezoelectric cable elements, in the piezoelectric coaxial cable the piezoelectric polymer acts as a polarized dielectric. Piezoelectric cables are extremely rugged and can withstand also heavy loads

The soft cellular polymers may only be used for applications in shallow water, although their dg product is quite large [11]. The reason is the heavy loading of the cellular material by the liquid in a water environment. Ferroelectrets are therefore much more promising for air-borne ultrasound applications, since their FOM is the highest of all materials available. Applications

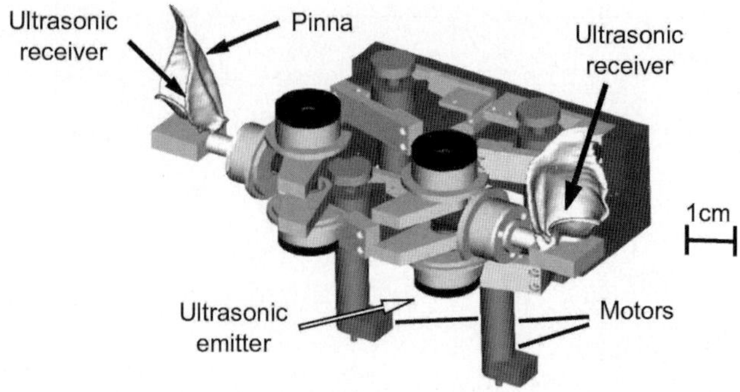

Fig. 6.9. Pulse echo location system based on cellular polypropylene ferroelectrets. The system is comparable to the ultrasound imaging systems of animals, like bats (after [44])

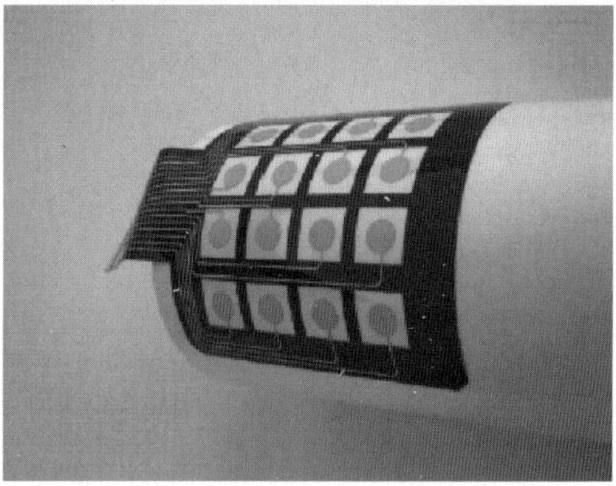

Fig. 6.10. Flexible keypad made from cellular polypropylene ferroelectrets. The keypad is vandal proved (taken from [15])

are available on the lab scale, an example being an ultrasound-based echo location system shown in Fig. 6.9 [44]. This echo location system is very similar to that of animals like bats.

The negligible longitudinal piezoelectric coefficients prevents applications where large deflections are required, an example is the piezoelectric bimorph. However, it is an advantage when only the thickness change is of interest, such as in switches. Applications of cellular polymer films are discussed in depth in a recent review of Wegener and Bauer [28], and on the homepages of several small and medium-sized companies in Finland [13–15]. Musical key pads based on cellular polypropylene are for example currently favored by many musicians. Figure 6.10 shows an example of a fully flexible, vandal proved keypad, successfully employed in the market [15].

Most noteworthy is the recent demonstration of a cellular polypropylene microphone by Hillenbrand and Sessler [45] with a sensitivity approaching that

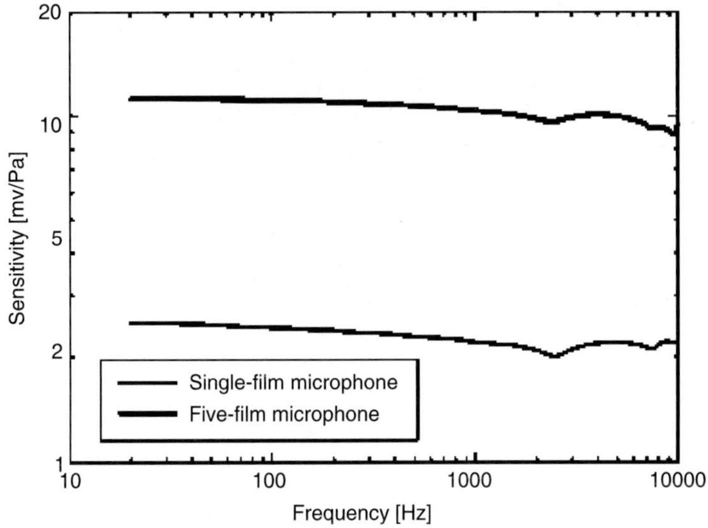

Fig. 6.11. Frequency response of a single layer and a five layer cellular polypropylene-based microphone (taken from [45])

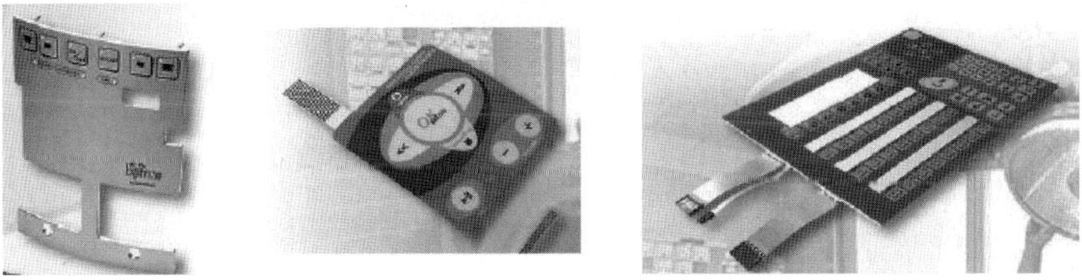

Fig. 6.12. Flexible keypads of different geometry, realized with piezoelectric laquer technology (taken from [6])

of conventional microphones. The microphone is based on cellular polypropylene and shows a flat response in the audio range, as shown in Fig. 6.11. In comparison to electret microphones, no micromachining is required for adjusting an air-gap. Therefore, cellular piezoelectrics may conquer a significant part of the microphone market.

Flexible keypads can be also realized with the piezoelectric laquer technique developed by ALGRA [6]. Many different forms and shapes can be easily developed, examples are shown in Fig. 6.12 for illustration. Currently, the most important advantage of the laquer technique is the high sensitivity achieved (approximately 1 V per 1 N applied force) and the large temperature range of operation in comparison to ferroelectrets.

6.4 Piezoelectric Ignition Systems

Piezoelectric materials are widely used as active elements in stress gauges used to provide nanosecond, time-resolved stress measurements of rapid impulsive stress pulses produced by impact, explosion or rapid deposition of radiation. In the earliest work, the gauges used crystalline sensors made of either x-cut quartz or various cuts of lithium niobate with thicknesses of many millimeters. The wave transit times through such sensors range from many tens of nanoseconds to a few microseconds. The upper response limits of the crystalline sensors are limited by either, or both, mechanical or electrical properties: dynamic yielding of the sensors or dielectric breakdown due to the large internal fields produced by the piezoelectric effect.

For using piezoelectric polymers in such applications, it is necessary to process them so that they display properties with reproducibility comparable to crystalline materials [46]. The specific poling cycling poling process, known as Bauer poling, is described in detail. The precise conditions during poling and the resulting highly reproducible samples allowed for an accurate description of the piezoelectric response of PVDF for both high-pressure shock compression and large volume compression. The behavior of PVDF is very well-characterized over a wide range of pressures using high-pressure shock loading with well-behaved and reproducible data for pressures up to 25 GPa in inert materials. Most important PVDF shows a response independent of the loading path up to 30 GPa [5].

The direct stress-derivative or stress-rate signals of a few nanoseconds duration from PVDF gauges, and the higher operating stress limits, provide capabilities not available with any other technique. The 10 ns transit time of a wave in PVDF achieves conditions in which the electrical behavior is described in the "thin mode". That is, the electrical effects as the wave propagates within the sensor are not resolved to first order.

PVDF gauges can be not only used for the characterization of materials under high pressure shock loading, they can be also used for Hugoniot measurements of porous explosives. Shock pressure profile measurements "in situ" porous H.E. in a detonation regime are also feasible. The measured shock pressure profiles with the PVDF gauge show a fast superpressure of a few nanoseconds followed by a pressure release down to a plateau level and then by a pressure decay.

6.5 Polarization Technique

For the preparation of piezoelectric polymers and copolymers, it is necessary to apply a high poling electric field to the essentially insulating material. The earliest electrical and most widely available process applied to pole PVDF films is the room temperature "corona poling" method, which produces large

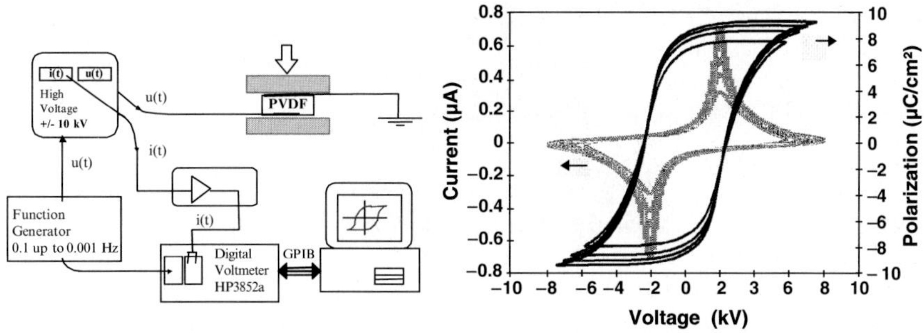

Fig. 6.13. Poling equipment for the cyclic Bauer poling process of PVDF, together with the evolution of the hysteresis curve. Gauges can be produced with a reproducibility of the material properties better than 2% (after [47])

area films for applications that do not require a very high electrical quality of the samples. In such a large volume production method, quality and reproducibility of data is not necessary or is it desired.

For applications and materials studies requiring high quality and reproducibility, individual samples must be produced with well-defined electrical properties in a process, which controls the amplitude, duration, and the history of the electric field. By controlling the electrical field applied to individual samples, the electric displacement, displacement current, remanent polarization, and homogeneity of polarization can be also controlled [46]. The Piezotech poling equipment, depicted in Fig. 6.13a, allows us to produce extremely well-defined piezoelectric polymer samples, as shown by the evolution of the hysteresis curves in Fig. 6.13b. With this Bauer cyclic poling process, reproducible remanent polarizations as large as $9\,\mu C\,cm^{-2}$ are routinely achieved in a commercial process. Because electric fields as high 5 MV per cm are utilized, inferior films are eliminated in the process. Each PVDF gauge, fabricated with this process is characterized with an individual poling history with well-defined electrical properties, better than 2%.

6.6 Piezoelectric PVDF Shock Response

The shock response of PVDF has been determined by several experiments. An experimental set-up, where very well pressure levels up to 25 GPa with uncertainties of 1% can be achieved is sketched in Fig. 6.14 [46]. The pressure is applied to the gauge by the impact of a precisely controlled impactor mounted on a projectile. The time scale of the experiment is on the order of 10 ns, the time stress waves need to propagate through the PDVF gauge. The short duration of the loading creates conditions where no heat is exchanged, so called "Hugoniot" conditions. A range of impact velocities from $250\,m\,s^{-1}$ to $1.8\,km\,s^{-1}$ is achieved with a powder gun for accelerating the projectile.

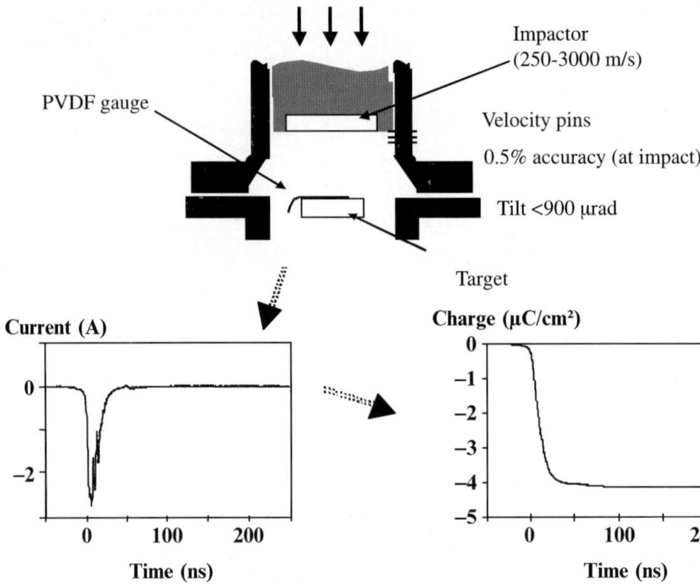

Fig. 6.14. Experimental arrangement for shock-wave measurements. An impactor mounted on a projectile impinges with a preselected velocity on the pressure gauge. Also shown are the current and charge response of a PVDF gauge (after [47])

With a two-stage light gas gun, impact velocities between 0.85 and 3 km s^{-1} are achieved. Symmetric impact, nonsymmetric impact, and reverse impact techniques have been utilized for investigating the shock response of the PVDF gauges. Also included in Fig. 6.14 is the current response of the PVDF shock gauge, and the released charge obtained by integrating the current response. Since the signal rise times are on the order of few nanoseconds, low inductance, GHz instrumentation must be used throughout, for example Tektronix TDS 684B digitizers.

From a series of symmetric and nonsymmetric impact experiments, the charge response of PVDF shock gauges is well documented up to pressure levels of 35 GPa. Figure 6.15 summarizes the experimentally obtained data and compares the data with calculations obtained from hydrocode DYNA (Livermore-Software Two Dimensional Hydrodynamic Finite Element Code) and SHYLAC (Hydrodynamic Lagrangian Simulation of Shocks) models. The experimental data show that the electrical charge response of the PVDF gauges is independent of the loading path [47]. Therefore, PVDF gauges can be used to study shock wave impacts under a large variety of conditions.

PVDF gauges are extremely versatile for investigating porous explosives under Hugoniot conditions [47]. An example of an experiment, where shock wave pressure signals are recorded within a porous H.E. Formex explosive is shown in Fig. 6.16. A PVDF gauge is mounted between the two H.E., two

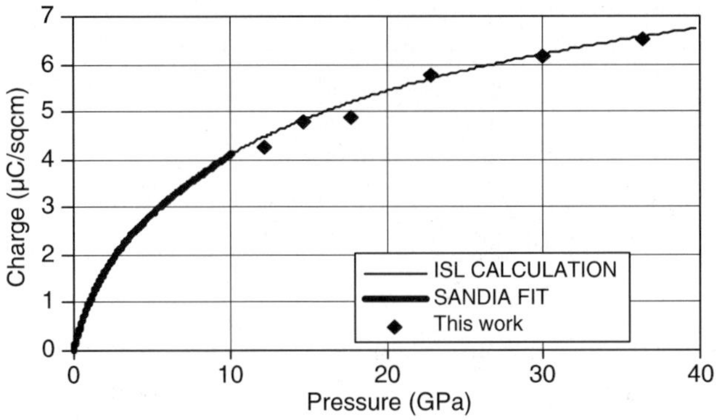

Fig. 6.15. Charge response of PVDF vs. pressure up to 35 GPa. The very well-documented charge response of PVDF allows for measurements of pressures during shock wave experiments (after [47])

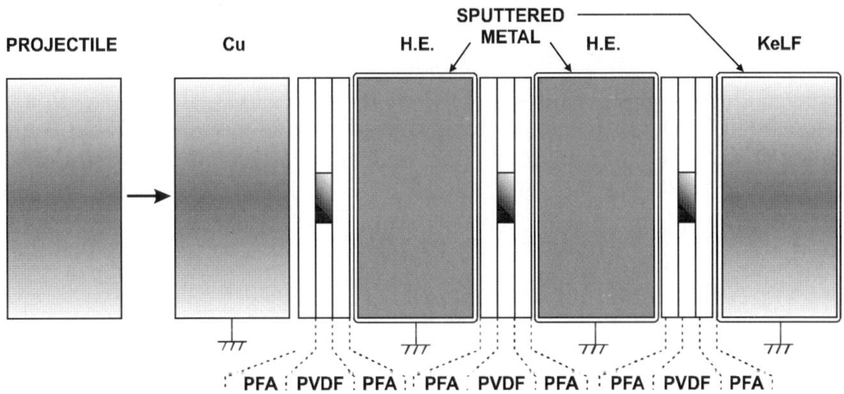

Fig. 6.16. Experimental arrangement for the Hugoniot investigation of porous high explosives. Several PVDF gauges are used to monitor the pressure during explosion, even within the explosive (after [47])

other gauges are placed in front and at the end of the H.E.'s. The shock wave response of the shock gauges in the front and between the H.E. Formex are shown in Fig. 6.17 for illustration. After the shock pressure jump, a plateau and a slight collapse is observed until the pressure reaches a maximum value of 9.7 GPa, following the reaction of the H.E. For the gauge in between the H.E. samples, a fast superpressure of a few nanoseconds duration is observed at a level of 14.5 GPa, followed by a pressure release down to a plateau level of 10.5 GPa and finally a pressure decay. The experiments reveal that precise pressure measurements are obtained by polymer shock gauges, so these gauges allow the study of materials and properties.

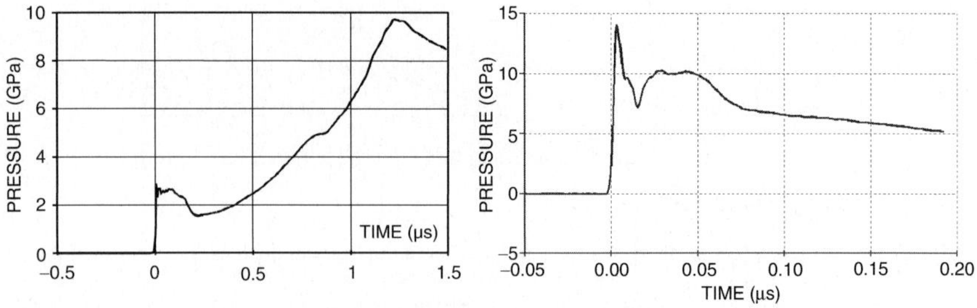

Fig. 6.17. Pressure levels during the shock-wave experiment with porous explosives (*left*), first PVDF gauge in Fig. 6.16, *right* second PVDF gauge in Fig. 6.16 (after [47])

6.7 Outline of Future Developments

The material class of piezoelectric electro-active polymers has been significantly broadened. It, therefore, seems most interesting, to combine electro-active polymers with flexible electronics. Flexible electronics may be based on amorphous silicon, a mature technology with a huge application market in current LCD flat screens, or on organic semiconductors, a field that is currently pursued by many research teams in academia and industry. The current field of flexible electronics is mainly driven by backplane applications for flexible displays, though developments of wearable electronics are also advancing rapidly.

By combining the worlds of electro-active polymers with flexible electronics, completely new electronic devices can be envisaged, possibly useful in nonvolatile memories, and in sensors for pressure, infrared radiation, etc. Memories were already reported in the literature [8, 48]. Sensors based on the combination of electro-active polymers and flexible electronic circuits are just emerging [49]. An example of such a sensor is shown in Fig. 6.18, where cellular ferroelectrets are combined with amorphous silicon matrices. Coupling piezoelectric laquers with such matrices would enable the production of fully flexible keypads with all the necessary electronics for read-out and signal amplification. Therefore, a huge potential exists for future research and application niches of electro-active polymers in this upcoming field.

Materials development may benefit from modern tools, like ab-initio studies of polarization and pyroelectricity in ferroelectric polymers [50], or density functional and molecular dynamics studies of electromechanical properties [51]. An example is the suggestion to synthesize and investigate BN-based polymers, which should display better ferro and piezoelectric properties [50]. So far, theoretical predictions of better piezoelectric properties of ferroelectric polymers await experimental verification. The thermally stable cellular polymers, available on the laboratory scale, await introduction to commercial processes for development.

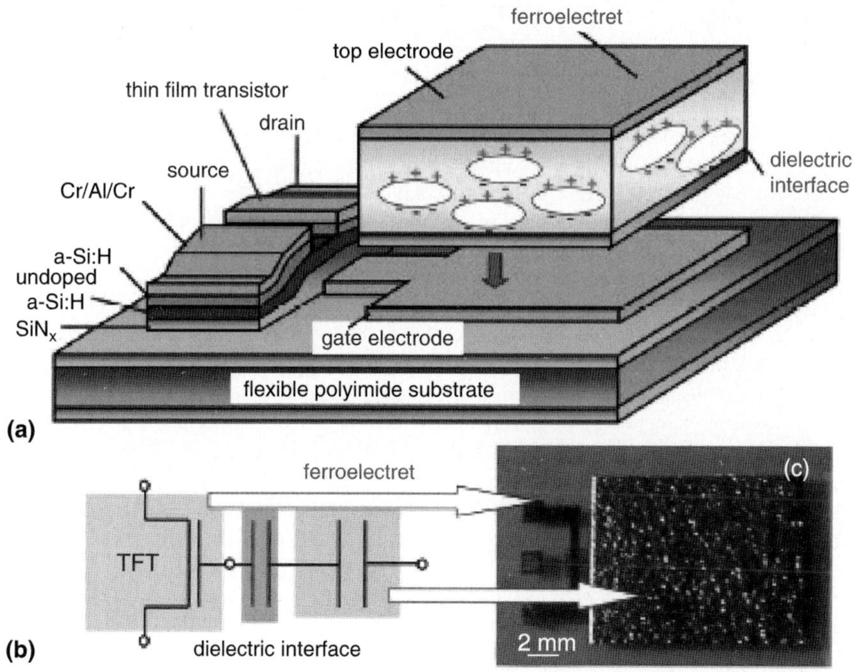

Fig. 6.18. Ferroelectret field effect transistor. The high impedance ferroelectret is directly coupled to an impedance matching thin film transistor, enabling the element to work as a force activated switch, a pressure sensor and a microphone (after [49])

6.8 Conclusions

Electro-active polymers such as ferroelectric PVDF, cellular polypropylene ferroelectrets, relaxor ferroelectric polymers, piezoelectric laquers, and multifunctional polymer-ceramic composites have useful piezoelectric properties that make these materials attractive for a large number of applications. Success stories for the research on such electro-active polymers are available, with many products having found their way into the market. The combination of electro-active polymers and flexible electronic systems promises new research opportunities in a fast developing, upcoming new field. The history and evolution of the PVDF pressure gauge has shown how the unique material properties of piezoelectric polymers make them the sensor of choice in the investigation of material properties under shock loading.

Acknowledgment

Financial support of the work in Austria from the Austrian Science Funds and the European Union (3 Plant) is gratefully acknowledged.

References

1. H.S. Nalwa (ed.), *Ferroelectric Polymers: Chemistry, Physics and Applications* (Marcel Dekker, New York, 1995)
2. http://www.meas-spec.com/myMeas/sensors/piezo.asp; http://piezotech.fr
3. see for example http://www.rp-acoustics.de/english/PVDF_e.htm
4. information of the company AVL, Austria, www.avlditest.com
5. F. Bauer, IEEE Trans. Ultrason. Ferroelectr. Freq. Contr., **47**, 1448 (2000)
6. http://www.algra.ch/Englisch/keyboards.htm
7. B. Ploss, B. Ploss, F.G. Shin, H.L.W. Chan, C.L. Choy, Appl.Phys. Lett. **76**, 2776–2778 (2000)
8. Q.M. Zhang, V. Bharti, X. Zhao, Science **280**, 2101 (1998)
9. R.J. Klein, F. Xia, Q.M. Zhang, F. Bauer, J. Appl. Phys. **97**, 094105 (2005)
10. A. Savolainen, K. Kirjavainen, J. Macromol. Sci. Chem. **A26**, 583 (1989)
11. S. Bauer, R. Gerhard-Multhaupt, G.M. Sessler, Phys. Today **57**(2), 37–43 (2004)
12. M. Paajanen, J. Lekkala, H. Välimäki, IEEE Trans. Diel. Electr. Insul. **8**, 629 (2001)
13. information of the company Emfit Ltd., Finland, www.emfit.com
14. information of the company B-Band, Finland, www.b-band.com
15. information of the company Screentec, Finland, www.screentec.com
16. A. Mellinger, M. Wegener, W. Wirges, R. Gerhard-Multhaupt, Appl. Phys. Lett. **79**, 1852–1854 (2001)
17. Z. Hu, H. von Seggern, J. Appl. Phys., **99**, 024102 (2006)
18. X. Zhang, J. Hillenbrand, G.M. Sessler, Appl. Phys. A **84**, 139–142 (2006)
19. S. Katzir, Arch. Hist. Exact. Sci. **57**, 61 (2003) and references therein
20. H. Kawai, Jpn. J. Appl. Phys. **8**, 975 (1969)
21. J.G. Bergman, J.H. McFee, G.R. Crane, Appl. Phys. Lett. **18**, 203 (1971)
22. K. Nakamura, Y. Wada, J. Polym. Sci. **A-29**, 161 (1971)
23. S.B. Lang, Phys. Today **58**(8), 31–36 (2005)
24. A.J. Lovinger, Science **220**, 1115 (1983)
25. B. Chu, X. Zhou, K. Ren, B. Neese, M. Lin, Q. Wang, F. Bauer, Q.M. Zhang, Science **313**, 334–336 (2006)
26. S.B Lang, S. Muensit, Appl. Phys. A 85, 125–134 (2006)
27. R. Gerhard-Multhaupt, IEEE Trans. Diel. Electr. Insul., **9**, 850–859 (2002)
28. M. Wegener, S. Bauer, Chem Phys Chem **6**, 1014 (2005)
29. S. Bauer, IEEE Trans. Diel. Electr. Insul. **13**, 953–962 (2006)
30. C. Z. Rosen, B.V. Hiremath, R. Newnham, *Key Papers in Physics: Piezoelectricity* (American Institute of Physics, New York, 1992)
31. J. Döring, V. Bovtun, J. Bartusch, A. Erhard, DACH Jahrestagung 2004 Salzburg, http://www.ndt.net/article/dgzfp04/papers/p11/p11.htm
32. M. Lindner, H. Hoislbauer, R. Schwödiauer, S. Bauer-Gogonea, S. Bauer, IEEE Trans. Diel. Electr. Insul. **11**, 255 (2004)
33. G.M. Sessler (ed.), Electrets, Vol. I, 3rd edn., (Laplacian Press, Morgan Hill, 1999)
34. technical information from the homepage http://www.piezotech.fr/
35. T. Furukawa, Phase Transitions **18**, 143 (1989)
36. G.S. Neugschwandtner, R. Schwödiauer, S. Bauer-Gogonea, S. Bauer, M. Paajanen, J. Lekkala, J. Appl. Phys. **89**, 4503–4511 (2001)
37. Q.M. Zhang, J. Shao, T. Shrout, N. Kim, L.E. Cross, A. Amin, B.M. Kulwicki, J. Appl. Phys. **77**, 2549–2555 (1995)

38. F. Bauer, E. Fousson, Q.M. Zhang, L.M. Fee, IEEE Trans. Diel. Electr. Insul. **11**, 293–297 (2004)
39. C. Huang, R. Klein, H. Li, Q.M. Zhang, F. Bauer, Z.Y. Cheng, IEEE Trans. Diel. Electr. Insul. **11**, 299–311 (2004)
40. V. Bobnar, B. Vodopivec, M. Kosec, A. Levstik, B. Hilczer, Q.M. Zhang, Macomolecules **36**, 4436–4442 (2003)
41. Q.M. Zhang, C. Huang, F. Xia, J. Su, in *Electroactive Polymer Actuators as Artificial Muscles*, Chapter 4, ed. by Y. Bar Cohen, (SPIE Optical Engineering Press, WA, 2004)
42. M. Rabuffi, G. Picci, IEEE Trans. Plasma Sci. **30**, 1939–1945 (2002)
43. R. Pechta, RP-acoustics, private communication, 2006
44. A. Streicher, R. Müller, H. Peremans, M. Kaltenbacher, R. Lerch, 2004 IEEE Ultrason. Symp., 1142–1145 (2004)
45. J. Hillenbrand, G.M. Sessler, J. Acoust. Soc. Am. **116**, 3267 (2004)
46. F. Bauer, R.A. Graham, Ferroelectrics **171**, 95–102 (1995) and references therein
47. F. Bauer, Mat. Res. Symp. Proc. **698**, 53–64 (2002)
48. R.C.G. Naber, C. Tanase, P.W.M. Blom, G.H. Gelinck, A.W. Marsman, F.J. Touwslager, S. Setayesh, D.M. de Leeuw, Nat. Mat. **4**, 243 (2005)
49. I. Graz, M. Kaltenbrunner, C. Keplinger, R. Schwödiauer, S. Bauer, S. Lacour, S. Wagner, Appl Phys. Lett. **89**, 073501 (2006)
50. S.M. Nakhmanson, M. Buongiorno Nardelli, J. Bernholc, Phys. Rev. Lett. **92**, 115504 (2004)
51. H. Su, A. Strachan, W.A. Goddard III, Phys. Rev. B **70**, 064101 (2004)

Part II

Applications and Innovations

7

Electromechanical Frequency Filters

W. Wersing and K. Lubitz

7.1 Introduction

Frequency filters select signals with a frequency inside a definite frequency range or band from signals outside this band, traditionally afforded by a combination of L-C-resonators. The fundamental principle of all modern frequency filters is the constructive interference of travelling waves. If a filter is set up of coupled resonators, this interference occurs as a result of the successive wave reflection at the resonators' ends. In this case, the center frequency f_c of a filter, e.g., set up of symmetrical $\lambda/2$-resonators of length l, is given by $f_c = f_r = v_{ph}/\lambda = v_{ph}/2l$, where v_{ph} is the phase velocity of the wave. This clearly shows the big advantage of acoustic waves for filter applications in comparison to electro-magnetic waves. Because v_{ph} of acoustic waves in solids is about 10^4–10^5 smaller than that of electro-magnetic waves, much smaller filters can be realised. Today, piezoelectric materials and processing technologies exist that electromechanical resonators and filters can be produced in the frequency range from 1 kHz up to 10 GHz. Further requirements for frequency filters such as low losses (high resonator Q) and low temperature coefficients of frequency constants can also be fulfilled with these filters. Important examples are quartz-crystal resonators and filters (1 kHz–200 MHz) as discussed in Chap. 2, electromechanical channel filters (50 kHz and 130 kHz) for long-haul communication systems as discussed in this section, surface acoustic wave (SAW) filters (20 MHz–5 GHz), as discussed in Chap. 14, and thin film bulk acoustic resonators (FBAR) and filters (500 MHz–10 GHz), as discussed in Chap. 15.

Although mechanical channel filters are no longer used in new communication systems due to the technological progress of integrated circuits and modern digital electronics, we decided to present this section not only because of historical reasons but also of the following:

- Electromechanical frequency filters represent a pioneering high-tech application of piezoelectricity in modern communication

- They have required a highly sophisticated tailoring of the electromechanical transducers, and many aspects of piezoelectric materials engineering have been developed to meet these challenges
- It underlines and illustrates the necessity of a close mutual cooperation of all disciplines (materials science, physics, electronics, and manufacturing technology) in high-tech innovations

7.2 Electromechanical Channel Filters

The development of frequency-division-multiplex (FDM) telephone systems created a need for very selective filters with tightly controlled pass bands in the vicinity of 100 kHz. The desired characteristics has been achieved with full-lattice type filters with incorporated coils, capacitors, and X-cut quartz crystals operating in flexure-mode. These filters with evolutionary design improvements but still quite voluminous due to the incorporated coils and capacitors were manufactured in large quantities until the 1970s.

Because of the steady growth of international telephone service, progressively more exacting demands for long-haul communication systems required increasing performance of frequency filters. To fulfil these requirements and at the same time to considerably decrease the filter size, a rather new filter concept has been chosen. The idea was to fulfil the required high performance by mechanical frequency filters merely based on mechanically coupled iron-nickel alloy resonators with extremely high mechanical Q (>15,000), and to perform the signal transformation (electrical to mechanical and vice versa) at the input and output using piezoelectric transducers with high electro-mechanical coupling to achieve the broad bandwidth (Fig. 7.1). These filters offered additional advantages such as superior stability, smaller bulk, and lower fabrication cost. Moreover, only mechanical channel filters are of sufficiently high quality to allow the frequency-division multiplexing of 12 telephone channels by premodulation to the internationally specified FDM base group between 60 and 108 kHz. This was an enormous advantage because only one single type of channel filter was required.

Figure 7.2 shows the structure of the 50 kHz channel filter of the first channel modem in the world to satisfy the 1/20 CCITT[1] specifications. It consists of 12 bending resonators with a resonance Q of >15,000, interconnected by means of a wire spot-welded to the base of each. The two transducer resonators consist of a bolt-shaped element with a piezoelectric platelet soldered to its levelled side. All resonators are fabricated from a special iron–nickel alloy. The temperature coefficients of the acoustic velocity of this alloy is set to $0 \pm 3\,\mathrm{ppm\,K^{-1}}$ by annealing [1].

The excellent results obtained with the first mechanical filter generation as well as advances in the know-how of piezoelectric ceramics and fabrication

[1] Comité Consultatif Internationale Téléphonique et Télégraphique.

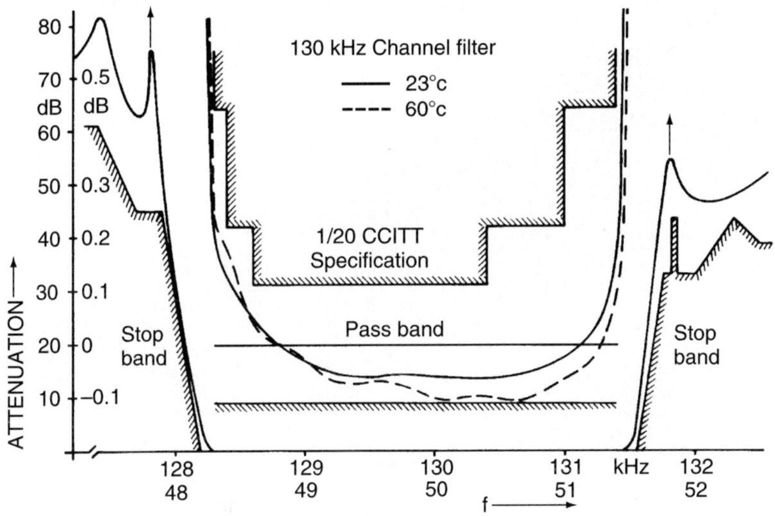

Fig. 7.1. Tolerance scheme for 50 and 130 kHz-channel filters as developed for telephone systems (1/20 of CCITT[1]-specification). It shows the extremely high performance required, i.e. a very wide bandwidth, very high attenuation and steep flanks in the stop band, and very low ripple in the pass band (*straight lines*). In addition, the response of a mechanical 130 kHz-filter is shown at 23°C (*full line*) and at 60°C (*broken line*)

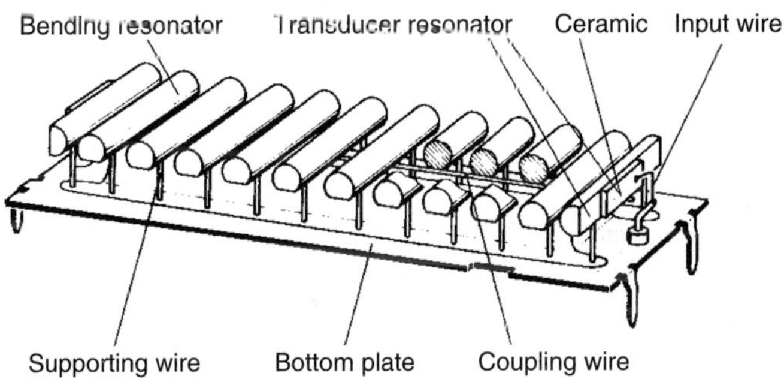

Fig. 7.2. Structure of a 50 kHz channel filter

technology made it very promising to develop a new channel filter generation with a reduction in bulk by almost 70% (Fig. 7.3). This was mainly possible due to the higher frequency level (130 kHz instead of 50 kHz) and the use of directly driven transducer resonators, obviated the need for coils at the input and output of the filter as discussed in detail in the next section.

Besides the broadband channel filters, narrow-band signal filters are also needed for the modems. These consist of two transducer resonators coupled either directly or via a third resonator.

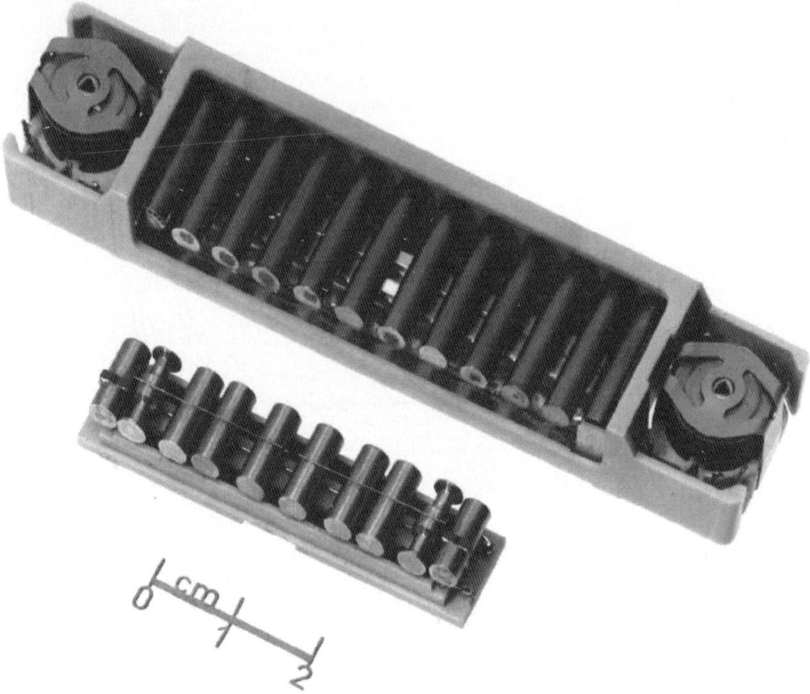

Fig. 7.3. Mechanical channel filters: 50 kHz-filter above and 130 kHz-filter below

7.3 Required Filter Ceramic Parameters

For the development of the first filter generation for 50 kHz channel modems, it was necessary to take into consideration the properties of suitable piezoelectric ceramics available at that time. On the basis of this the material specifications for 50 kHz-filters have been established (Table 7.1). Special emphasis was laid on the positive temperature coefficient of the frequency constant, this being required to compensate the temperature coefficient of the solder. Otherwise, the data of acceptor doped (hard) PZT ceramics as used for ceramic AM and FM radio intermediate frequency (IF) filters have been adopted. For comparison reasons, ceramic specifications for radio IF-filters are also shown in Table 7.1.

Still more stringent specifications had to be met by piezoelectric ceramics for channel filters for a new generation of 130 kHz channel modems. When the decision was reached to develop the new generation channel modems, it was clear that the goal to reduce the size of the channel filters by 70% without any release in performance (Fig. 7.1) could only be reached by a close mutual cooperation between ceramic materials experts, filter development engineers, and system experts.

Table 7.1. Specifications of piezoelectric ceramics for mechanical frequency filters

	Radio IF-filters	50 kHz-filters	130 kHz-filters
Permittivity	$\varepsilon/\varepsilon_0 \approx 600$	$\varepsilon/\varepsilon_0 = 900 \pm 40$	$\varepsilon/\varepsilon_0 = 1,360 \pm 40$
Coupling factor k_p	$k_\mathrm{p} \approx 0.35$	$k_\mathrm{p} = 0.44 \pm 0.04$	$k_\mathrm{p} = 0.54 + 0.02$
Coupling factor k_{31}	$k_{31} \approx 0.2$	$k_{31} = 0.25 \pm 0.02$	$k_{31} = 0.30 + 0.01$
Figure of merit	$(\varepsilon/\varepsilon_0)k_{31}{}^2 \approx 24$	$(\varepsilon/\varepsilon_0)k_{31}{}^2 = 56$	$(\varepsilon/\varepsilon_0)k_{31}{}^2 = 122$
Frequency constant	Not specified	Not specified	$N_1{}^\mathrm{E} = 1,680 \pm 10\,\mathrm{m\ s^{-1}}$
Resonance Q	$Q \approx 1{,}000$	$Q \geqslant 600$	$Q > 1,500$
Temperature coefficient[a]			
of permittivity	Not specified	$\mathrm{TC}\varepsilon \leqslant 3{,}000\,\mathrm{ppm\ K^{-1}}$	$\mathrm{TC}\varepsilon < 2,500\,\mathrm{ppm\ K^{-1}}$
of frequency constant	$\mathrm{TCF} < 60\,\mathrm{ppm\ K^{-1}}$	$\mathrm{TCF} = 80 \pm 50\,\mathrm{ppm\ K^{-1}}$	$\mathrm{TCF} = 80 \pm 10\,\mathrm{ppm\ K^{-1}}$
Aging of frequency constant per time decade	$\mathrm{AF} \leqslant 0.1\%$	$\mathrm{AF} \leqslant 0.1\%$	$\mathrm{AF} < 0.05\%$

[a]The temperature coefficient of a property p: $\mathrm{TCp} = (1/p)(\partial p/\partial T)$ is here defined as: $\mathrm{TCp} = (p(60°\mathrm{C}) - p(20°\mathrm{C})/p(20°\mathrm{C}) \cdot 40\,\mathrm{K}$

From Fig. 7.2, it can be seen that a considerable reduction in size can be achieved by avoiding the coils at the filter input and output. The function of these coils was to increase the bandwidth of the input and output transducers, which can otherwise only be achieved with a higher piezoelectric coupling factor of the transducer ceramic. However, the required coupling factor was estimated to be much higher than that achievable at best by any ceramic design. Thus, it was up to the system and filter designers to considerably reduce the demand on the coupling factor (to the value shown in Table 7.1). They proposed the following measures which have been realized the following way:

1. The center frequency was increased to 130 kHz to reduce the required relative bandwidth of the transducers (Fig. 7.1)
2. Directly driven transducer resonators utilizing the thickness coupling factor $k_{33}{}^2$ of the ceramic were used

The directly driven transducer resonators consist of two iron–nickel cylinders with two piezoelectric ceramic platelets of opposite polarity soldered in between (Fig. 7.3). The elimination of the input and output coils has become possible because of a higher mechanical coupling factor of the piezoelectric ceramic. This considerably reduces the length of the filter. The increased center frequency also reduces the resonator bolts' length and therefore the filter width, at least by a factor of two. Furthermore, by designing the new channel

[2] k_{33} is approximately two times larger than k_{31} utilized in transducer resonators of the 50 kHz-filters.

filter with attenuation poles on each side of its passband, it was possible to manage with fewer resonators, thereby again reducing the length of the filter and, above all, shortening its group delay [2]. Altogether, a bulk size reduction by almost 70% (Fig. 7.3) has been achieved.

To realize sufficiently low input and output impedances to allow the use of fully integrated amplifiers without impedance-matching transformers, the permittivity of the transducer ceramic had to be increased to about 1,350 and its temperature coefficient decreased to $TC\varepsilon < 2{,}500\,\mathrm{ppm\,K^{-1}}$.

More stringent specifications had to be met by the piezoelectric ceramic for the narrowband signal filters of the 130 kHz channel modems as well. Especially a very high resonance $Q > 1{,}800$ in combination with a very low temperature coefficient TCQ ($|\mathrm{TCQ/Q}| \leqslant 5\,\mathrm{ppm\,K^{-1}}$) was required to keep the insertion loss sufficiently independent of temperature (Sect. 7.4.5).

Finally, it should be stressed that not only improved ceramic properties but improved properties together with much lower fabrication tolerances had to be met for the 130 kHz-modems (Table 7.1).

7.4 Tailoring of Piezoelectric Ceramics for Frequency Filter Applications

The basic demand upon piezoelectric ceramics used in frequency filters is a high mechanical resonance Q that can be achieved with acceptor-doped ceramics. This doping also results in a high coercive force (Fig. 7.4), so that these materials are referred to as hard piezoelectric ceramics (Chap. 4).

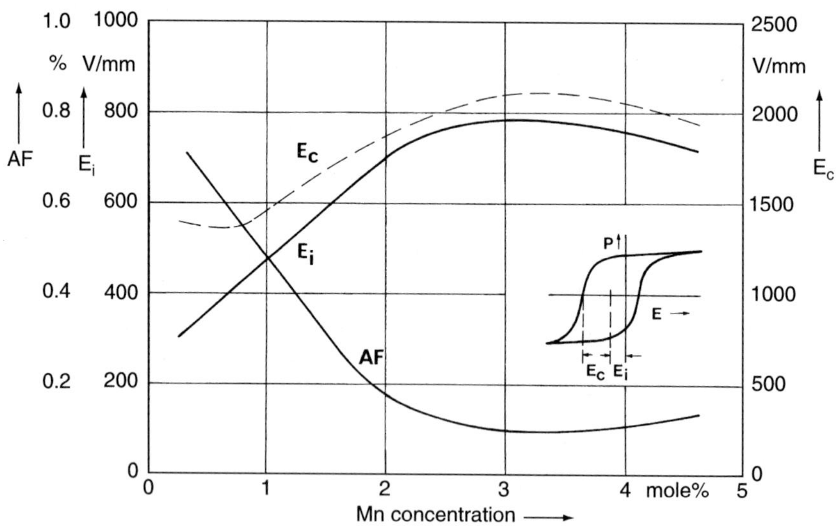

Fig. 7.4. Coercive field E_c, internal field E_i and aging rate of Mn-doped PZT ceramics at the morphotropic phase boundary vs Mn-Concentration

7.4.1 Manganese Doping of PZT

It has been found that the resonance Q of ceramics doped exclusively with acceptors such as Cr, Mn, and Fe is limited to about 800–1,000. This limitation results at least partly from the electrical conductivity[3] of these ceramics ($10^{-10} - 10^{-9}$ $(\Omega m)^{-1}$), which is far higher than that of ceramics doped with donors or modified with complex oxides (see later). The former sometimes show anomalously large dielectric losses [3] because of inhomogeneous high conductivities ($10^{-5} - 10^{-4}$ $(\Omega m)^{-1}$) at grain boundaries and as a result also decreased Q values of about 400.

Especially ceramics doped with Cr or Mn were found to have a frequency constant of very high time stability (Fig. 7.4) and thus already in the sixties the temperature stability of their frequency constant was also investigated to see whether they can be used for frequency filters [4].

The Cr or Mn doping stabilizes the domain configuration in the ceramic grains and thereby also the ferroelectric phase distribution within ceramics near the morphotropic phase boundary. Otherwise these ceramics would be generally unsuitable for filter applications because phase transitions induced by temperature variations would result in a frequency constant having a markedly nonlinear temperature coefficient with a superposed hysteresis loop. This is best demonstrated with the aid of samples having a composition fairly close to the phase boundary (Fig. 7.5). In the sample doped with only 0.4% Mn, the transition from the rhombohedral phase with a highly negative TCF to the tetragonal phase with a highly positive TCF is well pronounced, whereas in the sample doped with 4% Mn, the transition is seriously inhibited.

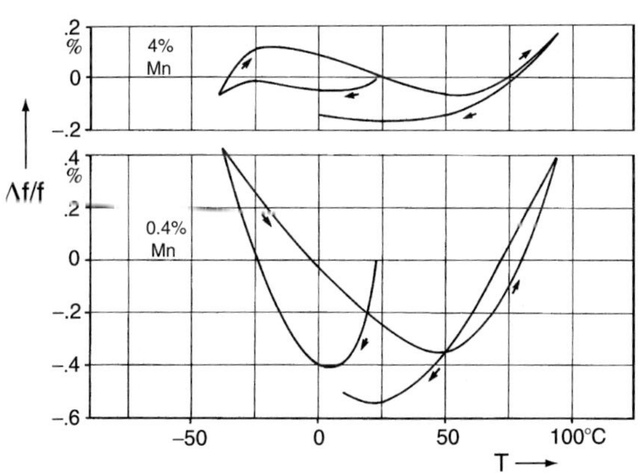

Fig. 7.5. Temperature response of frequency constant of Mn-doped PZT ceramics at the morphotropic phase boundary

[3] Because of the electromechanical coupling, the mechanical losses are coupled to the dielectric losses and, therefore, depend also on the conductivity of the ceramic [5].

For ceramics with a composition more in the tetragonal phase, e.g., ceramics with peak permittivity (poled ceramics) a purely positive TCF with nearly no temperature hysteresis is obtained.

The stabilization of the domain configuration of these ceramics becomes obvious from there hysteresis loops. Nonpolarized samples exhibit constricted and polarized samples exhibit displaced hysteresis loops (Fig. 7.4). This effect is described usually by introducing an internal stabilization field E_i (Fig. 7.4). This internal field E_i develops according to a logarithmic time law [6]. A similar logarithmic time law is experimentally observed for the aging of materials parameters such as permittivity, loss tangent, frequency constant, and resonance Q. This time law can be substantiated theoretically using a model that traces back the aging of ceramic parameters to a thermally activated transition from a domain configuration that is not in thermal equilibrium to a thermally balanced configuration. All symptoms described – high resonance Q, linearization of TCF, reduction of aging rate, appearance of an internal field – appear to have the same physical origin. Numerous models have been discussed in the literature for this, including bulk effects, domain wall effects, and grain boundary effects. In all these models, oxygen vacancies appear to be of critical significance, and it has been shown [6] that each of these effects has its justification, i.e., stabilization cannot be attributed to a single effect in the case of all materials.

7.4.2 Compensating-Valence Complex Oxide Substitutions

On the basis of these basic phenomena important for filter ceramics, it was possible to develop PZT ceramics solely doped with Mn that were able to fulfil the requirements for the first generation of channel modems (Table 7.1). We will now show how the properties of filter ceramics can be considerably improved by the combination of additives. As the mole fraction of an additive in PZT is increased, a point is soon reached, depending on the limit of the solubility of the additive, beyond which the ceramic properties cannot be further improved or even a marked degradation of the properties will set in. This situation was soon to lead to the use of coupled substitutions with ions of different valences, in other words, substitutions with complex oxides (Chap. 4). For filter ceramics various ternary ceramic systems such as $PbTiO_3$-$PbZrO_3$-$Pb(Mg_{1/3}Nb_{2/3})O_3$ have been investigated, which had to be doped additionally with a stabilizer such as Mn [7]. Takahashi et al. have shown that it is possible to use Mn directly as a constituent of a complex oxide [8]. They reported the presence of ceramics with favorable coupling factors and very high mechanical Q values in the system $PbTiO_3$-$PbZrO_3$-$Pb(Mn_{1/3}Nb_{2/3})O_3$ that recommend this system for filter ceramics [8].

Therefore, we studied this system in the vicinity of the morphotropic phase boundary varying the amount of the complex additive $(Mn_{1/3}Nb_{2/3})$ between 2 and 20%. The principal results – permittivity of poled samples, the planar

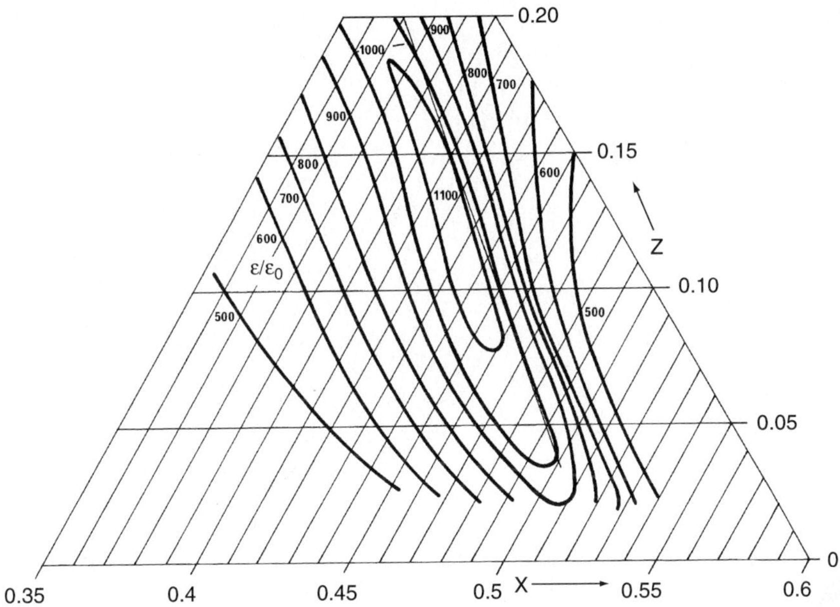

Fig. 7.6. Permittivity of poled $Pb[Zr_xTi_y(Mn_{1/3}Nb_{2/3})_z]O_3$ ceramics ($x+y+z=1$)

coupling factor and the temperature coefficients of permittivity and frequency constant – are shown in Figs. 7.6–7.9. The morphotropic phase boundary is continuously displaced with increasing ($Mn_{1/3}Nb_{2/3}$) concentration z toward smaller Zr concentration x. Thus due to the charge compensation of niobium, the bivalent Mn can be incorporated into the lattice far beyond its limit of solubility of ≈2.5% in PZT. As a result, the mechanical Q and the internal field strength are higher, and the aging rates and the conductivity are much lower than those of ceramics doped solely with Mn.

7.4.3 Temperature Coefficients of Frequency Constant and Permittivity

The strong correlation between the temperature coefficients TCε and TCF as well as their dependence on composition can be understood as a result of the strong dielectric anisotropy in the vicinity of the morphotropic phase boundary, as discussed in Chap. 3. This can be shown simply by calculating TCε and TCF of ceramic samples (slim bars) from the temperature coefficients of the single crystal permittivity and compliance components as derived with the aid of Devonshire's theory in Chap. 3.1. Because TCε and TCF anomalies in the vicinity of the morphotropic phase boundary likewise appear in unpoled samples (see Fig. 7.1 of [9]), one can calculate the temperature coefficients of unpoled ceramic samples, i.e., one can calculate them from the effective permittivity ε^{T*} (Equation 3.35 of Chap. 3.2) and from the isotropic fraction

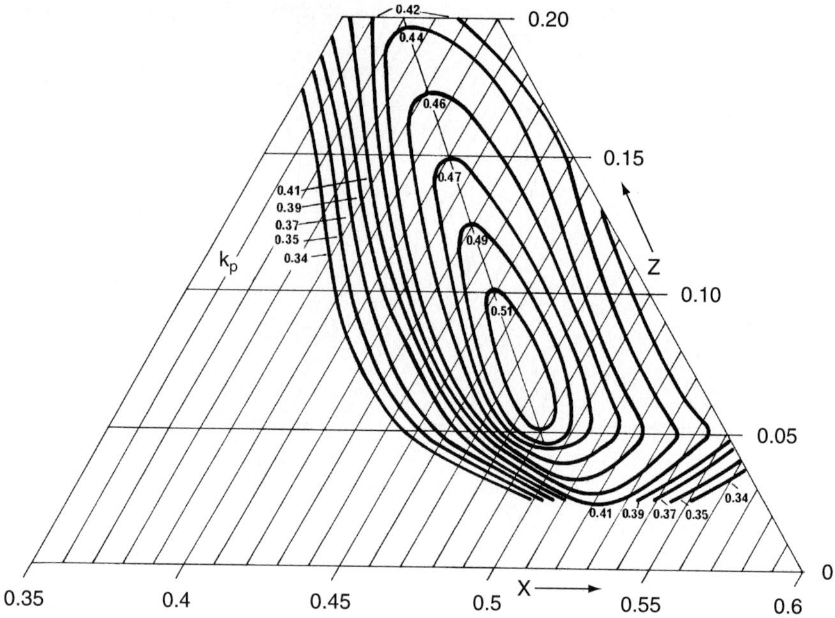

Fig. 7.7. Planar coupling factor of $Pb[Zr_xTi_y(Mn_{1/3}Nb_{2/3})_z]O_3$ ceramics $(x+y+z=1)$

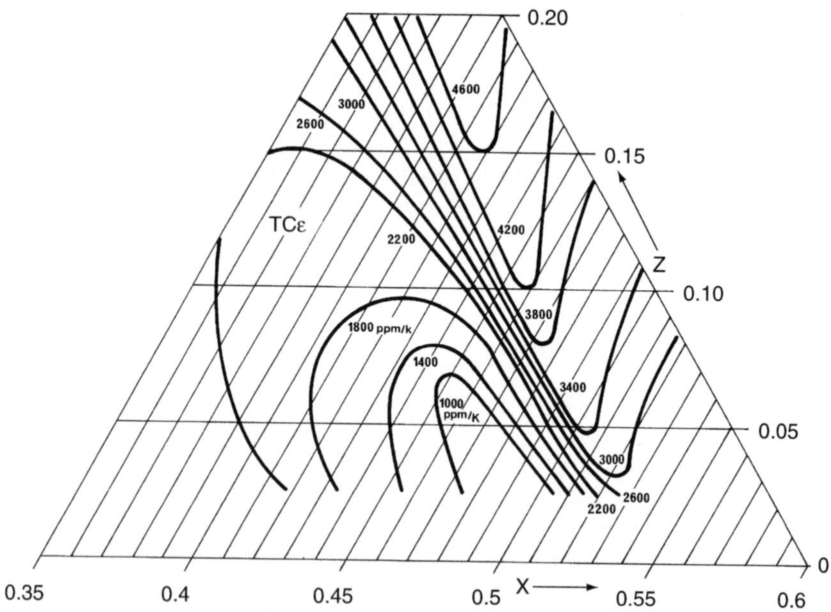

Fig. 7.8. Temperature coefficient of permittivity of $Pb[Zr_xTi_y(Mn_{1/3}Nb_{2/3})_z]O_3$ ceramics $(x+y+z=1)$

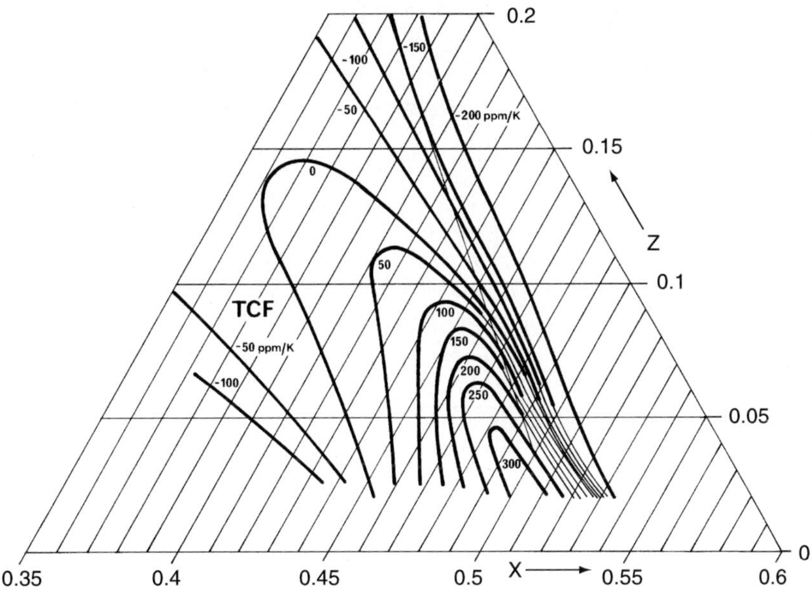

Fig. 7.9. Temperature coefficient of frequency constant of $Pb[Zr_xTi_y(Mn_{1/3}Nb_{2/3})_z]O_3$ ceramics ($x + y + z = 1$)

of the compliance matrix $s_{\alpha\beta}{}^{E}$ [10]. The advantage of this procedure is that the calculation of effective properties of unpoled ceramics is much simpler than that of poled ceramics. Of course, this procedure can only be used for single crystal properties, which do not depend on the direction of spontaneous polarization, i.e., for those that are described by tensors of even rank.

Let us first consider the temperature coefficient TCF. Because the TCF of discs is nearly identical to that of slim bars[4], we consider here the simpler case, i.e. we calculate the TCF of slim (unpoled) ceramic bars given by:

$$\mathrm{TCF} = -\tfrac{1}{2} \cdot \mathrm{TCs}_{11}^{D*} - \alpha_1 \qquad (7.1)^5$$

Calculating $s_{11}{}^{D*}$ from the isotropic fraction of the compliance matrix $s_{\alpha\beta}{}^{E}$

$$s_{11}^{D*} \approx \langle s_{11}\rangle^{D} = \tfrac{1}{5}\left(s_{11}^{E} + s_{22}^{E} + s_{33}^{E}\right) + \tfrac{2}{15}\left(s_{12}^{E} + s_{23}^{E} + s_{13}^{E}\right) + \tfrac{1}{15}\left(s_{44}^{E} + s_{55}^{E} + s_{66}^{E}\right) \qquad (7.2)$$

and using the form (3.31) for the $s_{\alpha\beta}{}^{E}$ components one obtains for the effective compliance $s_{11}{}^{D*}$

$$s_{11}^{D*} = \left\langle s_{11}^{D}\right\rangle + \kappa_1 P^{s2}\varepsilon_{11}^{T} + \kappa_3 P^{s2}\varepsilon_{33}^{T} \qquad (7.3)$$

[4] This is found experimentally and can also be derived theoretically.

[5] The thermal expansion coefficient α_1 vertical to P_r is typically 5–10 ppm K^{-1} and can be neglected for the following discussion.

with κ_1 and κ_3 are given for the

$$\text{tetragonal phase: } \kappa_1 = \tfrac{2}{15} Q_{44}^2 \quad \kappa_3 = \tfrac{4}{15}\left(3Q_{11}^2 + 8Q_{12}^2 + 4Q_{11}Q_{12}\right) \qquad (7.4)$$

$$\text{rhombohedral phase: } \kappa_1 = \tfrac{2}{45}\left(8Q_{11}^2 + 8Q_{12}^2 - 16Q_{11}Q_{12} + Q_{44}^2\right)$$

$$\kappa_3 = \tfrac{4}{45}\left(5Q_{11}^2 + 10Q_{12}^2 + 20Q_{11}Q_{12} + Q_{44}^2\right) \qquad (7.5)$$

The maximum in the compliance ${s_{11}}^{\mathrm{D}*}$ (minimum in frequency constants ${N_1}^{\mathrm{E}*}$ or ${N_\mathrm{p}}^{\mathrm{E}*}$, cf. Fig. 12 of Chap. 3) in the vicinity of the morphotropic phase boundary is caused by the strong dielectric anisotropy as can be seen from the second term of (7.3). As can be further seen from (7.3), the TCF is mainly controlled by the temperature coefficients $\mathrm{TC}({\varepsilon_{11}}^{\mathrm{T}} \cdot P^{\mathrm{s}2})$ and $\mathrm{TC}({\varepsilon_{33}}^{\mathrm{T}} \cdot P^{\mathrm{s}2})$ which are calculated from (3.25)–(3.28) to be:

$$\mathrm{TC}\left(\varepsilon_{33}^{\mathrm{T}} P^{\mathrm{s}2}\right) = 8\chi_0 b \varepsilon_{33}^{\mathrm{T}2} P^{\mathrm{s}4} \qquad (7.6)$$

$$\text{tetragonal phase: } \mathrm{TC}\left(\varepsilon_{11}^{\mathrm{T}} P^{\mathrm{s}2}\right) = 2\chi_0 \varepsilon_{33}^{\mathrm{T}} \left(\zeta_{112} - \zeta_{111}\right) \varepsilon_{11}^{\mathrm{T}} P^{\mathrm{s}4} \qquad (7.7)$$

$$\text{rhombohedral phase: } \mathrm{TC}\left(\varepsilon_{11}^{\mathrm{T}} P^{\mathrm{s}2}\right) = 2\chi_0 \varepsilon_{33}^{\mathrm{T}} \tfrac{4}{9} \left(\zeta_{111} - \zeta_{123}\right) \varepsilon_{11}^{\mathrm{T}} P^{\mathrm{s}4} \qquad (7.8)$$ [6]

Now we consider the temperature coefficient of the effective permittivity $\varepsilon^{\mathrm{T}*}$. It can be calculated from (3.35) to be:

$$\mathrm{TC}\varepsilon^{\mathrm{T}}* = \frac{1}{1+\nu}\left[\left(1 + \frac{1}{\nu} + \frac{4}{\nu}\frac{\varepsilon_{33}^{\mathrm{T}}}{\varepsilon_{11}^{\mathrm{T}}}\right)\mathrm{TC}\varepsilon_{11}^{\mathrm{T}} + \frac{4}{\nu}\frac{\varepsilon_{33}^{\mathrm{T}}}{\varepsilon_{11}^{\mathrm{T}}}\mathrm{TC}\varepsilon_{33}^{\mathrm{T}}\right] \qquad (7.9)$$

with $\nu = \sqrt{1 + 8\frac{\varepsilon_{33}^{\mathrm{T}}}{\varepsilon_{11}^{\mathrm{T}}}}$.

From $\mathrm{TC}\left(\varepsilon_{ik}^{\mathrm{T}} P^{\mathrm{s}2}\right) = \mathrm{TC}\varepsilon_{ik}^{\mathrm{T}} + \mathrm{TC}\left(P^{\mathrm{s}2}\right)$ follows with $TC\left(P^{s2}\right) = -2\chi_0 \varepsilon_{33}^{\mathrm{T}}$ (as calculated from (3.17)) the relation between $\mathrm{TC}({\varepsilon_{\mathrm{ik}}}^{\mathrm{T}} P^{\mathrm{s}2})$ and $\mathrm{TC}{\varepsilon_{\mathrm{ik}}}^{\mathrm{T}}$:

$$\mathrm{TC}\varepsilon_{ik}^{\mathrm{T}} = 2\chi_0 \varepsilon_{33}^{\mathrm{T}} + TC\left(\varepsilon_{ik}^{\mathrm{T}} P^{\mathrm{s}2}\right) \qquad (7.10)$$

Because of the strong dielectric anisotropy in the vicinity of the morphotropic phase boundary, the temperature coefficients of permittivity and frequency constant are mainly determined by (7.7) and (7.8). Thus the $\mathrm{TC}{s_{11}}^{\mathrm{D}*}$ growth to high negative values (TCF to high positive values) at the tetragonal side of the phase boundary and to high positive values (TCF to high negative values) at the rhombohedral side of the phase boundary[7]. However, on the one hand, the TCF peak on the tetragonal side is shifted to lower positive or even small negative values (cf. Fig. 12 in Chap. 3) by the third term in (7.3) because here κ_3 is about eight times higher than κ_1[8]. On

[6] We use here the same nomenclature as in Chap. 3. For simplicity the prime for principal axis components is omitted.

[7] $\zeta_{111} > \zeta_{112}$ and $\zeta_{111} > \zeta_{123}$, see Chap. 3.

[8] This is obtained by calculating κ_1 and κ_3 from electrostrictive coefficients as measured in [11].

the other hand, the third term in (7.3) has no effect at the rhombohedral side of the phase boundary because here κ_3 is even smaller than κ_1. Having in mind that at the morphotropic phase boundary both the tetragonal and the rhombohedral phase are coexisting, the jump of TCF from positive to large negative values at the phase boundary as obtained from (7.7) and (7.8) is smeared out and the experimentally found behavior is well described as can be seen in Fig. 12 of Chap. 3 and in Fig. 7.9. However, in Fig 7.9, the negative TCF peak is not fully visible within the plotted data set. The correlation between the temperature coefficient of permittivity and frequency constant can be seen easily from (7.10). The first term in (7.10) simply shifts the typical dependence of $\mathrm{TC}s_{11}{}^{\mathrm{D}*}$ on composition x mainly controlled by $\mathrm{TC}(\varepsilon_{11}{}^{\mathrm{T}}P^{\mathrm{s}2})$ to positiver values.

The theoretically found jump of the temperature coefficients at the phase boundary shows clearly that the measurement of TCF or TCε (or both) in the vicinity of the morphotropic phase is very useful to determine the exact phase boundary and the coexistence region by means of electrical measurements. Therefore, it can be noted from Figs. 7.8 and 7.9 that the phase boundary is shifted to lower Zr contents x, and the mixed phase region of the rhombohedral and tetragonal phases grows broader as the mole fraction of $Mn_{1/3}Nb_{2/3}$ is increased. In addition, however, a strong levelling off of the TCF maximum in the tetragonal phase is observed, where even the TCF for $z \geqslant 0.15$ is no longer positive at all. Furthermore, this lowering of the TCF peak is accompanied by its broadening and, above all, its displacement away from the phase boundary.

It is interesting to compare this behavior with that of dopants such as Nd which substitute Pb. Such dopants also lead to significant lowering of the TCF peak, while any large mole fraction of such dopants will even drive the TCF peak to negative values. Since, however, they will not result in any notable displacement of the peak away from the phase boundary, they can be used to lower the TCF peak of a ceramic system to a desired value without danger of any serious reduction of the coupling coefficient. Such an example is given in [9] where a complex ceramic system modified with $Ni_{1/3}Nb_{2/3}$ and with 4% $Mn_{1/3}Nb_{2/3}$ was additionally doped with Nd. Without any Nd doping a TCF peak of almost 300 ppm K^{-1} was obtained. With the increased addition of Nd, the TCF peak was sharply reduced and already assumes negative values as of about 4% Nd. Of course, the Nd additive causes a displacement of the phase boundary, but without any notable displacement of the TCF peak relative to the phase boundary.

The possibility to adjust the TCF peak to a value that is requested by the filter design is technologically of enormous significance. Only for a material close to the TCF peak, the TCF variation with composition is minimal and, therefore, its sensitivity to fabrication tolerances is low, i.e., it is possible to achieve a highly reproducible mass production of ceramics with tight TCF tolerances.

On the basis of the ceramic system studied so far we find a TCF value of about 80 ppm K^{-1} as required for channel filters for a composition with $x = 0.43$ and $z = 0.1$ (cf. Fig. 7.9). With this ceramic composition a TCF tolerance of ± 10 ppm K^{-1} is readily reproducible. Since the permittivity and coupling factor of this composition also vary only at a low rate, the too are easily reproducible. Furthermore, since their absolute values are higher than those required for 50 kHz channel filters, these ceramics were ideally suitable for the production of first-generation 50 kHz channel modems. However, the permittivity and coupling factor values of these ceramics could not fulfil the requirements for 130 kHz modems.

7.4.4 Noncompensating-Valence Complex Oxide Substitutions

In the ceramic system $Pb[Zr_xTi_{0.9-x}(Mn_{1/3}Nb_{2/3})_{0.1}]O_3$ and especially for the composition with $x = 0.43$, we obtained permittivity and coupling factor values of 1,050 and 0.46, respectively, considerably lower than required (for 130 kHz modems). However, all other important parameters (cf. Table 7.1) were met excellently. Therefore, it would have been obvious to improve permittivity and coupling factor simply by adding of a softener such as Nd or Nb because the addition of a softener to PZT is known to improve these parameters considerably (Chap. 4). However, since on the one hand, practically all additives lower the Curie temperature and, on the other hand, the Curie point of filter ceramics has to exceed 300°C substantially to avoid a degradation of the ceramics by soldering sufficiently small, the total mole fraction of additives has to be kept as small as possible. Thus it just was logical to induce a softening effect in the system $Pb[ZrTi(Mn_{1/3}Nb_{2/3})]O_3$ by reducing the ratio of Mn to Nb. This procedure was found to allow the realization of a continuous series of ceramics extending from hard to soft (exclusive Nb doping). It is obvious that this knowledge has become crucial for many other applications, too. Ceramics with optimally high permittivity and coupling factor on the one hand, and high resonance Q and low aging rates on the other, were obtained by choosing a Mn/Nb ratio of approximately 1/5. Figure 7.10 shows the principal material parameters of such ceramics vs. the Zr content. Ceramics with particularly favorable material parameters with respect to the requirements of 130 kHz modems (cf. Table 7.1) were realized with Zr concentrations x between 0.43 and 0.44.

We would like to emphasize that the noncompensating-valence substitution influences the TCF peak similarly like three-valent dopants such as Nd substituting Pb, as discussed earlier. With this substitution, the TCF peak was considerably reduced, but without any notable displacement of the TCF peak away from the phase boundary, i.e., the TCF peak remains quite close to the permittivity peak (cf. Fig. 7.10).

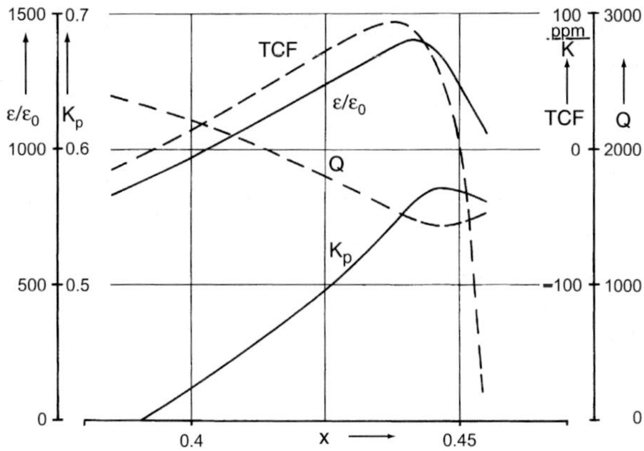

Fig. 7.10. Properties of PZT ceramics modified by Mn and Nb in the ratio 1/5

7.4.5 Temperature Coefficient of Resonance Q

As already described in Sect. 7.3 very stringent specifications had to be met by the piezoelectric ceramic for narrowband signal filters of the 130 kHz channel modems. Especially a very high resonance $Q > 1,800$ in combination with a very low temperature coefficient TCQ ($|\mathrm{TCQ}/Q| \leqslant 5\,\mathrm{ppm\,K^{-1}}$) was required to keep the insertion loss sufficiently independent of temperature.

It is well agreed that the electrical and mechanical losses in soft piezoelectric ceramics are mainly caused by vibrating domain walls and their interaction with lattice defects [12]. The question was whether the losses in stabilized ceramics are still dominated by domain wall losses. The losses in ferroelectric ceramics are known to depend on the amplitude of the measurement field strength. The presence of a well defined threshold field strength proportional to the internal field strength E_i has been noted, and it has been found that the small signal losses occurring below these threshold field strength are not hysteresis losses [13]. Since these small-signal losses have about the same frequency response in both the ferroelectric and the paraelectric state of the material [13], they have been attributed to relaxations that have no bearing on the ferroelectricity of the material. This frequency dependence of the observed small-signal losses in stabilized ceramics is quite different from the only slight frequency dependence found for hysteresis losses [12]. The physical origin of these small-signal relaxation losses is not really known. They are, however, obviously related to lattice defects, possibly oxygen vacancies, because a characteristic dependence on doping is observed.

Figure 7.11 shows the mechanical losses (measured at about 180 kHz) of Pb[ZrTiMnNb]O_3 ceramics with various Mn/Nb ratios compared with a soft PZT ceramic. Each Mn modified ceramic shows at least one relaxation peak, which decreases in frequency as the Mn content is increased. The soft ceramic (not doped with Mn) gives the impression that at about −50°C a

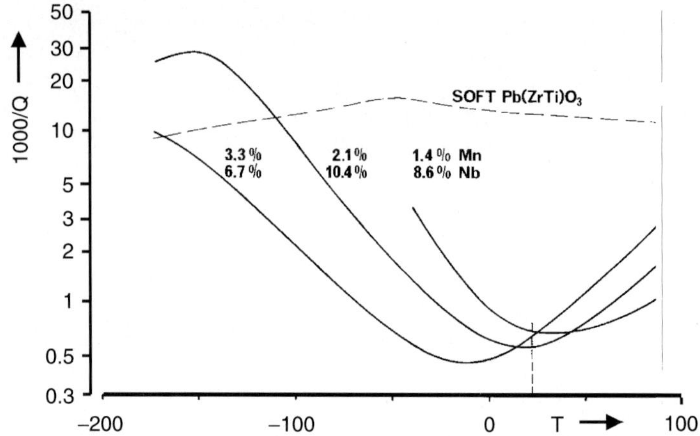

Fig. 7.11. Temperature dependence of mechanical losses of $Pb(ZrTiMnNb)O_3$ ceramics

relaxation appears superposed on the hysteresis losses. Further measurements have shown that the relaxation of small signal losses of stabilized ceramics do not depend on the Mn/Nb ratio, but solely on the Mn mole fraction. A similar relaxation effect has been found in Fe-doped ceramics [14]. The dependence of the loss relaxation on doping is of major significance for realizing ceramics with low TCQ as required for signal filters. As can be seen from Fig. 7.11, it is even possible to reduce TCQ to nearly zero within the temperature interval from 20 to 60°C. The stringent specifications for narrowband signal filters of the 130 kHz channel modems, i.e. resonance $Q > 1,800$ in combination with $|\mathrm{TCQ/Q}| \leqslant 5\,\mathrm{ppm\ K}^{-1}$ is also fulfilled by the material system with a Mn/Nb ratio of 1/5, the parameter of which have been shown in Fig. 7.10.

7.5 Production of Piezoelectric Ceramics for Frequency Filters

As already mentioned above, the electro-mechanical channel filter represents a pioneering step toward all modern high-tech applications of piezoelectric ceramics. For this innovation comprehended not only the basic idea, the design, the systems adaptation, its demanding specifications meant a challenge for technology as well. All the factors of reproducibility, reliability, competitiveness, and costs had to be met simultaneously. Thus it makes sense to describe briefly the decisive technology steps. They have remained the basis for many novel high-tech applications.

The piezoelectric ceramic development shown in the last section was performed in small batches (one mole) using a standard mixed oxide powder processing. Nevertheless, also advanced preparation and sintering techniques developed for the production of transparent ferroelectric lead zirconate titanate ceramics doped with lanthanum (PLZT) for optical applications have been tested for the preparation of filter ceramics. In general, advanced sintering techniques such as vacuum sintering and hot pressing allow the production of ceramics that are totally nonporous. However, nonporous piezoelectric ceramics are only required for applications at frequencies upwards of about 50 MHz, where significant losses occur due to the scattering of ultrasonic waves at pores. The extent to which filter ceramics can be improved by vacuum sintering has been investigated. It was found that the density could be increased to almost its theoretical value and additionally, the permittivity could be increased as well [15]. However, the coupling factor was not significantly improved over its optimal value obtained by firing in oxygen.

Therefore, due to economical reasons, it had been decided to keep the mixed oxide powder processing for the production of filter ceramics. It does not lead to the highest possible values of coupling factor and permittivity for ceramics close to the morphotropic phase boundary as a result of the only medium homogeneity obtainable in mass production. However, it is cost effective and leads to stable and reproducible results even under normal processing fluctuations.

The main fabrication steps of transducer platelets were the following:

- Powder preparation and granulation using spray drying
- Pressing of blocks ($100 \times 60 \times 20\,\text{mm}^3$) and sintering
- Cutting sintered blocks into large platelets with final transducer thickness
- Annealing of large platelets at high temperature, deposition of electrodes, poling, and spot checking
- Cutting to final size and final testing

This fabrication method allows the production of transducer platelets with a yield of more than 95%. However, an extensive process development work was necessary for several fabrication steps to achieve a high yield.

Especially important was the introduction of an annealing step at high temperature. After sawing the platelets were stacked and heated up to about 1,050°C for a short time. Thereby micro cracks due to sawing were healed, the platelet surfaces were perfectly cleaned and thus the adhesion of electrodes considerably improved. Furthermore, the high temperature annealing homogenizes the distribution of Pb and lattice defects within the platelets leading to extremely precise and reproduceable material parameters.

The high resonance Q of the transducer resonators is decisive for the function of mechanical frequency filters. It critically depends on the perfect connection, i.e., perfect soldering between ceramic platelets and resonator bolts. This can be achieved with an electrode system deposited by sputtering. First a thin chromium adhesion layer is deposited to obtain a good chemical bonding to the ceramic material. Then a thin Pt diffusion barrier layer is deposited followed by the deposition of the gold electrode. The barrier layer avoids any inter-diffusion (alloying) between the adhesion layer and the gold electrode during the soldering process. To distinguish the polarization direction of the transducer platelets in the final transducer resonator assembly (cf. Fig. 7.3), an additional very thin Pd layer is deposited on one side of the large ceramic platelets.

To achieve a sufficient domain orientation in highly stabilized ceramics by poling an electric field of rather high strength ($>3\,\mathrm{kV\ mm^{-1}}$) must be applied at elevated temperature ($\approx 150^\circ$C). Consequently, poling must be performed in a suitable isolation medium to avoid any sparking. Usually silicon oil is used to solve this problem. However, the perfect cleaning obtained by the high temperature annealing step would be completely ruined by the oil. Therefore, we developed a special poling procedure using SF_6 gas as isolation medium. In this way an optimum soldering is maintained.

After poling the large ceramic platelets can be tested reliably by performing small signal resonance measurements (Chap. 18). This preceding testing of each batch can be very useful because it allows to adjust the coupling factor precisely to a (little bit lower) target value using an additional annealing step at a typical temperature between 100 and 200°C.

Finally, the large ceramic platelets were cut by using precise wafer sawing to transducer platelets. The fabrication of transducer platelets for 130 kHz modems was completed by finally 100% testing of the transducer platelets.

In spite of a high sophisticated specification and tight tolerances, it was possible by the above-described fabrication method to achieve a yield of more than 95% over a fabrication period of more than one decade.

References

1. H. Albert, Feinwerktechnik **72**, 244 (1968)
2. A.E. Günther, H. Albsmeier, K. Traub, Proc. IEEE **67**, 102 (1979)
3. W. Wersing, Ferroelectrics **22**, 813 (1978)
4. H. Banno, T. Tsunooka, Jpn. J. Appl. Phys. **6**, 954 (1967)
5. R. Truell, C. Elbaum, B.B. Chick, *Ultrasonic methods in Solid State Physics* (Academic Press, New York, 1969) pp. 341–343
6. K. Carl, K.H. Härdtl, Ferroelectrics **17**, 473 (1978)
7. H. Schichl, NTZ **4**, 299 (1976)
8. M. Takahashi, N. Tsubouchi, M. Yonezawa, T. Ohno, T. Akashi, Nec Res. Develop. Jpn. **35**, 57 (1974)
9. W. Wersing, Ferroelectrics **37**, 611 (1981)
10. I.M. Lifschitz, L.N. Rosenzweig, J. Exp. Theoret. Phys. **16**, 967 (1946)
11. G. Zorn, W. Wersing, H. Göbel, Jpn. J. Appl. Phys. **24** suppl. 24–2, 721 (1985)
12. J.O. Genter, P. Gerthsen, N.A. Schmidt, R.E. Send, J. Appl. Phys. **49**, 4485 (1978)
13. H.-J. Hagemann, J. Phys. C. Solid State Phys. **11**, 3333 (1978)
14. K.H. Härdtl, Ferroelectrics **24**, 75 (1980)
15. K. Möhring, H. Schichl, Ber. Deutsch. Keram. Ges. **53**, 200 (1976)

8

Ultrasonic Imaging

W. Wersing and R. Lerch

8.1 Introduction

Ultrasonic medical imaging provides a low cost and non-invasive tool to look inside the human body. It is one of the most popular imaging tools, second only to conventional X-ray imaging in the number of clinical procedures performed. It has various advantages over conventional X-ray, computed tomography and magnetic resonance imaging, because it does not utilise harmful electromagnetic radiation, is non-invasive, easy to perform, cost effective, capable of providing images in real time, and provides good soft tissue differentiation [1].

Ultrasonic imagers for medical diagnosis are to be categorized as piezoelectric sonar sensors. These are sensors that emit ultrasound and detect ultrasound after it has been reflected from objects or passed through media. They are used for the identification of object structures, for the measurement of distances or for the measurement of medium or object properties. Ultrasonic transducer for sonar sensors differ considerably, depending on whether sound propagation happens in gas, liquid or bulk. For measurements of the amount of a liquid and its flow speed, special sonar techniques have been developed.

Ultrasonic imaging technology has come a long way from its beginnings in the 1970s. The image quality of medical ultrasound has improved vastly with linear and phased array transducer designs replacing the mechanical scanners used 2–3 decades ago. Further improvements have been achieved with the optimisation of the relevant properties of piezoelectric materials and array processing techniques. These advanced features are mandatory for modern imaging modalities, whether this is compounded imaging, second harmonic imaging, contrast agents, colour flow Doppler or 3D image reconstruction to enhance the image quality. Modern ultrasonic imaging has been and continues to be a field of intense research and development where many innovations from first ideas to final commercial products are achieved in rather short periods. New imaging modalities are continually developed and adapted to existing imaging systems. Many of those are recognized not only as useful but often as indispensable diagnostic aids. These improvements require the development of imaging transducers with specific properties and of high performance being capable of operating in a wide range of modes. These performance requirements have a strong impact on the complexity of transducer fabrication processes and piezoelectric materials' performance.

8.2 Principle of Ultrasonic Imaging Systems

Ultrasonic imaging is based on the phenomenon of the reflection and modified transmission of sound waves at the interface of two media with different characteristic acoustic impedances. The characteristic acoustic impedance Z of a medium such as biological tissue is

$$Z = \rho v, \tag{8.1}$$

where ρ is the density and v is the sound velocity of the medium. When the acoustic waves generated by the piezoelectric transducer encounter an impedance discontinuity, some of the energy is reflected and the remaining is transmitted ahead (see Fig. 8.1). When reaching the transducer, the reflected echo generates a voltage that is proportional to the acoustic impedance mismatch at the interface between the tissues. The transmitted part of the sound wave can serve to detect other impedance discontinuities that may be present at farther distances (Fig. 8.1). The relative small impedance mismatch between soft tissues allows for the ultrasonic waves to propagate across several interfaces, providing imaging capability up to large depth. The penetration depth depends on the tissue type and the ultrasound frequency f and reaches typically 10–15 cm at a frequency of 5 MHz. The lateral resolution is approximately three times the wavelength, that is at 5 MHz about 1 mm.

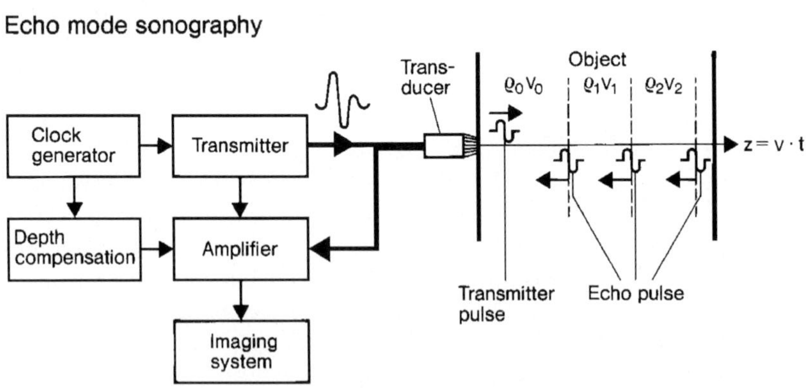

Fig. 8.1. Principle of ultrasonic imaging systems. Mean values for soft biological tissue are sound velocity $v = 1{,}540\,m\ \mathrm{s}^{-1}$, characteristic acoustic impedance $Z = \rho v = 1.5 \times 10^6\ \mathrm{Pa\ s\ m^{-1}}$, sound wavelength at $f = 3.5\,\mathrm{MHz}$ $\lambda = 0.44\,\mathrm{mm}$

The received voltage pulses deliver two kinds of information: first their amplitudes are proportional to the acoustic impedance mismatch at the position of reflection and second, the position itself can be calculated exactly from the delay time between pulse transmission and reception, provided the sound wave velocity is known. Because of the fact that biological tissue mainly consists of water, only longitudinal sound waves with a typical sound wave velocity between 1,500 and $1,600\,\mathrm{m\ s}^{-1}$ can propagate in biological tissue.

The simplest imaging mode is named A-scan (after amplitude). Here the amplitudes of the received echo pulses are simply displayed vs. the propagation path, that is penetration distance.

If the amplitudes of the echo pulses are transformed into brightness signals, the echo positions along the propagation path or penetration distance d can be depicted directly as brightness modulated dots alongside the picture line, usually vertical on the screen. To obtain a complete picture, many (at least 100) of these picture lines are lined up scanning along the body (span s) using mechanical or electronic scan techniques. By this means a sectional grey scale image (d vs. s) of the investigated tissue is obtained. This allows anatomy to be observed in a wide range of organs, for example in obstetrical or cardiological imaging. The brightness modulated sectional view is usually named B-scan (after brightness). Today, B-scan imaging techniques completely outweigh other imaging modalities.

There exists a third imaging mode, which is called M-scan (after motion). To obtain this mode the picture lines are not lined up from different transducer positions s but from the same position at different times t. Thus, the obtained grey scale image (d vs. t) shows the motion of the sounded tissue. On the contrary, this explains that for sectional grey scale imaging, it is essential to guarantee good real time capability to avoid artefacts caused by movements of the organs.

However, the movement of blood can also be used for estimating its flow velocity with high local resolution. This estimate is then superposed to the grey scale image using colour coding. The flow measurement itself is based on the physical effect, discovered by Doppler, that the sound frequency of a source is shifted due to its movement. In the case of blood flow measurement, the reflecting blood acts as sound source. The frequency f of the received signal backscattered from the blood particles is shifted in relation to the transmitter frequency f_0 by the Doppler frequency

$$\Delta f = f - f_0 = 2 f_0 \frac{v_{\mathrm{blood}}}{v} \cos\varphi, \tag{8.2}$$

with v the sound velocity, v_{blood} the velocity of the reflecting blood particles and φ the angle between sound direction and direction of blood flow. The detection of the weak backscattered signals from the blood particles puts forward highest demands upon the dynamics of signal processing. These performance requirements have direct consequences on those of the piezoelectric transducer materials.

There are many possibilities to scan and to focus ultrasonic images using specific ultrasonic transducers often also named ultrasonic antennas, as discussed in the following section.

8.3 Ultrasonic Antennas and Scanning Modalities

The simplest ultrasonic antenna is the classical single-element transducer, which is based on a piezoelectric disc metallised on both surfaces and poled along its thickness direction. An electric pulse applied to the metallised faces excites an acoustical resonance in the piezoelectric half-wavelength disc-resonator and generates pressure waves on both, the front- and the backside of the transducer disc. To adapt the acoustic impedance of the ceramic transducer material ($Z_c \approx 30 \times 10^6\,\mathrm{kg\,m^{-2}\,s^{-1}}$) to that of biological tissue (close to that of water, i.e. $Z_t \approx 1.5 \times 10^6\,\mathrm{kg\,m^{-2}\,s^{-1}}$), the front side of the transducer element is covered with a plastic matching layer designed to optimise the transfer of energy from the active layer to the tissue and vice versa [2]. As a result, the resonant ringing of the transducer is reduced and the axial resolution enhanced. The thickness of this matching layer is typically around a quarter-wavelength and its acoustical impedance Z_m is in between that of the ceramic transducer (Z_c) and that of biological tissue (Z_t), the optimum would be $Z_m = \sqrt{Z_c Z_t}$. Modern transducer designs now utilise multiple matching layers with graded impedance to further improve the transducer performance. To focus the ultrasound beam, a suitable formed intermediate layer is often used as acoustic lens (Fig. 8.2). Often both functions matching and focusing

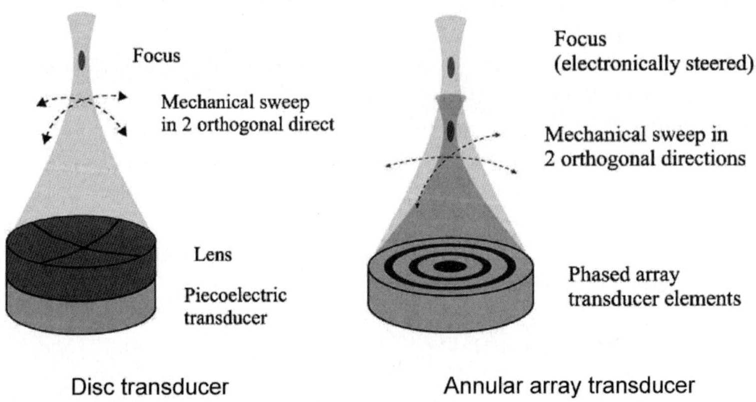

Fig. 8.2. Single element transducers, the classical disc transducer with fixed focus defined by the lens form (*left*) and annular array transducer consisting of acoustically isolated piezoelectric rings (*right*). By individually controlling the signal phases of the different rings, it is possible to vary the transducer's focus electronically. Matching layers and backing are not shown in the figure

are integrated into one layer or layer package. On the back side of the transducer, a thick backing layer is attached usually. It completely absorbs the ultrasonic power emitted in this direction so that no energy can be radiated back to the active layer, which would produce parasitic echo signals.

The image quality at a given ultrasound frequency[1] strongly depends on the axial and lateral resolution. Usually, because of a good acoustic matching, the axial resolution is better than the lateral one. Thus, improving of the image quality requires the improving of the lateral resolution. The lateral resolution of a single-element transducer is correlated with the beam profile and depends on the diameter of the transducer and the lens curvature (curvature radius R). Furthermore, it depends upon the degree of focusing, that is the ratio between the active transducer diameter $2a$ (aperture) and the focal distance $z_0 \approx R$. With increasing degree of focusing, the resolution increases within the focus range (Fig. 8.3); however, at the same time, the depth of focus decreases. Therefore, an optimal resolution requires for each focal distance z a different lens curvature. Furthermore, to keep the lateral resolution constant within the whole image depth, it would be necessary to increase the transducer aperture with increasing distance z so that $z/2a \approx$ constant.

This clearly shows that classical disc transducers cannot be focussed optimally, a compromise must be found between the degree of focusing and the depth of focus. To overcome this problem annular array transducer have been developed (Fig. 8.2). Such a transducer functions quite similarly to a classical single element disc transducer, merely the transducer disc is divided into concentric piezoelectric ring transducers (Figs. 8.2 and 8.4), each of which can be individually driven by a signal pulse. Focusing can be achieved by applying delays to the driving pulses in such a way that the acoustic pressure generated by each ring transducer arrives at the focal point at the same time, that is the driving signals applied to the inner rings must be more delayed than those applied to the outer rings. For a transmitted pulse the delays are fixed to a certain focal distance according to the desired image depth. However, for received echo signals, the delays can be adapted continuously to focus on the depths at which the echo signals were produced. This method is referred to

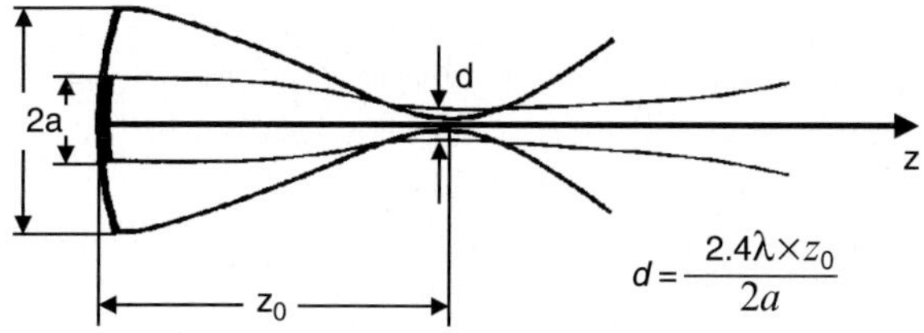

Fig. 8.3. Lateral resolution and depth of focus for a classical single element transducer. The beam profiles for two transducers with $2a = 15$ mm and $2a = 40$ mm are compared, the focal distance is $z_0 = 60$ mm

[1] The ultrasound frequency is usually limited by the necessary penetration depth.

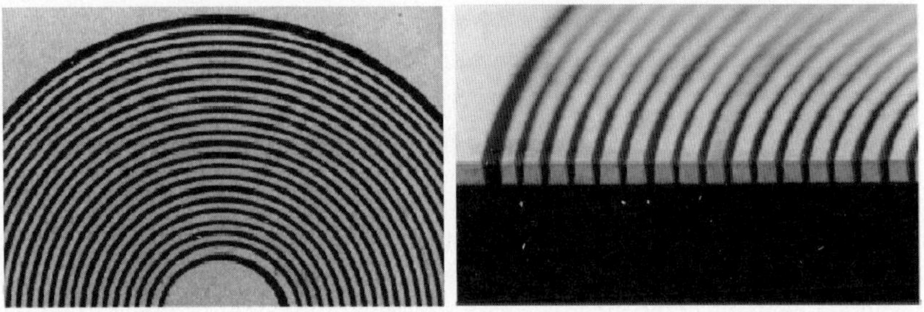

Fig. 8.4. Ultrasonic annular array transducer, top view (*left*) and side view of an array cut through (*right*)

as dynamic focusing. Compared to simple single element transducers, much larger depth ranges can be imaged with good lateral resolution.

Another dynamic approach, the so called time gain compensation, is used to improve imaging performance over large depth ranges. This method compensates for the strong attenuation that an ultrasound echo impulse experiences along its path through the biological tissue by dynamically increasing the amplification factor of the receiving amplifier. Because of the fact that attenuation increases exponentially with increasing distance or travelling time, it is necessary that the amplification gain increases exponentially too, typically within a short time period of 100 μs by more than 100 dB.

As already discussed in the last section, mechanical or electronic scan techniques are required to obtain a complete sectional grey scale image. In the beginning of the ultrasonic diagnostics, solely mechanical scanning techniques have been used. In the simplest case, a single element transducer (disc or annual array transducer) is guided over the body region under consideration. Via a mechanical scan arm the position of the transducer is measured by electronic means. Furthermore, motor driven real-time-scanners have been developed with parallel image and sector image scanning modes. Common to all these systems was the use of only one single element (disc or annual array) transducer. In mechanical linear scanners two or more single element transducers have also been used; however, only one being active at a time, always one after the other. In the seventies, mechanical ultrasonic scanners have been more and more replaced by electronic array scanners. Depending on the acoustic window, two different transducer configurations are used usually, the linear and the phased array (Fig. 8.5).

When the imaging window is large, linear arrays can be used. Linear array scanners are operated by applying voltage pulses to a small group of piezoelectric elements in succession, as shown in Fig. 8.5. The beam can start from one end of the transducer and slowly move towards the other end by a sequential excitation of the piezoelectric elements in the row, that is the radiating aperture is shifted pitch by pitch (centre to centre distance of adjacent elements). By doing so, the linear array can mimic the actual movement of a

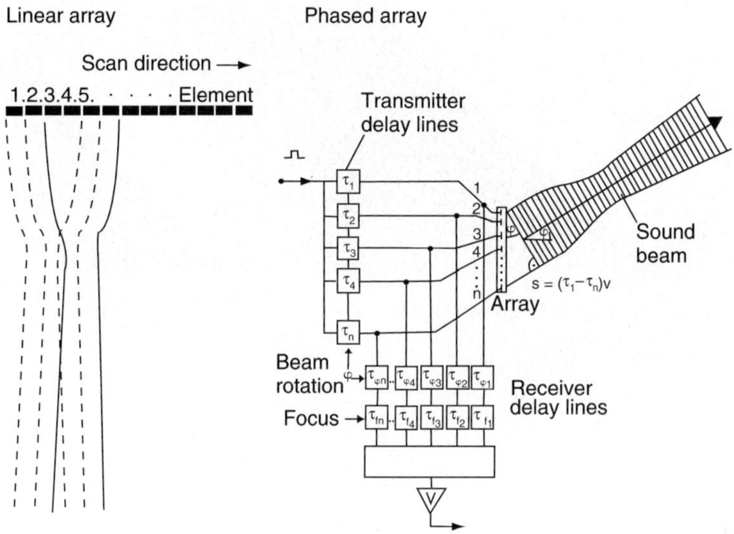

Fig. 8.5. Principle function of ultrasonic electronic array scanners, linear array (*left*) and phased array (*right*)

single element transducer mechanically moved along the scanning direction. If the repetition rate at which the scanning is done is at least 30 frames per second, a real time image with a rectangular cross section can be obtained. Focusing methods such as dynamic focusing are similarly applied as described for annular arrays to achieve large depth ranges with good resolution. Focusing vertical to the scanning direction is usually achieved using a mechanical lens with fixed focus.

Linear scanners are usually large, having a width around 1 cm and a length up to 10 cm, with 128–512 piezoelectric elements. Figure 8.6 shows the design of a linear 2 MHz array of the first generation[2] with a view of the ceramic thickness transducer elements. On the back side of the transducer, a thick backing layer has been attached, which completely absorbs the ultrasonic power emitted in this direction. To adapt the acoustic impedance of the ceramic transducer material ($Z_c \approx 30 \times 10^6\,\mathrm{kg\,m^{-2}\,s^{-1}}$) to that of tissue ($Z_t \approx 1.5 \times 10^6\,\mathrm{kg\,m^{-2}\,s^{-1}}$), the transducer elements are covered with a plastic matching layer (not shown in the figure). To further enhance beam sensitivity, modern transducer designs utilise multiple matching layers with graded impedance and integrated lens. Since the width of the single array elements is larger than their thickness (Fig. 8.6, top), the elements generate transverse vibrations with low frequencies that are also emitted and received. Because tissue hardly absorbs low frequency sound, artefacts occur resulting

[2] It shows details much better than modern linear arrays.

Fig. 8.6. View of the piezoelectric transducer elements (without matching layer) of a linear ultrasonic array of the first generation (*top*) and array with transducer elements divided into sub-elements (*bottom*)

in a pronouncedly reduced image quality [3]. These disturbing vibrations can be suppressed by fine sawing of the single elements into sub-elements with a width to thickness ratio <1 (Fig. 8.6, bottom).

When the acoustic window is small, large linear arrays cannot be used. For example, when images have to be taken of the heart, then the acoustic beam is required to fit in between the ribs. Under these circumstances, another scanning method based on signal delay has been developed using so-called phased arrays. In a phased array, for each line of the image all elements together are utilised at each sound impulse. The ultrasonic beam generated can be both focused at different points along the axial direction and also steered (by typically $\pm 45°$) by controlling the phase delay sequence of transmitted and received electrical signals (Fig. 8.5). Phased arrays must be much smaller (typically $\lambda/2$) than linear arrays due to the effect that with increasing aperture the achievable steering angle decreases. Therefore, fine sawing techniques must be employed for fabrication. Phased array transducers are very similar in construction to linear arrays (Figs. 8.7 and 8.8), they are about 1 cm wide and 1–3 cm long, and have fewer elements (64–256). The width of a single transducer element is only about half of its thickness and, therefore, transverse vibrations are completely prevented. However, because of the small area of the single elements, problems with crosstalk between adjacent elements occur. Furthermore, special piezoelectric materials of high performance are required to achieve a good sensitivity, a high signal-to-noise ratio, and an electrical impedance matching with the signal transmission lines (see Sect. 8.5).

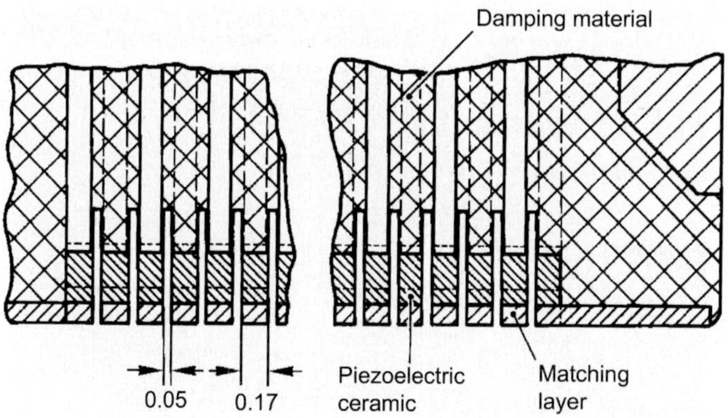

Fig. 8.7. Cross section of an ultrasonic phased array

Fig. 8.8. Ultrasonic phased array antenna

Finally, Fig. 8.9 shows a commercial ultrasonic imaging system with ultrasonic probes for different medical investigations.

The ultrasonic antennas discussed so far have one-dimensional (1D) array scanners to generate two-dimensional (2D) sectional grey scale images. By manually moving the transducer head vertical to the scanning direction, it is possible to obtain three-dimensional (3D) ultrasonic images sequentially.

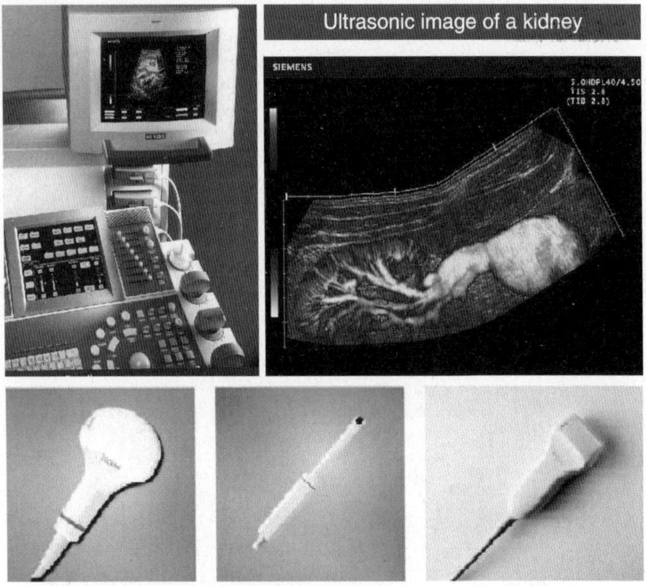

3.5 MHz Convex array 6.5 MHz Prostata probe 13 MHz 1.5 D array antenna

Fig. 8.9. Modern ultrasonic imaging system together with imaging probes for different applications in medical diagnostics

However, 3D ultrasonic real-time imaging is not possible this way. To make this possible a real 2D array antenna is requested, that is a transducer element matrix that can be addressed individually using line and row switching electronics.

Although, 2D-phased antennas have been in development for a long time, such arrays have not yet reached any commercial significance by far. However, new materials and array technologies promise a commercial breakthrough to occur in the future (see the following sections).

In principle, a 2D-phased antenna consists of several thousand transducer elements, in each direction typically 64–256 elements, that is in all 4,096–65,536 elements. The technical requirements, to some extend unforeseen, are identical to those of linear phased arrays, that is large scanning angle, low crosstalk, high sensitivity and high signal-to-noise ratio. Because of the possibility to address each transducer element individually, an ultrasound beam steering in any direction can be realized. The beam can be focused similarly as in the case of linear phased arrays. The cutting of such a 2D array antenna is shown in Fig. 8.10. Its application for 3D ultrasonic imaging systems requires real time imaging, that is at least 25 images per second.

Fig. 8.10. Cutting of a 2D ultrasonic phased array antenna. The transducer elements have a base area of $150 \times 150\,\mu m^2$ and a height of about 1 mm (matching layers included)

8.4 Material and Array Fabrication Aspects

In this section, general aspects concerning materials and array fabrication are explained, whereas in Sect. 8.5 we discuss all features related to both the state of the art piezoelectric ceramics used in ultrasonic imagers (or imaging equipment) and to piezoelectric materials specially designed for this application.

The fabrication of standard 1D linear and phased arrays usually starts with thin ceramic strips of length l and wideness w according to the array dimensions, typically 1 cm wide and 1–10 cm long. The thickness t of those strips is $\lambda/2$[3]. This corresponds to a thickness between 0.15 and 1 mm for arrays with a frequency between 15 and 2.5 MHz, respectively. These stripes cut from sintered ceramic blocks are processed very similar to the filter ceramic platelets discussed in Chap. 7. This concerns similarly the deposition of electrode systems, which usually consist of a thin adhesion layer, a thin diffusion barrier layer and a gold electrode. These ceramic stripes are connected with the backing block on the one side (rear side) and with the matching layers (up to three $\lambda/4$ layers) and a protection layer including the lens on the other side (front side) (Fig. 8.7). The sides of the backing usually serve to contact the transducer elements via fixed or deposited transmission lines. Contacting of the transducer elements is achieved using soldering or glueing.

The backing on the rear face of the transducer array is usually the thickest layer. It serves as the mechanical support of the antenna beside its absorber

[3] λ is the wave length of sound in the piezoelectric ceramic.

function (acoustic swamp). For this layer, a composite made of a high loss polymer (high visco-elastic attenuation) and embedded heavy metal particles such as tungsten particles [4] is used. Via the concentration and grain size of the embedded heavy particles, this composite with (0–3) connectivity[4] can be optimised concerning acoustic impedance and sound attenuation scattering, respectively. Very high attenuations together with a well defined acoustic impedance can be achieved using a mixture of different particle types. Additionally, the thickness of the backing can be reduced considerably to realize miniaturized antennas.

Matching layers are also fabricated from particle loaded polymer composites. However, their attenuation must be kept as low as possible. Therefore, a single type of well defined particles is added in an amount to achieve the desired acoustic impedance.

8.5 Piezoelectric Materials for Ultrasonic Imager

As in the past, also in future, the optimisation of piezoelectric material properties and processing techniques will continue to play a decisive role in the field of ultrasonic imaging.

8.5.1 General and Specific Requirements

Piezoelectric materials for ultrasound imaging transducers usually have to operate as both emitting and receiving transducers. Therefore, on the one hand, some conditions have to be fulfilled in common for both the receive as well as the transmit mode and, on the other hand, other conditions are contradictory and suitable compromises have to be found.

In general, the piezoelectric transducer elements should have a good acoustic and electrical impedance matching and a high effective electromechanical coupling coefficient k_{eff}, which is the capability of the transducer element to convert electrical energy into acoustical energy and vice versa. This coupling factor depends not only on the piezoelectric material, but also on the specific design of the transducer element. In ultrasonic imaging devices the vibration displacements are in the thickness direction (poling direction). Therefore, for single element transducers of large area (lateral dimensions $\gg$ thickness) the relevant coupling factor is $k_{33\mathrm{t}} \equiv k_{\mathrm{t}}$[5] $< k_{33}$ and for small area elements of arrays (e.g. bars or rods with lateral dimensions $\ll$ thickness as in Fig. 8.10) it is k_{33}. For transducer elements of phased arrays (Figs. 8.7 and 8.8) with one lateral dimension $<$ thickness and the other lateral dimension $\gg$ thickness, the effective coupling factor is intermediate k_{t} and k_{33}.

---

[4] Regarding the connectivity definitions, see Chap. 16.

[5] In this case the transducer element is laterally clamped at the thickness resonance frequency. For definitions of coupling factors, see Chap. 18.

Specific requirements for the receiving mode can be derived by considering an ultrasonic transducer element as pressure (sound pressure) sensor (Fig. 8.18), which is basically characterized – as every sensor – by its sensitivity (voltage sensitivity) and its signal-to-noise ratio. The figures of merit for these characteristics are derived in Appendix A ((8.13) and (8.14)) to be

$$F_{\mathrm{VS}} = \frac{d_{33}}{C_{\mathrm{E}} + C_{\mathrm{line}} + C_{\mathrm{input}}}, \tag{8.3}$$

with $C_{\mathrm{E}} \sim \varepsilon_{33}{}^{\mathrm{T}}$

$$\text{and } F_{\mathrm{SNR}} = \frac{d_{33}}{\sqrt{\varepsilon_{33}^{T} \tan\delta}}. \tag{8.4}$$

From (8.3) and (8.4) it can be seen that in the receiving mode the transducer material should have a high piezoelectric coefficient d_{33}, a low permittivity $\varepsilon_{33}{}^{\mathrm{T}}$ and a low loss factor tan δ. The general requirement of a high coupling factor k_{33} or k_{t} is usually in line with a high piezoelectric coefficient d_{33}. However, the requirement of a high d_{33} and a low permittivity $\varepsilon_{33}{}^{\mathrm{T}}$ is contradictory for MPB[6]-piezoelectrics and thus a suitable compromise must be found. Using the relation between k_{33} and d_{33} shown in (18.21) of Chap. 18 $d_{33} = k_{33}\sqrt{(\varepsilon_{33}{}^{\mathrm{T}} s_{33}{}^{\mathrm{E}})}$, it is recognized that $d_{33} \sim \sqrt{C_{\mathrm{E}}}$. Thus, both the voltage sensitivity and F_{SV} increase with increasing C_{E} (respectively, $\varepsilon_{33}{}^{\mathrm{T}}$), they assume a maximum value at $C_{\mathrm{E}} = C_{\mathrm{line}} + C_{\mathrm{input}}$ (electrical matching) and then decrease again. Assuming a transmission line of length 1 m, the line capacitance[7] C_{line} of a line with $Z_{\mathrm{line}} = 80\,\Omega$ is typically 88 pF. To achieve electrical matching with a standard phased array transducer element of wideness $w = 0.17$ mm (Fig. 8.7), length $l = 10$ mm and thickness $t = 0.55$ mm (3.5 MHz), a ceramic permittivity $\varepsilon_{33}{}^{\mathrm{T}}/\varepsilon_0 > 6,500$ would be required. Usually a soft PZT ceramic is chosen with k_{33} as high as possible and with $2,000 < \varepsilon_{33}{}^{\mathrm{T}} < 3,000$. But, of course, moderate dielectric and mechanical losses ($\approx$0.02) must be accepted with these kind of materials. Some mechanical losses can even be useful to damp internal mechanical resonances of the active layer. However, it should be noted that within complex PZT-based ceramic systems doped with Mn, it is possible to obtain materials with high d_{33}, medium $\varepsilon_{33}{}^{\mathrm{T}}$ and low losses (Chaps. 4, 7, and 9).

For high frequency devices and for arrays, that is in all cases where fine sawing techniques are required for structuring, fine-grained ceramics without any large pores or other defects (such as second phases like ZrO_2) are necessary. Furthermore, sawing parameters must be optimally adapted to the piezoelectric material, for example when benefiting from ferroelastic properties to avoid any flaw.

---

[6] PZT ceramics with a Zr/Ti ratio in the vicinity of the morphotropic phase boundary (MPB), see Chap. 3.

[7] The typical capacitance lining of a low capacitance transmission line with $Z_{\mathrm{line}} = 150\,\Omega$ is 27 pF m^{-1}.

8.5.2 Piezoelectric Ceramics, Polymers and Single Crystals

Lead zirconate titanate (PZT)-based ceramics have been used traditionally for ultrasonic antennas because of high coupling coefficients, wide range of dielectric constants and moderate dielectric losses. However, they have several disadvantages, including a high acoustic impedance. Therefore, great efforts have been made to improve these materials according to the requirements discussed in the last section. This has been achieved by the modification of PZT with complex oxides such as $(Ni_{1/3}Nb_{2/3})O_3$, by using raw materials of high quality and by optimising ceramic processing.

Non-ceramic materials such as piezoelectric polymers polyvinylidene fluoride (PVDF) and copolymers P(VDF-TrFE) have been investigated for ultrasonic antennas as well because they have an acoustic impedance close to biological tissue. However, they are very limited for use as transducers because of their low dielectric constant (<10), high dielectric losses, and low coupling coefficient (0.2–0.3). They can only be used in high frequency applications because their properties decrease with increasing thickness, due to poling difficulties.

In recent years, the development of piezoelectric single crystals such as lead zinc niobate–lead titanate (PZN-PT) and lead magnesium niobate–lead titanate (PMN-PT) provide features for major improvements in transducer performance. They have very high coupling coefficients (k_{33} up to 0.9) and also provide a range of dielectric constants (1,000–4,000) for optimum electrical impedance matching [5]. These single crystals compositions promise transducers with an improved sensitivity and bandwidth (see Chap. 9). The broad bandwidth of these transducers can enhance the image quality of present technologies, for example in low frequency Doppler flow measurements combined with high resolution imaging or in compounded images (deep penetration and high resolution) or in harmonic imaging with contrast agents and imaging at higher frequencies (harmonic imaging). Their price is still high but decreasing [6]. However, it is necessary that the increase in performance is proved to be reproducible under industrial production conditions [7].

8.5.3 Piezoelectric Composites

At the end of the 1970s, researchers at Pennsylvania State University showed that by combining high effective piezoelectric ceramics and soft polymers, it is possible to design a variety of composites with property combinations that cannot be achieved with any other piezoelectric material [8]. Therefore, in the 1980s, many efforts were started to develop composites with promising properties for ultrasound transducers. Tailoring of properties can be achieved by changing the connectivity[8] of the phases and the volume fraction and spatial

[8] New processing techniques for composites are shown in Chap. 16, regarding the connectivity definitions, see also Chap. 16.

distribution of the active ceramic phase [9, 10]. Different approaches based on traditional ceramic processing techniques such as tape casting, injection moulding, dicing and such have been proven in making composites with different but simple connectivity. However, these techniques do not permit the fabrication of structures with complex internal hierarchy and symmetry due to their rectilinear mode of operation. To overcome this limitation, a new manufacturing/prototyping technique the so-called Solid Freeform Fabrication (SFF) is emerging. It provides very high design flexibility due to an integrated way of manufacturing 3D components directly from computer-aided design (CAD) files (see Chap. 16).

It is interesting to notice that linear and phased arrays constructed from alternating layers of piezoelectric ceramic and polymer can be considered as ceramic–polymer composite with 2–2 connectivity, manufactured with the traditional dicing and filling technique. Highly porous ceramics can also be considered as composites, ceramic–air-composites with 3–0 and 3–3 connectivity. They can be fabricated quite inexpensively using different processing methods and show rather interesting properties [11]. Recently, a further more traditional technique has been developed for ceramic–polymer-composites with 1–3 connectivity, well suited for ultrasonic imaging. It is based on a fine structured piezoelectric ceramic (Fig. 8.11) obtained with a lost wax method in combination with a precision slip-casting process [12]. A high thickness coupling factor k_t, close to the ceramic's k_{33}, can be obtained even for relatively low ceramic contents. So that it is possible to reduce the acoustic impedance $Z_{composite}$ to about $Z_c/4$.

Fig. 8.11. Fine structured piezoelectric ceramics for 1–3 composites, fabricated by a lost wax method in combination with a precision slip-casting process

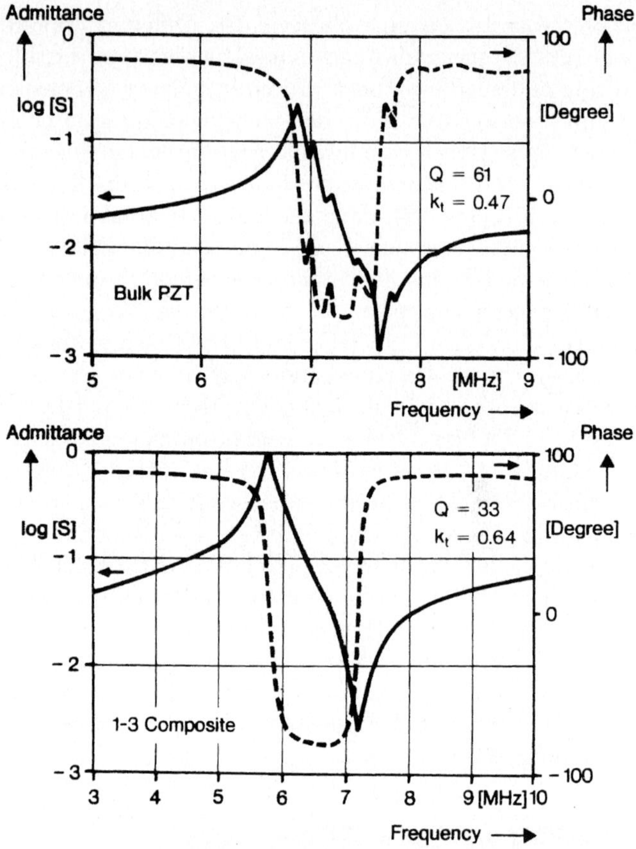

Fig. 8.12. Properties of thickness transducers made from bulk ceramics (*top*) and 1–3 composites fabricated by a lost wax method (*bottom*)

The advantage of such 1–3 composites can be seen in Fig. 8.12, where a thickness transducer (resonator) made of a bulk ceramic is compared to one made of such a 1–3 composite. From the admittance (amplitude and phase) vs. frequency measurements, shown in Fig. 8.12, the effective coupling factors and resonance Q values have been obtained. The advantage of the composite transducer is clearly to be seen. Its k_{t} is about 95% of the ceramic's k_{33}, $Z_{\mathrm{composite}} \approx 0.4 Z_{\mathrm{c}}$, and now spurious resonance artefacts due to lateral overtones are observed.

8.6 New Technologies for Ultrasonic Imager

8.6.1 Ultrasonic Antenna Arrays Based on Multilayer Ceramics

A general problem of piezoelectric actuation arises from the relative high electrical field strength (1–2 kV mm^{-1}) that is required to realize the largest

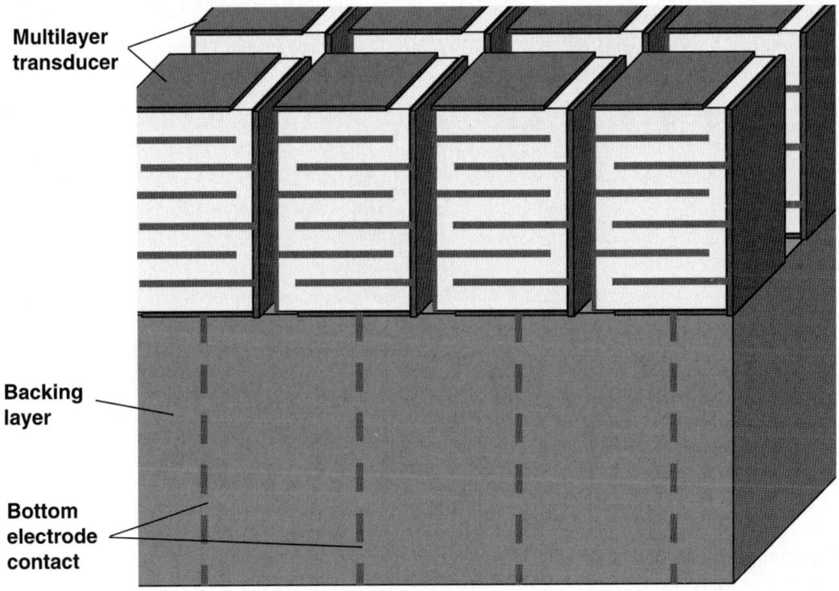

Fig. 8.13. Ultrasonic multilayer array antenna

possible deformations. Strictly speaking, piezoelectric ceramics are poorly matched to conventional amplifiers based on standard CMOS technology as used in present-day integrated electronics. In the case of piezoelectric actuators, this problem was solved by developing piezoelectric multilayer ceramics (see Chap. 13). A similar problem arises for ultrasound transducers in the transmitting mode. Therefore, it is not surprising that efforts have been undertaken to apply the multilayer ceramic technology to ultrasonic antennas as well. As can be seen from Fig. 8.13, the capacitance of a multilayer transducer consisting of m layers is by the factor m^2 higher than that of a single layer transducer of the same thickness (or resonance frequency). The amplitude of the driving impulse in the transmitting mode can be reduced by the factor $1/m$ and the electrical impedance matching of a transducer element to the transmission line impedance (see discussion in Sect. 8.5.1) will be even possible now for 2D arrays as shown in Fig. 8.10 [13]. This feature in combination with the use of transmitting pulses of low amplitude is especially important for miniaturized ultrasonic antennas that can be inserted into the human body, for example the antenna of a Trans-Esophagial-Endo (TEE). Figure 8.14 shows a TEE multilayer antenna that allows to image the heart through the oesophagus.

A further application for multilayer antennas is the realization of multi-spectrum transducers. They enable both imaging modalities, that is transmitting and receiving in the fundamental resonance mode as well as in the second harmonic mode, as shown in Fig. 8.15. This new ultrasonic imaging technique is called Multi-Spectrum Technology (MST).

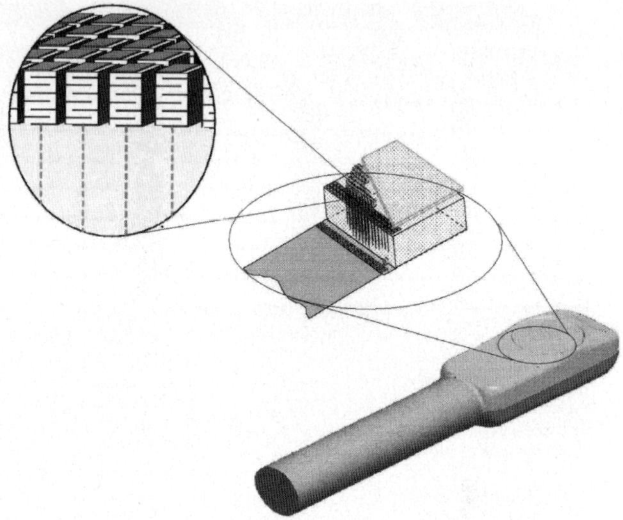

Fig. 8.14. Trans-esophagial-endo (TEE) imaging transducer based on an ultrasonic multilayer array antenna

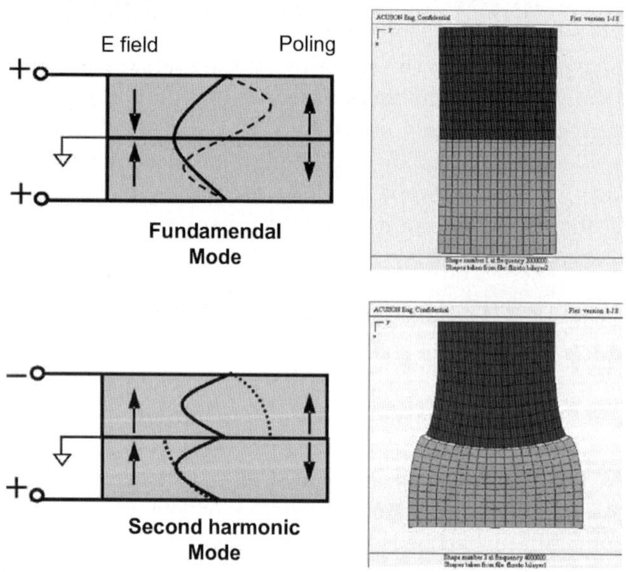

Fig. 8.15. Multi-spectrum technology based on ultrasonic multilayer array antennas

Finally, Fig. 8.16 gives an impression what can be achieved in 3D imaging with these array antennas. It shows the image of a foetus in the womb after the tenth week of pregnancy taken with a mechanically moved 1D linear multilayer antenna.

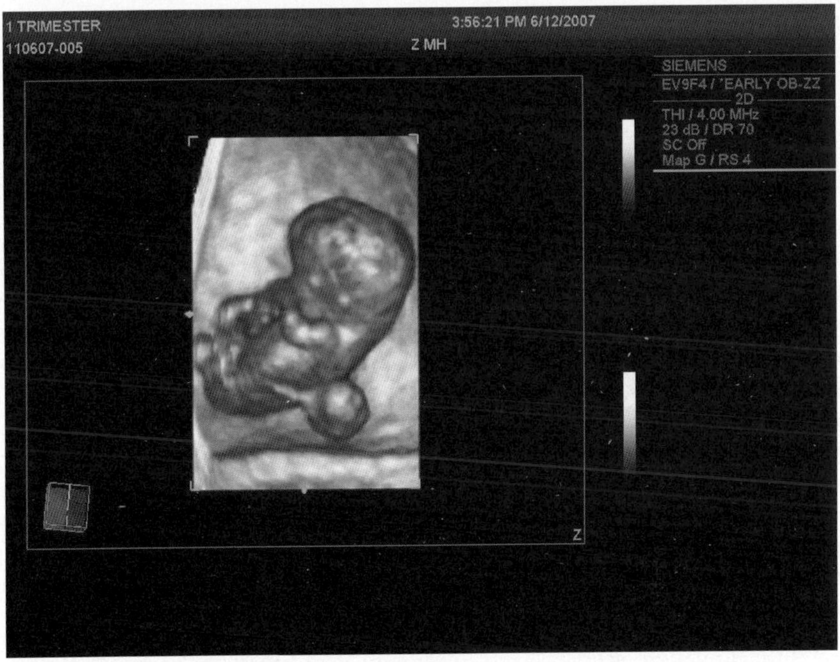

Fig. 8.16. Ultrasonic image of a foetus (tenth week pregnancy) taken with a mechanically moved 1D linear multilayer antenna

8.6.2 Piezoelectric Micro-machined Ultrasonic Transducers (P-MUT)

The fabrication of 2D transducer matrices for 3D imaging as shown in Fig. 8.10 using standard piezoelectric materials and transducer processing techniques becomes with increasing complexity and frequency more and more a technical challenge, because the ceramic material cannot be diced down to very small dimensions. Therefore, the micro-electro-mechanical systems (MEMS) technology is emerging for high frequencies, 3D real time imaging and low cost applications. Piezoelectric micro-machined ultrasonic transducers (p-MUT) and capacitive micro-machined ultrasonic transducers (c-MUT) are in competition for these applications.

In c-MUTs, the actuation is based on the coulomb force F_e between an electroded membrane and the bulk silicon wafer separated by an air gap r. This force is given by the derivative in respect to electrode distance of the electrostatic energy

$$F_e = \frac{\partial}{\partial r}\left(\frac{1}{2}CV^2\right), \tag{8.5}$$

with $C\,(r) = \varepsilon_0 A/r$ (A is effective capacitor area). When V is increased, r varies between the initial gap r_0 and 0. Assuming an elastic restoring force

$k \cdot (r_0 - r)$, the following non-linear relation between displacement and applied voltage can be derived

$$V = \sqrt{\frac{2k}{\varepsilon_0 A}} r \sqrt{r_0 - r}. \tag{8.6}$$

When V increases r decreases until $V(r)$ reaches its maximum at $r = 2r_0/3$. Then a collapse of the membrane occurs and the gap closes abruptly. A linear small signal response of the membrane can be achieved by applying a bias voltage to the membrane. If the bias voltage is chosen to be just below the collapse voltage (pull down voltage), a very high response can be achieved, equivalent to nearly 1,000 times the d_{33} coefficient of a typical piezoelectric ceramic. Of course, the dynamic is then very low, that is there is a trade-off between sensitivity and dynamics. A decisive advantage of c-MUTs is that they can be manufactured by surface micro-machining using standard CMOS technologies (Fig. 8.17).

In p-MUTs, the actuation is based on a piezoelectric bending transducer, which is obtained by depositing a thin (typically 1 μm) piezoelectric layer together with the necessary electrodes on top of a membrane (e.g., poly-Si) thermally matched to silicon. The p-MUT technology offers the following advantages:

- The thin piezoelectric layers, although very small, have a low electrical impedance and, therefore, can be easily matched to the 50 Ω coaxial transmission lines usually used.

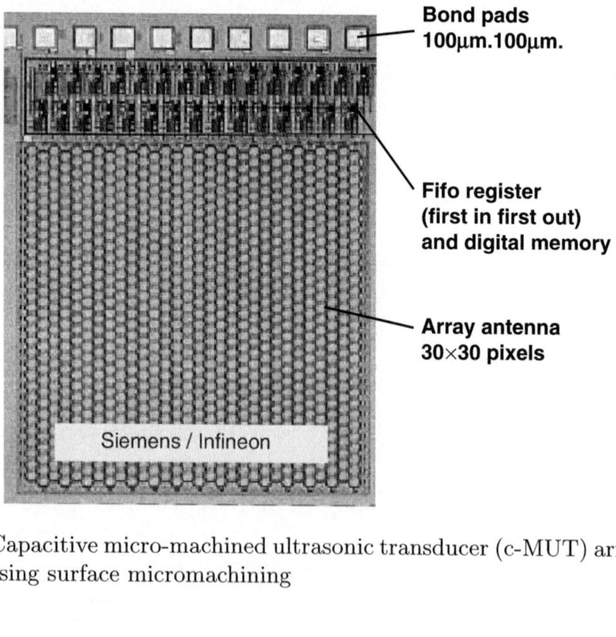

Fig. 8.17. Capacitive micro-machined ultrasonic transducer (c-MUT) array antenna fabricated using surface micromachining

- The general property of a piezoelectric bending transducer, relatively large free deflection and low blocking force (compared to thickness transducers), that is its low acoustic impedance can be used for ultrasonic array antennas because (due to its small size) its resonance frequency falls into the interesting region between 5 and 50 MHz.
- Although the effective coupling factor of the bimorph structure (${k_{\mathrm{eff}}}^2 < 0.15$) of a p-MUT is much smaller than that of a c-MUT near its pull down voltage (theoretical ${k_{\mathrm{eff}}}^2 \leqslant 1$), the acoustic energy $W_{\mathrm{a}} = {k_{\mathrm{eff}}}^2 W_{\mathrm{e}}$ obtained from a p-MUT can be much larger than that of a c-MUT. This is the consequence of the high electrostatic energy W_{e} that can be injected into the p-MUT due to the extremely high field strength applicable to the piezoelectric thin film.

Recently, the p-MUT technology has been adapted for droplet ejectors used for applications such as high quality ink jet printers and biomedical drug or micro-sample delivery systems [14].

Appendix Derivation of Figure of Merit for Sensor Material Properties in the Receiving Mode

The dependence of the basic sensor property's sensitivity and signal-to-noise ratio on material properties can be discussed using the equivalent circuit diagram for the receiving operation mode shown in Fig. 8.18. The received sound pressure signal p is transformed by the matching layers to the stress signal $T_3 = np$ acting on the transducer element in thickness direction (3-direction). The transformation ratio n follows from the impedance transformation of the matching layers, for optimal matching ($Z_{\mathrm{c}} = n^2 Z_{\mathrm{t}}$) follows $n = \sqrt{20} \approx 4.5$. In the transducer element the stress signal generates the charge signal $Q_{\mathrm{s}} = d_{33} T_3 A$, where d_{33} is the piezoelectric coefficient in thickness direction and A is the active transducer area. Therefore, the voltage signal V_{input} at the input of the imager processing electronic can be calculated to be

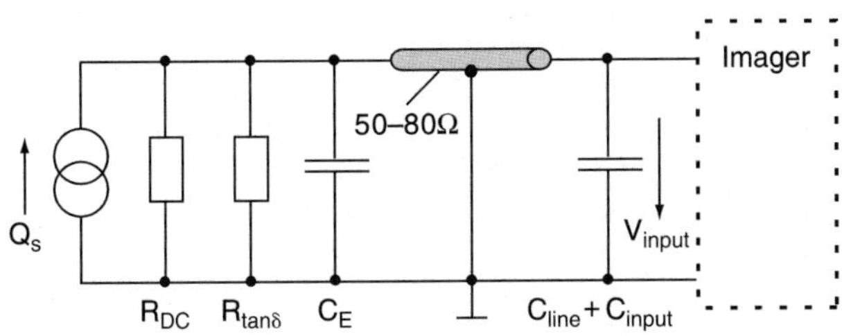

Fig. 8.18. Equivalent circuit of a piezoelectric transducer element of an imaging array operated in the receiving mode, that is as a pressure sensor element

$$V_{\text{input}} = \frac{d_{33}An}{C_{\text{E}} + C_{\text{line}} + C_{\text{input}}} p, \tag{8.7}$$ [9]

where C_{E} is the capacitance of the transducer element, C_{line} that of the transmission line and C_{input} the input capacitance of the processing electronic.

From (8.7) the voltage sensitivity of the antenna element is obtained to be

$$S_{\text{VS}} = \frac{V_{\text{input}}}{p} = \frac{d_{33}An}{C_{\text{E}} + C_{\text{line}} + C_{\text{input}}}. \tag{8.8}$$

The characterization of sensor response is insufficient without analyzing also its noise generation, which limits the smallest input signal that can be detected. Therefore, we evaluate the total noise measured at the output of the read-out amplifier. It is related to the noise generated in the matching layers, the transducer element and the read-out pre-amplifier. However, since a general discussion of the pre-amplifier noise is out of the scope of this chapter, only the noise is considered that can be measured at the input of the signal processing unit (sensitivity is also related to this input). The main noise voltage at this input results from a fluctuating current $<{I_n}^2>$ within a certain frequency band $\Delta f = f_2 - f_1$ often denoted as noise bandwidth

$$\langle I_n^2 \rangle = \int_{f_1}^{f_2} w_I(f)\,\mathrm{d}f, \tag{8.9}$$

which is due to polarization fluctuations in the piezoelectric (ferroelectric) transducer element. In this case, the spectral noise power density $w_{\text{I}}(f)$ follows, as in the case of pyroelectric sensor elements [15], from the Johnson noise generated by the conductance $1/R_{\tan\delta} = \omega C_{\text{E}} \tan\delta$ (Fig. 8.10)

$$w_I(\omega) = 4kT\omega C_E \tan\delta. \tag{8.10}$$

The signal-to-noise ratio (S/N) can be expressed by the noise equivalent sound pressure NESP, which is defined as the minimum pressure that can be detected. This pressure generates at the input of the signal processing unit a voltage signal that is exactly as high as that generated from the tan δ noise

$$V_{\text{input}}^{\text{noise}} = \frac{\sqrt{\langle I_{\text{N}}^2 \rangle}}{C_{\text{E}} + C_{\text{line}} + C_{\text{input}}} \Delta t = \frac{\sqrt{8\pi kTC_{\text{E}}\ \tan\delta\ \Delta f f}}{C_{\text{E}} + C_{\text{line}} + C_{\text{input}}} \Delta t. \tag{8.11}$$

In (8.11) Δt is given by the ultrasound impulse length, the noise bandwidth Δf is appropriately be chosen to correspond with the wideness of the impulse spectrum (typical 1–7 MHz) and f is the centre frequency of the spectrum (typical 3.5 MHz, resonance frequency of the transducer element). Using (8.11) the noise equivalent sound pressure can be calculated to be

[9] Equation (8.7) holds under the assumption that optimal acoustic matching and backing is achieved so that transducer resonance effects can be neglected in the receiving mode.

$$\mathrm{NESP}_{\tan\delta} = \frac{V_{\mathrm{input}}^{\mathrm{noise}}}{S_{\mathrm{VS}}} = \frac{\sqrt{8\pi kTC_{\mathrm{E}}\ \tan\delta\ \Delta ff}}{d_{33}An}\Delta t. \tag{8.12}$$

The figures of merits for comparison and selection of piezoelectric materials for ultrasonic imaging antennas are obtained by extracting the relevant material parameters from the voltage sensitivity (8.8) and the noise equivalent sound pressure (8.12). These are

$$F_{\mathrm{VS}} = \frac{d_{33}}{C_{\mathrm{E}} + C_{\mathrm{line}} + C_{\mathrm{input}}}, \tag{8.13}$$

with $C_{\mathrm{E}} \sim \varepsilon_{33}{}^{\mathrm{T}}$

$$\text{and } F_{\mathrm{SNR}} = \frac{d_{33}}{\sqrt{\varepsilon_{33}^{\mathrm{T}} \tan\delta}}. \tag{8.14}$$

References

1. K.K. Shung, IEEE Eng. Med. Biol. **15**, 18 (1996)
2. C.S. Desilets, J.D. Fraser, G.S. Kino, Ultrasonics **25**, 115 (1978)
3. J. Borburgh, I. Feigt, P. Hini, V. Zurinski, Siemens Forschungs-u. Entwickl. Ber. **9**, 116 (1980)
4. T.N. Nguyen, M. Lethiecq, F. Levassort, L. Pourcelot, IEEE Trans. Ultrason. Ferroelect. Freq. Contr. **43**, 640 (1996)
5. T. Kobayashi, S. Shimanuki, S. Saitoh, Y. Yamashita, Jpn. J. Appl. Phys. Part 1R **36**, 6035 (1997)
6. C.G. Oakley, M.J. Zipparo, IEEE Ultrason. Symp. Proc. **1**, 1157 (2000)
7. S. Saitoh, T. Takeuchi, T. Kobayashi, K. Harada, S. Shimanuki, Y. Yamashita, IEEE Trans. Ultrason. Ferroelect. Freq. Contr. **46**, 414 (1999)
8. R.E. Newnham, D.P. Skinner, L.E. Cross, Mat. Res. Bull. **13**, 525 (1978)
9. W. Wersing, in *Proceedings of 6th IEEE International Symposium on Application of Ferroelectrics*, p. 212, 1986
10. W.A. Smith, Proc. SPIE **1733**, 2 (1992)
11. W. Wersing, K. Lubitz, J. Mohaupt, Ferroelectrics **68**, 77 (1986)
12. U. Bast, D. Cramer, A. Wolff, in *Proceedings of 7th CIMTEC International Meeting on Modern Ceramic Technology*, p. 2005, 1990
13. R.L. Goldberg, S.W. Smith, IEEE Trans. Ultrason. Ferroelect. Freq. Contr. **41**, 761 (1994)
14. G. Percin, A. Atalar, F.L. Degertekin, B.T. Khuri-Yakub, App. Phys. Lett. **72**, 1397 (1998)
15. W. Wersing, R. Bruchhaus, in *Ferroelectric Film Devices*, ed. by M.H. Francombe, Handbook of Thin Film Devices, vol. 5, (Academic Press, New York, 2000) p. 143

9

High Effective Lead Perovskite Ceramics and Single Crystals for Ultrasonic Imaging

Y.(J.) Yamashita and Y. Hosono

9.1 Introduction

Some complex lead perovskites, namely $Pb(B'B'')O_3$ (B' = Mg, Zn, Ni, Fe, Sc, In, and Yb; B'' = Nb, Ta, and W), are relaxors with the rhombohedral or monoclinic structure and tetragonal structure with lead titanate $PbTiO_3$ (PT); these perovskites exhibit a morphotropic phase boundary (MPB) and large electromechanical coupling factor and dielectric constants near the MPB [1–8]. Table 9.1 shows the Curie temperature (T_c,), maximum dielectric constant (K), crystal structure, ferroelectric (F) or antiferroelectric (AF) phase, $PbTiO_3$ mol(%) at the MPB, and T_c at the MPB for various relaxor materials, along with their abbreviations Figure 9.1 shows the MPB (I) in PZT and the MPB (II) in $Pb(B'B'')O_3$-$PbTiO_3$ and the MPB (III) in $Pb(B'B'')O_3$-PZT. The locations of the MPB (II) and (III) vary when different B' or B'' ions are selected as shown in Fig. 9.1. As regards practical piezoelectric applications, little attention was paid to MPB (II) and (III) compositions until 1980, due to the poor piezoelectric properties of the ceramics, notably their low electromechanical coupling factors compared to those of PZT ceramics, and the high cost of the raw materials. Initially, lead magnesium niobate, $Pb(Mg_{1/3}Nb_{2/3})O_3$ (PMN), lead nickel niobate, $Pb(Ni_{1/3}Nb_{2/3})O_3$ (PNN), and lead zinc niobate, $Pb(Zn_{1/3}Nb_{2/3})O_3$ (PZN), were considered to be the three best relaxors.

The present authors have reported large electromechanical coupling factors of $k_p > 0.7$, $k_{33} > 0.75$, and $k_t > 0.5$ near the MPB in $Pb\{(Sc_{1/2}Nb_{1/2})_{0.58} Ti_{0.42}\}O_3$ (PSNT 58/42; in accordance with the standard expression for $Pb(ZrTi)O_3$ (PZT) composition) ceramic samples [6, 7]. We speculated that the selection of B' and B'' ions, which tend to have a disordered structure, and a large PT content in the MPB composition would be necessary to realize a large k_p [8]. We also proposed a new model, which focused on the correlation of the electromechanical coupling factor and the molecular mass of B-site ions in $Pb(B,Ti)O_3$-type perovskite [9,10]. In this model, a low molecular mass for B-site ions, such as those of Sc, Mg, and Ni, is effective for obtaining materials

Table 9.1. Relaxor materials, high T_c materials, and their MPBs with lead titanate

Relaxor materials	Abbrev.	T_c K (°C)	ε_{max}	Structure	Ferro.	T_i @ MPB (mol%)	T_c @ MPB K (°C)
$Pb(B^{2+}{}_{1/3}B^{5+}{}_{2/3})O_3$							
$Pb(Cd_{1/3}Nb_{2/3})O_3$	PCdN	273 (0)	8,000	PC	F	28	653 (380)
$Pb(Zn_{1/3}Nb_{2/3})O_3$	PZN	413 (140)	22,000	R	F	9–10	448 (175)
$Pb(Mg_{1/3}Nb_{2/3})O_3$	PMN	263 (−10)	18,000	PC	F	30–33	428 (155)
$Pb(Ni_{1/3}Nb_{2/3})O_3$	PNN	153 (−120)	4,000	PC	F	28–33	403 (130)
$Pb(Mn_{1/3}Nb_{2/3})O_3$	PMnN	153 (−120)	4,000	PC	F	30–35	403 (130)
$Pb(Co_{1/3}Nb_{2/3})O_3$	PCoN	175 (−98)	6,000	M	F	33	523 (250)
$Pb(Mg_{1/3}Ta_{2/3})O_3$	PMgT	175 (−98)	7,000	PC	F	30?	373 (100)
$Pb(B^{3+}{}_{1/2}B^{5+}{}_{1/2})O_3$							
$Pb(Yb_{1/2}Nb_{1/2})O_3$	PYbN	553 (280)	150	M	AF	50	633 (360)
$Pb(In_{1/2}Nb_{1/2})O_3$	PIN	363 (90)	550	M	F	37	593 (320)
$Pb(Sc_{1/2}Nb_{1/2})O_3$	PSN	363 (90)	38,000	R	F	42	533 (260)
$Pb(Fe_{1/2}Nb_{1/2})O_3$	PFN	385 (112)	12,000	R	F	7?	413 (140)
$Pb(Sc_{1/2}Ta_{1/2})O_3$	PST	299 (26)	28,000	R	F	45	478 (205)
Others							
*$PbZrO_3$	PZ	513 (240)	3,000	O	AF	47	633 (360)
(Pb, La)(Zr, Ti)O_3	PLZT	<623 (< 350)	30,000	R, T	F, AF	35–47	<623 (<350)
$BiScO_3$	BS	>673 (>350)	<1,000	R	F	64	723 (450)
$BiInO_3$	BIn	>973 (>700)	<1,000	R	F?	70	843 (570)
$PbTiO_3$	PT	763 (490)	9,000	T	F		

Abbreviations: *C* Cubic, *M* Monoclinic, *O* Orthorhombic, *PC* Pseudocubic, *PY* Pyrochlore, *R* Rhombohedral, *T* Tetragonal, *AF* Antiferroelectrics, *F* Ferroelectrics

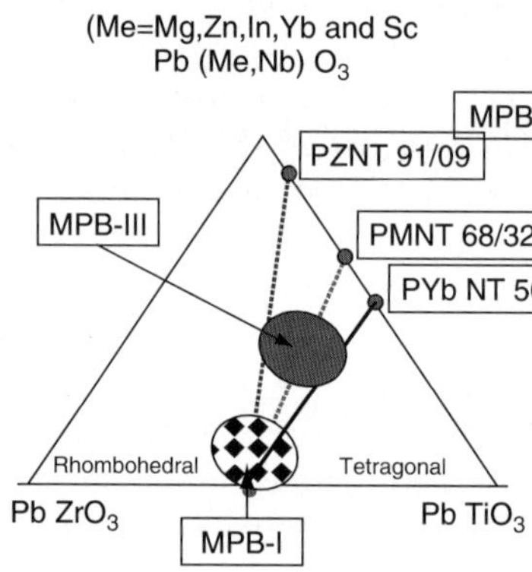

Fig. 9.1. Location of morphotropic phase boundary (MPB)-I, II, and III of relaxor-PT system

with large electromechanical coupling factors. In our most recent papers, we reported that $(Pb_{0.965}, Sr_{0.035})\{(Sc_{1/2}Nb_{1/2})_{0.08}(Mg_{1/3}Nb_{2/3})_{0.16}Pb(Ni_{1/3}Nb_{2/3})_{0.26}Zr_{0.135}Ti_{0.365}\}O_3$ piezoelectric ceramic materials have high dielectric constants of $\varepsilon_{33}{}^{\mathrm{T}}/\varepsilon_0 = 7,200$ and a large planar-mode electromechanical coupling factor of $k_{\mathrm{p}} = 69\%$ with $T_{\mathrm{c}} = 135°\mathrm{C}$ [11,12].

However, it is a difficult and time-consuming task to find an exact MPB composition with a proper T_{c} if the material consists of multicomponent complex lead perovskite structure compounds as shown above. We proposed a smart material design procedure that can predict the T_{c} and TiO_2 content of the material system by a simple calculation. High-performance piezoelectric materials with high dielectric constants of $\varepsilon_{33}{}^{\mathrm{T}}/\varepsilon_0 >$ 5,000, a large planar-mode electromechanical coupling factor of $k_{\mathrm{p}} > 70\%$, and a large piezoelectric constant of $d_{33} > 750\,\mathrm{pC\,N^{-1}}$ were developed by using the smart material design [13].

In addition, since the discovery of PZT, numerous $Pb(B'B'')O_3$ and $Pb(B'B'')O_3$-$PbTiO_3$ relaxor materials similar to PZT and new kinds of piezoelectric single crystals have been investigated [1–19]. In the last 15 years, the main focus of attention has been the potential industrial applications of these single crystals of relaxors.

In this chapter, we deal with large piezoelectric constant d_{33} materials based on lead perovskites. The materials consist of two groups: ceramics (part A, 3–4) and single crystals (part B, 5–6).

In part A, a novel guiding principle for discovering excellent piezoelectric materials is introduced, namely, the presence of low molecular mass B-site ions that can enter the lead–perovskite and enhancement of disordering of lead perovskite $Pb(B'B'')O_3$ structure. To obtain high dielectric constant $\varepsilon_{33}{}^{\mathrm{T}}/\varepsilon_0 >$ 5,000, a large electromechanical coupling factor (electromechanical coupling factor) planar mode $k_{\mathrm{p}} > 68\%$, and piezoelectric constant $d_{33} > 800\,\mathrm{pC\,N^{-1}}$ materials, the low-molecular-mass B-site ion model is applied.

In part B, high-quality piezoelectric single crystals, such as $Pb(Zn_{1/3}Nb_{2/3})O_3$-$PbTiO_3$ (PZNT) and $Pb(Mg_{1/3}Nb_{2/3})O_3$-$PbTiO_3$ (PMNT), have been investigated. Since their piezoelectric properties are greatly superior to those of PZT ceramics, they have been used for certain transducer applications since the early 2000s. The present situation and future prospects for these relaxor-PT single crystals are summarized. In this review, some possible high $d_{33} > 1,500\,\mathrm{pC\,N^{-1}}$ and high $T_{\mathrm{c}} > 180°\mathrm{C}$ single crystals are also introduced. Single crystals of the $Pb(Mg_{1/3}Nb_{2/3})O_3$-$Pb(In_{1/2}Nb_{1/2})O_3$-$PbTiO_3$ (PMN-PIN-PT) ternary system and their manufacturing process and electrical properties are reported. Application of these high d_{33} materials for the medical transducer application is also summarized.

9.2 Material Design for High $E_{33}{}^{t}/E_0$ and Large Electro-Mechanical Coupling Factor with a Certain T_c

Nakamura and coworkers proposed a novel concept in which the mass inequality of the A-site and Ti ions correlates with ferroelectricity, namely, it results in a large dielectric constant [14,15]. Figure 9.2 shows perovskite parameter vs. average mass of A ions for perovskite titanate $ATiO_3$ and $(A'_{1/2}A''_{1/2})TiO_3$. Solid circles (asterisk) and open squires (plus) indicate quantum paraelectric and ferroelectrics, respectively. Trial candidate ferroelectric substances are indicated by solid (hash). "QP region" and "FE region," respectively, donate quantum paraelectric and ferroelectric regions. According to their explanation, titanate perovskites, $ATiO_3$, and $(A'_{1/2}A''_{1/2})TiO_3$ are characterized by the following properties:

1. Their structure is three-dimensional, composed of oxygen shared among the $[TiO_6]$ octahedra, comprising the $[TiO_3]_\infty$ network.
2. The size of each Ti ion is suitable for rattling at the center of the individual $[TiO_6]$-octahedron.
3. The mass of each Ti ion (AW = 47.9 a.u.) is approximately equal to that of three oxygen ions (3 × O AW = 48.0).
4. A strong local electric field located on each Ti ion is coupled to collinear oxygen ions OI, according to Slater's correction of the Lorenz field. These properties give the perovskite titanates a high dielectric permittivity and the potential to undergo a displaceable ferroelectric phase transition.

If the above concept is applicable to the $Pb\text{-}(B', B'')O_3$ lead perovskite, we hypothesize that the low molecular mass B-site ions are important in realizing high ferroelectricity, namely, a large dielectric displacement, large dielectric constant, and large piezoelectric constant. Of course, the piezoelectricity of the perovskite is strongly influenced by many other factors. However, as shown in Fig. 9.3, clear correlations of electromechanical coupling factor k_p and B-site molecular mass of several lead perovskite compounds near the MPB were reported by the present authors [6–13]. To obtain a high-dielectric-constant piezoelectric material, one should pay attention not only to the room-temperature dielectric constant but also to the peak dielectric constant at Curie temperature (T_c). In addition, the phase change temperature of rhombohedral and tetragonal phases T_{rt} also plays an important role in increasing the room-temperature dielectric constants. Figure 9.4 shows two major approaches for obtaining modified PZT materials with large dielectric constant. A-site modification by La, Ba, or Sr for Pb in PZT decreases T_c of pure PZT and shows a large dielectric constant. However, it is difficult to obtain $\varepsilon_{33}{}^{T}/\varepsilon_0 > 7{,}000$ with the A-site modification process due to lower peak dielectric constant. In contrast, B-site modification by relaxors such as $(Mg_{1/3}Nb_{2/3})$ or $(Sc_{1/2}Nb_{1/2})$ also decreases T_c of pure PZT and shows a large dielectric constant of around $T_c = 150°C$ as well as at room

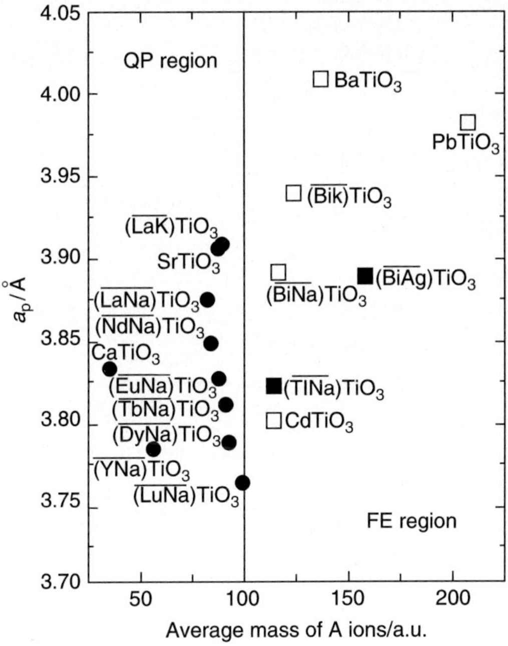

Fig. 9.2. Effect of ionic size and molecular mass of A-site ions of various $ATiO_3$ on the ferroelectric properties (after Nakamura et al. [14])

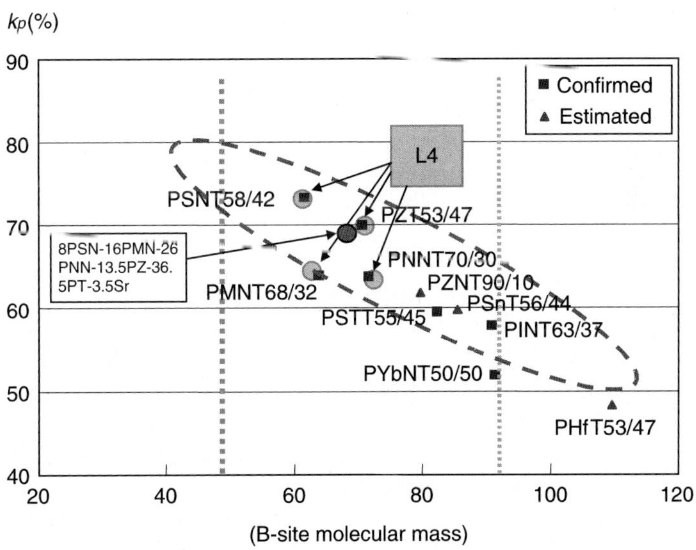

Fig. 9.3. Electromechanical coupling factor k_p and B-site molecular mass of several lead perovskite compounds near the MPB [9]

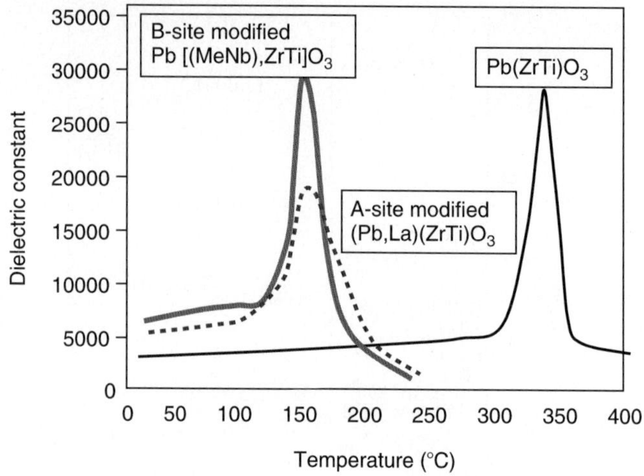

Fig. 9.4. High-dielectric-constant piezoelectric materials design for A-site and B-site modification of PZT

Table 9.2. Low-B-site-mass perovskite compounds for high-performance piezoelectric material

Perovskite compounds	Abbreviation	B-site M	Phase	T_c (°C)
$PbTiO_3$	PT	47.9	Tetragonal	490
$Pb(Sc_{1/2}Nb_{1/2})O_3$	PSN	68.93	Rhombohedral	90
$Pb(Mg_{1/3}Nb_{2/3})O_3$	PMN	70.07	Pseudocubic	−10
$Pb(Mg_{1/6}Ni_{1/6}Nb_{2/3})O_3$	PMNN	75.80	Pseudocubic	−70
$Pb(Ni_{1/3}Nb_{2/3})O_3$	PNN	81.52	Pseudocubic	−150
$PbZrO_3$	PZ	91.22	Rhombohedral	230
$Pb(In_{1/2}Nb_{1/2})O_3$	PIN	103.85	Pseudocubic	90

temperature. In addition, in view of the difficulty of obtaining a pure phase of perovskite structure due to the instability of the relaxor compounds even with a PZT composition of nearly 50 mol%, in mass production one should pay attention to stabilizing the perovskite structure after sintering to obtain a high-dielectric-constant and large-electromechanical coupling factor piezoelectric material based on the relaxor-PZT system. To overcome these shortcomings, substitutions of the A-site Pb with Ba and Sr and combination of relaxor substitution for B-site were applied. Kanai et al. reported that the decrease in T_c by the addition of Ba and Sr is 8.5 and 10.5°C $(mol\%)^{-1}$, respectively, for PZT 70/30 [16]. Thus, when 4.8 mol% Ba and Sr substitution is carried out, T_c decreases by 48°C compared with the unsubstituted relaxor-PZT composition. In addition, Ba or Sr substitution shifts MPB composition to the rhombohedral side by about 0.5–1.0 mol% mol^{-1}. Table 9.2 shows low molecular mass perovskite compounds and their T_c for high-performance piezoelectric mate-

Table 9.3. Electrical properties of low and high B-site molecular mass perovskite compounds of MPB composition [9]

Perovskite compounds	MPB composition	B-site M	k_p(%)	$\varepsilon_{33}{}^{T}/\varepsilon_0$	T_c (°C) Measured, (calculated)
$Pb\{(Sc_{1/2}Nb_{1/2})Ti\}O_3$	PSNT 58/42	60.1	72	2,500	260, (258)
$Pb\{(Mg_{1/3}Nb_{2/3})Ti\}O_3$	PMNT 68/32	63.0	65	5,500	150, (151.7)
$Pb(Zr, Ti)O_3$	PZT 53/47	70.9	70	2,200	380, (352.2)
$Pb\{(Ni_{1/3}Nb_{2/3})Ti\}O_3$	PNNT 71.4/28.6	71.9	–	20,000	33, (33.0)
$Pb\{(Zn_{1/3}Nb_{2/3})Ti\}O_3$	PZNT 90/10	80.2	(60)	2,200	175
$Pb\{(Sc_{1/2}Ta_{1/2})Ti\}O_3$	PSTT 55/45	83.7	60	2,200	205
$Pb(Sn, Ti)O_3$	PSnT 56/44	85.4	60	2,000	220
$Pb\{(In_{1/2}Nb_{1/2})Ti\}O_3$	PINT 63/37	87.1	58	2,000	320
$Pb\{(Yb_{1/2}Nb_{1/2})Ti\}O_3$	PYbNT 50/50	90.4	53	1,800	360
$Pb(Hf, Ti\}O_3$	PHfT 53/47	117.1	(50)	(1,800)	340
$(Pb_{0.965}Sr_{0.035})\{(Sc, Mg, Ni, Nb)_{0.5} Zr_{0.135} Ti_{0.365}\}O_3$	8PSN-16PMN-26PNN-13.5PZ-36.5PT	67.7	69	7,200	135, (139)
$(Pb_{0.952}Ba_{0.024}Sr_{0.024})\{(Sc, Mg, Ni, Nb)_{0.5} Zr_{0.13} Ti_{0.37}\}O_3$	16PSN-17PMN-17PNN-13PZ-37PT	65.6	72	5,500	153, (152.8)

rials used in this study. Three relaxor compositions are selected: PSN, PMN, and PNN have low molecular mass, from 68.93 to 81.52. In Table 9.2, the $T_c = -150°C$ of PNN was taken from a recent report [13]. Table 9.3 shows low molecular mass perovskite compounds of the MPB composition and their T_c's. In addition, the MPB composition of PNNT was determined as follows. Many researchers reported that the 0.50PNN-0.34PT-0.16PZ composition showed the highest electromechanical coupling factor $k_p = 70\%$. If the composition is regarded as a combination of MPBs of PZT 53/47 and PNNT ##/##, the MPB of PNNT is calculated as PNNT 71.4/28.6. Then, the MPB of PNNT 71.4/28.6 was applied to further calculations.

Figure 9.5 shows MPB composition lines in the PSN-(PMN-PNN)-PZ-PT system. In this quaternary pyramid system, a combination of $Pb\{(Mg_{1/3}Nb_{2/3})_{0.68}Ti_{0.32}\}O_3$ (PMNT 68/32, T_c = 150°C) and $Pb\{(Ni_{1/3}Nb_{2/3})_{0.714}Ti_{0.286}\}O_3$ (PNNT 71.4/28.6, T_c = 33°C) is indicated as $Pb\{((MgNi)_{1/3}Nb_{2/3})_{0.70}Ti_{0.30}\}O_3$ (PMNNT 70/30, T_c = 92°C) to locate four lead perovskite compounds in the corner of the pyramid. The MPB composition lines of the ternary and quartery systems are determined by the binary system MPB compositions. A plane in the pyramid shows an MPB composition. The MPB composition in Fig. 9.5 is an electromechanical coupling factor k_p peak composition. A composition corresponding to a peak in dielectric constant is located at about 1.5 mol% in the TiO_2-rich tetragonal side of the electromechanical coupling factor k_p-peak composition as shown in previous papers [5, 7, 11].

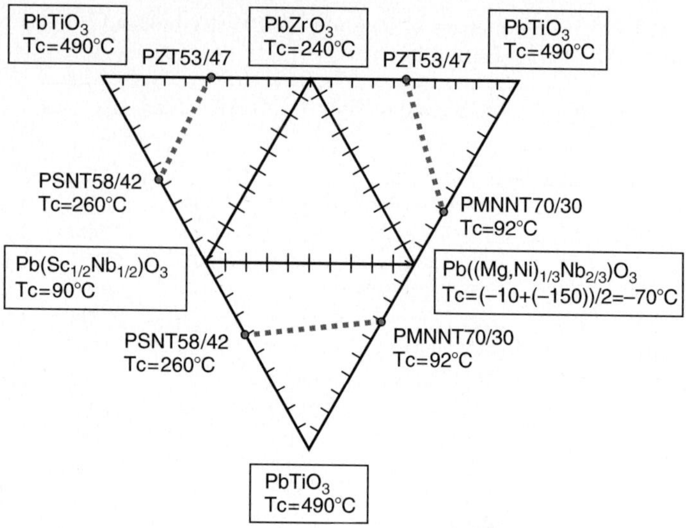

Fig. 9.5. MPB composition lines in $Pb(Sc_{1/2}Nb_{1/2})O_3$-$Pb(MgNi_{1/3}Nb_{2/3})O_3$-$PbZrO_3$-$PbTiO_3$ (PSN-(PMN-PNN)-PZ-PT) system

Figure 9.6 shows $T_c = 200°C$ composition lines in the PSN-(PMN-PNN)-PZ-PT system. The $T_c = 200°C$ composition lines of the ternary and quartery systems are determined by the $T_c = 200°C$ of each binary system composition. However, one portion of the plane has been eliminated because there is no $T_c = 200°C$ composition in the PZT binary system. Since the decreases in T_c induced by Ba and Sr substitution for A-site Pb are -8.5 and $-10.5°C\ mol^{-1}$, respectively [16], the $T_c = 155°C$ plane is obtained when 4.8 mol% Ba and Sr are substituted for A-site Pb.

Figure 9.7 shows $T_c = 200°C$ and MPB composition lines and planes in the PSN-PMN-PNN-PZ-PT system. The line on which there are two crossover planes is the MPB with the $T_c = 200°C$ composition in the multicomponent system. When calculating the MPB $TiO_2(x)$ content of $(Pb_{0.962}, Ba_{0.024}Sr_{0.024})\{(Sc_{1/2}Nb_{1/2})_{0.16}(Mg_{1/3}Nb_{2/3})_{0.17}Pb(Ni_{1/3}Nb_{2/3})_{0.17}Zr_{0.5-x}Ti_x\}O_3$ (PSN-PMN-PNN-PBSZT), the x content is determined by calculating the sum of the TiO_2 contents of the MPBs. In addition, the MPB shift caused by Ba and Sr was estimated to be 1.0 mol% mol^{-1}. Therefore, at the calculated peak electromechanical coupling factor k_p, the TiO_2 content is $TiO_2(mol\%) = (-1\times2.4)+(-1\times2.4)+(0.42\times0.16)+(0.32\times0.17)+(0.286\times0.17)+(0.47\times0.5) = 0.357$.

The peak-dielectric-constant composition is estimated to be at the position of the 1.5 mol% shift to the tetragonal side, namely, $TiO_2(mol\%) = 0.357 + 0.015 = 0.372$. For example, when calculating the T_c of $(Pb_{0.962}, Ba_{0.024}Sr_{0.024})\{(Sc_{1/2}Nb_{1/2})_{0.16}(Mg_{1/3}Nb_{2/3})_{0.17}Pb(Ni_{1/3}Nb_{2/3})_{0.17}Zr_{0.13}Ti_{0.37}\}O_3$, T_c is determined by calculating the sum of the T_c of the perovskite compounds. In

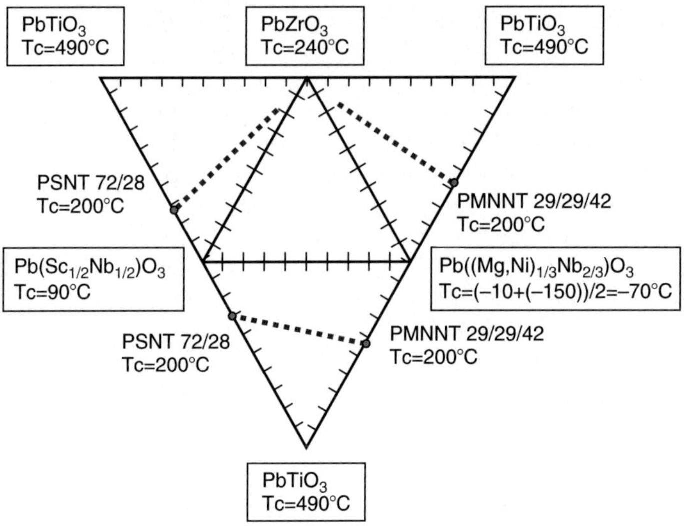

Fig. 9.6. Curie temperature (T_c) = 200°C composition lines in $Pb(Sc_{1/2}Nb_{1/2})O_3$-$Pb(MgNi_{1/3}Nb_{2/3})O_3$-$PbZrO_3$-$PbTiO_3$ (PSN-(PMN-PNN)-PZ-PT) system. Note: there is no T_c = 200°C composition in the PZT binary system

addition, T_c shifts caused by Ba and Sr were estimated as 8.5 and 10.5°C, respectively [16]. Calculated $T_c = (-8.5 \times 2.4) + (-10.5 \times 2.4) + (90 \times 0.16) + (-10 \times 0.17) + (-150 \times 0.17) + (230 \times 0.13) + (490 \times 0.37) = 152.8°C$.

These smart procedures can be used to find the MPB composition of the multicomponent lead perovskite piezoelectric material with any T_c within a reasonable time frame.

9.3 Processing of High-Performance Ceramics

The selected composition is $(Pb_{0.962}, Ba_{0.024}Sr_{0.024})$ $\{(Sc_{1/2}Nb_{1/2})_{0.16}(Mg_{1/3}Nb_{2/3})_{0.17}Pb(Ni_{1/3}Nb_{2/3})_{0.17}Zr_{0.5-x}Ti_x\}O_3$ (PSN-PMN-PNN-PBSZT). Ten samples of $TiO_2(x)$ contents from 0.33 to 0.37, which were calculated by the procedure, were investigated. The preparation process for the ceramic samples was described in detail in previous papers [7, 11]. All ceramic samples were prepared by a conventional solid-state-reaction method. All raw materials, PbO, $BaCO_3$, $SrCO_3$, Sc_2O_3, $MgCO_3$, NiO, Nb_2O_5, ZrO_2 and TiO_2, were wet-mixed with a ball mill, dried, and calcined at 850°C for 4 h. Firing was performed at 1,250–1,300°C for 2 h in a closed MgO crucible. The sample dimensions used for this experiment were 20 mm in diameter and 1.0 mm in thickness. A gold electrode was applied on both surfaces by sputtering. The samples were poled in a 20 kV cm^{-1} electric field at room temperature for 15 min. These samples were measured 24 h after poling. The electromechanical coupling factors were measured on the basis of the Electronic Materials

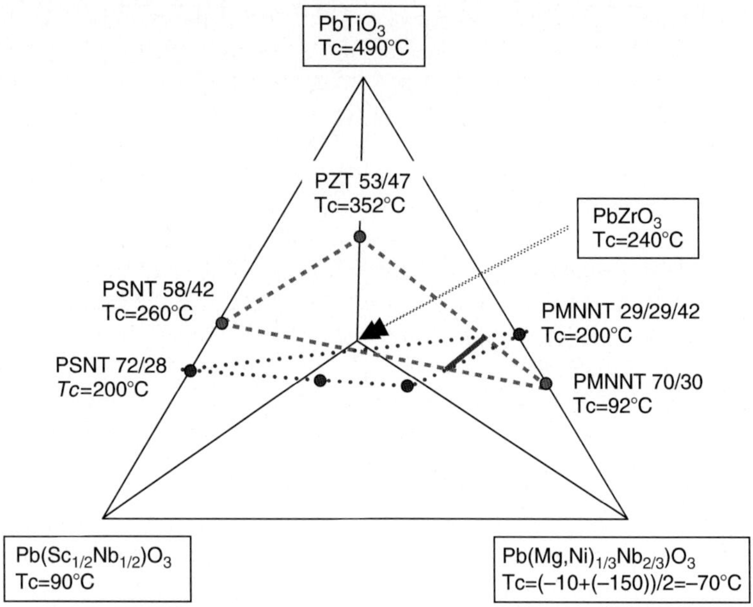

Fig. 9.7. Curie temperature (T_c) = 200°C and MPB composition lines and planes in $Pb(Sc_{1/2}Nb_{1/2})O_3$-$Pb(MgNi_{1/3}Nb_{2/3})O_3$-$PbZrO_3$-$PbTiO_3$ (PSN-(PMN-PNN)-PZ-PT) system. Note: the line that cross over two planes is the MPB with T_c = 200°C composition in the multicomponent system. The T_c = 155°C plane is obtained when 4.8 mol% Ba and Sr are substituted for A-site Pb

Manufacturing Association (EMAS)-6100 procedure of Japan for measuring piezoelectric ceramic elements. To calculate planar-mode electromechanical coupling factor k_p, the following equation was used:

$$1/(k_p)^2 = 0.398\{f_r/(f_a - f_r) + 0.579\}, \tag{9.1}$$

where f_r is the resonant frequency and f_a is the antiresonant frequency. The piezoelectric charge constant d_{33} was measured using a d_{33} meter (Institute of Acoustics Academia Sinica Model ZJ-3D).

9.4 Physical and Electrical Properties of High-Performance Ceramics

All ceramic specimens showed high densities of more than 8.00 g cm^{-3}, more than 98% of the theoretical densities, with an average grain size of 5 μm. The fired PSN-PMN-PNN-PBSZT samples have a complete single-phase perovskite structure with a mixture of rhombohedral and tetragonal phases. One of the shortcomings of relaxor-PT piezoelectric ceramic materials is the pyrochlore phase formation after sintering due to the instability of the perovskite phase. The stability of the perovskite structure near the MPB in the PSN-PMN-PNN-PBSZT ceramics is noteworthy. One of the reasons for this stability is that the PSN-PMN-PNN-PBSZT MPB, which is a very stable perovskite structure, contains a large amount of PT (more than 33 mol%) and PZ (more than 13 mol%). In addition, the 4.8 mol% Ba and Sr that replaces Pb in the A-site also plays an important role in stabilizing the perovskite structure [17].

Table 9.3 summarizes the electrical properties of the low and high molecular mass B site perovskite compounds of the MPB composition. In Table 9.3, the k_p value of the PZNT 90/10 ceramic is estimated from the result of PSN-PMN-PT and PSN-PZN-PT ternary system ceramics. Two k_p values of the PSnT 56/44 and PHfT 53/47 are shown: one is from Jaffe's data [1] and the other is an estimated one. The estimation of k_p values of the PSnT 56/44 and PHfT 53/47 was carried out in the next steps described in the following. First, the k_p of a high density PZT ($k_p = 70\%$) and that ($k_p = 40\%$) of a low density PZT in the 1950s were compared. Next, the k_p's of a high-density PSnT 56/44 and PHfT 53/47 were estimated by considering the dielectric constant values ($\varepsilon = 1,200$ and 800) and k_p values ($k_p = 28\%$ and 38%) of these materials as reported by Jaffe [1]. In all other relaxor – PT perovskites – the k_p values near the MPB are valid data, which have been confirmed using the same measurement procedures in the same laboratory [9].

Figure 9.3 and Table 9.3 show the e/m coupling factor, k_p, and molecular mass of B site ions of various lead–perovskite MPB compounds in ceramic form. The niobium-based PSNT58/42 ($k_p = 72\%$) has a larger k_p than the tantalum-based $Pb(Sc_{1/2}Ta_{1/2})O_3$-PT (PSTT 55/45) ($k_p = 60\%$). The indium-based PInNT 63/37 ($k_p = 58\%$) also has a larger k_p than the ytterbium-based PYbNT 50/50 ($k_p = 53\%$). These B site mass values for the four MPB composition compounds are 60.1, 83.7, 87.1, and 90.4, respectively. Although the PMNT 68/32, which has a low-B site mass of 63 and a slightly low k_p value of 64% compared to an expected $k_p = 70\%$, shows some deviation from the general trend, all the other lead–perovskite compounds show a clear correlation between k_p and B site mass. The lower k_p value of PMNT 68/32 ceramics may be affected by a small pyrochore structure presence in the ceramics. These results imply that a low B site mass or large difference in molecular mass between A-site (Pb) and B-site ions (Amass-Bmass) plays an important role in bringing about large e/m coupling factors for all lead–perovskites. Although, the high molecular mass A-site model and the low molecular mass B-site model are still not fully accepted by several scientists from material science viewpoint, it is worthy that the model can well explain the reason of the highest electromechanical coupling factor $k_p = 72\%$ of PSNT 58/42 than those of PZT 53/47, PMNT 68/32, or $Pb(Sc_{1/2}Ta_{1/2})O_3$-PT (PSTT 55/45) [9,10].

Figure 9.8 shows electromechanical coupling factor k_p as a function of the $TiO_2(x)$ content of $(Pb_{0.962}, Ba_{0.024}\ Sr_{0.024}))\{(Sc_{1/2}Nb_{1/2})_{0.16}(Mg_{1/3}$

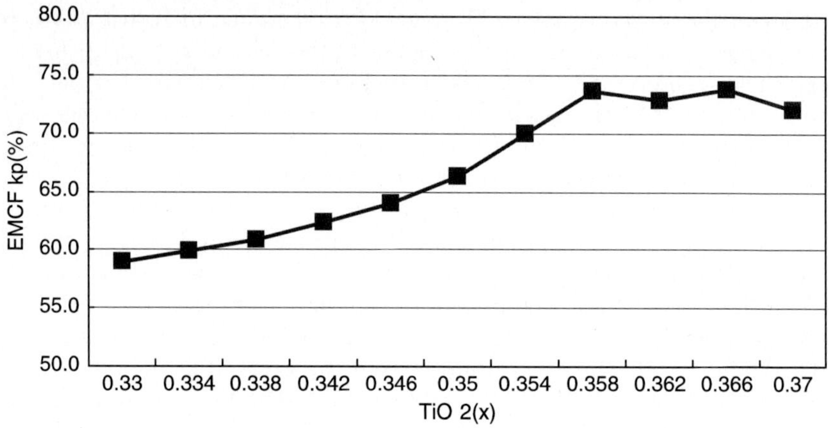

Fig. 9.8. Electromechanical coupling factor k_p as a function of TiO_2(x) content of $(Pb_{0.962}, Ba_{0.024}Sr_{0.024})$ $\{(Sc_{1/2}Nb_{1/2})_{0.16}(Mg_{1/3}Nb_{2/3})_{0.17}Pb(Ni_{1/3}Nb_{2/3})_{0.17}$ $Zr_{(0.5-x)}Ti_x\}O_3$ ceramics

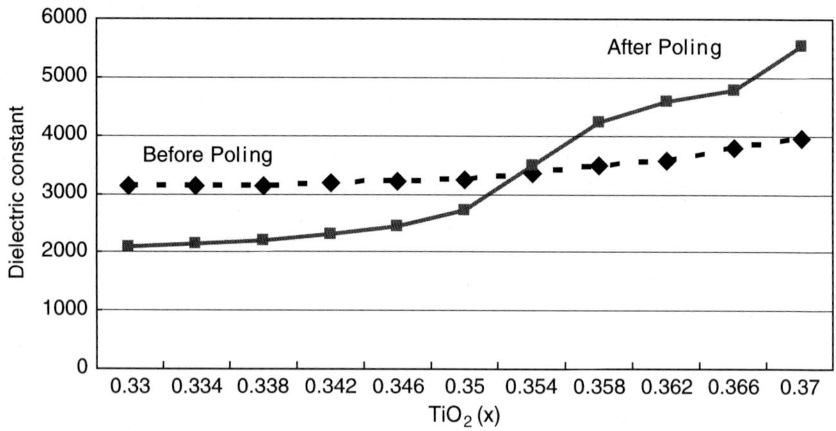

Fig. 9.9. Dielectric constant before and after poling as a function of $TiO_2(x)$ content of $(Pb_{0.962}, Ba_{0.024}Sr_{0.024})\{(Sc_{1/2}Nb_{1/2})_{0.16}(Mg_{1/3}Nb_{2/3})_{0.17}(Ni_{1/3}Nb_{2/3})_{0.17}$ $Zr_{0.5-x}Ti_x\}O_3$ ceramics

$Nb_{2/3})_{0.17}(Ni_{1/3}Nb_{2/3})_{0.17}Zr_{0.13}Ti_{0.37}\}O_3$ ceramics. A peak of a planar-mode $k_p = 73.6\%$ was found at $x = 0.36$, which is very close to the calculated value of $x = 0.362$. The k_p is very large for low-T_c piezoelectric materials.

Figure 9.9 shows the dielectric constant before and after poling as a function of the $TiO_2(x)$ content of PSN-PMN-PNN-PBSZT ceramics. For $x < 0.35$, the dielectric constant after poling is low and for $x > 0.35$, it is high. The same result has also been reported for PZT [1] and PSNT [7].

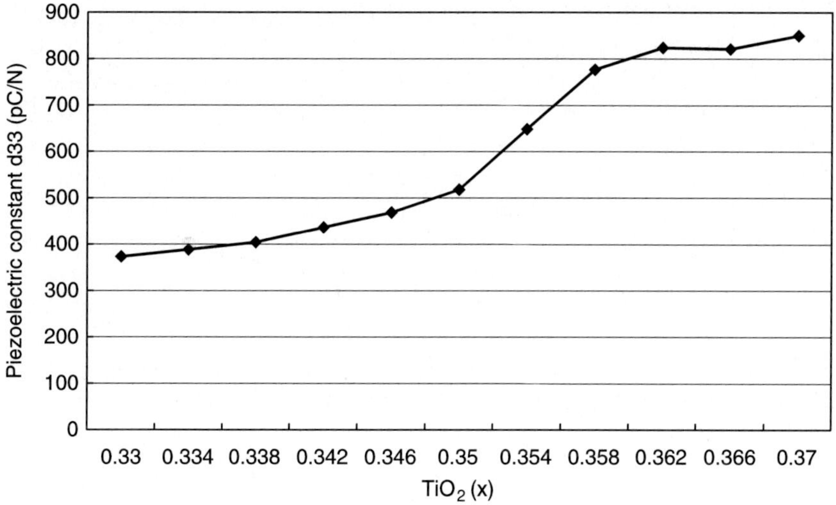

Fig. 9.10. Piezoelectric constant d_{33} measured by d_{33} meter as a function of $TiO_2(x)$ content of $(Pb_{0.962}, Ba_{0.024}Sr_{0.024})\{(Sc_{1/2}Nb_{1/2})_{0.16}(Mg_{1/3}Nb_{2/3})_{0.17}(Ni_{1/3}Nb_{2/3})_{0.17}Zr_{0.5-x}Ti_x\}O_3$ ceramics

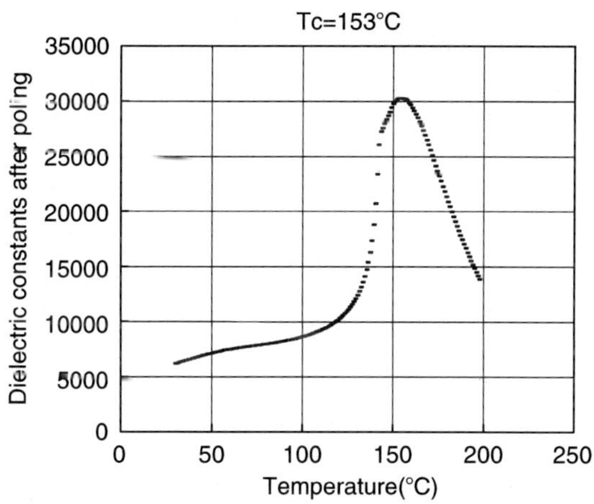

Fig. 9.11. Temperature dependence of dielectric constant after poling of $(Pb_{0.962}, Ba_{0.024}Sr_{0.024})\{(Sc_{1/2}Nb_{1/2})_{0.16}(Mg_{1/3}Nb_{2/3})_{0.17}Pb(Ni_{1/3}Nb_{2/3})_{0.17}Zr_{0.13}Ti_{0.37}\}O_3$ ceramic

Figure 9.10 shows piezoelectric constants d_{33} measured by the d_{33} meter as a function of the $TiO_2(x)$ content of $(Pb_{0.962}, Ba_{0.024}\ Sr_{0.024})\{(Sc_{1/2}Nb_{1/2})_{0.16}(Mg_{1/3}Nb_{2/3})_{0.17}(Ni_{1/3}Nb_{2/3})_{0.17}Zr_{0.13}Ti_{0.37}\}O_3$ ceramics. Piezoelectric charge constants of more than 800 pC N^{-1} were obtained when $x = 0.36$–0.37 for the PSN-PMN-PNN-PBSZT ceramics.

Figure 9.11 shows the temperature dependence of the peak dielectric constant after poling the $(Pb_{0.962}, Ba_{0.024} Sr_{0.024}) \{(Sc_{1/2}Nb_{1/2})_{0.16}(Mg_{1/3} Nb_{2/3})_{0.17}(Ni_{1/3}Nb_{2/3})_{0.17}Zr_{0.13}Ti_{0.37}\}O_3$ ceramic. The ceramic showed a maximum dielectric constant peak of 30,500 at T_c = 153°C. The obtained T_c (153°C) and calculated T_c (152.8°C) of the composition showed good agreement. The T_c and MPB composition of relaxor-type piezoelectric materials are more or less influenced by many other factors, such as raw material purities, particle size, reactivity, calcining temperature, firing temperature, grain size, firing density, crystal structure (perovskite/pyrochlore), order/disorder of the ions, domain structure, polarization process, and measurement frequency. However, as shown in Figs. 9.8–9.11 and Table 9.3, the material design of the T_c and TiO_2 content of the MPB in the multicomponent lead–perovskite PSN-PMN-PNN-PBSZT ceramic system shown in this study is a useful tool for obtaining piezoelectric materials with better performance [13].

9.5 Lead-Based Relaxor Single Crystal Materials

9.5.1 Large-Size Single Crystal Growth

The first large-size PZNT piezoelectric single crystal by flux growth for medical transducer applications was reported by Kobayashi in 1997 [18]. However, in recent years, single crystal growth has been performed by the solution Bridgman or direct Bridgman process. Using these processes, it is possible to control evaporation of PbO at high temperature during crystal growth. The single crystal growth by the Bridgman process started at about the same time in Japan, the USA, and China [19]. The Bridgman process furnace has a steep vertical temperature gradient, more than 20°C cm^{-1}, at around solid/liquid interfaces determined by the phase diagram. Figure 9.12 shows the large size, 80 mm in diameter PMN-PIN-PT single crystal boule produced by the Bridgman process [20]. PMNT and PZNT single crystals of almost the same size have been successfully grown by other research teams [21–23].

9.5.2 High Curie Temperature Single Crystals

One of the disadvantages of PZNT and PMNT single crystal is that they have a low Curie temperature T_c and a low rhombohedral to tetragonal phase change temperature (T_{rt}) compared to those of PZT ceramics (Table 9.4). Unlike PZT ceramics, PZNT and PMNT exhibit a phase change temperature T_{rt} of around 50–120°C. Although the T_{rt} values shift depending on poling conditions, fairly large decreases of k_{33} and d_{33} are observed above the T_{rt} [19]. Therefore, in practice, the operating temperature range of these single crystal devices is limited by T_{rt}, not by T_c. The T_c's and T_{rt}'s of PZNT 94/6 to PZNT 89/11 lie in the ranges of T_c = 161–178°C and T_{rt} = 107–47°C, and those of PMNT 70/30 to PMNT 67/33 in the ranges of T_c = 140–155°C and

$T_{rt} = 87$–$47°C$. These T_{rt} and T_c's are sufficiently high for naval or medical applications where the operating temperatures are usually below 60°C. However, they are unsuitable for high-temperature applications such as actuators for outdoor use. Therefore, new material systems with $T_c > 180°C$ and $T_{rt} > 100°C$ are required for high-temperature applications. Using the binary system of xPb($In_{1/2}Nb_{1/2}$)O_3-$y$$PbTiO_3$ (PINT $100x/100y$) and the ternary system of xPb($In_{1/2}Nb_{1/2}$)O_3-yPb($Mg_{1/3}Nb_{2/3}$)O_3-$z$$PbTiO_3$ (PIN-PMN-PT $100x/100y/100z$), attempts have been made to grow single crystal of sufficient size. Yasuda et al. reported that PINT single crystals of fairly good quality, more than $10 \times 10 \times 3\,mm^3$ in size, were obtained by the flux method. The PINT single crystal shows a T_c of 260°C, which is 80°C higher than that of the PZNT 91/9 single crystal. A T_{rt} is observed near 100–150°C. The PSN-PMN-PT single crystal shows a peak dielectric constant maximum at $T_c = 200°C$ at 1 kHz, which is 50°C higher than that of the PMNT 67/33 single crystal. A small peak of dielectric constant is observed near 120°C, which is considered to be the T_{rt} from rhombohedral to tetragonal phase. The PINT shows excellent temperature dependence of piezoelectric properties [19].

Recently, the present authors reported piezoelectric properties of xPb($In_{1/2}Nb_{1/2}$)O_3-yPb($Mg_{1/3}Nb_{2/3}$)O_3-$z$$PbTiO_3$ (PIN-PMN-PT $100x/100y/100z$) single crystal [21]. Figure 9.13 shows the dielectric constants before poling of the PIN-PMN-PT 16/51/33 and 24/51/33 single crystals. Small peaks of the dielectric constant are observed at 50–90°C, corresponding to the phase-

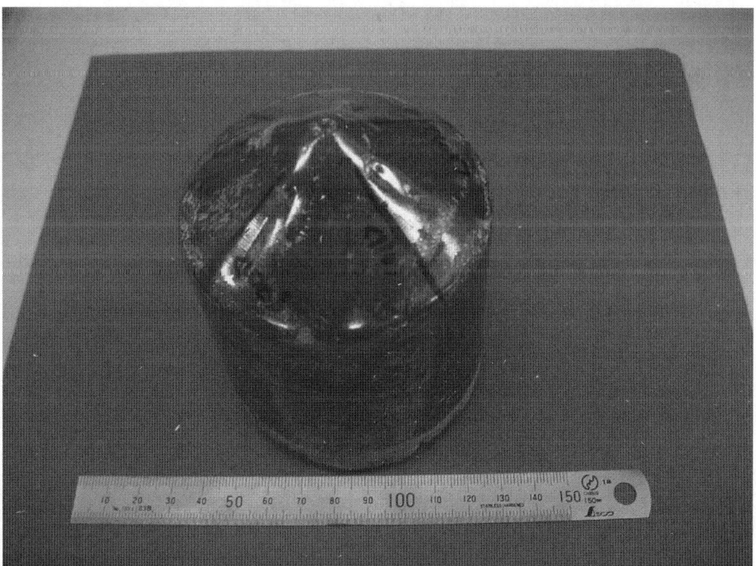

Fig. 9.12. Large size, 80 mm in diameter PMN-PIN-PT single crystal produced by the Bridgman process (Courtesy of JFE Minerals Co. of Japan, see [20])

Table 9.4. Comparison of several relaxor-PT single crystals and PZT type ceramics for transducer applications

Materials	Ti (mol%)	T_c (°C)	K_{33}	k_{33}(%)	k_{31}(%)	d_{33} (pC/N)	d_{31} (pC/N)	g_{33} (10^{-3} Vm N^{-1})	Size (mm)
PZNT	5–10	150–175	4,000	95	81	2,400	−1,700	60	50
PMNT	28–33	135–155	4,500	93	61	2,500	−1,000	50	70
PSNT	40–44	240–260	2,000	(94[a])	(80[a])	(1,500)	(−1,000)	(>40)	<10
PINT	33–38	300–320	2,000	(85[a])	(>50)	(1,000)	(−700)	(>40)	<15
PIN-PMN-PT	30–35	180–250	3,100	90	60	2,200	−1,000	60	25
PZT 5H[b]	40–47	210	3,200	75	40	700	−350	20	>100
Relaxor-PZT high K[b]	35–38	135–155	5000–7200	78	42	800–900	−380	20	>100

[a]Predicted value
[b]Ceramics

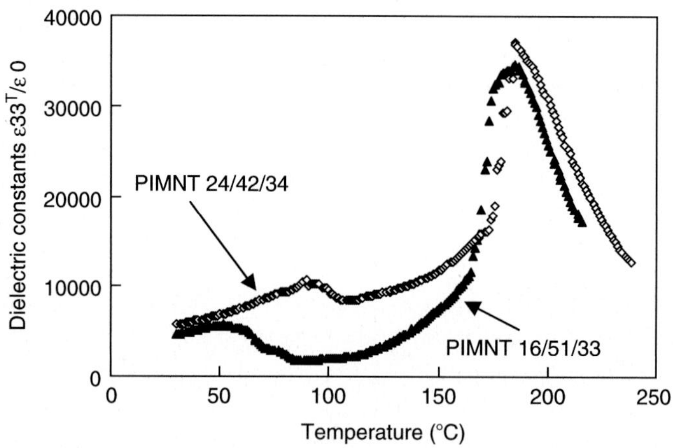

Fig. 9.13. Dielectric properties of PIN-PMN-PT (PIMNT) 24/42/34 and PIN-PMN-PT (PIMNT) 16/51/33 single crystals [24]

transition temperature, T_{rt}. The T_{rt} and T_c of PIN-PMN-PT 24/42/34 and PIN-PMN-PT 16/51/33 single crystal were 89 and 184°C and 52 and 181°C, respectively. The T_{rt} of PIN-PMN-PT 24/42/34 single crystal was about 40°C higher than that of PIN-PMN-PT 16/51/33 single crystal, whereas the T_c of PIN-PMN-PT 24/42/34 single crystal was almost the same as that of PIN-PMN-PT 24/42/34 single crystal.

Table 9.5 lists the dielectric and piezoelectric properties of the PZNT 93/07, PMNT 70/30, PIN-PMN-PT 24/42/34, and the ideal single crystal for medical array transducers. The dielectric constant after poling, ${\varepsilon_{33}}^T/\varepsilon_0$, of the PIN-PMN-PT 24/42/34 was 4,900 at room temperature, which is slightly higher than that of PIN-PMN-PT 16/51/33 single crystal. High dielectric constant of the piezoelectric material is very important for the ultrasonic

Table 9.5. Electrical properties of the obtained and ideal single crystal for medical array transducers

	PZNT 93/07	PMNT 70/30	PIN-PMN-PT 24/42/34	Ideal crystal
Free Dielectric constant $\varepsilon_{33}{}^{\mathrm{T}}/\varepsilon_0$	4,000	4,500	4,900	>4,000
Clamped dielectric constant $\varepsilon_{33}{}^{\mathrm{s}}$	600	1,000	1,000	>1,000
Phase-transition temp. T_{rt} (°C)	92	78	89	>100
Curie temperature T_{c} (°C)	165	155	184	>180
Coupling factor k_{31} (%)	58	49	70	<50
Coupling factor k_{33} (%)	95	93	(90)	>88
Piezoelectric const. d_{33} (pC/N)	2,400	2,000	2,200	>1,800
Coercive field E_{c} (kV cm^{-1})	5	4.6	6.8	>6
Coupling factor k'_{33} (%)	82	82	80	>80
Frequency const. $N'_{33\mathrm{r}}$ (Hz m)	900	1,180	1,200	>1,180
Single crystal size (mm)	50	70	25	>100

medical array transducer application. Recently, the capacitance of the array transducer has been decreasing, because the element size of the array transducer has been decreasing as the number of channels increases. To match electrical impedance between a tiny piezoelectric element and a coaxial cable, higher dielectric constant, which means larger capacitance, is required. A large piezoelectric constant, $d_{33} = 2,200\,\mathrm{pC\ N^{-1}}$, was observed in the PIN-PMN-PT 24/42/34 single crystal, which is almost the same as that of the PIN-PMN-PT 16/51/33 single crystal. The coupling factors $k_{33}{}'$, which are important for the phased array probe, were measured using sliver transducers obtained by the dicing mentioned above. Observation by microscopy revealed no serious cracking and chipping on the sides of the sliver transducers diced from the plane transducer. The transducer showed very clear impedance behavior and high $k_{33}{}'$ value of around 80%, which is almost the same as those of the PZNT 91/9 and PMNT 67/33 single crystal.

Next, the temperature dependence of $k_{33}{}'$ was measured using the sliver transducers of the PIN-PMN-PT 24/42/34 and PIN-PMN-PT 16/51/33 single crystals as shown in Fig. 9.14. The phase-transition temperature T_{rt}, the Curie temperature T_{c}, and the coupling factor $k_{33}{}'$ at room temperature of the PIN-PMN-PT 24/42/34 and PIN-PMN-PT 16/51/33 single crystals used for the measurement were 89°C, 184°C, 80%, and 52°C, 181°C, 82%, respectively. The T_{rt} and T_{c} values were measured using poled plate transducers before dicing to the sliver transducers. The $k_{33}{}'$ values are almost the same. Regarding the temperature dependence of $k_{33}{}'$ of the PIN-PMN-PT 24/42/34 and PIN-PMN-PT 16/51/33 single crystals, $k_{33}{}'$ decreased in a step-like manner from 50 to 80°C, which is similar to the behavior of PZNT single crystals we reported. Because the transducer may be heated up to around 70°C, thermal stability of piezoelectric properties is very important in the medical transducer application. Although $k_{33}{}'$ of the PIN-PMN-PT

16/51/33 single crystals decreased at about 50°C, $k_{33}{}'$ of the PIN-PMN-PT 24/42/34 single crystals hardly changed until 70°C. This result indicates that the PIN-PMN-PT 24/42/34 single crystals had not only excellent piezoelectric properties but also that those properties had good thermal stability due to its high T_{rt}. Moreover, it had good voltage-proof nature due to its high T_{c} and large coercive fields, E_{c}. Therefore, the PIN-PMN-PT 24/42/34 single crystal may become a candidate for the ideal piezoelectric material for the medical array transducer application. In particular, the large coercive fields and high-frequency constants make the PIN-PMN-PT 24/42/34 single crystal useful for high-frequency transducer and multilayer transducer applications [24].

9.5.3 New Single Crystal Manufacturing Process

The Bridgman process and top seeded solution growth (TSSG) process are well established single crystal growth processes [20–23]. However, these processes have several shortcomings, such as Pt crucible cost and composition homogeneity. To overcome these drawbacks, a new approach has been studied. Solid state crystal growth (SSCG) and templated grain growth (TGG) processes have many advantages. The reasonable size and quality of SSCG process for lead-relaxor materials was reported by researchers in Korea and USA [19]. Since these processes have many advantages compared to the conventional single crystal process, they are expected to lead to the reduction in the cost of relaxor single crystals for mass production in the near future.

9.6 Applications

Medical echo diagnostic ultrasound systems use ultrasonic array transducers for their applications. Frequencies used for these applications are 2–20 MHz.

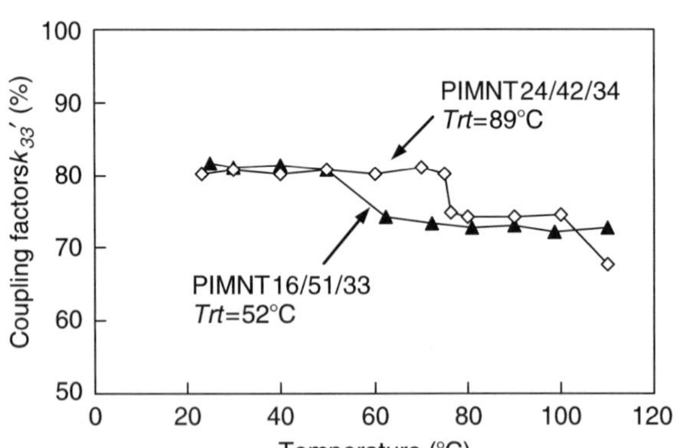

Fig. 9.14. Temperature dependence of the $k_{33}{}'$ of PIN-PMN-PT 24/42/34 (PIMNT) and PIN-PMN-PT (PIMNT) 16/51/33 single crystals [24]

The typical cardiac transducer operates at a center frequency of 2–4 MHz and makes use of a piezoelectric material plate with a size of $15 \times 25\,\mathrm{mm}^2$. Recently, medical echo ultrasound has been greatly improved. The color Doppler technique and tissue harmonic imaging are frequently used for functional diagnosis as well as for ultrasonic tomography (B mode). However, clearer images are required for early and reliable diagnosis of human organs with small tumors. This requires a wider bandwidth of probes as well as improved sensitivity. The key for achieving this is the performance of the piezoelectric material which transmits and receives the echo ultrasound. High dielectric constant type PZT5H ceramics have been mostly used for this application in recent decades. However, recent research has focused on obtaining greater sensitivity and broader bandwidth, and new transducers such as piezoelectric composites have been studied. Single crystals of PZNT 91/9 and PMNT 70/30 have a large d_{33} (>2,000 pC N^{-1}) and k_{33} (>92%), $k_{33}{}'$ (>80%), and g_{33} ($>60 \times 10^{-3}$ Vm N^{-1}), values far larger than those obtained for any PZT ceramic specimens. This is one reason that intensive research on medical transducers has been conducted in various parts of the world during the last 15 years. After the initial reports and presentations by Toshiba in the early 1990s, several organizations in Korea and the USA entered the field [25, 26]. Saitoh et al. reported the first images of B-mode and Doppler-mode obtained by using PZNT 91/09 single crystal for cardiac transducers operating at 3.5 MHz with 96 channels [27]. The echo amplitude of the PZNT probe is about 6 dB higher than that of the PZT probe, and the frictional bandwidth is 30% wider. This means that both the penetration and the resolution of the PZNT probe are superior to those of conventional PZT probes. Large coupling factor $k_{33}{}'$ (>82%), dielectric constant after poling ($K > 3{,}000$), and low acoustic impedance $Z_{33} < 24 \times 10^6\,\mathrm{kg\,m^{-2}s^{-1}}$ make PZNT 91/9 single crystals an excellent transducer material for medical diagnostic applications. At present, the size of PZNT and PMNT single crystal is sufficient for various applications. However, the quality and uniformity of these crystals within and between the wafers, as well as among lots and among manufacturers, are still insufficient [19]. Although it is usually easy to grow PZNT and PMNT single crystals in [111] or [110] directions, it is very difficult to grow them in [001] direction [28]. Since [001] wafers are used for medical transducers, there is always some TiO_2 distribution within wafers. This is one of the causes of large scattering of k_{33}, d_{33}, and dielectric constant within wafer. In addition, the manufacture of high-frequency single crystal transducers, that is, more than 7 MHz, requiring thin plate (less than 150 μm), is very difficult because of the low-frequency constant and low mechanical strength of single crystal. The present cost of single crystal wafers is more than 10 times that of PZT ceramics. Other drawbacks are temperature instability, poor mechanical properties, chipping bydicing, nonuniform bonding strengths between matching layers, electrodes, and crystal surfaces, low clamped dielectric constants, low E_c, depoling effects, low production yield, etc. So, at present, although single crystals have properties which are very attractive compared to those of

PZT ceramics, the supply of single crystal medical array transducers is limited to only a few probe manufacturers [25, 26].

Figure 9.15 shows the first commercialized 2.6 MHz PMNT single crystal medical probe manufactured by Humanscan Co. of Korea. The 6 dB bandwidth of the single crystal probe is about 100%, which is larger than that of conventional PZT probe of 80% [25].

9.7 Conclusions

1. A smart material composition design for realizing superior dielectric and piezoelectric properties in lead-based perovskite compounds has been proposed. The key to the composition design is the selection of T_c and the exact MPB composition of TiO_2 content in the complex lead–perovskite system ceramics $Pb(Sc_{1/2}Nb_{1/2}))O_3$-$Pb(Mg_{1/3}Nb_{2/3})O_3$-$Pb(Ni_{1/3}Nb_{2/3})O_3$-$(Pb, Ba, Sr)(Zr, Ti)O_3$. On the basis of this composition design, a proper composition of complex lead–perovskite system ceramics PSN-PMN-PNN-PBSZT near the MPB was proposed. A small amount of Ba and Sr, less than 5 mol%, for the Pb site was replaced to stabilize the perovskite structure of these ceramics. The $(Pb_{0.962}, Ba_{0.024}\ Sr_{0.024})\ \{(Sc_{1/2}Nb_{1/2})_{0.16}(Mg_{1/3}Nb_{2/3})_{0.17}Pb(Ni_{1/3}Nb_{2/3})_{0.17}Zr_{0.13}Ti_{0.37}\}O_3$ ceramics showed a pure perovskite structure with an average grain size of 5 μm and a density of 8.00 g cm^{-3}. A dielectric constants of $\varepsilon_{33}{}^{T}/\varepsilon_0 = 5,500$, a planar-mode electromechanical coupling factor of $k_p = 72\%$, $d_{33} = 850\ pC\ N^{-1}$ (measured using a d_{33} meter), and $T_c = 153°C$ were obtained. Thus, an effective approach for the smart material composition design of the complex lead–perovskite structure for high-performance piezoelectric materials was verified.

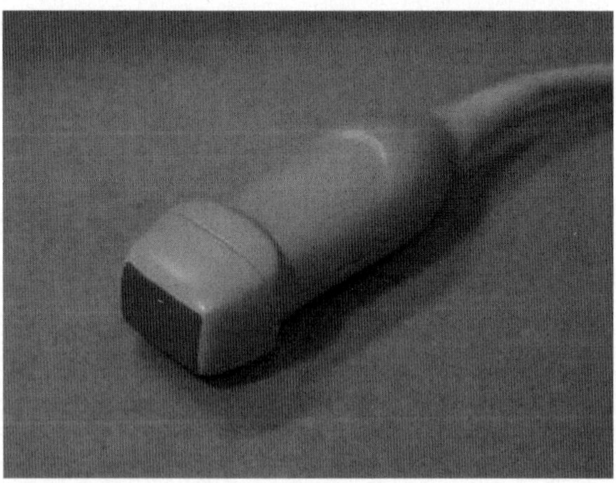

Fig. 9.15. The first commercialized 2.6 MHz PMNT single crystal medical probe (Courtesy of Humanscan Co. of Korea, see [25])

2. Sufficiently large PZNT and PMNT piezoelectric single crystals, that is, 25–80 mm in diameter, for various applications, such as medical transducers, actuators, sonar, and hydrophones, are available.
3. Process capabilities, such as polishing, electroding, dicing, lead-attaching, and the assembling of 2–4 MHz medical cardiac single crystal transducers, are almost established. High quality images on screens of medical echo systems have been confirmed by several echo equipment manufacturers.
4. Quality and uniformity level between and within wafers, lot to lot and among manufacturers of single crystal made by the (solution) Bridgman process are still inferior to those of PZT ceramics.
5. Design and manufacture of high-frequency single crystal array transducers, that is, higher than 7 MHz, is very difficult due to the low frequency constant and mechanical strength of single crystal.
6. New high T_c single crystals of In-doped PMNT, PIN-PMN-PT 24/42/34, would be very promising for the ideal single crystal if single crystals of sufficient size and quality were grown.
7. Some other drawbacks of single crystal array transducers have been pointed out. So, at present, single crystals have yet to win acceptance in the mass production of medical array transducers except at a few probe manufactures. Nevertheless, it is quite possible that single crystals and/or single crystal-like ceramics may become the predominant materials for sophisticated, performance-oriented piezoelectric products in the near future.
8. Regarding future tasks, research on methods of evaluating single crystal wafer quality and uniformity using nondestructive processes is urgently required. Improvement in the uniformity of capacitance, electromechanical coupling factors k_{33}, and piezoelectric constant d_{33} within and between wafers is also a major task for crystal growers. Detailed basic research to determine the mechanisms accounting for the scattering of electrical and acoustical properties should be performed from the viewpoints of composition, domain structure, and defect chemistry. Finally, research on new single crystal materials and single crystal growth technology to solve all the present problems is essential. From the viewpoint of applications, a new dicing process capable of preventing cracking of fragile single crystal transducer is also necessary to make fine pitch array transducers.

References

1. B. Jaffe, *Piezoelectric Ceramics* (Academic Press, London, 1971) p. 200
2. J. Kuwata, K. Uchino, S. Nomura, Jpn. J. Appl. Phys. **21**(9), 1298 (1982)
3. S.-E. Park, T. R. Shrout, IEEE Trans. Ultrason. Ferroelect. Freq. Contr. **44**, 1140 (1997)
4. H. Wang, B. Jiang, T. R. Shrout, W. Cao, IEEE Trans. Ultrason. Ferroelect. Freq. Contr. **51**, 907 (2004)

5. Y. Hosono, Y. Yamashita, H. Sakamoto, N. Ichinose, Jpn. J. Appl. Phys. **42**, 535 (2003)
6. Y. Yamashita, Jpn. J. Appl. Phys. **32**, 5036 (1993)
7. Y. Yamashita, Jpn. J. Appl. Phys. **33**, 4652 (1994)
8. Y. Yamashita, Jpn. J. Appl. Phys. **33**(9B), 5328 (1994)
9. Y. Yamashita, K. Harada, Y. Hosono, S. Natsume, N. Ichinose, Jpn. J. Appl. Phys. **37**(9B), 5288 (1998)
10. Y. Yamashita, Y. Hosono, K. Harada, N. Yasuda, IEEE Trans. Ultrason. Ferroelect. Freq. Contr. **49**, 184 (2002)
11. Y. Yamashita, Y. Hosono, Jpn. J. Appl. Phys. **43**, 6679 (2004)
12. Y. Hosono, Y. Yamashita, IEEE Trans. Ultrason. Ferroelect. Freq. Contr. **52**, 1823 (2005)
13. Y. Yamashita, Y. Hosono, Jpn. J. Appl. Phys. **44**, 7046 (2005)
14. T. Nakamura, Y.J. Shan, P. H. Sun, Y. Inaguma, M. Itoh, Ferroelectrics **219** (1–4), 71 (1998)
15. T. Nakamura, Y. J. Shan, M. Miyata, K. Kobashi, Y. Inaguma, M. Itoh, Kor. J. Ceram. **5**(1), 82 (1999)
16. H. Kanai, O. Furukawa, K. Abe, Y. Yamashita, J. Am. Ceram. Soc. **77**(10), 2620 (1994)
17. O. Furukawa, Y. Yamashita, M. Harata, T. Takahashi, K. Inagaki, Jpn. J. Appl. Phys. **24**(suppl 24–23), 96 (1985)
18. T. Kobayashi, S. Shimanuki, Y. Yamashita, Jpn. J. Appl. Phys. **36**, 6035 (1997)
19. S. Trolier McKinstry, L. E. Cross, Y. Yamashita, *Piezoelectric Single Crystals and Their Application*, Memorial Book of the Late Seung Eek (Eagle) Park, The Materials Research Lab, PSU, PA, 2004
20. http://www.jfe-mineral.co.jp/e_mineral/seihin/eseihin01.html
21. http://www.hcmat.com/
22. http://www.sinocera.net/en/crystal_pmn.asp
23. http://www.microfine-piezo.com/
24. Y. Hosono, Y. Yamashita, K. Hirayama, N. Ichinose, Jpn. J. Appl. Phys. **44**, 7037 (2005)
25. http://humanscan.co.kr/e_index/e_index.htm
26. http://www.medical.philips.com/main/products/ultrasound/products/technology/purewave.html
27. S. Saitoh, T. Takeuchi, T. Kobayashi, K. Harada, S. Shimanuki, Y. Yamashita, Jpn. J. Appl. Phys. **38**, 3380 (1999)
28. N. Setter, *Piezoelectric Materials in Devices.* (EPFL Swiss Federal Institute of Technology, Lausanne, Switzerland, 2002) ISBN-2-9700346-0-3, p. 455

10

High-Power Ultrasound Transducers for Therapeutic Applications

D. Cathignol

10.1 Introduction

It is not possible to examine all the innovations relating to the generation of high-power sound waves using composite materials, because of the multitude of applications (sonar, welding, cleaning, sonochemistry, etc.) and the extremely wide range of frequencies used (50 kHz–10 MHz). In this chapter, we will focus on the medical field. Nonetheless, a large number of problems or concepts raised here can be extended and adapted to industrial or military applications. After demonstrating the advantages that the composite material provides for the design of ultrasonic sources, we will examine the specific problems linked to continuous and pulsed ultrasound sources.

10.2 Piezocomposite Technology: Its Contribution to Ultrasound Power Generation

Piezocomposite materials ("piezocomposites") with 1–3 connectivity are made up of small, fine columns of piezoelectric ceramic embedded in a polymer material [1]. This structure enhances the electromechanical performance across the thickness of the plate. The physical properties of piezocomposites, particularly their dielectric, elastic and piezoelectric characteristics, can be modelled [2,3]. For a 1–3 piezocomposite, the coupling coefficient obtained is higher than that of the bulk ceramic "piezoceramics", thus resulting in better energy conversion. It is known that the mechanical losses of composite 1–3 materials are larger than those of the piezoceramics generally used in this type of application. Thus, the electroacoustical conversion efficiency of a transducer produced by piezocomposite technology is in theory – and particularly in the simplest cases – lower than that of a transducer produced by piezoceramic technology. However, in many applications, the efficiency of the transducer is not the only objective to be satisfied and piezocomposite technology allows these other objectives to be achieved. The particularly interesting properties of 1–3 piezocomposites are as follows:

- The lower acoustic impedance, around 8–12 MRayl, facilitates the transfer of energy to water or tissue.
- The higher value of the thickness mode coupling coefficient k_t (0.55–0.65) increases the energy conversion and widens the bandwidth of the emitting signal.
- The addition of a front layer is possible if this layer is designed in such a way that its thermal dilatation coefficient is close to that of the piezocomposite material. This layer can fulfil different functions, such as acoustic matching, electric insulation, chemical resistance of the transducer's radiating surface, and especially the dissipation of heat.
- Judicious selection of the polymers' thermomechanical properties allows the conditioning of the active surface and facilitates the production of transducers of complex shape.
- The very strong anisotropy of piezocomposite materials allows the reduced propagation of parasitic modes other than the thickness mode. This property facilitates the design of array transducers since, in numerous cases, simple patterning on the electrode at the surface of the same substrate allows the elements of the array to be acoustically independent. The hypothesis that all the points at the surface of the transducers vibrate in piston mode, generally supposed for modelling, is true in the case of piezocomposites in contrast with piezoceramics.

Thus, piezocomposite technology offers a set of solutions to the transducer designer for the development of high-power ultrasonic sources. The development of transducers for ultrasonic therapy applications using piezocomposite technology requires much research on polymer materials in terms of the stability of their mechanical characteristics over a wide range of temperatures. It also requires the design of multilayered structures (front layer/piezocomposite/backing layer) to facilitate acoustic energy transfer and thermal draining. In the following paragraphs, we review theoretical and experimental results obtained in the field of continuous waves ("cw") and pulsed sources.

10.3 High-Power cw Sources

10.3.1 Generation of High-Intensity Ultrasonic Waves

Recent progress achieved in the design of high-power transducers permits piezocomposite transducers to deliver acoustic power levels compatible with therapeutic applications (mean intensity of up to 15–20 W cm^{-2} on the front face of the transducer) with efficiency of around 60–70%. Excessive overheating constitutes the first source of limitations. To reduce this effect, it is necessary to minimise the losses in the transducer, by either optimising the efficiency of electrical energy conversion into acoustic energy at the working frequency or by draining the thermal energy to an exterior environment. Recent tests performed at the National Institute of Health in France and at the University of Michigan, on piezocomposite technology transducers manufactured by IMASONIC (Besançon, France), showed efficiencies of around 60% for transmitted acoustic intensity of the order of $10\,\mathrm{W\ cm^{-2}}$ [4]. These efficiency and acoustic power levels are retained for the duration of the excitation period (several minutes) and are reproducible.

10.3.2 Predictability of Radiation Pattern

The usual models used to predict the field radiated by the active surface of a transducer propose the hypothesis that the latter vibrates in a homogeneous fashion, i.e., the amplitude and phase of displacement are the same at any point of the emitting surface. This hypothesis is not valid in the case of a piezoceramic. In fact the vibrating behaviour in thickness mode of such transducers is disturbed by the existence of the plate and the surface waves propagating from the edges of the ceramic layer. For example, Fig. 10.1 shows that the on- axis pressure distribution differs greatly from the prediction of the Rayleigh integral in the case of a spherical shell made of piezoceramic material. In contrast, it is very close to the theoretical curve when the piezocomposite transducer is used. Chapelon et al. in [5] demonstrated the predictability of the radiation pattern emitted by a truncated annular array-type piezocomposite transducer where the acoustic field on the axis was measured for several electronic focusing configurations and compared to a theoretical simulation. Despite the probable effects of several factors that were not taken into account in the simulation (nonlinear effects of the propagation, directivity of the hydrophone, etc.), good radiation predictability is evident, allowing for the development of transducers with complex geometries.

10.3.3 Generation of Power in a Wide Frequency Range

In contrast to piezoceramics, piezocomposites allow high-intensity ultrasound beams with wide frequency bandwidths to be generated. This property may be used when designing phased array probes to reduce the grating lobes in annular or linear arrays while maintaining the number of elements constituting the probe [6]. A second benefit of widening the frequency bandwidth of acoustical waves is illustrated in the cavitation effects. The formation of air bubbles by cavitation causes degradation in the efficiency of a treatment and can bring about unwanted lesions outside the zone to be treated. The use of a signal phase modulated by a pseudorandom code, instead of a single-frequency emission, reduces the rectified diffusion effect and so the increase in the bubble size which is responsible for the inertial cavitation. The results are impressive: cavitation was successfully reduced by a factor of 50 for a mean acoustic

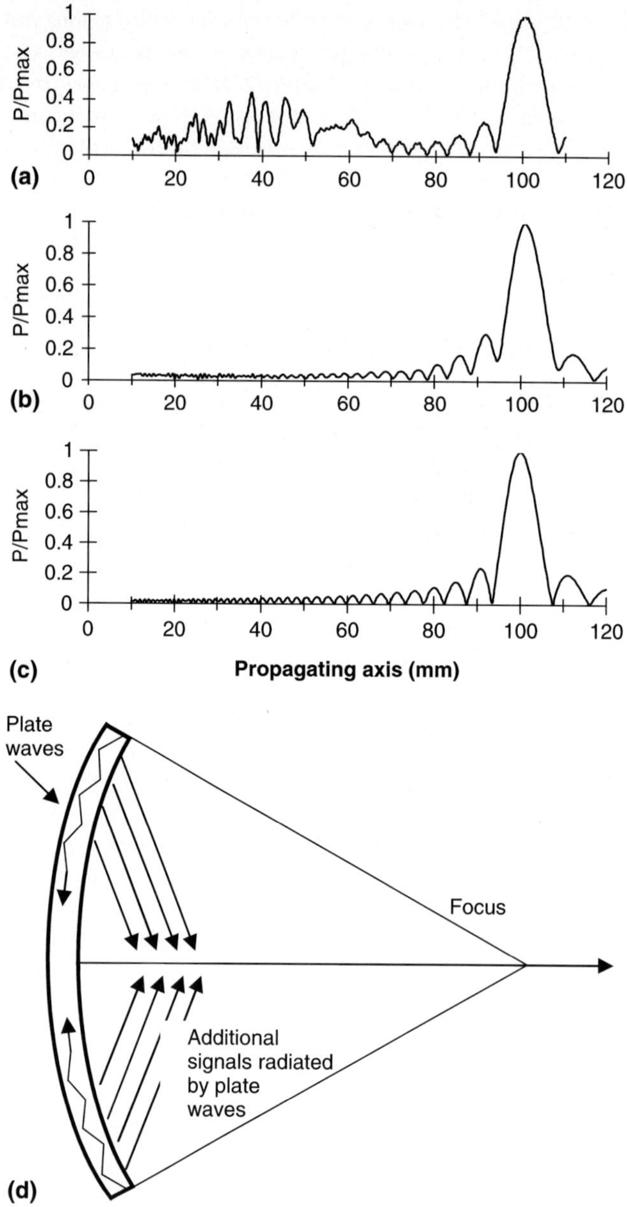

Fig. 10.1. Comparison of the pressure distribution along the axis between piezoceramic (**a**) and piezocomposite (**b**). The former differs greatly from the theoretical value (**c**). Plate waves and surface waves propagating from the edges are responsible for the grating lobes in **a**

intensity of 4.6 W cm^{-2}. The intensity limitation due to the cavitation threshold is thus relaxed, and high-intensity focused ultrasound treatments can be performed with greater precision in terms of aim and better protection of the surrounding tissues [7].

10.4 Progress in High-Pressure Pulse Sources

The second part of this chapter deals with high-pressure sources based on piezocomposite materials. The main application is lithotripsy but other applications are currently being investigated, including the treatment of tendonitis and the destruction of the non-tumour tissue. Nowadays new clinical concepts in lithotripsy demand small shock heads while actual ultrasonic shock heads are very cumbersome (50 cm in diameter) [8]. Reducing their size is possible only if the pressure generated at the surface can be increased so that the pressure at the focus remains the same. Two areas for possible advances are currently being investigated: the identification of new materials, and the development of new methodologies.

10.4.1 Experimental Identification of a Piezoelectric Material for High-Pulse Pressure Generation

Comparative trials conducted on piezoceramic and piezocomposite materials have shown that the latter allow much greater surface pressure to be generated prior to destruction [9]. The transducers tested had a diameter of 20 mm and a resonance frequency of 720 kHz: a frequency close to that generated by electrohydraulic lithotripters. In order to avoid constraints resulting from welding connection wires, power supply was ensured by means of flexible springs placed around the edges of the transducers. The pressure generated was progressively increased by increments of 0.5 MPa after 10^6 shocks had been successfully applied to the transducer. The piezoceramic transducers, whether hard or soft materials, are destroyed by being broken into several pieces for surface pressures of around 2 MPa (Fig. 10.2a). By contrast, piezocomposite transducer materials are destroyed by piercing the material and generating an electric arc (Fig. 10.2b) for pressures of at least 4 MPa. The cause of destruction was essentially judged to be surface waves in the case of solid materials, and localised overstressing which destroyed the active plots in the case of piezocomposite materials [10].

10.4.2 New Methodologies for the Generation of High Surface Pressure

10.4.2.1 Stacked Transducers

Although it has been suggested and demonstrated experimentally that the use of composite materials makes it possible to generate more significant surface

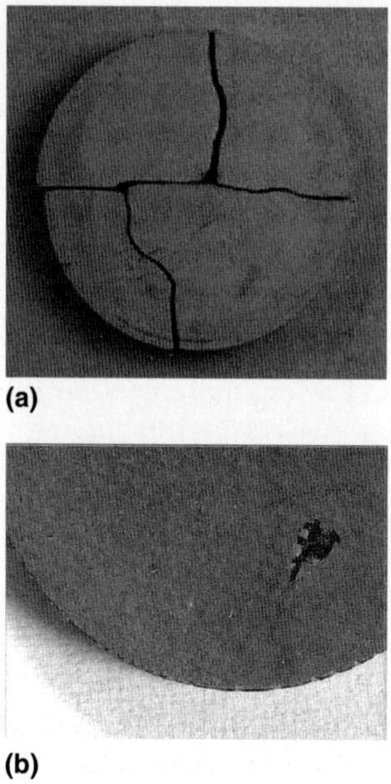

Fig. 10.2. Piezoceramic and piezocomposite transducers after aging tests. Piezoceramic brakes in several pieces while piezocomposite are destroyed by piercing

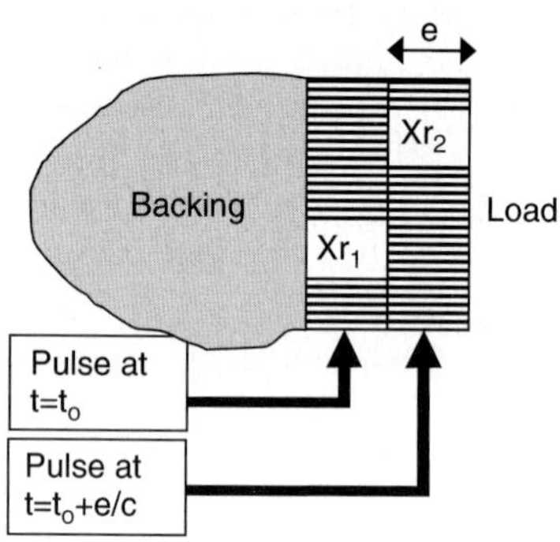

Fig. 10.3. Principle of a sandwiched transducer

pressures because of the reduced significance of the radial modes, the maximum pressure does not exceed ∼3 MPa for long ageing tests. However,these values still appear to be too weak for the creation of a shock-wave generator of very small diameter. For this reason, another method was investigated: stacking the transducers [11]. When excited at particular time intervals, the pressure waves generated by each element of the stacked transducers reinforce each other at the transducer-to-load interface. For example, in Fig. 10.3, two stacked piezoelectric transducers are shown. If the first transducer is excited at time t_0, the pressure wave reaches the transducer-to-load interface at time

$t_0 + e/c$, where e is the transducer thickness and c the ultrasound velocity. At this instant, this wave is transmitted into the load. If the second transducer is excited at this same instant $t_0 + e/c$, then the resultant pressure wave is instantaneously transmitted into the load and the two pressure waves reinforce each other. Feasibility was demonstrated using prototypes consisting of two stacked plates, each 30 mm in diameter, acoustically matched by a quarter-wavelength plate. Two high-voltage impulse generators are used to drive the stacked transducers. Each transducer is charged by its neighbour, which in turn is charged by the output impedance of the generator that controls it. In these conditions, a loss of energy leads to a reduction in pressure. A theoretical study showed that this reduction in pressure is greatest when the generator's output impedance is equal to that of the transducer, but never exceeds 10%. Using this concept, it is therefore possible to generate a number of limited pressure shocks (10^6) of around 6–7 MPa. However, endurance tests show that placing them next to one another limits the effective pressure to around 4 MPa. Following this work, the Richard Wolf GmbH company designed a compact shock head, thanks to the use of stacked transducers. The latest results indicate that a shell with a diameter of 28 cm has an in vitro fragmentation capacity equivalent to that of a traditionally designed shell with a diameter of 50 cm [12].

10.4.2.2 Electrical Prestrain

An alternative with the same aim of increasing the transducers' surface pressure was proposed, called prestrain. When a piezoelectric material is subjected to an electrical field, its thickness varies. When the field is in the same or opposite direction to the polarisation, a compressive or an extended wave is generated in the adjacent medium. For shock-wave generation, because a very strong compression wave has to be generated, the piezoelectric material is subjected to a major extension. Piezoelectric material obtained by sintering, which constitutes the active piezocomposite plots, withstands compression very well but extension very poorly. The principle of prestraining the material [13] is as follows (Fig. 10.4). First, an electric field in the opposite direction to that of the polarisation vector is applied, causing the material to contract, and second, a transient field in the same direction as the polarisation vector is superimposed, generating the compression wave. The resulting extension, when the transient field is applied, compared with what would have been obtained using conventional excitation, is reduced by the value of the compression induced by the first field applied. The prestrain may be applied in three different manners as illustrated in Figs. 10.5a–c. Applying an electrical field in the opposite direction of the polarisation vector leads to a depolarisation of the material, even in hard materials. Thirdly, the voltage step function is symmetric with respect to the zero voltage and returns to the base line with a large time constant. This part should contribute to repole the transducer and to allow an increase of the electrical prestrain. The compressive wave

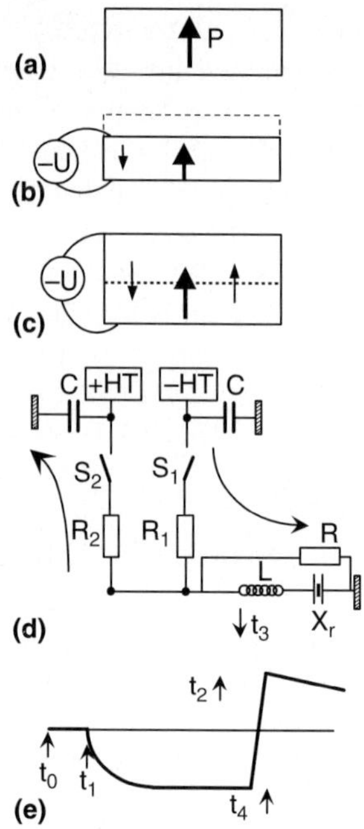

Fig. 10.4. Principle of the electrical prestrain (**a**–**c**). The electrical circuit (**d**) and the chronogram (**e**)

is obtained by superimposing a pseudo-step voltage function. The electrical driving circuit includes positive and negative high-voltage power supplies.

We shall now explain the working principle. At t_0 the two switches S_1 and S_2 are switched off and no voltage is applied to the transducer. At t_1, S_1 is closed and the voltage of the positive power supply is applied on the transducer with the time constant R_1CX_r, where CXr is the free capacitance of the transducer. As the resulting electrical field is in the opposite direction to the dipole orientation, a compression takes place inside the transducer. At t_2, S_1 is reopened and a few microseconds later, at time t_3, S_2 is switched on for a duration equal in length to the time of flight (1.4 μs) in the material. The optimal ramp t_3–t_4 is obtained using the correct values for R, C and L. An extensive stress is simultaneously generated in the material. After reopening S_2 at t_4, the transducer voltage returns to zero with the time constant RCX_r.

The acoustic pressure generated for the three methodologies in function of the applied voltage is given in Fig. 10.5. In the first two, the piezoelectric material depoles for an electrical voltage of about 2,000 V. In the third, the pressure increases linearly as a function of the voltage, up to a maximum of 6,000 V. At this value it is impossible to increase the prestrain voltage. Nevertheless, it is possible to generate more pressure by increasing the positive voltage. Comparative studies into ageing have shown that prestressing allows

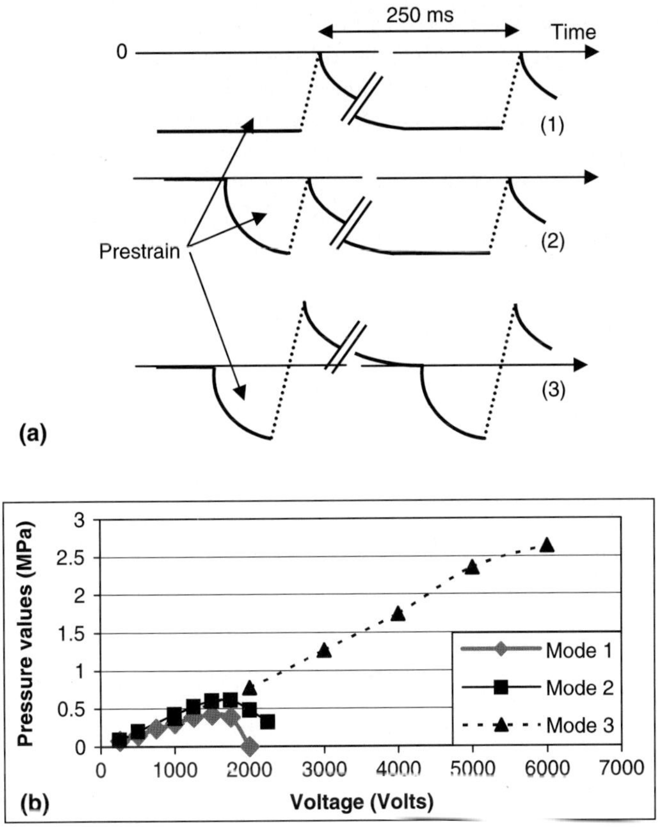

Fig. 10.5. The three different modalities tested (**a**). Only the third one does not depolarise the material. Maximum pressure values obtained for the three different modalities (**b**)

the surface pressure to be increased by around 40% prior to destroying the transducer.

10.4.2.3 Application to the Development of a Compact Shock Head

A spherical bowl was manufactured using piezocomposite material (Imasonic, France). This bowl, 12 cm in diameter, is focused at 12 cm and has a natural resonance frequency of 360 kHz. To generate a quasi-unipolar pressure wave, the backing has approximately the same acoustic impedance as the piezocomposite material and the matching layer has a thickness of $\lambda/2$ [14].

The acousto-electrical coefficient is $5\,\mathrm{kPa\,V^{-1}}$. For use over a long period, the voltage applied is maintained at a level equal to or less than 6 kV, corresponding to a surface pressure of 3 MPa. For this value, the pressure at the focus (measured with an optical fibre 100 μm in diameter and with a bandwidth of 30 MHz) is 60 MPa. The focal zone measured at −3 dB is 15 mm long in the propagating axis and ±1.5 mm in the off- axis. The disintegration efficacy was tested on standard kidney stone-mimicking plaster balls, 15 mm in diameter, made with calcium sulphate with hollow glass microspheres 100 μm in diameter (HMT, Kreuzlingen, Switzerland). Similar to the case of an electohydraulic generator, for a focal pressure of 60 MPa, the number of shocks to disintegrate the plaster balls is about 150 which is considered to be the golden standard.

10.5 Conclusion

The evolution of ultrasonic therapy and the growing interest that it arouses make it increasingly necessary to take into account the specific nature of each application. The applications of piezocomposite technology, which have been discussed, show the significance of this technology. This is particularly obvious if we consider all the needs that must be satisfied in the design of a transducer for therapeutic applications. The specific properties of composite materials and the combination of these properties bring to transducer design a flexibility and a degree of optimisation which are often lacking in designs based on ceramic materials. This flexibility and this degree of optimisation form the basis for new solutions, thus responding to the diverse requirements of therapeutic ultrasound.

References

1. W.A. Smith, B.A. Auld. IEEE Trans. Ultrason. Ferr. **38**(1), 40–47 (1991)
2. J.A. Hossack, G. Hayward,IEEE Ultrason. Ferr. **38**(6), 618–629 (1991)
3. W. Cao, M. Zhang, L.E. Cross, J. Appl. Phys. **72**(12), 5814–5821 (1992)
4. J.Y. Chapelon, D. Cathignol, C. Cain, E. Ebbini, J.U. Kluiwstra, O.A. Sapozhnikov, G. Fleury, R. Berriet, Ultrasound Med. Biol. **26**(1), 153–159 (2000)
5. J.Y. Chapelon, F. Prat, A. Arefiev, D. Cathignol, R. Souchon, Y. Theillère, IEEE Ultrason. Symp. Proc. 1265–1268 (1996)
6. F. Dupenloup, J.Y. Chapelon, D. Cathignol, O.A. Sapozhnikov, IEEE Ultrason. Ferr. **43**(6), 991–998 (1996)
7. J.Y. Chapelon, F. Dupenloup, H. Cohen, P. Lenz, IEEE Ultrason. Ferr. **43**(4), 623–625 (1996)
8. P.H. Thibault. Ann. Urol. **20**, 20–25 (1986)
9. J.P. Sferruzza, A. Birer, A. Matias, Y. Theillère, D. Cathignol, Sens. Actuators A Phys. **88**, 146–155 (2001)
10. E. Closset, P. Trompette, A. Birer, D. Cathignol, IEEE Ultrason Ferr. (to be published)
11. J.P. Sferruzza, A. Birer, D. Cathignol, Ultrasonics **38**, 965–968 (2000)

12. T. Dreyer, W. Krauss, E. Bauer, R.E. Riedlinger, in *Proceedings of IEEE Ultrasonics Symposium 2000*, San Juan, Puerto Rico, vol. 2, 2000, pp. 1239–1242
13. A. Birer, M. Ghohestani, D. Cathignol, IEEE Trans Ultrason Ferr. **51**(7), 879–886 (2004)
14. J.P. Sferruzza, F. Chavrier, A. Birer, D. Cathignol, Ultrasound Med. Biol. **49**(2), 177–183 (2002)

11

Piezoelectric Motors and Transformers

K. Uchino

11.1 Introduction

Piezoelectric ceramics forms a new field between electronic and structural ceramics [1–4]. Application fields are classified into three categories: positioners, motors, and vibration suppressors. From the market research result for 80 Japanese component industries in 1992, tiny motors in the range of 5–8 mm are required in large numbers for office and portable equipment; the conventional electromagnetic (EM) motors are rather difficult to produce in this size with sufficient energy efficiency, while Silicon MEMS actuators are too small to be used in practice. Piezoelectric ultrasonic motors whose efficiency is insensitive to size are superior in the millimeter motor area. The manufacturing precision of optical instruments such as lasers and cameras, and the positioning accuracy for fabricating semiconductor chips are of the order of 0.1 μm which is much smaller than the backlash of the EM motors. Vibration suppression in space structures and military vehicles also require compact but mighty piezoelectric actuators.

On the other hand, one of the bulkiest components in information processing equipment (such as laptop computers) is the power supply component, especially the electromagnetic transformer used in the power supply. Losses such as skin effect, thin wire and core losses of the electromagnetic transformer increase rapidly as the size is reduced. Thus, high efficiency, small size, and absence of electromagnetic noise are some of the attractive features of piezoelectric transformers (PT), making them more suitable for miniaturized power inverter components used in lighting up the cold cathode fluorescent lamp behind a color liquid crystal display or generating high voltage for air-cleaners [5].

The advantages of piezoelectric devices over electromagnetic types are summarized (see Fig. 11.1):

(a) More suitable to miniaturization – As the stored energy density is larger than that of an EM type, is 1/10 smaller in volume, and weight can be achieved

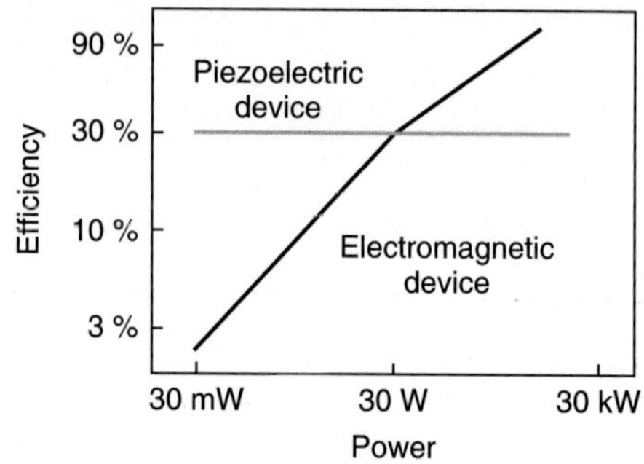

Fig. 11.1. Efficiency vs. power relation for electromagnetic and piezoelectric motors

(b) No electromagnetic noise generation – As magnetic shielding is not necessary, a compact design can be achieved
(c) Higher efficiency – As the efficiency is insensitive to the size, the piezo-device is more effective in the power range lower than 30 W
(d) Non-flammable – As the piezo-device is safer for the overload or the short-circuit at the output terminal

This article reviews recent developments of piezoelectric actuators and transformers in terms of new materials, device designs, drive/control techniques, modeling, and applications.

11.2 Transducer Materials

The primary research focus is still put on the enhancement of induced strain magnitude (real part). However, the research trend is gradually shifting to the improvement of strain hysteresis or loss (imaginary part) of the piezoelectric materials, which is more important for the actual commercialization of the piezoelectric motors. Due to the social demand, studies on Pb-free piezoelectric materials have also begun. Optomechanical coupling devices are a new trend, targeting the coming optical information age. Recent seven topics are reviewed here: high strain materials, high temperature materials, phase-change materials, polymers, high-power piezoelectrics, Pb-free materials, and photostrictive materials.

11.2.1 High Strain Materials

As the enhancement of the induced strain level is a primary target, single crystals with a better capability for generating larger strains are being used these days. In 1981, Kuwata et al. firstly reported an enormously large electromechanical coupling factor k_{33} = 92–95% and piezoelectric constant $d_{33} = 1,500\,\mathrm{pC\,N^{-1}}$ in solid solution single crystals between relaxor and normal ferroelectrics, $Pb(Zn_{1/3}Nb_{2/3})O_3$–$PbTiO_3$ [6, 7]. This discovery has not

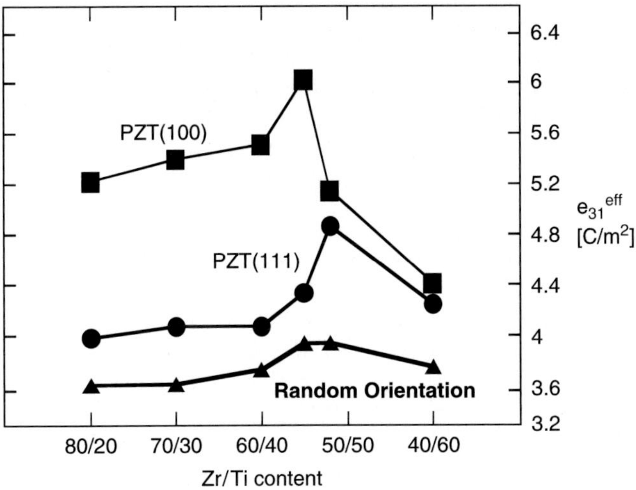

Fig. 11.2. Experimental results for the piezoelectric e_{31} constants in epitaxially grown PZT thin films with a rhombohedral composition 70/30 [11]

been used for more than 10 years until high k materials have gained attention in medical acoustics. These data have been reconfirmed, and improved data were obtained recently, aiming at medical acoustic applications [8, 9]. The strains as large as 1.7% can be induced practically for the PZN-PT solid solution single crystals. It is notable that the highest values are observed for a rhombohedral composition only when the single crystal is poled along the perovskite [001] axis, not along the [111] spontaneous polarization axis.

A series of theoretical calculations made on perovskite type ferroelectric crystals suggests that large d and k values in similar magnitudes to PZN-PT can also be expected in PZT. Crystal orientation dependence of piezoelectric properties was phenomenologically calculated for compositions around the morphotropic phase boundary of PZT [10]. The maximum longitudinal piezoelectric constant d_{33} (4–5 times enhancement) and electromechanical coupling factor k_{33} (more than 90%) in the rhombohedral composition were found to be at an angle of around 57°, canted from the spontaneous polarization direction [111], which corresponds roughly to the perovskite [100] axis. Figure 11.2 shows the experimental results for the piezoelectric e_{31} constants in epitaxially grown PZT thin films [11]. Note that the maximum e_{31} constant can be obtained in the rhombohedral phase near the morphotropic phase boundary, and in the [100] specimen (rather than in the [111] specimen). Damjanovic et al. reported similar supporting results [12].

11.2.2 High Temperature Materials

Applications such as diesel injection valve control require actuators used at an elevated temperature of around 150°C. This requirement urges to develop high Curie temperature piezoelectric materials. Shrout et al. reported the solid solutions of $BiScO_3$, $BiInO_3$ and $BiYbO_3$ with $PbTiO_3$ with the morphotropic phase boundary Curie temperature more than 450°C [13].

To the contrary, cryogenic actuator materials are required for space applications such as "Hubble" telescope adjustment. Mulvihill et al. reported PZT-based multilayer actuators for this purpose [14].

11.2.3 Phase-Change Materials

Concerning the phase-change-related strains, polarization induction by switching from an antiferroelectric to a ferroelectric state has been proposed [15]. In the field-induced strain curves for the lead zirconate stannate based $Pb_{0.99}Nb_{0.02}((Zr_xSn_{1-x})_{1-y}Ti_y)_{0.98}O_3$ system, the longitudinally induced strain reaches up to 0.4%, which is larger than that expected in normal polycrystalline piezostrictors or electrostrictors. A rectangular-shape hysteresis, referred to as a "digital displacement" because of the two on/off strain states, and a "shape memory" effect are observed in these samples. Once the ferroelectric phase has been induced, the material will "memorize" its ferroelectric state even under zero-field conditions, although it can be erased with the application of a small reverse bias field [16].

11.2.4 Polymer Actuators

Polyvinylidene difluoride-trifluoroethylene (PVDF-TrFE) copolymer is a well-known piezoelectric, which has been popularly used in sensor applications such as keyboards. Zhang et al. reported that the field induced strain level can be significantly enhanced up to 5% by using a high-energy electron irradiation onto the PVDF films [17].

11.2.5 High-Power Piezoelectrics

In order to obtain a large output mechanical power, the piezoelectric materials are driven under a high vibration level, namely at the mechanical resonance under a relatively large AC electric field. This causes heat generation as well as a significant degradation in piezoelectric properties. Therefore, the high-power devices such as ultrasonic motors and transformers require a very "hard" piezoelectric with a high mechanical quality factor Q_m, to suppress heat generation. The Q_m is defined as an inverse value of the loss factor, tan δ_m. It is also notable that the actual mechanical vibration amplitude at the resonance frequency is directly proportional to this Q_m value. The vibration velocity dependence of the piezoelectric constant, permittivity, elastic compliance and electro-mechanical coupling factor was studied for PZT based samples [18]. A significant decrease in Q_m is usually observed above a certain critical vibration level, which is the limit of the output vibration energy for that particular PZT. Even if the input electrical energy is increased further, the additional energy will be converted only into heat.

At present, the highest value $1\,\mathrm{m\,s^{-1}}$ of the maximum vibration velocity (defined at 20°C rise from room temperature) can be obtained in $\mathrm{Pb(Zr,Ti)O_3}$-$\mathrm{Pb(Mn_{1/3}X_{2/3})O_3}$ (X = Sb, Nb)-doped with rare-earth ions (Yb, Eu) [19, 20]. This velocity is more than three times larger than that in commercially available hard PZTs, leading to a power density capability more than ten times. The loss mechanisms in piezoelectrics have been gradually clarified [21]. The heat generation under off-resonance is basically originated from intensive dielectric loss, while under resonance it is caused by intensive mechanical loss, where both intensive (experimentally obtained apparent) dielectric and mechanical losses are inter-related with extensive (material's proper) dielectric and mechanical losses. Further, the resonance loss ${Q_\mathrm{m}}^{-1}$ at a small vibration velocity is mainly determined by the extensive mechanical loss, and with increasing vibration velocity, the extensive dielectric loss contribution significantly increases. Microscopically, under a small electric field the origin of loss will be primarily non-180° domain wall motion (extensive mechanical loss) and the contribution of 180° domain wall motion (extensive dielectric loss) will significantly increase above a certain vibration level.

Relating to the hysteresis in piezoelectrics, Smith proposed a computer simulation to evaluate the polarization and strain hysteresis under an electric field cycle, one example of which is shown in Fig. 11.3 [22].

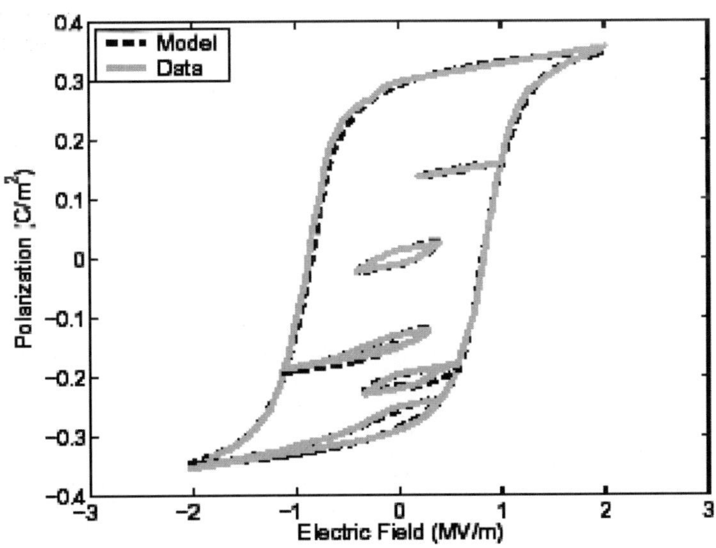

Fig. 11.3. Comparison among the model calculation (*dashed*) and experimental polarization (*solid*) as a function of applied electric field for PZT 5A [22]

11.2.6 Lead-Free Piezoelectrics

Japanese and European community may experience governmental regulation on the lead usage in these 20 years. Pb (lead)-free piezoceramics have started to be developed after 1999. Lead-free piezoelectric materials are classified into (1) piezoelectric single crystals, e.g., langasite ($La_3Ga_5SiO_{14}$) and lithium tetraborate ($Li_2B_4O_7$), (2) tungsten bronze-typed ferroelectric ceramics, (3) bismuth layer structured ferroelectric ceramics, (4) potassium–sodium niobate, $KNbO_3$–$NaNbO_3$ systems, and (5) other perovskite-typed ferroelectric ceramics. Figure 11.4 shows statistics of various lead-free piezoelectric ceramics. The share of the papers and patents for bismuth compounds (bismuth layered type and $(Bi, Na)TiO_3$ type) exceeds 61%. This is because bismuth compounds are easily fabricated in comparison with other compounds. The most recent report on giant strain in a single crystal $BaTiO_3$ is also noteworthy [23]. Refer to the most recent review paper by T. Takenaka [24].

11.2.7 Photostrictive Actuators

A photostrictive actuator is a fine example of an intelligent material, incorporating "illumination sensing" and self-production of "drive/control voltage" together with final "actuation." In certain ferroelectrics, a constant electromotive force is generated upon exposure to light, and a photostrictive strain results from the coupling of this bulk photovoltaic effect with inverse piezoelectricity. A bimorph unit has been made from PLZT 3/52/48 ceramic doped with slight addition of tungsten [25]. The remnant polarization of one PLZT layer is parallel to the plate and in the direction opposite to that of the other

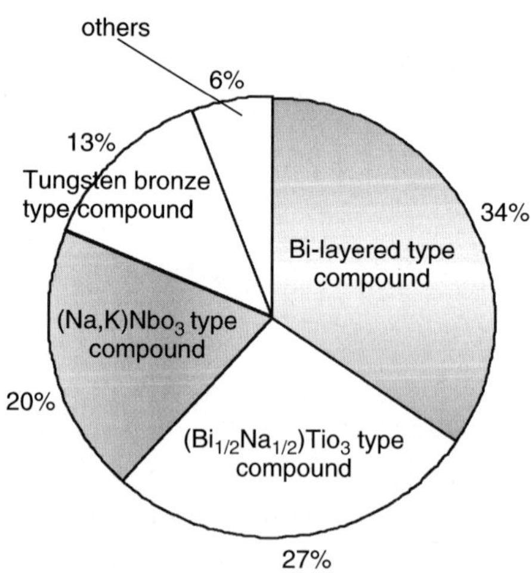

Fig. 11.4. Statistics of various lead-free piezoelectric ceramics (total number of patents and papers is 102)

Fig. 11.5. A photo-driven micro walking device, designed to begin moving by light illumination. It is simple in structure, having neither lead wires nor electric circuitry

plate. When a violet light is irradiated to one side of the PLZT bimorph, a photovoltage of 1 kV mm^{-1} is generated, causing a bending motion. The tip displacement of a 20 mm bimorph 0.4 mm in thickness was 150 μm, with a response time of 1 s.

A photodriven micro walking device, designed to begin moving by light illumination, has been developed (shown in Fig. 11.5) [26]. It is simple in structure, having neither lead wires nor electric circuitry, with two bimorph legs fixed to a plastic board. When the legs are irradiated alternately with light, the device moves like an inchworm with a speed of 100 μm min^{-1}.

In pursuit of thick film type photostrictive actuators for space structure applications, in collaboration with Jet Propulsion Laboratory, Penn State investigated the optimal range of sample thickness and surface roughness dependence of photostriction. 30 μm thick PLZT films exhibit the maximum photovoltaic phenomenon [27].

11.3 Actuator Designs

Amplification of displacements is one of the essential development issues in ceramic actuators. There are two categories: amplification mechanisms in terms of space (hinge lever, flextensional), and in terms of time (inchworm and ultrasonic motor).

11.3.1 Actuator Components

Two of the most popular actuator designs are multilayers and bimorphs. The multilayer, in which roughly 100 thin piezoelectric/electrostrictive ceramic sheets are stacked together, has advantages such as low driving voltage (100 V), quick response (10 μs), high generative force (100 kgf) and high electromechanical coupling. But the displacement in the range of 10 μm is not sufficient for some applications. This contrasts with the bimorph, consisting of multiple piezoelectric and elastic plates bonded together to generate a large bending displacement of several hundred μm, but the response (1 ms) and the generative force (100 gf) are low.

A multilayer actuator with interdigital internal electrodes has been developed by Tokin [28]. In contrast to the conventional electrode configuration, line electrodes are printed on piezoelectric green sheets, and are stacked so that alternating electrode lines are displaced by one-half pitch. This actuator generates motions at right angles to the stacking direction using the longitudinal piezoelectric effect. Long ceramic actuators up to 74 mm in length were manufactured, which could generate the displacement up to 55 μm. A three-dimensional positioning actuator with a stacked structure has been proposed by PI Ceramic, in which shear strain is utilized to generate x and y displacements [29].

A monomorph device has been developed to replace the conventional bimorphs, with simpler structure and manufacturing process. The principle is a superposed effect of piezoelectricity and semiconductivity [30]. A monomorph plate with 30 mm in length and 0.5 mm in thickness can generate 200 μm tip displacement, in equal magnitude of that of the conventional bimorphs. The "rainbow" actuator by Aura Ceramics [31] is a modification of the abovementioned semiconductive piezoelectric monomorphs, where half of the piezoelectric plate is reduced so as to make a thick semiconductive electrode to cause a bend.

A market research held in 1992 revealed that the customers' demands for actuators can be satisfied neither with multilayers nor bimorphs. The demands are 100 μm in stroke, 100 N in force with 100 μs in response speed. A composite actuator structure called the "cymbal" has been developed at Penn State to provide characteristics intermediate between the multilayer and bimorph actuators; this transducer exhibits an order of magnitude larger displacement than the multilayer, and much larger generative force with quicker response than the bimorph [32,33]. A Cymbal with a thickness 2 mm and a diameter 12 mm can generate a displacement up to 100 μm. The Cymbal has been applied to make a miniaturized laser beam scanner. Ring-morphs are another alternative to a displacement amplification mechanism [34]. By eliminating unnecessary ceramic center portion of the bimorph/unimorph, 30–40% enhancement of the field-induced displacement was realized.

11.3.2 Ultrasonic Motors

Modified from the original Sashida's surface wave type, Canon and Seiko Instruments have successfully commercialized camera automatic focusing mechanisms and date change mechanisms in a wrist watch, respectively.

However, a significant problem in miniaturizing this sort of traveling wave motor is found in the further miniaturization process down to 8 mm ϕ; without providing a sufficient buffer gap between the adjacent electrodes, the electrical poling process easily initiates cracks on the electrode gap due to the residual stress concentration. Uchino et al. proposed the following concepts for designing compact ultrasonic motors (a) simplification of the structure and reduction in the number of components, (b) the use of simple (i.e., uniform) poling configuration (i.e., *gentle to PZT*), and (c) the use of standing-wave type for reducing the number of drive circuit components.

A *Windmill* motor design with basically a flat and wide configuration (pancake type), using a metal–ceramic composite structure is the first good example [35], followed by a thin and long configuration, a metal hollow tube motor (needle type). It is composed of a metal tube and two rectangular PZT plates bonded on it (see Fig. 11.6) [36, 37]. When one of the PZT plates is driven by a sinusoidal voltage, the metal tube exhibits "wobbling" motion, like a hula-hoop or Chinese dish spinning. A no-load speed of 1,800 rpm and an output torque of more than 1 mNm were obtained at 80 V_{peak} for both directions for a motor 2.4 mmϕ in diameter and 12 mm in length. This significantly high torque was obtained due the dual stator configuration and the high pressing force between the stator and rotors made of metal.

Finite Element Analysis (FEA) is a useful tool to obtain the vibration mode simulation in a stator. The Penn State used an ATILA software code (developed by ISEN, and distributed by Micromechatronics, Inc., PA) for the motor stator calculation in linear and rotary motors, and found the resonance

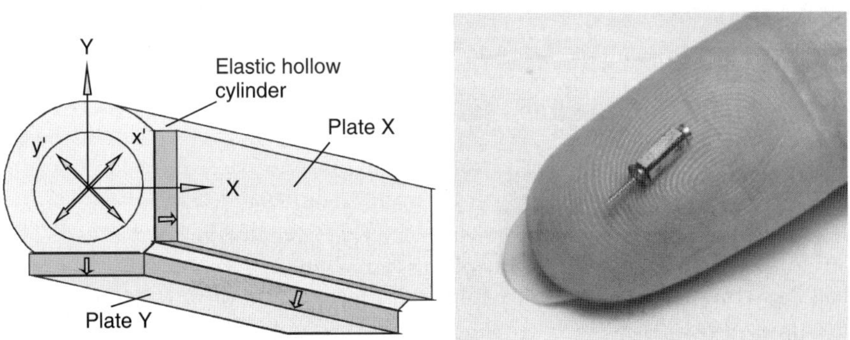

(a) (b)

Fig. 11.6. "Metal tube" motor using a metal tube and two rectangular PZT plates. (**a**) Schematic structure and (**b**) photo of a world-smallest motor (1.5 mmϕ) [37]

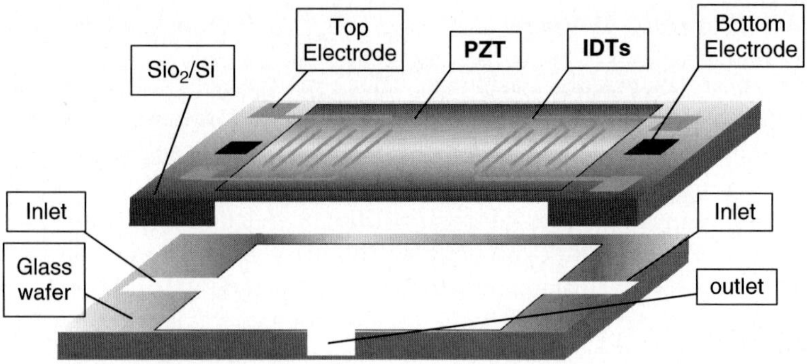

Fig. 11.7. Structure of a PZT/silicon MEMS device, blood tester [40]

frequency discrepancy to be less than 5% [38]. The design optimization trial for the ultrasonic motors using ATILA, in conjunction with the so-called genetic algorism, seems to suggest a future direction of the design algorism [39].

11.3.3 MEMS Devices

PZT thin films are deposited on a silicon wafer, which is then micro-machined to leave a membrane for fabricating micro actuators and sensors, i.e., micro-electromechanical systems (MEMS). Figure 11.7 illustrates a blood tester developed by The Penn State in collaboration with OMRON Corporation in Japan [40]. When a voltage is applied to two surface interdigital electrodes, the surface PZT film generates surface membrane waves, which soak up the blood and the test chemical from the two inlets, then mix them in the center part, and send the mixture to the monitor part through the outlet. FEA simulation was conducted to evaluate the flow rate of the liquid by changing the thickness of the PZT or of the Si membrane, inlet and outlet nozzle size, cavity thickness.

11.4 Drive Techniques

11.4.1 Pulse Drive Methods

Pulse drive of the piezoelectric/electrostrictive actuator generates very large tensile stress in the device, sometimes large enough to initiate cracks. In such cases, compressive bias stress should be employed on the device through clamping mechanisms such as a helical spring and a plate spring. Another solution to suppress this problem is to apply a suitable pseudo-step voltage; that is, the overshoot and ringing of the tip displacement is completely suppressed when the rise time is precisely adjusted to the resonance period of the piezo-device [41].

The rise time adjustment is also very important from the heat generation viewpoint. The temperature rise was examined in consideration of the pulse drive using a trapezoidal wave, as in the applications to diesel engine injection systems. The trapezoidal wave with a maximum voltage of 100 V, 60 Hz, and a duty ratio of 50% was applied to a commercialized multilayer actuator (product by Ceramtech). The temperature rise is surprisingly dependent on the step rise time, and this, in turn, is basically due to the actuator's vibration overshoot and ringing [unpublished].

11.4.2 Piezoelectric Transformers

One of the problems in wide commercialization of piezoelectric actuators and ultrasonic motors includes rather bulky and expensive drive circuitry that is inevitably required. Noticing that the largest components in the drive circuit are electromagnetic transformers, researchers at Penn State have been working on replacing these conventional EM transformers with piezoelectric transformers. The PT has input and output terminals fabricated on a piezoceramic plate, and the input/output voltage is changed through the vibration energy transfer.

Recent lap-top computers with a liquid crystal display require a very thin, no EM-noise transformer with a high efficiency as an inverter of a fluorescent backlight. This application has accelerated the development of the PT. NEC recently commercialized a multilayer rectangular type transformer in order to increase the voltage step-up ratio, using a third-order longitudinal mode, which is one of the methods to release the stress concentration and to increase its lifetime of the original Rosen type [42]. Disk type transformers with an asymmetrical electrode pattern which have a much higher voltage step-up ratio than the rectangular type and disk multilayer types with a floating layer were proposed successively recently [43, 44]. History of piezoelectric transformers in terms of power capability is illustrated in Fig. 11.8. Note that the multi-stacked disk type has a 100 W capability.

In parallel to the component developments, new applications of those PTs, that is, power supplies for piezoelectric actuators and ultrasonic motors have been proposed as schematically illustrated in Fig. 11.9. The PTs can reduce the total system volume and weight significantly.

Figure 11.10 compares the drive circuit sizes for the present EM transformer included type, Shinsei motor (a) and a new ultrasonic motor which integrates its drive circuit and a piezoelectric transformer (developed by Manispiya et al. [45]).

Vazquez Carazo et al. introduced a compact drive system with PTs for a multilayer piezoelectric actuator, aiming at active vibration control on a helicopter [46]. In this sort of military application, we need to realize a compact, light-weight and EM-noise free system while keeping quick response (minimum 200 Hz). Figure 11.11 summarizes the drive system for piezoelectric actuator control, using PTs. High AC voltage from the transformer was rectified, and

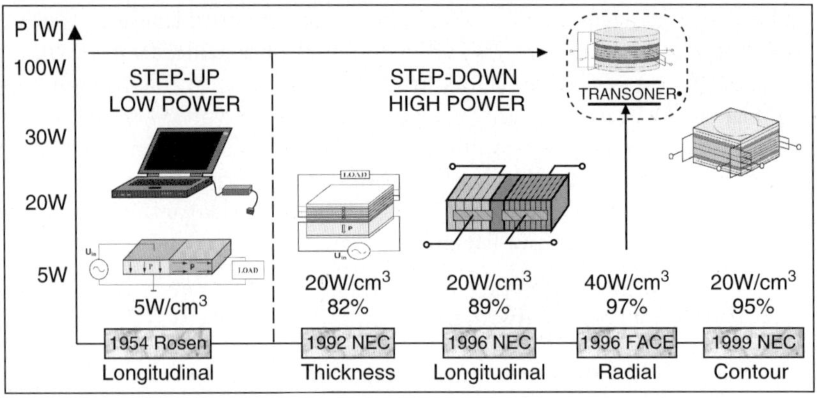

Fig. 11.8. Development history of piezoelectric transformers in terms of power capability

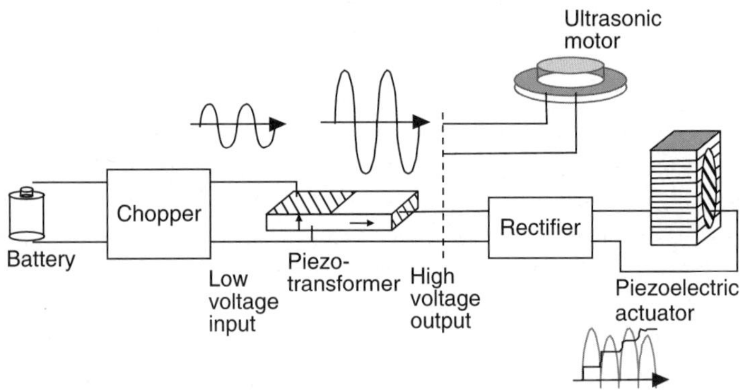

Fig. 11.9. Piezo-actuator drive system using a piezo-transformer

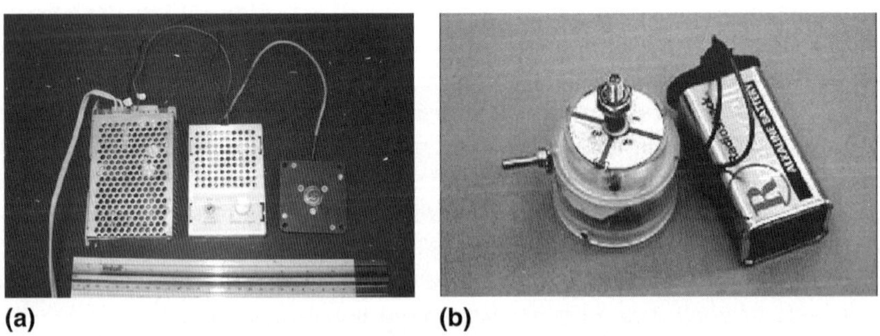

Fig. 11.10. (**a**) Commercialized Shinsei motor and its drive circuit and (**b**) piezo-transformer integrated ultrasonic motor (Penn State) [45]

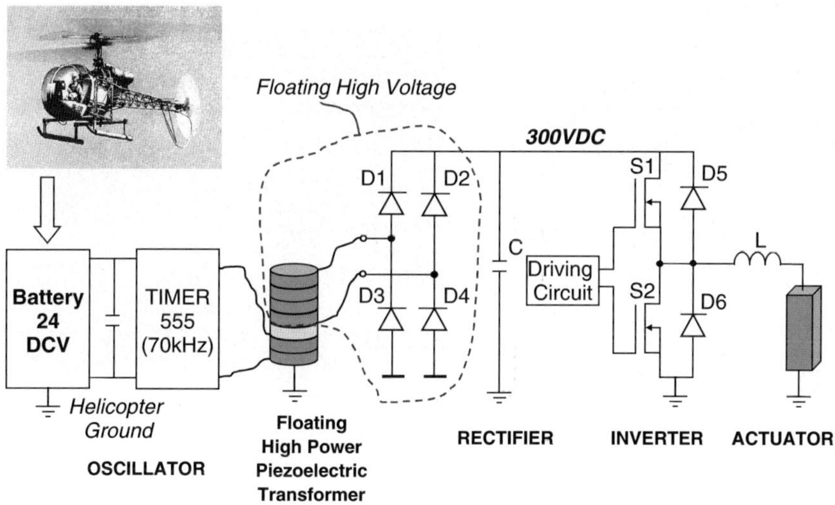

Fig. 11.11. Compact drive system for piezoelectric actuator control [46]

this DC voltage was applied to the actuator through a power amplifier to control the applied voltage. Among various power amplifier families, a Class-D switching amplifier was chosen, because it has advantages over the other switching and linear amplifiers, such as permitting amplitude and frequency control and realizing fast actuation response of the actuator by chopping a DC voltage. The displacement ±1.5 μm was controlled by ±20 V applied voltage. This drive system could be used at least up to 500 Hz, which is sufficient for an active vibration control on a helicopter. Note that the new drive system is compact, light-weight by 1/10 in comparison with conventional EM transformer circuits, and magnetic-noiseless.

11.4.3 Piezoelectric Energy Harvesting

One of the most recent research interests is piezoelectric energy harvesting. When a piezoelectric is adopted in a noise vibration system, cyclic electric field is excited. If this electric energy is consumed via a suitable resistor as Joule heat, mechanical noise vibration is significantly suppressed; that is, passive damper [47]. On the other hand, if the electric energy is accumulated into a rechargeable battery, a new energy harvesting system can be realised. Figure 11.12 shows an LED traffic light array system driven by a piezoelectric windmill developed by NEC-Tokin Corporation.

The Penn State group developed energy harvesting piezoelectric devices based on a "Cymbal" structure (29 mmϕ, 1–2 mm thick), which can generate electric energy up to 100 mW under an automobile engine vibration [48].

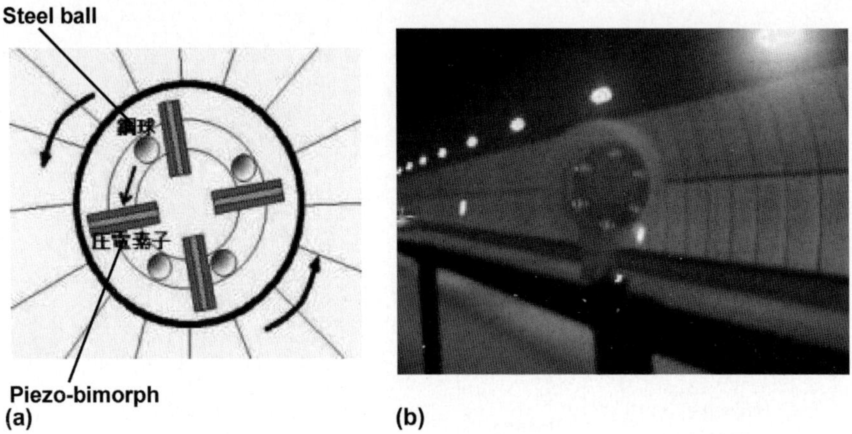

Fig. 11.12. Piezoelectric windmill (**a**) for driving an LEC traffic light array system (**b**) (courtesy by NEC-Tokin)

Table 11.1. Development history in piezoelectric actuators

Year	Key technology	Commercialization
1978	PMN electrostrictor (Penn State)	
1978	Cofired MLA (Penn State)	
1981	High d/k single crystal (TI Tech)	
1982	Ultrasonic motors (Shinsei)	
1987		Dot matrix printer (NEC)
1987		Camera auto focus (Canon)
1988	Linear π-type USM (Shophia U)	
1989		Camera shutter (Minolta)
1989	Monie/cymbal (Penn State)	Piezo suspension (Toyota)
1990		Headrest control (Toyota)
1995		Inkjet printer (Epson)
1999		Wrist watch (Seiko)
2000		Diesel injection (Siemens)
2000	Monie/cymbal (Penn State)	
2004		Phone camera (Samsung)

11.5 Applications

Commercialized products seem to come 5–7 years after the basic key technology is developed in the case of piezoelectric devices, which can be found in Table 11.1.

11.5.1 Inkjet Printer

Epson commercialized a piezoelectric inkjet printer, by utilizing a cofiring technique of PZT with ZrO_2 substrates (see Fig. 11.13). This cofiring

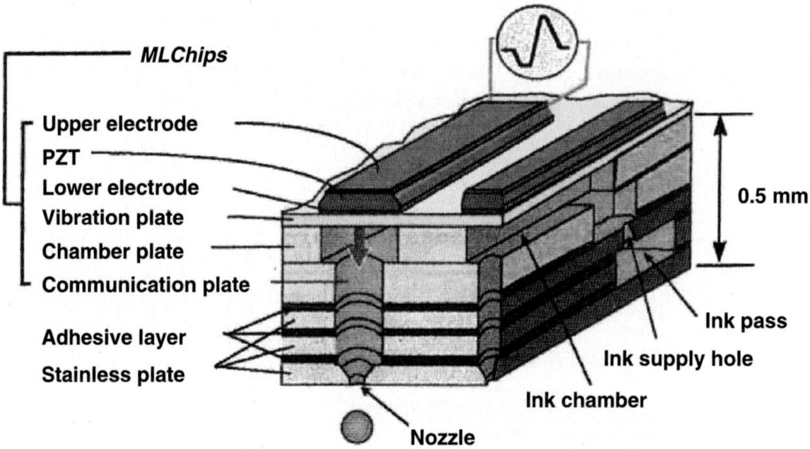

Fig. 11.13. EPSON MACH printer head structure [49]

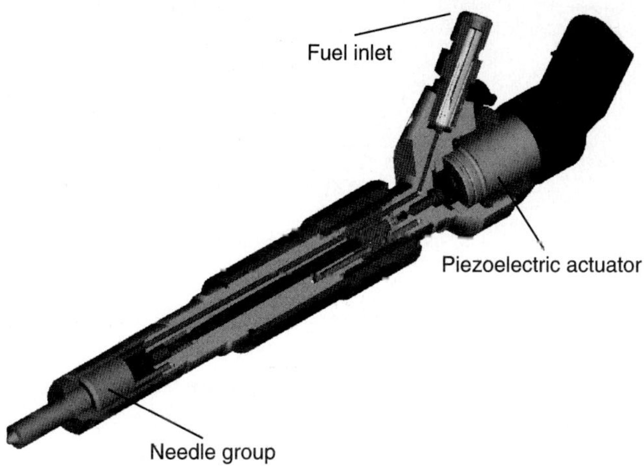

Fig. 11.14. Common rail type diesel injection valve with a piezoelectric multilayer actuator (News release by Bosch)

technique significantly improved stability in vibration for various inks during the patented pull–push–pull dynamic action of the piezoelectric chamber [49].

11.5.2 Diesel Injection Valve

Recent success in the multilayer piezoelectric actuators for diesel injection valve applications has verified the highest reliability of these devices. So-called common-rail type injection valves have been widely commercialised by Siemens and Bosch (Fig. 11.14).

11.5.3 Piezoelectric Pump

NEC recently news-released their piezoelectric pump water cooling module for a fuel cell power supply in a laptop computer (Fig. 11.15).

11.5.4 Ultrasonic Cleaner

Ultrasonic lens cleaner is commonly used in homes. Honda Electronics added another ultrasonic cleaner in conjunction with a washing machine. Figure 11.16 shows an L–L coupler horn to generate water cavitation for removing dirt on a shirt color.

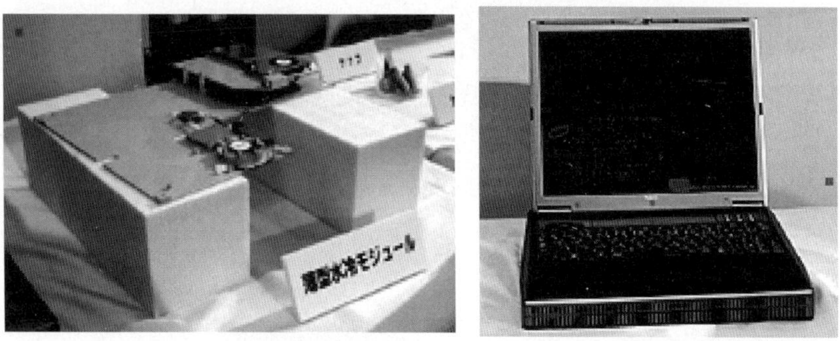

Fig. 11.15. A piezoelectric pump water cooling module (*left*) for a fuel cell in a laptop computer (Courtesy by NEC)

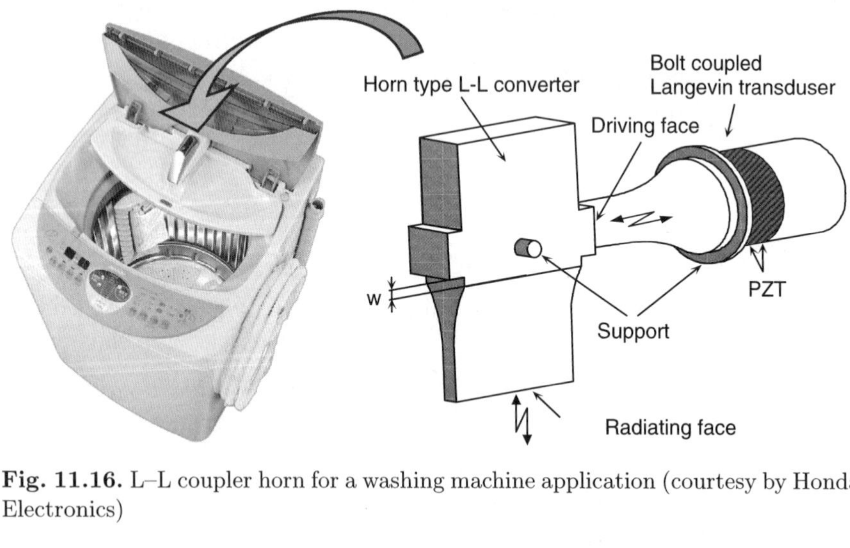

Fig. 11.16. L–L coupler horn for a washing machine application (courtesy by Honda Electronics)

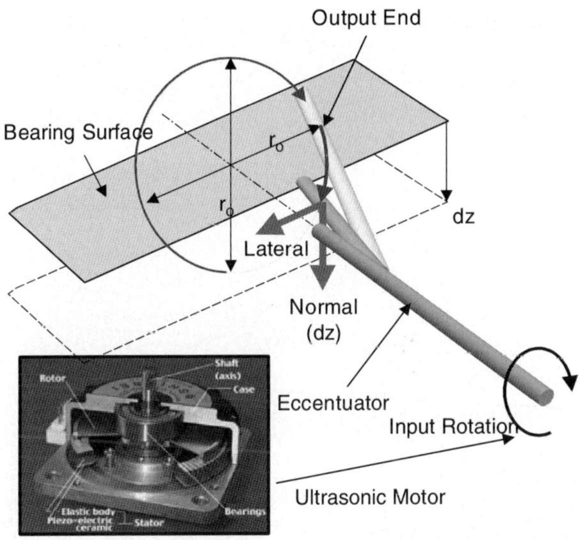

Fig. 11.17. Wing shape control with an ultrasonic motor and an eccentuator mechanism (courtesy by AFRL)

11.5.5 Smart Wing with Ultrasonic Motors

Shinsei ultrasonic motors have been adopted for the US Air Force Smart Wing Program. Coupled with an eccentuator mechanism, the USM functions to control the wing shape (morphing) (Fig. 11.17).

Seiko micro motors have already been used in wrist watches for silent alarm and perpetual function. The most recent application can be found in a microaerial vehicle. Figure 11.18 shows a micro flying robot, a helicopter of the size 13 cmϕ, 7 cm height and 8.9 g weight, in which the floating force is obtained by two piezoelectric motors and additional two motors provide the position/cant control.

The Penn State metal tube motors have been utilized in the world's smallest 4WD vehicle (Fig. 11.19) and the camera auto zooming and focusing mechanism in a Samsung cellular phone (Fig. 11.20).

11.6 Future Research Trend

Future research trends will be divided into two ways: up-sizing in space structures and down-sizing in office equipment. Further down-sizing will also be required in medical diagnostic applications such as blood test kits and surgical catheters. The development keywords will be *integration* and *hybridization.*

Piezoelectric thin films compatible with silicon technology will be much focused in MEMS systems. An ultrasonic rotary motor as tiny as 2 mm in

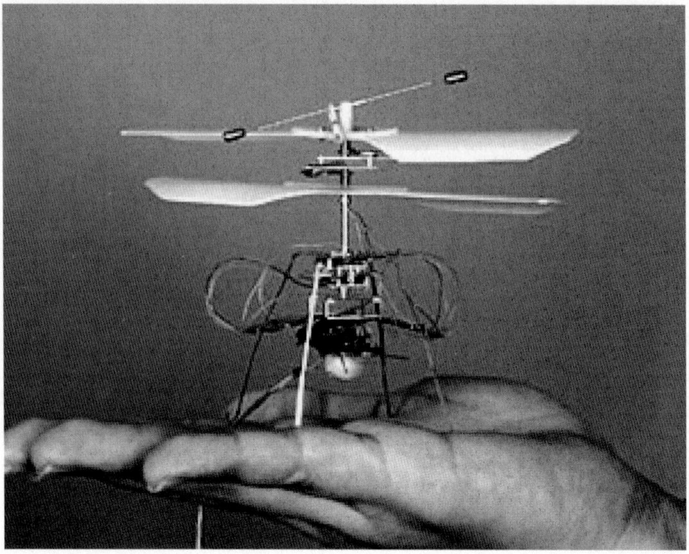

Fig. 11.18. Compact helicopter with micro ultrasonic motors (courtesy by Seiko–Epson)

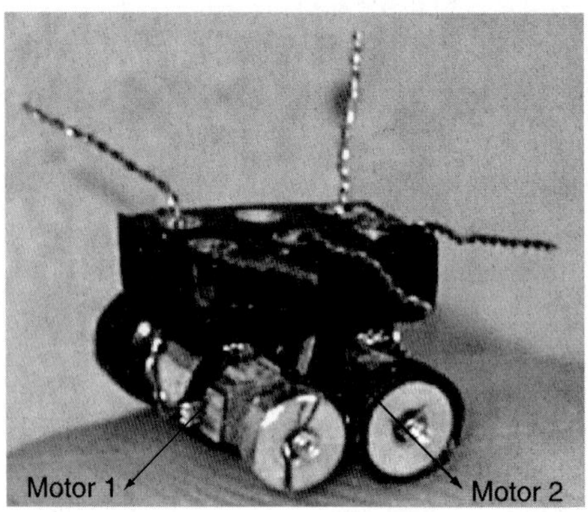

Fig. 11.19. World-smallest vehicle with two micro metal tube USMs. The size is $7 \times 7 \times 7\,\text{mm}^3$

diameter fabricated on a silicon membrane is a good example [50]. One ceramic multilayer component actuator proposed by Mitsui Chemical is very suggestive for predicting the future trend [51]. Only by the external connection, a combined vibration of the longitudinal L_1 and bending B_2 modes can be excited. Compact ultrasonic motors such as The Penn State's "windmill" and "metal tube" motors will expand the applications in medical diagnostic/surgical areas.

Fig. 11.20. Camera auto zooming/focusing mechanism with two metal tube USM's in a Samsung cellular phone

Miniaturization of the drive circuit will be also an important research area for the actual wide commercialization. Piezoelectric transformers may cause a breakthrough for this application.

With the expansion of the application field of ceramic actuators, the durability/reliability issue becomes more important. The final goal is, of course, to develop much tougher actuator ceramics mechanically and electrically. However, the reliability can be improved significantly if the destruction symptom of the actuator is monitored. Safety systems or health monitoring systems have been proposed with two feedback mechanisms: position feedback which can compensate for the position drift and the hysteresis, and breakdown detection feedback which can stop the actuator system safely without causing any serious damages onto the work, e.g., in a lathe machine [52]. Acoustic emission and internal potential measurements, and resistance monitoring of a strain-gauge type internal electrode embedded in a piezoactuator under a cyclic electric field drive are good predictors for the lifetime [53, 54].

Material-wise, Pb (lead)-free piezoceramics have started to be developed partly due to the governmental regulation expected in the near future. It is not soon, but barium titanate based or $(K, Na)(Ta, Nb)O_3$ crystals seem to come back after more than 60 years of their discovery and the sonar application.

References

1. K. Uchino, *Piezoelectric Actuators and Ultrasonic Motors* (Kluwer Academic, Norwell, MA, 1996)
2. K. Uchino, *Ferroelectric Devices* (Marcel Dekker, New York, 2000)
3. K. Uchino, MRS Bull. **18**(4), 42 (1993)

4. K. Uchino, in *Proceedings of the 4th International Conference Electronic Ceramics and Applications*, 1994, p.179
5. T. Yoshida, Tokin Techn. Rev. **24**, 1–9 (1997)
6. J. Kuwata, K. Uchino, S. Nomura, Ferroelectrics **37**, 579 (1981)
7. J. Kuwata, K. Uchino, S. Nomura, Jpn. J. Appl. Phys. **21**, 1298 (1982)
8. K. Yanagiwawa, H. Kanai, Y. Yamashita, Jpn. J. Appl. Phys. **34**, 536 (1995)
9. S.E. Park, T.R. Shrout, Mater. Res. Innovt. **1**, 20 (1997)
10. X. H. Du, J. Zheng, U. Belegundu, K. Uchino, J. Appl. Phys. Lett. **72**, 2421 (1998)
11. S. Kalpat, K. Uchino, J. Appl. Phys. **90**(6), 2703–2710 (2001)
12. D. Damjanovic, D.V. Taylor, N. Setter, in *Proceedings of Materials Research Society Symposium VIII*, 1999
13. R.E. Eitel, C.A. Randall, T.R. Shrout, P.W. Rehrig, W. Hackenberger, S.-E. Park, Jpn. J. Appl. Phys. **40**, 5999–6002 (2001)
14. M.L. Mulvihill et al., in *Proceedings of American Ceramic Society Electronics Division Marketing*, Greensville, Oct. 9–11, 2000, Journal of Electroceramics (Kluwer Acadmic, Boston, 2002)
15. K. Uchino, S. Nomura, Ferroelectrics, **50**(1), 191 (1983)
16. A. Furuta, K.Y. Oh, K. Uchino, Sensors Mater. **3**(4), 205 (1992)
17. V. Bharti, H.S. Xu, G. Shanti, Q.M. Zhang, K. Liang, J. Appl. Phys. **87**, 452 (2000)
18. K. Uchino, J. Zheng, A. Joshi, Y. H. Chen, S. Yoshikawa, S. Hirose, S. Takahashi, J.W.C. deVries, J. Electroceramics **2**, 33 (1998)
19. Y. Gao, Y.H. Chen, J. Ryu, K. Uchino, D. Viehland, Jpn. J. Appl. Phys. **40**, 687–693 (2001)
20. Ryu, J., H.W. Kim, K. Uchino, J. Lee, Jpn. J. Appl. Phys. **42**(3), 1307–1310 (2003)
21. K. Uchino, J. Zheng, Y.H. Chen, X.H. Du, J. Ryu, Y. Gao, S. Ural, S. Priya, and S. Hirose, J. Mater. Sci. **41**, 217–228 (2006)
22. R. Smith, in *Proceedings of ONR Transducer Workshop*, 2002
23. X. Ren, Nat. Mater. Lett. Published online: 11 Jan. 04; doi:10.1038/nmat1051 (2004)
24. T. Takenaka, Bulletin Ceram. Soc. Japan, **40**(8), 586–597 (2005)
25. K. Uchino, Mat. Res. Innovat. **1**, 163 (1997)
26. K. Uchino, J. Rob. Mech. **1**(2), 124 (1989)
27. P. Poosanaas, K. Tonooka, K. Uchino, J. Mechatronics **10**, 467–487 (2000)
28. J. Ohashi, Y. Fuda, T. Ohno, Jpn. J. Appl. Phys. **32**, 2412 (1993)
29. A. Bauer, F. Moeller, in *Proceedings of 4th International Conference on New Actuators* (AXON Technology Consultants, GmbH, Bremen, 1994), p.128
30. K. Uchino, M. Yoshizaki, K. Kasai, H. Yamamura, N. Sakai, H. Asakura, Jpn. J. Appl. Phys. **26**(7), 1046 (1987)
31. Aura Ceramics, Inc., Catalogue "Rainbow"
32. Y. Sugawara, K. Onitsuka, S. Yoshikawa, Q.C. Xu, R.E. Newnham, K. Uchino, J. Am. Ceram. Soc. **75**(4), 996 (1992)
33. A. Dogan, K. Uchino, R.E. Newnham, IEEE Trans. UFFC **44**, 597 (1997)
34. S. Dong, X.H. Du, P. Bouchilloux, K. Uchino, J. Electroceramics **8**, 155–161 (2002)
35. B. Koc, P. Bouchilloux, K. Uchino, IEEE Trans. Ultrason. Ferr. **47**(4), 836–843 (2000)

36. B. Koc, S. Cagatay, K. Uchino, IEEE Ultrason. Ferr. Trans. **49**(4), 495–500 (2002)
37. S. Cagatay, B. Koc, K. Uchino, IEEE Trans. UFFC, **50**(7), 782–786 (2003)
38. P. Bouchilloux, K. Uchino, in *Proceedings of the SPIE: The International Society for Optical Engineering*, USA, vol. 4693, 2002
39. P. Bouchilloux, K. Uchino, J. Intell. Mater. Syst. Struct. **14**(10), 657–667 (2002)
40. S. Kalpat, Ph.D. Thesis, Penn State University, Fall, 2001
41. K. Uchino, J. Giniewicz, *Micromechatronics* (Marcel Dekker, New York, 2003)
42. S. Kawashima, O. Ohnishi, H. Hakamata, S. Tagami, A. Fukuoka, T. Inoue, S. Hirose, in *Proceedings of IEEE International Ultrasonic Symposium 1994*, France, 1994
43. B. Koc, Y. Gao, K. Uchino, Jpn. J. Appl. Phys. **42**(2A), 509–514 (2003)
44. P. Laoratanakul, A.V. Carazo, P. Bouchilloux, K. Uchino, Jpn. J. Appl. Phys. **41**, 1446–1450 (2002)
45. S. Manuspiya, P. Laoratanakul, K. Uchino, Ultrasonics **41**(2), 83–87 (2003)
46. A.V. Carazo, K. Uchino, J. Electroceramics **7**, 197–210 (2001)
47. K. Uchino, T. Ishii, J. Jpn. Ceram. Soc. **96**(8), 863–867 (1988)
48. H.-W. Kim, A. Batra, S. Priya, K. Uchino, D. Markley, R.E. Newnham, H.F. Hofmann, Jpn. J. Appl. Phys. **43**(9A) 6178–83 (2004)
49. N. Kurashima, in *Proceedings of Machine Technical Institute Seminar* (MITI, Tsukuba, Japan, 1999)
50. A.M. Flyn, L.S. Tavrow, S.F. Bart, R.A. Brooks, D.J. Ehrlich, K.R. Udayakumar, L.E. Cross, J. Microelectromech. Syst. **1**, 44 (1992)
51. H. Saigo, in *15th Symposium on Ultrasonic Electronics*, No.PB-46, Nov. 1994, p. 253
52. K. Uchino, J. Ind. Educ. Soc. Jpn. **40**, 28 (1992)
53. H. Aburatani, K. Uchino, A. Furuta, Y. Fuda, in *Proceedings of 9th International Symposium on Application Ferroelectrics*, 1995, p. 750
54. H. Aburatani, K. Uchino, Jpn. J. Appl. Phys. **37**(Part 1, No. 1), 204 (1998)

12

Piezoelectric Positioning

S. Arnold, P. Pertsch, and K. Spanner

12.1 Introduction

Piezoelectric materials change shape under the influence of electric fields. Known as the inverse piezoelectric effect, this phenomenon is the principle upon which a large class of motion-producing devices is based.

The piezoceramic elements in today's piezo actuators are optimized to offer sufficient displacement to make them suitable for certain motion tasks. They are used in high-frequency oscillators in ultrasonic wave generators, in ultrasonic motors and as valve actuators in vehicles. Piezo actuators, however, can also be designed for ultra precise motion and positioning of objects in quasi-static operation, i.e., for nanopositioning.

The displacement of a piezo actuator upon exposure to an electric field is caused by ionic shifts and ferroelectric reorientations in the piezoceramics. The motion is virtually friction free and the resolution practically unlimited. Piezo actuators are especially well suited for positioning tasks requiring accuracies in the one-nanometer range over travel ranges of a few tens of microns (Table 12.1).

Modern nanopositioning systems offer motion in six degrees of freedom (X, Y, Z and the three corresponding rotation axes) with several hundred micron travel and resolutions in the realm of one nanometer. Ideally, they are designed as parallel-kinematics systems with non-contacting position sensors for all motion directions, each measuring the moved platform against the stationary baseplate directly, and the whole connected to digital signal processing and servocontrol electronics.

While in the early 1970s, positioning with such accuracies was reserved for exotic applications in university research labs, by the beginning of the twenty-first century, nanopositioning had found its way into a number of key industrial processes – processes which would be virtually impossible without piezoelectric technology. In commercial high-resolution optical, confocal and scanning-probe microscopes, piezo actuators have become standard and are playing an important role in research and quality-assurance inspection systems [1].

Table 12.1. Comparison of piezoactuators and other piezoelectric drives

Actuator/drive type	Max. displacement (μm)	Max. force (N)	Max. velocity (mm s^{-1})
Bender	2,000	1	–
Transversal multilayer	50	50	–
Longitudinal stack	300	50,000	–
Longitudinal stack with displacement amplification	1,000	10,000	–
Shear stack	10	100	–
Stepping drive/walk drive[a]	∞	1,000	20
Inertia drive[a]	∞	10	20
Ultrasonic motor[a]	∞	10	1,000

[a]Typical values, actual force and velocity are interdependent

The demand for nanopositioners experienced dramatic growth during the telecom boom around the year 2000. Optical fibers and micro-electro-mechanical systems (MEMS) needed to be aligned with extremely high accuracy and then physically joined. At the time of this writing, it is the semiconductor industry that dominates industrial demand for precision-motion piezo actuators. Increasing demands on performance, size, power consumption and cost of integrated circuits make necessary smaller and smaller feature sizes on semiconductor wafers. A lithographic process is used to transfer these features, and nanopositioning plays a role both in their creation and subsequent inspection.

The requirements of industrial applications are, however, not limited to position accuracy, but increasingly focus on long lifetime, short cycle time and reliability.

In the following sections, the requirements and solutions of nanopositioning technology will be introduced, and the path from technological novelty to industrial-class, high-precision nanopositioner traced.

12.2 Nanopositioning Actuator Technology

In nanopositioning, the piezoceramics used are optimized for travel range and lifetime. Two types of actuators are especially important for nanopositioning: stack actuators and shear actuators.

The induced mechanical deformation of a piezoceramic – and hence the travel range, ΔL, of an actuator – depends on the following interrelated parameters: electric field E, electrical potential difference U, actuator length L, mechanical stress T, the force applied F, and the properties of the piezoelectric material used. The material properties can be described by the piezoelectric strain coefficients d_{ij}, which describe the relationship between an applied electric field and the mechanical strain produced.

Piezo actuators are usually operated well below their resonant frequencies. Keeping a position will led to high DC electric field loads over long time. Therefore actuator lifetime is largely a function of the actuator's resistance to electric DC-fields, temperature and humidity. Another important positioning related property is the thermal expansion behavior.

12.2.1 Stack Actuators (Linear Actuators)

The active portion of a stack actuator consists of a stack of ceramic disks separated by metallic electrodes (Fig. 12.1). The maximum operating voltage is proportional to the thickness of the disks. Most high-voltage actuators use ceramic disks of 0.4–1 mm thickness. To obtain the nominal operating field of 1–2 kV mm^{-1}, these actuators must be operated at voltages of up to 1,000 V.

In foil-technology multilayer actuators, the ceramic layers and the electrodes are co-fired to form a monolithic block. There, the active layers are between 25 and 100 μm thick. The advantage of actuators with a larger number of layers lies in the lower operating voltage of 100 V which is sufficient to obtain the maximum displacement from the thinner ceramic layers.

Stack actuators can withstand high compression forces and have the highest stiffness of all piezoactuator designs. Standard versions are available with load capacities of up to 100.000 N, and preloaded actuators can also operate under tension.

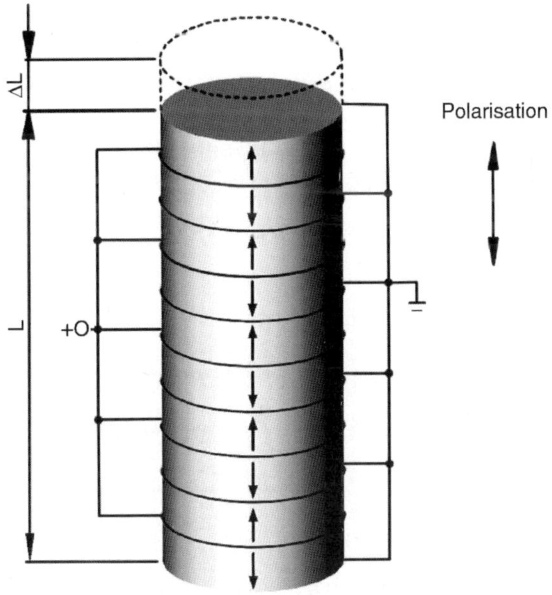

Fig. 12.1. Electrical design of a stack translator

Typical stack actuators achieve induced displacements of up to 0.2% of their lengths. The following equation can be used to estimate the obtainable displacement, ΔL:

$$\Delta L \approx d_{33} n U, \tag{12.1}$$

where ΔL = max. nominal displacement without external force or restraint, d_{33} = longitudinal piezoelectric strain coefficient, n = number of ceramic layers, U = operating voltage.

Modern piezoceramic materials for actuator applications exhibit longitudinal strain coefficients (d_{33}) of 400–650 pm V^{-1}. The coefficients describing the transverse effect, d_{31} – i.e., contraction perpendicular to the field and the polarization direction, have values from −180 to −250 pm V^{-1}. These figures only apply to the raw material at room temperature under small-signal conditions, i.e., with fields up to 10 V mm^{-1}.

The maximum allowable field for piezo actuators is between 1 and 2 kV mm^{-1} in the polarization direction and up to 0.5 kV mm^{-1} in the opposite direction. The maximum voltage that can be applied depends on the type of ceramic and the insulation materials used.

Exceeding the maximum voltage can lead to depolarization, dielectric breakdown or disintegration of the piezoceramics.

12.2.2 Shear Actuators

Shear actuators have the highest displacement coefficients. The effective value of d_{15} with a field strength of 0.5 kV mm^{-1} can be up to 1,200 pm V^{-1}, almost double the longitudinal coefficient at the same field strength. A further advantage of shear piezos is their suitability for bipolar operation, whereby the mid-position corresponds to a drive voltage of 0 V. In shear mode, unlike in the other modes, the electric field is applied perpendicular to the polarization direction (see Fig. 12.2). To prevent repolarization, the nominal field is limited to 0.5 kV mm^{-1} [2].

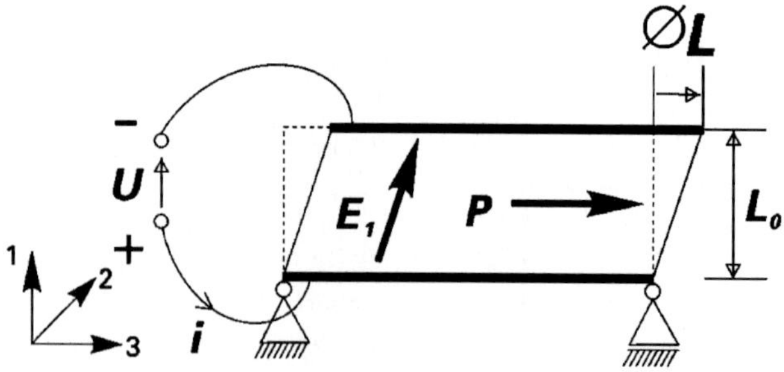

Fig. 12.2. Material deformation in a shear actuator

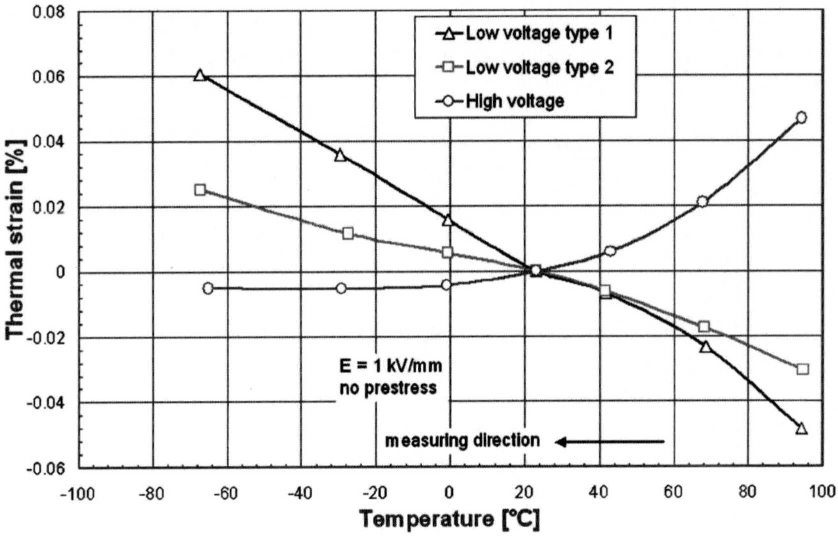

Fig. 12.3. Linear thermal expansion of different actuator types. The negative thermal expansion coefficient results from thermal depolarization of the piezoceramic

12.2.3 Temperature Effects

The thermal expansion of piezoceramics is less than that of most metals. Figure 12.3 shows the behavior of different actuator types.

Piezoceramics must be poled to exhibit the piezo effect. A poled ceramic may depole when heated; the degree of depolarization depends on the Curie temperature of the material. Modern piezoceramics have Curie temperatures of 350°C and can be operated at up to 150°C [3].

12.2.4 DC Humidity Degradation

The degradation of dielectric components under DC driving conditions is related to the strength of the electric field, the temperature and the humidity. For multilayer actuators, the dependency of the lifetime t_{L} on the applied electric field is usually described by a power law: $t_{\mathrm{L}} \sim E^k$ where k is between 2.5 and 3. However, the most severe condition is humidity, which can easily reduce the lifetime by several orders of magnitude. Increased humidity is not unusual in actuator applications. Clean rooms in the semiconductor industry, one of the major application areas of nanopositioning equipment, have artificially humidified air to avoid electrostatic discharge. In conventional actuator designs, the high humidity-DC-signal combination is most critical for the open-electrode configurations, where the internal electrodes are separated only by the active layer thickness and protected by a polymer coating.

Because the polymer coating is not able to prevent moisture penetration, water is attracted by the field and ionized by electrolysis. In the following different processes can occur. Among others Ag^+ ions can move as Ag(OH) from the anode through cracks, voids, dissolved secondary phases at the grain boundaries and, especially, on the surface to the cathode. There, it is reconverted to metallic silver. As a result, silver dendrites grow from the cathode to the anode. First they lower the resistance, later they can cause breakdown. Dissociated protons can also destroy the material directly. That's why DC humidity degradation is not just seen with Ag/Pd internal electrodes; even actuators with platinum electrodes tend to degrade at high fields [4].

12.2.5 Ceramic-Insulated Multilayer Actuators

Figure 12.4 shows a ceramic-insulated actuator design. On the surface with no terminations, it has a thin ceramic protection layer. This inorganic coating can not be penetrated by humidity. The layer is pressed on the surface while the ceramic is still in the green state and is made of the same PZT ceramic material as the active layers. In contrast to the buried design, where the insulation width has to be at least 200–300 μm for tolerance reasons, here it has the same uniform thickness as the active layers. Therefore, it is penetrated to some extent by the scattered field of the internal electrodes (Fig. 12.4) and, consequently, it is partially poled and expands during operation. Hence, stress and crack formation is limited and lifetime is dramatically improved (Fig. 12.5).

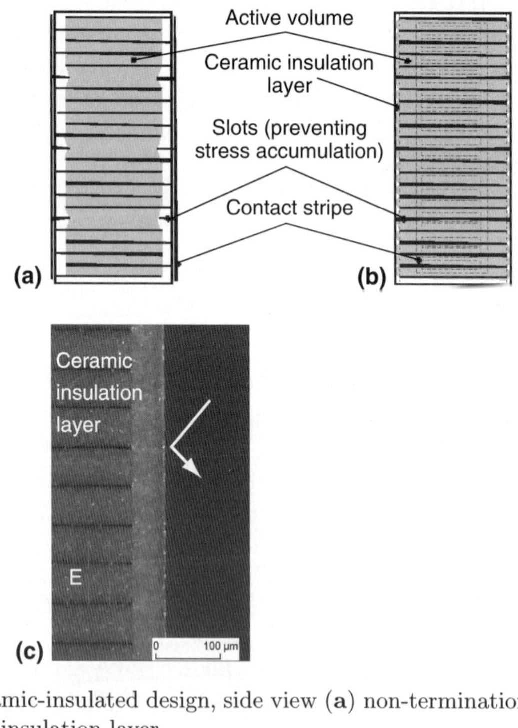

Fig. 12.4. Ceramic-insulated design, side view (**a**) non-termination side, (**b**) termination side, (**c**) insulation layer

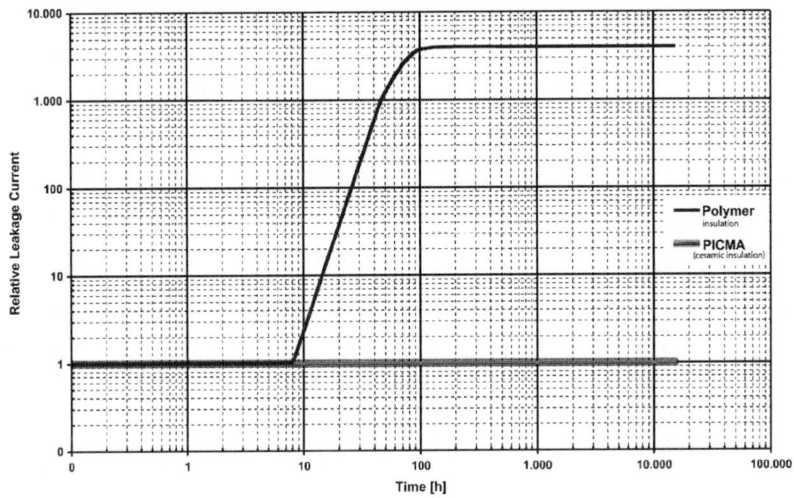

Fig. 12.5. Ceramic-insulated piezoactuators (*lower curve*) compared with conventional multilayer piezoactuators with polymer insulation under test conditions of $U = 100\,\text{VDC}$, $T = 25°\text{C}$, relative humidity = 70%, ceramic layer thickness = 60 µm. In conventional actuators, the leakage current begins to rise after only a few hours – an indication of degradation of the insulation and reduced life expectancy

12.3 Mechanical Coupling

The mechanical coupling between the load and the piezo actuator is of great importance. It is ideally designed to compensate for the undesirable features of piezo actuators: low pulling force capacity limiting dynamic operation, low lateral load capacity and limited displacement. In addition to compensating for these properties, the most modern mechanics designs include elements of parallel kinematics, which is superior to the conventional serial arrangements in many respects.

12.3.1 Push, Pull and Lateral Loads

Piezo actuators can withstand high compressional stresses of up to 250 MPa before being physically destroyed. Because depolarization effects occur, however, the compression forces should be kept below a maximum of 30 MPa.

The tensile load capacity of piezoceramic actuators is limited to only 5–10% of the compressive load capacity. Lateral and bending forces must be avoided completely, to avoid cracking the actuator at, for example, adhesive joints. Tensile forces also result from the inertia of loads attached to the actuator when moved dynamically.

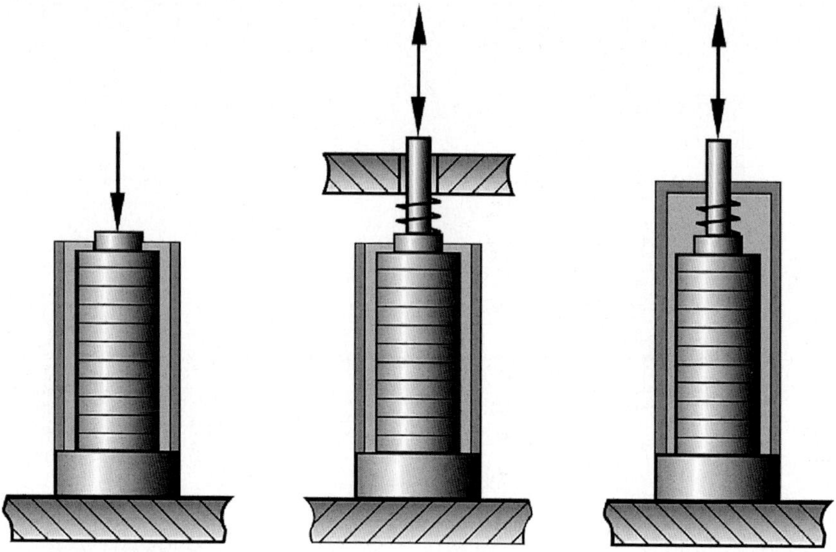

Fig. 12.6. To avoid damage, pulling forces are only applied to preloaded actuators; the preload is either applied externally, or integrated in the actuator housing

For sine wave motion, the maximum forces can be estimated as follows:

$$F_{\mathrm{dyn}} = \pm 4\pi^2 m_{\mathrm{eff}} \left(\frac{\Delta L}{2}\right) f^2, \tag{12.2}$$

where F_{dyn} = dynamic force (N), m_{eff} = effective mass: ~1/3 the weight the piezoceramics plus the moved mass (g), ΔL = displacement (peak–peak) (m), f = drive frequency (Hz).

The maximum permissible forces must be considered when choosing an operating frequency.

One way to extend the dynamic operation capacity is the use of a preload, for example by compressing the ceramic with a spring (Fig. 12.6). However, it must not be forgotten that when operating an actuator against a spring load, a part of the travel range is lost due to the elasticity of the piezo element.

Ball tips or flexures can be used to decouple lateral and bending forces from the actuator (Fig. 12.7).

12.3.2 Flexures

Decoupling a piezo actuator from its environment with flexures does more than just protect the piezoceramic from undesirable outside forces. Flexures can also serve as guidance elements, preventing motion components perpendicular to the actuator's displacement direction from being transferred to the moved mass.

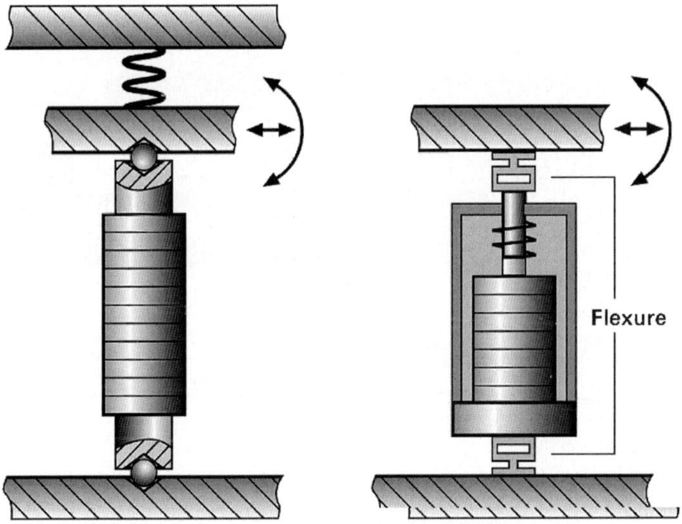

Fig. 12.7. Ball tips or flexures to decouple lateral and bending forces

A flexure is a frictionless, stictionless device based on the elastic deformation (flexing) of a solid material (e.g., steel). Sliding and rolling are entirely eliminated. The advantages over conventional guiding systems are high stiffness and freedom from friction and play. Flexures can be fabricated from non-magnetic materials and require no lubricants or other consumables. Unlike air cushion bearings, which are also virtually friction-free, flexures are suitable for vacuum operation.

Parallelogram flexures exhibit excellent guidance characteristics. Depending on complexity and tolerances, they have straightness/flatness values in the nanometer range or better. Basic parallelogram flexures cause arcuate motion (travel in an arc) which introduces an out-of-plane error of about 0.1% of the travel range (Fig. 12.8).

More complex guiding systems with multilink flexures can compensate for arcuate motion and make possible straightness/flatness in the nanometer or microradian range (Fig. 12.9).

12.3.3 Flexures with Integrated Lever Amplification

The displacement of piezo actuators in nanopositioning systems can be increased many-fold with integrated lever amplification mechanisms. To maintain dynamic performance and subnanometer resolution with the increased travel range, the lever system must be extremely stiff as well as backlash- and friction-free. Flexures are again ideally suited as linkage elements.

Compared with direct-drive systems, lever-amplified designs offer as advantages longer travel ranges, smaller form factors and reduced electrical

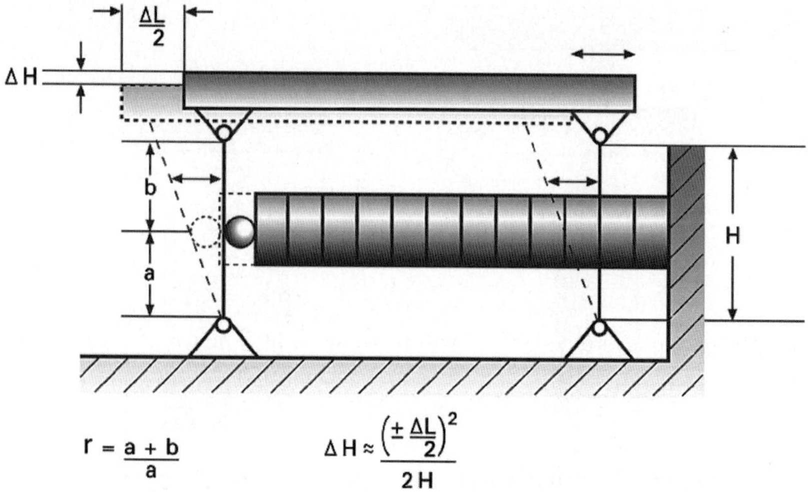

Fig. 12.8. Basic parallelogram flexure guiding system with motion amplification. The amplification r (transmission ratio) is given by $(a+b)/a$. The system displacement, ΔL, equals the lever amplification, r, times the displacement of the actuator

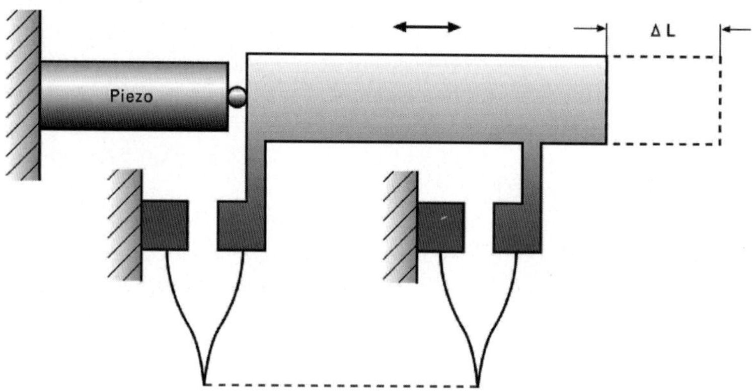

Fig. 12.9. Zero-arcuate-error multilink-flexure guiding system

capacitance. However, penalties must be paid in terms of stiffness and thus lower resonant frequencies and dynamic performance values.

The behavior of an ideal lever with ratio r can be described by the following equations:

$$k_{\mathrm{sys}} = \frac{k_0}{r^2}$$
$$\Delta L_{\mathrm{sys}} = \Delta L_3 r \,, \tag{12.3}$$
$$f_{\mathrm{res-sys}} = \frac{f_{\mathrm{res-0}}}{r}$$

where r = lever ratio, ΔL_0 = travel of the primary drive element, ΔL_{sys} = travel of the lever-amplified system, k_{sys} = stiffness of the lever-amplified system, k_0 = stiffness of the primary drive system (actuator stack and joints), $f_{res-sys}$ = resonant frequency of the amplified system, f_{res-0} = resonant frequency of the primary drive system (actuator stack and joints) (Fig. 12.8).

In practice, a balance between mass, stiffness and cost must be found.

The coupling between the actuator stack and the lever system is crucial. It must be very stiff in the motion direction but should be soft in all other degrees of freedom to avoid damage to the ceramics. If, for example, the stiffness of each of the two lever interfaces is only as high as that of the actuator stack alone, the overall system stiffness will be reduced by a factor of three. In many piezo-driven systems, the stiffness of the ceramic is thus not the limiting factor in determining the stiffness of the mechanism as a whole.

12.3.4 Parallel Kinematics

There are two basic ways to design multiaxis positioning systems: serial kinematics and parallel kinematics. Serial kinematics (Fig. 12.10) are easier to design and build and can be operated with simpler controllers. In a multiaxis, serial kinematics system, each actuator is assigned to exactly one degree of freedom. If there are integrated position sensors, they are also each assigned to one drive and measure only the motion caused by that drive and in its direction of motion. All undesired motion (guiding error) in the other degrees

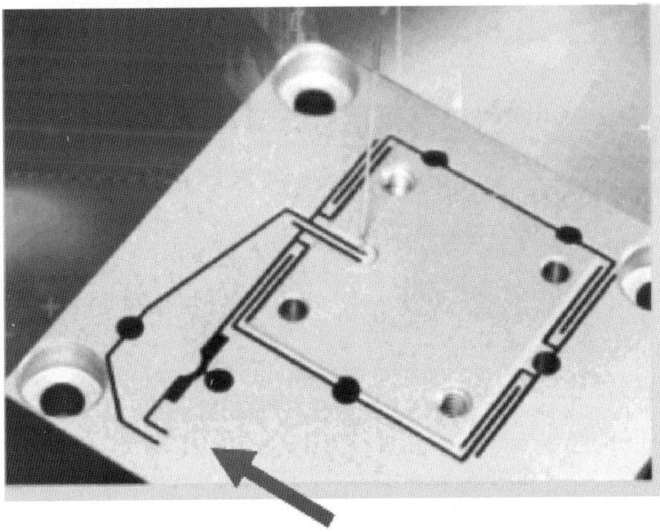

Fig. 12.10. Fabrication of flexures in a nanopositioning system using a wire-EDM cutting process. Actuator interfaces at arrow point with displacement in arrow direction

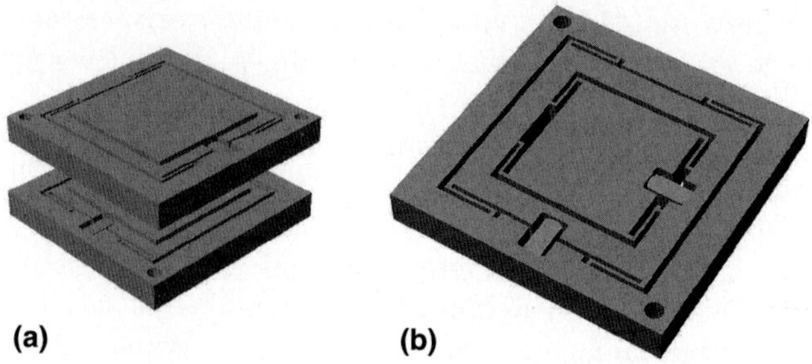

Fig. 12.11. Working principle of serial-kinematics, XY piezo stages: (**a**) stacked (**b**) nested

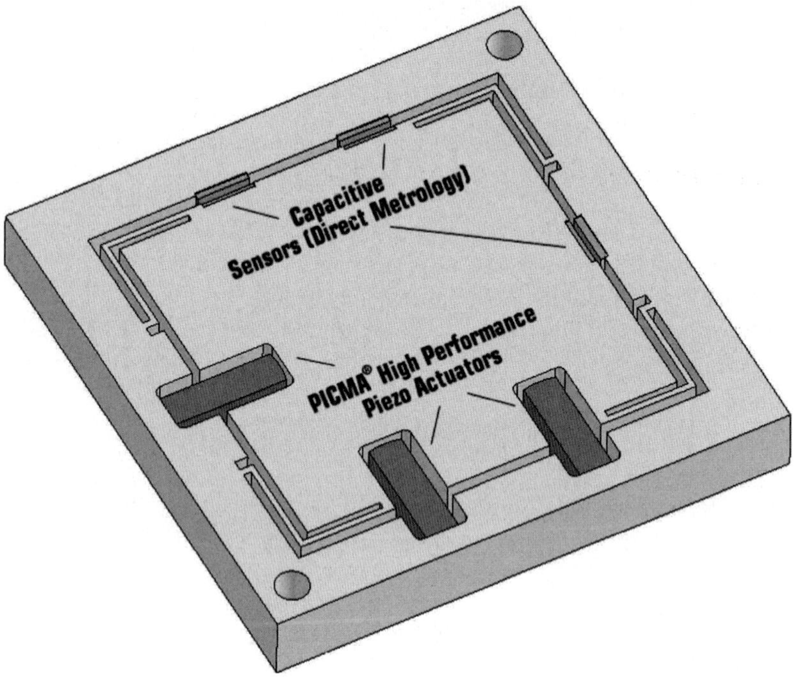

Fig. 12.12. Working principle of an XY parallel-kinematics piezo stage

of freedom is not seen and hence cannot be corrected in the servo-loop, a situation which leads to cumulative error (Fig. 12.11).

In a parallel-kinematics multiaxis system, all actuators act directly on the same, central, moving platform (Fig. 12.12). In this way, the same resonant frequency and dynamic behavior is obtained for both the X and Y axes. Parallel kinematics also facilitates integration of parallel metrology, with its direct measurement of guiding errors at the moving platform. A parallel metrology sensor sees all motion in its measurement direction, not just the motion of one actuator, so runout from all actuators can be compensated for in real-time. The results are better trajectory control, repeatability and flatness [5].

12.4 Sensors and Servocontrol

With servocontrolled operation using capacitive position sensors, nonlinear effects like hysteresis and drift are eliminated, and the basis for repeatable positioning in the nanometer range is established (Fig. 12.13). In addition, digital processing and control algorithms allow further optimization of system dynamics and linearity.

12.4.1 Capacitive Position Sensors

Capacitive position sensors use a probe excited with a high-frequency signal to sense the distance to the moving object or piezo stage platform (target). The gap between the probe and the target is inversely proportional to the

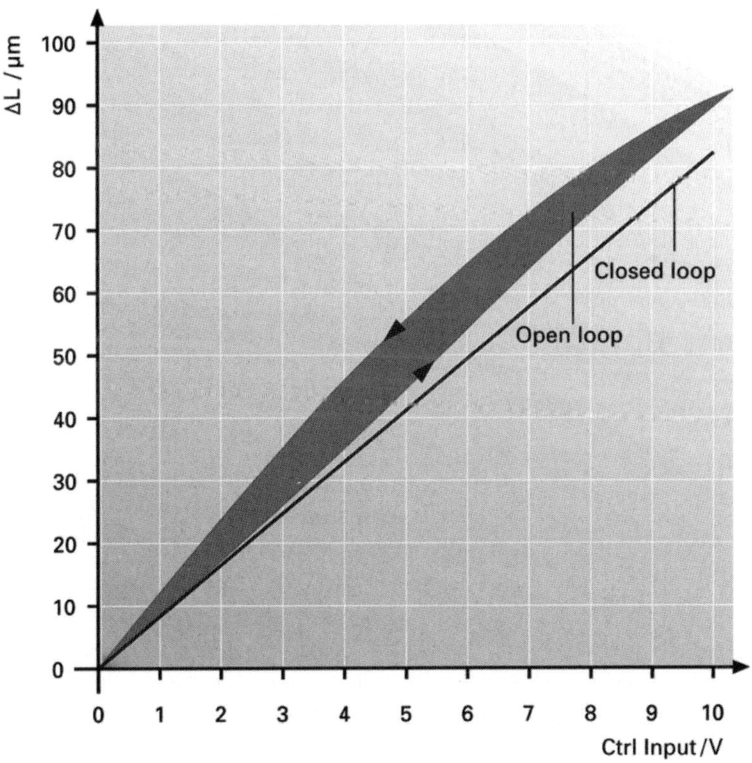

Fig. 12.13. Open-loop vs. closed-loop linearity of a typical piezotranslator

capacitance of the capacitor they form, so measuring the capacitance permits determining the distance (Fig. 12.14). Capacitive sensors can attain resolutions and repeatabilities of better than 0.1 nm. They can furthermore detect motion frequencies of up to 10 kHz.

The design of the nanopositioning system must ensure parallelism of the probe and the target, a further reason for preferring flexure guidance. The effect of residual tip/tilt error can be greatly reduced by applying additional linearization techniques to the sensor signal (Fig. 12.15).

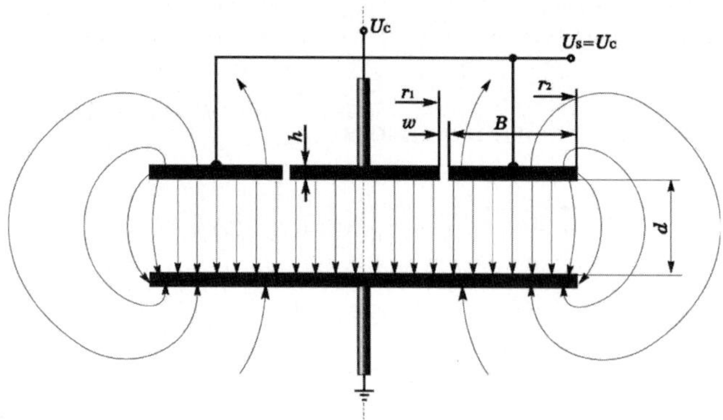

Fig. 12.14. Working principle of a two-plate capacitive sensor

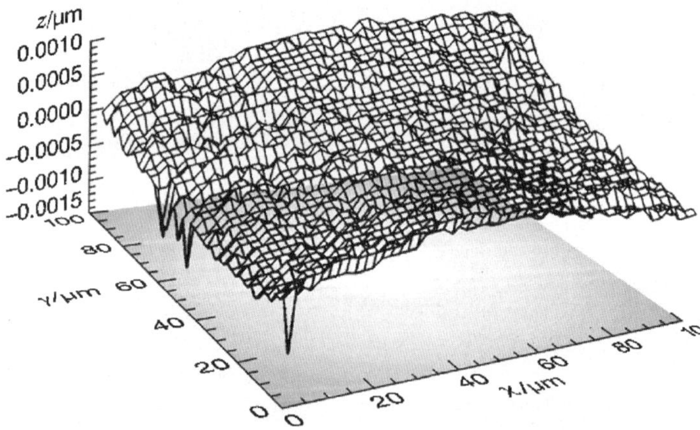

Fig. 12.15. Flatness (Z-axis deviation) of a $100 \times 100\,\mu m$, XY-plane scan with a six-axis nanopositioning system with active trajectory control. The moving portion of this parallel-metrology positioner is equipped with precise, parallel-metrology capacitive sensors for all six degrees of freedom. The sensors are continually measuring the actual platform position against the stationary reference. Any deviations that occur in the four out-of-plane directions are corrected in real-time through the corresponding actuators

12.4.2 Improving Dynamics and Linearity

The dynamic behavior of positioning systems depends on the stiffness of the system, on the position sensor and the servocontroller used. Simple servo-controllers have a limited closed-loop tracking bandwidth of about 10% of the mechanical resonant frequency. A number of techniques can be used to greatly improve the dynamic performance of piezomechanics systems.

Dynamic signal preshaping techniques can be used for optimizing periodic motion, such as scanning ellipses, grids or other shapes. These procedures require no external sensors or hardware; instead, an algorithm uses the position information from the sensors integrated in the piezomechanics to calculate an optimized control signal. The result is an improvement in linearity and tracking accuracy of up to three orders of magnitude (Figs. 12.16 and 12.17).

12.5 Extended Travel with Stepping Drives

Classical piezo-based nanopositioning systems are subject to severe limitations as regards the increased travel from lever amplification. In particular, system stiffness is reduced in proportion to the square of the lever amplification factor (see Sect. 12.3).

The higher the lever ratio is made in order to obtain a longer travel range, the softer the system becomes. This makes control more difficult and substantially increases settling times. At the same time, the resolution of the system is worsened proportional to the amplification, because mechanical noise such as that due to the finite resolution of the control signal is also amplified by the lever.

The actuators usually cannot be designed large and stiff enough to dispense with lever amplification, because the space available in integrated industrial applications is limited. To obtain a travel range of 1 mm, for example, a one-meter-long actuator would be required.

The demand for travel ranges of several millimeters with the stiffness and high resolution of classical piezoactuators and systems, is leading the nanopositioning technology down new paths. Here are the main features of three types of solutions: ultrasonic piezomotors, inertial piezo drives, and piezo stepping drives – of which the last is best able to fulfill the requirements mentioned.

Ultrasonic piezomotors are very fast – they reach speeds of up to 1,000 mm s^{-1} – and their travel ranges are virtually unlimited. In a piezomotor, a high-frequency piezoceramic driven at resonance drives a moving runner using a friction tip on the resonator. Because of the way they work, piezomotors do not generally have the high resolution of a piezo actuator. This could be remedied by driving the piezoceramic element in an analog mode, but the required control system would be complex, the voltages high and the utility of the additional fine adjustment in such a low-force system questionable.

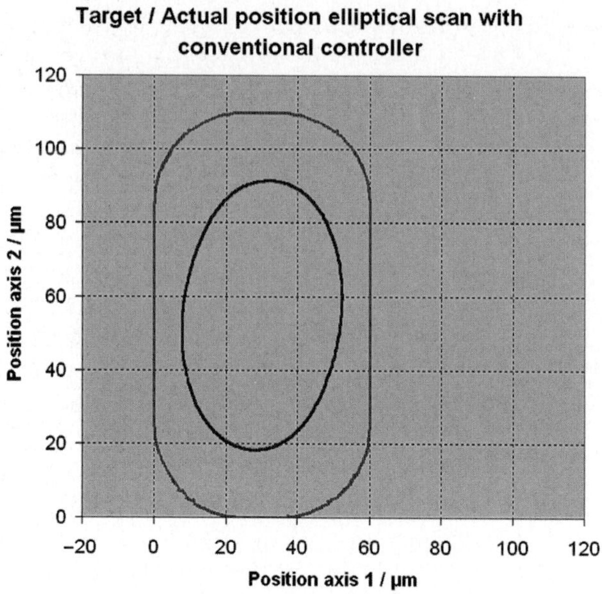

Fig. 12.16. Elliptical scan in a laser micro-drilling application with *XY* piezoscanning stage and conventional PID controller. The outer ellipse describes the target position, the inner ellipse shows the actual motion of the stage

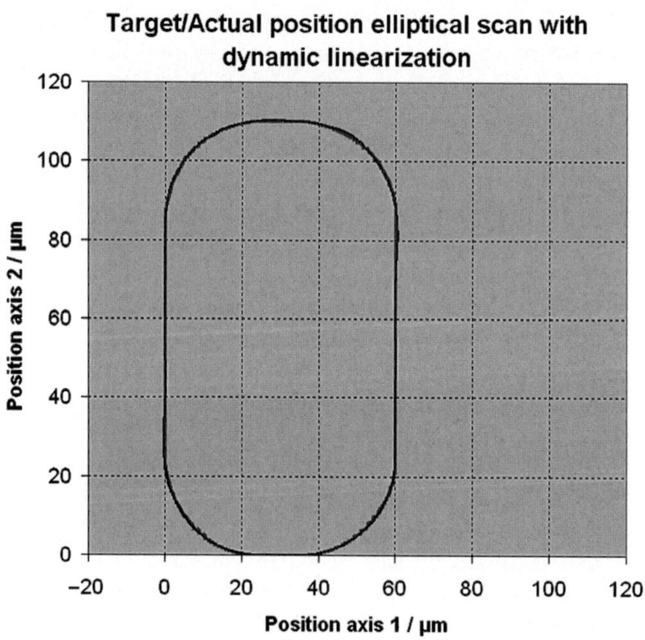

Fig. 12.17. Same scan as before but using dynamic digital linearization, a highly effective preshaping algorithm. Target and actual data can hardly be discerned. The tracking error has been reduced to a few nanometers

Inertial Piezo Drives also offer theoretically unlimited travel. They use a special controller and exploit the difference between static and dynamic friction to push a runner in the motion direction. Because of the role of friction between the runner and stator, the resulting force generation capacity and system stiffness is limited. The range over which high-resolution motion can be obtained is a few tens of microns; the speed up to a few tens of microns per second.

The third possibility is a drive in which multiple piezo elements execute a complicated, stepping motion sequence. Here too, the travel range is practically unlimited. The resolution, however, is not limited to the step size, because an analog mode allows high-resolution motion over the range of one step – a few tens of microns – and that with excellent dynamics. Speeds of up to several millimeters per second can be obtained.

The step-mode motion sequence (Fig. 12.18) includes phases in which the piezo element lifts off the runner and others where it drives the runner in the direction of motion. The clamp/unclamp and drive motions can be implemented with bender actuators. For applications with large moved masses, or where the dynamics require extremely high stiffness of the drive, the piezo elements can consist of two differently polarized segments: one for the clamping motion and one for the shear motion. The shear elements are responsible for the driving force, while the longitudinally polarized clamping elements control contact with the runner in the different step phases.

In analog mode, where only the shear segments are active, motion with subnanometer resolution can be carried out over a travel range of about one step (Fig. 12.19). Analog mode exhibits excellent dynamic performance, making these drives suitable for tasks as demanding as active vibration control. Piezoelectric nanopositioners with stepping drives can generate forces of up to 1,000 N and offer resolutions of better than 100 pm (Fig. 12.20). High-resolution incremental encoders, which provide very high accuracy over ranges of many millimeters, are used as position control sensors.

Specially designed electronics ensures the coordination of the shear and clamping motion of the individual piezos. Stepping drives represent a

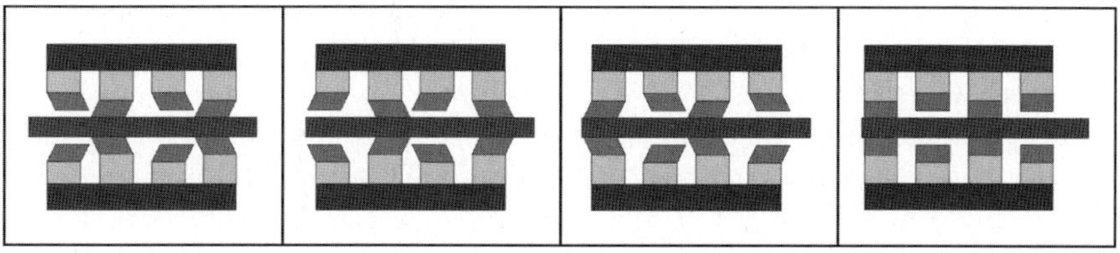

Fig. 12.18. Motion sequence in a piezo stepping drive with shear and clamping element segments. While one set of piezo elements clamps and drives, the other set takes a new bite

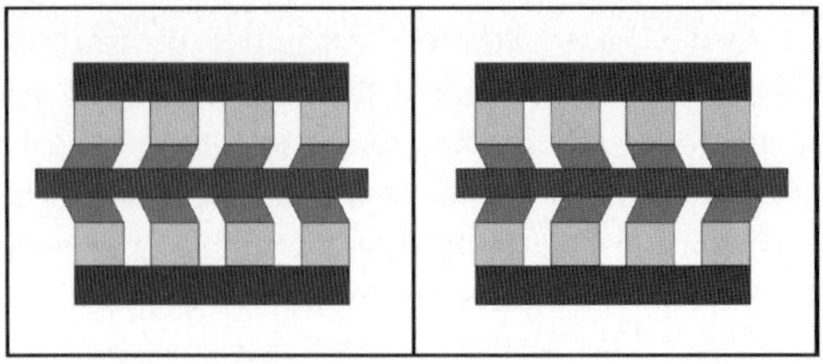

Fig. 12.19. In analog mode only the shear segments are active and both sets of elements are clamped to the runner

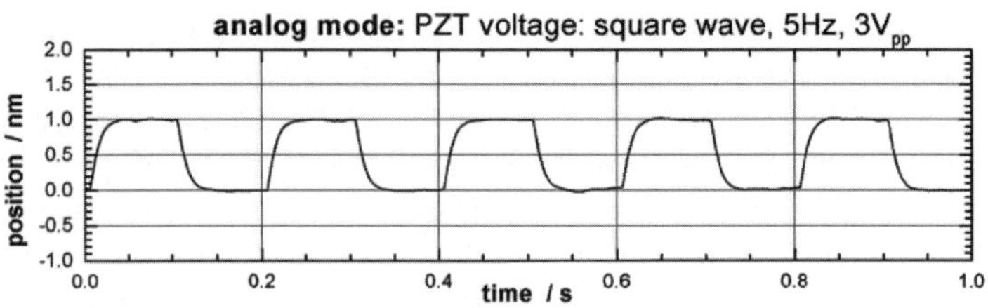

Fig. 12.20. Train of open-loop 1 nm moves in analog mode, as performed by a stepping drive with shear and clamping piezos

forward-looking technology for ultra-high-precision adjustment tasks, including the active vibration compensation of heavy machine components. There are already a large number of these drives in service in industrial applications [6].

12.6 Future Prospects

As mentioned earlier, in the semiconductor industry, lithographic techniques are used to produce feature widths of under 100 nm on silicon wafers. As development proceeds to even smaller features, it will be possible to manufacture higher-performance electronic components, and new manufacturing technologies will become available. This makes the semiconductor industry the driving force for the further development of nanopositioning where piezo actuators already became the key technology.

In biological research, high-resolution positioning is also used for the identification, localization, monitoring and control of molecules in cellular processes. Should molecular biological techniques find a place in early detection and therapeutic procedures, then nanopositioning could experience an upsurge through increased use in medical equipment.

Microscopy and other image-producing processes increasingly turn to nanopositioning. Examples are adjusting the focal plane when modeling samples in three dimensions, probe manipulation in atomic force microscopy, or rapid scanning of CCD arrays at the pixel level to increase the effective resolution (dithering). The widening use of such systems in digital cameras designed for long-distance aerial photography, in astronomy or in microscopy-based inspection systems and medical equipment will continue to pose new challenges for nanopositioning technology.

While taking a position within a few nanometers and holding it for a certain time is the central claim of nanopositioning, the ancillary conditions will become more and more demanding. Longer and longer travel ranges with no degradation in resolution or stiffness will be required. Harsh environments, such as vacuum, helium atmosphere or intense magnetic fields will challenge reliability. The manufacturers of nanopositioning equipment will find it increasingly necessary to examine new technologies like stepping drives. They will also have to look for more efficient manufacturing processes because industrial applications are generally cost-sensitive and the use of piezoelectric drives in an industrial process must not reduce the throughput, but rather increase it. Not all limitations have their solution in improved actuator design or other hardware changes. Dynamic signal reshaping for faster settling or better trajectory control, filtering techniques or new types of control algorithms are also part of the development of nanopositioning technology. Nonetheless, one important direction is the research into new active materials and technologies, with the object of obtaining greater induced mechanical displacement, lower hysteresis and improved linearity.

References

1. S. O. R. Moheimani, Rev. Sci. Instr., **79**(7), pp. 071101 (2008)
2. P. Pertsch, D. Rössger, E. Hennig, W. Plötner, A. Bauer, in *Proceedings of 8th ACTUATOR 02*, Bremen, June 2002, Axon, pp. 45–48
3. P. Pertsch, H. Marth, P. Anger, A. Feltz, in *Proceedings of 7th ACTUATOR 00*, Bremen, June 2000, Axon, pp. 41–44
4. P. Pertsch, S. Richter, D. Kopsch, N. Krämer, J. Pogodzik, E. Hennig, in *Proceedings of 10th ACTUATOR 06*, Bremen June 2006, Axon, pp. 527–530
5. R. Glöss, P. Pertsch, Mechatronik F&M **107**(9), pp. 64–68 (1999)
6. K.-D. Müller, H. Marth, P. Pertsch, R. Glöss, X. Zhao, in *Proceedings of 10th ACTUATOR 06*, Bremen, June 2006, Axon, pp. 149–153

13

Piezoelectric Injection Systems

R. Mock and K. Lubitz

13.1 Technical Motivation for Direct Injection Systems

The origin of direct injection can be doubtlessly attributed to Rudolf Diesel who used air assisted injection for fuel atomisation in his first self-ignition engine. Although it became apparent already at that time that direct injection leads to reduced specific fuel consumption compared to other methods of fuel injection, it was not used in passenger cars for the moment because of its disadvantageous noise generation as the requirements with regard to comfort were seen as more important than a reduced specific consumption.

Especially the cold start Diesel knock proved to be unacceptable for a long time. Although one knew at least on a theoretical basis that pilot injections would be the appropriate countermeasure, one was not able to realise them with the available injectors. This situation changed fundamentally in 1989 when Audi launched the first TDI engine with Diesel direct injection in which the problem of pilot injection was solved. Nonetheless, the combustion noise could not yet be reduced to the level of a gasoline engine. The reason for this was that using the pump-injector principle, the ratio of the dosed fuel amount as well as the timing between pilot and main injection could not be adjusted with the required flexibility.

It was not before Diesel high pressure pumps became available that the way was paved for the "common rail" technology. In 1997, the high volume production of "common rail" injectors was launched using electromagnetic Diesel injectors, in 2000 the first vehicles left the production plant which were equipped with piezoelectric "common rail" injectors [1–4] (see Fig. 13.1). Above all, it was the high flexibility of fuel dosing which permitted a variability of fuel injection in form of several (up to seven) split quantities – something which had not been considered as possible till then. This technical feature along with the high dosing precision and spray quality broke new ground for a triumphant success of Diesel power train technology which at least in Europe has by far outnumbered the Gasoline vehicles viewing the number of newly registered cars.

Fig. 13.1. Photographic view of the first generation PCR injector

13.2 Description of Advanced Direct Injection Systems

13.2.1 Diesel: Common Rail Injectors (2,000 Bar)

The first generation of piezoelectric "common rail" Diesel injectors (abbrev.: PCR) which in September 2000 was employed for the first time in a series production vehicle uses a servo-hydraulic principle instead of a direct drive by the piezo actuator to move the injector's nozzle section (Fig. 13.2).

For that purpose, the high pressure Diesel fuel ($0 < p < 1{,}500\,\text{bar}$) acts in two ways on the nozzle section. By directly feeding the Diesel fuel to the bottom part of the nozzle, a constant force is exerted on it. The "control piston" which forms the upper part of the nozzle section is pressurised by the fuel which is fed via two throttles into a control chamber which is closed by a servo valve. As long as the servo valve is closed, the pressure in the control chamber is equal to that acting on the lower end of the nozzle. As the cross section area of the control piston is slightly larger than that of the lower end of the nozzle, a resulting force is generated which presses the nozzle into its seat.

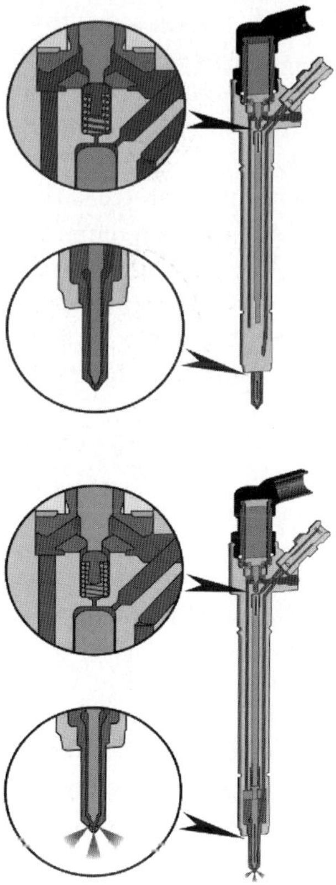

Fig. 13.2. PCR Injector: pressure distribution in the closed state (*upper view*) and in the open state (*lower view*) of the servo-hydraulic injector drive

To trigger an injection, an electric voltage is applied to the piezoelectric actuator. Via a tappet the expanding actuator lifts the servo valve off its seat and a ring-shaped gap opens through which the fuel inside the control chamber can escape into the low pressure drain ($p = 1\,\text{bar}$). As a consequence, the pressure above the control piston drops almost instantaneously. As the pressure on the lower end of the nozzle remains constant, the direction of the resulting hydraulic force on the nozzle is reverted so that the nozzle is lifted off its seat and thereby the nozzle outlets are released through which the fuel is injected into the combustion chamber. In addition it is atomised because of the high system pressure of up to 1,500 bar.

A spring between the control piston and the nozzle part on the one hand acts as an impact damper, on the other hand it keeps the nozzle in the closed state as long as no rail pressure is present – as it is, e.g. the case when the engine is turned off.

When the voltage across the piezo actuator is switched off, it contracts. Driven by a spring force, the servo valve is moved back to its seat. The ring-shaped gap is closed again so that via the two throttles the full system pressure is built up again inside the control chamber. As soon as the full pressure is reached, the resulting force acting on the nozzle and control piston again points towards the seat. The nozzle moves into its stationary position and the nozzle outlets are closed.

Via the diameter ratio of the two throttles, the opening and closing velocities of the nozzle can be adapted to the requirements of the respective engine.

That such a servo principle is used instead of, for instance, a direct drive is due to the fact that the mechanical work that has to be applied in order to open the nozzle exceeds the capabilities of the actual piezo actuator by one order of magnitude. This circumstance is discussed in more detail in Chap. 13.3.1.

13.2.2 Gasoline High Pressure Direct Injectors (200 Bar)

A fuel atomisation, the quality of which being sufficient for a good combustion, is reached in the case of gasoline at pressures of about 200 bar. As a consequence, it is feasible to realise gasoline high pressure injectors actuating the nozzle by a direct drive (Fig. 13.3).

In the case of the injector for which series production has been started in 2006 [5–8], the functional unit of piezo actuator, the nozzle and hydraulic compensator is mounted into a double-walled, cylindric housing. The fuel is fed to the nozzle via the gap between the inner and the outer tubes, thus acting as a coolant for the inner injector parts, mainly the actuator unit.

The top cap of the piezo actuator is supported by the hydraulic compensator, whereas the bottom cap is in contact with the valve needle. During the mounting procedure, the needle is loaded by a spring which generates a force of about 200 N by which the needle is pressed into the seat. After the installation and the application of the electric contacting to the piezo actuator, the hydraulic compensator unit is mounted into the injector housing and loaded by a force of about 100 N against the piezo actuator. Thereby the force between the needle and the seat is reduced by just this number so that 100 N remains as a resulting force which retains the nozzle in its seat against a maximum fuel pressure of 200 bar.

The fuel enters the gap between the needle and the cartridge at the lower end of the injector housing across two boreholes in the inner tube of the housing. To prevent the piezoactuator from being wetted by the fuel, a bellow between the upper end of the cartridge and the needle acts as a flexible seal in the axial direction. Its effective pressure-loaded cross section area is dimensioned such that the hydraulic forces acting on the bellow and the needle compensate in essence.

As the piezo actuator is electrically charged, it expands against the forces by the compensator and the needle. As the compensator is conceived such that it forms a hard bearing for a fast switching motion, the valve lifts off

Fig. 13.3. Photographic view and CAD cross section of a piezoelectric high pressure gasoline injector

due to the electric excitation of the actuator after the closing force has been overcome. A ring-shaped gap opens between the needle and the seat through which the fuel is injected and atomised.

As soon as the desired fuel quantity is reached, the piezo actuator is electrically discharged and as a consequence contracts. The preload spring then pulls the needle back to the seat. The ring-shaped gap is closed, and the injection cycle is complete.

13.2.3 Gas: CNG/Hydrogen Injectors (<200 Bar)

Considering the CO_2 problem and the global warming, gaseous fuels increasingly become the focus of attention for developers. They either exhibit a greatly improved CO_2 balance than gasoline or Diesel, or like hydrogen they can completely be generated by means of regenerative energy sources and accordingly are CO_2 neutral.

At present, the development focuses on injectors which take into account the specific requirements of the injection of gaseous fuels [9]. The fundamental difference to liquid fuels is the completely different flow characteristics of gases. To dose a gaseous fuel which contains the same amount of energy as liquid fuels, one has to inject a considerably larger volume. At a pressure of, e.g. 20 bar the injector has to deliver a throughput quantity which exceeds that of gasoline by a factor of fifty in order to obtain the same engine power. The injection timing and the gas pressure, however, are defined by system requirements so that the larger volume throughput has to be realised by an increased cross section of the orifice. As the diameter of the employed nozzle can be increased only to a limited amount, there remains only an increase of the needle lift to a value of 300 μm to guarantee the throughput quantity. However, using the present piezo actuators, a stroke of 50 μm cannot be exceeded. Therefore a hydraulic transformer is mounted between the actuator and the nozzle which at the same time acts as a thermal compensator. The working principle of this injector including the hydraulic transformer element is shown in Fig. 13.4.

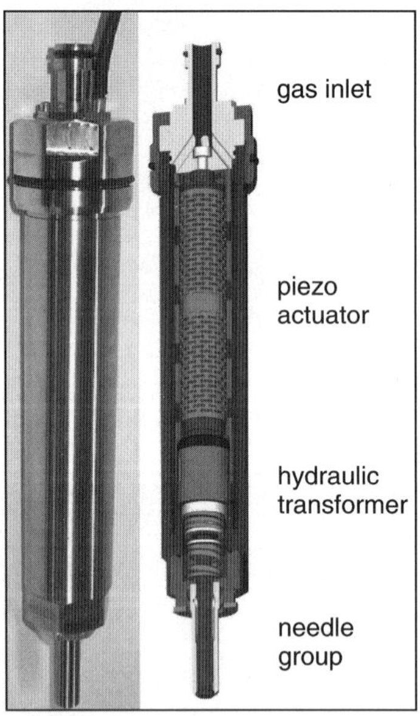

Fig. 13.4. Photographic view and CAD cross section of a piezoelectric CNG injector

13.2.4 Future Perspectives of Piezo Injectors

Unlike their electromagnetic counterparts, piezoelectric injectors provide a rather well defined sensor signal. This is simply due to the fact that a piezoelectric material always shows the piezo sensor and the actuator effect at once. Monitoring the time dependent voltage and the current of a piezo actuator, which is employed in an injector, yields a lot of information about the inner states of the injector as well as the surroundings of the injector. Thus using model-based control strategies which treat the injector as a part of a mechatronic control system it is possible to detect important information about the switching process such as opening and closing time of the servo valve in the PCR by recreating the inner state variables which describe the effect under consideration. Moreover, it appears to be possible to even employ a piezoelectric injector to determine quantities related to the combustion process just by isolating their impact on the injector function by a model-based control approach. At present, this is work in progress and the first substantial success in improving the existing products with respect to injection precision, environmental friendliness and the reduced system costs proves that the mechatronic system approach is one which is worth to be pursued over the next years.

13.3 Technical Specifications of Injection Systems

13.3.1 Diesel: Common Rail Injectors (2,000 Bar)

Due to the different working principles of the three injector concepts presented before, the technical requirements which have to be established for a suitable piezo actuator greatly differ. Nevertheless, the basic procedure by which the mechanical and electric characteristics are specified is the same in all the three cases. For the PCR, this procedure will be explained in some detail.

By the functional principle of the PCR, it becomes obvious that the opening and closing event only occurs when the pressure inside the control chamber has reached its respective end value. Particularly during the opening event, the pressure drop inside that chamber has to be effected sufficiently fast so that the movement of the control piston and the valve needle begins after a minimum time delay. As this pressure drop is substantially influenced by the flow characteristic of the servo valve, the piezo actuator has to fulfil two tasks in this case:

1. It has to provide sufficient stroke so that the servo valve is fully unthrottled in the opened condition (i.e. only the fixed throttles in the high pressure feed determine the flow rate)
2. It has to generate enough force to open the servo valve sufficiently fast against the pressure in the control chamber

To guarantee the first requirement, the flow cross section area released by it has to be significantly larger than that of the two fixed throttles which act in series with it. In case of the PCR the flow cross section at a stroke of 35 μm is roughly twice as large as that of the largest fixed throttle – therefore just sufficient to fulfil the requirement.

The force on the servo valve built up by the chamber pressure can be calculated from the pressure loaded surface of the valve at a given pressure. Taking into account also the force of the closing spring, the actuator has to generate a force of about 400 N at maximum system pressure before the servo valve is moved at all. Detailed simulations show that in order to obtain a sufficient dynamics of this process one indeed needs about 1,000 N!

The piezo actuator's maximum force, however, is not available during the complete actuation process. Due to the limited stiffness of the mechanical connections (weldings, etc.), part of the mechanical energy generated by the actuator is consumed by the mechanical work needed to expand the housing and its components. Provided that the stiffness values of housing and mechanical connections are optimised, the actuator has to exhibit a blocking force of about 2,100 N – at a blind lift of 35 μm this corresponds to a spring stiffness of 60 N μm^{-1} – to guarantee a net blocking force of 1,100 N for the switching cycle.

From the electric point of view, the second requirement means that the actuator has to be electrically energised with a rise and fall time of 100 μs to the maximum field strength of 2 kV mm^{-1}. Regarding technical boundaries and system costs, these electric characteristics have to be realised at moderate voltages (<200 V) and currents (<30 A). The consequence of this is that the piezo actuator has to be conceived as a multilayer stack with an appropriate number of piezoelectric layers.

13.4 Material Aspects

Motor injection systems in the past were realised solely with electromagnetic drives. In order to replace them by the completely new concept of a piezoelectric drive, the latter must have comparable production costs, life time and reliability and in addition a remarkably better functionality. It is described in the following section how a specific material design, i.e. matching of construction form, material composition and fabrication technology can lead to such a solution.

If we consider the functional principles of Diesel injection valves shown in Sect. 13.2.1, we can summarise the specifications for the piezoelectric drive itself by the following conditions: Drive voltage of up to 200 V DC, drive frequency of up to 100 Hz, piezoelectric elongation of 40 μm, blocking force of about 2,000 N, pulse length without rise and fall times 50–100 μs, operating temperatures from −40 to +160°C, where the maximum temperature is given by the external motor heating and by self-heating, and finally a minimum life time of $>10^9$ cycles.

The solution is a piezoceramic prismatic stack with dimensions of $7 \times 7 \times 30\,\text{mm}^3$. This stack is manufactured as a monolithic component in multilayer technology and has an internal structure as shown in Chap. 4.5, p. 123. On the base and top of this part, where force and elongation are transferred, always a 1–2 mm thick piezoelectrically inactive zone is incorporated in order to avoid friction and mechanical tensile stresses by piezoelectric transverse contraction. The active part of the stack consists of some hundreds of 80–100 μm thick ceramic sheets lying in between thin AgPd-electrodes having alternating polarity. They are driven by a voltage of 200 V or an electric field strength of $2\,\text{kV mm}^{-1}$. In combination with the effective d33* the active height of the stack determines the elongation and the cross section the force of the actuator.

During the rise of a pulse the actuator mainly generates force and after the opening of the valve, a no-load deflection. The specification values require an effective d33* of $750\,\text{pm V}^{-1}$ at $2\,\text{kV mm}^{-1}$. Such large values are only achieved by soft-PZT. A further boundary condition, the maximum operation temperature, in addition demands a material with Curie temperatures >300°C. In consequence only low doped soft PZT can be used, for example with Rare Earths on A-place or Nb on B-place of the perovskite.

Using standard ceramic fabrication technology such materials have sintering temperatures of 1,250–1,300°C. The manufacturing of multilayer stacks with AgPd-70/30 inner electrodes, however, needs a reduction of the sintering temperatures to values <1,130°C. This can be achieved by the application of high energy milling whereby the primary particle size before sintering can be remarkably reduced [10]. Particularly the sintering temperature can be reduced by small additions of acceptor elements like K or Ag (see Chap. 4.3 and 4.5) by which the soft PZT does not change its properties like low mechanical quality factor or high large signal d33. Looking at Figs. 4.22 and 4.23 of Chap. 4.5 we can see that using AgPd inner electrodes the Pd is forming a PbPd interface between PZT and electrode and that Ag may diffuse into the ceramic grain volume. In combination with a fine adjustment of the doping concentrations, the reduction of the sintering temperature can be achieved which is necessary for the application of the multilayer technology.

A first piezoelectric multilayer stack was realised already in 1983 as a driving element of a needle printer, admittedly using a low sintering complex ceramic with a Curie temperature of 150°C [11]. The first multilayer stack functioning in an injection valve with AgPd inner electrodes, low doping and a Curie temperature of 320°C however was presented not until the end of the nineties [12, 13].

The stack technology is similar to that described in Chap. 4.5 but requires two essential modifications: The first one relates to the polarisation in combination with the driving conditions, the second one concerns the method of contacting the stack.

If a free stack is driven by an above described pulse sequence, it splits after some hundreds of cycles. Although the lowest resonance frequency of the actuator is about 50 kHz and electrical pulses of 100 μs imply a quasistatic driving condition, in this dynamic mode during the falling pulses mechanical tensile stresses occur which by undercritical and critical crack growth destroy the stack very fast. Therefore a constant compressive stress must act on the stack in such a way that on no account tensile stresses can occur within the sample. That will be realised by tube springs in which the stack is welded under a compressive stress of 10 MPa [14].

The polarisation takes place within the tube spring. The superimposed compressive stress remarkably reduces the remanent relative elongation compared with a free poling. The number of domains oriented parallel to the poling field is lowered due to the ferroelastic influence of the mechanical stress. During driving and under an electrical field of 2 kV mm^{-1} many domains are switched in field direction and the stack elongation reaches the necessary amount.

This kind of polarisation has a further beneficial effect. In Fig. 13.5, the thermal expansion coefficients are plotted of unpoled and poled piezoceramics parallel and perpendicular to the poling direction [15]. In poling direction, against all normal experience the coefficient is negative. This is due to a decrease of the spontaneous polarisation and of the anisotropic lattice distortion with increasing temperature. The stack polarisation under mechanical compression stress creates a domain structure similar to that of an unpoled sample. Thus the thermal expansion coefficient is positive and comparable to that of the surrounding metallic tube spring.

The most important point which at all has made possible the application of piezo stacks as injection valve drives and which guarantees life time and reliability, is the application of quite specific electrical contacts. They must control and eliminate the detrimental influence of *in principle* originating

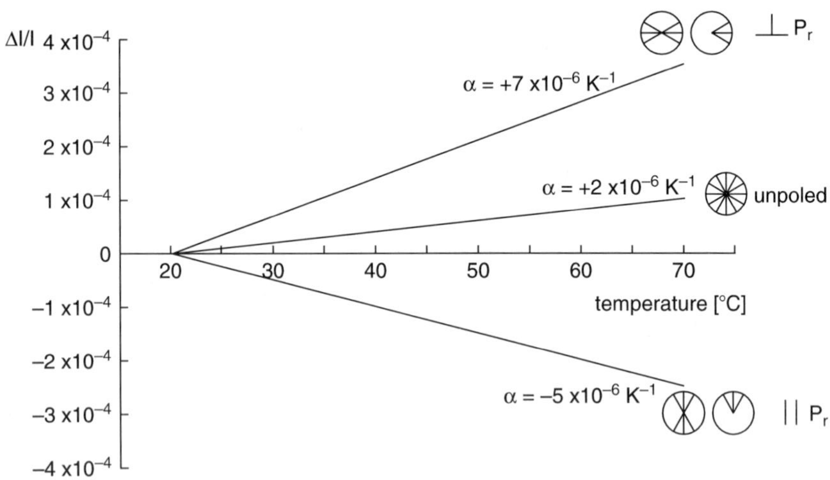

Fig. 13.5. Thermal expansion coefficients of unpoled and poled samples parallel and perpendicular to the remanent polarisation

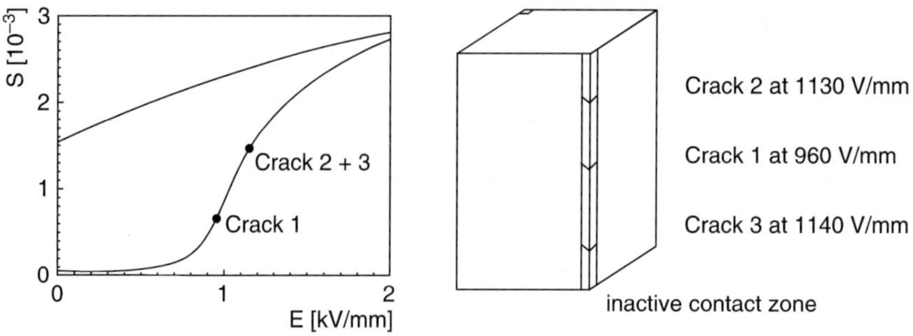

Fig. 13.6. Formation of delamination cracks during poling inside the inactive contact areas

cracks. Within the contact areas of the stacks (see Fig. 4.21, Chap. 4.5) even during polarisation and beginning operation tensile stresses arise although the active stack area is under compression stress by the tube spring. These stresses can be calculated, are highest in the middle of the stack and lead to delamination cracks in the contact area in general between PZT and the electrode. In Fig. 13.6, the evolution of such cracks during poling is measured by sound emission analysis and also their position is shown.

These fissures act as mechanical stress relievers so that their number after short operation time remains constant, depending on design, technology and especially on the strength of the interface PZT – inner electrodes. The crack tips stop permanently within the transition region to the active stack area which is under compression stress. Also some design variations are known where a series of artificial slits within the contact areas of the stack is incorporated during the fabrication having the same stress relief function [16, 17]. During operation in the region of the relief cracks a large stroke can be observed corresponding to the total elongation of the adjacent undamaged stack segment. Locally relative elongations of more than 100% may occur which, after short operation time, unravel the originally continuous metallic contact strip. In order to provide all these single stack elements with electrical energy, a multiple contact design is necessary. That can be realised for example by a wire harp or by soldered metal wool.

With the sum of the described measures, it succeeds to manufacture reliable stacks with life times of $>10^9$ cycles. Meanwhile, they are in use in cars a millionfold. It is remarkable that the absolute stack length as well as the elongation remain constant within narrow limits during life time. This means that both the remanent domain structure and the number and kind of domains switched up to saturation during each pulse remain unchanged. This also shows the future potential of piezoceramic stacks in the area of actuator applications if material and technology are modified specifically.

References

1. D. Baranowski, W. Kluegl, D. Schoeppe, in *Simulation and Design Optimization of a Common Rail Piezo Injector for Passenger Car DI Diesel Engines Fuel Injection System*, London, GB, Dec 1–2, 1999 (IMechE Seminar Publication, London, 1999)
2. K. Egger, D. Schoeppe, Diesel Common Rail II – Einspritztechnologie fuer die Herausforderungen der Zukunft Fortschritt-Berichte VDI Reihe 12, Nr. 348, Band 2
3. R. Mock, P.-C. Eccardt, B. Gottlieb, A. Kappel, K. Niederer, in *Modellierung der dynamischen und akustischen Eigenschaften eines piezoelektrischen Common-Rail-Diesel-Hochdruckinjektors mit ANSYS XXVI.* FEM-Kongress, Baden-Baden, D, 15–16 Nov. 1999
4. K. Egger, J. Warga, W. Klügl, Motortechnische Zeitschrift. **63**(9), 14–17 (English), 696–704 (German) (2002)
5. E. Achleitner, S. Berger, H. Frenzel, M. Klepatsch, V. Warnecke, Benzin-Direkteinspritzsystem mit Piezo-Injektor für strahlgeführte Brennverfahren MTZ (Motortechnische Zeitschrift) 05/2004
6. E. Müller, R. Otte, Potenzial der Doppeleinspritzung beim DI-Ottomotor MTZ (Motortechnische Zeitschrift) 09/2006
7. A.E. Funaioli, E. Achleitner, H. Baecker, in *Direct Injection Systems for Otto Engines SAE World Congress and Exhibition*, April 2007, Detroit, MI, USA, Session: Direct Injection SI Engine Technology (Part 3 of 3), 2007-01-1416
8. W. Mährle, C. Luttermann, M. Krauss, N. Klauer, High-Precision Injection in Combination with the New BMW Twin-Turbo Petrol Engine MTZ Worldwide 03/2007
9. A. Koch, G. Lugert, E. Magori, G. Bachmaier, S. Eisen, in *Key Components for Gaseous Fuels: Actual Demands and Future Concepts 1st International Symposium on Hydrogen Internal Combustion Engines*, University of Graz (Austria) September 2006
10. K. Lubitz, H. Hellebrand, D. Cramer, I. Probst, in *Euro-Ceramics II, Proceedings ECerS'91*, September 11–14 (1991), Augsburg, 1995
11. S. Takahashi, A. Ochi, M. Yonezawa, T. Yano, T. Hamatsuki, I. Fukui, Ferroelectrics **50**, 181 (1983)
12. C. Schuh, K. Lubitz, T. Steinkopff, A. Wolff, in *Piezoelectric Materials in Advances in Science, Technology and Applications*, ed. by C. Galassi et al. (Kluwer Academic, Dordrecht, 2000), pp. 391–399
13. K. Lubitz, C. Schuh, T. Steinkopff, A. Wolff, in *Proceedings of the 7th International Conference on Actuators,* Actuator 2000, June 19–21, Bremen, Germany, pp. 58–61 (2000)
14. Patent EP 1053568 B1, 2000
15. K. Lubitz, C. Schuh, T. Steinkopff, A. Wolff, in *Piezoelectric Materials in Devices*, ed. by N. Setter (edited and published, Lausanne, 2002), pp. 183–194
16. Patent US 6307306 B1, 2001
17. Patent EP 0844678 B1, 2002

14

Advanced RF Signal Processing with Surface Acoustic Waves on Piezoelectric Single Crystal Substrates

R. Veith, K. Wagner, and F. Seifert

14.1 Introduction

Today's rapidly growing proliferation of wireless services such as radio and TV broadcasting, cordless phones, mobile communication, wide and short range data communication, satellite based navigation systems and the growing number of users require an unprecedented increase of information that needs to be transmitted. Frequencies which are predominantly used in actual wireless transmission systems are limited to the radio frequency (RF) range from about 100 MHz to a few GHz. Today a big portion of the EM spectrum available for transmission is already allocated (Fig. 14.1, EM spectrum allocation) and today's services already require an efficient utilization of the allocated channels. In the future, with more services to be provided, the efficiency of utilization has to be further improved.

According to Shannon's Theory of Communication [1], the information capacity of transmission channels is proportional to the product of their bandwidth and signal to noise ratio (in dB). Thus, in order to provide more information capacity, the allocated frequency spectrum has to be continuously expanded to higher frequencies, any free space between allocated frequency bands has to be minimized and wider bandwidth channels have to be provided for higher data rate services. However, as information capacity depends on bandwidth as much as on signal to noise ratio, this signal to noise ratio must be maintained at its theoretical upper limit set by the signal power and the thermal noise present in the channel. Therefore filters are needed that select signals within the desired channel and prevent inter-channel interference from increasing the noise level.

The growing need for efficient utilization of the EM spectrum for wireless transmission poses challenging requirements on filter technology. Filters must operate at higher frequencies up to the GHz region, with higher bandwidth and low losses in the pass band, high out of band rejection and narrow transition bandwidth between pass band and stop band. As mobility is required for many services, miniaturization and low power consumption becomes another

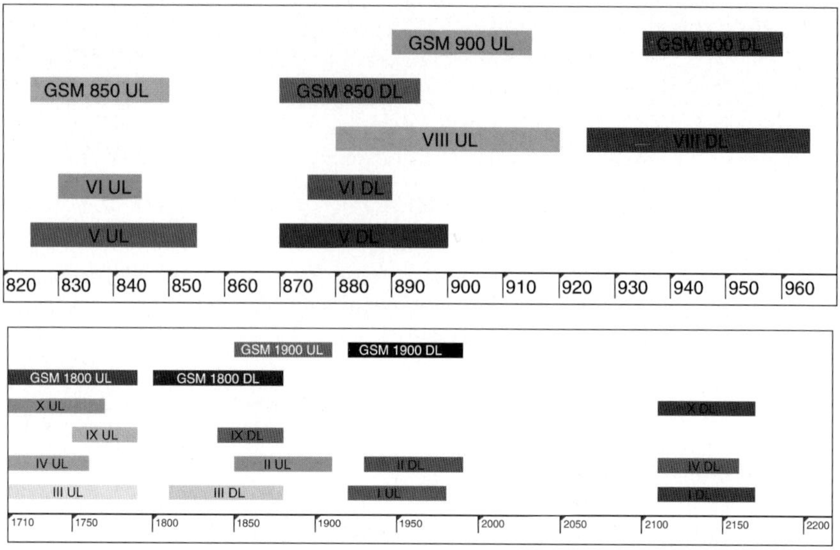

Fig. 14.1. EM spectrum allocation: Part of the spectrum allocated for GSM and WCDMA mobile communication in the 900 and 2,000 MHz regions (UL: uplink, DL: downlink)

challenging requirement for filters. Last but not least high volume consumer applications require reproducibility at high volume and economical cost, while others such as automotive applications require highest reliability in harsh environments.

With piezoelectric materials, mechanical waves in solids can be employed to perform filter functions for electrical signals. The low propagation velocity of mechanical waves provides miniaturization by a factor of 10^{-5} compared to realization of filters with EM waves in the free space. Further more, mechanical waves provide low propagation losses and high stability. Bulk wave resonators made up of thin quartz plates have been used since long for frequency control and are still the component of choice for precisely controlling frequencies in the range of 1–10 MHz. Their application to filtering has been limited as this technology is bound to a low frequency range and does not render itself easily to integration of multiple coupled resonators.

With the advent of the Interdigital Transducer (IDT), an electrode structure that can efficiently launch waves propagating along the surface of piezoelectric materials, higher frequencies in the range 10 MHz to 10 GHz became accessible together with almost unlimited design capabilities in conjuncture with submicron photolithographic planar patterning technology.

Though basically powerful for application to filter design, this technology required the identification of optimized piezoelectric materials in order to meet stringent system requirements. Single crystals proved to be most suitable because of their inherent homogeneous and highly reproducible structure.

As the ideal material with high electromechanical coupling, which is necessary to achieve wide bandwidth, and temperature stability were not available, a variety of materials covering ranges of coupling factors and temperature stability had to be developed – mainly quartz, lithium tantalate (LT) and lithium niobate (LN). Making use of the anisotropic properties of these crystals, special orientations of surfaces and propagation direction have been identified offering an optimum of segregation of coexistent waves, propagation losses, electromechanical coupling and temperature stability.

Together with other achievements in the areas of design, high resolution ultra precise micro patterning and miniaturized packaging, Surface Acoustic Wave (SAW) devices providing unrivalled filter solutions have been developed and are now widely used in today's wireless communication systems. They have especially contributed to enabling an efficient utilization of the EM spectrum for information transmission and to the miniaturization of mobile terminals.

The application of SAW on piezoelectric materials can be extended to other areas such as radio-frequency identification devices (RFIDs) where SAW devices can provide full transponder functionality with the advantage of a passive device without power consumption. Making use of piezoelectric materials, sensitivity against temperature, mechanical stress and interaction with adjacent gaseous or liquid media, SAW passive transponders can also be used as RF readable sensors, e.g. providing information out of harsh environments not easily accessible by other means.

In the future, the need to provide more information capacity, wider bandwidth channels and narrower spacing will continue requiring filters with even wider bandwidth and higher temperature stability. New piezoelectric material systems will have to be developed in order to meet these requirements. Layered structures offer good opportunities for achieving advanced filter performance and further miniaturization.

14.2 Wireless Communication Transceivers: Basic Structure, Filter Functions and Requirements

In modern wireless communication systems, transmitters have to produce a clean signal spectrum limited to the allocated frequency band and receivers have to segregate this spectrum from any other signal without degradation. The necessary signal processing functions, essentially frequency generation and filtering, are being implemented in various stages of the transmitter and receiver circuits. Signal processing becomes most challenging in two way radios such as mobile phones where both strong transmit (Tx) and weak receive (Rx) signals have to processed in close vicinity since one antenna is used for both. These two way radio systems usually have different frequency bands allocated for Tx and Rx signals. Both Tx and Rx bands are subdivided into a larger number of channels, allowing multiple users to use the system simultaneously with each user occupying one designated pair of transmit and receive channels.

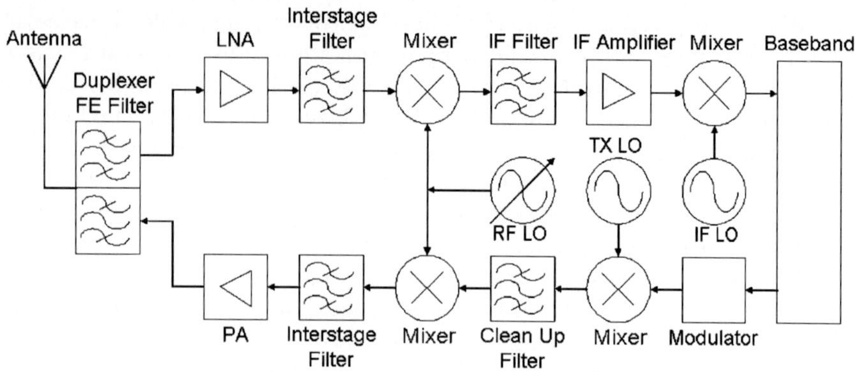

Fig. 14.2. Basic building blocks of traditional radio transceiver

Figure 14.2 shows the basic building blocks of a transceiver circuit, so designated as it combines Tx and Rx functions. Actual transmitters and receivers in any wireless communication system may consist of more or less complex architectures but they are essentially made up of the same basic building blocks. On the Tx path the Tx signals formed in the base band processor are up converted to the designated Tx channel carrier frequency by mixing with the carrier frequency generated in the corresponding RF local oscillator, passed through an RF Tx interstage filter, amplified in the RF power amplifier to the required output power level and finally passed through an RF Tx front end filter before reaching the antenna. On the Rx path, RF signals received from the antenna are first passed through an RF Rx front end filter, amplified in the low noise amplifier (LNA), passed through an RF Rx interstage filter, down converted to an intermediate frequency (IF) by mixing with the frequency generated in the corresponding RF local oscillator, passed through an IF filter, amplified and finally down converted to the base band by mixing with the IF frequency generated in the corresponding local oscillator.

The basic transceiver structure of Fig. 14.2 uses three different categories of filters. The first category, RF front end filters directly connected to the antenna, are used in the Tx path to remove any spurious components of the transmitted signal spectrum outside the allocated frequency band and in the Rx path to reject any unwanted signal outside the allocated frequency band in order to prevent interference with the wanted receive signal. Depending on the system and allocated frequencies, RF filter center frequencies and bandwidths typically range from 400 MHz to 5 GHz respectively from several 100 KHz to 100 MHz. RF front end filters in particular have to meet two critical requirements. On the one hand, front end filters need to provide minimum insertion attenuation in the pass band. Any loss of the wanted receive signal power would degrade the signal to noise ration at the LNA input and thus reduce the receiver sensitivity. Likewise any loss of the transmit signal power would be a waste of output power from the power amplifier and, at least for mobile devices, a waste of battery power reducing e. g. a mobile phone's talk time. On the other hand, front end filters need to withstand the high input power levels required for transmit signals which can reach up to 1–2 W in mobile phones. The transceiver shown in Fig. 14.2 uses the same antenna for both Tx and Rx paths. This configuration requires a duplexer, i.e. a device combining both Tx and Rx front end filters together with a special matching network that provides good matching of both Tx and Rx filters to the antenna and good isolation between Tx and Rx signals at the same time.

The second category, RF interstage filters located before the power amplifier in the Tx path respectively after the LNA in the Rx path, basically perform similar filtering functions as front end filters. As requirements on minimum insertion attenuation in the pass band and power handling capability are less stringent for interstage than for front end filters, interstage filters can be optimized for steep filter skirts in order to improve the rejection of unwanted signals adjacent to the allocated frequency bands.

IF filters, the third filter category, are used to segregate the designated channel from all other channels within the allocated frequency band and to prevent interference between channels. Typical center frequencies and bandwidths for IF filters range from 30 to 400 MHz respectively from 25 kHz to 25 MHz. IF filters have to be highly selective as they have to reject neighboring channels which may be received at much higher power levels than the designated channel. 50 dB and more adjacent channel rejection are typically required. As IF filters are also used to precisely shape the spectrum of the received signals, their pass band characteristics have to accurately match predetermined specifications in amplitude and phase for many applications. In analogue TV broadcasting, e.g., the picture and sound quality largely depends on the TV receiver's IF filter pass band matching the specified characteristics with an accuracy of a few 0.1 dB.

Analogue channel filtering at an IF frequency above 10 MHz offers the best overall suppression of interference between channels and hence the best preservation of channel information capacity. In a trade off between information capacity and miniaturization, the IF frequency can be set to zero or a value close to zero and channel filtering can be integrated into the digital base band processor semiconductor integrated circuit. This possibility becomes even more attractive for multifunctional mobile devices such as multi band and multi mode mobile phones where a multitude of channel filters with different bandwidths can be implemented in one digital signal processor. Today's mobile phone handsets therefore generally employ zero IF respectively low IF receiver architectures whereas mobile phone base stations with less stringent space and weight constraints employ analogue IF channel filtering in order to optimize channel information capacity.

Apart from filtering, frequency generation is the other key signal processing function critical to the system performance. Frequencies necessary to up convert and down convert base band signals are generated in local oscillators (LO). Local oscillators typically employ a low Q value voltage controlled oscillator (VCO) tied to a lower frequency, typically around 10 MHz, highly stable quartz crystal resonator. Comparing the local oscillator frequency to the quartz crystal resonator reference and controlling it accordingly provides the necessary long term stability of the LO frequency. However, this can not remove the high frequency fluctuations or noise generated in the low Q value VCO. In order to reduce this noise and improve channel information capacity, highly stable, high Q value and high frequency resonators to be used for direct generation of LO frequencies or as a reference are required.

In order to fit into mobile devices, where space and weight constraints are becoming more severe with every new more complex generation offering expanded functionalities, signal processing components for filtering and frequency control must meet a number of challenging general requirements in addition to equally challenging requirements on electrical characteristics. In particular, application in mobile devices requires continuous miniaturization and reduction of power consumption. Last not least, in view of the growing number of mobile devices produced every year, continuous improvement of the economy in fabrication of signal processing components is necessary.

14.3 Acoustic Waves in Piezoelectric Solids – Attractive for Filter Solutions – Bulk Waves, Surface Acoustic Waves (Rayleigh Waves), Surface Skimming Bulk Waves

Historically filters and resonators have been first realized with lumped element inductors (L) and capacitors (C) with one LC pair forming a resonator and several resonators forming higher order filters with steeper skirts and wider bandwidth. While such filters had been widely used at frequencies below 100 MHz, the lower quality factors (Q) of inductors and capacitors at higher frequencies were typically not sufficient to meet system requirements. In addition LC filters and frequency control elements were lacking the reproducibility and stability required for today's wireless communication systems and they were by far too bulky to be considered for mobile devices.

At higher frequencies, beyond 1 GHz, electromagnetic waves resonating in metal enclosed cavities provided the necessary Q values to implement high performance resonators and filters with excellent reproducibility and stability. However, with the dimensions of these devices ranging from half a wavelength to several wavelengths, the propagation velocity of EM waves in free space of 3×10^8 m s^{-1} and the corresponding wavelength at 1 GHz amounting to 0.3 m, these cavity filters were even more bulky than their LC counterparts and their application thus limited to large size Microwave and Radar transceiver installations.

The next step in miniaturization was to use electromagnetic waves in dielectric materials in order to slow down the propagation velocity by a factor proportional to the square root of the relative dielectric constant K of the material and shorten the wavelength at a given frequency accordingly. While materials have been developed for dielectric resonators approaching the high Q values and the excellent temperature stability of cavity resonators, the relative dielectric constant K of these materials is not higher than 50 and the maximum achievable miniaturization is thus limited to a factor of about 7. Despite this limitation, dielectric filters and resonators are still widely used at the front end of high power transceivers, e.g. in mobile phone base stations, because of their unmatched power handling capabilities.

Acoustic waves in solids propagating at about $3 \times 10^3\,\mathrm{m\,s^{-1}}$ offer further miniaturization by a factor of 10^5. Piezoelectric materials provide simple means to couple these waves to the electrical signals and piezoelectric single crystals in particular offer the low propagation loss, high Q values, high reproducibility and excellent stability needed for signal processing applications. The earliest and today still dominant application of these principles is in quartz crystal resonators generally used for frequency control. Quartz crystal resonators utilizing shear waves vibrating in a thickness resonant mode provide well defined frequencies, typically within a fraction of 1 ppm, stable within a few ppm over temperature and time. For practical reasons, however, because crystal plates thinner than a few hundred μm are difficult to prepare and handle, the restriction to bulk waves and thickness mode operation limited the achievable fundamental resonance frequencies to a maximum of about 10 MHz.

In 1965 White and Voltmer demonstrated for the first time that surface acoustic waves propagating on piezoelectric crystals could be applied for RF signal processing purposes [2]. With this they did not only overcome the frequency limitations of bulk acoustic wave devices depending on thickness vibrations, they also opened up the door for a host of new more complex RF signal processing functions employing acoustic waves in piezoelectric materials. Before the challenging requirements of wireless communication systems could be met, however, a thorough understanding of acoustic waves in piezoelectric crystals, their excitation, propagation and interaction had to be achieved [3,4].

As any solid material, piezoelectric crystals support three bulk acoustic waves in any direction – one longitudinal wave with displacements predominantly parallel to the direction of propagation and two shear waves with displacements predominantly perpendicular to the direction of propagation and orthogonal to each other for the two shear waves. The longitudinal wave has the highest propagation velocity – about 1.6–1.8 times higher than the two shear waves which in general differ in propagation velocities and are being nominated as fast respectively slow shear wave accordingly. While bulk acoustic waves are typically not employed in surface acoustic wave devices for signal processing, they have to be considered and accounted for as spurious waves that can transmit spurious signals or cause propagation losses to surface acoustic waves radiating part of their energy into bulk acoustic waves.

The surface of a piezoelectric crystal in general supports three different kind of SAWs.

Rayleigh waves, predicted in 1865 by Lord Rayleigh to exist on the free surface of isotropic solids [5], exhibit elliptical particle motion – a combination of longitudinal and shear vertical displacements – with amplitude decreasing exponentially with depth under the surface. With the propagation velocity at the free surface being lower than in the bulk the free surface is functioning as a waveguide. Rayleigh waves thus keep their energy concentrated in a layer about one wavelength deep under the surface, have the slowest propagation velocity and do not couple to any other acoustic wave. Rayleigh waves can be easily coupled to electric fields induced by metal electrodes on the surface and thus provide excellent means for realizing various signal processing functions. Because of their relatively weak electromechanical coupling – i.e. a relatively small fraction of their mechanical energy stored in electrical form – Rayleigh waves render themselves for applications that require precise filtering or frequency control but not strong conversion from electrical to mechanical energy and vice versa.

The first leaky surface acoustic wave (LSAW) consists of a shear horizontal wave propagating along the surface at a propagation velocity in between those of the fast and the slow shear bulk wave. These waves are named leaky as they are coupled to the slower bulk shear wave, continuously radiating energy into the slower wave and thus subject to propagation losses. Propagation losses, however, can be kept to a minimum for certain propagation directions and boundary conditions given by the metallization on the surface. LSAW generally exhibit stronger electromechanical coupling, faster propagation velocity and less concentration of acoustic energy at the surface than Rayleigh waves. LSAW therefore are most suitable for application in high frequency, high power and low loss signal processing functions in the front end of mobile phones.

The second LSAW exhibits predominantly longitudinal displacements and is therefore nominated as longitudinal leaky surface acoustic wave (LLSAW) [6]. LLSAW propagate at a velocity close to that of longitudinal bulk waves, almost twice as fast as Rayleigh waves. This is an attractive feature as it would allow expanding the fundamental frequency range accessible with electrode patterns of limited resolution. In order to reduce normally excessive LLSAW propagation losses to reasonable levels, thick and heavy electrode gratings have to be applied to the crystal surface. Essential LLSAW characteristics, such as electromechanical coupling, propagation velocity and propagation losses are thus becoming quite sensitive to electrode grating dimension variations during fabrication. LLSAW therefore have not found much practical use so far.

Piezoelectric single crystals are highly anisotropic and this has been used to identify directions of SAW propagation for practical use with optimum sets of desirable characteristics such as high electromechanical coupling, high temperature stability, low propagation losses and high suppression of spurious waves. The variation of the propagation velocity with direction can also be used to minimize diffraction of a narrow beam of surface acoustic waves. Provided the variation of propagation velocity in the propagation plane around the main propagation direction is symmetrical and the propagation velocity in the main propagation direction is at minimum, diffraction effects are reduced and the SAW beam stays more focused in the main direction as compared to propagation in an isotropic material. On the other hand, in the general case with the variation of propagation velocity around the main propagation direction being asymmetrical, the direction of power flow will not coincide with the direction of phase velocity and the SAW beam propagation will be diverted from phase velocity vector by the power flow angle. This effect of beam steering has then to be considered and accounted for in the design of SAW devices.

14.4 SAW Filters: Principle and Basic Elements

Although an astonishingly wide variety of signal processing functions can be achieved using SAW technology, all devices are built from only four basic elements which realize acoustic or acousto-electric functions: IDT, reflector, multistrip coupler (MSC), and acoustic waveguide. Whereas the latter two are needed infrequently, IDTs and reflectors set up the basis for all modern devices.

Interdigital transducers (IDTs) are used for generating and receiving surface waves. This transducer was first used by White and Voltmer in 1965 [2], and this date is widely considered as the birthday of surface wave technology, although it had already been referred to in earlier patents [7, 8]. The IDT consists of a set of finger-shaped metallic electrodes connected – in a regular way – to two larger connection electrodes called bus-bars. When an oscillatory voltage is applied to the bus-bars, the transducer generates a periodic electric field in the piezoelectric medium below it, thereby generating a surface wave. Efficient wave excitation takes place, if the IDT period p_{I} is equal to the wavelength λ of the SAW, since the contributions of the different periods superimpose constructively. Thus, even though dependent on the piezoelectricity of the substrate material below the IDTs the surface wave excited by each IDT period can be weak, a reasonable electroacoustic transduction can be obtained.

The configuration of Fig. 14.3a, which consists of fingers connected alternatingly to the bus-bars, is called single-electrode-type or normal-finger IDT. In it, the distance or "pitch" of finger centers is $\lambda/2$. Another very important form is the double-electrode-type or split-finger IDT of Fig. 14.3b, which has four fingers per IDT period. Whereas Bragg reflection occurs in the normal finger IDT for $\lambda \approx p_{\mathrm{I}}$, in the split finger IDT the Bragg reflection can be suppressed, since the period of mechanical reflection centers given by the electrode

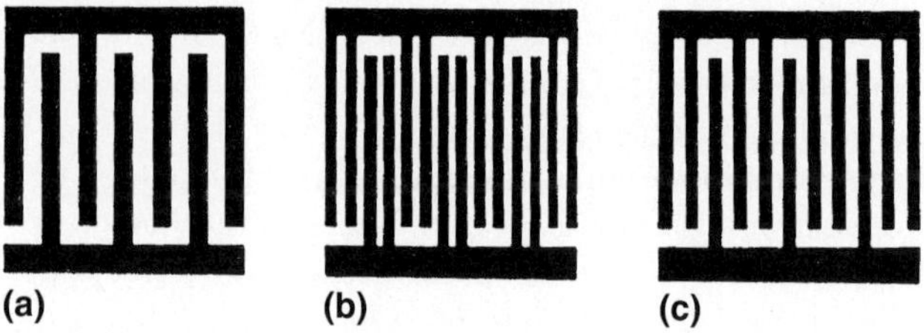

Fig. 14.3. Different types of IDTs. (**a**) Single electrode type, (**b**) double electrode type, (**c**) $\lambda/6$-finger-type

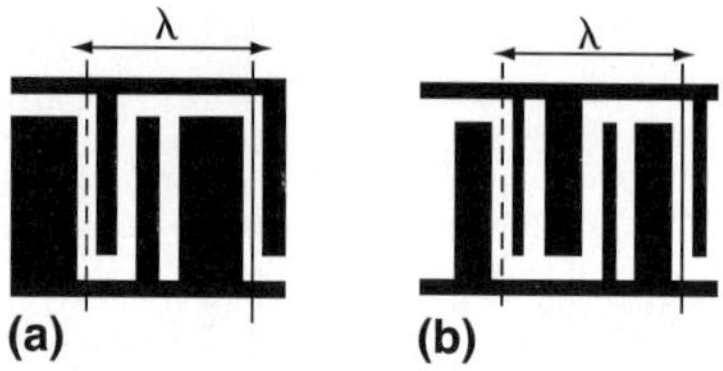

Fig. 14.4. Different types of SPUDT cells. (**a**) EWC SPUDT; (**b**) Hanma-Hunsinger-SPUDT

fingers is $\lambda/4$. Consequently, this type of IDT is used, whenever there are stringent requirements on the group delay characteristics of the SAW component.

Owing to their symmetry, the IDTs described above generate surface waves of equal amplitude in both the forward and reverse directions, so that it is called bidirectional. If an IDT can excite a SAW predominantly in one direction only, we call it a unidirectional IDT (UDT). Unidirectionality can be realized in three different ways: (1) using multi-phase electrical inputs, e.g., three-phase input signals with a relative phase shift of 120°; (2) using single-phase electrical input signals applied to comparatively complex, spatially asymmetric IDT cells, which rely on internal reflection to shift the "center of gravity" of excitation in the cell relative to the "center of gravity" of reflection of the cell such that unidirectionality is achieved; this transducer type is called single-phase unidirectional trandsducer (SPUDT); (3) using substrates with so-called natural unidirectionality, for which – due to crystallographic asymmetry – the center of reflection is displaced from the geometrical center of the electrode, where the center of excitation is located; this form is generally named NSPUDT.

Of all these different forms, only the SPUDTs have gained practical importance. There is a significant number of different realizations of SPUDTs: probably the most-used is the electrode-width-controlled SPUDT (EWC-SPUDT) and the Hanma-Hunsinger-SPUDT (HH-SPUDT) of Fig. 14.4, which consists of cells with three or four electrodes of unequal widths (Fig. 14.5).

Placing two IDTs next to each other and using one as the input transducer, which converts the electrical signal into an acoustic wave, the other one as the output transducer, which regenerates an electrical signal from received

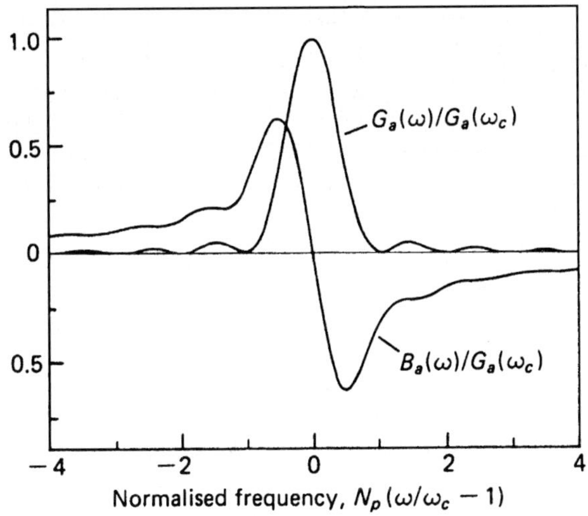

Fig. 14.5. Admittance $Y_a = G_a + jB_a$ of uniform reflection-less IDT of N_p finger pairs with center frequency ω_c

acoustic waves, the most simple filter configuration is obtained (Fig. 14.5). The best achievable insertion loss and the fractional 3 dB-bandwith are both directly proportional to the electro-acoustic coupling factor of the substrate material.

14.4.1 The Acoustic Reflector

For a guided surface wave propagating along a piezoelectric crystal, each electrode finger constitutes a discontinuity of the propagation medium, giving rise to a certain amount of wave reflection and scattering. The reason for this reflection is not only the mechanical perturbation, but also the shortening of the electric field, which both result in a change of propagation velocity. For a wave incident on a periodic array of electrodes, a particularly strong reflection occurs if the period of the grating, p, is equal or close to the half of the wavelength λ, since then the reflected waves are all in phase and interfere constructively. This is the Bragg condition $\lambda = 2p$, which is known for all types of waves in periodic media.

Even if the reflectivity of each electrode is small, total reflection of an incident wave can be reached if only the number of reflecting electrodes is sufficiently large. This fact has – since the early days of SAW technology – been exploited to realize one- or two-port SAW resonators by placing one or two SAW transducer between two reflector arrays. Reflector arrays consist of electrically open or shorted, i.e. interconnected, metallic strips. Also mixed forms (called open-short-reflectors) have been used with advantage.

For sake of completeness, it should be mentioned that – as an alternative to reflectors built of conductive electrode strips – arrays of nonconductive strips applied on or grooves etched into the substrate can be used. For economic reasons and ease of processing, however, metallic strips are usually preferred.

The choice of materials has important impact on resonators' performance. The reflectivity of each strip and, consequently, the width of the stop band are determined by the material properties of both the piezoelectric substrate and those of the strips as well as their geometry. Reflectivity is increased (1) by increased piezoelectric coupling of the substrate, (2) by increased mass density of the electrode material, and (3) increased thickness of the electrodes.

From the application's point of view, on one hand a high strip reflectivity is desirable, because using highly reflecting electrodes, less strips and consequently less die area is needed to attain total reflection. On the other hand a high strip reflectivity results in a higher sensitivity of frequency with respect to fabrication tolerances.

Finally, the maybe most relevant of the secondary properties of a resonator is its frequency behavior with respect to temperature. It is defined by the thermal expansion of the substrate α as well as by the variation of surface acoustic wave phase velocity with temperature $\mathrm{TCV_p}$: $\mathrm{TCF} = \mathrm{TCV_p} - \alpha$. To first order, TCF is defined by the substrate crystal alone. To second order, also the material and height of electrodes influence TCF considerably.

14.4.2 Design Techniques for RF Filters

From the early 1990s, the development of mobile communication technology provided an unprecedented impetus for research in Surface Acoustic Wave technology, at a time when many first generation researchers had already migrated to other research areas, because SAW seamed to be a well established, ripe technology.

Mobile communication equipment development brought about requirements previously almost unknown. It was no longer the precise shaping of a filter characteristic with minimum amplitude and group delay ripple or minimum deviation from a specified form; especially for RF-filters, the new, decisive criteria were minimum pass band insertion attenuation for TX and RX filters, durability to sustain transmit powers of 1–2 W, and minimum package and die size, to name a few only. Because of intensive research in materials as well as filter architectures and design techniques, SAW filters step by step mastered these challenges and replaced other filter technologies, like microwave ceramics, in mobile communication applications almost entirely.

The need for filters with insertion attenuations of less than, e.g. 3–4 dB, over a relative pass-band width of 2–4% expedited the development of new techniques optimized for fulfilling these requirements.

From network theory it is known that networks consisting of discrete reactance elements (i.e. capacitors and inductors) can be designed to implement filter functions. The basic idea is to design a network which approximates zero reflection at the electrical port over the desired pass-band frequency range. In the desired stop-band ranges, on the other hand, the port impedances are

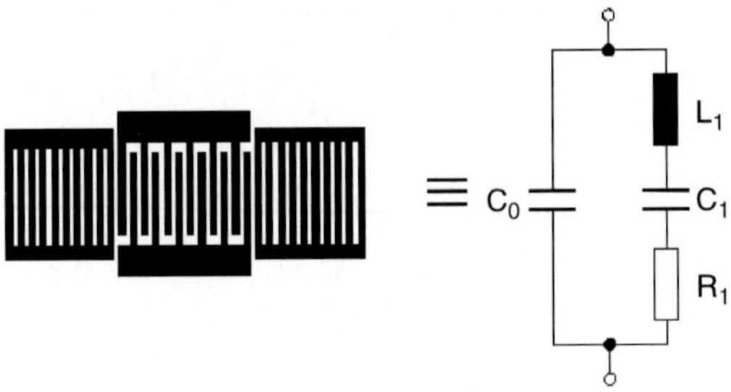

Fig. 14.6. Equivalent circuit of a one-port SAW resonator

deliberately mismatched so that all signals are reflected at the input port and the transmission is minimized. Since impedances of acoustic resonators have strong frequency dependence and can be controlled within certain limits by choice of materials and design measures, the objective was to use them instead of discrete reactance elements.

There are several methods from network synthesis that aim at finding a circuit that implements a given transfer characteristic. Nearly all these methods result in a net-list consisting of ideal inductance (L) and capacitance (C) elements. Since a SAW one-port resonator's electrical characteristics can be well approximated by an equivalent circuit consisting of L and C elements only, it is possible to construct a filter circuit with SAW resonators as building blocks, if the network topology is adequate (so-called ladder type topology) and the element values are realizable.

The equivalent circuit of a one-port SAW resonator consisting of an IDT and two shorted reflectors is shown in Fig. 14.6. It comprises of four elements: a static capacitance C_0, the dynamic capacitance C_1, the dynamic inductance L_1, and a small series resistor R_1. The resistor models the losses that exist in a realistic SAW resonator, and can be neglected for simplified considerations. Figure 14.7 shows a plot of the admittance function of an on-port resonator. There are two characteristic frequencies:

- The (series) resonance frequency f_s, where the admittance has a pronounced maximum ("pole")
- The parallel resonance frequency f_p or anti-resonance frequency f_a, where the admittance function approaches zero

With fixed metallization height and ratio, the series resonance frequency of an IDT is inversely proportional to its finger period. Off the resonances, the admittance function is dominated by the static capacitance C_0.

Figure 14.8 shows the basic section of a ladder-type filter. It consists of a series-resonator and a shunt resonator, whose anti-resonance frequency $f_{\mathrm{a},2}$ coincides with the series resonance frequency $f_{\mathrm{s},1}$ of the series resonator. This basic section has a band-pass characteristic: At $f = f_{\mathrm{s},2}$, the parallel resonator provides a short circuit for the input signal, at $f = f_{\mathrm{a},1}$, the series resonator

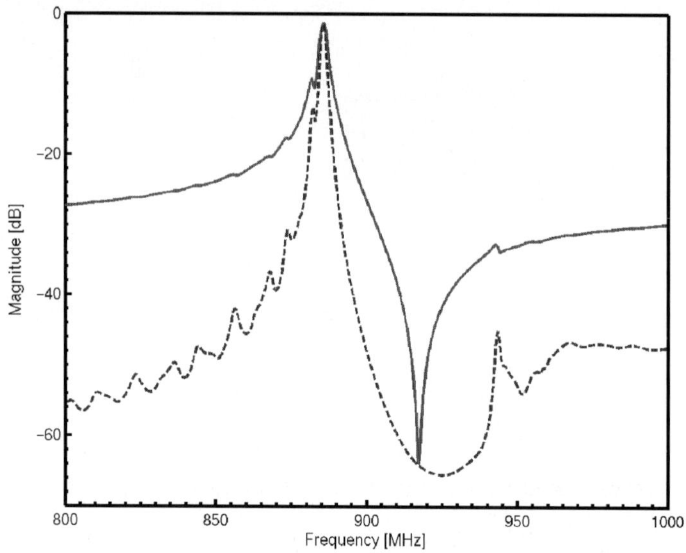

Fig. 14.7. Absolute value (*red solid line*) and real part (*dashed blue line*) of admittance Y11 of a one-port resonator in logarithmic scale

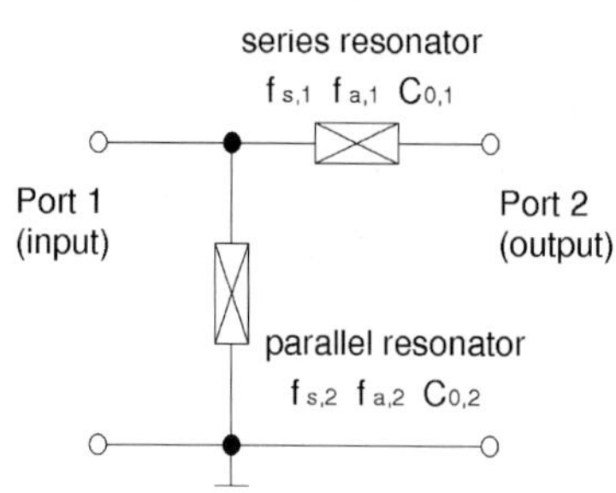

Fig. 14.8. Basic section of a ladder type filter

provides an open circuit for the input signal. So in both situations, the transmission approaches zero. At $f = f_{s,1} = f_{a,2}$, the parallel resonator provides an open circuit and the series resonator a short circuit, so transmission from input to output port is maximal (see Fig. 14.9).

The selectivity of a single basic section is rather low. For practical applications, therefore typically up to nine basic sections are cascaded in order to form a filter characteristic with high selectivity, on the cost of some additional insertion attenuation.

Whereas in the so-called image parameter design, identical basic sections and their mirror images are combined, in most state-of-the-art filter designs the individual sections are tuned individually to reach optimum electrical performance. The common name of this filter structure is LTF) or Impedance Element Filter (IEF).

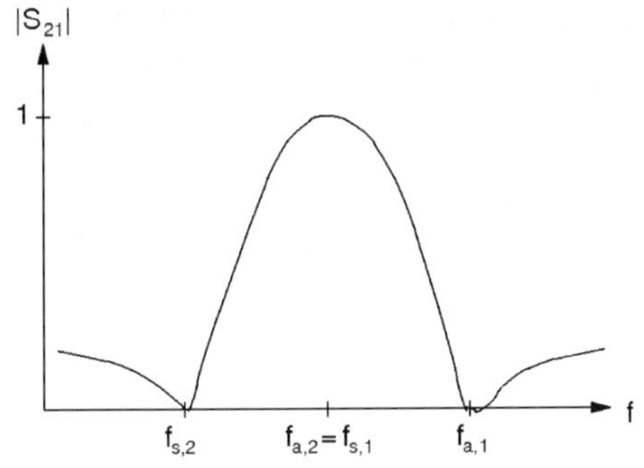

Fig. 14.9. Transfer characteristic of ladder type basic section

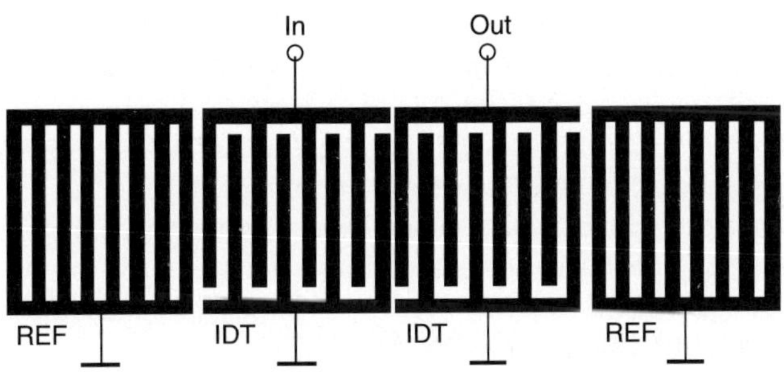

Fig. 14.10. Schematic of a one-track two-IDT DMS-filter

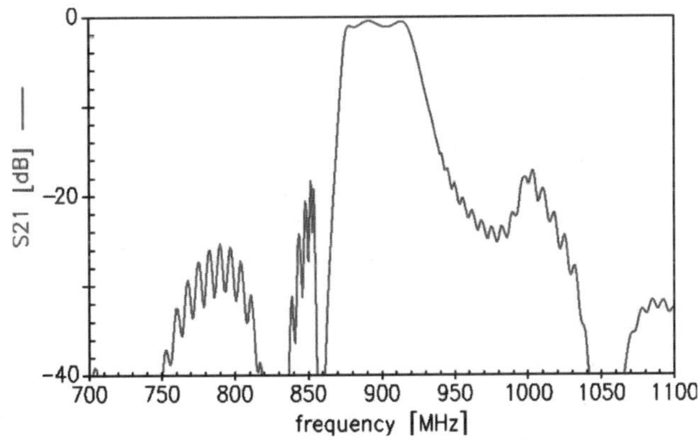

Fig. 14.11. Transfer characteristic (S21) of a one-track two-IDT DMS-filter

An alternative filter configuration to the LTF described above is the so-called Double Mode SAW-filter (DMS-filter) or Longitudinally Coupled Resonator Filter (LCRF). The basic design idea here is to allow for more than one resonance in a single acoustic track. Fig. 14.10 depicts schematically a 1-track-2-IDT DMS filter. The corresponding transfer function is depicted in Fig. 14.11. In the simplest case, the IDTs and reflectors all have the same electrode period, building a homogenous grating, whose resonance frequency determines the left skirt of the pass-band. By introducing a suitable additional gap between the IDTs, a second resonance, named cavity-resonance, is created due to constructive interference of waves reflected at left-hand and right-hand IDT-Reflector cascades, respectively. Both IDT- and cavity-resonance are exploited to achieve a band-pass filter characteristic.

14.5 Piezoelectric Single Crystal Materials for SAW Filters

In order to fully exploit the signal processing capabilities of SAW devices the optimum material system, i.e. the combination of piezoelectric substrate, plane and direction of SAW propagation and metal electrode material, has to be identified for each application. Specific weight, elastic, plastic, dielectric and piezoelectric material properties and the resultant factors such as SAW propagation velocity, electromechanical coupling coefficient K^2, propagation losses, temperature coefficient of delay TCD, spurious wave levels and tensile strength directly affect the achievable limits for miniaturization, center frequency, bandwidth, insertion loss, transition bandwidth, out of band rejection and power handling capability of SAW devices.

Aluminum (Al) has become the most preferred metal electrode material for SAW devices. Al provides sufficient electric conductivity and the low specific weight causes a minimum mass loading at the surface of the piezoelectric substrate with a corresponding minimum effect on SAW propagation velocity and minimum sensitivity of center frequency to dimensional variations of the Al electrodes during fabrication.

The low tensile strength of pure polycrystalline Al, however, is not sufficient for high power applications such as SAW front end filters connected to the antenna of mobile phones. Here an RF input power level to the SAW filter of up to 30 dBm and the resulting acoustic power density at the surface will induce stress levels to the metal electrodes exceeding the tensile strength of pure polycrystalline Al and lead to acoustic migration within the metal electrode, a gradual deterioration of the SAW device performance and finally to a complete failure due to whisker growth and short cuts between electrodes. Copper (Cu) doped Al electrodes [9], AlCuAl sandwich structures [10] and single crystal Al electrodes [11] have been used to provide increased tensile strength metal electrodes for SAW devices that can withstand 30 dBm input power.

As for the piezoelectric substrate, piezoelectric single crystals have become the preferred material for high performance SAW devices. Single crystals provide the best reproducibility of material properties and in contrast to polycrystalline materials, single crystal materials do not suffer from SAW propagation losses due to scattering at grain boundaries.

In the race for SAW filters with wide pass band, minimum insertion loss and narrow transition bandwidth, the ideal piezoelectric material would have a high electromechanical coupling coefficient K^2 and a zero temperature coefficient of delay TCD. As such a material does not exist as a single crystal, SAW devices, depending on the requirements for specific applications, use a number of different piezoelectric crystals. The vast majority of today's SAW devices are thus using any of the following three piezoelectric single crystals, quartz, LT, and LN, of which quartz offers the highest temperature stability with lowest K^2, LT offers moderate temperature stability and K^2 and LN combines the highest K^2 with the lowest temperature stability. Other piezoelectric crystals such as $Li_2B_4O_7$, $Bi_{12}Ge_4O_{20}$ and $La_3Ga_5SiO_{14}$ have been tried for special applications but not found widespread usage so far.

Piezoelectric crystals are highly anisotropic and this fact has been used to search for and identify optimum planes and directions for SAW propagation offering high electromechanical coupling K^2 and low temperature coefficient of delay TCD. In combination with the boundary conditions at the surface, basically the metallization employed, further optimizations have been achieved by identifying planes and directions for low SAW propagation losses, reduced SAW diffraction and reduced excitation of spurious bulk waves.

The identification of optimum SAW propagation directions depends on the type of wave to be employed. The Rayleigh wave, having the slowest propagation velocity and the lowest propagation loss, offers the best potential for miniaturization of high performance SAW devices but electromechanical coupling K^2 is limited to about 5% within the range of most frequently used piezoelectric crystals (Table 14.1). Rayleigh waves are therefore employed for frequency control SAW resonators, narrow band low loss front end filters and highly selective IF band pass filters of all bandwidth ranges.

Rayleigh waves on ST-cut quartz, propagating in the X-direction are being used for frequency control devices and narrow band filters. ST-cut quartz

Table 14.1. Piezoelectric single crystals for Rayleigh wave devices

Crystal	Crystal cut	SAW axis	Velocity (m s^{-1})	K^2 (%)	TCD[a] (ppm K^{-1})
Quartz	ST	X	3,158	0.11	$0.034\ (T-T_0)^2$
$LiTaO_3$	X	112° rot Y	3,295	0.80	18
$LiNbO_3$	Y	Z	3,488	4.1	92
$LiNbO_3$	128° rot Y	X	3,980	5.5	72
$Li_2B_4O_7$	45° rot X	Z	3,401	1.0	$0.230\ (T-T_0)^2$
$Bi_{12}Ge_4O_{20}$	Z	X	1,720	1.5	110
$La_3Ga_5SiO_{14}$	48.5° rot Y	26.6° rot X	2,400	0.32	1.1

[a]Quartz ST-cut has a parabolic temperature coefficient of delay (TCD) with a turnover temperature T_0 of 22.5°C, $Li_2B_4O_7$ has zero TCD at turnover temperature as quartz ST-cut but less temperature stability

offers a parabolic temperature dependence of delay with zero TCD at the turnover temperature T_0. T_0 depends on cut-angle and metallization and can thus be centered with respect to the intended operating temperature range by choosing the corresponding combination. Variation of center frequency over a temperature range from −20 to +80°C can thus be limited to about 85 ppm. AT-cut quartz plates, vibrating in shear bulk wave thickness mode and widely used as low frequency (up to several MHz) control elements, offer a third order temperature dependence of resonant frequency with higher temperature stability and frequency variations of a few ppm only over the same temperature range. Much effort has been spent therefore to find quartz cuts with similar temperature stability for SAW devices. One of these cuts, the so called LST-cut for a leaky SAW, offers as low as 10 ppm frequency variations over the −20°C to +80°C temperature range [12] but has not found widespread use because of high propagation velocity of 3,960 m s^{-1} and high sensitivity to metallization thickness and cut angle as well as strong temperature dependence of leaky propagation losses [13].

For medium bandwidth IF filters, Rayleigh waves on X-cut LT propagating in 112° rotated Y direction are being used. This cut offers a higher K^2 of 0.8% and a moderate TCD of 18 ppm K^{-1}. For this cut, the SAW propagation direction is not normal to the slowness surface and a power flow angle of 1.1° has to be accounted for in the design of SAW devices using this substrate.

High-bandwidth IF filters with reasonable insertion loss can be realized only on LN. Y-cut Z-propagating LN offers a low curvature slowness surface along the propagation direction and consequently substantially reduced diffraction of Rayleigh waves. This can be used to minimize diffraction related deteriorations in the frequency response of high performance SAW filters. However, as this cut also exhibits strong bulk wave excitation, an MSC has to be applied in order to separate SAW and bulk wave propagation. 128° rot Y-cut X-propagating LN offers not only higher K^2 and lower TCD compared to Y-cut Z-propagating LN but also a good suppression of shear bulk wave excitation eliminating the need for MSC application and providing substantial miniaturization potential. Most of the SAW TV IF filters being produced in high volumes are fabricated on this substrate.

$Li_2B_4O_7$ has been used for a few special applications as it offers electromechanical coupling like LT and a parabolic temperature dependence of delay like quartz. With temperature stability much less than quartz and the material being water soluble, this material has become less attractive. $Bi_{12}Ge_4O_{20}$ offers the lowest SAW propagation velocity and the best potential for miniaturization, but has found only limited applications because of the relatively high TCD. $La_3Ga_5SiO_{14}$ offers an attractive combination of low propagation velocity, moderate K^2 and low TCD, and has been used to produce IF band pass filters with extremely narrow transition bandwidth. Limited availability and economy of this material, however, has restricted its widespread use.

The medium to high bandwidth low loss filters needed in the front end of mobile phones have to employ first LSAW on LT or LN with K^2 and

Table 14.2. Piezoelectric single crystals for leaky surface acoustic wave (LSAW) devices

Crystal	Crystal cut	SAW propagation	Velocity ($m\ s^{-1}$)	K^2 (%)	TCD ($ppm\ K^{-1}$)
$LiTaO_3$	36° rot Y	X	4,100–4,212	4.7–7.6	28–35
$LiTaO_3$	42° rot Y	X	4,100–4,212	4.7–7.6	28–35
$LiNbO_3$	64° rot Y	X	4,478	11.3	81
$LiNbO_3$	41° rot Y	X	4,379	17.2	80

TCD ranging from about 5 to 17%, respectively 32–81 ppm K^{-1} (Table 14.2). The critical issue for employing LSAW, however, is to find planes and directions for LSAW propagation with minimum propagation loss. For LT rotated Y-cut X-propagating LSAW, leaky propagation losses exhibit a parabolic dependence on both cut-angle and metallization height. Thus minimum leaky propagation loss is observed for zero metallization height on 36° rot Y LT and for a more practical Al metallization height h/λ of about 10% on 42° rot Y LT. The majority of mobile phone front end filters with relative bandwidth about 5% are made on 42° rot Y LT substrates. For wider bandwidths, LN has to be used, which offers a range of rotated Y-cuts with low LSAW propagation losses and K^2 of 11–17%. However, higher temperature shifts of frequency due to the TCD of about 80 ppm K^{-1} has to be taken into account in the design of SAW filters on these substrates.

Before SAW devices could find their widespread application in wireless communication, much efforts had to be spent for the development of crystal growth and wafer fabrication processes for the economical production of piezoelectric single crystal substrates with the necessary low defect densities, compositional variances and dimensional tolerances.

Quartz crystals are grown in a hydrothermal process in autoclaves at temperatures between 300 and 400°C and pressures between 700 and 1,500 bar by immersing seed plates into aqueous solutions of quartz with sodium hydroxide or carbonate acting to increase quartz solubility [14]. Growth rates are mainly controlled by the temperature difference between quartz growing and quartz supply regions and are typically ranging between 0.4 and 1 mm day^{-1} (Z-direction). Critical defects for SAW devices are voids and dislocations that can originate from the seed solution interface and propagate from there through the crystal in the course of crystal growth. X-ray topography has been used to analyze such defects and provide valuable hints about their origin as well as effective ways to prevent them [15]. With careful control of seed surface quality and growth parameters, especially in the initial growth phase, state of the art growth processes allow to grow quartz crystals almost free of such defects (Fig. 14.12).

LT and LN crystals are grown from the melt using the Czochralski method by first dipping a crystal seed rod into the melt and then pulling the rotating seed crystal rod upward. With precise control of temperature gradients, rate

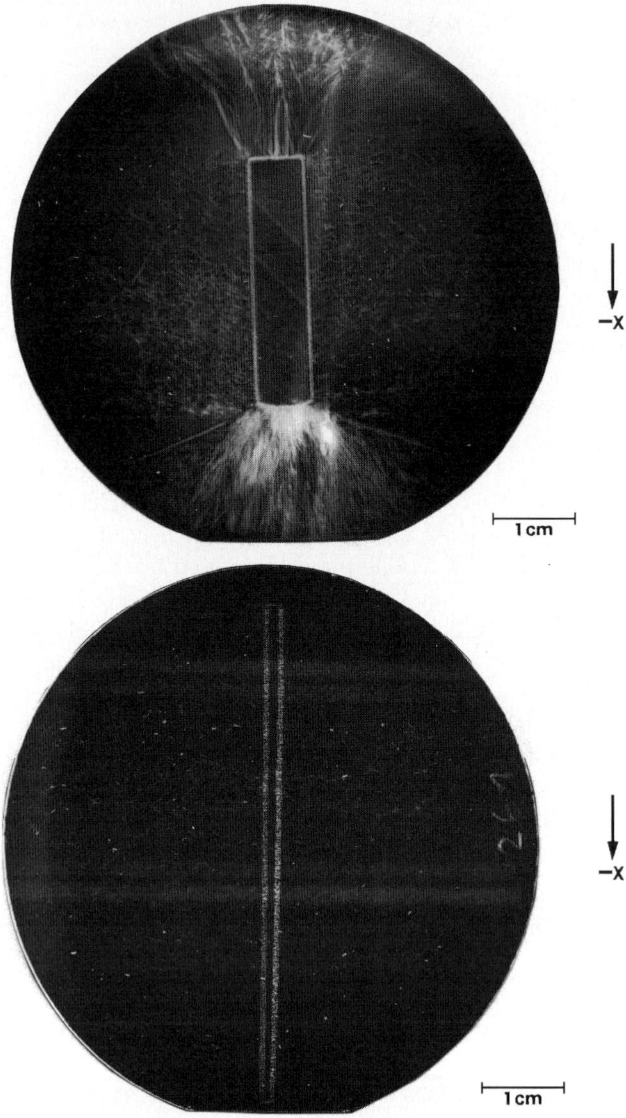

Fig. 14.12. X-ray topography of 4 in. quartz wafers with (**a**) and without (**b**) crystal defects originating from the seed interface [16]

of pulling and speed of rotation, it is possible to extract large, cylindrical, single-crystal boules from the melt. The high melting temperature of LT of 1,650°C necessitates high melting metal such as Iridium crucibles to contain the melt and the application of inductive heating whereby the heat is dissipated by a strong RF magnetic field inducing eddy currents in the crucible and the melt. The lower melting temperature of 1,253°C of LN allows boules

Fig. 14.13. X-ray topography of $LiNbO_3$ boule cross sections with defects originating from the seed and the cone segment [16]

to be pulled from Platinum crucibles and the application of resistance heating which provides means for lower as well as better controlled thermal gradients. Critical crystal defects such as clusters of dislocations and small angle grain boundaries can originate from the seed or the cone shaped top of the boule where the crystal gradually grows from the smaller seed rod to the larger boule diameter. Mechanical stress built into the crystal during and after growth can lead to cracking of boules or substrates fabricated from such boules. As in the case of quartz X-ray topography has been successfully applied to identify such defects, their origin and effective ways to prevent them [17]. Extensive optimization of the complex set of seed in conditions, thermal gradients, pull rates and rotation speeds was necessary to produce large diameter boules with sufficient low defect densities and mechanical stress levels needed for SAW devices (Fig. 14.13). LT is a brittle material with a considerable risk of breakage in crystal, substrate or SAW device production. It has been shown, however, that the breakage strength can be substantially improved by Fe doping resulting in the so-called yellow LT [18].

In order to avoid compositional variations in LT and LN boules, the growth process has to be started with exactly the congruent melt composition. The congruent melt is slightly off the stoichiometric 50/50 mol% Li/Ta respectively Li/Nb ratio to the lower Li content side and only with this congruent melt the composition of the solidified crystal will exactly match that of the melt. Any off congruent melt composition will thus result in Li content variations throughout the growth process and consequently in SAW propagation velocity variations across the boule. The true congruent composition has been determined for both LN [19] and LT [20] and with proper control, compositional variations can thus be kept within 0.05 mol% Li for both materials. However, as LN SAW propagation velocity is less sensitive to compositional variations, the corresponding variations of SAW propagation velocity are lower as compared to LT, which partially compensates for the higher temperature dependence of LN.

The economy of photolithographic SAW device fabrication processes requires large substrate wafer diameters. While the semiconductor industry is moving to 200 and 300 mm silicon wafers for the most complex devices, the majority of far less complex SAW devices are currently being produced on 100 mm wafers. In response to the future trend to increasingly complex SAW devices, up to 150 mm wafers have been made available for the major piezoelectric crystals quartz, LT and LN. Wafer fabrication processes such as slicing, lapping and polishing have to be kept under tight control to maintain orientation tolerances between ±5 and ±15 arc min for the SAW propagation surface and direction. SAW propagation surfaces are polished to residual roughness of around 0.3 nm Ra. This small residual roughness still causes some dispersion of surface acoustic waves, i.e., a frequency dependent propagation velocity that has to be properly accounted for in the design of SAW devices. The back side of SAW substrates is typically left as lapped with a roughness around 300 nm Ra in order to suppress spurious bulk wave reflection. The stress resulting from different propagation surface and back side roughness has to be removed by suitable stress relieve etching in order to prevent wafers from warping. Subsequent SAW device photolithographic and assembly processes require tight wafer thickness control. State of the art wafers achieve total thickness variations (TTV) of less than 5 μm and local thickness variations (LTV) of about 1 μm over 5 mm.

In contrast to quartz, which is piezoelectric, ferroelectric LT and LN are both piezoelectric and pyroelectric. Pyroelectric charge build up due to temperature changes during SAW device fabrication or operation can lead to destruction of metal electrodes or reversal of ferroelectric domains and thus cause SAW devices to fail. As a countermeasure, chemically reduced LT and LN wafers have been developed that provide conductivity in the order of $2–5\times10^{-5}\,\Omega^{-1}\,\mathrm{cm}^{-1}$, enough to prevent pyroelectric damages, without changing other material properties.

14.6 Fabrication of SAW Filters: Metallization, Micro Patterning, and Packaging

While the fabrication of SAW devices basically utilizes the same micro patterning and packaging technologies as the fabrication of semiconductor integrated circuits (ICs), both areas need to emphasize different aspects. Both areas depend on high resolution submicron micro patterning technologies to assess high frequencies; however, semiconductor IC need precise registration of 25 and more topographically structured layers, whereas the active area of SAW devices typically consists of a single layer of metallization on a flat wafer surface. To achieve tight frequency control, SAW device micro patterning has to put emphasis on tight control of line width, metallization height and even the profile of metallization edges. In packaging, SAW devices in contrast to semiconductor IC's need a cavity to protect acoustically active areas and allow free SAW propagation.

With I-line 5× projection printing wafer steppers 300 nm line width resolution can be achieved with the photolithographic micro patterning process corresponding to an upper limit of 3 GHz for the center frequency of LSAW RF filters on LT. With proper control of all relevant process parameters standard deviations of critical dimensions can be reduced to 20 nm for the line width and to 0.5% for the relative metallization height. The sensitivity of SAW devices, mainly their frequency characteristics, to such dimensional deviations depends on many factors such as piezoelectric material, propagation direction, type of wave, metallization material, line width to gap ratio, metallization height as a fraction of the acoustic wavelength and even the metallization edge profile [21]. In case of a 1.9 GHz LSAW filter with 150 nm Al-Cu metallization on LT, 20 nm line width variations correspond to about 75 ppm and 450 ppm variation of the lower and higher frequency filter skirt frequencies, respectively, and 0.5% variation of relative metallization height correspond to about 290 ppm variation of center frequency. In another example of a 500 MHz SAW filter with 120 nm Al metallization on ST-cut quartz the dimensional standard deviations mentioned above translate into about 30 ppm standard deviation of center frequency for each of the line width and relative metallization height variations. Together with about 30 ppm standard deviation caused by subsequent processes including mechanical stresses introduced during packaging, the total center frequency standard deviation of the completed device thus amounts to 52 ppm or 26 KHz.

After the metallization of the acoustically active areas, subsequent metallization processes are needed to enhance the metallization height of bus bars and bond pads for improved electrical conductivity and bonding ability and to provide the multilayer under bump metallization, typically a Ti, Ni, Au sequence, necessary for forming reliable solder ball joints in flip chip device assembly. Passivation of SAW devices for improved reliability, i.e., protection against corrosion and electrical shorts by conductive particles, can be achieved by applying thin layers of insulating SiO_2 or SiN, electrolytic enhancement of the Al oxide layer or the fabrication of micro cavities on top of the acoustically active areas consisting of two photolithographically structured organic layers. Further wafer level processes comprise the application of organic acoustic absorbers for SAW IF filter devices and the application of solder bumps for subsequent flip chip bonding to ceramic substrates.

Singulation of wafers into individual chips is performed by diamond blade dicing for wafers between 500 and 250 nm thick. If thinner chips are required for low profile packages, a dicing before the grinding process has to be applied whereby a thicker wafer of, e.g., 350 nm is first half-diced from the top surface to a depth of 130 nm and then ground from the back surface to achieve the desired 130 nm chip thickness.

Encapsulation of SAW chips is to protect against environmental influences and to provide electrical interconnection between bond pads on the chip surface and corresponding contact areas on the application circuit board. Packages have to conform to the technology used for mounting components on

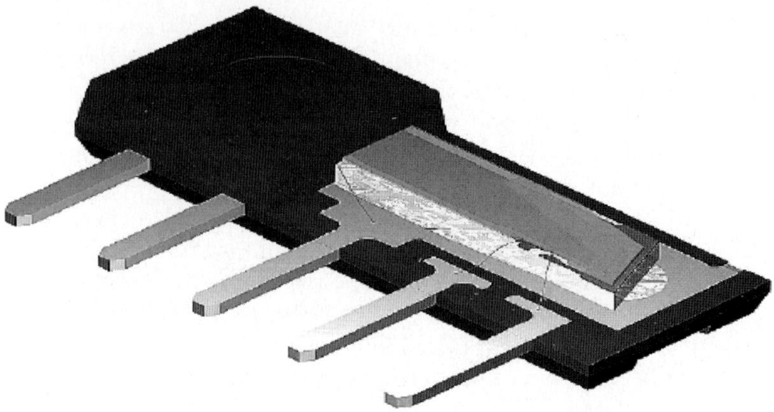

Fig. 14.14. SIP5D SAW TV-IF filter package – cross section

the application circuit board, i.e., through the hole (TTH) or surface mount technology (SMT). The development of special packaging technologies has been the main driver for the miniaturization of SAW devices.

For application in TV sets or tuners, lead frame packages for TTH mounting occupying the minimum of precious real estate on the circuit board are required. A five pin single in line package has thus become the standard for SAW TV-IF filters. In the package shown in Fig. 14.14 the SAW chip is bonded to the lead frame with organic glue, serving also as an absorber for acoustic bulk waves radiated towards the back side of the chip, and electrically connected by wire bonding. The acoustic active area of the chip is protected by an organic two layer structure forming a cavity allowing the SAW filter to be encapsulated by molding with a standard thermosetting mold compound. The outer dimensions of this package are $13.7 \times 4.8 \times 2.4\,\mathrm{mm}^3$. World wide production volume for polymer encapsulated lead frame packaged SAW filters amounts to about 500 million pieces per year.

For mobile applications, low profile SMT packages are required. First mobile phones used SAW filters packaged in leadless ceramic chip carriers with wire bonded SAW chips and soldered or seam-welded metallic lids (Fig. 14.15). The outer dimensions of these packages range from $19 \times 6.5 \times 1.6\,\mathrm{mm}^3$ to $3 \times 3 \times 1.1\,\mathrm{mm}^3$. While today's mobile phones require much smaller filter packages, the hermetically sealed ceramic packages are still in use for high reliability applications such as automotive keyless entry systems and for SAW IF filters in mobile digital radios and TV tuners. World wide production volume for ceramic SMT packaged SAW devices amounts to about 200 million pieces per year.

Further miniaturization as required by more complex multifunctional mobile phones was achieved with chip-sized packages (CSP) utilizing SAW chips flip chip bonded to a high temperature co-fired alumina ceramic substrate and covered with a polymer laminate (Fig. 14.16). The outer dimensions of CSP

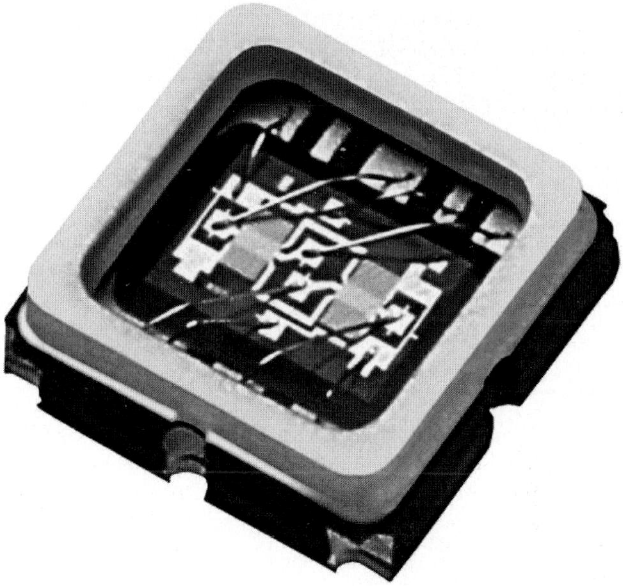

Fig. 14.15. SAW filter packaged in leadless ceramic chip carrier

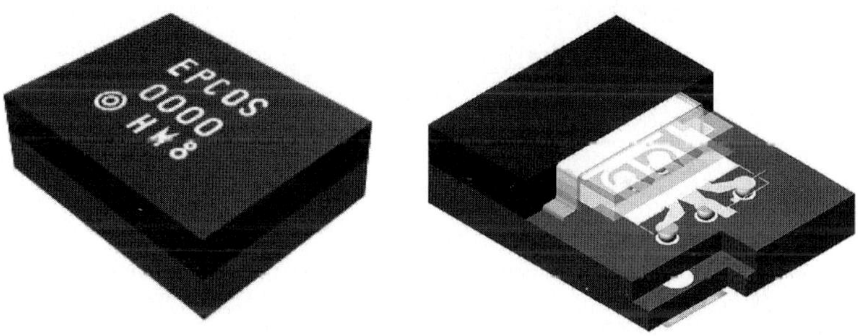

Fig. 14.16. Saw filter in CSP package (CSSP3)

SAW filters range from $2 \times 2.5 \times 0.74\,\mathrm{mm}^3$ to $1.4 \times 1.1 \times 0.4\,\mathrm{mm}^3$. Functionally integrated devices with more then one filter function may need larger packages. With multilayer low temperature co-fired ceramic (LTTC) substrates high Q matching elements, as required for duplexers, can be integrated into the package. With 1 billion mobile phones produced per year the world wide production volume for CSP SAW devices amounts to about 4 billion pieces per year.

The last step in miniaturization could be the wafer level packaging with two organic or inorganic layers on top of the piezoelectric crystal chip forming a cavity over the active area, a ball grid array of solder bumps on top of the cavity and three-dimensional interconnect lines connecting bond pads on the chip surface with the corresponding under bump metallization on top of the cavity. The outer dimensions of single RF SAW filters could thus shrink to $0.8 \times 0.6 \times 0.35\,\mathrm{mm}^3$.

14.7 SAW Filters for TV-IF Applications

TV-IF filters have been the first application of SAW devices in a consumer mass market. Highly selective SAW TV-IF filters with precisely shaped pass band characteristics have become a necessary prerequisite for adjacent TV channel occupation in cable and terrestrial TV as well as for higher resolution pictures on flat panel displays.

World wide TV programs are broadcast in frequency channels spaced 8, 7 or 6 MHz apart, depending on the standard, over various very high frequency (VHF) and ultrahigh high frequency (UHF) bands covering the frequency range from 54 to 862 MHz. In classical analog TV broadcasting the transmitted spectrum of a TV program comprises picture, color and sound carrier frequencies modulated with the respective signals (Fig. 14.17). The TV-IF filter required in the receiver to select the wanted program and reject all others has three distinct features. A Nyquist slope at the picture carrier, i.e., a filter skirt perfectly symmetrical about the picture carrier required to completely restore one side band after demodulation from the vestigial side band transmitted, a sound shelf to suppress the sound carrier preventing interference with the other carriers and to provide a flat pass band in the vicinity of the

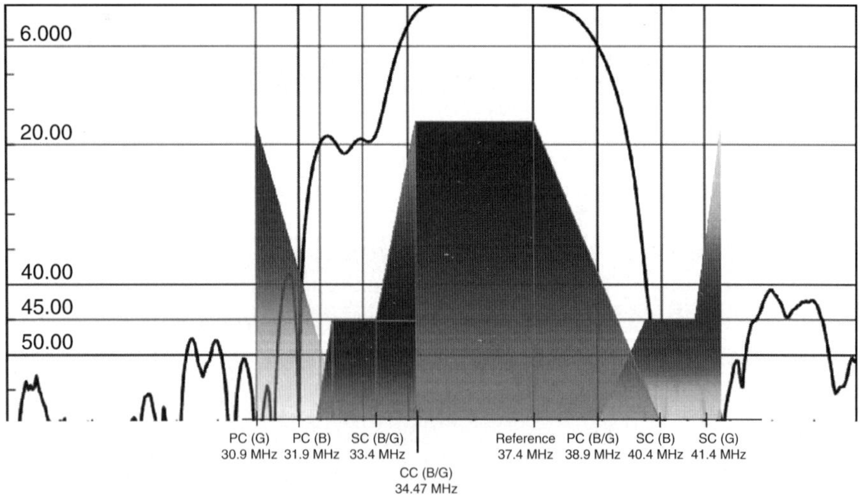

Fig. 14.17. Analog TV broadcasting signal frequency spectrum and IF filter characteristic (BG standard spectrum, BG intercarrier filter)

sound carrier for good sound quality and deep traps to reject strong adjacent channel carrier frequencies. In analog TV broadcasting any distortion in the TV-IF filter pass band or any insufficient rejection of neighboring channels will cause visible and audible deteriorations of picture and sound quality and SAW TV-IF filters therefore have to fulfill tight specifications.

Digital TV broadcasting allows for a substantially more efficient utilization of the EM spectrum. While analog TV needs one full 8–6 MHz wide channel to broadcast one program, digital modulation allows transmitting one HDTV program with six times the resolution or the equivalent number of standard programs simultaneously over the same channel. Over the next decade, analog TV broadcasting will therefore be replaced by digital broadcasting. In digital TV broadcasting, the transmitted spectrum is evenly distributed over the channel bandwidth (Fig. 14.18). The characteristic features of the IF filter are a flat pass band, steep skirts and high rejection of adjacent channel signals. In digital TV broadcasting, any distortion in the TV-IF filter pass band or any insufficient rejection of neighboring channels will cause deterioration of the bit error rate and channel information capacity and SAW TV-IF filters for digital applications therefore have to fulfill similar tight specifications as their analog counterparts, though sensitivity of degradation to certain parts of the specifications may be different in both cases.

Both analog and digital TV broadcasting need IF filters with 10–20% relative bandwidth, precise pass band shaping, narrow transition bandwidth and high out of band rejection. The exact relative bandwidth required depends

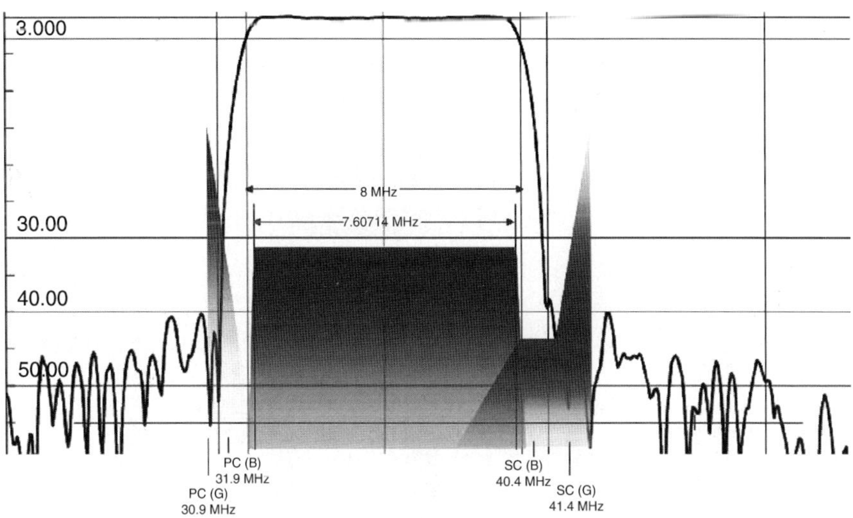

Fig. 14.18. Digital TV broadcasting signal frequency spectrum and IF filter characteristic (DVBT spectrum with analogue adjacent channels, DVBT filter)

on the standard. While European standards, using the widest bandwidths and lowest IF center frequencies of about 36 MHZ, require the largest relative bandwidth and US based standards with lower bandwidth and IF center frequencies of about 44 MHz require medium relative bandwidth, Japanese standards with lower bandwidth and IF center frequencies of about 57 MHZ require the lowest relative bandwidth. SAW TV-IF filters are being implemented as transverse filters with Rayleigh waves on X-cut 112° rot Y-propagating LT or 128° rot Y X-propagating LN. The advantages of LT are a lower SAW propagation velocity and a higher temperature stability allowing realization filters with steeper skirts on a given chip length and narrower transition bandwidths. The low electromechanical coupling factor of LT, however, limits applications to lower relative bandwidths and the matching inductors, needed to reduce insertion losses, add to pass band distortions that are difficult to control. SAW TV-IF filters on LT are therefore used where emphasis is on steep skirts, narrow transition bandwidths and low relative bandwidths, i.e. predominantly in Japan and for a minor part of US based standard applications. LN has the advantage of a high electromechanical coupling factor with no need for matching inductors, better control of pass band characteristics and lower coupling to bulk waves with lower spurious responses and higher out of band rejection. SAW TV-IF filters on LN are therefore used where emphasis is on precise pass band control, high out of band rejection and high to medium relative bandwidths, i.e. predominantly for European based standard applications and for the major part of US based standard applications. Figure 14.19 shows typical SAW TV-IF filters in five pin single in line lead frame packages.

Fig. 14.19. Typical SAW TV-IF filters in five-pin single in line lead frame packages

14.8 SAW Filters and Resonators for Short Range Remote Control Applications

A typical application of SAW devices for short range remote control is in keyless car entry systems. Here a narrow bandwidth signal is transmitted from the car key (Fig. 14.20). The carrier frequency of the transmitted signal is stabilized by a SAW resonator with fundamental resonance at the nominal carrier frequency of 433.92 MHz. The receiver in the car uses a front end SAW filter connected to the antenna to cut off noise and interference from other signals in order to improve receiver sensitivity and selectivity. Apart from the signal spectrum, the filter bandwidth has to accommodate frequency variations due to fabrication tolerances and temperature changes of both the SAW resonator and the SAW filter. With both SAW resonator and SAW filter implemented on ST-cut quartz, center frequency tolerances specified as ±75 KHz and center frequency changes over a temperature range from −20 to +80°C of ±35 KHz, the SAW filter needs to have a minimum usable bandwidth of 220 KHz to account for the SAW resonator frequency variations and an actual bandwidth of 440 KHz to also account for the filter's frequency variations. Figure 14.21 shows SAW resonators and the SAW filters, respectively, packaged in $3{\times}3\,\text{mm}^2$ and $5 \times 5\,\text{mm}^2$ footprint hermetically sealed ceramic packages.

The system performance could be further improved with respect to receiver sensitivity and selectivity as well as power consumption in the transmitter with narrower filters having a bandwidth closer to that of the signal

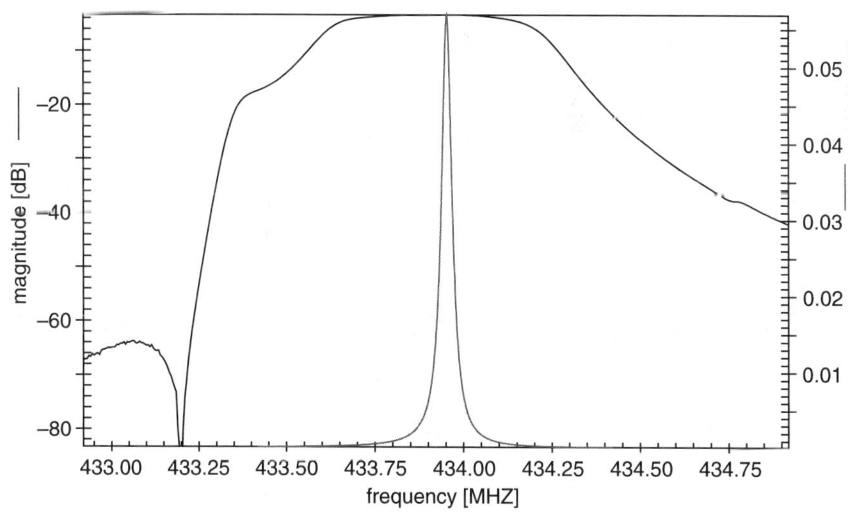

Fig. 14.20. SAW devices for remote keyless car entry systems (R 900 SAW resonator: frequency response of conductance: B3575 SAW filter: frequency response S_{21})

Fig. 14.21. SAW resonator and SAW filter for remote keyless car entry system (Photo R900, B3575)

spectrum. To provide SAW resonators and filters with tightened fabrication tolerances and improved temperature stability that are required for such an improvement, ongoing activities aim to develop frequency trimming processes for individual SAW devices and to identify SAW propagation directions in quartz crystals with lower temperature coefficient of delay.

14.9 SAW Filters for Mobile Phone Front End Applications

One major application of SAW filters are front end duplexers in frequency division duplex (FDD) and CDMA/WCDMA cellular phones. A duplexer allows simultaneous receiving of signals from an antenna and transmitting signals to it, with very high isolation between the Tx and Rx ports [22]. For this, two RF-filters, with well-separated pass bands, are integrated on a ceramic substrate (low temperature co-fired ceramics, LTCC, or high temperature co-fired ceramics, HTCC). The transmit signal is passed through the Tx filter, while signals entering the antenna are received by the Rx filter, whose pass band has a higher frequency than that of the Tx filter. Both pass bands are separated by a frequency gap (duplex spacing). For US-PCS and UMTS cellular phones, both operating at about 2 GHz, the pass bandwidths are 60 MHz. The duplex spacing in the case of the US-PCS amounts to only 20 MHz, while for the UMTS case it is 130 MHz.

The requirements for these filters are very stringent. Besides low insertion attenuation, a high rejection in the passband of the respective other filter is required. In particular, high Tx–Rx isolation is needed. In order to provide sufficient frequency selection sufficient steepness of the passband edges is necessary, in particular at the frequency gap, where temperature shifts and fabrication tolerances limit the available space for filter roll-off. A further important requirement is the high power durability of the Tx filter. The choice of material system and design technique to fulfill these requirements is exemplified now in US-PCS and UMTS 2 GHz filters.

In general RF-front-end filters are built on LSAW substrates. First this is due to their increased penetration depth with respect to SAW substrates [23]. In this way the energy density at the surface is reduced and power durability increased. Second and even more important, since LSAW waves have higher phase velocities and in turn larger wavelengths than SAW waves, for a given minimal structure size operation frequencies of about 60% higher than on SAW crystal cuts can be realized. The most common substrates are rotated *YX*- LT and *YX*-LN cuts (Table 14.2). For a duplex spacing of 20 MHz, 36°–46° *YX*- LT are chosen since the coupling strength of the *YX*-LN cuts is too large. Furthermore, 36°–42° *YX*- LT exhibit a smaller TCF. However, as will be discussed below for the US-PCS system, an even smaller TCF of below 20 ppm K^{-1} is needed.

Both DMS and ladder type filter designs are possible for RF duplexers. The ladder type technique has the advantage of higher power durability and a high rejection in the close upper stop band (about 10 MHz outside pass band), while DMS filters have steeper skirts and a high rejection in the remote stop band region [22].

Figures 14.22 and 14.23 show examples of UMTS and US-PCS duplexers, which have been realized on 39°– and 42°–*YX*-LT, respectively, employing the ladder type design technique.

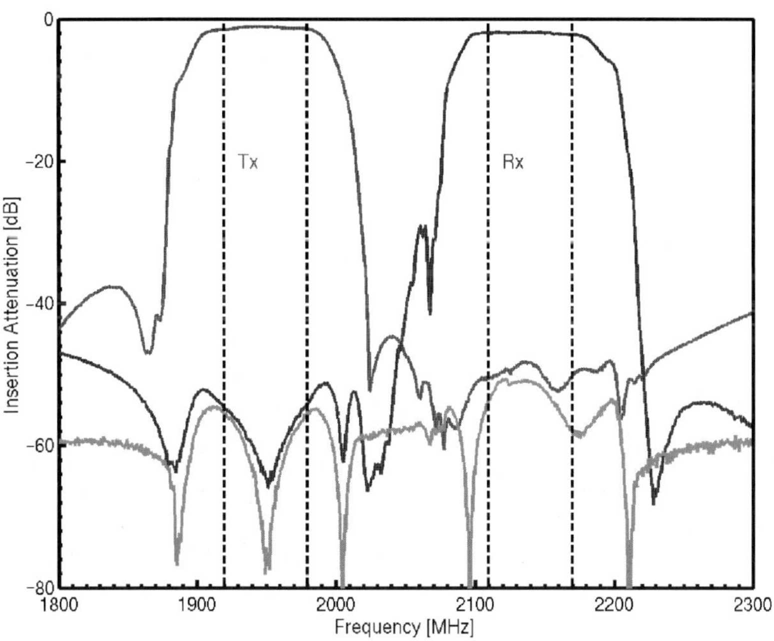

Fig. 14.22. Transfer characteristic of UMTS Duplexer: TX-path (*red*), RX-path (*blue*) and isolation TX–RX (*green*)

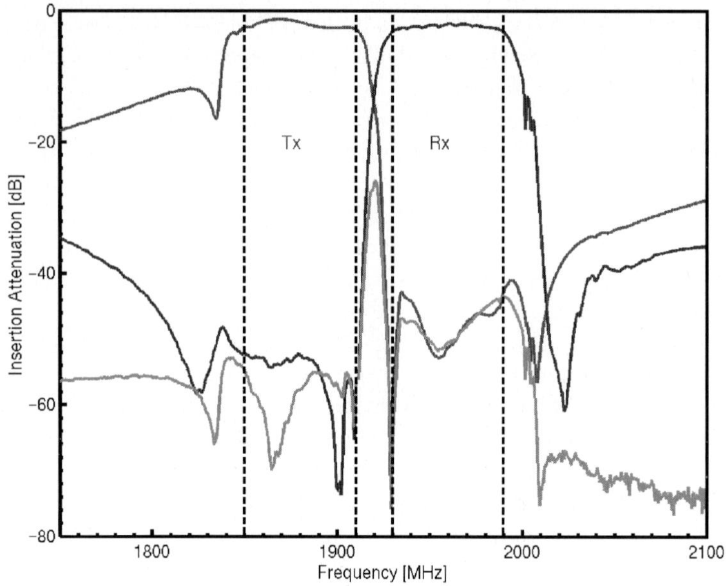

Fig. 14.23. Temperature compensated US-PCS Duplexer: TX-path (*red*), RX-path (*blue*) and isolation TX–RX (*green*)

To achieve a good mutual rejection of the Tx and Rx bands of the UMTS duplexer (Fig. 14.22), electromagnetic poles are sometimes introduced in the stop band regions of the filters. This can be achieved by mounting the duplexer chips on an LTCC package with integrated coils. An alternative approach in this case is to employ a substrate of higher coupling factor [24]. In this approach the antiresonance of the Tx series resonator is placed in the Rx passband, while the resonance of the parallel resonator of the Rx band is placed in the Tx passband (compare Fig. 14.9). In this way, by avoiding additional design measures, a high mutual rejection as well as lower insertion attenuation is achieved. This method is, however, only applicable if no steep passband edge is required.

Owing to the small duplex spacing of only 20 MHz, the US-PCS system is in particular challenging. For the 39° *YX*- LT the TCF amounts to $-45\,\text{ppm K}^{-1}$ at antiresonance and $-35\,\text{ppm K}^{-1}$ at resonance [25]. For the required operating temperature range from -30 to $+85$°C this results in a frequency shift of 9–10 MHz. The remaining 10 MHz is insufficient to allow for the fabrication tolerances of mass production and the roll-off of the filter edge.

It is therefore necessary to lower the TCF. A promising approach is to coat the substrate and electrodes with an SiO_2 layer [25–27]. The idea is to compensate the negative TCF of the substrate with the positive TCF of SiO_2. With increasing SiO_2 layer thickness, however, the wave type evolves from LSAW into a Love wave, a significant contribution of which is confined in the layer. This effect decreases the electroacoustic coupling. Therefore a trade-off of acceptable coupling and TCF reduction has to be made. Figure 14.23 shows a temperature-compensated US-PCS duplexer on a 39° *YX*- LT substrate with a 20% relative SiO_2 coating thickness. The TCF values amount to only -17 and -11 ppm K^{-1} at the right skirt of the Tx and the left skirt of the Rx filters, respectively.

14.10 SAW Devices for RFID and Sensor Applications

Within the last 20 years a rapid growth of the field of wirelessly readable RFIDs has taken place [28]. This is mainly due to the cheap and comprehensive availability of complementary metal-oxide silicon integrated circuits (CMOSIC). But also SAW devices gained an important sector for reliable industrial applications of RFID and remote sensors. Here unique wirelessly operating systems were developed, based on special piezoelectric devices.

All RFID systems work with a two-way (up–down) communications protocol: An interrogator (reader) transmits an RF request signal over an antenna to the RFID device. After processing this request in the transponder it answers with its ID and/or sensor measurement value (code). Two types of transponders are usually distinguished: The passive one gets its energy for the response signal from the reader by RF continuous wave (CW) or the request signal. The active tag has a power source on board. Since the standardized limit of the reader signal is rather low (10–30 mW), the achievable interrogation distance for passive transponders is much smaller than for active tags. On the other hand, passive tags have advantages in cost, reliability, lifetime, weight and volume. So, for low-range RFID systems (e.g., admission control for persons) passive silicon systems are used. However, for systems requiring higher range and mechanical and jamming resistance (industry and traffic), the special integrating signal processing possibilities of piezoelectric SAW devices compensate for the cost advantage of silicon RFIDs.

The first SAW RFID system was invented by NYSEN, SKEIE and ARMSTRONG [29] and implemented for cattle flock counting. As sketched in Fig. 14.24a, the RFID tag is a beam substrate of an appropriate piezoelectric material (LN, LT, etc.) with crystallographic orientation for SAW propagation parallel to its axis. The simplified sketch of Fig. 14.24a shows an exciting interdigital transducer IDT0, on the upper surface of the substrate connected to the tag antenna. We explain the operation of this SAW RFID assuming an interrogation by a short RF burst at the centre frequency *fm* of the IDTs [30,31]. This excites a SAW burst propagation towards the other reflecting IDT1, IDT2, IDT4, etc. Each of the IDTs in the SAW path on the surface of the tag reflects the SAW burst back to the antenna IDT0. This radiates at the burst frequency *fm* to the interrogator, so that we get the response pattern of Fig. 14.24b. The time intervals are given by twice the tag's length divided by the SAW velocity v_s. Typically, we have 1 µs time between request

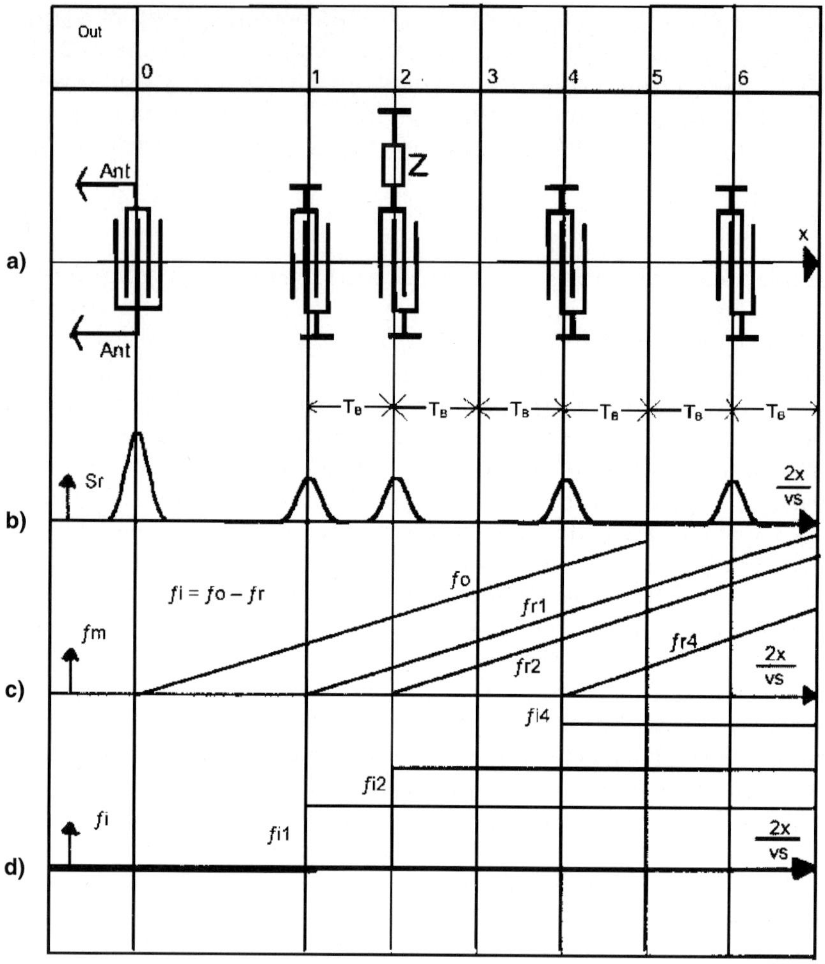

Fig. 14.24. Sketch of SAW RFID structure and LFM read out signal scheme. (**a**) Substrate with reflective IDTs. Antenna is connected to IDT0. External electrical impedance Z is connected to IDT2. (**b**) RFID response requested by a short burst. (**c**) Frequency dependence of response signal due to a LFM request chirp. (**d**) Audio frequencies originating by self interference of LFM request chirp with response chirps

and first response impulse for switching the reader between transmission Tx and reception Rx. This duplexer works at request impulse powers of typically 10 mW compared to many kilowatts of a radar Tx impulse. Here we should note cautiously that special IDTs (e.g. single phase unidirectional) with shifted locations are necessary to achieve such distinct responses as in Fig. 14.24b.

The location of the reflecting IDTs determines the tag's binary identity number (Fig 14.24b shows the code 1,1,0,1,0,1). For a road toll system in Norway, a 22 bit code was used providing 2^{20} ($= 1{,}024^2 > 1$ million) possible identity numbers plus a start and a stop bit [32]. In this car toll system the SAW tag together with its antenna is mounted behind the car's windshield. No electrical connection is necessary from the tag to the battery. The interrogating reader antenna is mounted on a steel frame above the highway. A pioneer of SAW technology, Clinton Hartmann, has carefully investigated a possible global SAW ID tag system in a 40 MHz range of the 2.4 GHz ISM band [33]. Making use of the signal processing capabilities given by the precise position of the reflecting IDTs and corresponding possibilities of pulsed signal matched filtering, the design of tag devices with capacities up to 256 bits is described, showing high signal accuracy and low losses. For usual tagging systems 25–30 bits can be estimated to suffice, so that a large redundancy may be exploited for error correction, as done in communications.

In the sketch of the passive ID tag signal response Fig. 14.24b we observe long time intervals between the response bursts, determined by the tag geometry only. This allows low noise signal processing after decay of the request burst. So, additionally to response burst amplitude detection, applied to most silicon ID tags, using RF interrogation bursts with stabilized burst period T_B and RF frequency fm, also phase measurements are possible for each response burst of the SAW ID tag. This is accomplished using a quartz stabilized master clock for T_B and fm over many periods T_B and integrating over this long time the difference in amplitude and phase of each burst. Therefore very accurate measurements of small shifts of the response bursts are feasible. Since SAW tags at constant temperature and constant mechanical clamping are crystallographically and electrically stable, no difference in burst response is observed for these. But changes of tag temperature and/or mechanical strain (e.g., acceleration) cause observable substrate length effects on the measurement results, i.e. mainly the relative phase of the response burst. These simple wirelessly readable ID tag strain and temperature sensors could be realized only with piezoelectric single crystal SAW substrates.

Additionally to this remote mechanical strain and temperature sensors the SAW tag of Fig. 14.24 together with accurate signal processing creates the possibility of wirelessly readable sensors for electrical impedance. As shown in Fig. 14.24a an electrical impedance Z can be connected as electrical load to one of the reflecting IDTs. Here the Z-loaded IDT2 is situated between the shorted IDT1 and IDT4, which serve as reference of SAW amplitude and phase in the response signal. As described by Hashimoto [34], change of electrical load Z of an IDT causes a definite change of its acoustical reflectivity in amplitude and phase. Thus, by observing the response of a loaded SAW-RFID we can remotely measure resistance R, inductance L and capacity C contained in Z.

Two examples may be mentioned: (1) As changing capacitive load the abrasion of a piston bearing ring of a working natural gas pump was measured [35]. (2) Magnetic field and/or electric current were measured wirelessly by using a giant magneto impedance effect wire as load Z [36]. Its impedance is very sensitive to a magnetic field yielding great effect in the corresponding SAW sensor response burst [37].

In most industrial applications SAW RFIDs are interrogated by a linearly or stepwise frequency modulated continuous wave (LFMCW) microwave sweep. This technique has been in use for 50 years in airplane landing systems to measure the plane's distance to the ground: An LFMCW is radiated to the ground and superposed with the reflected wave to yield a low frequency interference signal whose frequency is proportional to the ground distance. Applying this technique to a SAW RFID, each reflecting IDT in Fig. 14.24a gives a certain delay of the LFM and therefore a certain interference frequency *fi* as shown in Fig. 14.24c. So, a Fourier analysis of the superposition product yields the ID-code of the tag. Figure 14.24d should demonstrate two facts: (1) It is very important that the interrogating LFM is strictly linear, to keep all modulation noise out of *fi*. (2) Extending the time of interrogation yields the same extension of the signal-to-noise ratio of the resulting low frequency *fi* [32]. This is fundamental in all spread spectrum modulation systems: A signal, extended in frequency to a bandwidth B and in time to a length T has a process or correlation gain $G = TB$ compared with the single pulse with $TB = 1$. The gain G also describes the system's resistance against jamming, other RF-noise and eavesdropping [38]. Since the frequency bands standardized for RFIDs more and more get crowded with users, the spread spectrum readability of SAW IDs is very important. In the environment of a tire producing factory court, only FMCW read SAW tags allowed to identify tires of cars driving in this court. All other tire identifying systems failed because of EM pollution. Also for railway and metropolitan subway trains, identification SAW tags have been used successfully for many years. Many other industrial applications in harsh environment require the reliability of piezoelectric crystal RFIDs.

The advantage of not needing a battery for passive RFIDs is connected with the much shorter range R of readability of passive tags with the standardized interrogation RF power (typically 10 mW). Considering radiation from a point source yields an R^{-2} range attenuation law for active tags and an R^{-4} law for passive ones, since the RF energy is attenuated by R^{-2} on its way from the interrogator to the tag and its way back. Recently, a clever invention has made possible longer readability ranges with an R^{-2} law for energy autonomous piezoelectric RFID sensors [39].

The SAW structure of these tags (Fig. 14.24a) is extended by further technological elements on the substrate's surface, which advantageously can be fabricated together with the IDTs. The task of these elements is to extract energy from the tag's environment and to transform the energy into an exciting SAW burst which, after passage of the ID structure, is radiated as ID-coded electromagnetic wave to the interrogator (attenuated by an R^{-2} law). Explaining this principle we consider a door to be monitored by the reader system. When the door is opened or closed, it activates a tiny knocking device on the substrate, which produces a high electrical pulse voltage to excite a spark in a nanotechnology spark gap at the input of the tag's ID path. Likewise, a high voltage generated by some pyroelectric generator can excite a voltage spark in a SAW ID temperature sensor. So, in such autonomously powered SAW ID systems, the tag signals its own ID to the base station once the parameter exceeds a certain limit.

So, for radio transmission we have the link budget with the reading range of active IDs – some 100 m in the open – but the battery replaced by environment energy. Using additional piezoelectric parameters of SAW RFIDs we get rid of battery replacement and leakage problems and achieve a great reading range. In the future, similar clever applications of piezoelectric properties will hopefully be exploited by means of nanotechnology and physics.

High-temperature piezoelectric materials are urgently needed in aircraft, automotive, metalcast, glass, ceramic and nuclear industry to be implemented as temperature and pressure sensors at elevated temperatures. Within the last two decades, research and development concentrated on gallium orthophosphate ($GaPO_4$) [40], but also langasite ($La_3Ga_5SiO_{14}$) and other lanthan and strontium compounds were crystal grown and investigated. For temperatures above 300°C, the well-established aluminum (Al) IDTs on these crystals are not usable owing to heat, electro- and stress-migration effects, causing destruction (droplets) of the thin Al films. A solution of this problem was found by fabrication of platinum (Pt) IDTs supported by a thin (20 nm) surface adherence layer of titanium (Ti) or zirconium (Zr). These structures resisted temperatures of 600–800°C up to 144 h.

In a recent investigation targeting wireless SAW RFID sensors for temperatures up to 1,000°C, Fachberger et al. [41] have compared the usefulness of three different piezoelectric materials: Y-cut LN, 48.5° rot Y-cut langasite and 5° rot Z-cut $GaPO_4$. The result of this work was, that for $GaPO_4$ and langasite, the electromechanical coupling constants K^2 of 0.3% and 0.44%, respectively, were too small and the f^2 losses were very high at the 2.4 GHz ISM band for wirelessly readable passive RFIDs at elevated temperatures, whereas with LN substrates, measurements of temperature and ID were possible up to a temperature of about 420°C. Employing langasite and $GaPO_4$, however, SAW resonators can be fabricated as described in Sect. 14.4 and used as sensors. Buff and his team [42] have measured the resonance frequency of such devices in dependence of temperature up to 700 or 800°C near a center frequency of 435 or 415 MHz respectively.

14.11 Outlook: Future Development of System and Filter Requirements, and New Piezoelectric Material Solutions

Today, SAW devices fabricated on piezoelectric single crystal materials are widely used in wireless communication systems. With their advanced RF signal processing capability, they have played an important part in allowing more efficient utilization of the EM spectrum and, as a consequence, the provision of more information capacity for a rapidly growing number of wireless communication services and users. They have also substantially contributed to the comfort and usability of mobile terminals owing to the miniaturization achieved through their design and their packaging and because of their passive, i.e. non-power-consuming, nature.

In the future, the trend for more wireless communication services to be offered and more users taking advantage of them will continue. With such anticipated growth of wireless communication systems and the fact that the available frequency spectrum remains limited, the necessity to further increase the efficiency of EM spectrum utilization will become even more pressing. The trend to allocate higher frequencies, broader frequency channels and bands as well as the trend to place these channels and bands more closely to each other will continue and the signal processing capabilities of SAW devices will be further challenged accordingly. Furthermore, the trend to combine telephone, radio, TV reception, GPS navigation, and much more in multimedia mobile terminals will drive the complexity of transceiver architecture and further miniaturization and integration of all components.

The signal processing capabilities of SAW devices on piezoelectric single-crystal materials are, however, approaching their limits. This is due to the fact that none of the available single crystal materials can provide the high electromechanical coupling necessary for achieving wide bandwidths and the low temperature coefficient of frequency necessary for achieving an extremely narrow effective transition bandwidth between the low-loss pass band and the high-rejection stop band. Miniaturization of SAW devices on piezoelectric single crystals is also approaching limits, as packages for these devices have to provide for a cavity to protect the acoustic waves propagating along the crystal surface.

Present limitations of SAW devices on piezoelectric single crystals can be overcome, however, with layered structures, combining the negative temperature coefficient of frequency of a suitable surface of a high coupling single crystal such as LN with the positive temperature coefficient of a layer such as SiO_2 deposited on top of this surface and an array of heavy electrodes, e.g. made of Cu instead of Al, at the interface of the single crystal and the layer deposited on top of it. In this configuration, the piezoelectric single crystal provides the coupling between electrical signals and acoustic waves; the array of heavy electrodes provides guidance of the acoustic waves along an inner surface and the top layer itself provides compensation of the single crystal substrate's temperature coefficient of frequency. Layered structures thus have a good potential for the realization of wideband RF filters with low losses and narrow transition bandwidth as required by system demands. Also, miniaturization can be driven to new borders as acoustic waves guided along inner surfaces do not need to be protected by cavities when devices are packaged.

The new benefits of layered structures will not come without new challenges. With the ongoing trend to safe information capacity wherever possible, layered structure devices will have to meet stringent requirements in terms of losses, steepness of filter skirts and fabrication tolerances affecting the center frequency. To this end, first of all, processes have to be developed for the deposition of highly reproducible top layers and electrode gratings with low acoustic and electric losses. Second, the sensitivity of layered structure devices against dimensional variations of heavy electrodes, both electrode height and electrode width, will become much larger than in the case of relatively light Al electrodes on piezoelectric single crystal surfaces. As the dimensional control of Al electrodes of SAW filters on piezoelectric single crystal materials has already reached its physical limits, same or lower fabrication tolerances for layered structure devices can only be achieved with appropriate trimming processes that are still yet to be developed. Innovative approaches may become necessary in order to provide means for trimming structures on buried inner surfaces.

With these new challenges being met, signal processing with acoustic waves propagating on or in piezoelectric materials will continue to play an important part in further developments of wireless communication systems. While new materials, material combinations and processes will have to be developed in order to meet growing requirements resulting from the growing need for efficient utilization of available RF spectrum, the state of the art achieved so far for SAW signal processing on piezoelectric single crystals does provide a solid base on which these new technologies can be built.

References

1. C.E. Shannon, Bell Syst. Tech. J. **27**, 379–423, 623–656 (1948)
2. R.M. White, F.W. Voltmer, Appl. Phys. Lett. **7**, 314–316 (1965)
3. B.A. Auld, *Acoustic Waves and Fields in Solids*, vol. 1 (Wiley, New York, 1973)
4. K.-ya Hashimoto, *SAW Devices in Telecommunications* (Springer, Berlin 2000), pp. 1–22
5. Lord Rayleigh, Proc. Lond. Math. Soc. **7**, 4–11 (1885)
6. T. Makkonen, Appl. Phys. Lett. **82**, 3351–3353 (2003)
7. W.S. Mortley, British Patent 988 102, May 1963
8. J.H. Rowen, U.S. Patent 3 289 114, Dec. 1963
9. W. R. Shreve, in *Proc. 1977 IEEE Ultrasonic Symposium*, 1977, pp. 857–861
10. T. Nishihara, H. Uchishiba, O. Ikata, Y. Satoh, Jpn. J. Appl. Phys. **34**(Part 1, 5B), 2688–2692 (1995)
11. A. Sakurai, H. Nakanishi, Y. Yoshino, Jpn. J. Appl. Phys. **34**(Part 1, 5B), 2674–2677 (1995)
12. Y. Shimizu, M. Tanaka, T. Watanabe, in *Proc. 1985 IEEE Ultrasonic Symposium*, 1985, pp. 233–236
13. C.S. Lam et al., *First Symposium on Piezoelectric Acoustic Waves and Device Applications*, Dec. 14–17, 2004, Ningbo, Zhejiang, China, download from http://www.txc.com.tw/en/d_support/03.html
14. G. Johnson, J. Foise, Encyclopedia Appl. Phys. **15**, 365 (1996)
15. H.R. Stocker, R. Veith, E. Willibald, G. Riha, in *Proc. 1981 IEEE Ultrasonic Symposium*, 1981, 78–82

16. E. Willibald, Dissertation TU-München, 1983
17. B.C. Grabmaier, E. Willibald-Riha, E. Born, Siemens Forschungs- und Entwicklungsberichte **17**(3), 159–164 (1988)
18. Yamaju Ceramics Co. Ltd., European Patent Application EP 1741809A1
19. P.F. Bordui et al., J. Cryst. Growth **113**(1–2), 61–68 (1991)
20. J. Kushibiki, Y. Ohashi, IEEE Trans. Ultrason. Ferroelectr. Frequ. Control **53**(2), 385–392 (2006)
21. M. Solal et al., *Second International Symposium on Acoustic Wave Devices for Future Mobile Communication Systems*, 3–4 March 2004, Chiba University, Japan, download from http://www.usl.chiba-u.ac.jp/~ken/Symp2004/PDF/3A2.PDF
22. Y. Satoh, O. Iktata, in *Advances in Surface Acoustic Wave Technology, Systems and Applications*, ed. by C. Ruppel, T. Fjeldly, vol. 1 (World Scientific, Singapore, 2000)
23. C.K. Campbell, *Surface Acoustic Wave Devices* (Academic, New York, 1998)
24. M. Kadota, T. Nakao, K. Nishiyama, S. Kido, M. Kato, R. Omote, H. Yonekura, N. Takuda, R. Kita, in *Third International Symposium of Acoustic Wave Devices for Future Mobile Communication Systems*, March 2007 (Chiba University, Japan, 2007), pp. 143–147
25. G. Kovacs, W. Ruile, M. Jakob, U. Rösler, E. Maier, U. Knauer, H. Zottl, in *IEEE Ultrasonic Symposium*, 2004, pp. 974–977
26. M. Kadota, T. Nakao, N. Taniguchi, E. Takata, M. Mimura, K. Nishiyama, T. Hada, T. Tomura, in *IEEE Ultrasonic Symposium*, 2003, pp. 2105–2109
27. M. Kadota and T. Kimura, *IEEE Ultrasonic Symposium*, 2006, pp. 2305–2309
28. K. Finkenzeller, *RFID Handbuch (Transl. RFID Handbook)* (Hanser, München, Wien, 2002)
29. P.A. Nysen, H. Skeie, D. Armstrong, US Patent 4725841, 1986
30. L. Reindl, A. Pohl, in *SAW*, ed. by K. Finkenzeller. RFID Handbuch (Hanser, Muenchen, Wien 2002), pp. 153–163
31. A. Pohl, *IEEE Trans. UFFC* **47**(2), 317–332 (2000)
32. Technology Focus, Microw. RF Eng. 42–43 (1989)
33. C.S. Hartmann, P. Brown, J. Bellamy, in *Proc. 2. International Symposium on SAW Devices for Future Mobile Communication. Systems*, Chiba Universtiy, Japan, March, 2004, http://www.idtechex.com
34. K.-ya Hashimoto, *SAW Devices in Telecommunictions* (Springer, Berlin, 2000), pp. 74–79
35. R. Steindl, A. Pohl, F. Seifert, *IEEE Trans. MTT* **47**(12), 2625–2629 (1999)
36. H. Hauser et al., IEEE Trans. Instr. Meas IM **49**(3), 648–652 (2000)
37. R. Steindl et al., Sens. Actuators **85**, 169–174 (2000)
38. R.C. Dixon, *Spread Spectrum Systems* (Wiley, New York, 1998)
39. W.-E. Bulst, F. Schmidt, O. Sczesny, US Patent 2006/0037 380A 1, 2006
40. P. Krempl, G. Schleinzer, W. Wallnöfer, Sens. Actuators **A61**, 361–363 (1997)
41. R. Fachberger et al., IEEE Trans. UFFC **51**(11), 1427–1437 (2004)
42. W. Buff et al., in *IEEE Ultrasonic Symposium*, 2003, pp. 187–191

15

Piezoelectric Films for Innovations in the Field of MEMS and Biosensors

P. Muralt

15.1 Introduction

Microelectromechanical systems (MEMS) were born as a new technological discipline during the 1980s (for an introductory textbook, see, for instance [1]). The idea of the pioneers was to enlarge capabilities of integrated circuits based on silicon beyond pure electronics by adding mechanical elements, which were made of silicon and further materials of semiconductor technology. The addition of mechanics extended the application range of silicon technology to motion sensors, pressure and force sensors, small actuators, and a number of acoustic and ultrasonic devices, most importantly resonators for signal treatment. In order to profit from the symbiosis with electronics, those mechanical elements should, of course, be controlled by electronic signals. Evidently, this new silicon technology makes sense only for small, miniaturized devices. The technical advantage comes from the fact that powerful thin-film deposition and patterning techniques as used for semiconductor fabrication allow unprecedented precision of mechanics in the nano- to micrometer range. As a large number of devices are produced in parallel on the same wafer (batch processing), the cost level is acceptable in spite of expensive fabrication tools, at least at high production volumes. Concerning processing, the chemistry of silicon turned out to be very helpful: high etching rates of anisotropic wet etching in a base solution (as, e.g., KOH) and anisotropic deep silicon etching in a plasma reactor are crucial issues in efficiently tailoring silicon. Over the last 20 years, MEMS technology has became a proven and mature technology with many applications. While "MEMS" is still taken as a standing brand name for the field, the actual MEMS field has become much wider than stipulated by the notion of electromechanics, including thermal, optical, magnetic, chemical, biochemical, and further functional properties. Also, the main material of the device is not necessarily silicon, but may be glass or plastics, especially for biomedical applications.

It is expected that piezoelectric materials with their intrinsic electromechanical coupling should and will play an important role in MEMS. Indeed, piezoelectric thin film devices based on ZnO were among the first ones to be demonstrated [2]. In piezoelectric materials, a deformation occurs upon application of an electric field (converse effect) whereby the obtained strains depend, to first approximation, linearly on the electric field. The opposite effect works as well: upon application of a stress, charges develop on the surfaces of the piezoelectric body (direct effect). The most frequently applied bulk material is PZT, a solid solution with formula $PbZr_xTi_{1-x}O_3$, which exhibits outstanding piezolectric properties at the morphotropic phase boundary at $x = 0.53$. Considering the applications of bulk materials (including thick films), we identify many typical features reflecting their unique characteristics, which may be also unique on the micro- or nanoscale and also useful for applications at small dimensions. The use of piezoelectrics in scanning microprobe microscopy is justified by the high rigidity prevailing during deformation and enabling sub-nanometer precision in all directions. Piezoelectrics are used to generate and sense ultrasonic waves as needed in ultrasonic imaging at frequencies of 1–50 MHz and in nondestructive testing at similar frequencies by virtue of the large efficiency in energy conversion between electrical and mechanical energy. In signal filtering, piezoelectric resonators play a dominant role in the range 0.1–10 GHz thanks to their high quality factors, outperforming by far LC resonators. In injection valves for combustion motors, piezoelectrics are applied thanks to large forces and high operation speeds. Without discussing dimensional details, one can summarize the advantages as follows:

- Strong forces or, alternatively, large excursions in bending structures
- Low voltage because of high dielectric constants
- High efficiency in energy conversion, and equivalently low noise detection
- Possibility of high-speed and high-frequency operation
- High acoustic quality of many piezoelectrics (not true in multidomain ferroelectrics)
- Last but not least, a more or less linear behavior.

For thin-film applications, the same principles remain valid; however, one has to keep in mind that smaller dimensions lead, on the one hand, to smaller voltages and, on the other, to weaker forces, smaller excursions, and higher resonance frequencies. Given the large number of degrees of freedom in mechanics, the outstanding features of piezoelectricity will lead to a plethora of applications and devices with application frequencies in a large range between a few hertz and some gigahertz.

Piezoelectricity is not the only possible choice, though. There are also other electromechanical phenomena available in the micro world, and not necessarily the same as in the macro world. There is first of all electrostatic interaction across an air or vacuum gap of a capacitor [3,4] or between comb fingers [5,6]. Even with the disadvantage of nonlinearity, electrostatic interaction can well be used for actuators and, with a bias voltage, also for sensors and transducers. Not so practical in the macro world (very high voltages needed), electrostatics can be a smart solution in the micro world. In theory, electrostatic devices may even reach an electromechanical coupling coefficient (or conversion efficiency) k^2 of 100% in the small-signal limit [4], a value that hardly exceeds 60% in piezoelectrics. A further competing effect is the thermomechanical bimorph effect. It can be used for actuation if a micro heater is embedded in the device. The heat dissipation, and thus power consumption, of such a device is naturally higher and the application frequency lower than with a piezoelectric drive. The same holds for shape memory alloy actuators. Further possibilities are offered by electromagnetic interactions and magnetostrictive materials. Disadvantages are, in general, resistive losses in coils. On the other hand, magnetic interaction can be an advantage if remote power feeding is required [7].

The scientific community was always aware of the importance of piezoelectric MEMS. Interestingly, PZT thin films (and other ferroelectric thin films) and their integration were developed in parallel to general MEMS processing. Both fields profited from a general improvement of physical and chemical deposition techniques during the 1980s and dry etching techniques during the 1990s. However, the main momentum in ferroelectric films came from memory applications (ferroelectric random access memory, FERAM [8,9]). While PZT FERAMs have now found their niche markets and appear to be more and more established in low-power/low-voltage nonvolatile memories, PZT MEMS have not yet established a place in MEMS. The implied industry is much smaller, and even though physical criteria speak for PZT MEMS, the availability of high-quality material and its price play important roles. Whereas AlN has become readily available because of numerous companies and institutes working on thin-film bulk acoustic wave (BAW) RF filters – the first piezo-MEMS success – the ferroelectric $Pb(ZrTi)O_3$ thin film is hardly available on a commercial basis. This is not only due to the very limited number of experienced producers (producing mostly for themselves only), but also due to the integration scheme that is often not well mastered and which imposes its rules to the design (see, e.g., [10]). As a result, PZT MEMS is considered only if there is no other less demanding solution. It seems now that there is at least one good mass application justifying the effort in PZT MEMS, the ink jet printer.

15.2 Operation Principles

Thin-film materials are in various aspects quite different from bulk materials. First of all, they have to be grown on the device in production. As the possible growth temperature is usually smaller than that applied in bulk material synthesis, differences in microstructure and purity may lead to modifications of their properties. If we do not deal with single-crystalline growth on single crystals (epitaxy), the first step in growth is the nucleation of the crystalline material on the substrate. In MEMS, this substrate is usually a buffer layer for growth on insulators or a metal (including oxide conductors) layer for growth on an electrical conductor. The nuclei grow first in all directions until they touch each other. Afterwards, the grains grow in columns perpendicular to the film plane (z-direction), as indicated in Fig. 15.1a. Each column is a single crystal. If one specific crystalline axis stays perpendicular to the film in each grain, meaning that each column grows in the direction of this specific crystalline axis, one deals with a so-called textured film. A (001)-textured film thus means that the (001) planes are parallel to the film plane, and that the direction of the columns (the z-direction) coincides with the c-axis of the crystalline lattice (denoting the indices in the film system by i, and the one in the grain-crystal system by i', we can state $3 = 3'$, and random relation between 1 and $1'$). The control of the film texture is the only means to control the orientation of the anisotropic piezoelectric materials. It follows that only polar materials can be used as thin-film piezoelectrics (which excludes quartz). In case of ferroelectric materials, a net polarization can be obtained by reorienting domains in an external electric field (poling process). In case of non-ferroelectric polar materials, such as the hexagonal wurtzites ZnO and AlN, the growth process must not only provide for a film texture, but in addition also for alignment of the polar direction (c-axis in this case). This is indeed possible. In most cases films are deposited in (001) texture, or tilted (001) texture (see later). Considering (001)-textured, polycrystalline columnar film slab as a free body, we would find a thickness mode d_{33} that is equal to the $d_{3'3'}$ of the single-crystal column. The in-plane ordering is random, resulting in a cylindrical symmetry. This gives the same symmetry of the piezoelectric tensor as for hexagonal or tetragonal crystal symmetry. It follows that in case of wurtzites, $d_{31} = d_{32}$ would be equal to $d_{3'1'}$, and that the shear mode coefficients d_{15} or d_{24} coincide as well. In perfectly (001)-textured films of ZnO or AlN, as schematically drawn in Fig. 15.1a, we thus find single-crystal properties again for the relevant piezoelectric coefficients.

In ferroelectrics, the situation is in general more complicated. Uniaxial ferroelectrics such as the tungsten-bronze $BaSrNb_2O_5$ are research topics [11, 12] but not (yet) established as thin-film materials. Much better established are ferroelastic perovskite structures such as PZT. The polarization direction is not uniquely defined, but may point along different crystallographic axes, subdividing the material into domains. The poling of such a film leads to an average polarization direction along the external electric field, which in the case of a planar capacitor will lead to a situation as shown in Fig. 15.1a, where the arrows now indicate the average polarization direction. We obtain the same symmetry as a ceramic, i.e., cylindrical symmetry with one infinite rotation axis, usually denoted as axis 3, and find again that the free film slab should show the three coefficients d_{33}, d_{31}, and d_{15} as in a ceramic slab. In PZT thin films, no trivial link to single-crystal values is possible. First, the single-crystal values are not as well known, and second, domain sizes, grain sizes, and textures are different from those of bulk ceramics. A single column of a PZT film does not possess the average property of a PZT ceramic, simply because it is a monocrystal, usually small compared to a ceramic grain, and

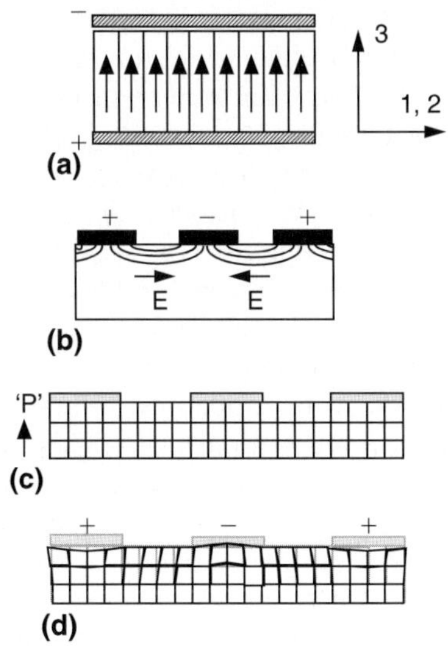

Fig. 15.1. Electrode systems for driving piezoelectric films: (**a**) Planar capacitor structure with top and bottom electrodes. (**b**) Electric field with interdigital electrodes grown on insulator. (**c**) Case of non-ferroelectric, polar film with uniform perpendicular polarization. (**d**) Schematic drawing of strains induced by d_{15} between the electrodes, and d_{33}, d_{31} below the electrodes for the situation in (**c**)

has much more surface and interface contributions as compared to a large grain in the bulk ceramic.

In addition to the planar capacitor structure, one can also make use of interdigitated (ID) electrode structures. Also, in this case one has to differentiate between wurtzite and perovskite ferroelectrics. The polar direction in wurtzites is, in case of net piezoelectricity, perpendicular to the film plane, whether there is a bottom electrode or not. If there is no bottom electrode (see Figs. 15.1b–d), the transverse electric field between the electrodes excites a shear deformation through d_{15}, and the longitudinal component excites a transverse strain (S_1 or S_2) below the electrodes, as indicated in Fig. 15.1d. Such modulated strain patterns are good for exciting surface acoustic modes on, for instance, AlN-coated sapphire samples. When using a ferroelectric, the film needs first poling (except if sputter-deposited in the poled state [13, 14]). Poling without a bottom electrode is hardly imaginable. An interesting configuration is nevertheless obtained when growing ferroelectric films on insulators and using ID electrodes. The poling obtained by using directly the ID electrodes leads to a completely different operation mode. Electric field and average polarization are everywhere parallel (or antiparallel), meaning that everywhere the longitudinal effect is at work. The result is a positive strain in the film plane between the electrodes. This mode is an alternative to the planar capacitor structure, with the difference that high voltages are needed (actuator) or delivered (sensor) and that the film is in compression instead of tension when working at high fields parallel to the polarization. Such structures make very much sense in energy harvesting applications [15], which require higher voltages than for sensors. Actuator and transducer structures were recently investigated [16], showing promising results. Theoretically, the effective piezoelectric coefficients should be larger than in case of a parallel plate structure [16,17]. When combining the ID electrodes with a continuous bottom electrode, the transverse effect (d_{31}) is manly employed in non-ferroelectric polar materials. The field direction changes sign between neighboring electrodes, and can be used again for SAW excitation with wurtzites. In the case of ferroelectrics, this structure makes no sense without prior homogeneous poling of the film. This might be again a poling by the deposition process, arc poling, or an extended top electrode that is removed after poling.

Another important difference between thin-film and bulk materials lies in the fact that thin films are used in a composite structure where the total elastic properties are often dominated by the passive part of the structure. This passive part may be a silicon cantilever, or a silicon oxide or nitride membrane, for instance. The interaction with the substrate is very anisotropic. At the interface along the in-plane directions (indices 1, 2), the piezoelectric thin film and the substrate have identical strains. Perpendicular to the film plane, the thin film is usually free to move, i.e., $T_3 = 0$. As a consequence, the complete equation of state needs to be analyzed. For illustration, a thin film clamped to a much thicker substrate is considered (i.e., a piezoelectric laminated plate). When a field is applied, the strains in the plane stay at zero ($S_1 = S_2 = 0$); in-plane stresses and an out-of-plane strain are developed. The equation of state using compliance tensor and d_{ij} coefficients reads now as (for the average polar direction parallel to E-field: no shear components):

$$\begin{aligned} S_1 &= (s_{11}^{\mathrm{E}} + s_{12}^{\mathrm{E}})T_1 + d_{31}E_3 = 0, \\ S_3 &= 2s_{13}^{\mathrm{E}}T_1 + d_{33}E_3. \end{aligned} \tag{15.1}$$

The first line of (15.3) allows derivation of the in-plane stresses, that is, an effective $e_{31,\mathrm{f}} = -T_1/E_3$. The second line leads to the definition of an effective $d_{33,\mathrm{f}} = S_3/E_3$:

$$\begin{aligned} e_{31,\mathrm{f}} &= \frac{d_{31}}{s_{11}^{\mathrm{E}} + s_{12}^{\mathrm{E}}}, \\ d_{33,\mathrm{f}} &= d_{33} - \frac{2s_{13}^{\mathrm{E}}}{s_{11}^{\mathrm{E}} + s_{12}^{\mathrm{E}}}d_{31}. \end{aligned} \tag{15.2}$$

These two coefficients can be used to describe piezoelectric laminated structures using a mixed variable set of (S_1, S_2, T_3) in cases where the shear strains can be neglected [18]. The clamping of the ferroelectric film on a substrate affects piezoelectric properties in two ways. There is first the reduction of $d_{33,\mathrm{f}}$ as given by (15.2). A second problem arises from the elastic clamping of

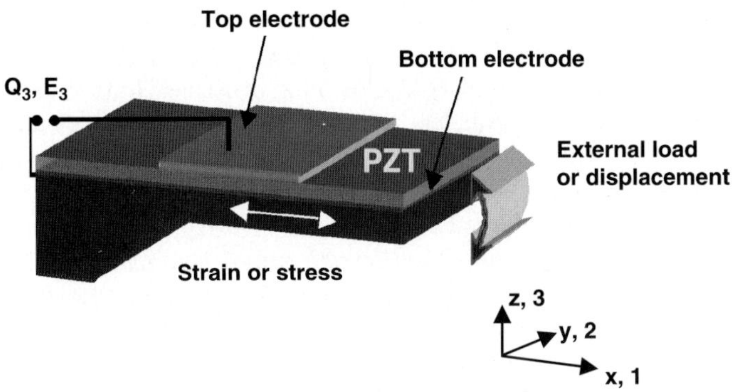

Fig. 15.2. Input (E_3, load) and output parameters (displacement, Q_3) in actuator and sensor applications for piezoelectric laminated plates

ferroelastic domain walls, with the consequence of a reduced displacement for a given field, and thus a smaller piezoelectric response. This will be discussed in Sect. 15.4.

Both effective coefficients can be measured directly (see [19] for $d_{33,\mathrm{f}}$, and [20–22] for $e_{31,\mathrm{f}}$). It is obvious that $d_{33,\mathrm{f}}$ is always smaller than d_{33} and that the absolute value of $e_{31,\mathrm{f}}$ is always larger than that of e_{31}. A typical MEMS operation mode of a piezoelectric laminated cantilever is shown in Fig. 15.2. In the actuator mode, application of a voltage $V_3 = E_3 t_\mathrm{p}$ leads to a piezoelectric in-plane stress T_1 causing a deflection of the structure. In the sensor mode, in-plane strains ($S_1 + S_2$) create the piezoelectric charges that record the deformation of the flexible structure. In addition, the film is sensitive to the out-of-plane stress T_3. Attention has to be paid to the fact that such stress is difficult to apply without bending and that it causes in addition in-plane strain by the Poisson effect (see discussions in [23, 24]). A cantilever structure can be modeled by analytical methods [25]. In the actuator mode, a constant curvature is established at a given voltage. The deflection at the end of the beam is proportional to the square of the beam length. Excursions of 10–20 µm with 500 µm long beams have been obtained [26]. More complicated structures must be analyzed by means of finite element calculations [27].

15.3 Materials, Deposition Processes, and Integration

The thin-film materials are in principle the same as the known bulk materials. However, as they have to be synthesized on the device, various limitations exist related to thermal and chemical compatibility. Such limitations are certainly a problem for ferroelectrics with their many fast diffusing and volatile species such as K, Na, Pb, etc. For this reason, integration needs to be studied as intensively as the growth itself.

15.3.1 AlN

There is at least one material that can be synthesized better as a thin film than as a bulk single crystal: the wurtzite structure AlN (see Fig. 15.3). It appears to be very difficult to synthesize single crystals with good piezoelectric properties [28], whereas excellent piezoelectric properties are obtained at films processed by reactive sputter deposition at very moderate temperatures of 300°C or even below [29]. The growth mode leading to high piezoelectric values can be ascribed to enhanced mobility of atoms (adatoms) at the growing surface promoted by ion bombardment, and is typical for sputter-deposited films under certain temperature–pressure/RF-bias conditions (Zone T growth mode according to Thornton [30]). Since ion bombardment leads also to compressive stress, the process needs to be carefully tuned to balance the piezoelectric coefficient and stress [31]. A summary of properties is given in Table 15.1.

15.3.2 PZT and Further Lead-Based Ferroelectrics

The best investigated ferroelectric for piezoelectric MEMS is quite naturally PZT. At the beginning of PZT thin-film development, ferroelectric memories (see [36, 37]) were the driving force. MEMS applications were much less intensively investigated (see [38] for a review). Much effort has been invested in growing PZT thin films with all types of deposition methods (see reviews in [39, 40]). Successful growth and integration of ferroelectric thin films comprise three essential steps: (1) Deposition of a suitable substrate layer (many

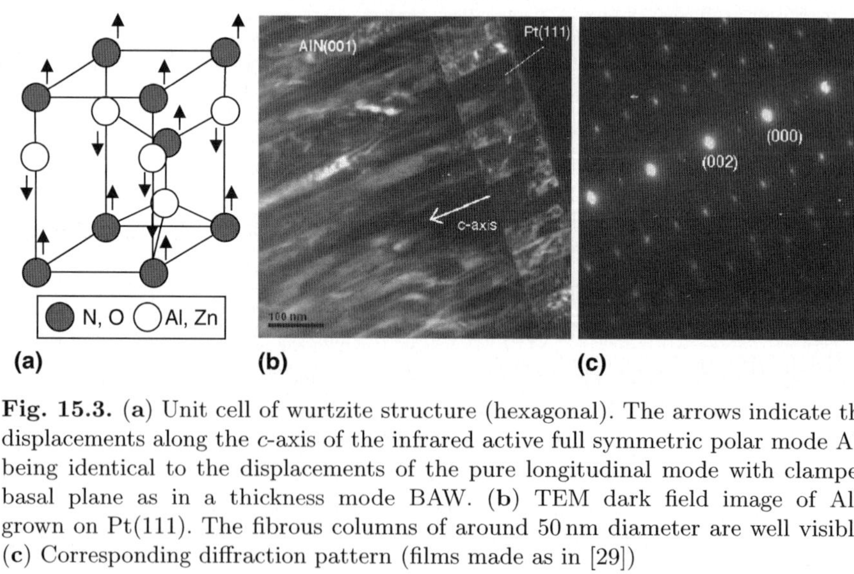

Fig. 15.3. (**a**) Unit cell of wurtzite structure (hexagonal). The arrows indicate the displacements along the c-axis of the infrared active full symmetric polar mode A1, being identical to the displacements of the pure longitudinal mode with clamped basal plane as in a thickness mode BAW. (**b**) TEM dark field image of AlN grown on Pt(111). The fibrous columns of around 50 nm diameter are well visible. (**c**) Corresponding diffraction pattern (films made as in [29])

Table 15.1. Materials constants of AlN from different sources

	e_{33} (C m^{-2})	e_{31} (C m^{-2})	d_{33} (pm V^{-1})	d_{31} (pm V^{-1})	c^{E}_{33} (GPa)	c^{E}_{13} (GPa)	c^{E}_{11} (GPa)	c^{E}_{12} (GPa)	ε_{33} $\varepsilon_{33,f}$	$d_{33,f}$ (pm V^{-1})	$e_{31,f}$ (C m^{-2})	k^2_t (%)
Epi AlN/ sapphire [32]	**1.55**	**−0.58**	*5.53*	*−2.65*	**395**	**120**	**345**	**125**	**9.5**	*3.92*	*1.37*	*6.5*
Ab initio [33]	*2.0*	*−0.47*	**6.72**	**−2.71**	395	120	345	125				
AlN/Pt [34]	*2.0*	*−0.68*	*7.0*	*−3.2*	395	120			**10.2**	**5.1**	**1.3**	*11*
Resonators	*1.67*											**7.8**

Standard data from epitaxial films on sapphire are compared to ab initio calculations and data from polycrystalline thinfilms (bold: primary data, italics: derived data) (from [35])

times an electrode) on a buffer/adhesion layer that is chemically and mechanically stable enough to survive PZT deposition without too much recrystallization (no hillock formation allowed) and without delamination; (2) Nucleation of the perovskite on the bottom electrode, or other substrate, in such a way as to obtain columnar growth and enable texture control; and (3) Continuation of growth avoiding pores (important for $e_{31,f}$ applications) and cracks. The combination of pore and crack-free films becomes a problem for films thicker than about 3 μm. The most common electrode system is passivated Pt(111)/TiO_2/Ti. There is no space here to treat all the growth issues. Just one item is picked up that is common to all deposition methods: nucleation. The perovskite structure is sufficiently complex to exhibit a so-called nucleation-limited growth – at least if we do not deal with homoepitaxy. Hence, substrate structure and chemistry as well as chemical composition of the first monolayers plays an essential role. It is, for instance, observed that nucleation on Pt(111) needs some lead excess, while no excess is needed for continuing growth on an existing PZT film. Another example is the orientation as obtained on Pt(111), which depends on the chemical composition of the first few monolayers. Starting a layer of $PbTiO_3$ with TiO_2 leads rather to (111) orientation, and starting with PbO leads rather to (100) orientation [41]. Such behavior allows texture control [42]. It was shown that the nucleation density is much larger on a thin TiO_2 layer than on bare Pt (Fig. 15.4). On very small TiO_2 islands, quasi-single-crystal platelets were obtained, allowing for a site controlled or registered growth of ferroelectric nanosized structures [43, 44].

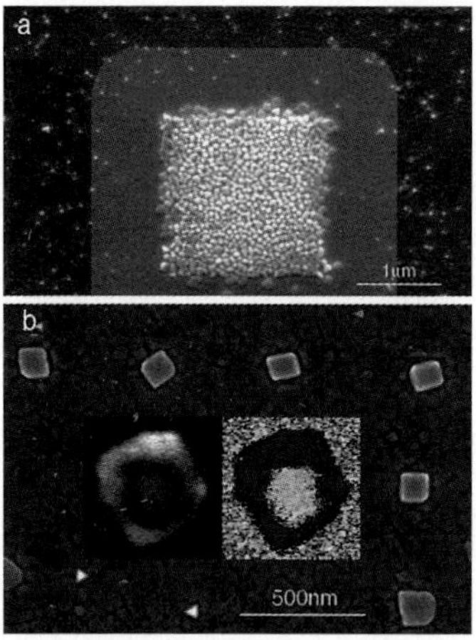

Fig. 15.4. Evidence for nucleation-controlled growth in experiments of in situ physical vapor deposition [43]: (**a**) A 2-nm thick TiO_2 square island of 2×2 μm is situated on an epitaxial Pt(111) film on a $SrTiO_3$(111) single crystal. The TiO_2 film provokes the nucleation of 60 times higher density than on Pt(111). (**b**) On very small islands (here 150 nm wide), single-crystal platelets are obtained after 20 nm PZT40/60 deposition. The inserts show images of amplitude and phase of the piezoelectric response of one of the structures, as measured by piezoelectric-sensitive AFM

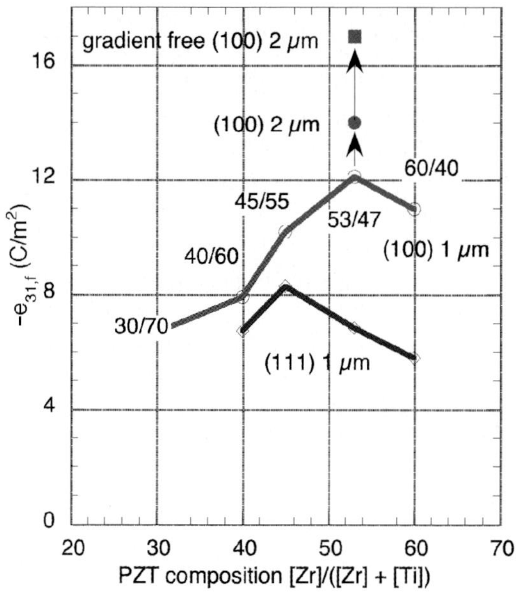

Fig. 15.5. Transverse piezoelectric coefficient $e_{31,\mathrm{f}}$ as a function of Zr/Ti ratio obtained by a sol-gel route. The full lines represent 1 μm thick films. (from [17] and [46])

Using a beam tool, the optimal PZT composition and orientation were identified as the morphotropic composition (53/47 for Zr/Ti) with (100) orientation [45] (Fig. 15.5). This is in fact the same composition as is used for standard PZT ceramics. This composition excels with a large amount of domain activities as well as peaking of the dielectric constant and all piezoelectric coefficients, especially d_{15}. It was observed that sol–gel films exhibit a compositional gradient that does not allow a very precise hitting of the maximum. This problem could recently be corrected by a preventive variation of the solution composition [47].

A good indicator of the progress is the published value of the effective in-plane transverse piezoelectric coefficient $e_{31,\mathrm{f}}$, which is very relevant for MEMS applications and indicates at the same time a good density, and is conveniently measured in the direct [20–22, 48] and converse mode [26]. In Fig. 15.6, the reported $e_{31,\mathrm{f}}$ is shown as a function of the year of publication. A constant progress is observed in the case of PZT thin films integrated on silicon. The major integration and texture control issues were solved in the 1990s. It is instructive to note that now integrated films on silicon outperform the epitaxial films, which were better initially, since in their case no integration and seeding issues had to be solved. The lead-based epitaxial films show always the same values in the range of -11 to $-13\,\mathrm{C\,m^{-2}}$, whether it is a PZT or a relaxor PMN-PT ($Pb(Mg,Nb)O_3$-$PbTiO_3$).

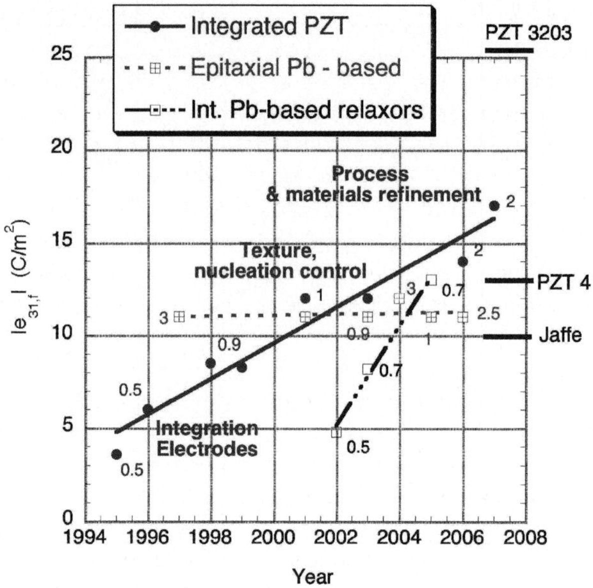

Fig. 15.6. The history of the transverse piezoelectric coefficient $e_{31,f}$ as an indicator of the advancement in the field. The films are in the 0.5–2 μm thickness range. PZT thin films integrated on silicon (data from [20, 21, 26, 45–47, 49]) are compared with several types of lead-based epitaxial films (PZT [50, 51]), $BiScO_3$–$PbTiO_3$ [52], $Pb(MgNb)O_3$–$PbTiO_3$ [53–55], relaxor type integrated thin films such as $Pb(Yb,Nb)O_3$–$PbTiO_3$ [56–58]. The values derived from ceramics data refer to early PZT ceramics (Jaffe), standard hard PZT 4 (product of Morgan Electroceramics), and modern optimized PZT ceramics, the 3203 of Motorola. The small numbers aside of the points are the film thickness in micrometers (from [17])

The difference between integrated and epitaxial films cannot be explained by a thickness effect, as both series contain values in the 2–3 μm range (Fig. 15.6). Thickness would be a relevant parameter since it was indeed observed that thicker films show more domain contributions ([59, 60], see Sect. 15.4). Nevertheless, the explanation has to do most likely with these contributions (if we exclude simple process quality reasons). A value of $-10\,\mathrm{C\,m^{-2}}$ is about the expected value for pure intrinsic piezoelectricity, meaning the response of a single *c*-domain (according to accepted values for electrostrictive, elastic, and dielectric constants of the lattice for PZT40/60 [61, 62]). So one could argue that the lattice in epitaxial films is perfect and that, however, the ferroelastic domains do not cooperate as well as in the best films grown on platinized silicon. There is a difference in domain configuration due to different thermal strain histories, because the substrates for epitaxy (MgO, $SrTiO_3$, etc.) exhibit larger thermal expansion coefficients than that of silicon. As a result, the *a*-domain volume fraction in tetragonal phases is much larger in a film on silicon than in the others. A further difference lies in the electrode: Pt as used for integrated films is a true metal that can screen (or compensate) the polarization charges on a very short range by free electrons. This is quite different when using an oxide electrode. Charge compensation involves a deeper zone, and might even involve oxygen vacancies. Furthermore, the nature of chemical bonds across the interface is different. Stress-driven Pt diffusion at the interface is known to happen at the growth temperature [63]. All such effects could lead to differences in dislocation densities and strain relaxation close to the interface, and thus eventually to stronger ferroelastic domain wall pinning near oxide electrodes, which would impede domain wall contributions to piezoelectricity at the small-signal response level.

Relaxor materials do not seem to give any advantage over PZT to date. This can be explained by the following: (1) Relaxors are more difficult to nucleate and to grow, thus requiring higher growth temperatures. This complicates integration tasks; (2) The still relatively low growth temperatures in comparison to bulk materials do not allow for a very homogeneous distributions of the disordered B-site ions, leading to a much lower dielectric constant (see discussion in [17,64]); (3) The piezoelectric e_{ij} coefficients express mechanical stress per applied electric field, and are thus proportional to the stiffness. The high strain response of relaxors does not equally translate in stress response because relaxors are softer.

Ignoring domain effects, well-oriented or textured films should exhibit superior properties than ceramics. This is indeed achieved comparing with pure PZT ceramics (such as the old material published by Jaffe and Berlincourt [65]) and with standard PZT 4 (Morgan Electroceramics). However, there are modern ceramics with elaborate substitutions and dopants that are still superior to the best PZT thin films (Motorola's 3203 in Fig. 15.6). One can thus argue that improvements are still possible.

15.4 Ferroelectric Domains

In ferroelectric perovskites, the ferroelectric phase is derived from a cubic high-temperature phase. The ferroelectric phase is, for instance, tetragonally distorted as in Ti-rich PZT, giving rise to ferroelastic domains. The polarization direction along the longer unit cell axes can point along all the former cubic axes. The so-formed 90° domains are separated by {110}-type planes (domain walls). Figure 15.7a shows the example of a tetragonal PZT(111) film showing a dense domain structure. In the (111) orientation, strain relaxation by such domains is less obvious than in PZT(100) films. Figure 15.7b schematically shows the polarization as it must behave theoretically. Vertical $(1,-1,0)$ domain walls are observed as well. It is an unsolved question whether these vertical domains walls will disappear during poling, or whether tail-to tail (if we consider poling in upwards direction in Fig. 15.7b) configurations are stabilized by defect charges (positive oxygen vacancies in this case). Ferroelastic as well as 180° domain configurations play a very important role

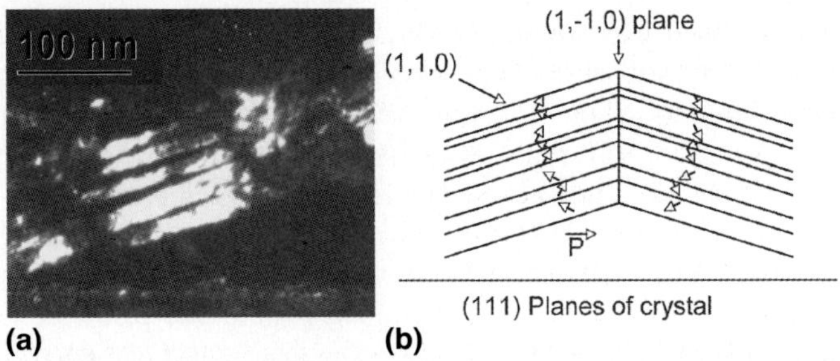

Fig. 15.7. (**a**) Ferroelastic domains in a (111)-oriented, tetragonal PZT film grown as epitaxial film by reactive sputter deposition on a $SrTiO_3(111)$ single crystal (process described in [66]), as revealed by dark field TEM images. (**b**) Schematic drawing of domain patterns corresponding to the observed ones

for the properties. On the one hand they may reduce properties because of antiparallel alignments; on the other, the strain changes by domain wall motions in an electric field may exceed the intrinsic effect of ionic displacements in the crystal lattice [67]. Furthermore, higher-order terms have to be added to the linear piezoelectric effect when domain wall motions become relevant [68].

In PZT ceramics, about 40–60% [67, 69] of the piezoelectric small-signal response is due to ferroelastic domain wall motions. In an electric field, domains change volumes depending on whether their potential energy $-\vec{E}\vec{P}$ is increased or reduced. In the switched volumes, the direction of the spontaneous strains change, yielding a net strain change in the ferroelectric body. In thin films, ferroelastic domains adapt to strain or stress imposed by the substrate in order to reduce the elastic energy [70]. In very thin films, such strains are due to epitaxial mismatch [71, 72]; in thicker films, the cause is the thermal mismatch upon cooling down from the growth temperature. Considering a tetragonal {100}-oriented film, a domain pattern is formed by compensating tensile stress by switching the c-axis into the plane (forming an a-domain), and compensating compressive stress by switching the c-axis out of plane (c-domain). The laminar $c/a/c/a$ domain pattern is a possible solution to the elastic problem [73–75]. The ratio of a- to c-domain volume fraction is given by substrate clamping and thermal mismatch history. However, the domain walls may still be free to move in an electric field compensating the increase in elastic energy. Analytical model calculations of the extrinsic piezoelectric effect ($d_{33,\mathrm{f}}$) in thin films are available for an ideal $c/a/c/a$ domain pattern [74].

For many years not much evidence of such piezoelectric domain contributions was found in thin films. It was thought that ferroelastic domain walls were very much pinned by defects. This idea was supported by hot-poling experiments of tetragonal {100} PZT15/85 films [76]. It was observed that hot poling helped to reduce the a-domain faction. This was attributed to unpinning from defects, mainly oxygen vacancies, since oxygen vacancies may move at the applied temperature of 150–200°C. But after such poling, domains were even more fixed than before (which is advantageous for pyroelectric detectors). Later, it was observed that more than 3 μm thick polycrystalline films deliver significant contributions from ferroelastic domain motions as indicated by considerable nonlinear effects in piezoelectricity [59, 60].

In recent years, much progress in hunting domain phenomena has been made thanks to piezoelectric-sensitive atomic force microscopy (AFM) measurements [77–79]. Local piezoelectricity can be measured, and even individual domains can be observed. When patterning a 200-nm thick PZT 40/60 film into dots with widths in the 100–200 nm range, therefore smaller than the height, an increasing piezo-response with increasing aspect ratio was found [66] (Fig. 15.8). As unclamping can account for at maximum a doubling of the response, the observed fourfold increase was attributed to unpinning of 90° domains and to the major domain reconfiguration taking place in addition to simple unclamping. Surprisingly, such a drastic increase took place just by patterning, i.e., without "additional help" such as, for example, annealing or poling, indicating that ferroelastic domain walls must be nevertheless quite mobile once the clamping is removed. Recently, a fivefold increase of the piezoresponse close to the corners of patterned features was observed when the film region was additionally subjected to poling before the measurement [80]. In a very recent paper, it was reported that the response at a 90° domain

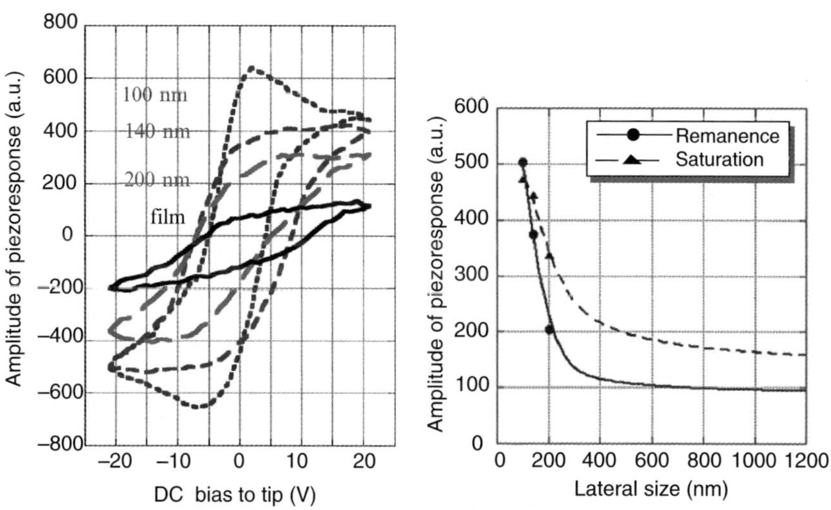

Fig. 15.8. Experimental elucidation of the clamping problem by AFM investigations at nanopatterned thin-film structures with aspect ratio of up to 2. These were etched out from a 200-nm thick epitaxial PZT 40/60 films deposited on top of a conductive $SrTiO_3$ crystal. The ferroelectric loop of the piezoelectric response was measured (*left*), showing a factor of 4–5 times increase in remnant piezoelectricity (*right*) (from [66])

wall boundary can be detected even on a continuous film [81]. It can be concluded that the mechanical clamping to a substrate leads to a very significant stiffening of ferroelastic domain wall motions.

15.5 Flexural Devices for Linear Microactuators and Sensors

With the exception of RF resonators, and possibly inertial sensors using T_3 as input parameter [23, 82], MEMS devices are based on flexural structures exploiting the piezoelectric stress generated in the film plane by either the transverse coefficient, $e_{31,\mathrm{f}}$, or the ID electrodes, as discussed in Sect. 15.2.

Many applications have already been demonstrated with beam-type or bridge-type structures as displayed in Fig. 15.2: movable mirrors and mirror arrays for video-projection [83], low-frequency microphones [84], accelerometers, active cantilevers for AFMs, micro active damping for MEMS devices [82]. The most promising device with respect to production volume and feasibility is most likely the ink jet printing head, or more generally speaking a droplet ejector in liquid delivery systems. First thin-film versions were developed with ZnO [85,86]. More recently, industry is investigating PZT thin-film versions for high-density ink jet printing. The role of the piezoelectric element is to exert a short pressure pulse (rise time typically 1 μs) to an ink cavity (see Fig. 15.9). The resulting pressure wave leads to a droplet ejection at a suitably placed orifice.

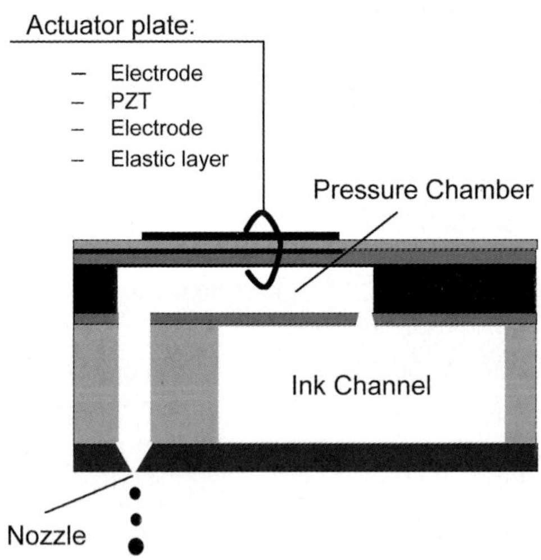

Fig. 15.9. Schematic drawing of an ink jet printing device based on piezoelectric thin films (after [106])

15.6 Flexural Devices for Ultrasonic Applications: Piezoelectric Micromachined Ultrasonic Transducers (pMUT)

If we drive flexural structures into resonance, we obtain a transducer. It is a device able to transmit and receive ultrasonic waves. Potential applications are ultrasonic imaging, nondestructive testing, and sensors using ultrasound. The current state of integrated PZT films on micromachined structures is illustrated in Fig. 15.10. Dense and crack-free PZT films of up to 4 μm thickness can be deposited by sol–gel techniques at the wafer level. The measured admittance curves allow for an evaluation of the electromechanical coupling factor, confirming very well the $e_{31,\mathrm{f}}$ of 12 C m^{-2} or even more measured at beams. pMUTs have attracted some interest in solving integration and precision issues in ultrasonic imaging at higher frequency than the most common 3 MHz, and for two-dimensional probe arrays. The membrane structures exhibit also much smaller acoustic impedances than bulk PZT, allowing for a direct coupling with water-like tissues or media. The inherent problem is, however, the small coupling coefficient due to the fact that the transverse coupling factor is used (which is twice smaller than the longitudinal one), and also that passive material must be used for getting the right operation mode (another factor of 2 is lost). This rule-of-thumb argument makes a forecast of 15% for best k^2, using optimal PZT. Best experimental results so far have reached 6% [88].

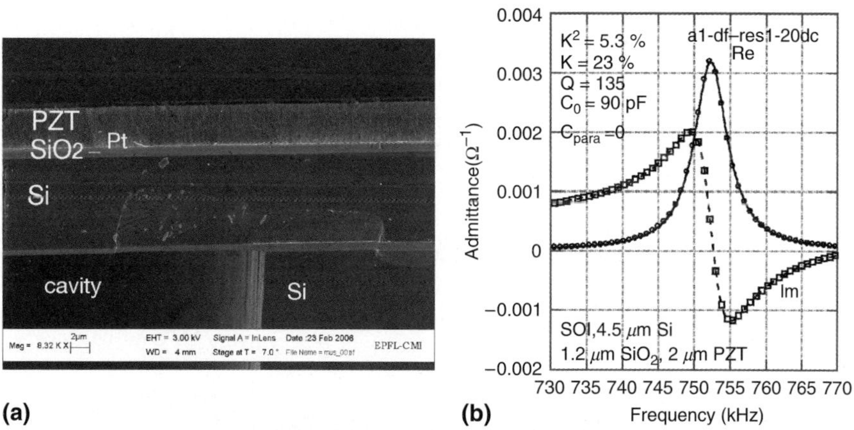

Fig. 15.10. Piezoelectric micromachined ultrasonic transducers: (**a**) SEM cross-section of a rectangular element (from [87]); (**b**) Admittance curve measured in air of a circular element. The solid line represents the equivalent circuit with fitted parameters given in the figure [88]

15.7 Bulk Acoustic Wave Devices

BAW devices are known for their efficient transformation of electrical power into mechanical power and vice versa. A longitudinal ultrasound wave is trapped in an electroded, half-wavelength-thick slab (thickness t_p) of piezoelectric material having its polarization along the wave propagation (see Fig. 15.11). In the bulk world this principle is very successfully applied in ultrasound probes for medical imaging and nondestructive testing at a few MHz. The antiresonance frequency $f_{\mathrm{a}} \approx v_{\mathrm{s}}/(2t_{\mathrm{p}})$ is inversely proportional to the thickness t_{p}. Thin-film BAW resonators (TFBARs) resonate at frequencies of a few gigahertz. The high sound velocity v_s of AlN of 11,000 m s^{-1} yields roughly 2 GHz at 2 µm (in reality somewhat less due to mass loading by electrodes). The typical film thickness range is thus ideal for RF filters in mobile communication. (The US-CDMA standard requires, for instance, a duplex filter centered at frequencies 1960 and 1880 MHz for the receive and the transmit line, respectively.) The elastic energy trapped in the TFBAR constitutes an energy tank of high quality as needed in RF signal processing. TFBARs have been investigated since the early 1980s, initially mainly motivated by applications in television filters. The required frequencies were, however, too small, meaning that the required film thickness was too large to be meaningful. The competing SAW devices were more easily produced at these low frequencies. In addition, sputter deposition was not yet the optimized technique with outstanding uniformity and reproducibility as it is today. The advent of mobile communication and the ever-increasing process control in semiconductor technology, together with the momentum of MEMS research, have led to the convergence of all requirements for a new product called TFBAR RF filter. Today, it is one of the most successful MEMS products. The present market of around 100 million dollars is believed to extend to above 1 billion dollars.

There are two key properties for the choice of the piezoelectric material in TFBARs: the coupling coefficient k_{t}^2 of the thickness mode and the material quality factor Q_{m}:

$$k_{\mathrm{t}}^2 = \frac{e_{33}^2}{c_{33}^{\mathrm{D}} \varepsilon_0 \varepsilon_{33}^{\mathrm{S}}}, \quad Q_{\mathrm{m}} = \frac{1}{\omega \tau_1}. \tag{15.3}$$

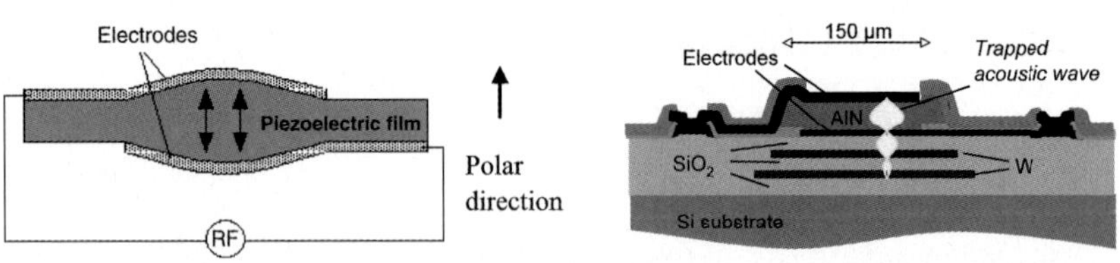

Fig. 15.11. BAW devices: (**a**) Principle of excitation in a half-wavelength slab of piezoelectric material; (**b**) Schematic drawing of BAW resonator used in RF filters for mobile phones at about 2 GHz (after R. Aigner [93])

Table 15.2. Relevant materials parameters for TFBAR applications for the two most used wurtzite thin films, as derived from standard literature data [89, 90]

	k_t^2 (%)	τ_1 (fs)	Q_m at 2 GHz	$k_t^2\ Q_m$ at 2 GHz	v_s (m s^{-1})
AlN	6.5	32	2,490	160	11,000
ZnO	9	45	1,770	160	6,100

Q_m can be related to the viscosity of the material and is written as in (15.3) to take account of the frequency dependence. τ_1 is a material constant. Both these key properties are given for AlN and ZnO in Table 15.2. The coupling constants are achieved in high-quality TFBARs [91–94]; however, the observed quality factor is usually smaller. The main reasons are losses by emission of parasitic waves and resistive losses in electrodes or further parasitic elements. It is interesting to note that the figure of merit for filter insertion loss, the product $k_t^2 Q_m$, is the same for both materials. As the bandwidth depends directly on k_t^2, ZnO looks more interesting at first sight. The fact, however, is that AlN is the favorite material at present. The coupling coefficient of AlN is sufficient to fulfill filter specifications. The decisive properties are higher thermal conductivity and larger breakdown field, which allow for better power handling capabilities with AlN, enabling the use of AlN TFBARs in the transmit line. While there is hardly any material performing better than AlN at frequencies above 2 GHz, one may ask whether at lower frequencies ferroelectrics could do the job. Ferroelectric TFBARs would be smaller and would add some tuning capabilities. Major problems are the acoustic losses due to domain wall relaxation. A review on this issue has recently been published [35].

A TFBAR requires some measures to achieve acoustic isolation. Two different designs have been established. In the first one, the piezoelectric capacitor structure is simply liberated from the substrate by bulk micromachining techniques, more recently also by surface micromachining, i.e., by removing a sacrificial intermediate layer [95]. The complete filter is usually composed of resonators situated on one membrane [92]. A second possibility consists in depositing an acoustic reflector below the resonator. Such a reflector is composed of several pairs of high- and low-acoustic-impedance $\lambda/4$ layers (as shown in Fig. 15.11b), which is able to reflect practically 100% of the longitudinal waves in a relatively large band [91]. Both types are currently fabricated and are on the market. It appears that industry has mastered the very stringent requirements on film uniformity and frequency tuning. The precise resonance frequency of TFBARs depends also on the electrode thickness, and one has to account carefully for all thickness changes during processing (see, e.g., [96] for a description of resonance shifting for filter fabrication).

Resonating piezoelectric bodies are good not only for imaging and signal filtering; piezoelectric resonators are also very sensitive to mass loading. This effect has been used for decades for measuring the evaporation rate by means of quartz crystal oscillators, and also in may types of biological sensors [97,98].

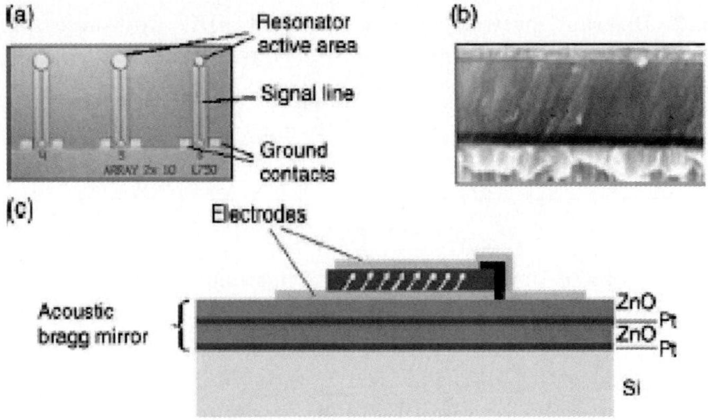

Fig. 15.12. Setup of the FBAR biosensor: (**a**) top view of the resonator and the electrical leads; (**b**) electron microscope picture of c-axis inclined ZnO; (**c**) schematic picture of the lateral structure comprising the resonator with two electrodes solidly mounted on an acoustic Bragg mirror (from [102])

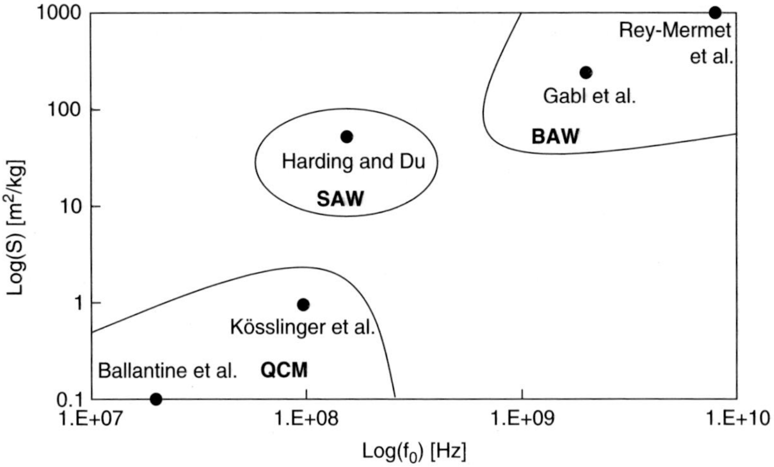

Fig. 15.13. overview on sensitivity results obtained with quartz micro balances (QCM) [97], SAW devices [98], and BAW devices [99, 101]

With the advent of TFBARs, one can imagine the realization of miniaturized biomedical, chemical, and environmental sensors with enhanced functionality, as their integration into MEMS fluidic systems follows quite naturally and the smaller size due to the higher frequency would additionally allow for large arrays with strong screening capabilities. Work in this field is already quite advanced (see Fig. 15.12).

Such sensors were first realized in versions operating with longitudinal waves [99, 100]. The increase of sensitivity with frequency is indeed observed, as shown in Fig. 15.13, which depicts the sensitivity reported in some selected works [101]. The BAW numbers given in this figure correspond to longitudinal modes. Such devices could be useful as gas detectors. A thin polymer layer on the resonator would serve as the gas-sensitive layer changing its mass and its elastic constant with absorption of the gas.

Detection in a liquid, as required for biomedical sensors, requires the use of shear mode BAWs. Longitudinal BAWs emit too much power into the liquid, causing excessive damping. Ideally, shear modes have no amplitude perpendicular to the film plane, and thus excite no emission in the liquid. The resonator is less damped and a good resolution is still possible [102–104]. The excitation of a shear wave requires an electric field that is inclined with respect to the polar axis of the film ($S_5 = d_{15}E_1$). This requires either the growth of polar films with a tilted polar axis in combination with a usual planar capacitor, or the use of ID electrodes. The growth of tilted grains requires first the nucleation of grains that are not (001) oriented, and furthermore an oblique flux of the arriving atoms from the sputter target in order to give a growth advantage for non-(001)-oriented grains [103,105]. It is not evident to achieve a homogeneous film on a large wafer. By tilting the wafer away from the usual face target situation, the deposition uniformity becomes very bad.

In general, the sensitivity S of such a device is defined as the relative frequency change $\Delta f/f_0$ of the resonance (originally at f_0) upon loading with a surface mass density μ (kg m^{-2}):

$$S = \frac{1}{f_0} \lim_{\mu -> 0} \frac{|\Delta f|}{\mu}. \tag{15.4}$$

If we consider the simple case when loading with a material of identical acoustic properties as the piezoelectric (same acoustic impedance, same sound velocity) and identical density ρ, one obtains the frequency $f(\mu)$ and sensitivity S as:

$$f(\mu) = \frac{v_s}{2}\left(\frac{1}{t_0 + \mu/\rho}\right), \quad S = \frac{v_s}{2 f_0 \rho}\left(\frac{1}{t_p}\right)^2 = \frac{2}{v_s \rho} f_0. \tag{15.5}$$

The relative sensitivity thus increases linearly with frequency. If we thus increase the frequency to the gigahertz range by means of a TFBAR, we can expect much more sensitive sensors than those operating at a few megahertz! However, if we have to figure out the detectivity, or in other words the signal-to-noise ratio, of such sensors, we have to argue in a different way. We have to consider the absolute signal size. We do not measure the mass loading directly, but we observe it through its manifestation in the electrical signal. The part in the admittance (or impedance) that is due to acoustic interaction is proportional to the coupling factor k^2. The evaluation of the frequency shift can be carried out only to a precision that must be proportional to the width of the resonance peak (FWHM). We can thus define a detectivity parameter D of the form:

$$D = \frac{k^2}{\mathrm{FWHM}} \lim_{\mu->0} \frac{|\Delta f|}{\mu} \approx \frac{k^2 Q v_\mathrm{s}}{2 f_0 \rho} \left(\frac{1}{t_\mathrm{p}}\right)^2 = \frac{2k^2 Q}{v_\mathrm{s} \rho} f_0. \tag{15.6}$$

Formally, we still see the proportionality to the frequency. Unfortunately, we have to consider that in any material Q decreases with frequency and we can arrive at an expression that includes only material constants.

$$D \approx \frac{k^2}{\pi \tau_1 v_\mathrm{s} \rho} \propto \frac{k^2}{\tau_1 c^{1/2} \rho^{1/2}}$$

Of course, there are many other parameters influencing losses and noise levels, certainly the electronics. A real thin-film resonator is not of a homogenous material, but contains electrodes and passive layers. A simple figure of merit as D is only part of the story. In Weber et al. [102], detectivities were determined by experiments. They showed that their shear-mode TFBAR was indeed better than a quartz microbalance, but not by a factor 100 as the sensitivity would suggest, but only by a factor 4. The reason lies in the increase of noise level with frequency. Since TFBAR sensors are new, one can nevertheless expect that some improvements are still possible.

To conclude this chapter, it is certainly well justified to recognize the large progress in piezoelectric thin-film MEMS over the last two decades. Materials have been very much improved, property–process relations are now much better understood, and integration in micromachined structures has been well mastered. Whereas AlN TFBARs have achieved industrial validation, PZT and other perovskite materials still need to prove their reliability in industrial or consumer applications.

References

1. N. Malouf, Artech House **265** (2000)
2. K. Uozumi, K. Ohsone, R.M. White, Appl. Phys. Lett. **43**, 917–919 (1983)
3. H.T. Soh, I. Ladabaum, A. Atalar, C.F. Quate, B.T. Khuri-Yakub, Appl. Phys. Lett. **69**, 3674–3676 (1996)
4. A.S. Ergun, G.G. Yaralioglu, B.T. Khuri-Yakub, J. Aerospace Eng. **16**, 76–84 (2003)
5. V.P. Jaecklin, C. Linder, N.F. de Rooij, J.-M. Moret, Sensors Actuators A **39**, 83–89 (1993)
6. P. Cheung, R. Horowitz, R.T. Howe, ASME Micormech. Sens. Actuators Syst. **32**, 269–278 (1992)
7. E. Quandt, A. Ludwig, Sens. Actuators **81**, 275–280 (2000)
8. J.F. Scott, C.A. Paz de Araujo, Science **246**, 1400–1405 (1989)
9. O. Auciello, J.F. Scott, R. Ramesh, Phys. Today **51**, 22–27 (1998)
10. H. Raeder, F. Tyholdt, W. Booij, F. Calame, N.P. Ostbo, R. Bredesen, K. Prume, G. Rijnders, P. Muralt, J. Electroceram. **19**, 357–362 (2007)
11. A. Infortuna, P. Muralt, M. Cantoni, N. Setter, J. Appl. Phys. **100**(10), 104110 (2006)

12. A. Infortuna, P. Muralt, M. Cantoni, A. Tagantsev, N. Setter, J. Eur. Cer. Soc. **24**, 1573–1577 (2004)
13. R. Bruchhaus, D. Pitzer, R. Primig, W. Wersing, Y. Xu, Integr. Ferr. **14**, 141–149 (1997)
14. S. Hiboux, P. Muralt, Integr. Ferr. **36**, 83–92 (2001)
15. Y.B. Jeon, R. Sood, J.-H. Jeong, S.-G. Kim, Sens. Actuators A **122**, 16–22 (2005)
16. E. Hong, S. Trolier-McKinstry, R. Smith, S.V. Krishnaswamy, IEEE Trans. UFFC **53**, 697–706 (2006)
17. P. Muralt, J. Am.Ceram. Soc. **91**, 1385–1396 (2008)
18. P. Muralt, Integr. Ferr. **17**, 297–307 (1997)
19. A.L. Kholkin, C. Wüthrich, D.V. Taylor, N. Setter, Rev. Sci. Instr. **67**, 1935–1941 (1996)
20. M.-A. Dubois, P. Muralt, Sens. Actuators A **77**, 106–112 (1999)
21. J.F. Shepard, P.J. Moses, S. Trolier-McKinstry, Sens. Actuators A **71**, 133–138 (1998)
22. K. Prume, P. Muralt, F. Calame, T. Schmitz-Kempen, S. Tiedke, IEEE Trans. UFFC **54**, 8–14 (2007)
23. Y. Nemirovsky, A. Nemirovsky, P. Muralt, N. Setter, Sens. Actuators A **56**, 239–249 (1996)
24. A. Barzegar, D. Damjanovic, N. Ledermann, P. Muralt, J. Appl. Phys. **93**, 4756–4760 (2003)
25. Q. Meng, M. Mehregany, K. Deng, J. Micromech. Microeng. **3**, 18–23 (1993)
26. P. Luginbuhl, G.-A. Racine, P. Lerch, B. Romanowicz, K.G. Brooks, N.F.D. Rooij, P. Renaud, N. Setter, Sens. Actuators A **53**, 530–535 (1996)
27. T. Fabula, H. J. Wagner, B. Schmitt, S. Büttgenbach, Sens. Actuators A **41–42**, 375–380 (1994)
28. G. Bu, D. Ciplys, M. Shur, L.J. Schowalter, S. Schujman, R. Gaska, Appl. Phys. Lett. **84**, 4611–4613 (2004)
29. F. Martin, P. Muralt, M.-A. Dubois, A. Pezous, J. Vac. Sci. Techn. A **22**, 361–365 (2004)
30. A. Rodriguez-Navarro, W. Otano-Rivera, J.M. Garcia-Ruiz, R. Messier, J. Mater. Res. **12**, 1850–1855 (1997)
31. M.-A. Dubois, P. Muralt, J. Appl. Phys. **89**, 6389–6395 (2001)
32. K. Tsubouchi, K. Sugai, N. Mikoshiba. in *IEEE Ultrasonics Symposium*, 1981
33. T. Kamiya, Jpn. J. Appl. Phys. **8**, 4421–4426 (1996)
34. T. Shiosaki, T. Yamamoto, T. Oda, A. Kawabata, Appl. Phys. Lett. **36**, 643–645 (1980)
35. P. Muralt, J. Antifakos, M. Cantoni, R. Lanz, F. Martin. in *IEEE Ultrasonics Symposium 2005* (IEEE, Rotterdam, 2005)
36. C.A. Araujo, L.D. MacMillan, B.M. Melnick, J.D. Cuchiaro, J.F. Scott, Ferroelectrics **104**, 241–256 (1990)
37. R.E. Jones, P. Zürcher, P. Chou, D.J. Taylor, Y.T. Lii, B. Jiang, P.D. Maniar, S.J. Gillespie, Microelectronic Eng. **29**, 3–10 (1995)
38. P. Muralt, J. Micromech. Microeng. **10**(2), 136–146 (2000)
39. C.P.d. Araujo, J.F. Scott, G.W. Taylor (eds.), in *Ferroelectricity and Related Phenomena*, ed. by G.W. Taylor. vol. 10 (Gordon and Breach, Amsterdam, 1996)
40. P. Muralt, IEEE Trans. Ultrason. Ferr. **47**, 903–915 (2000)

41. S. Hiboux, P. Muralt, J. Eur. Cer. Soc. **24**, 1593–1596 (2004)
42. P. Muralt, T. Maeder, L. Sagalowicz, S. Hiboux, S. Scalese, D. Naumovic, R.G. Agostino, N. Xanthopoulos, H.J. Mathieu, L. Patthey, E.L. Bullock, J. Appl. Phys. **83**(7), 3835–3841 (1998)
43. S. Bühlmann, P. Muralt, S. VonAllmen, Appl. Phys. Lett. **84**, 2614–2616 (2004)
44. S. Clemens, T. Schneller, A. van der Hart, F. Peter, R. Waser, Adv. Mater. **17**, 1357–1361 (2005)
45. A. Seifert, N. Ledermann, S. Hiboux, P. Muralt, Mater. Res. Soc. Symp. Proc. **596**, 535–540 (2000)
46. N. Ledermann, P. Muralt, J. Baborowski, S. Gentil, K. Mukati, M. Cantoni, A. Seifert, N. Setter, Sens. Actuators A **105**, 162–170 (2003)
47. F. Calame, P. Muralt, Appl. Phys. Lett. **90**, 062907 (2007)
48. M.-A. Dubois, P. Muralt, D.V. Taylor, S. Hiboux, Integr. Ferr. **22**, 535–543 (1998)
49. F. Tyholdt, F. Calame, K. Prume, H. Raeder, P. Muralt, J. Electroceram. **17**, 311–314 (2007)
50. I. Kanno, S. Fujii, T. Kamada, R. Takayama, Appl. Phys. Lett. **70**, 1378–1380 (1997)
51. I. Kanno, Y. Yokoyama, H. Kotera, K. Wasa, Phys. Rev. B **69**, 064103 (2004)
52. J.C. Nino, S. Trolier-McKinstry, J. Mater. Res. **19**, 568–572 (2004)
53. T. Yoshimura, S. Trolier-McKinstry, J. Cryst. Growth **229**, 445–449 (2001)
54. J.-P. Maria, J.F. Shepard, S. Trolier-McKinstry, T.R. Watkins, A.E. Payzant, Appl. Ceram. Technol. **2**, 51–58 (2005)
55. S. Yokoyama, S. Okamoto, H. Funakubo, T. Iijima, K. Saito, H. Okino, T. Yamamoto, K. Nishida, T. Katoda, J. Sakai, J. Appl. Phys. **100**, 054110 (2006)
56. Q.F. Zhou, Q.Q. Zhang, S. Trolier-McKinstry, J. Appl. Phys. **94**, 3397–3402 (2003)
57. Q.Q. Zhang, Q.F. Zhou, S. Trolier-McKinstry, Appl. Phys. Lett. **80**, 3370–3372 (2002)
58. N. Bassiri Gharb, S. Trolier-McKinstry, J. Appl. Phys. **97**, 064106 (2005)
59. A. Kholkin, Ferroelectrics **238**, 235–243 (2000)
60. F. Xu, S. Trolier-McKinstry, W. Ren, B.M. Xu, Z.L. Xie, K.J. Hemker, J. Appl. Phys. **89**, 1336–1348 (2001)
61. N.A. Pertsev, V.G. Khukhar, H. Kohlstedt, R. Waser, Phys. Rev. B, **67**, 054107 (2003)
62. M.J. Haun, *Ferroelectrics* **99**, 45–54 (1989)
63. A. Kingon, P. Muralt, N. Setter, R. Waser, in *Ceramic Materials for Electronics*, ed. by R.C. Buchanan (Mercel Dekker, New York, 2004)
64. K. Brinkman, Y. Wang, D. Su, A. Tagantsev, P. Muralt, N. Setter, J. Appl. Phys. **102**, 044110 (2007)
65. D.A. Berlincourt, C. Cmolik, H. Jaffe. in *Proceedings of the IRE* (Institute of Radio Engineers, New York, 1960)
66. S. Bühlmann, B. Dwir, J. Baborowski, P. Muralt, Appl. Phys. Lett. **80**, 3195–3197 (2002)
67. Q.M. Zhang, H. Wang, N. Kim, L.E. Cross, J. Appl. Phys. **75**, 454–459 (1994)
68. G. Robert, D. Damjanovic, N. Setter, A.V. Turik, J. Appl. Phys. **89**, 5067–5074 (2000)
69. J.L. Jones, M. Hoffmann, J.E. Daniels, A.J. Studer, Appl. Phys. Lett. **89**, 092901 (2006)

70. A.L. Roitburd, Phys. Stat. Sol. A **37**, 329–339 (1976)
71. J.S. Speck, W. Pompe, J. Appl. Phys. **76**, 466–476 (1994)
72. J.S. Speck, A. Seifert, W. Pompe, R. Ramesh, J. Appl. Phys. **76**, 477–483 (1994)
73. G. Arlt, N.A. Pertsev, J. Appl. Phys. **70**, 2283–2289 (1991)
74. N.A. Pertsev, A.Y. Emelyanov, Appl. Phys. Lett. **71**, 3646–3648 (1997)
75. N.A. Pertsev, A.Yu. Emelyanov, Phys. Solid State **39**, 109–114 (1997)
76. M. Kohli, P. Muralt, N. Setter, Appl. Phys. Lett. **72**, 3217–3219 (1998)
77. M. Aplanalp, L.M. Eng, P. Günter, Appl. Phys. A Mater. Sci. Process **66**, S231 (1998)
78. A. Gruverman, H. Tokumoto, A.S. Prakash, S. Aggarwal, B. Yang, M. Wuttig, R. Ramesh, O. Auciello, T. Venkatesan, Appl. Phys. Lett. **71**, 3492–3494 (1997)
79. E.L. Colla, S. Hong, D.V. Taylor, A.K. Tagantsev, N. Setter, Appl. Phys. Lett. **72**, 2763–2765 (1998)
80. V. Nagarajan, A. Roytburd, A. Stanishevsky, S. Prasertchoung, T. Zhao, L. Chen, J. Melngailis, O. Auciello, R. Ramesh, Nat. Mater. **2**, 43–47 (2002)
81. G. Le Rhun, I. Vrejoiu, M. Alexe, Appl. Phys. Lett. **90**, 012908 (2007)
82. Y.V. Meyer, C.M. Collet, J. Baborowski, P. Muralt, Smart Mater. Struct. **16**, 128–134 (2007)
83. K.H. Hwang, Y.J. Song, S.G. Kim, Jpn. J. Appl. Phys. Part 1 **37**, 7074–7077 (1999)
84. N. Ledermann, P. Muralt, J. Baborowski, M. Forster, J.-P. Pellaux, J. Micromech. Microeng. Syst. **14**, 1650–1658 (2004)
85. G. Percin, T.S. Lundgren, B.T. Khuri-Yakub, Appl. Phys. Lett. **73**, 2375–2377 (1998)
86. G. Percin, B.T. Khuri-Yakub, Rev. Sci. Instr. **73**, 2193–2196 (2002)
87. B. Belgacem, F. Calame, P. Muralt, J. Electroceram. **17**, 369–373 (2007)
88. P. Muralt, N. Ledermann, J. Baborowski, A. Barzegar, S. Gentil, B. Belgacem, S. Petitgrand, A. Bosseboeuf, N. Setter, IEEE Trans. UFFC **52**, 2276–2288 (2005)
89. A. Ballato, J.G. Gualtieri, IEEE Trans. UFFC **41**, 834–844 (1994)
90. J.G. Gualtieri, A. Ballato, IEEE Trans. UFFC **41**, 53–59 (1994)
91. K.M. Lakin, K.T. McCarron, R.E. Rose. in *IEEE Ultrasonics Symposium* (IEEE, Seattle, 1995)
92. R.C. Ruby, P. Bradley, Y. Oshmyansky, A. Chien, J.D. Larson. in *IEEE Ultrasonics Symposium* (IEEE, Atlanta, 2001)
93. R. Aigner, J. Kaitila, J. Ellia, L. Elbrecht, W. Nessler, M. Handmann, T. Herzog, and S. Marksteiner. in *IMS'2003* (Philadelphia, 2003)
94. H. Heinze, E. Schmidhammer, C. Diekmann, and T. Metzger. in *IEEE Ultrasonics Symposium* (*IEEE*, Montreal, 2004)
95. M.-A. Dubois, J.-F. Carpentier, P. Vincent, C. Billard, G. Parat, C. Muller, P. Ancey, P. Conti, *Monolithic above-IC resonators technology for integrated architectures in mobile and wireless communication.* IEEE J. Solid-State Circuits **41**, 7–16 (2006)
96. R. Lanz, P. Muralt, IEEE Trans. UFFC **52**, 936–946 (2005)
97. E. Gizeli, in *Biomolecular Sensors*, ed. by E. Gizelli, C.R. Lowe (Taylor & Francis, London, 2002)
98. G.L. Harding, J. Du, Smart Mater. Struct. **6**, 716–720 (1997)
99. R. Gabl, H.-D. Feucht, H. Zeininger, G. Eckstein, M. Schreiter, R. Primig, D. Pitzer, W. Wersing, Biosens. Bioelectron. **19**, 615–620 (2004)

100. M. Benetti, D. Cannata, F. DiPietrantionio, Foglietti, Verana, Appl. Phys. Lett. **87**, 173504 (2005)
101. S. Rey-Mermet, R. Lanz, P. Muralt, Sens. Actuators B **114**, 681–686 (2006)
102. J. Weber, W.M. Albers, J. Tuppurainen, M. Link, R. Gabl, W. Wersing, M. Schreiter, Sens. Actuators A **128**, 84–88 (2006)
103. J. Bjurstrom, G. Wingqvist, I. Katardjiev, IEEE Trans. UFFC **11**, 2095–2100 (2006)
104. G. Wingqvist, J. Bjurstrom, L. Liljeholm, V. Yantchev, I. Katardjiev, Sens. Actuators B **123**, 466–473 (2007)
105. T. Yanagitani, M. Kiuchi, M. Matsukawa, Y. Watanabe, J. Appl. Phys. **102**, 024110 (2007)
106. E. Fujii, R. Takayama, K. Nomura, A. Murata, T. Hirasawa, A. Tomozawa, S. Fujii, T. Kamada, and H. Torii, IEEE Trans. UFFC **54**, 2431–2438 (2007)

16

Piezoelectric Composites by Solid Freeform Fabrication: A Nature-Inspired Approach

A. Safari and E.K. Akdoğan

16.1 Introduction

Piezoelectrics and electrostrictors are indispensable materials for use in transducer technology, as they inherently possess both direct (sensing) and converse (actuation) effects. A piezoelectric/electrostrictive sensor converts a mechanical input (displacement or force) into a measurable electrical output through piezoelectric/electrostrictive energy conversion. In the case of a piezoelectric, an applied mechanical force (stress) induces a voltage across the terminals of the transducer. On the other hand, an applied mechanical force induces a change in the capacitance of an electrostrictive transducer that could be electrically detected. Hence, the mechanical to electrical energy conversion is accomplished directly when a piezoelectric is used, while the same is obtained indirectly if the electroactive material of choice is an electrostrictor. Conversely, both piezoelectric and electrostrictive materials develop an elastic strain under an applied electric field. The said elastic strain is linearly proportional to the applied field in a piezoelectric, whereas electrostrictive coupling involves the second-order (quadratic) coupling of electric field with elastic strain. While piezoelectricity is possible only in noncentrosymetric point groups, electrostriction is observed in all solids, which make it a much more general solid-state phenomenon. Sensing and actuation functions can coexist in a given transducer by the intelligent use of such materials. Piezoelectrics and electrostrictors, therefore, constitute the backbone of modern transducer technology, as mechanical to electric energy (and vice versa) conversion can be accomplished with great efficiency in a way that is second to none among all phenomena known to date [1, 2].

The direct piezoelectric effect enables a transducer to function as a passive sound receiver or as a pickup by converting sound pressure into an electrical signal. Applications in which such electromechanical transducers are used include hydrophones for underwater low-frequency noise detection and microphones, among others. On the other hand, the converse piezoelectric effect permits a transducer to act as an active sound transmitter or a loudspeaker.

When used in the pulse-echo mode, a transducer can perform both active and passive functions in a quasi-simultaneous fashion. This mode of operation entails the emission of an acoustic wave into the medium of interest, and sensing the reflection (echo) back onto it. Such echoes are produced when sound waves are reflected back from an interface (boundary) between two substances of different acoustic impedances [3]. The magnitude of the echo is proportional to the acoustic impedance mismatch between the two materials forming the interface and the medium in which the wave travels. Electrostrictive materials, on the other hand, can be used as a high-displacement actuator since the strain is coupled to the applied field quadratically. Such materials have also been used as sensors where the change in permittivity with applied stress constitutes the mechanism for sensing.

The design and fabrication of composite materials opens new avenues to optimizing the electrical, magnetic, and mechanical properties for specific application. Properties that are otherwise impossible to achieve with constituent materials in single-phase fashion can be synthesized by the use of several connectivity patterns (see Sect. 16.3). Piezoelectric composites originated from this line of reasoning to tailor the input and output responses of an electromechanical transducer to a specific application. For instance, one may wish to maximize the piezoelectric sensitivity, minimize the density to obtain a good acoustic matching with water, and at the same time, make the transducer mechanically flexible to conform to a curved surface [3]. These properties are mutually exclusive; and Mother Nature does not provide us with single-phase materials that simultaneously satisfy such requirements. Thus, in many applications, one might meet conflicting design requirements by combining the most useful properties of two or more phases through the effective use of black and white symmetry – the essence of the composite approach.

What is elaborated herein appertains to the development of piezoelectric ceramic–polymer composites and novel actuators by solid freeform fabrication (SFF). This advanced fabrication method brings unique degrees of freedom to composite making, which is likely to set the stage for the prototyping of next-generation piezoelectric composites in the twenty-first century.

16.2 Connectivity Patterns: A Unifying Concept for Composites

The arrangement of the phases comprising a piezoelectric composite dictates the field patterns inside the composite, which in turn govern the electromechanical properties. A system that allows one to codify the manner in which the individual phases are self-connected (continuous) was first proposed by Newnham et al. on the basis of black and white symmetry principles [4,5]. There are 10 connectivity patterns for a two-phase (diphasic) system, in which each phase could be continuous in zero, one, two, or three dimensions as illustrated in Fig. 16.1.

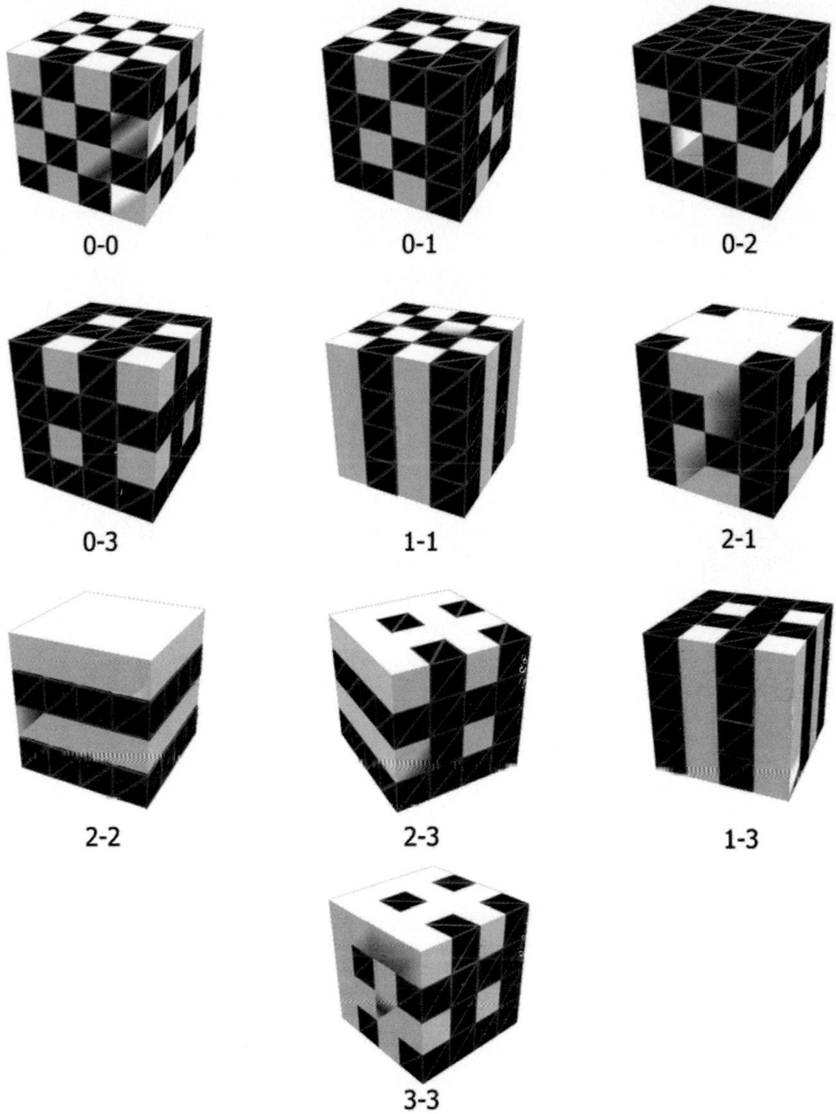

Fig. 16.1. The 10 connectivity patterns in a two-phase composite

The internationally accepted nomenclature to describe such composites is (0–0), (0–1), (0–2), (0–3), (1–1), (1–2), (2–2), (1–3), (2–3), and (3–3). The first digit within the parentheses refers to the number of degrees of freedom the piezoelectrically active phase possesses, while the latter designates the same for the inactive polymer phase. The polymer phase is virtually always chosen to be inactive since its role is to reduce the acoustic impedance and dielectric permittivity, and not to contribute to producing a mechanical response. Several piezoelectric ceramic/polymer composites have been developed on the basis of the connectivity concept, as shown in Fig. 16.2 [6–10]. The connectivity pattern formalism plays a similar role as what point groups represent for crystals within the framework of Neumann's principle, as well as space groups within the context of translational symmetry. On the basis of the aforementioned symmetry principles, a multitude of composites created using fabrication techniques such as dicing, tape casting, lost mold method, injection molding, and others is summarized elsewhere [3].

16.3 Solid Freeform Fabrication Fundamentals

While traditional ceramic processing techniques such as tape casting, injection molding, dicing, and so on have proven to be effective in making composites with simple connectivity patterns, none of them permits the fabrication of structures with complex internal hierarchy and symmetry because of their rectilinear mode of operation. From this limitation arose the idea to develop a new manufacturing/prototyping technique that ultimately would impart very high design flexibility to the fabrication of polymer, ceramic and metal structures, as well as their permutations. SFF, an emerging technology that provides an integrated way of manufacturing 3D components from computer aided design (CAD) files, seems to be the answer. It is an additive process by which an arbitrary 3D structure is made by depositing a given material in a layer-by-layer fashion, without using any hard tooling, dies, molds, or machining [11–17].

In the mid-1990s, several SFF methods have been developed to fabricate polymer, metal, or ceramic structures on a fixtureless platform, directly from a CAD file. Some of these techniques are designed to produce large parts with a fast output rate, and with modest surface finish. Some of SFF techniques, on the other hand, target markets where a very high resolution and a good surface finish are required. SFF or rapid prototyping methods that have found commercial success include stereolithography (SLA, 3-D Systems, Inc.) [18, 19], fused deposition modeling (FDM, Stratasys, Inc.) [12–16,20,21], selective laser sintering (SLS, DTM Corp.) [16], laminated object manufacturing (LOM, Helisys, Inc.) [22,23], 3D printing (3-DP, Soligen, Inc.) [24], robocasting (Sandia National Labs) [25, 26], and Sanders Prototyping (SP, Sanders Prototyping, Inc.) [27]. Most of these techniques are designed to manufacture net shape polymer parts for form-fitting applications and design verification. However, some are also capable of manufacturing metal or ceramic parts as exemplified by the Rutgers fused deposition of ceramics (FDC) and fused deposition of multimaterials (FDMM) processes [20, 21, 28].

All SFF techniques begin with a common approach. First, a CAD data description of the desired component is prepared. Second, a surface file (aka stl file) is created from the CAD file, which is later input into the manufacturing

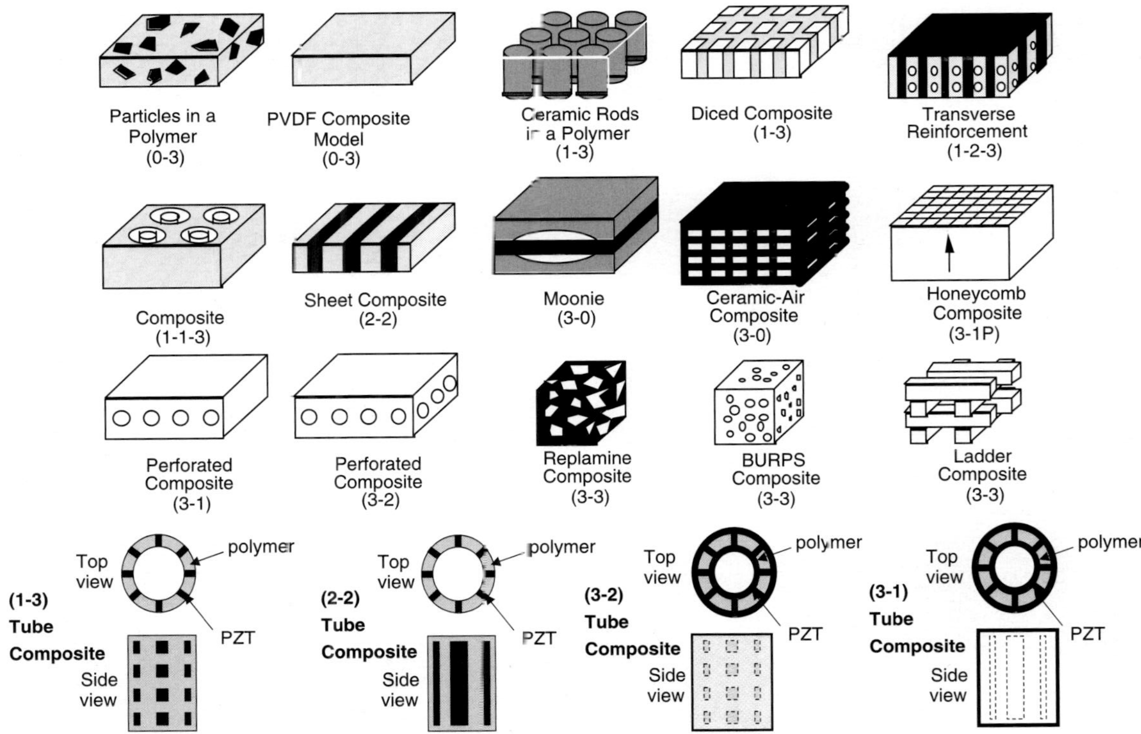

Fig. 16.2. Examples of various composites with reference to their connectivity pattern

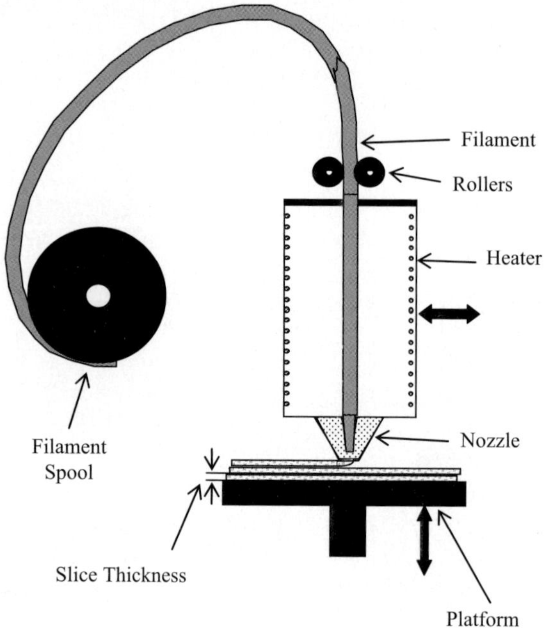

Fig. 16.3. Schematic of solid freeform fabrication by fused deposition of ceramic process

system. Third, the so-called stl file is converted into cross-sectional slices, or a slice file, in which each slice can be uniquely defined about its build strategy by varying the tool path. The slices collectively define the shape of the part. Finally, the information related to each slice (i.e., toolpath) is then transmitted in a layer-by-layer fashion to the SFF machine. In Fig. 16.3 a schematic representation of the FDC process is shown. The lowering of the fixtureless platform by one slice thickness in the z-direction follows the completion of each layer. Subsequent layers are built on top of the preceding one in a sequential manner, and the process is repeated until the whole part is finished. Each layer fuses onto each other, forming a solid continuum thereby. The process typically uses a filament feedstock made out of a thermoplastic polymer in which is dispersed a ceramic or metal, using the principles of colloidal chemistry. The sequential or layered approach of manufacturing a 3D object of arbitrary shape is indeed the quintessential feature of SFF. In other words, the way SFF works is to mimic Mother Nature's method of putting together a 3D crystal structure, which is one atomic plane at a time.

In the fused deposition of multiple materials (FDMM) process, the same approach is pursued as in the FDC process; however, a multihead assembly is used instead of a single-head assembly. The multihead assembly enables one to build 3D structures comprising two or more materials in a sequential fashion.

While the FDMM process offers great flexibility in the design and fabrication of multimaterial performs, co-densification is a major challenge owing to the differential shrinkage of each material. However, judicious materials selection as well as the fine tuning of the solids loading in the filament feedstock can circumvent this difficulty [6, 21, 22].

SFF techniques provide many advantages for the manufacturing of advanced functional components over traditional ones. A part normally takes weeks or months to fabricate because of the time spent in fabrication of the mold and tools, machining operations, and so on. In contrast, the same part can be prototyped in a few days' time using SFF, providing reduced lead times and costs in the development of new functional parts thereby. This technology is advantageous to use in cost-effective iterative design to produce components with optimum shape and desired properties.

The feasibility of the FDC process in making piezoelectric composites was demonstrated in the fabrication of various PZT/polymer composites with different connectivity patterns. The FDC process was used via two distinct processing routes: (1) direct FDC and (2) indirect technique or the lost mold process where polymeric molds were manufactured by FDM. In the direct processing route, FDC is used with a PZT-powder-loaded polymer filament as the feed material for a direct-layered manufacturing of the 3D green ceramic structure. The indirect route was used in fabricating a polymer mold in which the negative of the desired structure is formed via the FDM process using the Stratasys commercial filament (ICW-04) and Sanders prototyping. The mold is infiltrated with the PZT slurry and then placed in a vacuum oven to ensure complete filling of the voids. The samples are then dried in an ambient atmosphere (~2 h), followed by drying at higher temperatures (~70°C, 2 h). The initial slow drying step has to be included to reduce the possibility of cracking in the slip. The thermoplastic polymer mold is evaporated during the early stages of a specifically designed binder burnout cycle.

The optimum solids loading used in the production of filaments are typically 50–55 vol.%, which utilizes a four-component thermoplastic binder system containing a polymer, tackifier, wax, and plasticizer. It was found that powder pretreatment with an organic surfactant before compounding is critical in obtaining feedstock materials that can be extruded. Filaments of 1.78 mm in diameter are typically used in the process. Fused deposition of these filaments is accomplished using the 3D Modeler (Stratasys, Inc.) or an in-house-built FDMM system.

Previous research which harnessed the design guidelines provided by the connectivity pattern formalism has unequivocally shown that composites fabricated by SFF exhibit superior piezoelectric properties in comparison to single-phase piezoelectric ceramics as well as composites made by traditional methods. In the following section, we will present some examples of composites and actuators which were fabricated by SFF.

16.4 Piezoelectric Ceramic Polymer Composites

16.4.1 Example of Piezoelectric Composites by SFF

Piezocomposites with (2–2) connectivity can be readily made using dicing technique as well as by SFF. However, the traditional dicing technique poses severe difficulties when one wishes to introduce a volume fraction gradient (VFG) in the composite [3, 16]. That is so because the so-called kerf width has to be increased from the center to the edges, while a cutting blade with a single thickness is used in the dicing machine [5].

To overcome these limitations, Panda [6, 29] studied VFG (2–2) composites, which were fabricated with the indirect lost mold route using Sanders Prototyping. Many mathematical functions including regular, Gaussian, linear, and exponential gradients, and their effects on various performance metrics were studied. The electromechanical and acoustic properties of such VFG composites were compared to those having no VFG. In Panda's work [6], all VFG distributions were designed to have ∼60 vol. % ceramic in the center region of the composite to obtain a high acoustic pressure output at the center. However, no further increase in the ceramic content at the center of the composite was pursued, as that would also increase the acoustic impedance difference between the central region and the edges albeit further improving the output. As a result, very complex matching layers would have been needed to obtain good energy transfer across the whole composite. The ceramic content in such VFG composites was gradually decreased to approximately 20% at the edges for all the types of distributions studied to ensure that the edges would act as a good receiver of ultrasonic waves owing to the low dielectric constant of those regions. A further decrease in the total ceramic amount at the edges could further enhance the receiving sensitivity. However, it would concomitantly cause a decrease in the thickness-coupling coefficient because of interference from the lateral modes, for a given ceramic element width and spacing.

In this approach, an air gap of 135 μm between the walls of the mold was created. Such a design enables one to ascertain that ceramic walls maintain a uniform width after slurry infiltration, as well as after the subsequent heat treatment. Composites with wall thickness ranging from 95 μm at the center to 575 μm at the edges were successfully fabricated, in accordance with the respective mathematical function. The positioning and thickness of the mold walls were very accurate with an $X-Y$ deposition error of only ±5 μm. After infiltrating the PZT slurry into the mold and by subsequent heat treatment, the sintered structures were embedded in Spurr epoxy. The measurement of the dielectric properties of such linear, exponential, and regular (2–2) composites has shown that the dielectric constant was as high as 950 for the Gaussian and 430 for the exponential design, following the total volume percent of PZT in the composite as shown in Table 16.1 along with other relevant properties. All structures had well-discernable thickness mode resonances with $k_\mathrm{t} \sim 66\%$,

Table 16.1. Comparison of electromechanical properties of volume fraction gradient (2–2) PZT composites with traditional (2–2) composites

Gradient type	Ceramic wall width (μm)	Spurr epoxy width (μm) Center	Spurr epoxy width (μm) Edge	Ceramic volume (%)	K	k_t (%)	k_p (%)	d33 (pC N^{-1})
Gaussian	115	85	500	36	950	68	26	430
Linear	115	90	550	28	800	66	28	470
Exponential	115	90	570	25	430	66	20	410
Regular (2–2) composite	220	440	440	33	570	63	–	350

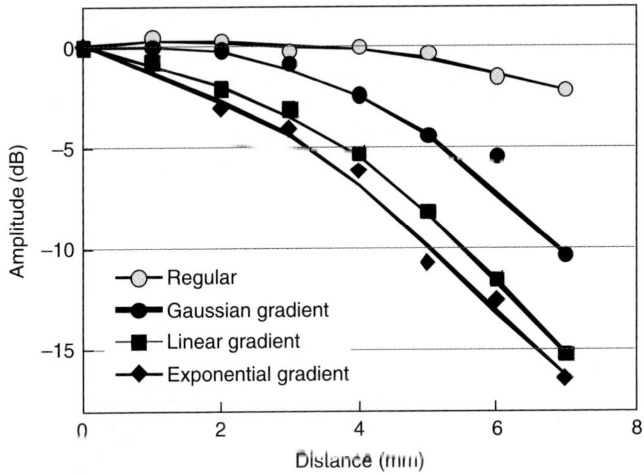

Fig. 16.4. Variation of acoustic amplitude with distance from the center in a volume fraction gradient (2–2) composite as a function of gradient type

and moderately low values of the k_p. The high k_t in such composites was attributed to the high aspect ratio of the ceramic elements (>6). The longitudinal piezocharge coefficient (d_{33}) of such composites was found to be greater than 400 pC N^{-1} (see Table 16.1).

The vibration profiles of the (2–2) composites with and without VFGs are shown in Fig. 16.4 as a function of the distance from the center. The pressure output of the VFG designs were compared with a regular (2–2) composite, which was obtained from a diced PZT-5H ceramic (wall width: 220 μm; Kerf: 440 μm). The regular design had ~33 vol. % ceramic with a d_{33} and k_t of 340 pC N^{-1} and 63%, respectively. As expected, the amplitude of the vibration for this composite did not vary much with the distance from the center. The small decrease in the amplitude at the edges was attributed to the clamping of the edges. On the other hand, all VFG distributions showed a decrease in the vibration amplitude as distance from the center increased (see Fig. 16.4). The Gaussian VFG was designed to have a very slow decrease in ceramic volume fraction with distance from the center. The vibration amplitude observed in Fig. 16.4 followed a similar trend, initially remaining constant and then dropping by only −4 dB at a distance of 5 mm. Hence, this distribution was found to be very different from the regular (2–2) composite at the same distance from the center. The linear gradient showed a very smooth decline in the vibration output with distance (see Fig. 16.4), linearly falling to about −8 dB at a distance of 5 mm. The exponential distribution had a vibration profile that was slightly lower in amplitude than the linear distribution. It should be noted that although the ratio of the total ceramic content in the center and the edges is constant for all VFG distributions, the pressure output ratios are not similar, which was attributed to the limitation in the measurement techniques [22]. The precise manipulation of the volume fraction within the same composite allowed for the control of the vibration amplitude profile to obtain the desired pressure output as seen in Fig. 16.5. The profile for the Gaussian vibration amplitude did not fall as fast, and was very similar to the regular distribution. The linear or the exponential gradient showed a very smooth decrease in the vibration from the center to the edges.

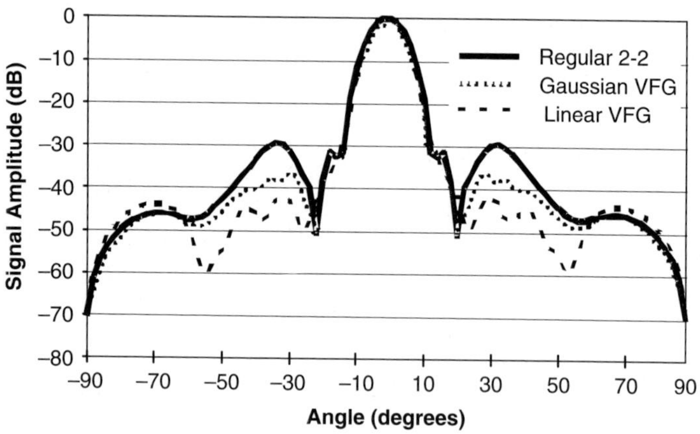

Fig. 16.5. Variation of signal amplitude with angle in a volume fraction gradient (2–2) composite as a function of gradient type

Turcu et al. [30, 31] studied oriented (2–2) soft PZT-Epoxy composites that were fabricated by FDC, which are depicted in Fig. 16.6. The inspiration for this study came from the modeling work of Nan et al. [32, 33], who theoretically investigated the effects of orientation of the ceramic phase relative to the poling directions in (0–3) and (1–3) piezocomposites using a Green's function formalism. Nan et al.'s modeling results showed that the dielectric and piezoelectric properties of the piezocomposites with (1–3) connectivity decreased when the poling direction deviated from the orientation of the ceramic phase. In Turcu's work, the orientation angle of the ceramic phase relative to the poling direction varied in the range of 0–75° with 15° increments (see

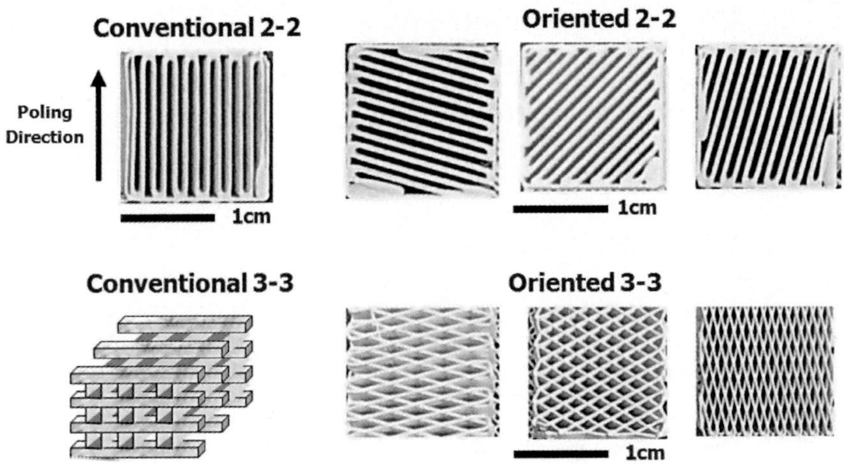

Fig. 16.6. Cross-sectional photographs of oriented (2–2) and (3–3) composites with various orientation angles

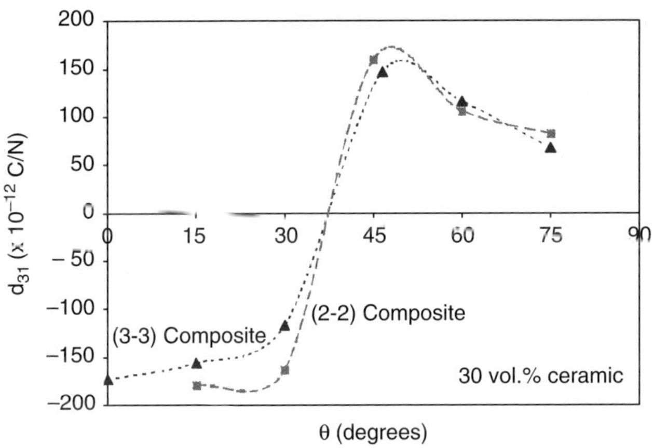

Fig. 16.7. Orientation angle dependence of effective transverse piezocharge coefficient in oriented (2–2) and (3–3) composites showing the critical angle for cross-over

Fig. 16.6) [30, 31]. The volume fraction of the ceramic phase in both composites was fixed at 0.30. It was found that the effective dielectric constant and effective d_{33} coefficients of both composites decreased with increasing orientation angle. However, the effective d_{31} coefficient became positive at 45°, reaching a value of 150 pC N^{-1} as shown in Fig. 16.7. The data showed a maximum in $d_h g_h$ at ∼45° as well, reaching a magnitude of 15,000 × 10^{-15} m^2 N^{-1} as depicted in Fig. 16.8. On the other hand, the d_{33} was ∼430 pC N^{-1} up to 45°, followed by a sharp decrease to 100 pC N^{-1} at 75°. Turcu et al. also studied oriented (3–3) soft PZT–epoxy composites by FDC (see Fig. 16.6). These composites exhibit an 85% increase in piezoelectric voltage coefficient

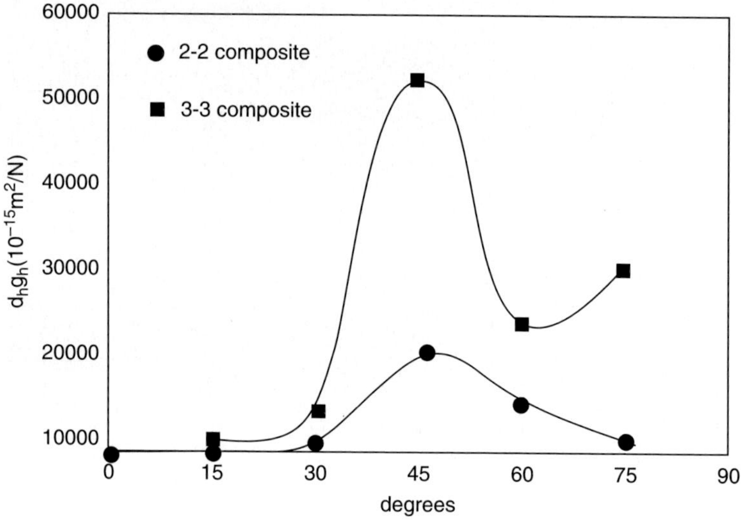

Fig. 16.8. Orientation angle dependence of the hydrostatic figure of merit for oriented (2–2) and (3–3) composites

g_{33} at an orientation angle of 45°. The FOM of the oriented (3–3) composite reached an extremely high value of ~50,000 × $10^{-15}\,m^2\,N^{-1}$, while is much higher than the obtained in (2–2) composites as shown in Fig. 16.8.

A composite with (3–3) connectivity could also be envisioned as two phases mutually penetrating each other, thereby forming two 3D self-connected networks in intimate contact. The feasibility of the FDC process for making piezoelectric composites was demonstrated in the fabrication of PZT/polymer 3D honeycomb and ladder composites with (3–3) connectivity patterns [6,10]. The ladder structures were built by using a raster-fill strategy with a fixed inter-road spacing. The consecutive layers were built at 90° to each other. The volume fraction of the ceramic phase in the structure is approximately 70%, and can be varied by changing the width and spacing between the ceramic roads. In this particular design, the ceramic roads are ~300 μm wide with ~800 μm center-to-center spacing. The mesoscopic structures created by FDC are very uniform with excellent unit cell repeatability, unequivocally demonstrating the superiority of the FDC process over traditional methods of composite making [6]. The indirect method was also used to fabricate 3D honeycomb composites. The scanning electron micrograph in Fig. 16.9 depicts a structure of a sintered ceramic prior to epoxy infiltration. The structure consists of a 3D lattice composed of "air pipes" defining the 3D piezoelectric skeleton (matrix). The "air pipes" were created upon burning out the sacrificial polymer. By virtue of the flexibility offered by the FDC process, the diameter of the holes, the spacing, and the volume fraction of the ceramic phase can be varied between the holes. In the example given in Fig. 16.9, the diameter of the holes is ~200 μm, with ~350 μm center-to-center uniform spacing, and the PZT content is ~25% by volume [6, 34–36].

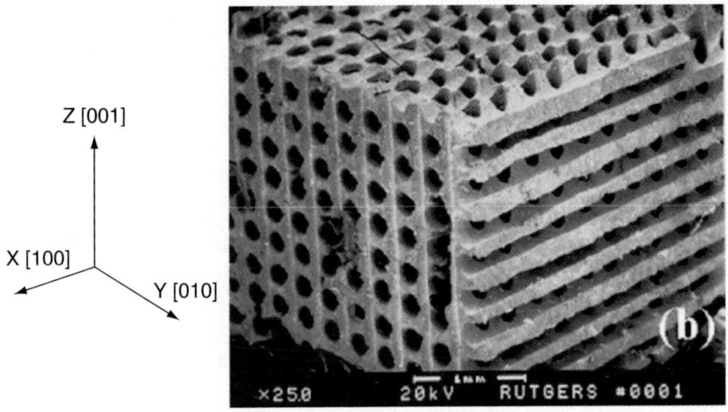

Fig. 16.9. Scanning electron micrograph of an air-pipe composite with (3–3) connectivity showing the ceramic phase prior to infiltration with polymer

Monolithic ceramic tubes, like their planar counterparts, suffer from similar deficiencies such as high weight, high capacitance, and low figure of merit $d_h g_h$. Until recently, the fabrication of functional piezoelectric composites with radially oriented connectivity has been difficult or impossible using conventional processing. Radial composites with four different connectivity patterns have been fabricated using FDC and the SP processes. These consist of (1–3), (2–2), (3–1), and (3–2) type connectivity as shown in Fig. 16.10 [37]. For the purpose of simplicity, cylindrical coordinate systems are being used to designate the connectivity of each composite design. For the (1–3) type composite, the ceramic phase is continuous in only the r direction, while the polymer phase is continuous in the diameter (r), height (z), and sweep angle (θ) directions (see Fig. 16.10). For (2–2) type composites, both the ceramic phase and the polymer phases are continuous in the r and z directions. In the case of (3–1) type composites, the ceramic phase is continuous in all three directions, while the polymer is continuous in only the z direction. Finally, for the (3–2) type composites, the ceramic phase is continuous in all three directions, while the polymer phase is continuous in the z and θ directions. In this structure, the inner and outer ceramic rings are approximately 1 mm in thickness, and the ceramic sheets connecting the inner and outer rings are approximately 1 mm by 1 mm.

16.4.2 Piezoelectric Actuators Fabricated by SFF

SFF has been used to fabricate novel actuators as well [38–46]. It has been demonstrated that improved electromechanical properties can be achieved by tailoring the geometry and dimensions of an actuator, which we call *form factor engineering*. Processing and properties of a few rapidly prototyped actuators will be reviewed with emphasis on the recent results on unimorph and multimaterial spirals and bending actuators.

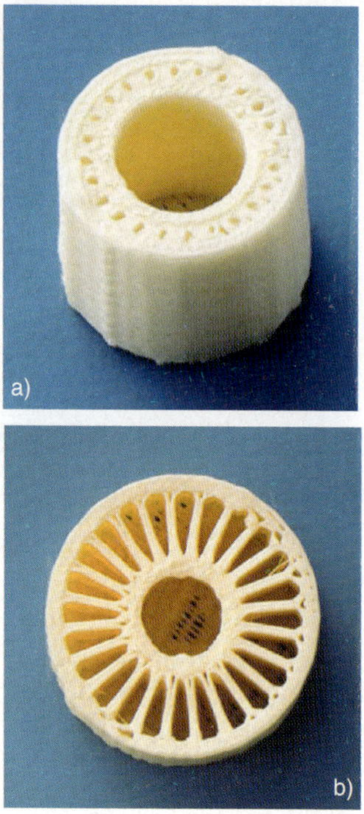

Fig. 16.10. Photographs of radial composites with (**a**) 3–1 and (**b**) 3–2 connectivity

Piezoelectric actuators generate field-induced displacements that can be used in motion systems such as micropositioners, ultrasonic motors, and vibration damping. It is desirable to enhance the displacement while maintaining a reasonable amount of load bearing capability. This goal has been achieved by manipulating the geometry of simple shape (rod or strip) actuators into more complex geometries [43, 47]. For example, the axial displacement of a solid piezoelectric cylinder (rod) can be improved by replacing it with a tube of similar height and diameter. Other techniques such as the addition of a stiff inactive layer (a metal shim in unimorph and bimorph actuators) have also been utilized to improve displacement [40,48]. The flexibility of FDC has been harnessed in developing novel actuators such as spiral [44], telescoping [45], dome [46], tube [47], curved, and flextensional transducers [48].

Spiral actuators with a compact geometry show large field-induced strains. The application of an electric field across the spiral width induces both tangential and radial displacements, the former being much larger than the latter. It has been shown that for a PZT spiral actuator with 32 mm diameter, the tangential movement reaches 1.9 mm at $\sim$11.6 kV cm^{-1} [49–51]. This is about 12 times higher than the displacement of a PZT strip with the same dimensions, which is an indication that displacement can be substantially enhanced by tailoring the geometry of the actuator.

In a curved geometry such as the spiral, there exists a slight gradient in the displacement of small elements across the thickness. Here, the strain of a volume element with outer diameter (r_o) is larger than that of an element in the inner diameter (r_i) (see Fig. 16.11). Such a strain creates a bending moment, which leads to rotation or large tangential displacement in the spiral. The large actuation of spirals is obtained at the cost of load-bearing capability, since the two are inversely related. For example, the blocking force for the above-mentioned spiral actuator, measured at various electric fields (1.3–3.3 kV cm^{-1}), is ~0.8 N. Despite low blocking force, spiral actuators are of interest due to large displacement and compact geometry.

To enhance the actuation of spiral actuators, unimorph PZT spirals have been investigated, and asymmetric electrodes/shims were implemented: i.e., a stainless steel shim is attached on the inner side of the spiral wall. In a conventional unimorph, the relative difference between the length of ceramic and metal strips under an electric field results in bending, and results in large tip displacement. The observed increase in displacement is analogous to the

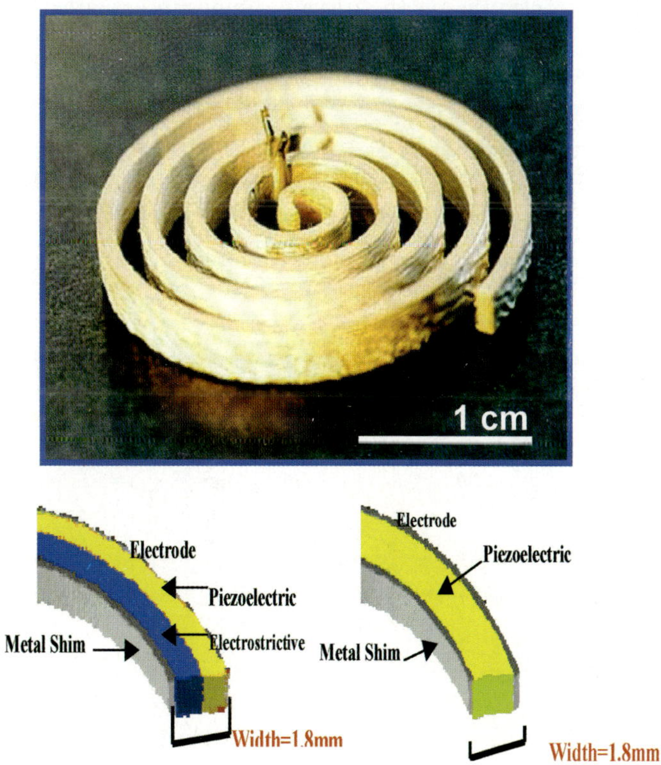

Fig. 16.11. Photograph of a spiral actuator (*top*), and schematic of monolithic multimaterial spiral actuator design compared with a single material of the same design

bending of a conventional unimorph. It was also determined that the hysteresis of the unimorph spiral was 6 times higher than the regular spiral actuator at 400 V bias.

In a series of experiments, the tangential displacement of spirals with dissimilar ceramic strips was investigated as depicted in Fig. 16.11 (bottom half). Multimaterial spiral actuators, composed of piezoelectric $0.65Pb(Mg_{2/3}Nb_{1/3})O_3 - 0.35PbTiO_3$ (PMN-PT 65/35 hereafter) and electrostrictive $0.90Pb(Mg_{2/3}Nb_{1/3})O_3 - 0.10PbTiO_3$ (PMN-PT 90/10 hereafter), were fabricated. The multimaterial spiral actuator was designed such that the outer side of the spiral wall was the piezoelectric 0.65PMN-0.35PT layer, while the inner side of the wall was the electrostrictive 0.90PMN-0.10PT layer. Such spirals were prototyped by FDC and co-fired successfully at 1250°C for 1 h. Such a monolithic bilayer configuration can be electrically biased to induce a strain differential between the piezoelectric and electrostrictive layers, which was found to enhance the bending moment in the spiral. Measurements have shown that the combined piezoelectric–electrostrictive composite spiral actuator yielded higher displacements than piezoelectric actuators of similar dimensions. For example, the piezoelectric–electrostrictive composite actuator has ~510 μm displacement at 500 V, as compared to ~200 μm displacement of the piezoelectric actuator at the same bias field magnitude as shown in Fig. 16.12.

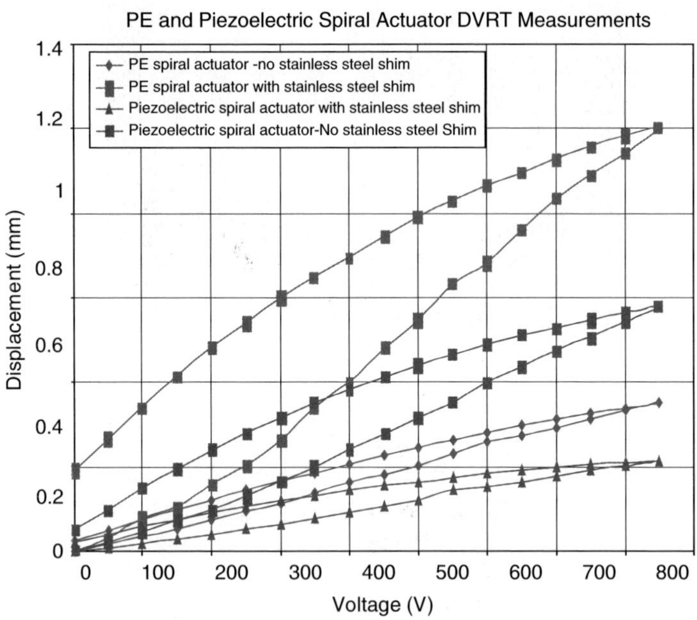

Fig. 16.12. Comparison of spiral actuators based on the monolithic multimaterial design with single material spiral actuators, showing the enhancement in tip displacement

The enhancement of tangential displacement was accompanied by large hysteresis of the multimaterial actuators similar to the unimorph spirals. The effect of the thickness ratio of piezoelectric and electrostrictive layers in monolithic multimaterial monomorphs, composed of varying ratios of piezoelectric PMN-PT 65/35 to electrostrictive PMN-PT 90/10, has also been studied [49–51]. As shown in Fig. 16.13, where the displacement is plotted as a function of thickness ratio, the displacement per unit length of the actuator (δ/L) reaches a maximum when the ratio is 5:1 piezoelectric to electrostrictive. The aforementioned comparison indicates that the tip displacement in such monomorph actuators can substantially be increased by clamping the transverse displacement of the piezoelectric with an electrostrictive layer that is lower in volume fraction. The electrostrictive layer also imparts lower hysteresis to the actuators as compared to traditional piezoelectric bimorphs, which is desirable for applications demanding good positioning accuracy.

Fundamentally, a piezoelectric–electrostrictive monolithic bilayer composite (hereafter PE-MBLC), such as the (PMN-PT 65/45)–(PMN-PT 90/10) bilayer, enables one to obtain asymmetry in the upward and downward displacement of the actuator, for which the field-induced differential transverse strain assumes the form (to a first approximation) [52]:

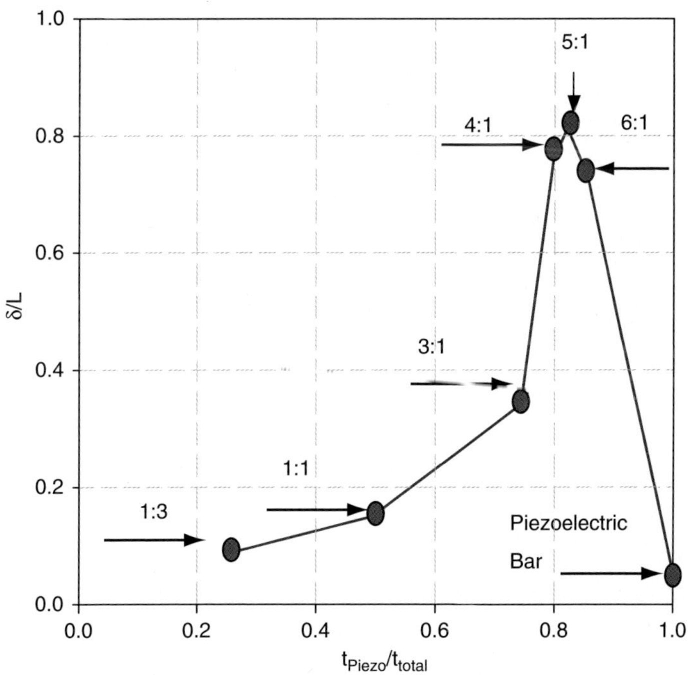

Fig. 16.13. Dependence of tip deflection (δ/L) per unit actuator length on piezoelectric–electrostrictive layer ratio in PMN–PT monolithic bilayer bending actuators

$$\Delta x_1 = d_{31}^{\mathrm{P}} E_3^{\mathrm{P}} + \left\{ Q_{12}^{\mathrm{P}} \left(E_3^{\mathrm{P}}\right)^2 - Q_{12}^{\mathrm{ES}} \left(E_3^{\mathrm{ES}}\right)^2 \right\}, \tag{16.1}$$

where the superscripts P and ES designate the piezoelectric and electrostrictive layers, respectively. Here, x, d, E, and Q are the strain, piezocharge coefficient, electric field, and electrostriction coefficient, respectively. The electric field distribution in the bilayer is computed from the solution of the Poisson equation in conjunction with the dielectric flux continuity condition [52]. In the bilayer configuration, a first-order effect (piezoelectricity) is combined with a second-order effect (electrostriction) as shown in Fig. 16.14. From a group theoretical viewpoint (Curie principle), such an MBLC should possess asymmetric field-induced strain response because the resultant composite property is due to a superposition of a symmetric effect (electrostriction) with respect to the polarity of the applied electric field, and an asymmetric effect (piezoelectricity) with respect to the same. When the applied electric field (E_3) is in the opposite direction of the spontaneous polarization (P_s), the transverse differential strain expression (Δx_1) becomes $\Delta x_{31} = |d_{31}| E_3^{\mathrm{P}} - \left|Q_{12}^{\mathrm{P}}\right| \left(E_3^{\mathrm{P}}\right)^2 + \left|Q_{12}^{\mathrm{ES}}\right| \left(E_3^{\mathrm{ES}}\right)^2$, while the same expression is $\Delta x_{31} = -|d_{31}| E_3^{\mathrm{P}} - \left|Q_{12}^{\mathrm{P}}\right| \left(E_3^{\mathrm{P}}\right)^2 + \left|Q_{12}^{\mathrm{ES}}\right| \left(E_3^{\mathrm{ES}}\right)^2$ when the polarity of P_s and E_3 is the same [52]. Hence, the tip deflection is greater for an antiparallel alignment of P_s and E_3 (< 0): i.e.: $\left|\Delta x_{31}^{E_3<0}\right| > \left|\Delta x_{31}^{E_3>0}\right|$. The same line of reasoning applies to complicated MBLC geometries as well, although the

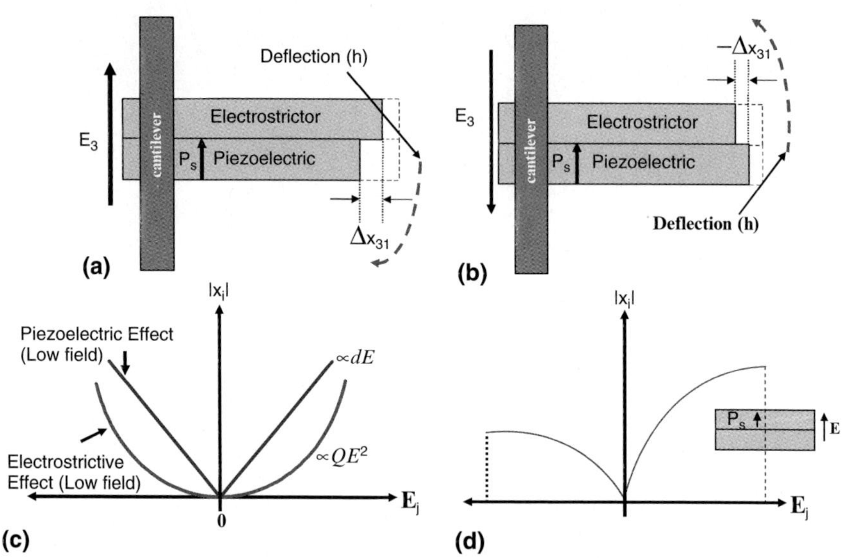

Fig. 16.14. Schematic showing the concept of asymmetric with respect to the polarity of the applied electric field in piezoelectric–electrostrictive bilayer bending actuators

analytical expressions for the characteristic differential strain would become very complicated. However, it suffices to mention that the aforementioned approach shows great promise in developing actuators with directional displacement properties.

Recently, an array of novel multimode piezoelectric transducers has been fabricated using the FDC process [53]. Figure 16.15 shows the photographs of some of the prototyped multimode transducers, which include flextensional transducers such as the Wagon Wheel, Class IV, Concave Class I, and Convex Class I, and vector sensors (projectors) [53]. These transducer designs possess internal displacement and/or stress amplification mechanisms that allow tuning of the resonance frequencies and optimization of the vibration modes. In addition, the monolithic nature of the transducers and the flexibility to drive each segment of the transducers separately or in tandem allow a broad bandwidth, unique acoustic radiation beam patterns, and directional response in

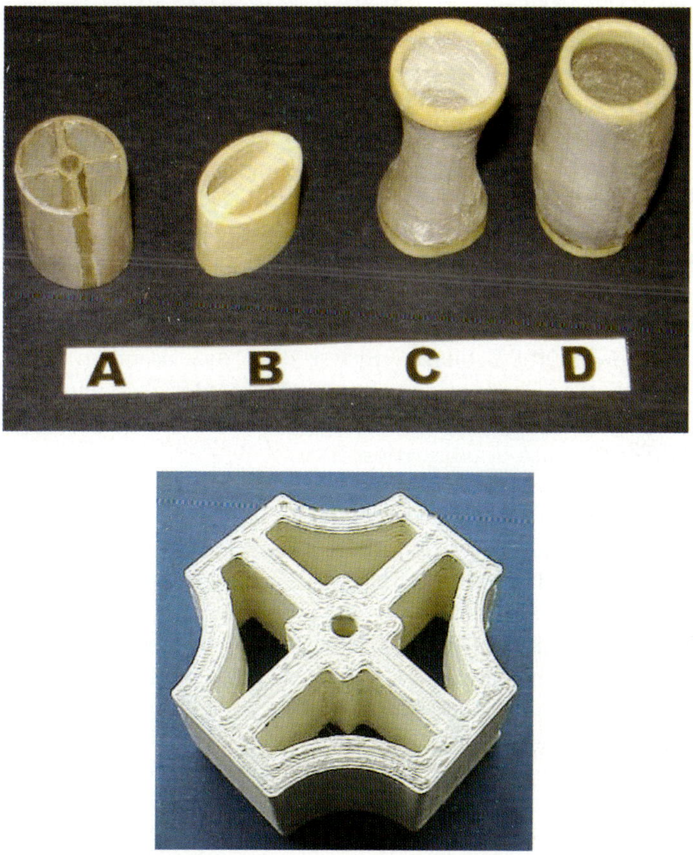

Fig. 16.15. Examples of multimode piezoelectric transducers prototyped by fused deposition of ceramic

an unprecedented fashion [53]. The advantage of using SFF emerges here once again, as such complex transducers can be prototyped within days, instead of weeks and months by using ceramic extrusion technology.

16.5 Summary and Concluding Remarks

The use of SFF in the development of novel piezoelectric composites was elaborated on; its effectiveness in developing composites with unmatched structural complexity was advocated. Through a series of examples, such as the "ladder" and "3D Honeycomb" composites with (3–3) connectivity as well as the oriented (2–2) and VFG (2–2) piezocomposites, it was shown that the properties of composites can be "engineered" on the basis of symmetry principles, thereby mimicking Mother Nature's way of forming natural crystals. In addition, it was demonstrated that composites with controlled phase periodicity, volume fractions, and micro- and macro hierarchy could be obtained in such structures, which are otherwise extremely difficult, if not impossible, to fabricate with traditional techniques. It was also shown that SFF can be used to make various novel actuators.

The flexibility of SFF allows for the rapid fabrication of parts, whereby one can go through a multitude of iterations and make frequent changes to the dimensions of such actuators to accomplish full-scale optimization. Therefore, SFF is ideally suited for what we have come to call "form factor engineering." As in the case of spiral actuators, the large tangential displacements (albeit moderate blocking force) are a consequence of the form factor used, onto which is superimposed the asymmetric deflection of the monolithic piezoelectric–electrostrictive PMN–PT bilayer. SFF is also very useful in accomplishing monolithic integration of various electroactive materials, which provides for multifunctionality as well as better reliability than layered structures made by the "glue-on" route of fabrication.

Earlier, we mentioned that SFF is a nature-inspired fabrication technique. A quick glance to the mode by which crystals form would indicate that growth takes place in an atomic-layer-by-atomic-layer progression. Of course, the scale of such a growth mode is much finer, which results in the macroscopic smoothness of crystal surfaces. Hence, one might visualize the vast range of material properties as consequence of the interaction among such atomic layers in a given crystal. Following this line of reasoning, SFF in principle should be able to provide the means to create manmade macro- and meso-crystals made out of different materials, enabling one to induce across the interfaces. Such an approach seems to have great potential in obtaining cross-coupling of coupling phenomena in a composite structure: e.g., ferroelectric–ferromagnetic coupling, magentostrictive–piezoelectric coupling [54–56], among others. Such novel designs based on the intelligent use of materials would surely supercede the traditional composite properties, since one can possibly induce interactions that are otherwise absent.

Cross-coupling of coupled phenomena, therefore, emerges as a new avenue of research using SFF, where the key concepts will be symmetry superposition, control of interaction cross-sections, and the scale of the interaction among the phenomena of interest. While SFF has already proven to be most effective in providing the answer to the symmetry superposition issue, cross-coupled cross-sections and tuning of the scale of interactions remain as challenges awaiting solutions. Currently, SFF technology can build features typically 25 μm in thickness, which needs to be reduced below 1,000 nm thickness in the years ahead to accomplish the aforementioned artificial cross-coupling. Therefore, advanced hardware as well as superior feedstock material with superb flow properties will be needed. As such, we expect major breakthrough in systems design and in rheological studies of materials in the immediate future.

Acknowledgement

The authors are thankful to the Office of Naval Research (ONR), the Defense Advanced Research Projects Agency (DARPA), New Jersey Commission on Science and Technology (NJCTS), and the Glenn Howatt Foundation at Rutgers University for the generous funding provided. The authors thank past graduate students and colleagues A. Bandhophadhyay, R.K. Panda, T.F. McNulty, S. Turcu, G. Lous, F. Mohammadi, B. Jadidian, A. Hall and M. Allahverdi.

References

1. K. Uchino, *Piezoelectric Actuators and Unltrasonic Motors* (Kluwer Academic, New York, 1996)
2. X. Yu, Ferroelectric Materials and Their Applications (North Holland, Amsterdam, 1991)
3. E.K. Akdogan, M. Allahverdi, A. Safari, IEEE Trans. Ultrason. Ferr. **52**(5), 746–775 (2005)
4. D.P. Skinner, R.E. Newnham, Mater. Res. Bull. **13**, 599–607 (1978)
5. S.M. Pilgrim, R.E. Newnham, L.L. Rohlfing, Mater. Res. Bull. **13**, 677–684 (1987)
6. R.K. Panda, Ph.D. Thesis, Rutgers University, New Brunswick, NJ, 1998
7. A. Safari, Ph.D. Thesis, The Pennsylvania State University, State College, PA, 1983
8. W. Schulze, Ferroelectrics **50**, 33–37 (1983)
9. T.R. Shrout, L.J. Bowen, W.A. Schulze, Mater. Res. Bull. **15**, 1371–1379 (1980)
10. A. Safari, Mater. Res. Bull. **17**, 301–304 (1982)
11. J.W. Halloran, Br. Ceram. Trans. **98**, 299–303 (1999)
12. H.L. Marcus, J.J. Beaman, J.W. Barlow, D.L. Bourell, R.H. Crawford, Rapid Prototyping Using FDM: A Fast, Precise, Safe Technology. Paper presented at Solid Freeform Fabrication Symposium, Austin, TX, 1992

13. H.L. Marcus, J.J. Beamen, J.W. Barlow, D.L. Bourell, R.H. Crawford, in *Rapid Prototyping Using FDM: A Fast, Precise, Safe Technology, Solid Freeform Fabrication Proceedings*, Austin, TX, 1992
14. H.L. Marcus, J.J. Beamen, J.W. Barlow, D.L. Bourell, R.H. Crawford, in *Fast, Precise, Safe Prototypes With FDM, Solid Freeform Fabrication Proceedings*, University of Texas at Austin, Austin, TX, 1991
15. H.L. Marcus, D.L. Bourell, Adv. Mater. Proc. **9**, 677–684 (1993)
16. H.L. Marcus, Mech. Eng. **117**, 62–66 (1995)
17. D.L. Bourell, J.J. Beamen, H.L. Marcus, J.W. Barlow, Solid Freeform Fabrication, An Advanced Manufacturing Approach. Paper presented at Solid Freeform Fabrication Proceedings, Austin, TX, 1990
18. P.F. Jacobs, Stereolithography and Other RP & M Technologies from Rapid Prototyping to Rapid Tooling (Society of Manufacturing Engineers, Dearborn, MI, 1995)
19. P.F. Jacobs, Rapid Prototyping and Manufacturing: Fundamentals of Stereo Lithography (Society of Manufacturing Engineers, Dearborn, MI, 1992)
20. M.A. Jafari, W. Han, F. Mohammadi, A. Safari, S.C. Danforth, N. Langrana, J. Rapid Prototyping **6**(3), 161–175 (2000)
21. T.F. McNulty, F. Mohammadi, A. Bandyopadhyay, D.J. Shanefield, S.C. Danforth, A. Safari, *Development of a Binder Formulation for Fused Deposition of Ceramics*, **4**(4), 144–150 (2001)
22. M. Feygin, B. Hsieh, in *Proceedings of the Solid Freeform Fabrication Symposium*, 1991, pp. 123–125
23. M. Feygin, U.S. Patent No. 5354414, 1994
24. J. Cesarano, T.A. Baer, P. Calvert, in *Proceedings of the Solid Freeform Fabrication Symposium*, vol. 8, 1997, 25–27
25. J. Cesarano, B.H. King, H.B. Denham, in *Proceedings of Solid Freeform Fabrication Symposium*, vol. 9, 1998, 697–699
26. J. Cesarano, P. Calvert, U.S. Patent No. 6027326, 2000
27. E.M. Sachs, M.J. Cima, P. Williams, D. Brancazio, J. Cornie, J. Eng. Ind. **114**, 481–485 (1992)
28. A. Bandyopadhyay, R.K. Panda, V.F. Janas, M.K. Agarwala, S.C. Danforth, A. Safari, J. Am. Ceram. Soc. **80**(6), 1366–1371 (1997)
29. A. Safari, S.C. Danforth, R.K. Panda, B. Jadidian, T.F. McNulty, G. Lous, in *Proceedings of 8th U.S.-Japan Conference on Composite Materials*, Baltimore, MD, 1998, pp. 526–535
30. S. Turcu, M.S. Thesis, Rutgers University, New Brunswick, NJ, 2002
31. S. Turcu, B. Jadidian, S.C. Danforth, A. Safari, J. Electroceram. **9**, 165–171 (2002)
32. C.W. Nan, L. Liu, L. Li, J. Phys. D **33**, 2988–2991 (2000)
33. C.W. Nan, G.W. Weng, J. Appl. Phys. **88**, 416–420 (2000)
34. A. Bandyopadhyay, R.K. Panda, V.E. Janas, M.K. Agarwala, S.C. Danforth, A. Safari, J. Am. Ceram. Soc. **80**, 1366–1372 (1997)
35. A. Safari, V.F. Janas, R.K. Panda, Fabrication of Fine Scale 1–3 PZT Ceramic/Polymer Composites Using a Modified Lost Mold Method. Paper presented at SPIE Symposium on Smart Structures and Materials, San Diego, CA (1996)
36. A. Safari, J. Cesarano, P.G. Clem, B. Bender, Fabrication of Advanced Functional Electroceramic Components by Layered Manufacturing (LM) Methods. Paper presented at Proceedings of the 13th IEEE International Symposium on Applications of Ferroelectronics, Nara, Japan, 2002

37. T.F. McNulty, Ph.D. Thesis, Rutgers University, Piscstaway, New Jersey, USA, 1999
38. F. Mohammadi, A. Kholkin, B. Jadidian, A. Safari, Appl. Phys. Lett. **75**, 2488 (1999)
39. Q.M. Wang, Q. Zhang, B. Xu, R. Liu, L.E. Cross, J. Appl. Phys. **86**, 3352 (1999)
40. A. Safari, S.C. Danforth, M.A. Jafari, M. Allahverdi, B. Jadidian, F. Mohammadi, N. Vankataraman, S. Rangarajan, in *Proceedings of the 9th European Conference on Rapid Prototyping and Manufacturing*, 2000, pp. 247–249
41. A. Safari, S.C. Danforth, in *Proceedings of the 11th IEEE International Symposium on the Applications of Ferroelectrics*, 1998, pp. 229–231
42. A. Bandyopadhyay, R.K. Panda, T.F. McNulty, F. Mohammadi, S.C. Danforth, A. Safari, Rapid Prototyping J. **4**, 37–42 (1998)
43. A. Safari, M. Allahverdi, Ceram. Eng. Sci. Proc. **22**, 473–475 (2001)
44. F. Mohammadi, Ph.D. Thesis, Department of Ceramic and Materials Engineering, Rutgers University, New Jersey, USA, 2001
45. F. Mohammadi, A. Kholkin, S.C. Danforth, A. Safari, in *Proceedings of the 11th IEEE International Symposium on the Applications of Ferroelectrics*, 1998, pp. 273–275
46. S.A. Wise, Sens. Actuators **A69**, 33–37 (1998)
47. Q.M. Zhang, H. Wang, L.E. Cross, J. Mater. Sci. **28**, 3962–3965 (1993)
48. Y. Sugawara, K. Onitsuka, S. Yoshikawa, Q.C. Xu, R.E. Newnham, K. Uchino, J. Am. Ceram. Soc. **75**, 996–998 (1992)
49. A. Hall, Ph.D. Thesis, Rutgers University, Piscataway, New Jersey, USA, 2006
50. A. Hall, E.K. Akdogan, M. Allahverdi, A. Safari, J. Eur. Ceram. Soc. **25**, 2991–2997 (2005)
51. A. Hall, E.K. Akdogan, M. Allahverdi, A. Safari, J. Electroceram. **15**(2), 143–150 (2005)
52. E.K. Akdogan, A. Hall, A. Safari (unpublished)
53. R.E. Newnham, D.C. Markley, R.J. Meyer Jr., W.J. Hughes, A.-C. Hladky-Hennion, J.K. Cochran Jr., Ceram. Trans. **150**, 427–443 (2004)
54. S. Dong, J.-F. Li, D. Viehland, IEEE Trans. Ultrason. Ferr. **50**(10), 1253–1261 (2003)
55. S. Dong, K. Uchino, L. Li, D. Viehland, IEEE Trans. Ultrason. Ferr. **54**(6), 1240–1249 (2007)
56. S. Dong, J.-F. Li, D. Viehland, IEEE Trans. Ultrason. Ferr. **54**(6), 1240–1249 (2007)

Part III

Characterisation Methods

17

Microstructural Analysis Based on Microscopy and X-Ray Diffraction

M.J. Hoffmann, H. Kungl, R. Theissmann, and S. Wagner

Characteristic features of ferroelectric materials are the transition in crystal structure from a cubic paraelectric phase to a lower-symmetry ferroelectric phase at the Curie temperature T_c and the possibility of changing the polarization of polycrystalline materials by means of electric fields. Both features give rise to challenges with respect to X-ray analysis and microscopy.

The paraelectric to ferroelectric transition results in a considerable anisotropic change of lattice constants. In order to reduce the mechanical stresses resulting from the structural transformation, the formation of domains is observed, which is typical for the microstructure of ferroelectrics. The domain formation is governed by minimizing the sum of elastic energy, electric field energy, and domain wall (interface) energy and can be described as mechanical twinning or as martensitic transformation. The investigation of the domain structure by microscopy compounds the domain configuration and the interfaces between domains, the domain walls. Studies of the domain configuration are related to stacking of domains within grains, domain size, and, recently, to nanodomains. The domain wall regions are subject to high mechanical stresses due to mismatch in lattice spacings and change of polarization direction of the adjacent domains. Domain walls, nanodomains, and regions with high microstresses are features of the real structure that affect peak shape and positions in diffraction patterns. The proper interpretation of the diffraction patterns, in particular with respect to the analysis of phase compositions at the morphotropic phase boundary (MPB), has to take into account these microstructural features and their effect on peak shapes. X-ray and microscopy techniques are complimentary techniques for structural investigations. Microscopy provides detailed insight on local features of the microstructure. From X-ray diffraction patterns information on the average structure is obtained. The challenge for structure analysis is to identify the features of real structure by microscopy, then calculate their theoretical effects on diffraction patterns, and finally to prove the overall relevance of the local structural features by matching the measured with calculated diffraction patterns.

By applying electric fields or mechanical stresses the polarization direction of domains in ferroelectric materials can be changed by domain wall movement, nucleation of new domains, and domain switching. At the same time, there are piezoelectric and elastic effects. While information on the orientation of the domains can be obtained from the relative intensities of specific peaks, the effects from piezoelectric and elastic strain result in a change of the lattice spacings. In situ X-ray studies can be applied to the poling process, electrical cycling, and mechanical loading of ferroelectrics. The challenge in the in situ investigations of ferroelectric materials under an electric field and mechanical load, therefore, is to provide information on the microstructural mechanisms that cause strain and polarization.

Most technically important ferroelectric materials crystallize in the perovskite structure ABO_3. The unit cell of this structure can be described by the A-side cations occupying the corners, the B-site cation the center, and the oxygen anions the faces of the pseudocubic cell. Discussion of X-ray patterns will focus on technically most relevant systems, which are barium titanate ($BaTiO_3$), the $KNbO_3$–$NaNbO_3$ (KNN) system and lead zirconate titanate (PZT) solid solutions. Barium titanate is the most important ceramic material for ceramic capacitors. State-of-the-art materials for actuators and sensors are doped PZT and related relaxor ferroelectric solid solutions $P(B_1,B_2)$–PT or –PZT. Materials based on a modified KNN system are one of the most promising lead-free ferroelectrics for substitution of PZT in these applications.

17.1 X-Ray Diffraction Analysis (XRD)

17.1.1 Structure

17.1.1.1 $BaTiO_3$

The paraelectric modification of $BaTiO_3$, which is stable in the temperature range between 1460°C and 130°C, is cubic. Below the Curie temperature of 130°C, $BaTiO_3$ transforms to a tetragonal modification ($P4mm$), which on further cooling turns into an orthorhombic structure (Cmm) at 0°C and to rhombohedral symmetry ($R3m$) at −90°C [1]. The tetragonal, orthorhombic, and rhombohedral structures are ferroelectric. There is considerable thermal hysteresis with respect to the phase transitions, which depend on the size of the crystallites and stress condition [2]. Polarization of the tetragonal phase is towards the <001> direction. The ratio between the c- and the a-axis, the spontaneous strain, is approximately 1.01 at room temperature and subject to variations with crystallite size [3]. Stress dependence of transition temperatures and the sensitivity of lattice parameters to crystallite size indicate the pronounced influence of microstructural features on crystallographic parameters in ferroelectric materials.

17.1.1.2 KNN

The solid solution system KNN shows a Curie temperature of approximately 400°C, which is only slightly dependent on composition (Fig. 17.1). When cooling the cubic structure, the KNN transforms first into a tetragonal structure and then at approximately 200°C into an orthorhombic structure. Pure $KNbO_3$ with the space group *Cm2m* as well as KNN up to an Na content of 0.9 mole are ferroelectric at room temperature, whereas $NaNbO_3$ with the space group *Pbma* shows an anti-ferroelectric behaviour. Furthermore, an MPB at 47.5 mole% $KNbO_3$ is observed. Additives can strongly influence the crystallographic structure of KNN. When substituting Na and K by Li, the orthorhombic structure transforms completely into the tetragonal structure at room temperature at an Li content of 6 mol%. At approximately 5 mol%, the existence of an orthorhombic/tetragonal MPB is proposed. At this MPB, 18 possible polarization directions exist, 12 in <110> directions for the orthorhombic system and 6 in <100> directions for the tetragonal phase. With additional doping of Li–KNN near the MPB with Ta and Sb, excellent piezoelectric properties can be observed [4].

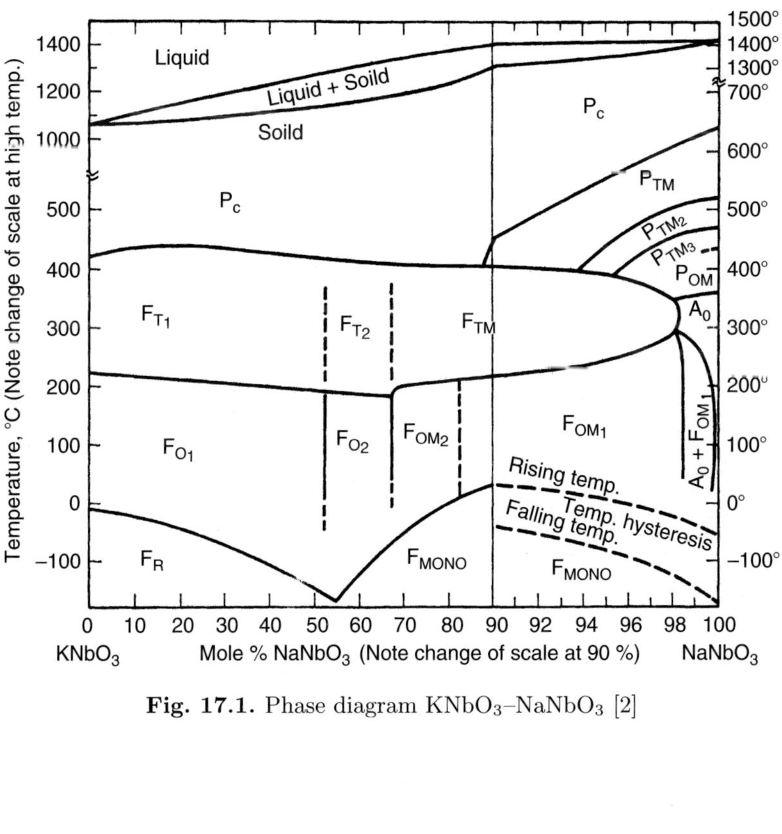

Fig. 17.1. Phase diagram $KNbO_3$–$NaNbO_3$ [2]

17.1.1.3 PZT

For the solid solution system $PbZrO_3$–$PbTiO_3$ the Curie temperature ranges from 495°C for lead titanate to 250°C for lead zirconate. The crystal structure for all compositions above the Curie temperature is cubic with space group *Pm*3*m*. Below T_c, the materials with high Ti content are tetragonal with the space group *P*4*mm* (Fig. 17.2 left side). Materials with very high Zr content are transformed to an anti-ferroelectric orthorhombic structure (*Pbam*) when cooling. A large part of the Zr-rich compositions, however, adopts a rhombohedral structure below T_c (Fig. 17.2 right side). Space groups of the rhombohedral materials are *R*3*c* and *R*3*m* depending on the composition. The compositional range with Zr contents between 50 and 60 mol% Zr is denoted as the MPB. The structural feature of these materials has been considered for many years as the coexistence of the tetragonal and rhombohedral phases [2]. However, the nature of the crystallographic structure and phase composition of these materials have been the subjects of ongoing scientific discussion.

Approaches to estimate lattice constants can be based on {200} reflections within a pseudocubic setting. For the tetragonal phase materials, the reflections split into the (002) and (200) reflections. The lattice distortion, given in terms of $\delta_t = (c/a) - 1 = (d_{002}/d_{200}) - 1$ (Fig. 17.2 right side), ranges from 6% for lead titanate to about 2% for tetragonal compositions close to the MPB. According to Randall et al. [54], lattice parameters are grain-size dependent, showing a decreasing lattice distortion δ_t with lower grain size. With increasing temperature, the c/a ratio diminishes by increasing a and decreasing c lattice spacings, thus approaching the cubic state. Rhombohedral lattice parameters can also be obtained from {200} reflections. Information on the rhombohedral lattice distortion are derived from the (111) and (−111) reflections and can be quantified in terms of $\delta_r = (9/8)(d_{111}/d_{-111} - 1)$ (Fig. 17.2 right side). The lattice distortion for the rhombohedral phase of approximately 0.7% is only slightly sensitive to composition and corresponds to a small angular distortion

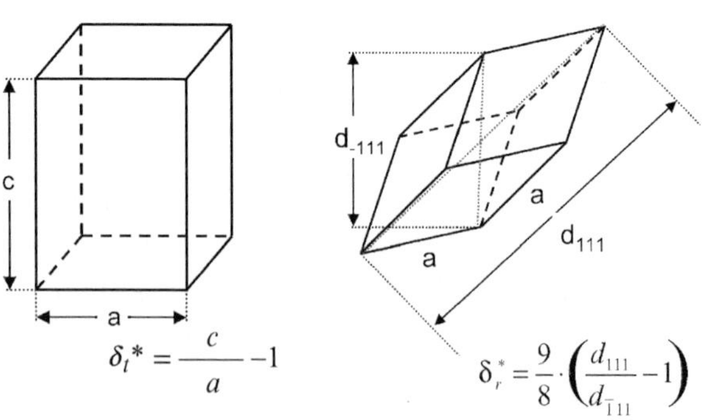

Fig. 17.2. Crystal structure and lattice distortion for the tetragonal (left) and the rhombohedral (right) phase of the PZT solid solution system

from cubic symmetry of 89.7°. For doped PZT (<5 mole%), there are only minor shifts with respect to the phase relations. Qualitatively similar dependence of phases and structures from composition prevail also for many Pb(B1,B2)–PT relaxor ferroelectric solid solutions.

17.1.2 Specific Features of XRD Patterns of Ferroelectrics

A specific feature of the XRD patterns of PZT and $BaTiO_3$ is a marked deviation of the peak shapes from the Gaussian–Lorentzian profiles (Fig. 17.3). Both tetragonal (002) and (200) reflections show a strong asymmetric distribution of the reflection intensity. The intensity of the (002) reflection is biased toward higher values of the 2θ diffraction angles, whereas the (200) peaks exhibit a asymmetry towards lower diffraction angles 2θ. In addition to the asymmetry of the peak shapes, intensity between the peaks has been observed. Approaches to explain the pronounced asymmetry are based on microstrains related to the domain structure [5, 6], for which strain has been predicted at 180° as well as 90° domain walls [7] (Fig. 17.4). In addition to the scattering from the strain-modified lattice spacings at the domain walls, effects such as diffuse scattering and coherent scattering may also occur [8, 9], giving rise to the intensity that cannot be attributed to the Bragg peaks. The intensity from these contributions depends on the microstructure close to the domain walls, such as domain wall width and straight or rough domain boundaries. Therefore, detailed information on the domain structure as well as the structure of the domain walls is important when considering the effects on the diffraction

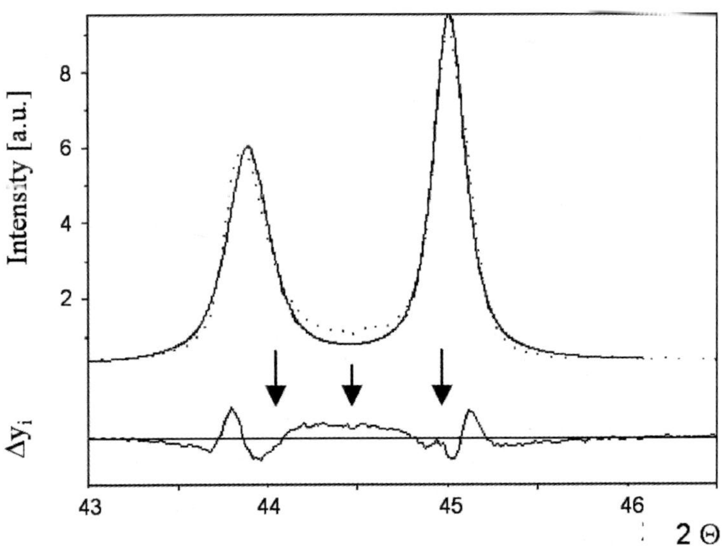

Fig. 17.3. Asymmetry of (002) and (200) reflection intensity between peaks in tetragonal PZT 50/50 1La2Sr

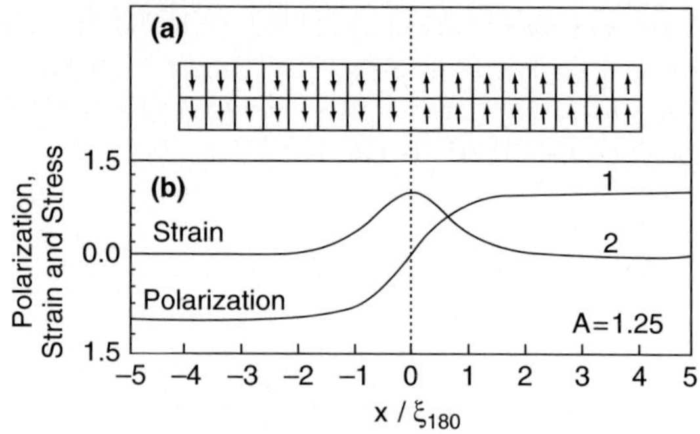

Fig. 17.4. Calculated strain profiles at 180° domain boundaries [7]

patterns. Approaches to model the effects on diffraction patterns have been made by using asymmetrical peak shapes, the introduction of a secondary (cubic) phase, and a distribution of lattice parameters of the t-phase [9–11]. Further effects from the domain structure on the X-ray pattern are detected by analyzing the full width half-maxima (FWHM) and evaluating them in a Williamson–Hall plot. Marked differences with respect to the line broadening is observed for the (*00l*) and (*h0h*) planes in $PbTiO_3$, which was attributed to strain [6]. The irregular line shape, peak broadening, and diffuse intensity between peaks therefore suggest that strain and microstructural features play an important role in the X-ray scattering properties of ferroelectric materials.

17.1.3 Analysis of X-Ray Patterns by Rietveld Method

In contrast to the ad hoc profile matching of specific reflections, Rietveld refinement is based on the simultaneous calculation of complete diffraction patterns based on structure data. By adjusting the calculated to the measured XRD pattern, information on the structure parameters is obtained. By means of this method, high-precision data for lattice parameters can be achieved. In addition to the lattice parameters, the atomic positions and temperature factors, which significantly influence the calculated diffraction pattern, are obtained. Displacement of Pb ions from ideal positions have been detected for tetragonal [12] and rhombohedral PZTs [13]. The shift in atomic positions has been suggested due to improved matching of calculated and measured patterns. In contrast to the adjustment of the patterns by anisotropic temperature factors, which may result in physically unrealistic results, the shift in atomic positions seems to be supported by first-principles calculations. The deviation of the line shapes may, however, be ambiguous. At the present stage, the Rietveld method provides only limited options to integrate features of the real structure such as domain walls into the analysis. Anisotropic broadening may be described by the phenomenological Stevens [14] parameter, but approaches to model asymmetric peak shapes from domain structure are not available so far.

17.1.4 Phase Analysis

Quantitative determination of phase contents for morphotropic PZT ceramics has been traditionally based on the evaluation of the {200} reflections using a structural model according to which the materials consist of a tetragonal and a rhombohedral phase (Fig. 17.5). Estimates of the tetragonal volume fraction χ_t were based on the ratio of the corresponding tetragonal (t) and rhombohedral (r) peak intensities

$$\chi_t = \frac{(I_{200,t} + I_{002,t})}{(I_{200,t} + I_{002,r} + I_{002,t})}.$$

Part of the intensity that is attributed to the rhombohedral phase may contain also contributions of the strained tetragonal phase. Results from this approach indicate an approximately linear dependence of phase content from the Zr/Ti ratio, with the dependence flattening out towards the single-phase regions [15–18]. The lattice distortion of the tetragonal phase decreases towards the MPB compositions (Fig. 17.6). In addition to the results on phase composition, a characteristic dependence of the FWHM can be observed by this approach (Fig. 17.7). The FWHM for the tetragonal peaks are almost constant with composition, whereas the rhombohedral 200 reflections show a sharp increase when lowering the Zr content and approaching the tetragonal side of the MPB [17, 18]. The pronounced broadening may indicate high strains, which could stabilize the coexistence of tetragonal and rhombohedral phases (Fig. 17.7) [17, 19]. Observations of Noheda et al. at reflection groups {200}, {220}, and {111} of undoped PZT 52/48 and 54/46 attribute the reflection intensity to a monoclinic phase [20–22]. The monoclinic reflections were more marked at low temperatures. The monoclinic displacements of the Pb sites as detected by Rietveld analysis correspond to the shifts that had

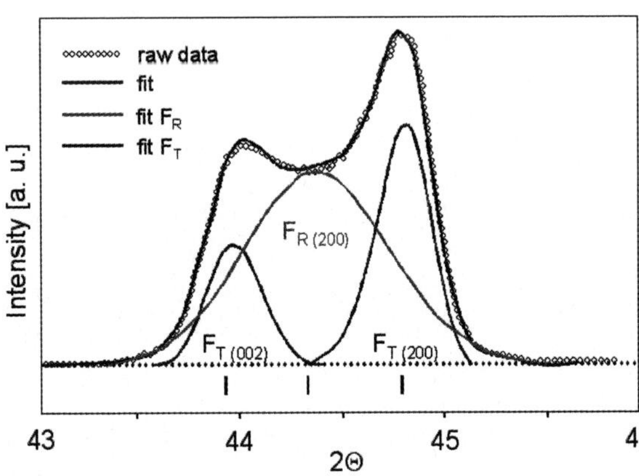

Fig. 17.5. Overlapping of the tetragonal (t) and rhombohedral (r) peaks in morphotropic PZT

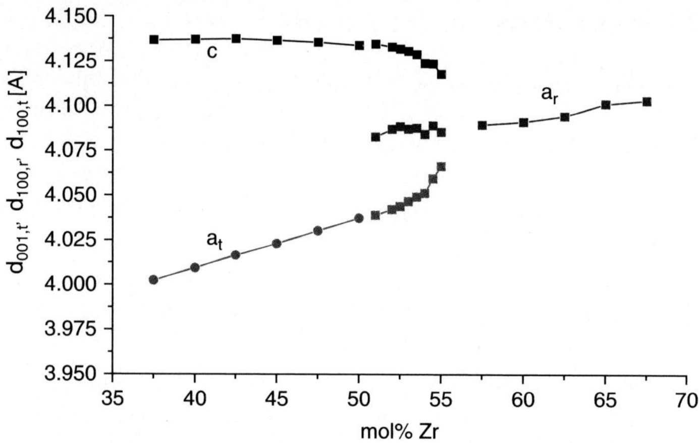

Fig. 17.6. Lattice constants in single phase and in morphotropic PZT 1La2Sr

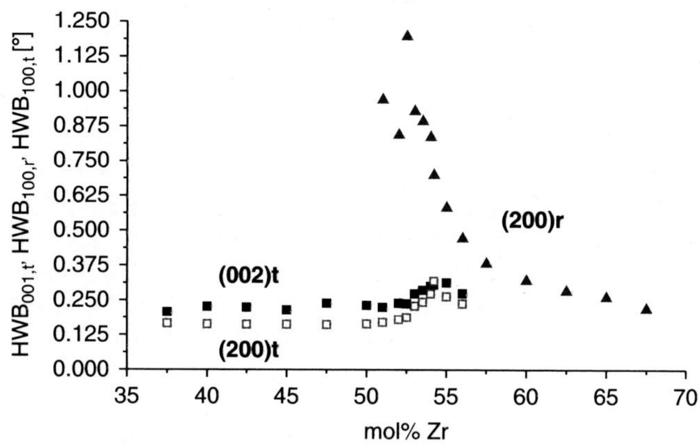

Fig. 17.7. Full width half-maximum (FWHM) in single phase and in morphotropic PZT 1La2Sr

been suggested for t- and r-phase material. Therefore, the monoclinic phase was considered as a structural bridge providing polarization rotation. Similar patterns were found in the MPB compositions of several relaxor ferroelectrics. In PZN–PT the monoclinic distortion and attribution to the monoclinic space group depend on the strain state [23, 24]. Differences were found between the surface and the inner parts of the material. From further investigations on undoped PZT 52/48 it turned out that the materials on the tetragonal side close to the MPB exhibited an average structure that appeared monoclinic in the diffraction patterns at room temperature [12]. Locally distorted crystalline regions in a tetragonal matrix structure were suggested to be the origin of the monoclinic reflections [25, 26]. The origin of the locally distorted regions is still under discussion. Several approaches have been presented. Local structures showing monoclinic distorted structures in the diffraction patterns may arise by (1) microstrain resulting from domain formation, (2) clustering inhomogenities of B-site cations [27], (3) disorder [25, 26], (4) coherent scattering of stacked domains [28], or (5) due to an adaptive phase consisting of nanoscale domains [28–30].

17.1.5 Texture Analysis

In the unpoled state the statistical probability of alignment corresponds to the multiplicity of crystallographically equivalent directions. For tetragonal materials the relative volume fractions of domains aligned parallel to a given direction with their *c*-axis is 1/3, whereas 2/3 of the domains should be oriented with the *a*-axis. For rhombohedral phase material, the relations between the volume fractions oriented <111> and <−111> are 1/4 and 3/4. The degree of alignment can be quantified from the intensities of the corresponding reflections by the parameters

$$f_t = \frac{I_{002,t}}{(I_{200,t} + I_{002,t})} \quad \text{and} \quad f_r = \frac{I_{111,r}}{(I_{111,r} + I_{-111,r})}.$$

Care has to be taken in the preparation of samples from the sintered material. Mechanical treatment during polishing provides a texture on the surface layers of the material [31, 32]. In order to provide a texture-free state, subsequent heat treatment is required. A large part of the texture can be removed by heat treatment at temperatures between 500 and 600°C for several hours. Application of higher temperatures may cause PbO evaporation and therefore changes in composition. Changes in texture are closely related to changes in polarization and strain. In PZT, polarization is related to the crystallographic <001> direction for tetragonal and to <111> direction for rhombohedral phase. From changes in the intensity ratio, f_{t}, the 90° switching of tetragonal phase, can be analyzed, whereas for rhombohedral phase material changes in f_{r} are related to 71 and 109° switching. No information on 180° switching is provided by the X-ray patterns, for an orientation of the polarization vector parallel and antiparallel to [001] or [111] can not be distinguished in the powder patterns.

17.1.6 In Situ High-Resolution XRD Measurements

Along with electric fields, two types of microstructural mechanisms are activated in ferroelectric ceramics. Strain is provided by the piezoelectric effect and domain switching. Both effects are coupled to elastic strains by stress–strain interactions of neighboring grains [38, 50]. In a similar way, mechanical pressure results in an elastic strain and strain by ferroelastic domain switching, both of them also involve stress–strain interaction. The microscopic features of the strain mechanisms are changes in lattice spacings for the piezoelectric effect and elastic strain, whereas ferroelectric and ferroelastic domain switching

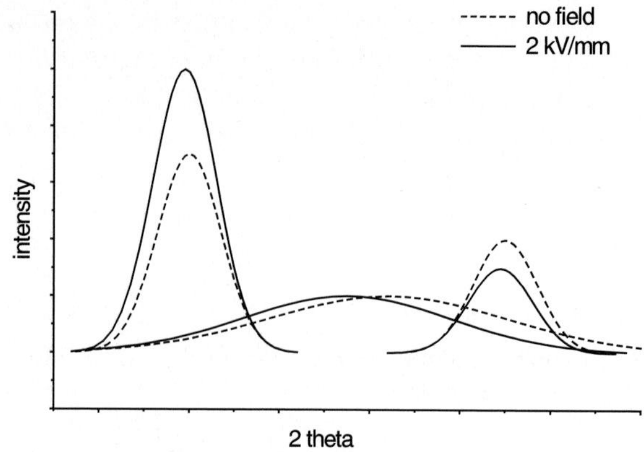

Fig. 17.8. Effects of electric field on the diffraction pattern (schematic)

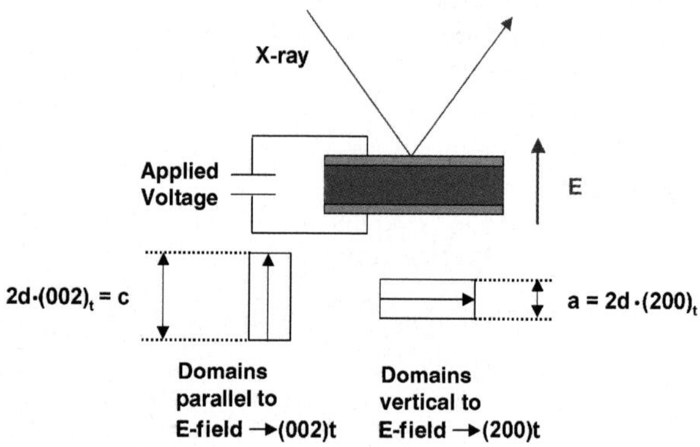

Fig. 17.9. Measurement setup for in situ X-ray measurements in reflection

changes the orientation of the polarization axis. The effects on the diffraction patterns are as follows: (1) Piezoelectric and elastic strain results in a shift of the peak positions; (2) Strain from 90°, 71° and 109° domain switching can be analyzed from changes in the intensities of particular diffraction peaks (Fig. 17.8). XRD therefore provides an option for the analysis of the microstructural mechanisms of strain in ferroelectrics during poling, with electric fields and mechanical loads. An experimental setup for an in situ experiment with electric field is sketched in Fig. 17.9. An experimental precondition when measuring in situ with an electric field or mechanical loads in reflection is that the sample is focused for the complete experiment. Parallel beam setup and suitable detectors as available in synchrotron beamlines are required for detection of the peak shift [32, 33]. Careful choice of electrode materials and balancing thickness of electrode layers are necessary. On the one hand absorption from the electrode material and generation of additional peaks has to be minimized; on the other, sufficient conductivity is required in order to

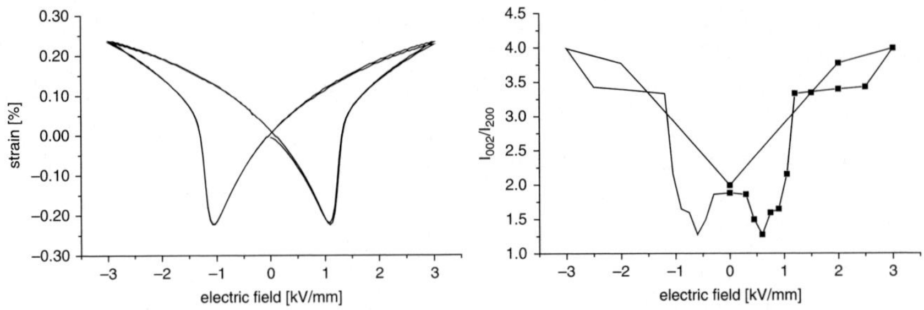

Fig. 17.10. Macroscopic strain and intensity ratio $I_{002,\mathrm{t}}/I_{200,\mathrm{t}}$ during bipolar electrical cycling of donor-doped morphotropic PZT

establish a homogeneous electric field. Measurements in transmission are favorable with respect to geometry, minimizing surface effects, and problems related to electroding, but they provide decreasing resolution related to lower wavelength resulting from the high-energy radiation required for transmission experiments.

Analyzing bipolar electrical cycles shows qualitatively similar reaction with respect to macroscopic strain and non 180° domain switching (Fig. 17.10). Application of low electric fields opposite to initial poling direction reduces strain and alignment of domains parallel to the field direction. With fields exceeding the coercive field, polarization reversal occurs. The degree of texture given in terms of f_t (or f_r) strongly increases up to maximum field level. Detailed description of the texture development with bipolar cycles are given by Saito [34], Ogawa [35], and Hoffmann et al. [36, 37]. Results on focusing the lattice strain during bipolar electrical cycling were presented by Reszat et al. [38]. The lattice strain consists of piezoelectric strain and elastic strain from stress–strain interactions in these cycles, as pointed out by Hall et al. [39].

The effects of electric field and mechanical load on switching behavior are dependent on the orientation of the polarization axis of the domain in the initial state. Orientation dependence of domain switching can be analyzed by means of high-energy transmission XRD or from measurements with a 4-axis diffractometer. Transmission XRD was used when studying texture and lattice strain during poling by Hall et al. [39–41]. From a corresponding micromechanical Eshelby-type model, it was concluded that domain switching and lattice strain depend on the cosine square of the angle towards the poling field.

Analysis of texture by X-ray was applied to poling by Wan and Bowman [42] and Jones et al. [43], to thermal depolarization by Chang et al. [44], and to mechanically loading PZT ceramics by Rogan et al. [45] and Li et al. [46].

Attempts were made to give a quantitative assessment with respect to the components of the macroscopic strain. Wersing et al. [47] approached the estimation of the piezoelectric effect and its anisotropy from X-ray data.

Their method was based on calculating the electrostrictive coefficients from lattice strain determined by in situ XRD measurements at high electric fields. Combining these results with data on spontaneous polarization and low-field measurement of dielectric constants, the piezoelectric strain was calculated by means of Devonshire theory. Recent research points to the contributions of strain from piezoelectric effect and domain switching. On the basis of the evaluation of texture data from transmission using a geometrical approach, Jones calculated the strain from domain switching and its contribution to total strain for PZT at 34% [48]. A method to estimate piezoelectric and domain switching strain based on XRD texture data and macroscopic strain measurements in the remanent state and at electric fields was developed by Kungl et al. [49, 50]. The data on texture and macroscopic strain are related by a semiempirical approach in this procedure. Investigations of morphotropic donor doped PZT suggested that high strain from domain switching is obtained from materials that show lower poling texture. These are materials with compositions on the tetragonal side of the MPB. In contrast to that, the piezoelectric strain increases towards the rhombohedral side of the phase boundary.

17.2 Microscopy

17.2.1 SEM

Microstructures of PZT, $BaTiO_3$, or KNN in terms of grain size, grain size distribution, porosity, and the presence of secondary phases can be analyzed using scanning electron microscopy (SEM). Furthermore, the type of secondary phases can be analyzed if the SEM is equipped with an energy dispersed X-ray (EDX) detector. The SEM technique requires polished and etched samples and, because the samples are insulators, sputtering with a thin conductive layer such as Au or Au/Pd is necessary. There exist two types of etching processes. The first one is a thermal etching at temperatures 100–150°C below the sintering temperature. Here the grain boundaries are mostly affected. More information is obtained by chemical etching at room temperature with a solution consisting of 95 ml distilled H_2O, 5 ml 0.1 molar HCl, and five drops of HF. Hereby the etching time depends strongly on the composition varying between 5 s and 4 min. Chemical etching affects not only the grain boundaries but also 90° domain walls can be visualized, because the etching attack of the domains depends on the polarization direction. On the contrary, 180° domains provide no contrast. Typical domain structures are lamellar patterns for tetragonal and wedge-shaped structures for rhombohedral PZT as shown in Fig. 17.11.

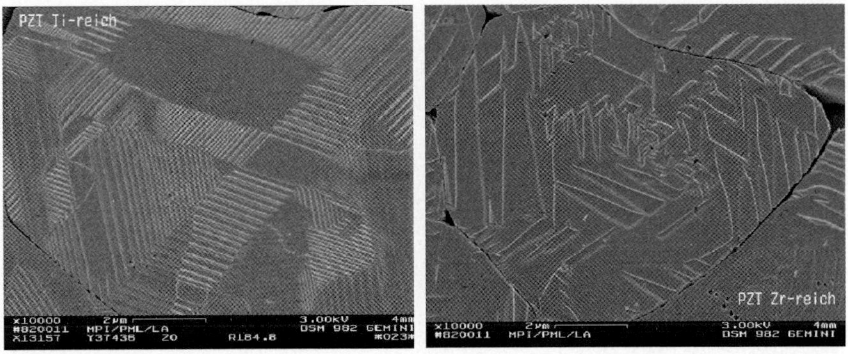

Fig. 17.11. Domain structures of tetragonal (left) and of rhombohedral (right) PZT as observed by SEM

17.2.2 TEM

The different techniques available in transmission electron micrsocopy (TEM) provide powerful tools for the structural investigation of ferroelectric materials. TEM closes the gap between macroscopic averaging structural investigation methods, such as X-ray diffraction, and spectroscopic methods that focus on point defects and their local environment. The main subject of TEM investigations is domain structures (conventional TEM), properties of domain walls (conventional TEM and high-resolution TEM (HRTEM)), and the investigation of the symmetry and polarization direction (convergent beam electron diffraction and holography). The development of C_s-corrected TEMs has opened up the opportunity to get insights into the local environment and the local ordering of solid solutions if they are combined with modern "exit wave reconstruction" software packages. Also, analytical methods are available in the field of TEM, such as electron energy loss spectroscopy (EELS) and energy dispersive X-ray analysis, but their practical use is inhibited in most cases because of lack of accuracy.

17.2.2.1 Conventional TEM Imaging and Electron Diffraction

Imaging of the domain structure by TEM bright field shows lamellar domains for the tetragonal ferroelectric materials (Fig. 17.12 left side). The domain configurations of rhombohedral PZT are described as domain bands, wedge shaped, or herringbone structures [51] (Fig. 17.12 right side). Owing to many variable operating modes, high resolution, and the combination of imaging with electron diffraction, additional details and characteristics can be analyzed and crystallographic orientations can be attributed to the domains. The correct representation of the microstructures in bulk materials by the microstructures observed in the thin specimens is, however, a point of ongoing discussion. Care has to be taken with respect to preparation of the samples

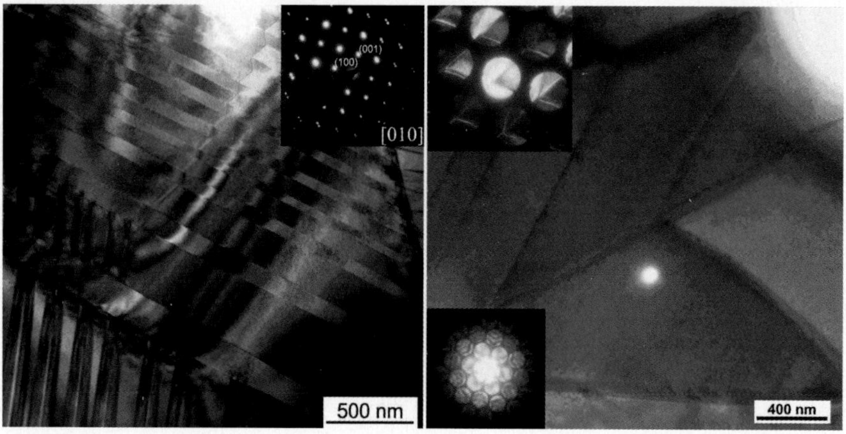

Fig. 17.12. The lamellar domain structure of a tetragonal PZT sample with the composition $Pb(Zr_{0.52}Ti_{0.48})O_3$ observed in the [010] direction (left) [52], and the herringbone structure of a rhombohedral PZT sample (right) with the composition $Pb(Zr_{0.60}Ti_{0.40})O_3$ observed along a [−111] direction (Courtesy R. Schierholz). The CBED pattern (bottom right) shows a mirror plane, but no 3-fold symmetry as expected for the [111] direction. The defocused CBED pattern (top right) gives a bright-field image (central spot) of the observed area, each diffraction spot containing the corresponding dark-field image. All domain walls which are *on-edge* (are within the crystallographic zone) can be attributed directly to the corresponding reflection

in order to avoid preparation-induced effects. In addition to the conventional methods preparing specimens by grinding thin layers followed by dimpling and ion beam thinning, new techniques such as focused ion beam (FIB) are available. As to the answer to the question whether a specific structure observed in TEM is representative for the bulk state, one has to take the conditions for its stability as well as the method of preparation into account. Many specific features of the micro- and domain structure have been worked out by TEM.

Domain Size

An immediate evaluation of the domain wall width perpendicular to the domain walls of the lamellar structure in tetragonal ceramics from the TEM image is possible only if the domain alignment is edge-on in the sample, which means that the <001> or <100> direction is perpendicular to the incoming beam. Evaluating tilted domains requires the measurement of the tilt angle. Care has to be taken with respect to specimen thickness, because the domain width is sensitive to the size in this dimension [53]. The domain size is related to grain size, whereby for PZT 52/48 a parabolic relationship was found for the range between 3 and 10 μm by Randall et al. [54]. These results correspond to the formula derived by Arlt [55], in which the domain size d is related to grain size g, domain wall energy σ_{90}, the stiffness coefficient $k \times c$, and a lattice distortion parameter β by

$$d = \sqrt{\frac{\sigma_{90} \cdot g}{2 \cdot k \cdot c \cdot \beta^2}}.$$

Recent investigations indicate that there might be a bimodal distribution in the domain wall width. TEM investigations of undoped PZT showed periodically alternating domain width of adjacent 90° domains [53].

Stacking Configuration

Varying grain size not only affects the domain width but also the stacking configuration of domains. As predicted from an energy minimizing approach considering elastic strain and interface domain wall energy [55], the domain configuration changes from a three-dimensional pattern to a two-dimensional structure with decreasing grain size. Reduction in the number of domain variants was found for grain sizes <1 μm in PZT 52/48 by Cao and Randall [56]. The critical grain size is specific for each material depending on lattice distortion, domain wall energy, and elastic modulus.

Strain at Domain Junctions

The microstructure and the stress conditions at a four-domain junction between two lamellar bands in tetragonal PZT 45/55 exhibiting a 3D domain structure have been studied by Mac Laren et al. [57]. Orientations of the domains at the junction were determined from Kikuchi patterns. In the lamellar regions at either side of the junction, the angles corresponding to 90°-δ domain boundaries were found, with 90°-δ being close to 2 $\tan^{-1}(a/c)$. Thus, the domains were shown to be not aligned strictly orthogonal, resulting in stresses. Additional strain was detected at the junction, which was shown to be equivalent to an array of disclination dipoles [58]. Local stresses at these disclinations – as calculated from the angular positions and assuming elastic accommodation – may reach values up to 1 GPa if not partially compensated by additional defects.

Nanodomains

In recent investigations of MPB compositions of undoped PZT, nanoscale domains were detected in addition to the different types of micron-scale domains [27,53,59]. The arrangement of the domains in PZT 50/50 was characterized as clustering [27]. In contrast to that in PZT 53/47, the nanodomains were found to be subdomains within a microscale domain. Approaches to explain the existence of these nanodomains were also contrary. Fluctuation of the B-cation environment, i.e., the local distribution of Zr and Ti cations, has been suggested by Woodward et al. [27]. The formation of the nanodomains has been also discussed in relation to martensitic theories of domain formation [53]. The theoretical work of Jin et al. predicts miniaturization of the domain structures in systems that exhibit low domain wall energies [29].

Grain Boundaries

In contrast to SEM imaging, in which grain boundaries can be observed only after etching, mapping of grain boundaries by TEM is possible without chemical or thermal treatment of these regions. Investigations of Sr-doped commercial PZT (EC65) by Tan and Shang [60] using Fresnel imaging and diffuse dark-field imaging detected a continuous layer of an amorphous phase between the grains of about 2 nm thickness. Corresponding EDX revealed a high content of lead with significant amounts of Si, Al, and Fe in the grain boundary phase. The presence of Si was attributed to impurities in the starting materials, which segregate during sintering. Grain boundaries of La-doped PZT at stoichiometric composition and with 3 mole% PbO excess were investigated by HRTEM and analytical STEM [31,32]. It was found that the planar grain boundaries between two grains were mostly free from grain boundary film in the stoichiometric materials, whereas the materials prepared with PbO excess had a grain boundary film with a thickness up to 4 nm. In the triple junctions, an amorphous PbO-rich secondary phase was found in both compositions.

Long- and Short-Range Ordering

Electron diffraction patterns can be used to obtain the Pb cation displacement in PZT of different phases [61]. From the presence of diffuse streaks joining Bragg points in a Kikuchi map, it was concluded that in rhombohedral compositions a highly oriented planar order with respect to the displacement of the Pb cations prevails. A long-range order has been suggested for the materials in the vicinity of the MPB. For the tetragonal phase, a short-range order has been indicated.

17.2.2.2 HRTEM Imaging of Domain Walls and Local Structure

HRTEM investigation of the local ordering in mixed crystal systems has become possible with the development of exit wave reconstruction software, but is often limited by practical aspects. The technique reconstructs the exit wave function on the basis of a series of images with varying defocus if the kinematic approximation is satisfied. In the case of lead zirconate titanate, the heavy lead atoms restrict the sample thickness to approximately two elementary cells along the beam direction. Other systems, e.g., $KNaNbO_3$, are more promising candidates for this kind of investigation. The imaging of domain walls is principally limited by the same restrictions. But still a number of questions can be answered using even thicker samples. The tilt of adjacent domains can be measured and correlated to the lattice parameters, the width of domain walls (the strain field) can be estimated, and the morphology of the domain wall can be seen (Fig. 17.13).

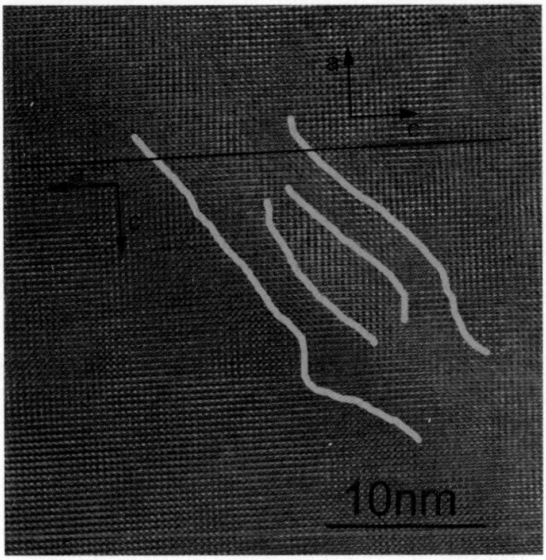

Fig. 17.13. Fourier-filtered high-resolution transmission electron microscopy image of a 90° domain wall in a PZT sample with the composition $Pb(Ti_{52.5}Zr_{47.5})O_3$. The two domains are tilted, 1.4(2)°, the angle can be seen with respect to the black line. The grey lines indicate the main contrast changes over the domain wall, and the thickness is estimated to be less that 10 nm (Courtesy R. Schierholz)

References

1. P.W. Forsbergh, Phys. Rev. **76**, 1187–1201 (1949)
2. B. Jaffee, W.R. Cook, H. Jaffee, *Piezoelectric Ceramics* (R.A.N. Publishers, Marietta, OH, 1971)
3. M.T. Buscaglia, V. Buscaglia, M. Vivani, J. Petzelt, M. Savinov, L. Mitioseriu, A. Testino, P. Nanni, C. Harnagea, Z. Zhao, M. Nygren, Nanotechnology **15**, 1113–1117 (2004)
4. Y. Saito, H. Takao, T. Tani, T. Nonoyama, K. Takatori, T. Homma, T. Nagaya, M. Nakamura, Nature **432**, 84–87 (2004)
5. G. Zorn, W. Wersing, H. Goebel, Jpn. J. Appl. Phys. **24**(Suppl.2), 721–723 (1985)
6. H. Boysen, Zeitschrift für Kristallographie **20**, 726–734 (2005)
7. W. Cao, L.E. Cross, Phys. Rev. B **44**, 5–12 (1991)
8. D.A. Bruce, J. Phys. C Solid State Phys. **14**, 5195–5214 (1981)
9. H. Boysen, Phase Transit. **55**, 1–16 (1995)
10. N. Floquet, C.M. Valot, M.T. Mesnier, J.C. Niepce, L. Normand, A. Thorel, R. Kilaas, J. Phys. III **7**, 1105–1128 (1997)
11. C.M. Valot, N. Floquet, P. Perriat, M.T. Mesnier, J.C. Niepce, Ferroelectrics **172**, 235–241 (1995)
12. J. Frantti, J. Lappalainen, S. Eriksson, V. Lantto, S. Nishio, M. Kakihana, S. Ivanov, H. Rundlöf, Jpn. J. Appl. Phys. **39**, 5697–5703 (2002)
13. D.J. Corker, A.M. Glazer, R.W. Whatmore, A. Stallard, F. Fauth, J. Phys. Condens. Matter **10**, 6251–6269 (1998)
14. P.W. Stevens, J. Appl. Crystallogr. **27**, 462 (1999)

15. L. Hanh, K. Uchino, S. Nomura, Jpn. J. Appl. Phys. **39**, 5697–5703 (1978)
16. W. Cao, L.E. Cross, Jpn. J. Appl. Phys. **31**, 1399–1402 (1992)
17. W. Wersing, W. Rossner, G. Eckstein, G. Tomandl, Silicates Industriels **3–4**, 41–46 (1985)
18. M.J. Hoffmann, H. Kungl, Curr. Opin. Solid State Mater. Sci. **8**, 51–57 (2004)
19. K.H. Härdtl, D. Hennings, Science of Ceramics **6**, VII/1 (1973)
20. B. Noheda, D.E. Cox, G. Shirane, R. Guo, B. Jones, E. Cross, Phys. Rev. B **63**, 14103-1–14103-9 (2000)
21. B. Noheda, J.A. Gonzalo, E. Cross, R. Guo, S.E. Park, D.E. Cox, G. Shirane, Phys. Rev. B **61**, 8687–8695 (2000)
22. R. Guo, E. Cross, S.E. Park, B. Noheda, D.E. Cox, G. Shirane, Phys. Rev. Lett. **84**, 5423–5426 (2000)
23. K. Ohwada, K. Hirota, P.W. Rehrig, P. Gehring, B. Noheda, Y. Fujii, S.E. Park, G. Shirane, J. Phys. Soc. Jpn. **70**, 2778–2783 (2001)
24. M.K. Durbin, C.J. Hicks, S.E. Park, T.R. Shrout, J. Appl. Phys. **87**, 8159–8164 (2000)
25. J. Frantti, S. Ivanov, J. Lappalainen, S. Eriksson, V. Lantto, S. Nishio, M. Kakihana, H. Rundlöf, Ferroelectrics **266**, 71–90 (2002)
26. J. Frantti, S. Ivanov, S. Eriksson, H. Rundlöf, V. Lantto, J. Lappalainen, M. Kakihana, Phys. Rev. B **66**, 64108-1–64108-15 (2002)
27. D.I. Woodward, J. Knudsen, I.M. Reaney, Phys. Rev. B **72**, 104110-1–104110-8 (2005)
28. K.A. Schönau, L. Schmitt, M. Knapp, H. Fuess, R. Eichel, H. Kungl, M.J. Hoffmann, Phys. Rev. B **75**, 184117 (2007)
29. Y.M. Jin, Y.U. Wang, A.G. Khachaturian, J.F. Li, D. Viehland, Phys. Rev. Lett. **91**, 197601-1–197601-4 (2003)
30. Y.M. Jin, Y.U. Wang, A.G. Khachaturian, J.F. Li, D. Viehland, J. Appl. Phys. **94**, 3629–3640 (2003)
31. M. Hammer, M.J. Hoffmann, J. Am. Ceram. Soc. **81**, 3277–3284 (1998)
32. M. Hammer, C. Monty, A. Endriss, M.J. Hoffmann, J. Am. Ceram. Soc. **81**, 721–724 (1998)
33. M. Knapp, C. Baehtz, H. Ehrenberg, H. Fuess, J. Synchrotron Radiat. **11**, 328–334 (2004)
34. Y. Saito, Jpn. J. Appl. Phys. **36**, 5963–5969 (1997)
35. T. Ogawa, Jpn. J. Appl. Phys. **39**, 5538–5541 (2000)
36. M.J. Hoffmann, M. Hammer, A. Endriss, D.C. Lupascu, Acta Materialia **49**, 1301–1310 (2001)
37. M.J. Hoffmann, H. Kungl, J.-Th. Reszat, S. Wagner, Polar Oxides – Properties, Characterization and Imaging, ed. by R. Waser et. al., Wiley (2005)
38. J.T. Reszat, A.E. Glazounov, M.J. Hoffmann, J. Eur. Ceram. Soc. **21**, 1349–1352 (2001)
39. D.A. Hall, A. Steuwer, B. Cherdirunkorn, T. Mori, P.J. Withers, J. Appl. Phys. **96**, 4245 (2004)
40. D.A. Hall, A. Steuwer, B. Cherdirunkorn, P.J. Withers, T. Mori, Mater. Sci. Eng. **409**, 206–210 (2005)
41. D.A. Hall, A. Steuwer, B. Cherdirunkorn, T. Mori, P.J. Withers, Acta Materialia **54**, 3075–3083 (2005)
42. S. Wan, K. Bowman, J. Mater. Res. **16**, 2306–2313 (2001)
43. J.L. Jones, E.B. Slamovic, K. Bowman, J. Appl. Phys. **97**, 034113-1–034113-6 (2005)

44. W. Chang, A.H. King, K.J. Bowman, Appl. Phys. Lett. **88**, 242901 (2006)
45. R. Rogan, E. Üstündag, B. Clausen, M.R. Daymond, J. Appl. Phys. **93**, 4104–4111 (2003)
46. J.Y. Li, R. Rogan, E. Üstündag, K. Bhattacharya, Nat. Mater. **4**, 776–781 (2005)
47. W. Wersing, K. Lubitz, J. Mohaupt, IEEE Trans. Ultrason. Piezoelectr. Freq. Control **36**, 424–433 (1989)
48. J.L. Jones, M. Hoffmann, J.E. Daniels, A.J. Studer, J. Appl. Phys. **98**, 092901-1–092901-3 (2006)
49. H. Kungl, Dehnungsverhalten von morphotropem PZT Dissertation, Universität Karlsruhe (2005)
50. H. Kungl, R. Theissmann, C. Baehtz, M. Knapp, H. Fuess, S. Wagner, T. Fett, M.J. Hoffmann, Acta Materialia **55**, 5780–5791 (2007)
51. J. Ricote, R.W. Whatmore, D.J. Barber, J. Phys. Condens. Matter **12**, 323–337 (2000)
52. R. Theissmann, L.A. Schmitt, J. Kling, R. Schierholz, K.A. Schönau, H. Fuess, M. Knapp, H. Kungl, M.J. Hoffmann, J. Appl. Phys. **102**, 024111 (2007)
53. L. Schmitt, K.A. Schönau, R. Theissmann, H. Fuess, H. Kungl, M.J. Hoffmann, J. Appl. Phys. **101**, 074107 (2007)
54. C.A. Randall, N. Kim, J. Kucera, W. Cao, T.R. Shrout, J. Am. Ceram. Soc. **81**, 677–688 (1998)
55. G. Arlt, J. Mater. Sci. **25**, 2655–2666 (1990)
56. W. Cao, C.A. Randall, J. Phys. Chem. Solids **57**, 1499–1505 (1996)
57. I. Mac Laren, L. Schmitt, H. Fuess, H. Kungl, M.J. Hoffmann, J. Appl. Phys. **97**, 094102-1–094102-8 (2004)
58. N.A. Pertsev, G. Arlt, Ferroelectrics **132**, 27 (1992)
59. P.G. Lucuta, V. Teodorescu, Appl. Phys. A: Solid Surf. **37**, 237 (1985)
60. X. Tan, J.K. Shang, Philos. Mag. A **82**, 1463–1478 (2002)
61. A. Glazer, P.A. Thomas, K.Z. Baba-Kishi, G.K.K. Pang, C.W. Tai, Phys. Rev. B **70**, 184123-1–184123-9 (2004)

18

Small Signal Resonance Methods

W. Wersing

18.1 Introduction

The properties of a piezoelectric material are defined by the elastic, dielectric, and piezoelectric tensor components which will be introduced in Sect. 18.2. Because of the inherent asymmetrical nature of piezoelectric materials and the fact that we want to predict different response directions, a tensor-based description of the properties of these materials is unavoidable. Generally these components are by far not constant, they depend on temperature, applied mechanical stress and electric field strength. Furthermore, they are amplitude-dependent and become non-linear or even non-reversible when the applied signal stress or signal field amplitudes exceed limits characteristic of the material. The dielectric, elastic and piezoelectric coefficients have not only effective (real, in-phase) components but also dissipative (imaginary, out-of-phase) terms which are of particular importance at high amplitudes.

The fact that we have to deal with the directionality of the response of piezoelectric materials and consequently with tensors complicates engineering to some extent, however, on the other hand, it offers a great design potential for engineers in the development of piezoelectric devices. Because, today, the design of new devices is done more or less by using computational tools, it is decisive for the engineer to have access to a complete set of the property coefficients (tensor components) of the materials applied.

Therefore, in this subchapter we will describe methods to precisely determine dielectric, elastic, and piezoelectric coefficients under small signal conditions. The determination of these properties under large signal conditions is discussed in Chap. 19. Because we consider small signal conditions, we need not take into account the non-linearities in the constants and the equations of state relating the electric and elastic variables can be written in the linear form derived in Sect. 3.1 from the thermodynamic potentials. Under small signal conditions the contributions to the piezoelectric response in ferroelectric materials are due to the intrinsic and the reversible extrinsic piezoelectric effect caused by the reversible movement of domain walls.

In general, static and quasi-static techniques for measuring piezoelectric coefficients are capable and have proven useful in certain special cases [1]. These techniques are not recommended here because dynamic methods are capable of greater accuracy and can be applied easily to a much wider range of crystal orientations and sample geometries. Therefore, small signal resonance methods have also proven its worth for testing piezoelectric transducers and devices in development and production.

18.2 The Electromechanical Equations of State

In Sect. 3.1, we derived from the Gibbs free energy the equations (3.11):

$$\begin{aligned} \Delta S_\alpha &= \alpha_\alpha^{\mathrm{E}} \Delta T + s_{\alpha\beta}^{\mathrm{E}} \Delta T_\beta + d_{i\alpha} \Delta E_i, \\ \Delta D_i &= p_i^{\mathrm{T}} \Delta T + d_{i\alpha} \Delta T_\alpha + \varepsilon_{ik}^{\mathrm{T}} \Delta E_k, \end{aligned} \qquad (18.1)^1$$

where $\alpha_\alpha^{\mathrm{E}}$ are the thermal expansion coefficients, ${s^{\mathrm{E}}}_{\alpha\beta}$ the elastic compliances at constant electric field, $d_{i\alpha}$ the piezoelectric constants, ${p^{\mathrm{T}}}_i$ the pyroelectric coefficients, and ${\varepsilon^{\mathrm{T}}}_{ik}$ the permittivities at constant stress. In addition, six other linear equations, i.e., three further sets of equations can be derived by use of the other thermodynamic potentials. The interrelationships among the mechanical, electrical, and thermal properties mediated by the different material constants can be well illustrated using a construct called a Heckmann diagram [4] shown in Fig. 18.1. The circles of the outer triangle represent the intensive variables (forces), the circles of the inner triangle represent the extensive variables (displacements).

In their isothermal form ($\Delta T = 0$) these four sets of equations are called basic piezoelectric equations or piezoelectric constitutive equations:

$$\begin{aligned} T_\alpha &= c_{\alpha\beta}^{\mathrm{D}} S_\beta + h_{i\alpha} D_i, \\ E_i &= -h_{i\beta} S_\beta + \beta_{ik}^{\mathrm{S}} D_k, \end{aligned} \qquad (18.2)$$

$$\begin{aligned} S_\alpha &= s_{\alpha\beta}^{\mathrm{D}} T_\beta + g_{i\alpha} D_i, \\ E_i &= -g_{i\beta} T_\beta + \beta_{ik}^{\mathrm{T}} D_k, \end{aligned} \qquad (18.3)$$

$$\begin{aligned} T_\alpha &= c_{\alpha\beta}^{\mathrm{E}} S_\beta - e_{i\alpha} E_i, \\ D_i &= e_{i\beta} S_\beta + \varepsilon_{ik}^{\mathrm{S}} E_k, \end{aligned} \qquad (18.4)$$

$$\begin{aligned} S_\alpha &= s_{\alpha\beta}^{\mathrm{E}} T_\beta + d_{i\alpha} E_i, \\ D_i &= d_{i\beta} T_\beta + \varepsilon_{ik}^{\mathrm{T}} E_k. \end{aligned} \qquad (18.5)$$

These piezoelectric constitutive equations are written in matrix form using Voigt's notation[2] and Einstein's summation convention of repeated subscripts (as in Chap. 3.1.4). Often the following matrix form is used were the matrices

[1] Notations are according to the IRE Standards [2, 3].
[2] $i,\ k = 1,\ 2,\ 3$ and $\alpha,\ \beta = 1,\ 2, \ldots,\ 6$.

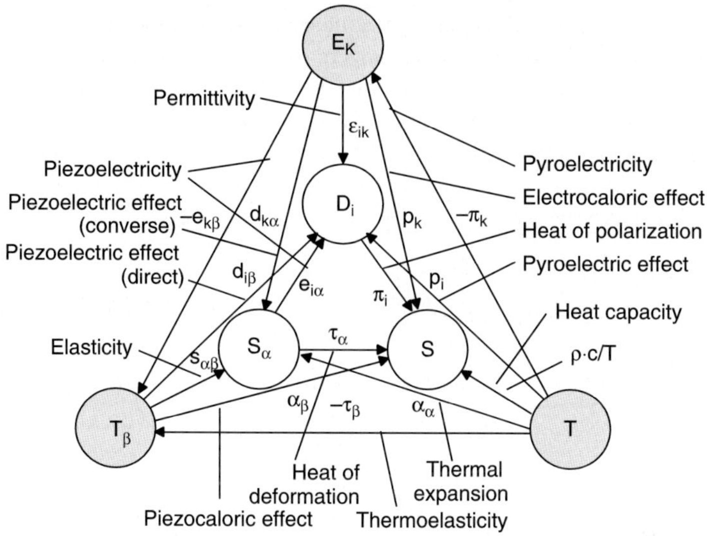

Fig. 18.1. Heckmann diagram showing the interrelationship among the mechanical, electrical, and thermal properties of materials mediated by the different material constants. The *circles* of the outer *triangle* represent the intensive variables (temperature T, stress components T_β, and electric field components E_k), the *circles* of the inner *triangle* represent the extensive variables (entropy S, strain components S_α, and displacement components D_i)

are symbolized by simple characters, sometimes of bold type. For example (18.5) read in this form

$$\begin{aligned} S &= s^{\mathrm{E}}T + d^{\mathrm{t}}E, \\ D &= dT + \varepsilon^{\mathrm{T}}E, \end{aligned} \tag{18.6}$$

where the superscript t denotes a transposed matrix.

For comparison reasons we write here (18.5) also in tensor form

$$\begin{aligned} S_{ij} &= s^{\mathrm{E}}_{ijkl}T_{kl} + d_{kij}E_k, \\ D_i &= d_{ikl}T_{kl} + \varepsilon^{\mathrm{T}}_{ik}E_k, \end{aligned} \tag{18.7}$$

where all variables and material properties are described by tensors of first (D_i, E_k) to fourth rank (s_{ijkl}) and again the summation convention applies.

The property matrices used in (18.2)–(18.6) have the following meaning:

c^{E} and c^{D} are the closed-circuit (constant E) and open-circuit (constant D) elastic stiffnesses,

s^{E} and s^{D} are the closed-circuit (constant E) and open-circuit (constant D) elastic compliances, ε^{T} and ε^{S} are the free (constant stress) and clamped (constant strain) permittivities, β^{T} and β^{S} are the free (constant stress) and clamped (constant strain) impermeabilities.

d, e, h, and g are called the piezoelectric compliances (the superscript t denotes transposed matrices). However, often the following specific names are used:

- d piezoelectric strain coefficients
- e piezoelectric stress coefficients
- g piezoelectric voltage coefficients
- h piezoelectric h coefficients (not specifically named)

Since each thermodynamic potential contains a complete description of the material of interest, we may deduce a large number of identities relating the property matrices. Some important examples are presented here. Naturally we have

$$\begin{aligned} \beta^{\mathrm{T,\ or\ S}} &= (\varepsilon^{\mathrm{T,\ or\ S}})^{-1}, \\ c^{\mathrm{E,\ or\ D}} &= (s^{\mathrm{E,\ or\ D}})^{-1}, \end{aligned} \tag{18.8}$$

$$\begin{aligned} d &= es^E = \varepsilon^T g \\ e &= dc^E = \varepsilon^S h \\ g &= hs^D = \beta^T d \\ h &= gc^D = \beta^S e \end{aligned} \tag{18.9}$$

$$\begin{aligned} s^E - s^D &= d^t g = g^t d \\ c^D - c^E &= e^t h = h^t e \end{aligned} \tag{18.10}$$

$$\begin{aligned} \varepsilon^T - \varepsilon^S &= ed^t = de^t \\ \beta^S - \beta^T &= hg^t = gh^t \end{aligned} \tag{18.11}$$

Crystal symmetry imposes restrictions on the form of the property tensors and consequently on the property matrices. In many cases the full pattern of symmetries imposed by each of the 32 point groups has been worked out [5] (see Table 8 of [6]). We show here in (18.12) as an example the detailed form of the matrices of (18.5) of poled ceramics[3] (class ∞mm) identical to those of crystals belonging to the hexagonal class 6 mm.

$$\begin{pmatrix} S_1 \\ S_2 \\ S_3 \\ S_4 \\ S_5 \\ S_6 \end{pmatrix} = \begin{pmatrix} s_{11}^E & s_{12}^E & s_{13}^E & 0 & 0 & 0 \\ s_{12}^E & s_{11}^E & s_{13}^E & 0 & 0 & 0 \\ s_{13}^E & s_{13}^E & s_{33}^E & 0 & 0 & 0 \\ 0 & 0 & 0 & s_{44}^E & 0 & 0 \\ 0 & 0 & 0 & 0 & s_{44}^E & 0 \\ 0 & 0 & 0 & 0 & 0 & 2\left(s_{11}^E - s_{12}^E\right) \end{pmatrix} \begin{pmatrix} T_1 \\ T_2 \\ T_3 \\ T_4 \\ T_5 \\ T_6 \end{pmatrix} + \begin{pmatrix} 0 & 0 & d_{31} \\ 0 & 0 & d_{31} \\ 0 & 0 & d_{33} \\ 0 & d_{15} & 0 \\ d_{15} & 0 & 0 \\ 0 & 0 & 0 \end{pmatrix} \begin{pmatrix} E_1 \\ E_2 \\ E_3 \end{pmatrix}$$

$$\begin{pmatrix} D_1 \\ D_2 \\ D_3 \end{pmatrix} = \begin{pmatrix} 0 & 0 & 0 & 0 & d_{15} & 0 \\ 0 & 0 & 0 & d_{15} & 0 & 0 \\ d_{31} & d_{31} & d_{33} & 0 & 0 & 0 \end{pmatrix} \begin{pmatrix} T_1 \\ T_2 \\ T_3 \\ T_4 \\ T_5 \\ T_6 \end{pmatrix} + \begin{pmatrix} \varepsilon_{11}^T & 0 & 0 \\ 0 & \varepsilon_{11}^T & 0 \\ 0 & 0 & \varepsilon_{33}^T \end{pmatrix} \begin{pmatrix} E_1 \\ E_2 \\ E_3 \end{pmatrix} \tag{18.12}$$

[3] The 3-axis is usually chosen to be the polar axis (parallel to the poling direction).

The elastic behavior of isotropic materials is usually characterized by Young's modulus Y, the Poisson ratio σ, and the shear modulus G. Because two of these parameters are sufficient to fully characterize an isotropic material there exists the relation

$$Y = 2(1+\sigma)G \tag{18.13}$$

between these parameters.

If we simplify the matrices of (18.12) to the isotropic case we naturally have vanishing piezoelectric coefficients, $\varepsilon_{11} \equiv \varepsilon_{33}$,[4] $s_{11} \equiv s_{33}$, $s_{13} \equiv s_{12}$, and $s_{44} \equiv s_{66} = 2(s_{11} - s_{12})$ and therefore we identify

$$Y = T_1/S_1 = 1/s_{11}, \tag{18.14}$$

$$\sigma = -Y\tfrac{S_2}{T_1} = -s_{12}/s_{11}, \tag{18.15}$$

and with (18.13)

$$G = \frac{1}{2(s_{11} - s_{12})} = 1/s_{66} = 1/s_{44}. \tag{18.16}$$

Because mechanical engineers are often more familiar with modules Y and G (or Y and σ) than with the compliance or stiffness matrices introduced above, these modules are also often used to characterize non-isotropic materials, especially piezoelectric ceramics. It is clear that in this case these modules can only characterize the material in a certain given direction. If we assume, for example, that the only effective stress component (normal stress component) in a piezoelectric ceramic is T_1 (compared to the axis system used for (18.12)) and all other components are zero, Young's modulus can be defined similarly as for isotropic materials (18.14).

In non-isotropic materials the directions of the cross contraction (vertical to the direction of the normal stress) are not equivalent and thus it is usual (and necessary) to define two different Poisson ratios

$$\sigma_{12} = -Y\frac{S_2}{T_1} = -s_{12}/s_{11} \tag{18.17}$$

and

$$\sigma_{13} = -Y\frac{S_3}{T_1} = -s_{13}/s_{11}. \tag{18.18}$$

18.3 Electromechanical Coupling Factors

Although each coefficient set in (18.2)–(18.5) as derived from a thermodynamic potential completely describes the properties of the material of interest,

---

4 We must no longer distinguish between ε^{T} and ε^{S} or s^{E} and s^{D} as can be seen from (18.10) and (18.11).

another property set is often used, especially in the case where small signal resonance methods are applied for characterizing material properties. This property set are the electro mechanical coupling factors k_{lm} which describe the strength of the electromechanical response. They are dimensionless measures of the strength of the piezoelectric effects and therefore, are very useful as a figure of merit. The coupling factors may also be considered as a measure of the efficiency of the conversation of electrical to mechanical or mechanical to electrical energy, i.e., for example when an electrical field is applied, k^2 measures the fraction of the electrical energy converted into mechanical energy. Therefore, in terms of k^2 we have:

$$k^2 = \frac{\text{electrical energy converted to mechanical energy}}{\text{input electrical energy}} \tag{18.19a}$$

or

$$k^2 = \frac{\text{mechanical energy converted to electrical energy}}{\text{input mechanical energy}}. \tag{18.19b}$$

Since the energy conversion is always incomplete, k^2 (and thus also k) is always <1.

Figure 18.2 shows as an example the conversion of mechanical to electrical energy. In the first step of the conversion cycle (A $\Rightarrow$ B), the mechanical energy is applied to the transducer material by stressing it, e.g., in Fig. 18.2 in the direction of the 3-axis of the material (e.g., a ceramic cylinder poled in the 3-direction). During this stress (pressure) increase up to $-T_{\max}$ the material sample is short-circuit, i.e., $E_3 = 0$ and the dielectric displacement growth to $-D_3$. Then the short-circuit is opened and the stress is removed (second step

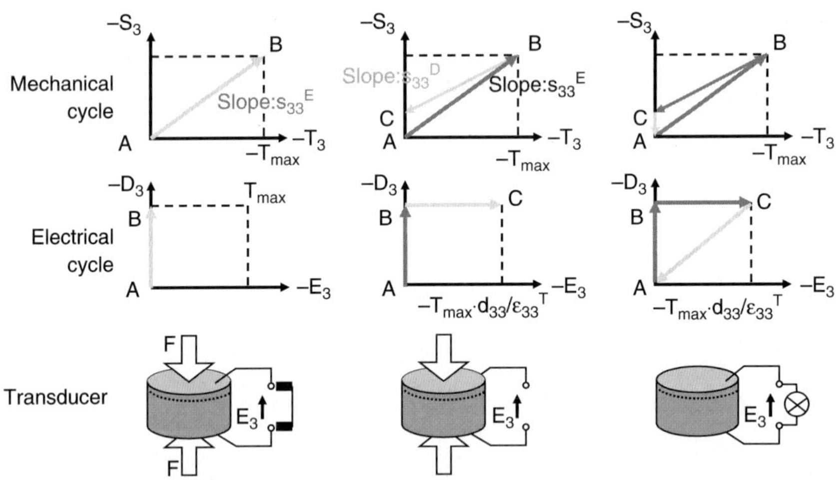

Fig. 18.2. Diagram showing a conversion cycle of mechanical to electrical energy to illustrate the relation between the electromechanical coupling factor k_{33} and d_{33}, $\varepsilon_{33}{}^{\mathrm{T}}$, and $s_{33}{}^{\mathrm{E}}$

of the cycle B $\Rightarrow$ C). Thereby the electric field across the transducer increases to $T_{\max} \cdot d_{33} \cdot \varepsilon_{33}{}^{\mathrm{T}}$. In the third step of the cycle (C $\Rightarrow$ A) the electrical energy stored in the transducer material is transferred to the external load and the transducer returns to the initial state. The mechanical input energy, i.e., the energy mechanically applied to the transducer in the step A $\Rightarrow$ B is

$$W_{\mathrm{AB}}^{\mathrm{mech}} = \tfrac{1}{2} s_{33}^{\mathrm{E}} T_{\max}^{2},$$

the energy mechanically released from the transducer in the second step B$\Rightarrow$C is

$$W_{\mathrm{BC}}^{\mathrm{mech}} = \frac{1}{2} s_{33}^{\mathrm{D}} T_{\max}^{2},$$

and thus, the electrical output energy transferred to the load reads

$$W_{\mathrm{CA}}^{\mathrm{elec}} = W_{\mathrm{AB}}^{\mathrm{mech}} - W_{\mathrm{BC}}^{\mathrm{mech}} = \frac{1}{2} \varepsilon_{33}^{\mathrm{T}} E_{\max}^{2} = \frac{1}{2} \frac{d_{33}^{2}}{\varepsilon_{33}^{\mathrm{T}}} T_{\max}^{2}.$$

Therefore, the coupling factor $k_{33}{}^{2}$ according to the definition (18.19b) is calculated to be:

$$k_{33}^{2} = \frac{W_{\mathrm{AB}}^{\mathrm{mech}} - W_{\mathrm{BC}}^{\mathrm{mech}}}{W_{\mathrm{AB}}^{\mathrm{mech}}} = \frac{s_{33}^{\mathrm{E}} - s_{33}^{\mathrm{D}}}{s_{33}^{\mathrm{E}}} = \frac{W_{\mathrm{CA}}^{\mathrm{elec}}}{W_{\mathrm{AB}}^{\mathrm{mech}}} = \frac{d_{33}^{2}}{\varepsilon_{33}^{\mathrm{T}} s_{33}^{\mathrm{E}}}. \tag{18.20}$$

Similarly the cycle for the conversion of electrical to mechanical energy results in

$$k_{33}^{2} = \frac{\varepsilon_{33}^{\mathrm{T}} - \varepsilon_{33}^{\mathrm{S}}}{\varepsilon_{33}^{\mathrm{T}}} = \frac{d_{33}^{2}}{s_{33}^{\mathrm{E}} \varepsilon_{33}^{\mathrm{T}}}. \tag{18.21}$$

Equations (18.20) and (18.21) define the electromechanical coupling factor k_{33} in terms of piezoelectric, dielectric, and elastic coefficients. Coupling factors defined in this way are called material coupling factors, they are a measure of the energy conversion efficiency of the piezoelectric material and not of a specific transducer. The relations between the elastic compliances at constant E and constant D and the permittivities at constant T and constant S also included in (18.20) and (18.21), respectively can already be seen from (18.10) and (18.11).

The general definition of a coupling factor given in (18.19) can be used, on the one hand, to characterize the energy conversion efficiency of a specific piezoelectric transducer[5] (transducer coupling factor) or, on the other hand, to characterize the energy conversion efficiency of a material of interest (material coupling factor). In the latter case different coupling factors must be introduced depending on the symmetry of the material considered. These coupling factors are directly related to the material property coefficients. In

[5] A transducer is often characterized by its effective coupling factor which is defined in (18.52) via its resonance frequencies (see Sect. 18.6).

the simplest case of the highest possible symmetry (class 6 mm) as considered for (18.12), besides the coupling factor k_{33}, the coupling factors

$$k_{31}^2 = \frac{d_{31}^2}{\varepsilon_{33}^{\mathrm{T}} s_{11}^{\mathrm{E}}} \tag{18.22}$$

and

$$k_{15}^2 = \frac{d_{15}^2}{\varepsilon_{11}^{\mathrm{T}} s_{55}^{\mathrm{E}}} \tag{18.23}$$

are required.

18.4 Piezoelectrically Excited Resonators

Elastic bodies show numerous mechanical resonances related to half wavelength standing elastic waves. The piezoelectric effect offers a simple method to excite these elastic waves electrically to permit observation of mechanical and electromechanical material properties. These relations can be clearly seen by studying the electrical behavior of a simple resonator as shown in Fig. 18.3.

Under circumstances where width w and thickness t are negligible compared to the length l of the bar ($l \gg w,\ t$; e.g., $l > 10w,\ w > 2t$), and when no external forces act on the bar, the following boundary conditions are fulfilled:

$$\begin{aligned} &T_1 = T_1(x_1), \quad T_2 = T_3 = T_4 = T_5 = T_6 = 0, \\ &E_1 = E_2 = 0, \quad \partial E_3/\partial x_1 = \partial E_3/\partial x_2 = \partial E_3/\partial x_3 = 0, \end{aligned} \tag{18.24}$$

i.e., we only have inertia stress along x_1 and an electric field component along x_3 which does not depend on x_1, x_2, and x_3.[6]

The equation of motion

$$\rho \frac{\partial^2 u_i}{\partial t^2} = f_i = \frac{\partial}{\partial x_k} T_{ik} \tag{18.25}$$

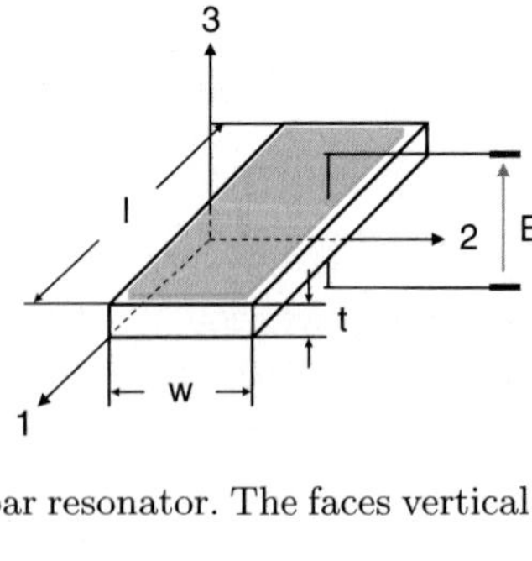

Fig. 18.3. Piezoelectric bar resonator. The faces vertical to the 3-axis are electroded

[6] This can be shown with the aid of Maxwell's equations for the quasi static case: $\mathrm{rot}\vec{E} = 0,\ \mathrm{div}D = 0$.

with mass density ρ, elastic displacement u_i,[7] force density f_i, and stress tensor T_{ik}[8] results together with the constitutive equation (18.7) under the boundary conditions (18.24) in the simple differential equation

$$\frac{\partial^2 u_1(x_1,t)}{\partial t^2} = v_s^2 \frac{\partial^2 u_1(x_1,t)}{\partial x_1^2} \quad \text{with } v_s = \frac{1}{\sqrt{\rho s_{11}^E}}. \tag{18.26}$$

Using complex amplitudes, e.g., $\bar{u}_1(x_1)$ defined by

$$u_1(x_1,t) = Re\left\{\bar{u}_1(x_1)e^{i\omega t}\right\} \tag{18.27}$$

to solve (18.26) for periodical functions one finds the following solution for the free vibrating bar ($T_1(x_1 = \pm l/2) = 0$), symmetrically clamped at $x_1 = 0$ ($v_1 = \partial u_1/\partial t = 0$ at $x_1 = 0$):

$$\bar{u}_1\left(x_1 = \frac{1}{2}l\right) = d_{31}\frac{\bar{U}}{t}\frac{1}{\omega\sqrt{\rho s_{11}^E}}\tan\left(\frac{1}{2}\omega l\sqrt{\rho s_{11}^E}\right) \qquad (18.28)^9$$

where $U(t) = \text{Re}\left\{\bar{U}e^{i\omega t}\right\}$ is the voltage applied on the resonator bar.

Calculating $D_3(x_1)$ using the second constitutive equation (18.7), integrating over x_1 and forming the time derivative gives the electrical current (that is calculated in (18.29))

$$\bar{I} = i\omega \cdot w \cdot \left(2 \cdot \frac{d_{31}}{s_{11}^E} \cdot \bar{u}_1\left(\frac{l}{2}\right) + \varepsilon_{33}^S \cdot \frac{l}{t} \cdot \bar{U}\right) \tag{18.29}$$

flowing through the resonator bar and with the aid of (18.28) the admittance of the free vibrating resonator bar

$$\bar{Y} = \frac{\bar{I}}{\bar{U}} = i\omega\frac{wl}{t}\varepsilon_{33}^S + i\frac{2w}{t}\frac{d_{31}^2}{s_{11}^E}\frac{1}{\sqrt{\rho s_{11}^E}}\tan\left(\frac{1}{2}\omega l\sqrt{\rho s_{11}^E}\right). \tag{18.30}$$

Together with the admittance, all coefficients are supposed to be complex to include losses in (18.30), i.e., we have:

$$\varepsilon_{33}^S = \varepsilon_{33}^{\prime S} - i\varepsilon_{33}^{\prime\prime S}, \quad s_{11}^E = s_{11}^{\prime E} - is_{11}^{\prime\prime E}, \quad d_{31} = d_{31}' - id_{31}''. \tag{18.31}$$

In the case of an ideal lossless material the resonator admittance is purely reactive and two sets of characteristic frequencies can be identified from

[7] Displacements are related to strain tensor components via $S_{ik} = \frac{1}{2}\left(\frac{\partial u_i}{\partial x_k} + \frac{\partial u_k}{\partial x_i}\right)$. These are related to matrix elements S_α : $S_1 = S_{11}$, $S_2 = S_{22}$, $S_3 = S_{33}$, $S_4 = 2S_{23} = 2S_{32}$, $S_5 = 2S_{13} = 2S_{31}$, $S_6 = 2S_{12} = 2S_{21}$.

[8] Tensor components T_{ik} are related to matrix elements T_α : $T_1 = T_{11}$, $T_2 = T_{22}$, $T_3 = T_{33}$, $T_4 = T_{23} = T_{32}$, $T_5 = T_{13} = T_{31}$, $T_6 = T_{12} = T_{21}$.

[9] Here t denotes the bar thickness, not to be confused with t used to denote time as for example in (18.27).

(18.30): $f_\nu^{(1)}$ for $\bar{Y} \to \mathrm{i}\infty$ and $f_\nu^{(2)}$ for $\bar{Y} = 0$. Thus $\omega_\nu^{(1)} = 2\pi f_\nu^{(1)}$ is obtained from (18.30) to be

$$\omega_\nu^{(1)} = \frac{1}{\sqrt{\rho s_{11}^{\mathrm{E}}}} \frac{\pi}{l}(2\nu - 1) = \omega_1^{(1)}(2\nu - 1), \quad \nu = 1, 2, 3 \ldots \tag{18.32}$$

and $\omega_\nu^{(2)} = 2\pi f_\nu^{(2)}$ can be obtained from

$$1 + \frac{d_{31}^2}{\varepsilon_{33}^{\mathrm{S}} s_{11}^{\mathrm{E}}} \cdot \frac{1}{\frac{1}{2}\omega_\nu^{(2)} l \sqrt{\rho s_{11}^{\mathrm{E}}}} \tan\left(\frac{1}{2}\omega_\nu^{(2)} l \sqrt{\rho s_{11}^{\mathrm{E}}}\right) = 0. \tag{18.33}$$

Replacing $\sqrt{\rho s_{11}^{\mathrm{E}}}$ in (18.33) with the aid of (18.32) and introducing $\Delta\omega_\nu = \omega_\nu^{(2)} - \omega_\nu^{(1)}$ it is possible to recast (18.33) to give:

$$\frac{d_{31}^2}{\varepsilon_{33}^{\mathrm{S}} s_{11}^{\mathrm{E}}} = \frac{k_{31}^2}{1 - k_{31}^2} = \frac{\omega_\nu^{(2)}}{\omega_\nu^{(1)}} \frac{\pi}{2}(2\nu - 1) \tan\left[\frac{\pi}{2}(2\nu - 1)\frac{\Delta\omega_\nu}{\omega_\nu^{(1)}}\right] \tag{18.34}$$

where (18.22) was used to write the left relation.

Equations (18.32) and (18.34) relate elastic and piezoelectric coefficients (or electromechanical coupling factors) to the characteristic frequencies of the piezoelectric resonator. These kinds of equations are fundamental for the determination of elastic and piezoelectric coefficients using resonance measurements.

Figure 18.4 shows the admittance and impedance circles of a real piezoelectric resonator with losses. In this case the resonator impedance is not purely reactive but complex. The dissipation present in real materials obscures the critical frequencies $f^{(1)}$ and $f^{(2)}$, instead associated frequencies $f_{\mathrm{m}}, f_{\mathrm{s}}, f_{\mathrm{r}}$ and

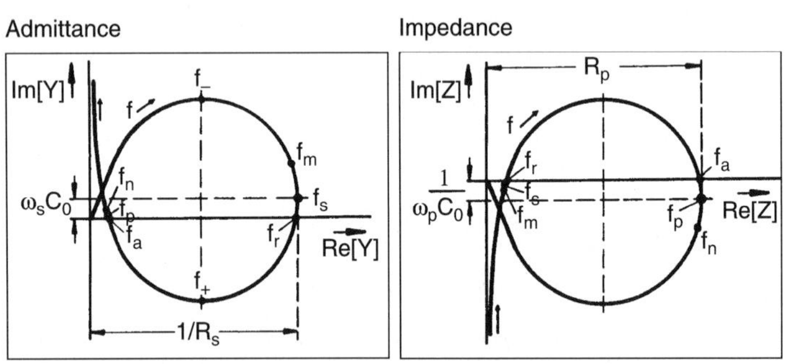

Fig. 18.4. Complex admittance and impedance circles in the vicinity of the characteristic frequencies $f^{(1)}$ and $f^{(2)}$ of a real lossy piezoelectric resonator. In a lossy resonator, there are three frequencies of interest near the admittance maximum (serial resonance) and, similarly, near the impedance maximum (parallel resonance). The three frequencies f_{m}, f_{s}, f_{r} associated to $f^{(1)}$ and f_{n}, f_{p}, f_{a} associated to $f^{(2)}$ correspond to maximum absolute admittance (impedance), maximum conductance (resistance), and zero susceptance (reactance), respectively

$f_\mathrm{n}, f_\mathrm{p}, f_\mathrm{a}$, respectively are of interest as explained in Fig. 18.4. It is clear that the resonance behavior of the complex admittance is caused by the tangent term in (18.30). Thus, to get a better understanding of the resonance behavior it is sufficient to use a complex ${s_{11}}^\mathrm{E}$ in the tangent argument and to consider ${s_{11}}^\mathrm{E}$ in the denominator and d_{31} as purely real. Furthermore, it is very useful to apply Cauchy's method to split a complex analytic function with infinite poles into partial fractions.[10] If this method is applied to the tangent in (18.30) we can write

$$\tan\left(\frac{1}{2}\omega l\sqrt{\rho s_{11}^\mathrm{E}}\right) = -\frac{1}{\frac{1}{2}\omega l\sqrt{\rho s_{11}^\mathrm{E}}}\sum_{\nu=1}^{\infty}\frac{2\omega^2}{\omega^2-\omega_\nu^2} \tag{18.35}$$

with ω_ν identical to ${\omega_\nu}^{(1)}$ as given in (18.32). However, one has to note that due to a complex ${s_{11}}^\mathrm{E}$ the frequencies ${\omega_\nu}^{(1)}$ are now also complex. In order to split $\omega_\nu^{(1)2}$ into its real and imaginary part we must, according to (18.32), split $1/{s_{11}}^\mathrm{E}$ into its real and imaginary part. Using (18.31) we can write:

$$\frac{1}{s_{11}^\mathrm{E}} = \frac{1}{s_{11}^{\prime\mathrm{E}}} + \mathrm{i}\frac{1}{s_{11}^{\prime\mathrm{E}}}\frac{1}{Q(\omega)} = \frac{1}{s_{11}^{\prime\mathrm{E}}} + \mathrm{i}\omega\frac{1}{s_{11}^{\prime\mathrm{E}}}\frac{1}{Q(\omega_\lambda)\omega_\lambda}, \tag{18.36}$$

where we have used that the resonator's mechanical losses (reciprocal resonator Q) are given by $1/Q = s_{11}^{\prime\prime\mathrm{E}}/s_{11}^{\prime\mathrm{E}}$ and that the losses are in general frequency dependent. Assuming that the losses are due to mechanical friction, they increase linearly with frequency, i.e., $Q(\omega)\cdot\omega$ is constant and the right side of (18.36) holds for any fixed ω_λ. With (18.36) $\omega_\nu^{(1)2}$ can be split into its real and imaginary part to give:

$$\omega_\nu^{(1)2} = \omega_\nu^{(1)2'} + \mathrm{i}\omega_\nu^{(1)2''} \cong \omega_\nu^{\prime(1)2} + \mathrm{i}\omega\omega_\nu^{\prime(1)2}\frac{1}{Q(\omega_\lambda)\omega_\lambda} = \omega_\nu^{\prime(1)2} + \mathrm{i}\omega\omega_\nu^{\prime(1)}\frac{1}{Q(\omega_\nu^{\prime(1)})}, \tag{18.37}$$

(18.37)[11]

where $\omega_\nu^{\prime(1)}$ is given by (18.32) with ${s_{11}}^\mathrm{E}$ replaced by ${s'_{11}}^\mathrm{E}$.

Using (18.35) together with (18.37), the complex admittance (18.30) of the bar resonator can be written in the following form:

$$\bar{Y} = \mathrm{i}\omega C_0 + N^2\left(-\mathrm{i}\omega\frac{1}{L^\mathrm{m}}\sum_{1}^{\infty}\frac{1}{\omega^2-\omega_\nu^{\prime(1)2}-i\omega\cdot\omega_\nu^{\prime(1)}/Q_\nu}\right) \tag{18.38}$$

[10] This method and thus also equations identical or equivalent to (18.35) can be found in many textbooks on the theory of complex analytic functions, e.g., in [7].

[11] We assume $\mathrm{Re}(\omega_\nu^{(1)2}) \cong (\mathrm{Re}(\omega_\nu^{(1)}))^2$ which is already for $Q > 5$ a good approximation.

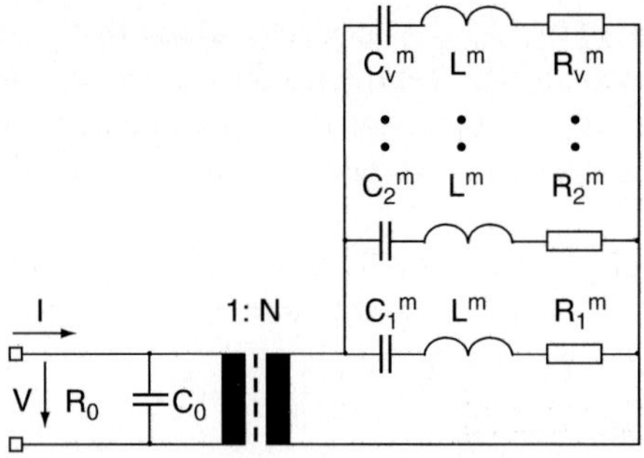

Fig. 18.5. Equivalent lumped-element circuit for the piezoelectrically excited bar resonator. The circuit elements C_ν^{m}, L^{m}, and R_ν^{m} symbolize mechanical resonator properties; they are proportional to compliance, mass of the resonator bar, and mechanical losses. The transformer symbolizes the transformation of the mechanical circuit elements on the secondary side of the transformer to real electrical circuit elements on the primary side via the piezoelectric effect. The electrical capacitor C_0 gives the clamped capacitance of the resonator

with $Q_\nu \equiv Q\left(\omega_\nu'^{(1)}\right)$,

$$C_0 = \frac{wl}{t}\varepsilon_{33}^{\mathrm{S}}, \tag{18.39}$$

$$L^{\mathrm{m}} = \frac{1}{8}\rho wlt, \tag{18.40}$$

and

$$N = w\frac{d_{31}}{s_{11}^{\mathrm{E}}}. \tag{18.41}$$

It can be easily verified that the equivalent circuit in Fig. 18.5 exactly describes the resonator admittance given in (18.38). Inspection of (18.38) for $\omega \to 0$ ($\omega \ll \omega_1'^{(1)}$) shows that the mechanical capacitors C_ν^{m} can be identified to be

$$C_\nu^{\mathrm{m}} = \frac{1}{L^{\mathrm{m}}\omega_\nu'^{(1)2}} = \frac{8}{\pi^2(2\nu-1)^2}\frac{l}{wt}s_{11}'^{\mathrm{E}}. \tag{18.42}$$

Inserting (18.42) into (18.38) for $\omega \to 0$ leads with $\sum\limits_{\nu=1}^{\infty} 1/(2\nu-1)^2 = \pi^2/8$ to

$$\bar{Y}(\omega \to 0) = \mathrm{i}\omega\left(C_0 + N^2\sum_{\nu=1}^{\infty} C_\nu^{\mathrm{m}}\right) = \mathrm{i}\omega\frac{wl}{t}\left(\varepsilon_{33}^{\mathrm{S}} + \frac{d_{31}^2}{s_{11}^{\mathrm{E}}}\right) = \mathrm{i}\omega\frac{wl}{t}\varepsilon_{33}^{\mathrm{T}} = \mathrm{i}\omega C. \tag{18.43}$$

Equation (18.43) shows that ${\varepsilon_{33}}^{\mathrm{T}}$ can be obtained by measuring the bar capacitance at a frequency much lower than the lowest resonance frequency of the bar.

Inspection of (18.38) in the vicinity of the resonance frequencies $\omega_\nu'^{(1)}$ shows that the motional resistors $R_\nu{}^{\mathrm{m}}$ symbolizing the resonator losses can be identified to be

$$R_\nu^{\mathrm{m}} = \frac{\omega_\nu'^{(1)} L^{\mathrm{m}}}{Q_\nu} = \frac{1}{Q_\nu}\sqrt{\frac{L^{\mathrm{m}}}{C_\nu^{\mathrm{m}}}}. \tag{18.44}$$

Furthermore, from (18.42), it can be seen that

$$\omega_\nu'^{(1)2} \equiv \omega_{s\nu}^2 = \frac{1}{L^{\mathrm{m}} C_\nu^{\mathrm{m}}}, \tag{18.45}$$

Similarly, it can be shown that

$$\omega_\nu'^{(2)2} \equiv \omega_{\mathrm{p}\nu}^2, \tag{18.46}$$

i.e., the real parts of the characteristic frequencies $f^{(1)}$ and $f^{(2)}$ are nothing but the associated serial resonance frequencies f_{s} and parallel resonance frequencies f_{p}, respectively.

For the determination of material property coefficients, piezoelectrically excited resonators are measured in the vicinity of isolated resonances, usually near the fundamental resonance. In this case, it is convenient to apply the traditional Butterworth-von Dyke equivalent circuit, a lumped-parameter equivalent circuit in the simplest form shown in Fig. 18.6. The electrical circuit elements C_1, L_1, and R_1 are obtained from the mechanical elements C_1^{m}, L^{m}, $R_1{}^{\mathrm{m}}$ (Fig. 18.5) by the transformation via the piezoelectric transformer in Fig. 18.5:

$$C_1 = N^2 C_1^{\mathrm{m}} = \frac{8}{\pi^2}\frac{k_{31}^2}{1-k_{31}^2} C_0, \tag{18.47}$$

$$L_1 = \frac{1}{N^2} L^{\mathrm{m}}, \tag{18.48}$$

$$R_1 = \frac{1}{N^2} R_1^{\mathrm{m}} = \frac{\omega_{\mathrm{s}1} L_1}{Q_1} = \frac{1}{Q_1}\sqrt{\frac{L_1}{C_1}} = R_{\mathrm{s}}. \tag{18.49}$$

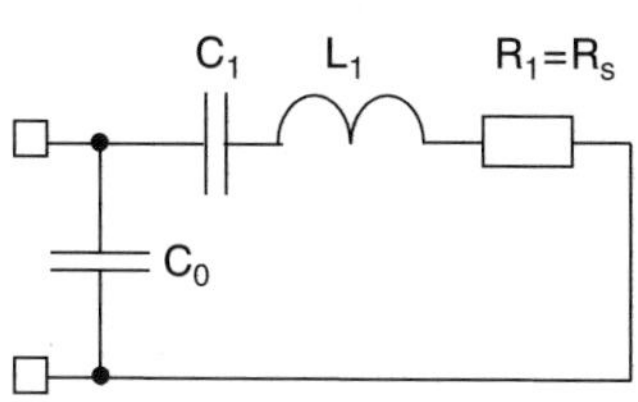

Fig. 18.6. Simplest form of an equivalent lumped-element circuit for a piezoelectrically excited bar resonator in the vicinity of the fundamental resonance. The electrical circuit elements C_1, L_1, and R_1 are the piezoelectrically transformed mechanical elements C_1^{m}, L^{m}, and $R_1{}^{\mathrm{m}}$. The electrical capacitor C_0 is the clamped capacitance of the resonator

In terms of the circuit elements of Fig. 18.6 the resonance frequencies read:

$$\omega_{\mathrm{s1}}^2 = \omega_{\mathrm{s}}^2 = \frac{1}{L_1 C_1} = \frac{1}{L^{\mathrm{m}} C_1^{\mathrm{m}}} = \omega_1^{\prime(1)2} = \frac{1}{\rho s_{11}^{\mathrm{E}}} \frac{\pi}{l}, \tag{18.50}$$

$$\omega_{\mathrm{p1}}^2 = \omega_{\mathrm{p}}^2 = \frac{1}{L_1}\left(\frac{1}{C_0} + \frac{1}{C_1}\right) = \omega_1^{\prime(2)2} = \omega_{\mathrm{s}}^2 \left(1 + \frac{8}{\pi^2} \frac{k_{31}^2}{1 - k_{31}^2}\right). \tag{18.51}$$ [12]

For the derivation of the lumped element equivalent circuits, we have assumed that only s_{11}^{E} in the tangent of (18.30) is complex. This approximation is, at least for piezoelectric ceramics, quite reasonable because many of these ceramics have dielectric, elastic, and piezoelectric losses in the range of 1–2% and, therefore, the imaginary part of the most left term in (18.34) cancels more or less leaving k_{31} real.

As can be seen from (18.50) and (18.51), the material coefficients ${s_{11}}^{\mathrm{E}}$ and k_{31} can be determined by measuring the serial and parallel resonance frequencies of a slim bar resonator. Measuring the bar capacitance at a frequency considerably lower than the resonance frequencies allows in addition the determination of ${\varepsilon_{33}}^{\mathrm{T}}$ and d_{31}. Similarly, it is possible to determine the material coefficients ${s_{33}}^{\mathrm{D}}$ and k_{33} from the serial and parallel resonance frequencies of a slim bar resonator with the electric field applied parallel to the length of the bar resonator. Together with ${\varepsilon_{33}}^{\mathrm{T}}$ the coefficients ${s_{33}}^{\mathrm{E}}$ and d_{33} can then also be determined.

18.5 Resonator Shapes Useful for Resonance Measurements

The length-extensional modes of bars have particular significance for the determination of material constants because these are the only simple modes of vibration for which the material can be an arbitrarily oriented crystal. However, the number of elastic compliances determinable from the measurements on bars decreases with increasing symmetry and measurements solely on extensional modes of bars are in no case sufficient to determine all the elastic compliance coefficients. Quite similar is the problem of determination of piezoelectric coefficients. Therefore, in addition to the length extensional modes of slim bars, radial modes and thickness-extensional modes of large thin disks, and thickness-extensional modes and thickness-shear modes of large thin quadratic plates are used. A detailed compilation of all the required formulas for these resonator types can be found in [6].

---

[12] This equation is equal to (18.34) within the approximation $\left(1 + \frac{4}{\pi^2}\frac{k_{31}^2}{1-k_{31}^2}\right)^2 = \left(1 + \frac{8}{\pi^2}\frac{k_{31}^2}{1-k_{31}^2}\right)$. This is excellently fulfilled even for the highest observed k_{31} values of $k_{31} \leqslant 0.35$.

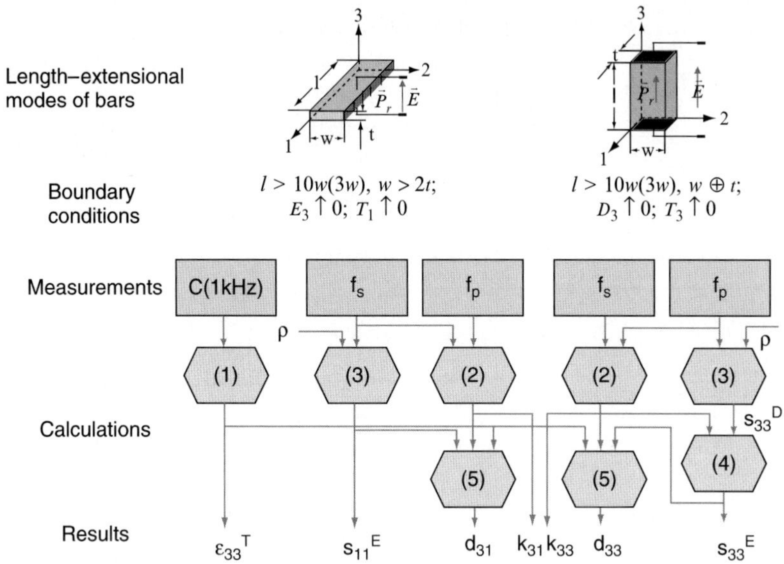

Fig. 18.7. Determination of elasticity and piezoelectric coefficients using length extensional modes of bar resonators in the vicinity of the fundamental resonances. $\varepsilon_{33}{}^{\mathrm{T}}$ is obtained from measuring the resonator's capacitance at 1 kHz. The *numbers in brackets* denote the applied formulas shown in Fig. 18.8

We show here, as an example, a detailed scheme for the determination of all the ten matrix coefficients (18.12) of poled ceramics[13] (class (∞ m)) identical to those of crystals belonging to the hexagonal class (6 mm). Figure 18.7 shows measurements and calculations required to obtain the material coefficients indicated as results (coefficients output) in the case of length extensional resonators. The formulas required are shown in Fig. 18.8. Some output coefficients of Fig. 18.7, namely $s_{11}{}^{\mathrm{E}}$, k_{31}, $s_{33}{}^{\mathrm{E}}$, and k_{33} are required ('Figs. 18.9 and 18.10) as input for the determination of the coefficients $s_{12}{}^{\mathrm{E}}$ and $s_{13}{}^{\mathrm{E}}$. These coefficients can be obtained either by measuring the resonance frequencies of the radial mode and the thickness-extensional mode of a disk resonator (Fig. 18.9) or by measuring the resonance frequencies of the thickness-extensional mode of a thin rectangular plate (Fig. 18.10).

For the determination of $s_{12}{}^{\mathrm{E}}$ according to Fig. 18.9, it is useful to calculate first k_{p} from the radial resonance frequencies using (2, Fig. 18.8) and σ^{E} and $s_{12}{}^{\mathrm{E}}$ using (6, Fig. 18.8). Then this σ^{E} can be used in (3, Fig. 18.8) to check whether both the output values of $s_{12}{}^{\mathrm{E}}$ in Fig. 18.9 are sufficiently equal. Admittedly, in the case of very high k_{p} values (>0.6) it might be necessary to calculate k_{p} not with the aid of Eq. (2, Fig. 18.8) but, in order to obtain a better accuracy, with (5) or Fig. 18.2 from [8]. For the determination of $s_{13}{}^{\mathrm{E}}$ according to Fig. 18.9, there exist also two possibilities: using solely the

[13] The 3-axis is usually chosen to be the polar axis (parallel to the poling direction).

Sample/mode

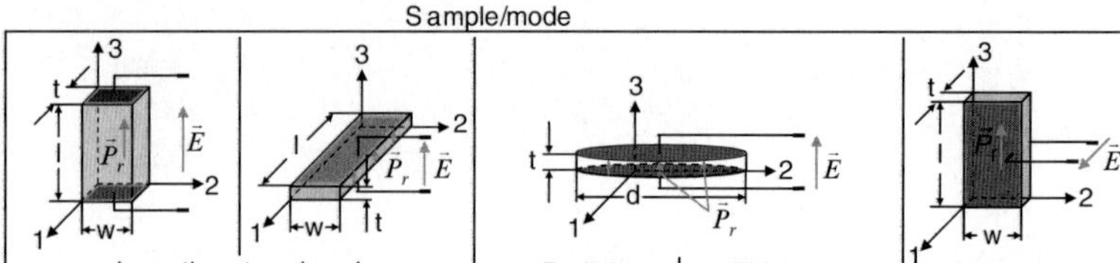

	Length-extensional	Length-extensional	Radial	Thickness	Thickness shear
Eq. (1)		$C=\varepsilon_{33}^{T}\cdot\frac{w\cdot l}{t}$	$C=\varepsilon_{33}^{T}\cdot\frac{\pi\cdot d^2}{4t}$		$C=\varepsilon_{11}^{T}\cdot\frac{w\cdot l}{t}$
Eq. (2)	$k_{33}^2 = \frac{\pi}{2}\frac{f_s}{f_p}\tan\left(\frac{\pi}{2}\frac{\Delta f}{f_p}\right)$	$k_{31}^2/(1-k_{31}^2) = \frac{\pi}{2}\frac{f_p}{f_s}\tan\left(\frac{\pi}{2}\frac{\Delta f}{f_p}\right)$	$k_p^2 \cong \frac{5}{2}\frac{\Delta f}{f_p}-\left(\frac{\Delta f}{f_p}\right)^2$	$k_t^2 = \frac{\pi}{2}\frac{f_s}{f_p}\tan\frac{\pi}{2}\frac{\Delta f}{f_p}$	$k_{15}^2 = \frac{\pi}{2}\frac{f_s}{f_p}\tan\left(\frac{\pi}{2}\frac{\Delta f}{f_p}\right)$
Eq. (3)	$\frac{1}{s_{33}^D}=4\rho\left(f_p\cdot l\right)^2$	$\frac{1}{s_{11}^E}=4\rho\left(f_s\cdot l\right)^2$	$s_{11}^E\left(1-\sigma^{E^2}\right)=\frac{\eta_1^2}{\pi^2\rho\left(f_s\cdot d\right)^2}$	$c_{33}^D=4\rho\left(f_p\cdot t\right)^2$	$c_{44}^D=\frac{1}{s_{44}^E}=4\rho\cdot\left(f_p\cdot t\right)^2$
Eq. (4)	$s_{33}^E=\frac{s_{33}^D}{\left(1-k_{33}^2\right)}$	$s_{11}^D=s_{11}^E\left(1-k_{31}^2\right)$		$c_{33}^E=c_{33}^D\left(1-k_t^2\right)$	$c_{44}^E=1/s_{44}^E=c_{44}^D\left(1-k_{15}^2\right)$
Eq. (5)	$k_{33}^2=\frac{d_{33}^2}{s_{33}^E\cdot\varepsilon_{33}^T}$	$k_{31}^2=\frac{d_{31}^2}{s_{11}^E\cdot\varepsilon_{33}^T}$	$k_p^2=\frac{2d_{31}^2}{\left(s_{11}^E+s_{12}^E\right)\varepsilon_{33}^T}$	$k_t^2=\frac{e_{33}^2}{c_{33}^D\cdot\varepsilon_{33}^S}$	$k_{15}^2=\frac{d_{15}^2}{s_{44}^E\cdot\varepsilon_{11}^T}$

Eq. (6)	$k_p^2=\frac{2}{1-\sigma^E}\cdot k_{31}^2$, $\sigma^E=\left\lvert\frac{s_{12}^E}{s_{11}^E}\right\rvert$
Eq. (7)	$s_{13}^{E^2}=\frac{1}{2}\left(s_{11}^E+s_{12}^E\right)\left(s_{33}^E-\frac{1}{c_{33}^E}\right)$
Eq. (8)	$k_t^2=\frac{\left(k_{33}-a\cdot k_p\right)^2}{\left(1-a^2\right)\left(1-k_p^2\right)}$, $a^2=2\cdot s_{13}^{E^2}/s_{33}^E\left(s_{11}^E+s_{12}^E\right)$

Fig. 18.8. Formulas required for the determination of material coefficients. η_1 is the lowest positive root of $(1+\sigma^{\mathrm{E}})J_1(\eta)=\eta J_0(\eta)$, J_0 and J_1 are Bessel functions of first kind and of zero and first order, respectively; $\eta_1=2.05$ for $\sigma^{\mathrm{E}}=0.31$

compliance and stiffness coefficients to calculate $s_{13}{}^{\mathrm{E}}$ from Eq. (7, Fig. 18.8), or measuring the thickness coupling factor $k_{\mathrm{t}}\equiv k_{33\mathrm{t}}$ and calculating $s_{13}{}^{\mathrm{E}}$ from Eq. (8, Fig. 18.8). Again it should be checked that both the output values of $s_{13}{}^{\mathrm{E}}$ in Fig. 18.9 are sufficiently equal. It should be mentioned that in case of piezoelectric ceramics, a fully electroded and poled ceramic disk is a very attractive test specimen, because it is easily prepared and shows a strongly excited fundamental radial resonance mode, free from interfering modes. However, one must make sure that the ceramic disk's microstructure and poling state are identical to those of all other resonator specimens used. In case of single crystals, it is not convenient to prepare disk specimens. In this case, one can use a thin rectangular plate to determine the coefficients $s_{12}{}^{\mathrm{E}}$ and $s_{13}{}^{\mathrm{E}}$ from the thickness extensional resonance frequencies, as shown in Fig. 18.10. To calculate the coefficients $s_{12}{}^{\mathrm{E}}$ and $s_{13}{}^{\mathrm{E}}$ from (6–8, Fig. 18.8), k_{p} in (8, Fig. 18.8) is substituted using (6, Fig. 18.8). Then the coefficients $s_{12}{}^{\mathrm{E}}$ and $s_{13}{}^{\mathrm{E}}$ can be calculated from ((7) and (8), Fig. 18.8).

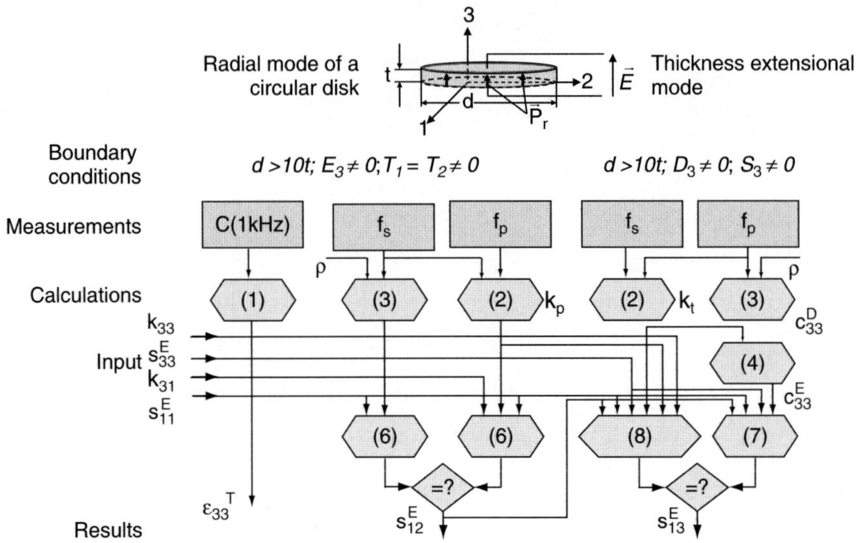

Fig. 18.9. Determination of elasticity coefficients $s_{12}{}^{\mathrm{E}}$ and $s_{13}{}^{\mathrm{E}}$ using radial modes and thickness extensional modes of a circular disk resonator. Again $\varepsilon_{33}{}^{\mathrm{T}}$ can be obtained from measuring the resonator's capacitance at 1 kHz. The *numbers in brackets* denote the applied formulas shown in Fig. 18.8

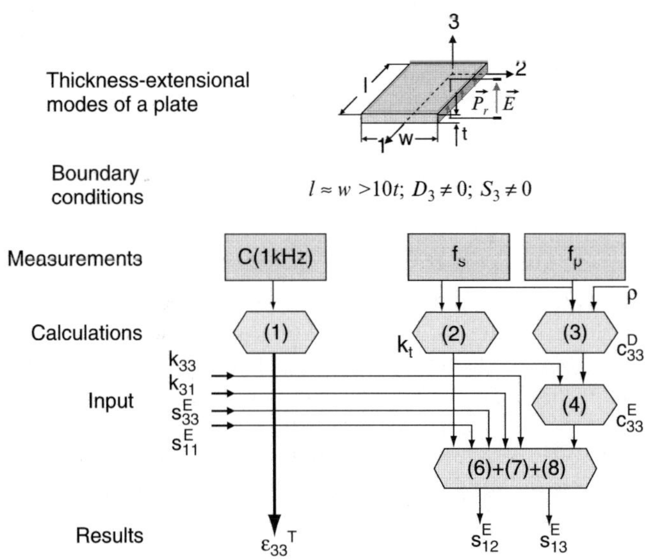

Fig. 18.10. Determination of elasticity coefficients $s_{12}{}^{\mathrm{E}}$ and $s_{13}{}^{\mathrm{E}}$ using the thickness extensional mode of a plate resonator. Again $\varepsilon_{33}{}^{\mathrm{T}}$ can be obtained from measuring the resonator's capacitance at 1 kHz. The *numbers in brackets* denote the applied formulas shown in Fig. 18.8

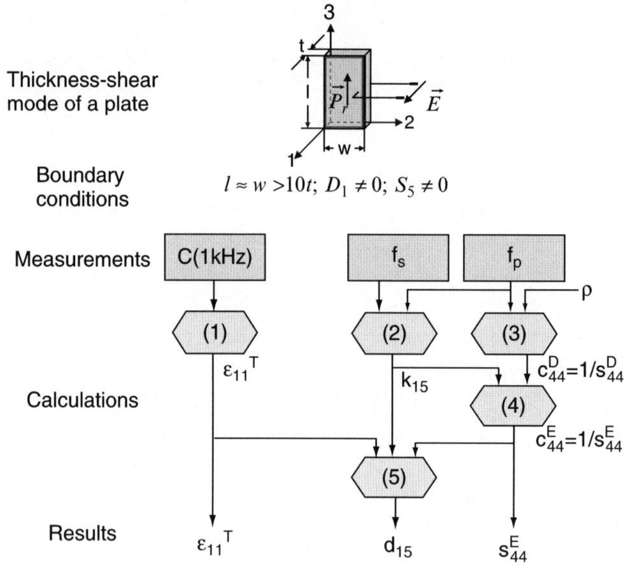

Fig. 18.11. Determination of shear coefficients $s_{44}{}^E$ and d_{15} using the thickness shear mode of a plate resonator. $\varepsilon_{11}{}^T$ can be obtained from measuring the resonator's capacitance at 1 kHz. The *numbers in brackets* denote the applied formulas shown in Fig. 18.8

The determination of the remaining coefficients permittivity $\varepsilon_{11}{}^T$, $s_{44}{}^E$, and d_{15} requires test samples with electrode configurations to allow the application a measuring field vertical to the *c*-axis or the polar axis, in case of piezoelectric single crystals, or vertical to the remnant polarization or poling direction, in case of ceramics. In the last case, it is important to make sure that by removing the electrodes used for poling and by depositing new electrodes, the poling state will not be changed. Figure 18.11 shows the required measurement and calculation steps for the determination of the coefficients.

To determine the material coefficients of crystals with lower symmetry, additional test specimens are required. For example, in the important case of crystals in the class (3 m) such as $LiNbO_3$ and $LiTaO_3$, one needs to determine six independent elastic, four piezoelectric, and two dielectric coefficients.

As mentioned, the length-extensional mode of slim bar resonators with the electric field applied vertical to the large faces of the bar resonator has two decisive advantages:

- The measurements are performed around the lowest resonance frequency. Thus, no overtones can disturb the results.
- The influence of parasitic inductivities and capacities of the measuring circuit is small.

In spite of these disadvantages, slim bar resonators with electrodes applied to its small faces and thin plate resonators cannot be avoided. If large crystals are available, one can select test specimens according to these advantages. All coefficients can be determined using slim bars with orientations[14] (ZX) or (ZY), (YX), $(YZw) \pm 45°$ and electrodes applied to their large faces, one (XZ) oriented bar with electrodes applied to its small faces, one (ZX) oriented thin plate with electrodes applied to the large faces, one (ZX) oriented thin rectangular plate with electrodes applied to the Y faces, and one (YZ) oriented plate with electrodes applied to the X faces [9]. Because none of these specimens is useful to determine $\varepsilon_{11}{}^{\mathrm{T}}$, a (XZ) or (YZ) oriented thin plate with electrodes applied to its large faces is additionally required [9].

In spite of the advantages of length-extensional resonators, there are situations where only small samples which are in addition often irregular in shape can be supplied. The ability to use such samples is of great importance for the development and evaluation of new materials. In this case the other way round may be useful for an adequate determination of material coefficients, i.e., not length-extensional but thickness resonators are used [10]. In principle it is possible to determine all the material coefficients in class (3 m) by use of thickness modes. However, since the thickness mode frequencies are not very sensitive to certain constants, in particular e_{31} and $c_{13}{}^{\mathrm{E}}$, it was found necessary to use at least one length-extensional resonator [10].

18.6 Measuring Circuits

The procedures used to determine the piezoelectric material coefficients with small signal resonance methods have already been described in the Standards on Piezoelectric Crystals, 1957 [11]. At that time the transmission method based on a scalar network analyzer was used for the determination of the resonance frequencies f_{m} at minimum impedance and f_{n} at maximum impedance as shown in Fig. 18.12. The resonator figure of merit

$$M = Q\frac{k_{\mathrm{eff}}^2}{1 - k_{\mathrm{eff}}^2} \tag{18.52}$$

has been defined to have a measure of the accuracy to what $\Delta \mathrm{f} = f_{\mathrm{p}} - f_{\mathrm{s}}$ can be approximated by $f_{\mathrm{n}} - f_{\mathrm{m}}$:

$$\Delta f = f_{\mathrm{p}} - f_{\mathrm{s}} \approx \frac{f_{\mathrm{n}} - f_{\mathrm{m}}}{\sqrt{1 + 4/M^2}}. \tag{18.53}$$

The effective coupling factor in (18.52)

---

[14] The notation for designating the orientation of crystalline bars and plates can be found in [6].

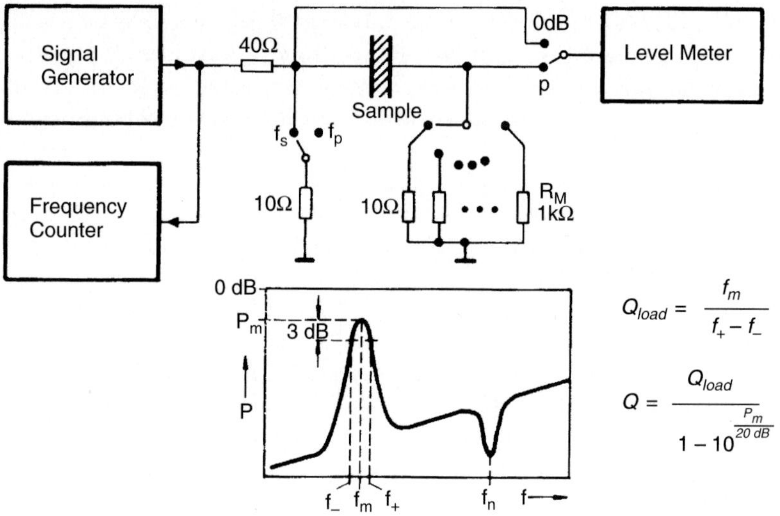

Fig. 18.12. Measuring equipment for the application of scalar small signal resonance methods

$$k_{\mathrm{eff}}^2 = \frac{f_\mathrm{p}^2 - f_\mathrm{s}^2}{f_\mathrm{p}^2} = \frac{C_1}{C_0 + C_1} \tag{18.54}$$

is defined solely via the resonance frequencies f_s and f_p of a resonator (or transducer) independently of its shape and its relation to the equivalent circuit elements C_0 and C_1 follows from (18.50) and (18.51).

The resonator quality factor Q can be determined with the measuring circuit of Fig. 18.12 either by measuring the impedance of the resonator $|Z_\mathrm{m}| \approx R_\mathrm{s}$ at $f_\mathrm{m} \approx f_\mathrm{s}$ or by determining the 3 dB frequencies f_- and f_+ (see Fig. 18.12). $|Z_\mathrm{m}|$ is usually measured by replacing the test resonator (at $f = f_\mathrm{m}$) with a resistor decade and adjusting it to the same level meter reading as obtained with the resonator. Q can then be calculated via (18.49). If the 3 dB frequencies f_- and f_+ are determined to find Q, one must be aware of the influence of the insertion level P_m. If P_m is not sufficiently small ($\ll$−20 dB) the directly estimated Q_load must be corrected to obtain Q (see Fig. 18.12).

Today, usually modern computer controlled vector analyzer systems are used to determine the piezoelectric material coefficients with small signal resonance methods. Here we have to distinguish between measuring systems capable of measuring impedances directly and systems which can only measure voltage levels, i.e., the level difference $P_{a-b} = 20\log(U_\mathrm{B}/U_\mathrm{A})$ and the phase difference $\varphi = \varphi_\mathrm{B} - \varphi_\mathrm{A}$ between two input signals A and B (Fig. 18.13). These differences directly allow the calculation of the complex impedance (or admittance) of the test resonator as shown in Fig. 18.13. Due to the fact that the material coefficients are dependent on the applied signal amplitude, particularly in ferroelectric materials, it is important to keep the voltage across the

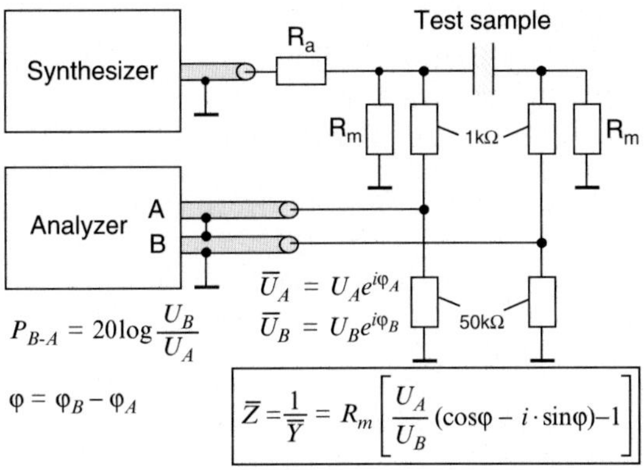

Fig. 18.13. Small signal measuring circuit using a vector level meter. R_m and R_a should be chosen to fulfil the conditions: $R_\mathrm{m} \ll R_\mathrm{s}$ (test sample's impedance at the series resonance) and $R_\mathrm{m} + R_\mathrm{a} = 50\ \Omega$ (characteristic wave guide impedance)

sample constant when the frequency is varied. Therefore, the circuitry used to excite the sample must have voltage source characteristics. This means the output impedance should be zero, or at least low with respect to the test sample's impedance at the series resonance R_s, i.e., $R_\mathrm{m} \ll R_\mathrm{s}$. Depending on the sample geometry, $R_\mathrm{m} < 1\ \Omega$ could be required. In critical cases, it can happen that the voltages to be measured by the level meter are so small that problems with noise occur. Then it is possible to increase R_m ($R_\mathrm{m} \approx R_\mathrm{s}$) and to control the voltage across the sample throughout the whole measurement by means of software tools.

A decisive advantage of computer controlled vector analyzer systems is their ability to perform piezoelectric measurements automatically with high precision and usually also with high speed. These features make small signal resonance methods very attractive for testing piezoelectric materials, transducers and even devices in development and production. Thereby, the effective piezoelectric performance of transducers is usually measured via the effective coupling factor as defined in (18.54).

Finally it should be emphasized that with modern computer controlled vector analyzer systems one is by far not limited to determine the material coefficients solely via the resonance frequencies. One measures the whole admittance or impedance cycle of a test sample and adjusts the related permittivity, elasticity, and piezoelectric coefficients so that the calculated impedance cycle fits exactly to the measured one. In other words one can use much more frequencies along the admittance or impedance cycle to determine the material coefficients with much higher accuracy. In this way, it is also possible not only to determine the real parts of the material coefficients but also all their imaginary parts with high precision [12].

References

1. W.P. Mason, H. Jaffe, Proc. IRE **42**, 921 (1954)
2. IRE Standards on Piezoelectric Crystals, Proc. IRE **37**, 1378 (1949)
3. IRE Standards on Piezoelectric Crystals, Proc. IRE **46**, 764 (1958)
4. G. Heckmann, Ergeb. Exakt. Naturwiss **4**, 100 (1925)
5. J.F. Nye, *Physical Properties of Crystals* (Clarendon Press, Oxford, 1964)
6. IEEE Standard on Piezoelectricity, ANSI/IEEE Std. **176** (1987)
7. A. Hurwitz, R. Courant, *Allgemeine Funktionentheorie und elliptische Funktionen* (Springer, Berlin, 1964)
8. IRE Standards on Piezoelectric Crystals, Proc. IRE **49**, 1169 (1961)
9. W.E. Köstler, Diploma Thesis, Technical University Munich, 1989
10. A.W. Warner, M. Onoe, G.A. Coquin, J. Acoust. Soc. Am. **42**, 1223 (1967)
11. IRE Standards on Piezoelectric Crystals, Proc. IRE **45**, 353 (1957)
12. J.G. Smits, IEEE Trans. Sonics Ultrason. **SU23**, 393 (1976)

19

Large Signal Resonance and Laser Dilatometer Methods

A. Wolff

19.1 Introduction

In many piezoelectric devices maximum deformations are desirable. They can only be achieved under high field strengths (1–2 kV mm^{-1}) and cause large mechanical stress amplitudes in the piezoelectric material. This calls for piezoelectric materials with a strong piezoelectric effect that are capable of withstanding both high electric field strengths and large mechanical stresses. Under these large signal conditions, the properties of piezoelectric materials are considerably non-linear and show hysteretic behaviour. As discussed in Chap. 3.3, the non-linear behaviour can be attributed to the motion of non 180° domain walls. Therefore, the non-linearity of the piezoelectric response and the accompanying hysteresis can be described with the aid of Rayleigh's law (Equations (3.48) and (3.49) in Chap. 3.3). The most important aspect of the Rayleigh-like behaviour is the fact that hysteresis and non-linearity in soft piezoelectric ceramics are essentially linked, as can be seen in Fig. 19.7, that is, both hysteresis and non-linearity are results of the same-domain wall pinning processes. Since the hysteretic response is closely related to the electrical and mechanical losses occurring in piezoelectric ceramics or ferroelectric perovskites, in general, the performance of power ceramics can be tested substantially by measuring either the dielectric losses at large field amplitudes or the mechanical losses at large mechanical stress amplitudes. However, to perform these measurements, it is not sufficient to extend simply the standard small-signal capacitance/loss tangent techniques according to IRE or IEEE standards [1] or the impedance measurement techniques as described in Chap. 18 to higher-signal amplitudes. For then the measurement results would be considerably falsified by an overheating of the test samples. With this in mind, pulsed test signals are mostly used [2].

Out of several possible large signal measurement methods, two are described in detail in the following. They have been selected because of their usefulness in the case of soft piezoelectric ceramics.

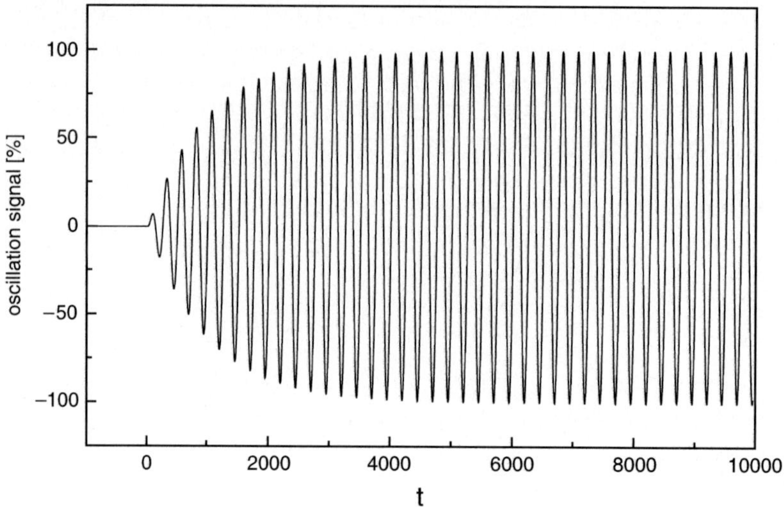

Fig. 19.1. Transient behaviour of the induced resonance oscillations in a test sample with a Q factor of 10

19.2 Large Signal Resonance Method Using Burst Pulse Excitation

Within this measuring method, in a piezoelectric sample, resonance oscillations are excited by the use of a sinusoidal burst-impulse. The time constant of the forced excitation can be measured directly through an assisting electrode or through the transient response of the load current or the resulting piezoelectric voltage (see Fig. 19.1).

The Q factor results from the number of oscillations within the time constant of the exponential increase in oscillation amplitude. The simplicity of determining the Q factor is very advantageous because, when measured with an assisting electrode, it is determined directly from an electrical voltage measurement. The loss factor results from the multiplicative inverse of the Q factor. This measurement procedure, however, is limited to the resonant frequencies of the sampling body and particularly to unclamped samples. The section below describes a measuring method that works below the resonant frequency of the test sample and enables the sample to be mechanically loaded.

19.3 Laser Dilatometer Methods

19.3.1 Force-Deflection Measurements

In this measurement set-up, force–distance measurements are performed on piezoelectric actuators under mechanical pre-loads. The deflection measurement is taken by two laser interferometers situated outside the clamp holders.

The clamp holders, the force sensor, and the mechanical load have a central opening through which the measuring beams can enter and exit. The measuring beams focused directly on the actuator surface are reflected there, allowing for a contactless, and thus, inertia-free distance measurement with a time resolution in the microsecond range and a precision of 0.1 µm. A suitable laser measuring system was chosen that is marked by its very high measuring speeds of up to 10 m s^{-1} maximum velocity of the actuator movement [3].

The main improvement of the measuring system is the introduction of a servo piezoelectric actuator – as opposed to a mechanical spring – that takes over the role of the mechanical load (see Fig. 19.2). This has a number of advantages:

- The stiffness of the load can be altered electrically, without requiring the measurement holders to be rearranged. This means that a constant static pre-stress of the actuator can be maintained throughout the measurement cycle.
- In contrast to real mechanical loads that show non-linear properties due to friction effects, etc., ideal linear loads (Hookean behaviour) can be realized with the servo piezoelectric actuator.
- The stiffness of the load can be set independently of the static pre-load.
- Realization of load stiffness settings ranging from negative to even infinite values that cannot be achieved with real passive loads.

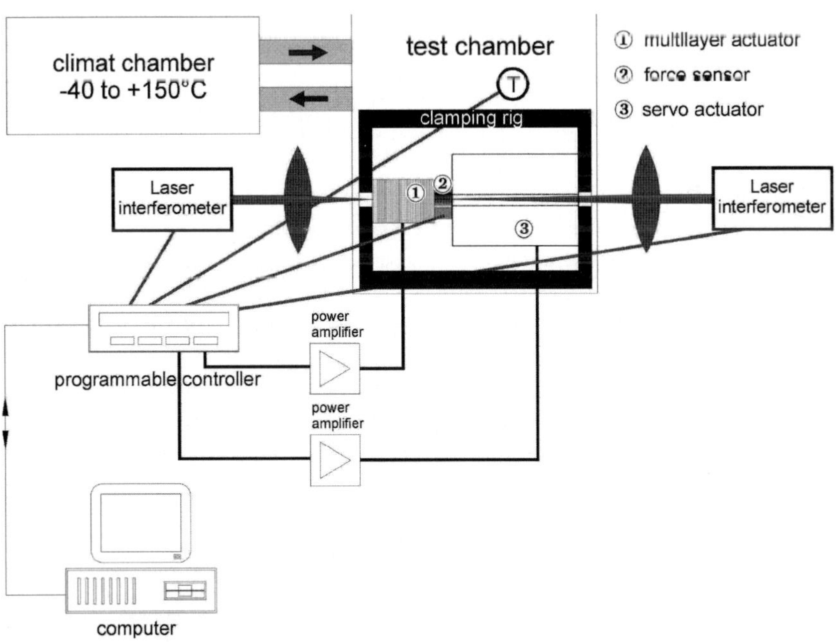

Fig. 19.2. Schematic set-up of a force-deflection measuring rig with a servo piezoelectric actuator as an active load

- Arbitrary non-linear force distance (F(x)) behaviour (mechanical limit) or processes influenced by hysteresis (friction) can be represented.
- Random load effects (change in the clamping force, change in stiffness) within a measurement cycle are possible.
- Random electrical and mechanical load combinations at frequencies up to 50 Hz can be set, and the resulting behaviour of the piezoelectric ceramic actuators can then be determined.
- The F(x) behaviour – and with it, the passive stiffness – can be determined during the clamping procedure.
- The stiffness of the load is no longer dependent on temperature due to the adjustment.
- The variation in clamping force due to various expansion coefficients can also be eliminated, when temperature cycles are run.

All of these advantages together make it possible to determine the optimum point of work, or rather, the optimal design of a piezoelectric actuator for a given application.

The programmable controller that controls the entire measuring procedure takes over the following tasks during measurement:

- Recording all measurement signals: Deflection of the piezoelectric stack (two-channel) mechanical force acting on the stack and, hence, also on the servo piezoelectric actuator voltage and current
- Generating the voltage pulse to trigger the piezoelectric actuator to be measured
- Generating the control voltage for the servo piezoelectric actuator
- PID controller for real time setting (pick-up rate of 20 kHz) of the clamping force, dependent on the measured variables of excursion and the piezoelectric stack force
- Optional temperature measurement

Figure 19.3 shows an example of a force-deflection measurement on a multilayer actuator with dimensions of $7 \times 7 \times 30\,\mathrm{mm}^3$ at 80 μm single-layer thickness. Between the five sinusoidal measuring pulses with 160 V maximum, corresponding to states 3–7, there is a time period of around 70 ms at a pulse duration of 50 ms.

For a measurement with a total period of 1 s – and assuming a nearly unloaded state at Point 1 – Points 2, 3, 2, 4, 2, 5, 2, 6, 2, 7, 2, and then again Point 1 are triggered in this sequence in a force–distance state space in one measuring cycle.

Individually, the state changes in Fig. 19.3 mean:

1–2 corresponds to accumulating a static pre-load of 1,000 N

2-x-2 with x = 3–7 Triggering the piezoelectric actuator with one of the total of five individual voltage pulses, where the servo actuator here forms load stiffness values of 0, 20, 40, 80, and 160 N $\mu\mathrm{m}^{-1}$ in sequence

2–1 Load discharge of the short-circuited piezoelectric stack from a force of 1,000 to 0 N

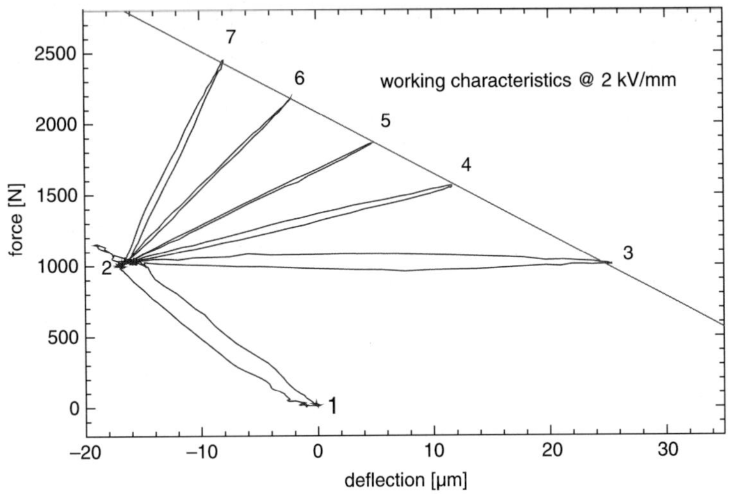

Fig. 19.3. Force-deflection diagram with load stiffness values of 0, 20, 40, 80, and 160 N μm^{-1} at a field strength of 2 kV mm^{-1} at room temperature

Points 3–7 are on the force–distance state curve for 2 kV mm^{-1} trigger field strength, which, during the rise of force regimen throughout the measurement, can be very closely described by a straight line. The slope of this line, whose absolute value depicts the actuator's stiffness, is −43 N μm^{-1}. The idle travel of 41.3 µm and the extrapolated blocking force of 1,800 N can also be seen in the force–distance diagram in Fig. 19.3.

It is worth noting the difference between the above aforementioned slope of the regression line of −43 N μm^{-1} and the slope described by a line through Points 1 and 2 of −59 N μm^{-1}. For a linear behaviour, the same stiffness values would be expected for both instances. Yet the difference can be explained by the different field strengths of 2 kV mm^{-1} in the first and 0 kV mm^{-1} in the second instance. Piezoelectric ceramics at a field strengths of 2 kV mm^{-1} show non-linear large signal properties that differ substantially from those at 0 kV mm^{-1}.

19.3.2 Determination of Losses from Force-Deflection Measurements

For this method [4], the piezo-ceramic sample is triggered quasi-statically with a uni-polar sinus pulse or a series of single sinusoidal pulses. The energy loss W_{loss} can be calculated from the measured voltage and current curves $U(t)$ and $I(t)$ (see Fig. 19.4) using the equation

$$W_{\mathrm{loss}} = \int_0^T U(t')I(t')\mathrm{d}t',$$

where T denotes the pulse length.

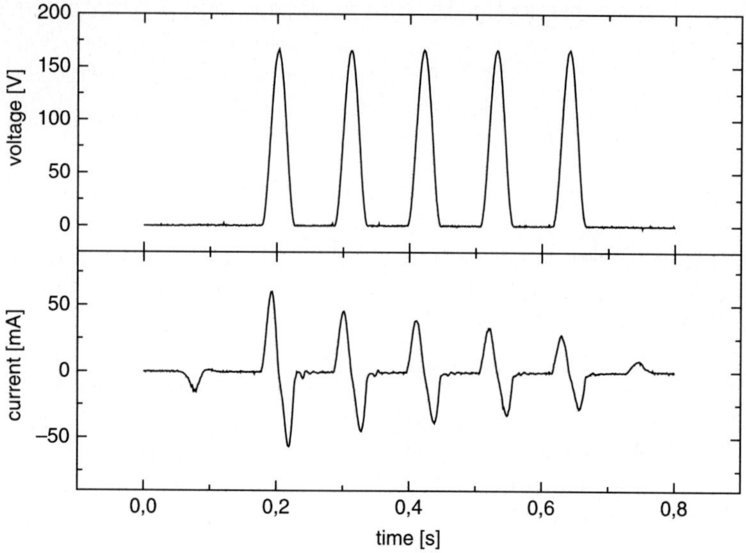

Fig. 19.4. Voltage–time and current–time diagram with load stiffness values of 0, 20, 40, 80, and 160 N μm^{-1} at a field strength of 2 kV mm^{-1} at room temperature

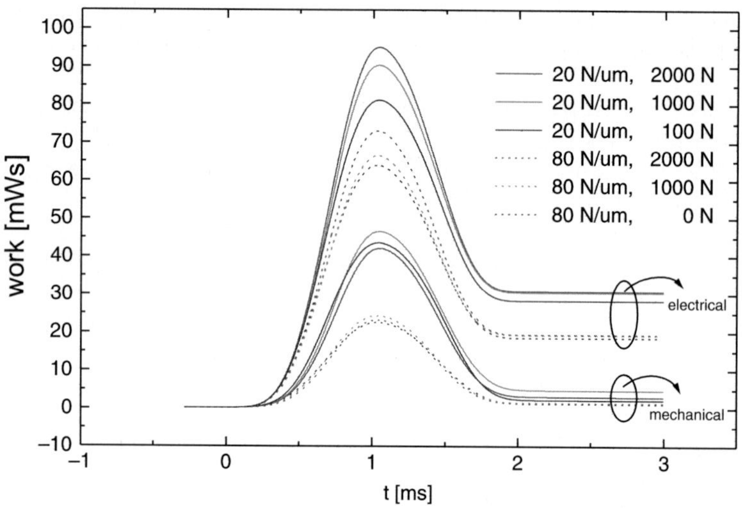

Fig. 19.5. Electrical energy–time curves for different mechanical load conditions of a piezoelectric actuator

This method of measuring energy is used as a component of the piezoelectric force-deflection measurement set-up (see Fig. 19.2).

Depending on the different pre-load forces and different load stiffnesses, different $W_{\mathrm{el}}(t)$ have been determined (see Fig. 19.5).

The electrical energy consumption is calculated using

$$W_{\mathrm{el}}(t) = \int_0^t U(t')I(t')\mathrm{d}t'$$

where t represents time.

The energy loss $W_{\mathrm{loss}} = W_{\mathrm{el}}(T)$ rises monotonously with the deflection of the piezoelectric actuator. A high load stiffness leads to a reduced deflection that is also expressed as a decreased loss energy. The influence of the clamping forces on the electrical losses, in contrast, is minor.

The advantage of this measuring method is the freely selected measurement frequency. Mechanical pre-loading is also permitted. The current measurement has to be very precise especially for small trigger frequencies because, small current offset errors lead to substantial energy measurement errors. Additional attention has to be drawn to the influence of pulse duration [3]. The transient behaviour of reversible domain processes (see Fig. 19.6) influences directly the observed deflection.

On applying only small voltages to the piezoelectric material, the intrinsic piezoelectric effect occurs (see Fig. 19.7). At high voltages, additional domain processes strongly increase the deflection. Thus, the deflection that can be obtained is composed of two physical contributions. One is due to the intrinsic piezoelectric effect and the second due to reversible domain processes.

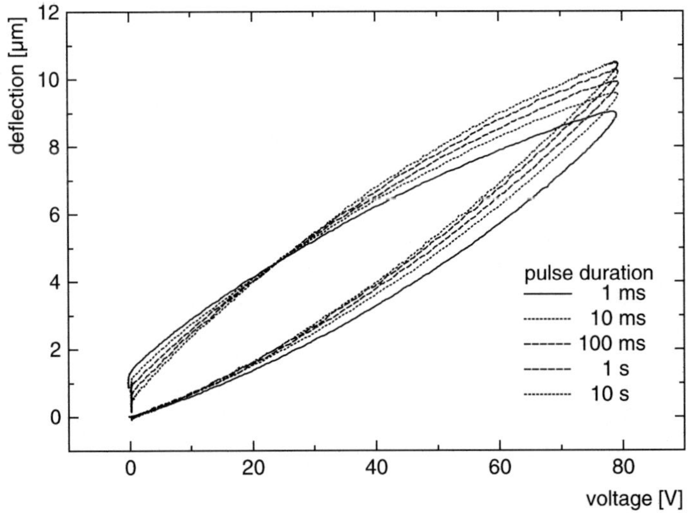

Fig. 19.6. Deflection-voltage diagram for different pulse durations (sinusoidal waveform, pulse amplitude 80 V, load condition 900 N, stiffness of load 20 N µm^{-1})

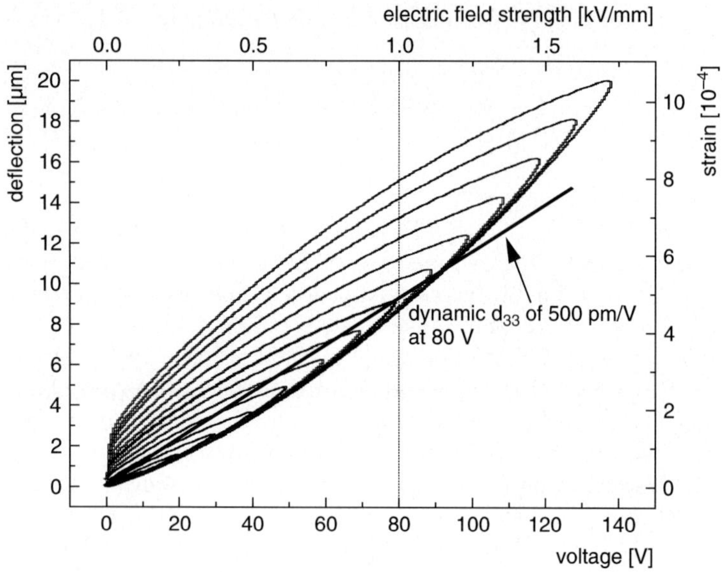

Fig. 19.7. Deflection-voltage diagram for different pulse amplitudes (sinusoidal waveform, pulse duration 1 ms, load condition 1,500 N, spring-loaded 20 N μm^{-1})

19.4 Determination of Thermal Expansion Coefficients Using a Laser Dilatometer

19.4.1 Challenges and Requirements

For quite a number of applications, the thermal expansion coefficient of the piezoelectric material plays an important role along with the piezoelectric coefficients d_{33}, d_{31}, d_{15}. Because of low thermal expansion of ceramics, high demands are placed on the precision of the distance measurements. Dilatometers with mechanical detection of the thermal expansion are the state of the art [5]. They require complex reference measurements to compensate for the thermal expansion of the measuring stamp. To overcome this problem, laser-supported dilatometer measurements are increasingly used [6] and, therefore, considered in the following.

Dilatometer measurements should characterize the thermal expansion behaviour in a temperature range from −40 to 150°C. A measurement precision of 0.1 µm is required for samples of several centimeters in length. The measurements should still be completed within a period of at most 3 h.

19.4.2 Realization and Experimental Results

The above-named requirements are best met using interference-based measuring systems [7]. In use here is a two-channel laser-Doppler measurement

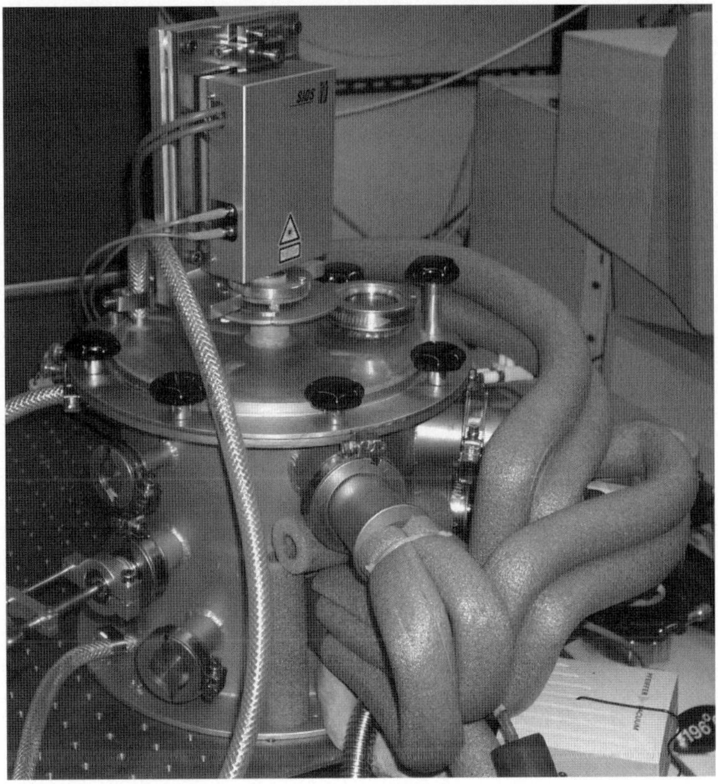

Fig. 19.8. Image of the laser-dilatometer measuring station

system in a low-drift configuration. The low drift is achieved, above all, by having two measuring beams simultaneously record the position of both sample ends. The change in length of the sample is calculated by the difference formed between the two individual distance measurements, where drift effects and other disturbances, such as air pressure variations, are nearly cancelled out. An only minimal portion of the measurement beam course goes through normal atmosphere, whereas the largest part of its course is through a vacuum. This test configuration (see Fig. 19.8) thus achieves drift velocities of less than $0.1\,\mu\mathrm{m\ h}^{-1}$. To achieve fast cooling and heating ramps, a combination of liquid nitrogen cooling with infrared heating in a vacuum oven was selected.

Figures 19.9 and 19.10 show a series of measuring cycles on the same sample with a cycle duration of 3 h each.

These thermal expansion measurements performed in direct succession show the influence of the polarization state on the sample's length. While piece of PZT ceramic with a sample length of 30 mm shrank by over 10 μm due to thermal depolarisation according to the first thermal measurement cycle (see Fig. 19.9), the remaining thermal depolarisation and the resulting remaining

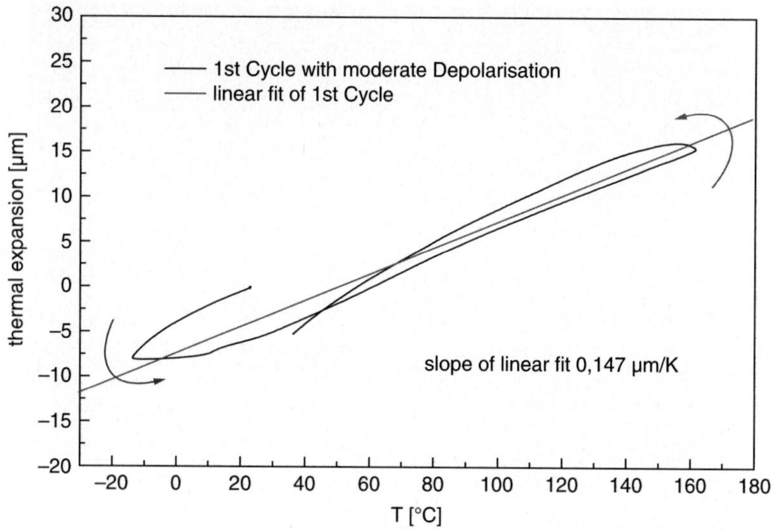

Fig. 19.9. Thermal expansion curve for a piezoelectric ceramic with thermal depolarisation (first thermal cycle)

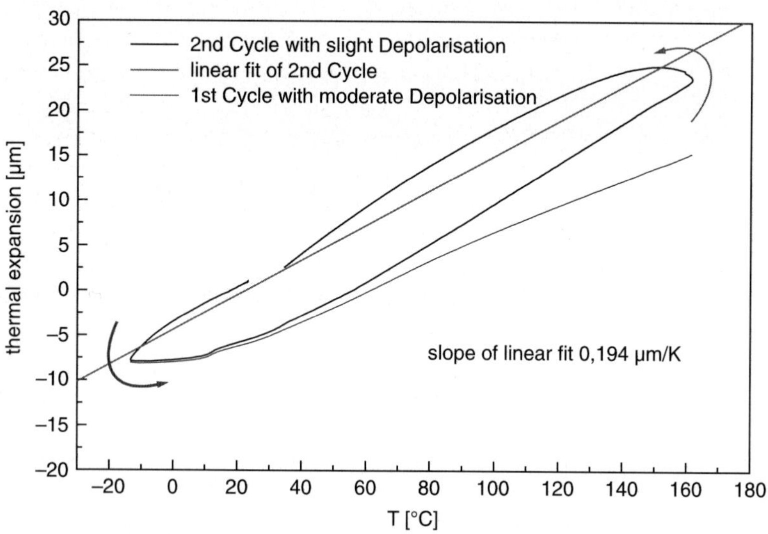

Fig. 19.10. Thermal expansion curve for a piezoelectric ceramic with slight depolarisation (second thermal cycle)

shrinkage of less than 5 µm in the second measuring cycle is considerably lower (see Fig. 19.10). Consequently, the corresponding expansion coefficients α ($4.9 \times 10^{-6}\,K^{-1}$ for a freshly poled ceramic and $6.5 \times 10^{-6}\,K^{-1}$ for a ceramic slightly depolarised by the first measurement cycle) show a clear dependence on the polarization state.

19.4.3 Accuracy of the Method

Because of the contactless measurement of length in a vacuum, unaffected measurement values that do not require a complex reference measurement on calibrated test samples are obtained . Measurement feasibility studies provide evidence for the high measurement accuracy of the system. An error in the heat expansion coefficient of $\leqslant 0.3 \times 10^{-6}\,\mathrm{K}^{-1}$ has been determined.

References

1. IEEE standard on piezoelectricity, E-ISBN: 0-7381-2411-7, (1988)
2. K. Lubitz, W. Wersing, Ferroelectrics, **40**, 237 (1982)
3. A. Wolff, D. Cramer, H. Hellebrand, I. Probst, K. Lubitz, in *Proceedings ISAF '94*, Pennsylvania State University, University Park, PA USA, 7–10 August 1994, p. 755
4. A. Wolff, D. Cramer, H. Hellebrand, C. Schuh, T. Steinkopff, K. Lubitz, in *Proceedings ISAF '96*, Rutgers University, East Brunswick, NJ, USA, 18–21 August 1996, p. 317
5. H. Walcher, *Winkel- und Wegmessung im Maschinenbau* (VDI-Verlag, Düsseldorf Germany, 1985)
6. R. Schödel, Proc. SPIE **5858**, (2005)
7. Sios Messtechnik GmbH, Anwenderdokumentation zum Miniaturinterferometer mitPlanspiegelreflektor, Ilmenau Germany, (1999)

20

Ferroelastic Characterization of Piezoelectrics

T. Fett

20.1 Young's Modulus Under Purely Elastic Deformation

The *elastic deformation* behavior of an isotropic material is governed by Young's modulus Y and Poisson's ratio σ. Knowledge of the Young's modulus, for instance, is necessary to split the total deformation S_i measured under load stress T_i into elastic S_i^{el} and "plastic" S_i^{pl} strain contributions

$$S_i = S_i^{\mathrm{el}} + S_i^{\mathrm{pl}} = \frac{T_i}{Y} + S_i^{\mathrm{pl}}, \tag{20.1}$$

where Y is not necessarily a constant, but may depend on the amount of nonelastic strains, as will be shown later. The methods for the determination of the Young's moduli, well-known from conventional ceramics and also applicable to unpoled piezoelectric materials, are much more complicated for nonisotropic materials due to the tensorial character of the modulus and the special electric boundary conditions [1]. The conditions of a constant electrical field, $E_{\mathrm{z}} \equiv E_3 = \mathrm{const}$, or a constant dielectric displacement, $D_{\mathrm{z}} \equiv D_3 = \mathrm{const}$, influence the elastic properties. This effect may be illustrated for a uniaxial stress state with $T_x \equiv T_1 \neq 0$ exclusively. In the case of a piezoelectric material, we are interested in the components

$$\begin{aligned} S_1 &= s_{11}^{\mathrm{E}} T_1 + d_{31} E_3, \\ D_3 &= d_{31} T_1 + \varepsilon_{33}^{T} E_3. \end{aligned} \tag{20.2}$$

The influence of the electrical boundary conditions on the Young's moduli[1] is mostly characterized by the interconnection or insulation of the two opposite electrodes. In the case of short-circuited electrodes, a disappearing voltage is obtained

[1] Concerning the definition of Y and σ in the case of non-isotropic materials see Chap. 18.

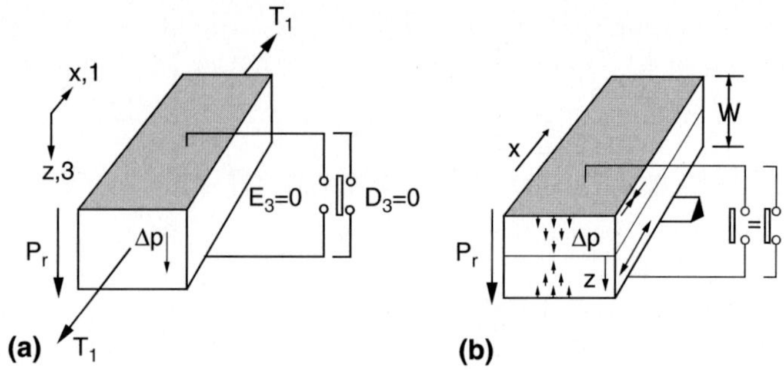

Fig. 20.1. Ceramic specimen, poled vertically to the applied stress T_1, **a** with short-circuited and insulated electrodes, **b** specimen, poled vertically to specimen's length axis, under bending load

$$U = \int_0^w E_z \, dz = 0, \tag{20.3}$$

with the specimen wideness w in z-direction. Only in the case of a homogeneous electrical field, this leads to the condition of a disappearing electrical field, that is $E_z \equiv E_3 = 0$. This case is illustrated in Fig. 20.1a for ceramic specimens poled perpendicular to the loading direction (closed switch). In this special case, insertion of $E_z \equiv E_3 = 0$ in (2) results in the x (or 1) component of strain S_x (or S_1) and electric displacement component D_z (or D_3).

$$\begin{aligned} S_1 &= s_{11} T_1, \\ D_3 &= d_{31} T_1. \end{aligned} \tag{20.4}$$

It follows from (4) that there is no piezoelectric coupling between the electrical and mechanical quantities. Consequently, the Young's modulus in x-direction is simply given by

$$Y_{11}^E = \frac{T_1}{S_1} = 1/s_{11}, \tag{20.5}$$

where superscript E denotes a constant electric field (especially $E = 0$).

Let us now consider the case of insulated electrodes. In Fig. 20.1a, this electrical condition is symbolized by an open switch, which prevents any flow of current. The real charges on the electrodes are constant. Consequently, D_z is constant, too. For $D_z = 0$, (2) yields

$$E_3 = -\frac{d_{31} T_1}{\varepsilon_{33}^T} \Rightarrow S_1 = \frac{T_1}{s_{11}^E} \frac{1}{1 - k_{31}^2}, \tag{20.6}$$

with the coupling factor

$$k_{31} = \frac{|d_{31}|}{\sqrt{s_{11}^E \varepsilon_{33}^T}}. \tag{20.7}$$

Finally, we obtain the Young's modulus under conditions of constant electric displacement as

$$Y_{11}^{\mathrm{D}} = \frac{1}{s_{11}^{E}} \frac{1}{1 - k_{31}^{2}} = Y_{11}^{E} \frac{1}{1 - k_{31}^{2}} . \tag{20.8}$$

In tests where the electrodes are not connected, we trivially obtain from the measurement Y_{11}^{D}.

In *bending tests*, the electrical boundary conditions are much more complicated. First, the bending bar is studied in flat position with the electrodes at the tensile and compression surfaces (see Fig. 20.1b). The change in polarization due to the piezoelectric effect (symbol Δ dropped in the following considerations) is

$$P_3 = d_{31} T_1 = d_{31} T_{1b}\, 2z/w \tag{20.9}$$

($T_{1\mathrm{b}}$ = outer fiber bending stress) and, consequently, antisymmetric to the center line of the bar (note that z is an odd function).

At the two electrodes, polarizations with opposite signs are generated due to the bending stress. Between the two electrodes, no electrical voltage can appear, that is, $U = 0$. Consequently, the switch position is of no importance. No electrical current is possible and the Young's modulus obtained by the measurements always is Y_{11}^{D}.

This effect is not quite trivial and rather somewhat surprising. From these considerations, it can be concluded that the results from the tests with homogeneous stress states are much more trustworthy than the results obtained under nonhomogeneous stresses, for example, in bending tests.

20.2 Young's Modulus Under Nonelastic Deformation

The compressive deformation of piezoelectric materials shows an interesting change of Young's modulus with increasing strain. From loading–unloading diagrams given in literature [2–5], a strongly increased Young's modulus is obtained for the unloading phase, where the unloading modulus is about 2–3 times the initial value Y_0 (for example, see Fig. 20.2).

Measurements of the stress–strain behavior in compression tests with superimposed partial unloading were performed in [6] for three poling states. Short-circuited electrodes were used, thus satisfying the condition of a disappearing electric field in the specimens. The Young's modulus obtained from partial unloading is represented in Fig. 20.3. In the undeformed state, the Young's moduli are ranked as $Y_{\parallel} < Y_{\perp} < Y_{\mathrm{unpoled}}$.

Compression tests were performed with two pairs of cross-wisely attached strain gauges that allowed the strain to be determined in stress direction (S_1) and normally to the stress (S_2). The Poisson's ratio during partial unloading, defined by the elastic strain response as $\sigma = -\mathrm{d}S_2/\mathrm{d}S_1$, is plotted in

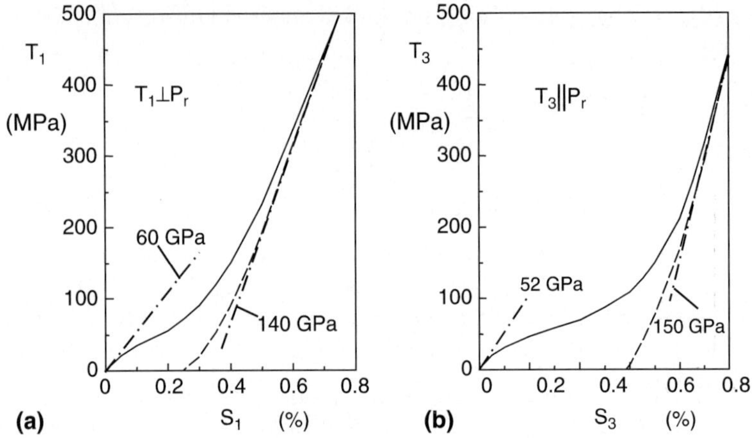

Fig. 20.2. Compression tests on soft PZT ceramics. **a** and **b** Cao and Evans [2] (poled perpendicular $\perp$ and parallel $\|$ to the specimen length axis)

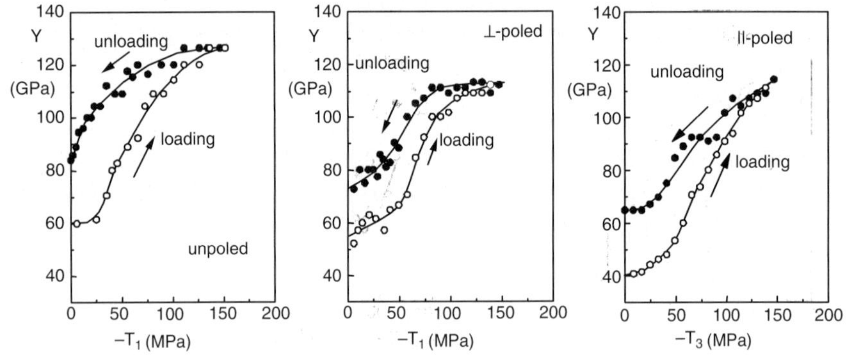

Fig. 20.3. Young's modulus determined from partial unloading tests for the three poling states during loading up to 150 MPa and during unloading

Fig. 20.4a vs. the applied stress for unpoled material. Starting from $\sigma \approx 0.39$, the Poisson's ratio decreases and approaches an asymptotic value of about $\sigma \rightarrow 0.3$. During unloading, σ is nearly the same as under increasing load and reaches $\sigma \approx 0.37$–0.38 after total unloading. The poled PZT ceramics show a more pronounced hysteresis effect. Figure 20.4b represents the results for the $\|$-poled material modification. In the case of the $\perp$-poled material, the transverse strain S_2 was measured in the poling direction (resulting in $\sigma_{\perp}$) and perpendicularly to the poling direction ($\sigma_{\|}$). In Fig. 20.4c, the average value $\sigma_{\text{aver}} = {}^1\!/_2(\sigma_{\|} + \sigma_{\perp})$ responsible for the volume change is plotted. Also this average value tends towards 0.3.

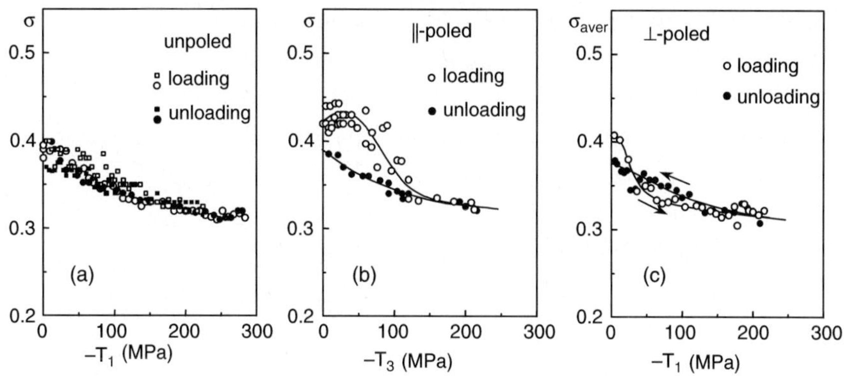

Fig. 20.4. Poisson's ratio obtained from a partial unloading test **a** unpoled PIC 151 (two specimens), **b** $||$-poled PZT, **c** $\perp$-poled PZT average $\nu_{aver} = {}^1\!/_2(\nu_{||} + \nu_{\perp})$

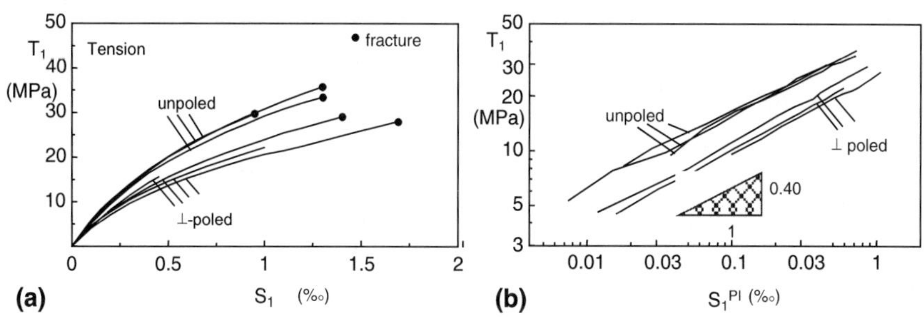

Fig. 20.5. Tensile tests performed on unpoled and poled PZT PIC 151. Fractured specimens marked by circles, **a** stress vs. total strain, **b** stress vs. plastic strain

20.3 Nonelastic Deformation

In the literature, the *nonelastic deformation* behavior of piezoelectric ceramics is investigated mainly under uniaxial loading, predominantly in compression tests. Nonlinear stress–strain curves were observed and interpreted by domain switching processes.

In tensile and compression tests on soft PZT ceramics, a *nonsymmetric deformation* behavior in tension and compression was observed. Plastic tensile strains in the unpoled material are significantly larger than in compression at the same stress level. An extremely sensitive method to assess differences in the plastic deformation behavior under tensile and compressive loading is based on strain measurements carried out in bending tests [7].

Some results of tensile tests are shown in Fig. 20.5a. In the poled specimens, the polarization was perpendicular to the applied load. Some specimens were unloaded before failure. Figure 20.5b shows the stress vs. the plastic strain $S_i^{\rm pl}$. A comparison of the results for tension and compression is given in Fig. 20.6 for

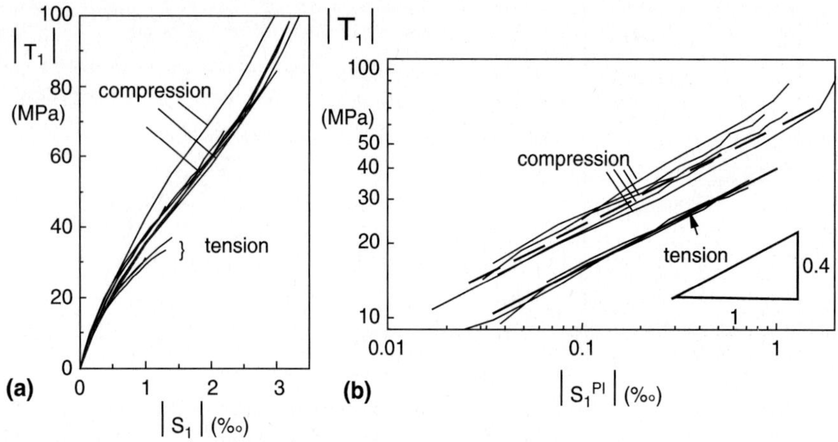

Fig. 20.6. Stress–strain curves (unpoled PZT): **a** linear plot T_1 vs. S_1; **b** represented as $\log(T_1) - \log({S_1}^{\mathrm{pl}})$ diagram

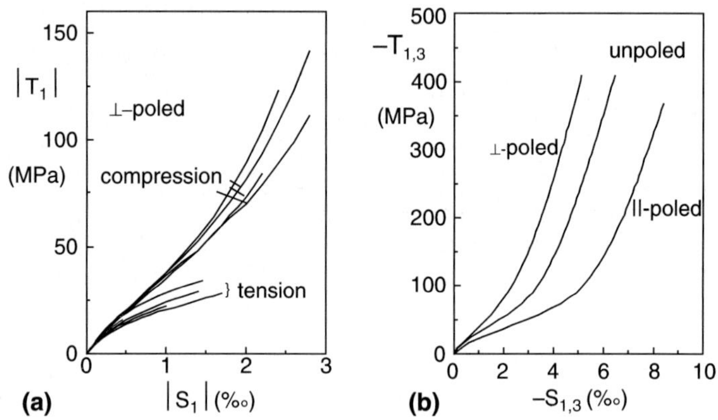

Fig. 20.7. Linear stress–strain curves **a** ⊥-poled PZT **b** compression tests on PIC 151, influence of the poling state

the unpoled and in Fig. 20.7a for the poled specimens. The tensile specimens show larger strains at the same stress level for unpoled and ⊥-poled PZT. In the case of ||-poled PZT, the compressive strains are significantly increased.

The plastic strains obtained from the data of Fig. 20.5a are plotted in Fig. 20.5b vs. the stress in a logarithmic representation. The experimental results, represented by the lines, can be expressed for $S_i^{\mathrm{pl}} < 1\%$ as

$$T_i = A^+ S_i^{\mathrm{pl}n}, \tag{20.10}$$

with $A^+ = 635\,\mathrm{MPa}$ and $n = 0.40$ for the unpoled material and with $A^+ = 545\,\mathrm{MPa}$ and $n = 0.43$ for the poled material.

Also at the compressive plastic strains, a straight linear behavior appears in the $\log(T_i) - \log(S_i^{\text{pl}})$ plot up to $S_i^{\text{pl}} \cong 1\%$ with the same slope as for the tensile curve. Therefore, the plastic strains in compression can be expressed as

$$T_i = A^- |S_i^{\text{pl}}|^n, \tag{20.11}$$

with $n = 0.4$ and $A^- = -940\,\text{MPa}$ for the unpoled and $n = 0.43$ and $A^- = -975\,\text{MPa}$ for the $\perp$-poled PZT.

Apart from the poling state, the applied electrical field will also influence the stress–strain curves. Results for PZT PIC 151 can be seen in [8]. As a singular result, the compression strain is shown in Fig. 20.8a for the unpoled material under a variable transverse electric field.

The *nonsymmetry* in the deformation behavior must have consequences on the stress distribution in nonhomogeneously loaded components, for example, in a bending bar. The information given in Fig. 20.6a for the unpoled material is sufficient to compute the stresses in a bending bar. A result is given in Fig. 20.8b for the unpoled material, where the stress distribution is shown for various elastically computed bending stresses T_1^{el}. The resulting stresses become strongly nonlinear. A significant stress reduction at the tensile surface is visible. While the stresses are reduced at the tensile surface, the compressive stresses increase at the compressive surface. This deviation from the linear-elastic behavior affects the strength [9] and toughness data obtained in bending tests.

Finally, it may be concluded that:

- PZT shows a nonsymmetric stress–strain behavior under tension and compression.

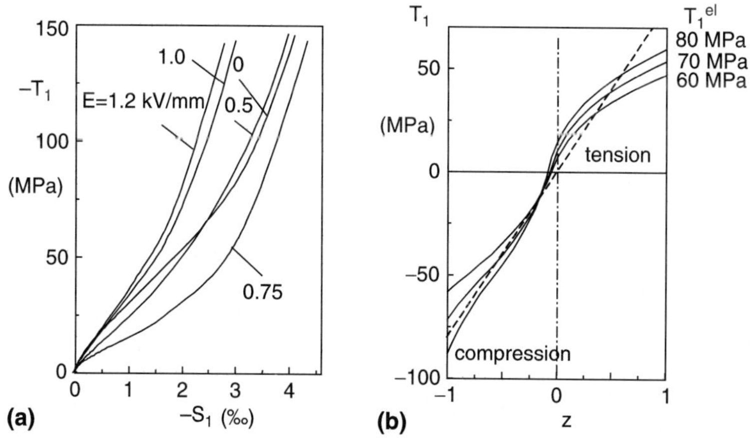

Fig. 20.8. **a** Influence of an electric field perpendicular to the stress direction in unpoled PIC 151, **b** stress state in a bending bar as a consequence of the nonsymmetry in the plastic deformation behavior (unpoled material). Linear distribution (*dashed*): elastically computed for a bending stress of 80 MPa

- Plastic strains of unpoled PZT at the same stress in tension exceed those in compression by a factor of about 2.
- PZT, poled perpendicularly to the external load, shows a stronger plastic deformation than unpoled material.

20.4 Multiaxial Deformation

Generally, measurements of the deformation behavior are made in uniaxial tests. For fracture mechanics problems in particular, where multiaxial stresses occur at crack tips, knowledge on the *multiaxial deformation behavior* is indispensable. Deformation under biaxial loading, for instance, can be studied on pressurized tubes [10] or in torsion tests carried out on thin-walled tubes [11]. Such tests allow the nonsymmetric yield condition to be determined for the onset of plastic deformations.

As an example of multiaxial loading, the stress vs. plastic strain curves measured in torsion tests (τ = shear stress) are plotted in Fig. 20.9a, together with the stress–strain behavior obtained from tensile and compression tests. Different symbols indicate different tests.

As obvious from the deformation curves presented above, plastic strains occur at lowest stresses already. To describe an "onset of plastic deformation," a limit value of the effective plastic strain S_0 of 0.01% is assumed formally as the yield condition, where

$$S_0 = \frac{\sqrt{2}}{3}\left[(S_1 - S_2)^2 + (S_1 - S_3)^2 + (S_2 - S_3)^2\right]^{1/2} . \tag{20.12}$$

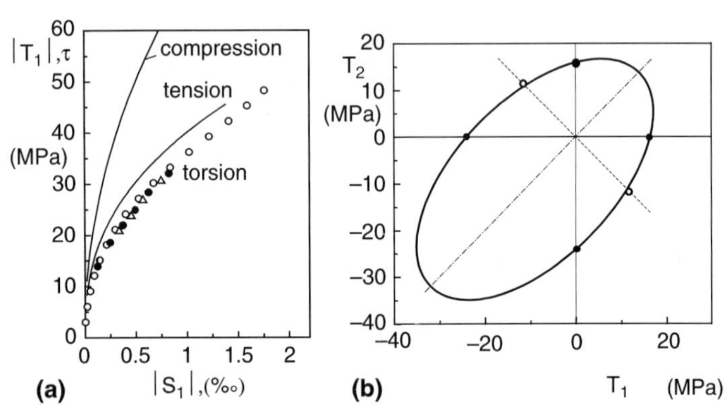

Fig. 20.9. **a** Stress vs. plastic strain curves for PIC 151 under uniaxial and biaxial loading, stresses plotted vs. S_1 (tension, torsion) and $-S_2$ for compression, **b** $T_1 - T_2$ representation of the onset of plastic deformation (defined as an effective strain $S_0 = 0.01\%$). Tensile and compression test: *filled circles*, torsion test: *open circles*, curve: (15)

For torsion tests with $S_2 = -S_1$ and $S_3 = 0$, this yields $S_0 = 2S_1/\sqrt{3}$.

This plastic strain seems to be sufficiently small when compared to the elastic strains. The related tensile and compression stresses are shown in Fig. 20.9b as filled circles. The torsion results are given as open circles.

It is a clear a priori that a Tresca or von Mises yield criterion can never describe the experimental results. Application of a Drucker–Prager yield criterion [12] seems to be much more promising. Here, the stress sign enters via the hydrostatic stress T_h. For $T_3 = 0$, it reads

$$T_\mathrm{h} = \tfrac{1}{3}(T_1 + T_2). \tag{20.13}$$

In terms of the first invariant J_1 of the stress tensor and the second invariant J'_2 of the stress deviator, the yield condition is given by

$$\alpha J_1 + \sqrt{J'_2} = \kappa, \tag{20.14}$$

with the material parameters α and κ. For plane stress conditions, $T_3 = 0$, it reads

$$\alpha(T_1 + T_2) + \frac{1}{\sqrt{3}} \left[T_1^2 + T_2^2 - T_1 T_2\right]^{1/2} = \kappa. \tag{20.15}$$

From the values of T_t for uniaxial tension and T_c for the compression test, it results

$$\alpha = \frac{1}{\sqrt{3}} \frac{|T_\mathrm{c}| - T_\mathrm{t}}{|T_\mathrm{c}| + T_\mathrm{t}}, \kappa = \frac{2}{\sqrt{3}} \frac{|T_\mathrm{c}| T_\mathrm{t}}{|T_\mathrm{c}| + T_\mathrm{t}}. \tag{20.16}$$

For $S_0 = 0.01\%$, $T_\mathrm{t} = 16.0$MPa, and $|T_\mathrm{c}| = 23.8$ MPa. From these stresses, $\alpha = 0.113$ and $\kappa = 11.05$ MPa are obtained. The yield condition, resulting in an ellipse, is plotted in Fig. 20.9b. The torsion results show only slight deviations from the ellipse.

20.5 Time-Dependent Deformation

PZT ceramics show *time-dependent deformation* effects under mechanical loading. Figure 20.10a displays the total strain as a function of time after load application in a tensile test, and Fig. 20.10b after load removal [13].

Figure 20.11a shows some tensile creep curves in a logarithmic representation. Straight lines that are characteristic of primary creep behavior are observed for a short number of times. The stress dependence is shown in Fig. 20.11b. A power law description of creep is given by

$$S_1^\mathrm{c} = B T_1^n t^m, \tag{20.17}$$

with $m = 0.27$, $n = 1.52$, and $B = 2.7 \times 10^{-4}$ (for T in MPa, t in s), where the exponent n results from the plot of the creep strain at $t = 120$ s vs. the applied stress. Similar creep effects are also visible in compression loading and under electric fields (see, e.g., [14, 15]).

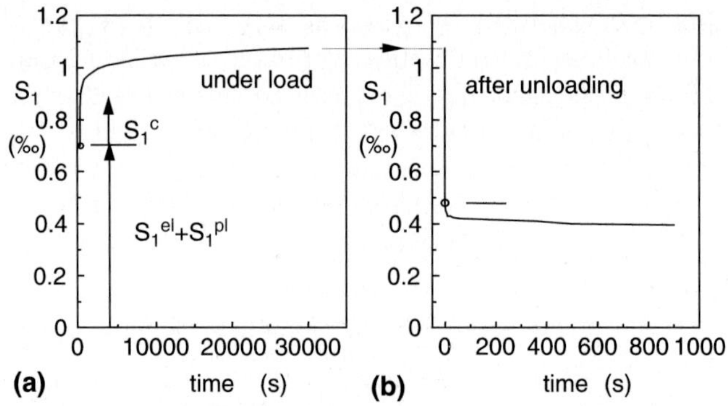

Fig. 20.10. Tensile creep curve of PZT PIC 151 at $T_1 = 25\,\mathrm{MPa}$, **a** deformations under load, **b** back deformations after unloading (unpoled material)

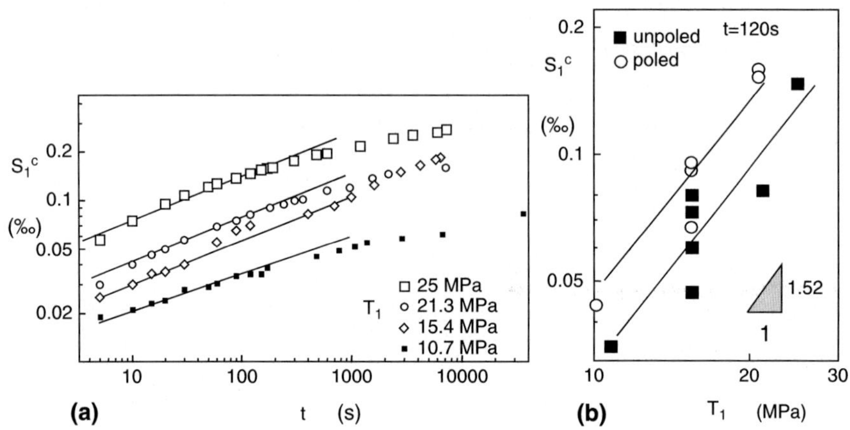

Fig. 20.11. Creep curves in tension, **a** dependence on time t, **b** dependence on stress for $t = 120\,\mathrm{s}$ [13]

While constant load is applied in the creep tests, the load depends on time in most practical applications. Under such conditions, the creep strains can be obtained from the so-called hardening rules, similar to the high-temperature creep of conventional ceramics. Two well-known rules are the time and strain hardening rules (see, e.g., [7]).

References

1. T. Fett, D. Munz, J. Test. Evaluat **28**, S.27 (2000)
2. H. Cao, A.G. Evans, J. Am. Ceram. Soc. **76**, 890 (1993)
3. A. Schäufele, *Ferroelastische Eigenschaften von Blei-Zirkonat-Titanat-Keramiken*, Fortschrittsberichte VDI, Reihe 5, Nr. 445, (VDI-Verlag Düsseldorf, 1996)

4. A. Schäufele, K.H. Härdtl, J. Amer. Ceram. Soc. **79**, 2637 (1996)
5. T. Fett, S. Müller, D. Munz, G. Thun, J. Mater. Sci. Lett. **17**, 261 (1998)
6. T. Fett, D. Munz, G. Thun, Ferroelectrics **274**, 67 (2002)
7. D. Munz, T. Fett, *CERAMICS, Failure, Material Selection, Design* (Springer-Verlag, März, 1999)
8. T. Fett, D. Munz, G. Thun, Ferroelectrics **297**, 83 (2003)
9. T. Fett, D. Munz, G. Thun, J. Mater. Sci. Lett. **18**, 1899 (1999)
10. W. Chen, C. Lynch, J. Eng. Mat. Technol. **123**, 169 (2001)
11. T. Fett, D. Munz, G. Thun, J. Am. Ceram. Soc. **86**, 1427 (2003)
12. W. Prager, *An Introduction to Plasticity* (Addison-Wesley, Amsterdam, 1959)
13. T. Fett, G. Thun, J. Mater. Sci. Lett. **17**, 1929 (1998)
14. D. Zhou, M. Kamlah, Acta Mat. **54**, 1389 (2006)
15. D. Zhou, M. Kamlah, J. Appl. Phys. **98**, 104 (2005)

Part IV

Multiscale Modelling

21

First-Principles Theories of Piezoelectric Materials

R.E. Cohen

Piezoelectrics have long been studied using parameterized models fit to experimental data, starting with the work of Devonshire in 1954 [1]. Much has been learned using such approaches, but they can also miss major phenomena if the materials properties are not well understood, as is exemplified by the realization that low-symmetry monoclinic phases are common around morphotropic phase boundaries, which was missed completed by low-order Devonshire models, and can only appear in higher-order models [2]. In the last 15 years, a new approach has developed using first-principles computations, based on fundamental physics, with no essential experimental input other than the desired chemistry (nuclear charges). First-principles theory laid the framework for a basic understanding of the origins of ferroelectric behavior [3–7] and piezoelectric properties [8–11]. The range of properties accessible to theory continues to expand as does the accuracy of the predictions. We are moving towards the ability to design materials of desired properties computationally. Here, we review some of the fundamental developments of our understanding of piezoelectric material behavior and the ability to predict a wide range of properties using theoretical methods. This is not meant as a review of the literature. Comprehensive reviews of the literature of theoretical studies of ferroelectrics are given by Resta [12] and Rabe and Ghosez [13].

Most first-principles calculations for piezoelectrics are based on the density functional theory (DFT) [14], and most are within the local density approximation (LDA) [15]. The energy for any configurations of atoms is computed by solving a set of effective Schrödinger equations with an effective potential that includes many-body contributions like those of a uniform electron gas at each point in space. Forces, phonon frequencies (via the dynamical matrix), effective charges, dielectric constants, elastic constants, piezoelectric constants, and polarization are all directly computable for the static lattice (zero temperature) for ordered structures.

To obtain properties for finite temperatures, it is necessary to use the primary first-principles results to parameterize an effective Hamiltonian or potential model, which can then be used to study the effects of temperature and simulate disordered materials.

A number of major advances in our understanding of piezoelectric materials are due to first-principles studies. First is the importance of hybridization. Ferroelectricity is due to the competition between long-range forces, which favor off-centering, and short-range forces that favor the high-symmetry centric phase [5]. Unlike the venerable Slater rattling ion model [16], the key element in oxide ferroelectrics is covalency, or hybridization between the cation and its oxygen neighbors that allows the cation to move off-center. This concept is now widely used in experiments and for development of new piezoelectric materials. Second is the concept of polarization rotation, which is responsible for the giant electromechanical coupling seen in relaxor ferroelectrics such as $Pb(Mg_{1/3}, Nb_{2/3})O_6(PMN) - PbTiO_3$ (PT) [17]. Third is the relationship between cation ordering and polar nanoregions in relaxors [18, 19]. Fourth is the prediction of a morphotropic phase boundary in pure PT at high pressure, with huge electromechanical coupling in the transition region, indicating that the main effect of the relaxor PMN or PZN, for example, is to tune the transition to zero pressure, rather than something intrinsic to relaxor behavior [11]. Fifth is the discovery of re-entrant ferroelectricity, with ferroelectricity reappearing at very high pressures, indicating the possibility of whole new classes of ferroelectric materials [20].

First-principles studies of ferroelectrics have also given rise to major advances in theoretical methods. The development of the modern theory of polarization [21–23] was motivated entirely to understand piezoelectrics, and has other broader implications as well [24, 25]. The theory of insulators under applied electric fields was also developed to understand piezoelectric materials [26, 27].

The goal of computational research on piezoelectrics is threefold: to help understand experiments, to help guide experiments, and to make predictions for new materials. An example of piezoelectric materials by design is exemplified by the predictions of interesting and tunable properties of ferroelectric superlattices [28–31].

21.1 The Origin of Piezoelectricity and Ferroelectricity: What We Have Learned from First-Principles Studies

Any insulating crystal with two or more sublattices that has the appropriate symmetry, i.e., is a member of one of the 20 polar crystal classes, will have a macroscopic polarization and be piezoelectric, but to have a large piezoelectric effect the crystal should contain ions with large effective charges and they should easily move as a result of lattice strains. All ferroelectrics are piezoelectric, but in addition must have a spontaneous polarization that is switchable by an applied electric field, and generally have a phase transition with increased temperature to a nonpolar, paraelectric state. Many perovskites are ferroelectric, and this review will mainly consider the oxide perovskites. Ferroelectricity in the $LiNbO_3$ structure is very similar in origin, but the structure allows uniaxial ferroelectricity only (the soft mode is nondegenerate and the polarization must be along the c-axis), which significantly changes the dynamical behavior.

Ferroelectrics research was a very hot topic in the 1960s and earlier, driven by experimental studies and phenomenological models. Starting in the 1990s, the field was rejuvenated with the introduction of first-principles methods along with a new generation of fundamental experimental studies.

21.1.1 First-Principles Methods

First-principles methods are so called because they require no experimental data as input, and materials properties are computed using fundamental physics starting with electrons and nuclei and the Coulomb interactions among them. The exact solution of the Schrödinger equation (or its relativistic version the Dirac equation) is impossible for many electron systems because it is a $3N$-dimensional differential equation, where N is of the order of 10^{23} for a bulk material. Almost all first-principles studies of ferroelectrics (and other materials as well) have used DFT, which reduces the problem to solution of a three-dimensional differential equation to find the charge density and total energy and its derivatives. The derivatives are used to compute elastic and piezoelectric constants, as well as vibrational frequencies. In principle, DFT is exact, but the exact functional is not known explicitly, so approximations are used. A typical and commonly used approximation is the LDA. In the LDA, the complicated many-body interactions among the electrons at each point in space in a material are modeled as being the same as the interactions in an electron gas of the same density as that density at that point. This is a surprisingly good approximation. Many tens of thousands of studies have been performed using these methods since the 1960s, and we now have a very good idea of their accuracy and reliability. Volumes are typically within a few percent of experiment, phase stability and phase transitions can be estimated reasonably, and many physical properties can be computed with reasonable accuracy, in some cases rivaling that of experimental studies. The small error in volume is not greatly significant in many materials, but for ferroelectrics it is crucial. This is because ferroelectrics are extremely sensitive to pressure (volume). Quite modest pressures destroy ferroelectricity in $BaTiO_3$ and $PbTiO_3$ corresponding to volume compressions of 2%, e.g., so a small error in volume is very significant. As a result, most first-principles studies of ferroelectrics have used the experimental volume, or include a small shift in pressure of -1 to -5 GPa. New density functionals may remove this need to rely on experimental data for the experimental volume [32].

An important advance in our theoretical understanding of polar solids came through the development of the first-principles theory of polarization [21–23]. Contrary to what one finds in many textbooks, the polarization is not the dipole moment in a unit cell of a material, which in fact depends on the choice of cell. The polarization is rather computable as a Berry's phase from the wave functions, and can be thought of as a current flow from one structure to another. Thus, it is the change in polarization that is computable. If the reference state is centrosymmetric, the polarization is the charge flow from that state to the final state of interest.

Another important advance was the development and implementation of the density functional perturbation or linear response theory, which allows direct computation of phonon frequencies as well as Born transverse effective charges [33, 34], and very recently elastic and piezoelectric constants [35]. Such linear response methods effectively solve directly for derivatives of the Kohn–Sham equations to obtain derivatives of the energy with respect to perturbations in the potential. The dynamical matrix can be obtained directly in this way and diagonalized, giving the quasiharmonic phonon frequencies, which give insight into underlying crystal instabilities, can be compared with vibrational spectroscopy from Raman or Infrared spectroscopy experiments, and to inelastic neutron and X-ray scattering. A huge amount of information is contained in the phonon dispersion, particularly when coupled with the phonon eigenvectors, which are not generally available experimentally.

Linear response also allows the direct computation of the Born transverse effective charges, Z^* [36]. The effective charges are important for understanding piezoelectric behavior in a material, as well as for quantitative computations of piezoelectric constants. The effective charges give another clear indication of the importance of hybridization or atomic polarization. The transverse effective charges are related to the amount of charge that moves when a nucleus is displaced; it is the change in polarization with displacement of a nucleus, $V\,\mathrm{d}\mathbf{P}/\mathrm{d}r_i$. The effective charge is a tensor, because the amount of charge that moves depends on the direction of the displacement. The transverse effective charges are also obtainable from optical experiments from the oscillator strengths or LO–TO splitting [37, 38].

Many important problems in piezoelectrics involve temperature or complex solid solutions and cannot be treated efficiently using only first-principles methods directly. Effective Hamiltonians and potential models, both fitted to first-principles computations, allow the computation of anharmonic dynamical properties and nonequilibrium phenomena such as ferroelectric switching, as well as thermal properties for complex chemically disordered or heterogeneous materials. First-principles-based effective Hamiltonians have been exploited especially well for ferroelectrics, starting with the pioneering work of Rabe and Joannopoulos [7]. In effective Hamiltonian models, only the lowest energy modes are considered and a Hamiltonian is written in terms of the normal-mode coordinate, the strain, and the strain-mode coupling. First-principles computations are then used to obtain the parameters of the Hamiltonian, and energy minimization, Monte Carlo, and/or molecular dynamics can then be used to obtain materials properties [39–41].

21.1.2 Applications to Ferroelectrics

21.1.2.1 Perovskite Ferroelectrics

First-principles calculations of the soft-mode potential surface in perovskite $BaTiO_3$ and $PbTiO_3$ in 1989 showed that the LDA can predict ferroelectric behavior (Fig. 21.1) [4,5]. The total energy is a local maximum for the perfect perovskite structure. For a cubic lattice, the energy is a minimum along [111] but tetragonal strain makes the minimum along [001] (Fig. 21.2). The strain coupling is critical for the tetragonal phases, and of course is primary for the piezoelectric effect. Strain has little effect in the rhombohedral phase under zero field, and to a good approximation the lattice remains cubic through the transition to a rhombohedra phase.

One of the long-standing questions was the origin of the ferroelectric effect. Slater [16] presented a model for ferroelectricity in perovskite known as the "rattling ion model." The idea is simply that the B-cation, e.g., Ti in $BaTiO_3$, is too small for the octahedral cage, and thus moves off-center due to the electrostatic forces from the other ions, the local Lorentz force. The problem with this classic model is that it does not explain why almost all ferroelectric perovskites have B-cations which have d_0 electronic configuration, i.e., their lowest unoccupied conduction band orbital has d character, and no d-states are nominally filled. Examples of d_0 cations in B-sites in ferroelectrics are Ti^{4+}, Zr^{4+}, Nb^{5+}, Ta^{5+}, and Sc^{3+}.

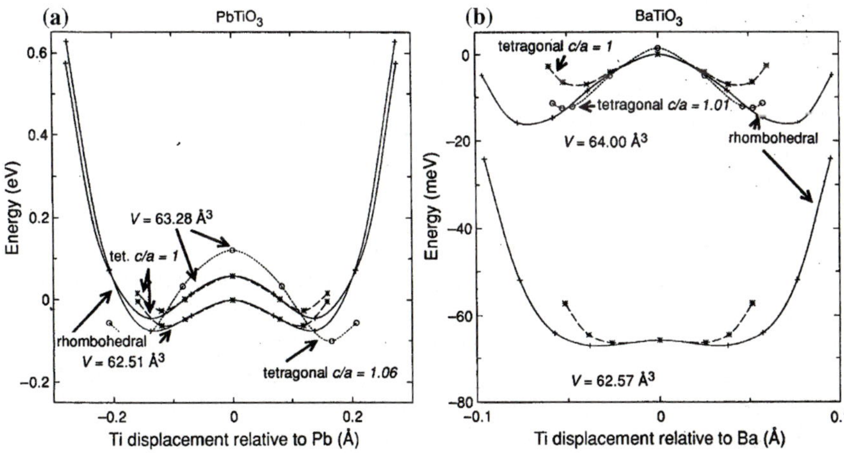

Fig. 21.1. Energy vs. phonon displacement for (**a**) $PbTiO_3$ and (**b**) $BaTiO_3$ showing the multiple-well potential surfaces that underlie ferroelectric behavior. Also illustrated is the great sensitivity to volume and shear strain (from [5])

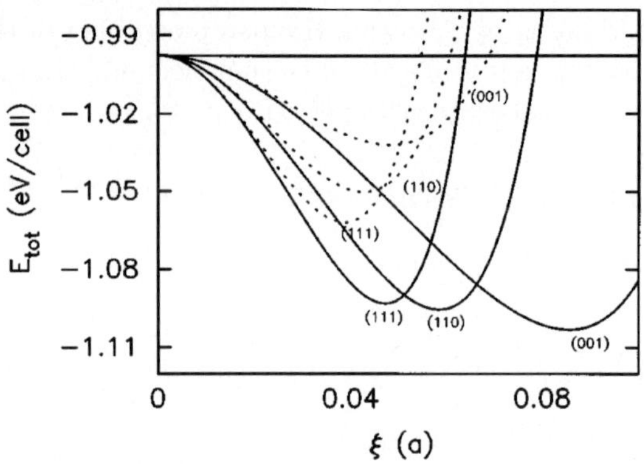

Fig. 21.2. Energy vs. mode coordinate for $PbTiO_3$ for the unstrained, cubic lattice (*dotted*), and relaxed tetragonal lattice (*solid*). Note that the energy is a minimum along the (111), rhombohedral, direction for the unstrained lattice and along the (001), tetragonal, direction for the strained lattice. The great energetic importance of the lattice strain is also evident. This behavior contrasts with that of $BaTiO_3$, which would show (111) always most stable, which is why $BaTiO_3$ has a series of three-phase transitions at zero pressures, whereas $PbTiO_3$ has only 1 (from [41])

First-principles calculations show clearly the following picture of the ferroelectric instability in oxide perovskites. Long-range Coulomb (Madelung) forces favor off-centering; there is a maximum in the ionic electrostatic energy for the atoms in the ideal perovskite positions. Short-range repulsive forces favor the ideal structure, where the atoms are as far apart as possible. Hybridization between the B-cation d-states and oxygen p-states reduces the repulsion and allows the atoms to move off-center. The Pb "lone-pair" 6s-states also lead to intrinsic off-centering of the Pb ions. The same picture holds in uniaxial ferroelectrics such as $LiNbO_3$ and $LiTaO_3$ [42], and may operate in many other ferroelectrics. Theoretical evidence for the above picture comes from the studies of charge densities and electronic densities of states for the ferroelectric and centrosymmetric structures. A clear indication is also the fact that if one removes the d-state degrees of freedom from the basis, the ferroelectric state is no longer stable [4, 5].

Another matter of long-standing interest is the nature of the ferroelectric phase transition, whether it should be considered primarily as an order/disorder or as a displacive phase transition. In the classical displacive picture, in the high-temperature phase, the high-symmetry structure is considered as stable. As the phase transition temperature is approached from above, the soft-mode frequency decreases. In a first-order transition, the polarization, which is the primary order parameter, suddenly acquires a nonzero value at the transition temperature, and the atoms move off-center. Simultaneously, the soft mode hardens and increases in frequency with further reduction in temperature. In a second- or higher-order transition, the order parameter varies continuously from zero to a nonzero value at the critical temperature T_c. Ferroelectrics tend to be weakly first order. In a pure order–disorder transition, the off-center displacements persist above T_c, the main distinction being whether the displacements are ordered or disordered. Whether a transition will be displacive or order–disorder depends on the well depths relative to the temperature, and the coupling strength between displacements in one cell and its neighbors.

Experiments have long suggested that ferroelectric phase transitions are not simple displacive transitions. For example, the soft mode in $BaTiO_3$ or $KNbO_3$ in the cubic high-temperature phase is triply degenerate, and at the transition to tetragonal one mode hardens, while the other two degenerate modes continue to soften as if nothing had happened. At the orthorhombic phase boundary, one of these modes hardens and the other continues to soften. Only at the orthorhombic to rhombohedral boundary, all of the modes do harden below the transition (Fig. 21.3) [43]. Further experimental evidence that the cubic phase is not "simply cubic" comes from X-ray diffraction; instead of the simple Bragg diffraction spots for the cubic perovskite structure,

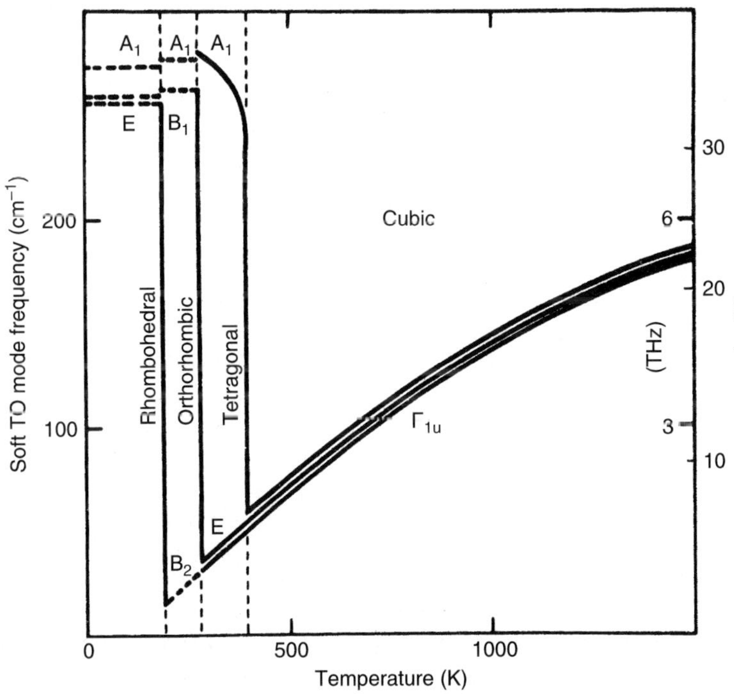

Fig. 21.3. Soft-mode behavior in $BaTiO_3$. One mode hardens at each transition, and the mode frequencies do not go to zero. Only the orthorhombic to rhombohedral transition looks similar to a conventional soft mode, even though it should be understood as having a large order–disorder character (from [43])

a complex streaky pattern is observed [44]. Furthermore, EXAFS shows that bond distances do not vary across T_c, as would be expected in a displacive transition [45]. On the other hand, it has been known that the transition is not a pure order–disorder transition, because the entropy of transition is much reduced from what it would for a pure order–disorder transition [46].

The first-principles static energy calculations from the beginning showed very deep wells along the [111] directions, the eight cube diagonals, giving rise to the eight-site model, and the order–disorder character was emphasized as had been proposed by Comes [44]. In the high-temperature phase, all eight sites are occupied, four are occupied in the tetragonal phase, two in orthorhombic, and one in rhombohedral, which is an ordered structure.

In the pure displacive model, the soft mode is the lowest excitation related to the phase transition. In the pure order–disorder model, there is no soft mode, and the dynamics are dominated by hopping, giving rise to low-frequency relaxational response. Dynamical simulations and experiments show that the true nature of the ferroelectric phases and phase transitions are intermediate between order–disorder and displacive. There is a soft mode and there is also a low-frequency relaxational response. In different materials, the relative intensity and importance of the relaxational and soft-mode response varies, and it varies in a single material as a function of temperature and pressure.

Linear response lattice dynamics calculations for the cubic structure show the underlying instabilities, and how they differ among ferroelectrics. Figure 21.4 shows the computed dispersion curves for the cubic structure of $BaTiO_3$, $PbTiO_3$, and $PbZrO_3$ [47]. This type of information is completely unobtainable experimentally. Experiments give renormalized phonon frequencies (the actual dressed modes) but first-principles theory can be used to obtain the harmonic potential surfaces which illustrate the underlying instabilities in the structure. The frequencies on these plots actually represent the square root of the curvature of the potential surface as a function of wave vector. Unstable (imaginary) frequencies are plotted as negative numbers. Note that the ferroelectric instability forms a flat band over most of the Brillouin zone for $BaTiO_3$ and $PbTiO_3$, but in the former it stiffens toward the R point, indicating strong stability against rotations and octahedral distortions in $BaTiO_3$. Both $BaTiO_3$ and $PbTiO_3$ have the most unstable mode at Γ, indicating that the ferroelectric distortions are lower energy than antiferroelectric ones. $PbZrO_3$, on the other hand, shows the most unstable points at R and M, i.e., antiferroelectric or rotational instabilities, consistent with its antiferroelectric ground state structure with octahedral tilts. The interplay of ferroelectric and rotational distortions has been studied in detail in $PbZrO_3$ and PZT [48–51].

One of the great advances in understanding ferroelectrics was the development of the first-principles theory of polarization [21–23, 52]. Previously, it was thought that polarization could be thought of as the dipole moment per unit cell divided by the unit cell volume. However, this quantity depends on the choice of unit cell. On second thought, many considered polarization

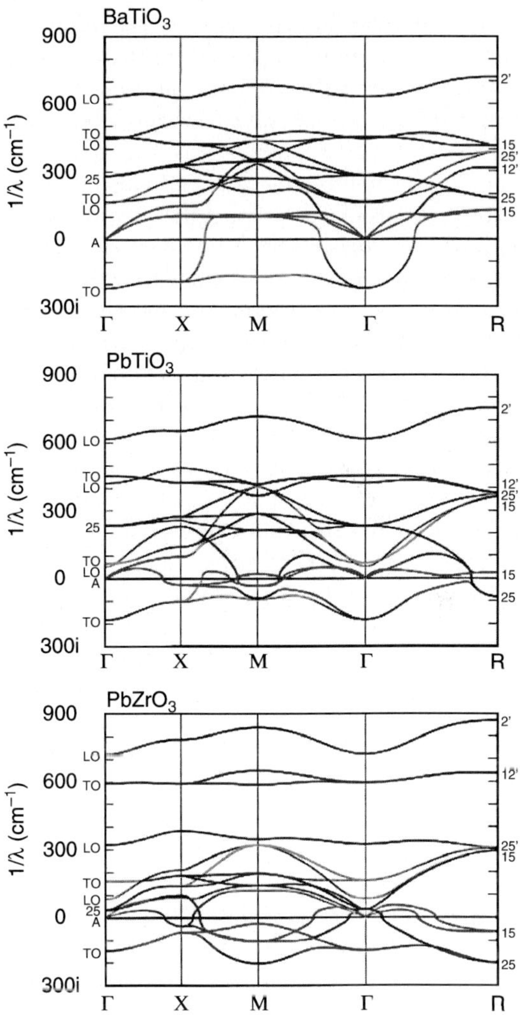

Fig. 21.4. Computed phonon dispersion curves for cubic $BaTiO_3$, $PbTiO_3$, and $PbZrO_3$ (from [47])

to be determined as a difference between the dipole moment of the unit cell in a polar phase and the dipole moment of the unit cell in a centrosymmetric phase. This also does not work in general; one would have to know how the charge flows as the atoms displace from the nonpolar to polar positions. In the case of a localized model, such as a shell model or localized Wannier functions, one still obtains a lattice of polarizations rather than a single-value polarization. The polarization lattice has a spacing of $e\mathbf{R}/V$, where $\mathbf{R}$ are the lattice vectors. These formal polarizations are not the quantities that are generally measured in experiments, which are always polarization differences;

generally, a current is measured in response to switching the polarization, for example. If one switches the polarization by 180° in a ferroelectric, the measured current flow corresponds to $2P$. It is particularly counterintuitive that a centrosymmetric crystal can have a polarization lattice of $(\pm\mathbf{n}+1/2)e\mathbf{R}/V$, where $\mathbf{n}$ are integers.

The reason for this behavior is that in periodic boundary conditions, polarization is a Berry phase of the wave function; it depends on the *phase* of the wave function as a function of wave vector $\mathbf{k}$, rather than the charge density, which depends on the square of the wave function. The polarization computed using the Berry's phase approach is the electronic part of the polarization, to which the atomic ion core contributions must be added. The claim that a localized model gives an unambiguous polarization [53] is not correct; even the completely localized Clausius–Mossotti model produces a lattice of compatible polarization in periodic boundary conditions.

Vanderbilt and King-Smith [22] showed that the polarization is related to the surface charge, and the different compatible formal polarizations correspond to different cell choices that can be consistent with different surface charges (modulo $e\mathbf{S}/A$, where $\mathbf{S}$ is a surface lattice vector and A is the surface cell area). The centrosymmetric cases that have no $\mathbf{P}$ compatible with $P = 0$ are where one cannot make a stoichiometric crystalline slab with the unit cell composition that has no dipole. For example, consider cubic $BaTiO_3$ with nominal charges, Ba^{2+}, Ti^{4+}, and O^{2-}. One can make a slab with BaO and TiO_2 surfaces with composition $BaTiO_3$ that has no dipole, and thus the lattice of polarizations includes $P = 0$. However, cubic $KNbO_3$ with nominal charges, K^+, Nb^{5+}, and O^{2-} gives one $(KO)^{1-}$ surface and one $(NbO_2)^{1+}$ surface; thus the polarization lattice is symmetric about 0, but the smallest values are $(\pm 1/2)e\mathbf{R}/V$. Real $KNbO_3$ would have defects or impurities to pacify the huge naked dipole across a perfect slab. Whether this difference in behavior is reflected in different surface, film, or bulk properties of ferroelectrics with paraelectric phases with formal polarization of $(\pm\mathbf{n}+1/2)e\mathbf{R}/V$ vs. those with formal polarization $(\pm\mathbf{n})e\mathbf{R}/V$ is an interesting question.

Usually, the derivatives of the polarization are of most interest. The Born transverse effective charges, defined as

$$Z^*_{s,\alpha\beta} = \frac{V}{e}\frac{\partial P_\alpha}{\partial u_{s,\beta}},$$

are greatly enhanced in oxide ferroelectrics, and the large coupling to electric fields is partly due to the softness of the potential surface, and partly due to the enhanced effective charges, since the field E couples with displacements d via Sum $Z^* d$.

In ferroelectrics, the effective charges are often much larger than their nominal values, contrary to chemical intuition that they would be lower than their nominal charges due to covalency, as would be static Mulliken charges, for example. Tables 21.1 and 21.2 show computed effective charges for $PbTiO_3$ [8] and rhombohedrally ordered PMN [54]. The effective charges can be greatly

Table 21.1. Computed effective charges for $PbTiO_3$

Atom	Z_{xx}	Z_{yy}	Z_{zz}
Pb	3.7	3.7	3.5
Ti	6.2	6.2	5.2
$O_{(0,0.5,0.5)}$	−2.6	−5.2	−2.2
$O_{(0.5,0.5,0)}$	−2.2	−2.2	−4.4

Note that they are greatly enhanced due to hybridization over the nominal charges of +2, +4, and −2 for Pb, Ti, and O, respectively (from [8])

Table 21.2. Computed effective charges for 1:2 ordered PMN with symmetry C2 (from [54])

Atom type	WP	Z^*_{xx}	Z^*_{yy}	Z^*_{zz}	Z^*_{xy}	Z^*_{xz}	Z^*_{yx}	Z^*_{yz}	Z^*_{zx}	Z^*_{zy}
Pb	*1a*	3.34	3.93	3.35	−0.29	−0.24	−0.50	−0.50	−0.23	−0.28
Pb	*2c*	3.57	3.79	3.50	−0.13	0.00	0.09	0.58	0.24	0.53
Mg	*1b*	2.89	2.87	2.88	0.04	0.14	0.02	0.02	0.14	0.04
Nb	*2c*	6.39	6.41	5.70	0.03	−0.39	0.11	−0.26	−0.13	−0.08
O1	*2c*	−3.39	−2.74	−2.32	−0.11	0.43	0.05	0.12	0.43	0.03
O2	*2c*	−1.95	−2.19	−4.03	−0.06	−0.19	−0.04	0.03	−0.19	−0.10
O3	*2c*	−2.39	−3.69	−2.34	−0.20	0.16	−0.35	0.19	0.15	0.33
O4	*2c*	−2.08	−2.27	−4.79	−0.03	−0.56	−0.10	−0.19	−0.14	0.12
O5	*1a*	−2.09	−5.37	−2.07	−0.13	0.31	0.20	0.19	0.31	−0.15

WP Wyckoff position

enhanced, so that Z^* for oxygen can reach magnitudes of above 5. The fact that these enhanced charges come from hybridization was shown in the same way that the importance of hybridization was shown for the energies. When the d-states are removed from the basis, the effective charges regain nominal values [55].

Enhanced effective charges can also be considered to arise from atomic polarization, as can be seen from the shell model. The shell model is a mechanical model for materials, where atoms are modeled as charged "cores" and charged "shells" connected by springs. Atomic polarization is modeled by displacement of an atom's shell from its core. Shell models can reproduce large effective charges through the "knock-on" effect. For example, when an oxygen core is displaced, its shell is displaced. The shell interacts with other ionic shells through its short-range interactions (springs) and charge, and the other shells move as well. This can give a much greater change in polarization of the crystal than that would occur in an isolated ion, and thus an enhanced effective charge. Effective charges can also be determined experimentally, from LO–TO splitting and from oscillator strengths. In fact, the enhancement in effective

Table 21.3. Piezoelectric (in C/m^2) and elastic (in GPa) constants for $PbTiO_3$ at the experimental lattice computed using density functional perturbation theory (DFPT) and frozen strains (FS) [11] compared with experiment (from [11])

Method	e31	e33	e15	c11	c12	c13	c33	c44	c66
DFPT	2.06	4.41	6.63	230	96.2	65.2	41.9	46.6	98.8
FS	2.07	4.48	6.66	229	95.6	64.3	41.2	47.2	98.6
Experiment	2.1	5.0	4.4	237	90	70	66	69	104

charges over their nominal charges was observed first experimentally [37,38], although its significance was not widely recognized.

Piezoelectric constants are derivatives of the polarization with respect to strain:

$$e_{\alpha\beta\delta} = \frac{\partial P_\alpha}{\partial \varepsilon_{\beta\delta}} = \sum_{s\mu} Z^*_{s,\alpha\mu} \frac{du_{s\mu}}{d\varepsilon_{\beta\delta}}..$$

The piezoelectric constants can be computed using a frozen phonon method, by either computing the **P** using the Berry phase method as a function of strain, or both first computing the effective charges and then relaxing the atomic positions as functions of strain to find the strain derivatives $du/d\varepsilon$ [8–10]. Wu and Cohen [11] applied the new density functional perturbation theory [34] for elasticity and piezoelectricity [35] to $PbTiO_3$ and obtained the elastic and piezoelectric constants as functions of pressure (Table 21.3). We found a set of transitions under pressure from tetragonal, to monoclinic, and then rhombohedral, like a morphotropic phase region seen in complex solid solutions like PZT ($PbZrO_3$–$PbTiO_3$) in relaxor ferroelectrics. Furthermore, we found that the piezoelectric constants peaked in the transition region, giving electromechanical coupling larger than those found in any known material (Figs. 21.5 and 21.6). This strongly suggests that large coupling seen in solid solutions with $PbTiO_3$ is simply due to chemical tuning of this transition to zero pressure rather than any complex structural or electronic effect.

21.1.2.2 Relaxor Ferroelectrics

Relaxor ferroelectrics are solid solutions between ferroelectrics and relaxor endmembers. A ferroelectric has a clear phase transition between a paraelectric and a ferroelectric state, possibility with additional ferroelectric phase transitions as temperature is lowered. In the absence of an applied electric field, a relaxor does not have a phase transition. Instead, it behaves more like a glass, with a broad freezing transition that gives a broad frequency-dependent maximum in the dielectric response. Under field cooling, a ferroelectric phase transition is generally observed. As composition is varied and a ferroelectric component is added, a relaxor to ferroelectric phase transition occurs. Near the transition, the ferroelectrics retain some of the characteristics of relaxors such as dispersion or frequency dependence in properties. Relaxor ferroelectrics

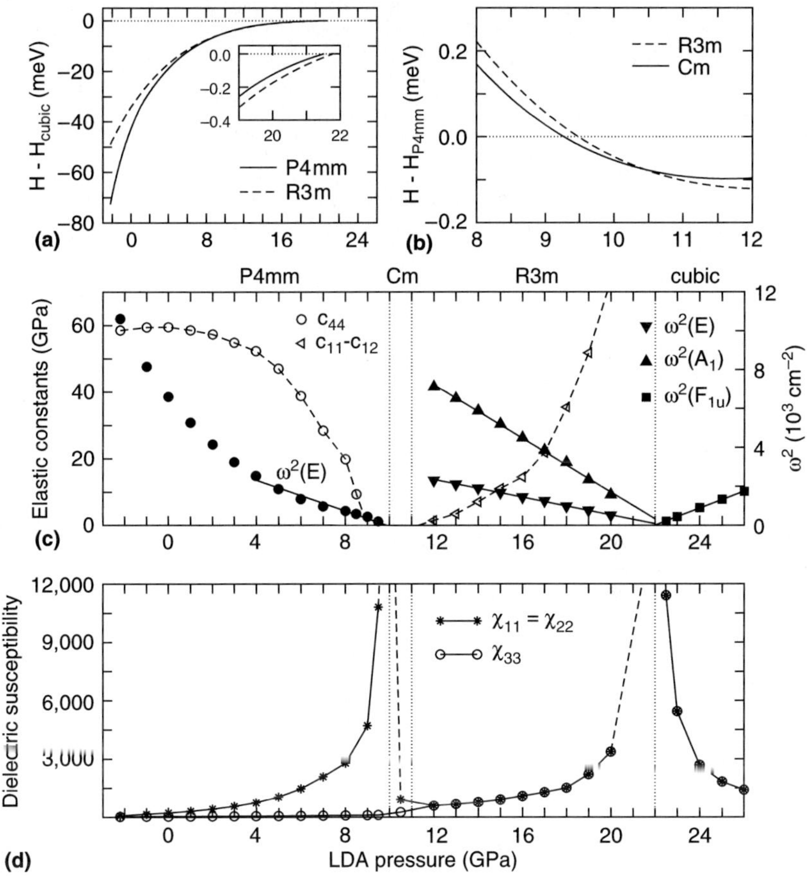

Fig. 21.5. Stability of $PbTiO_3$ phases and properties vs. pressure. (**a**) Enthalpy difference with respect to the cubic (C) phase for the tetragonal (T) and rhombohedral (R) phases. (**b**) Enthalpy difference with respect to the T phase for the R and monoclinic (M) phases. (**c**) Elastic constants for the T and R phases, and square of the lowest optical phonon frequencies ω^2 for the T, R, and C phases. (**d**) Dielectric susceptibility (from [11])

have been given much attention since the discovery of huge electromechanical coupling in single crystals such as $PbMg_{1/3}Nb_{2/3}O_3$–$PbTiO_3$ (PMN–PT) and $PbZn_{1/3}Nb_{2/3}O_3$–$PbTiO_3$ (PZN–PT) [56]. These crystals are rhombohedral at the relaxor end of the phase diagram and tetragonal at the $PbTiO_3$ side, with a morphotropic phase boundary and intermediate monoclinic or orthorhombic phase in-between. The largest piezoelectric response occurs near the morphotropic phase boundary in the rhombohedral phase, with the polarization along the [111] directions, and an applied electric field along the [001] direction. The large response has been understood as due to polarization rotation [57].

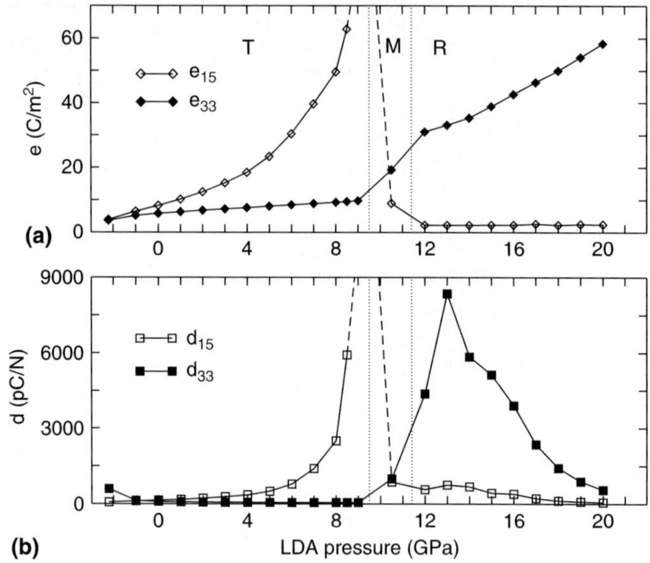

Fig. 21.6. Pressure dependence on piezoelectricity. (**a**) Piezoelectric stress coefficients. (**b**) Piezoelectric strain coefficients. The piezoelectric coefficients peak in the phase transition region and are larger than any known materials (from [11])

The polarization rotation effect can give much higher coupling than the normal collinear effect, in which the polarization and electric field are parallel to each other. The polarization along [111] is compatible with a rhombohedral structure, along [001] with tetragonal, and along [011] with orthorhombic. Intermediate directions like $[xxz]$ or $[x0z]$ are monoclinic. When the polarization is rotated with an applied field, the response is like that of a field-driven phase transition (Figs. 21.7 and 21.8). The large response in PMN–PT, for example, can be understood by considering pure PT. PT has a large tetragonal strain of 6%. If PT had a rhombohedral phase at zero pressure, one could apply a field to the rhombohedral phase driving the structure toward tetragonal. The net result would be a huge 6% strain. However, PT does not have a rhombohedral phase at zero pressure, but a solid solution with PMN, PZN, or PZ does have a rhombohedral phase that allows this rotation effect. The resulting strains are smaller than in pure PT, but still much larger than the collinear effect. The huge electromechanical coupling predicted in pure PT under pressure is due to polarization rotation as well. Ahart et al. recently confirmed these theoretical predictions using cryogenic in situ microRaman and high resolution synchrotron X-ray experiments [57B].

The polarization rotation effect has been verified now by numerous experiments. It leads to intermediate low-symmetry monoclinic or orthorhombic phases between rhombohedral and tetragonal phases on the phase diagram [58–63]. Rotation has also been seen in situ measurements under applied electric fields [64–66].

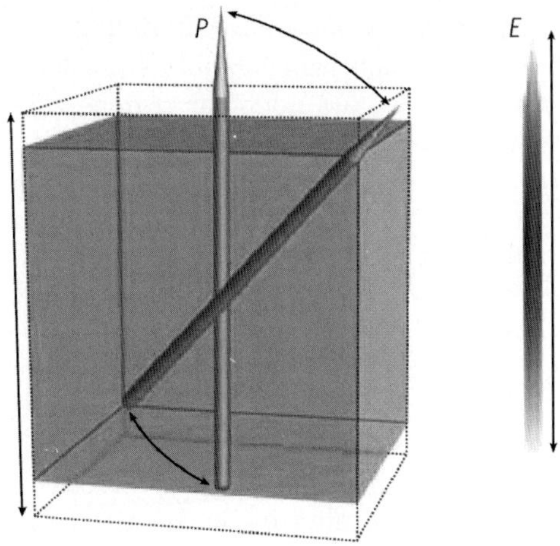

Fig. 21.7. Polarization rotation effect. An applied field along the cubic [001] direction rotates the polarization from [111], the rhombohedral direction, toward [001], through intermediate monoclinic states (from [57])

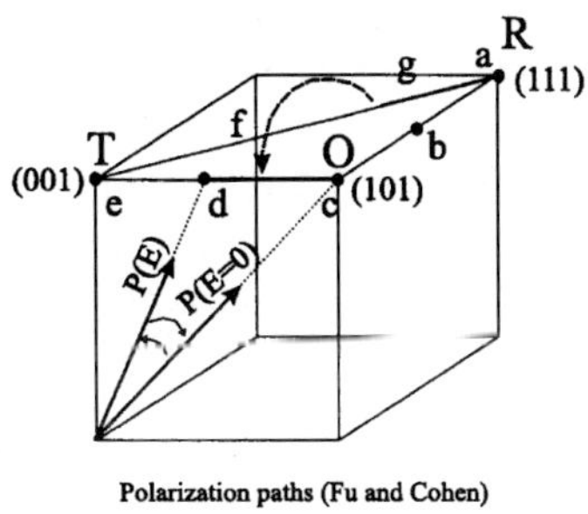

Fig. 21.8. Representation of polarization directions: (**a**) rhombohedral symmetry $P = [xxx]$, (**b**) monoclinic, spacegroup symmetry Pm, $P = [x0z]$, (**c**) orthorhombic symmetry, $P = [x0x]$, (**d**) monoclinic space group Pm, $P = [0yz]$, (**e**) tetragonal symmetry, $P = [00z]$, (**f**) monoclinic spacegroup Cm, $P = [xxz]$, and (**g**) triclinic $[xy1]$

In addition to the polarization rotation, which gives rise to the large electromechanical coupling, relaxor ferroelectrics display a wealth of intriguing phenomena, especially in the relaxor endmember or close to it. These relaxors do not freeze into an ordered ground state, but instead display a broad frequency-dependent maximum in the dielectric response. The generally accepted paradigm for these materials was outlined by Burns and Dacol [67], who attributed nonlinearity in the refractive index with temperature below a certain temperature as the onset of polarization in small nanoregions. Different polar nanoregions have polarizations pointing in different directions, and there is no long-range order among nanoregions which would give a phase transition. According to many views, these polar nanoregions are embedded in a nonpolar matrix, but some data suggest that it might be better to regard the polarization correlation, which would show long tails rather than a clear separation between polar and nonpolar regions. Theoretical models for disordered relaxors will be considered further below.

Relaxors can display two types of disorder. One type of disorder is the polarization disorder mentioned above. Another is chemical disorder. PMN is not generally found in a long-range ordered state, but there is evidence for small chemically ordered domains [68–70]. Other relaxors such as PST ($Pb(Sc_{0.5}Ta_{0.5})O_3$) can be chemically ordered by annealing, so that the effects of chemical order on their relaxor properties can be studied experimentally. In the ordered state, PST has a normal ferroelectric phase transition but shows relaxor behavior when chemically disordered [71, 72]. Statistical mechanical calculations show that the PST ground state is ordered [73–75]; the disordered state occurs from lack of equilibrium in the B-cation distribution caused by the slow diffusion times of the highly charged cations. The question remains if the chemical ordered state will always be ferroelectric in a relaxor, but it seems probable. Tinte et al. have shown in simulations that polar nanoregions form around chemical ordered nanoregions, and this leads to many of the observed phenomena of relaxor ferroelectrics [19].

First-principles calculations for PMN do show a polar, ordered ground state with monoclinic C2 symmetry with 1:2 stacking of Mg and Nb along (111) planes [76] or 1:1 symmetry [77, 78]. If cubic PMN did form nanoscale monoclinic domains, it would be difficult to determine definitively using diffraction data since it is very difficult to deconvolute the diffraction data from a sample with 12 twin variants. Nanodomains of the twin variants would provide static random fields that would contribute further to the disorder of the material and possibly some of the frequency-dependent relaxor behavior.

The effective charges in relaxors may contribute to the low-symmetry ground state found for PMN. Computations of effective charges in PMN show them to be very anisotropic. Local fields in the polar phase will lead to oxygen displacements in oblique directions, giving rise to a monoclinic ground state.

21.1.2.3 $LiNbO_3$ and $LiTaO_3$

The lithium niobate structure is closely related to perovskite and can be considered to be formed from perovskite by rotation of the octahedra, but the rotation is quite large the properties of ferroelectric lithium niobate and lithium tantalite are quite different from perovskite as a result of the different structure. The symmetry of the ferroelectric phase is R3c and thus they are uniaxial ferroelectrics. The soft mode is thus one dimensional and polarization rotation is not possible in this structure, greatly simplifying understanding of the ferroelectric phase transition. Complicating the picture is that Li is quite light, and thus can tunnel between the two wells in the double well potential surface. However, first-principles calculations show the wells to be quite deep (Fig. 21.9) and T_c is high, so classical hopping dominates. Before the first-principles calculations were performed, fitting to experiments suggested

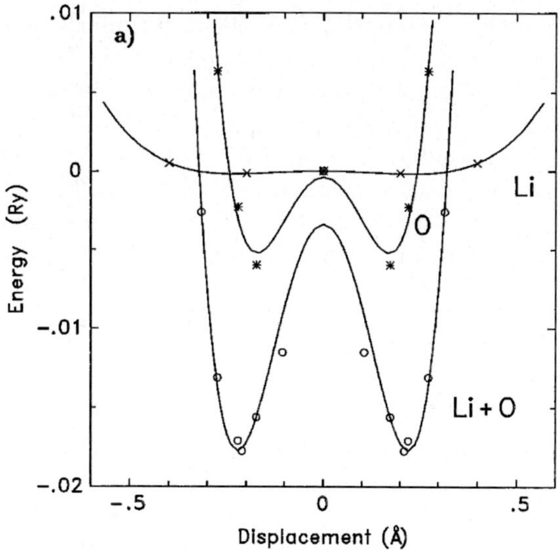

Fig. 21.9. Computed energy surface for $LiTaO_3$ for displacing Li only, O only, or Li and O along the soft-mode coordinate. Clearly, the instability is not due to Li alone, and the concerted motion of Li and O is important. The results for $LiNbO_3$ are very similar (from [42])

a triple well potential surface, which is not consistent with the first-principles results. Frozen phonon calculations showed that the double well is largely due to hybridization between the d_0 Nb^{5+} and Ta^{5+} ions with oxygen, much as for the oxide perovskite, whereas previous work had considered the ferroelectricity to be mainly due to hopping of the Li. The Li does hop, but mainly in response to the changing fields from the off-centering of the Nb or Ta. Linear response computations gave the phonon frequencies of the ground state ferroelectric state [79]. Modeling of the double well potential as an anharmonic oscillator gave good agreement with experiments. One outstanding problem is a first-principles understanding of the difference of T_c of $LiNbO_3$ (1,480 K) and $LiTaO_3$ (950 K) since they have almost identical potential surfaces.

21.1.2.4 GeTe and Other IV–VI Rocksalt Chalcogenides

One of the first first-principles studies of a ferroelectric phase transition was the work of Rabe on GeTe [7]. This work gave the formation of an effective Hamiltonian, and the statistical mechanics necessary to compute T_c and other thermal properties were developed. This landmark work has not yet been fully developed in more complex ferroelectrics, where instead efforts have focused on understanding of polarization and the use of molecular dynamics and Monte Carlo simulation techniques rather than renormalization group theory. Many of the important aspects of ferroelectrics were introduced in this study, including the importance of strain coupling, although the words "ferroelectric" or "piezoelectric" do not appear in the paper. GeTe remains a material with the simplest possible structure for a ferroelectric phase transition. There are only two atoms per unit cell and the paraelectric phase is rocksalt structured. Waghmare et al. [80] revisited GeTe and other chalcogenides, using first-principles techniques developed in the last 20 years that were unavailable in the 1980s, including linear response and Barry's phase techniques. They found large effective charges and large optical dielectric constants, and showed that the ferroelectric instabilities are partly due to lone-pair activity.

21.1.2.5 KDP

KH_2PO_4 (KDP) is a complex so-called "hydrogen-bonded" ferroelectric. Although it is very hard to understand how it works, KDP can easily be grown in large crystals suitable for nonlinear optic applications. There is a large theoretical and experimental literature which is impossible to review here, but very briefly, there are tetrahedral phosphate groups that are bonded to each other with hydrogen bonds. The groups can librate and the hydrogen bonds obey the Slater ice rules [81,82]. The hydrogen orientations order below a T_c of 122 K, all being bound to the top ($+c$) or bottom ($-c$) tetrahedral edges, giving rise to a net polarization. There is a huge negative isotope effect, with the deuterated form of KDP (DKDP) having a T_c of 229 K. Deuteration changes the proton (deuteron) dynamics, possibly changing tunneling probabilities, and changes the crystal structural parameters. Koval et al. [83] performed a detailed DFT study of KDP and shed some light on the nature of the ferroelectric phase transition in KDP and many long-standing issues are discussed, including the origin of the polarization, the relative importance of proton motions and the heavy ions, and the importance of tunneling. Contrary to many studies, they concluded that tunneling is not of major importance in KDP and that the large isotope effect is primarily due to changes in the structural geometry on deuteration, due to changes in the zero point motion of the protons (deuterons) but not due to tunneling among the different wells. KDP is a good example of how first-principles methods can help clear up long-standing mysteries that remained elusive using only experimental data and phenomenological models.

21.1.2.6 Multiferroics

Multiferroics are magnetic polar materials, which have crosscoupling between magnetism and polarization [84, 85]. Usually, magnetism and ferroelectricity are incompatible, because ferroelectricity is usually driven by hybridization between O 2p-states and d-states that are the lowest unoccupied states, giving a d_0 configuration, whereas magnetism depends on partially occupied d (or f) states. One material which has received much attention as a magnetic ferroelectric is $YMnO_3$ [86, 87], which is an improper ferroelectric antiferromagnet. This material is a theoretical challenge because conventional DFT gives a metallic band structure, and it is necessary to use self-interaction corrections (SIC) [86], LDA + U [87], or other technique which includes local Coulomb interactions to obtain the proper insulating state. $YMnO_3$ is probably not a route to a useful multiferroic, as it has a small polarization of only $6.2\,\mu C\ cm^{-2}$ and is an antiferromagnetic, yet it points out several interesting materials properties. It is an improper ferroelectric, in that it is a zone boundary instability that is the primary order parameter. It is also a good case example of how theory can be useful in explaining ambiguous experimental data. Contrary to most primary ferroelectrics, $YMnO_3$ does not have enhanced transverse effective charges and there is little or no displacement of the Mn relative to its neighboring oxygens. The small polarization is primarily due to displacements of the Y ions in response to octahedral tilts.

Theory is on the forefront of the multiferroic field and one candidate material, $BiFeO_3$, has received much experimental and theoretical attention [88–90]. Theory has been indispensable in understanding experimental data. Particularly, important was the theoretical determination of the magnitude of polarization in $BiFeO_3$, which has been discrepant between thin film and bulk measurements. Another important and crucial theoretical contribution was an understanding of how magnetization direction could be switched by switching the polarization direction, a result that had incorrectly appeared to be unallowed by symmetry. Experiments continue on bismuth ferrite [91] and theory will be needed to explain, for example, the phonon anomalies observed at the Néel temperature. The observed complete absence of a Raman active soft mode at the transition should also be examined theoretically. Singh [92] studied the multiferroic $PbVO_3$ and showed the origin of its insulating state, magnetism, and polar ground state. Very recently, Fennie and Rabe [93] designed a new multiferroic, $EuTiO_3$, using first-principles theory, bringing us to the final topic of "Materials by Design."

21.1.2.7 Materials by Design

One of the key goals of first-principles theory on piezoelectrics is to develop understanding and methods for carrying out computational materials by design. Someday, it will be possible to design materials of desired properties computationally, which then can be produced in the laboratory for testing, conformation, and development.

One example of such work is that of Singh's work on new materials. Singh et al. [94] used first-principles methods applied to a variety of oxide perovskite ferroelectrics to come up with a design plan for new materials. Such an approach shows a broader use of first-principles methods to gain deep understanding for materials design, as opposed to predicting properties only for certain candidate materials. Multiferroics also provide an important example of materials by design. The prediction of Bi_2FeCrO_6 as a large polarization multiferroic [95] is very exciting and an example of modern materials by design at its best. Other examples have been given above. The prediction of ferroelectric superlattices and their realization is another major advance in the "Materials by Design" concept [28–31].

21.1.3 Summary

First-principles theory has contributed significantly to our understanding of piezoelectrics and ferroelectrics. This is a rapidly growing field. We can expect theoretical methods to continue to get more robust and accurate, and to address more complex properties of piezoelectric materials. Key advances that came from theory are (1) understanding the role of hybridization and covalency in ferroelectric instability, (2) the role of large effective charges in electromechanical coupling, (3) a fundamental understanding of macroscopic polarization, (4) the role of polarization rotation in the single-crystal relaxor ferroelectrics, (5) the concept of polar ferroelectric superlattices, (6) an understanding of the requirements for materials to be multiferroics, and (7) design paths for new materials.

Acknowledgments

This work is supported by the U.S. Office of Naval Research. Thanks to M. Ahart, B. Burton, P. Ganesh, R.J. Hemley, H. Krakauer, K. Rabe, R. Resta, D. Singh, D. Vanderbilt and Z. Wu for helpful discussions.

References

1. A.F. Devonshire, Philos. Mag. **3**(10), 85 (1954)
2. D. Vanderbilt, M.H. Cohen, Phys. Rev. B **63**, 94108 (2001)
3. L.L. Boyer et al., Ferroelectrics **111**, 1 (1990)
4. R.E. Cohen, H. Krakauer, Phys. Rev. B **42**(10), 6416 (1990)
5. R.E. Cohen, Nature **358**, 136 (1992)
6. R.E. Cohen, Nature **362**, 213 (1993)
7. K.M. Rabe, J.D. Joannopoulos, Phys. Rev. B **36**(12), 6631 (1987)
8. G. Saghi-Szabo, R.E. Cohen, H. Krakauer, Phys. Rev. Lett. **80**(19), 4321 (1998)
9. G. Saghi-Szabo, R.E. Cohen, H. Krakauer, Phys. Rev. B **59**, 12771 (1999)
10. Z. Wu et al., Phys. Rev. Lett. **94**(6), 069901 (2005)
11. Z. Wu, R.E. Cohen, Phys. Rev. Lett. **95**, 037601 (2005)
12. R. Resta, Model. Simul. Mater. Sci. Eng. **11**(4), R69 (2003)
13. K.M. Rabe, P. Ghosez, in *Physics of Ferroelectrics: A Modern Perspective*, ed. by C.H. Ahn, K.M. Rabe, J.M. Triscone (Springer, New York, 2006)
14. C. Fiolhais, F. Nogueira, M. Marques (eds.), *A Primer in Density Functional Theory*. Lecture Notes in Physics, vol. 620 (Springer, New York, 2003), p. 256
15. W. Kohn, L.J. Sham, Phys. Rev. A **140**, 1133 (1965)

16. J.C. Slater, Phys. Rev. **78**(6), 748 (1950)
17. H. Fu, R.E. Cohen, Nature **403**, 281 (2000)
18. B.P. Burton, E. Cockayne, U.V. Waghmare, Phys. Rev. B **72**(6), 064113 (2005)
19. S. Tinte et al., Phys. Rev. Lett. **97**(13), 137601 (2006)
20. I.A. Kornev et al., Phys. Rev. Lett. **95**(19), 196804 (2005)
21. R.D. King-Smith, D. Vanderbilt, Phys. Rev. B **47**(3), 1651 (1993)
22. D. Vanderbilt, R.D. King-Smith, Phys. Rev. B **48**(7), 4442 (1993)
23. R. Resta, Rev. Mod. Phys. **66**, 899 (1994)
24. R. Resta, Phys. Rev. Lett. **96**(13), 137601 (2006)
25. R. Resta, J. Phys.: Condens. Matter **14**(20), R625 (2002)
26. I. Souza, J. Iniguez, D. Vanderbilt, Phys. Rev. Lett. **89**, 117602 (2002)
27. I. Souza, J. Iniguez, D. Vanderbilt, Phys. Rev. B **69**(8), 851061 (2004)
28. C. Bungaro, K.M. Rabe, Phys. Rev. B **6522**(22), 4106 (2002)
29. M.P. Warusawithana et al., Phys. Rev. Lett. **90**, 036802 (2003)
30. M. Dawber et al., Phys. Rev. Lett. **95**(17), 1 (2005)
31. S.M. Nakhmanson, K.M. Rabe, D. Vanderbilt, Phys. Rev. B **73**(6), 1 (2006)
32. Z. Wu, R.E. Cohen, Phys. Rev. B **73**(23), 235116 (2006)
33. X. Gonze, D.C. Allan, M.P. Teter, Phys. Rev. Lett. **68**(24), 3603 (1992)
34. X. Gonze, C. Lee, Phys. Rev. B **55**(16), 10355 (1997)
35. D.R. Hamann et al., Phys. Rev. B **71**(3), 035117 (2005)
36. P. Ghosez, J.P. Michenaud, X. Gonze, Phys. Rev. B **58**(10), 6224 (1998)
37. J.D. Axe, Phys. Rev. **157**, 429 (1967)
38. J.F. Scott, Phys. Rev. B **4**(4), 1360 (1971)
39. W. Zhong, D. Vanderbilt, K.M. Rabe, Phys. Rev. Lett. **73**(13), 1861 (1994)
40. K.M. Rabe, U.V. Waghmare, Phys. Rev. B **52**, 13236 (1996)
41. U.V. Waghmare, K.M. Rabe, Phys. Rev. B **55**, 6161 (1997)
42. I. Inbar, R.E. Cohen, Phys. Rev. B **53**, 1193 (1994)
43. Y. Luspin, J.L. Servoin, F. Gervais, J. Phys. C: Solid State Phys. **13**, 3761 (1980)
44. R. Comes, M. Lambert, A. Guinier, Solid State Commun. **6**, 715 (1968)
45. N. Sicron et al., Phys. Rev. B **50**(18), 13168 (1994)
46. G.A. Rossetti Jr., J.P. Cline, A. Navrotsky, J. Mater. Res. **13**(11), 3197 (1998)
47. P. Ghosez et al., Phys. Rev. B **60**, 836 (1999)
48. D.J. Singh, Phys. Rev. B **52**, 12559 (1995)
49. M. Fornari, D.J. Singh, Phys. Rev. B **63**(9), 921011 (2001)
50. M. Ghita et al., Phys. Rev. B **72**(5) (2005)
51. M.D. Johannes, D.J. Singh, Phys. Rev. B **71**(21), 1 (2005)
52. R. Resta, D. Vanderbilt, in *Physics of Ferroelectrics: A Modern Perspective*, ed. by C.H. Ahn, K.M. Rabe, J.M. Triscone (Springer, New York, 2007)
53. L. Boyer, H. Stokes, M. Mehl, Phys. Rev. Lett. **84**(4), 709 (2000)
54. N. Choudhury, R.E. Cohen, E.J. Walter, Comp. Mater. Sci. **37**, 152 (2006)
55. M. Posternak, R. Resta, A. Baldereschi, Phys. Rev. B **50**, 8911 (1994)
56. S.E. Park, T.R. Shrout, J. Appl. Phys. **82**(4), 1804 (1997)
57. R.E. Cohen, Nature **441**, 941 (2006)

57B. M. Ahart et al., Nature **451**, 545 (2008)

58. D. La-Orauttapong et al., Phys. Rev. B **65**(14), 4101 (2002)
59. K. Ohwada et al., J. Phys. Soc. Jpn **70**(9), 2778 (2001)
60. B. Noheda et al., Phys. Rev. B **61**(13), 8687 (2000)
61. B. Noheda et al., Ferroelectrics **237**(1–4), 541 (2000)

62. R. Guo et al., Phys. Rev. Lett. **84**(23), 5423 (2000)
63. B. Noheda et al., Appl. Phys. Lett. **74**(14), 2059 (1999)
64. K. Ohwada et al., Phys. Rev. B **6709**(9), 4111 (2003)
65. B. Noheda et al., Phys. Rev. B **6522**(22), 4101 (2002)
66. B. Noheda et al., Phys. Rev. Lett. **86**(17), 3891 (2001)
67. G. Burns, F.H. Dacol, Phys. Rev. B **28**(5), 2527 (1983)
68. S. Wakimoto et al., Phys. Rev. B **65**, 172105 (2002)
69. P.K. Davies, M.A. Akbas, J. Phys. Chem. Solids **61**(2), 159 (2000)
70. T. Egami et al., Ferroelectrics **206–207**(1–4; 1–2), 231 (1998)
71. L. Farber et al., J. Am. Ceram. Soc. **85**(9), 2319 (2002)
72. L. Farber, P. Davies, J. Am. Ceram. Soc. **86**(11), 1861 (2003)
73. B.P. Burton, R.E. Cohen, Ferroelectrics **151**, 331 (1994)
74. B.P. Burton, R.E. Cohen, Phys. Rev. B **52**(2), 792 (1995)
75. B.P. Burton, R.E. Cohen, Ferroelectrics **164**, 201 (1995)
76. N. Choudhury et al., Phys. Rev. B **71**(12), 125134 (2005)
77. S.A. Prosandeev et al., Phys. Rev. B **70**(13), 134110 (2004)
78. M. Suewattana, D.J. Singh, Phys. Rev. B **73**(22), 224105 (2006)
79. M. Veithen, P. Ghosez, Phys. Rev. B **65**(21), 214302 (2002)
80. U.V. Waghmare et al., Phys. Rev. B **67**(12), 125111 (2003)
81. J.C. Slater, J. Chem. Phys. **9**(1), 16 (1941)
82. R. Kind et al., Phys. Rev. Lett. **88**(19), 195501 (2002)
83. S. Koval et al., Phys. Rev. B **71**(18), 1 (2005)
84. N.A. Hill, Annu. Rev. Mater. Res. **32**, 1 (2002)
85. C. Ederer, N.A. Spaldin, Curr. Opin. Solid State Mater. Sci. **9**(3), 128 (2005)
86. B.B. Van Aken et al., Nat. Mater. **3**(3), 164 (2004)
87. C.J. Fennie, K.M. Rabe, Phys. Rev. B **72**(10), 100103 (2005)
88. J.B. Neaton et al., Phys. Rev. B **71**(1), 014113 (2005)
89. C. Ederer, N.A. Spaldin, Phys. Rev. B **71**(6), 060401 (2005)
90. P. Baettig, C. Ederer, N.A. Spaldin, Phys. Rev. B **72**(21), 1 (2005)
91. R. Haumont et al., Phys. Rev. B **73**(13), 132101 (2006)
92. D.J. Singh, Phys. Rev. B **73**(9), 094102 (2006)
93. C.J. Fennie, K.M. Rabe, Phys. Rev. Lett., **97**(26), 267602 (2006)
94. D.J. Singh et al., J. Phys. IV **128**, 47 (2005)
95. P. Baettig, N.A. Spaldin, Appl. Phys. Lett. **86**(1), 012505 (2005)

22

Thermodynamic Theory

G.A. Rossetti, Jr.

22.1 Introduction

The classic work of Devonshire [1, 2] in developing the phenomenological theory of barium titanate represented a significant advance in the study of the physical properties of piezoelectric crystals and ceramics based on the perovskite-structured ferroelectric oxides. Although various theories of ferroelectricity in $BaTiO_3$ had been put forward by other workers during the period, the phenomenological theory as formulated by Devonshire provided the first unified description of its changes in crystal form and of the variations of the thermal, dielectric, and piezoelectric properties observed on cooling through the sequence of cubic → tetragonal → orthorhombic → rhombohedral ferroelectric phase transitions. By taking account of the multicomponent nature of the ferroelectric polarization in the perovskite oxides, the Devonshire theory allowed the full matrix of tensor components describing the properties of the monodomain single-crystalline state to be assembled, even within the constraints of a comparatively limited set of available experimental data. The influence of the Devonshire theory on contemporary workers in the field is evident in the early investigations of Slater [3] into the origins of ferroelectricity in $BaTiO_3$, in that of Kittel [4] who predicted the possibility of antiferroelectricity in crystals of the perovskite type before the effect had been experimentally verified, and in that of Cross [5] in his treatment of ferroelectricity and antiferroelectricity in $NaNbO_3$–$KNbO_3$ and related perovskite systems.

Though it has been more than 50 years since the Devonshire papers appeared, phenomenological theories continue to play an important role in connection with the study of piezoelectric materials, and there are several reasons why this has been the case. Nearly all of the materials of technological importance for piezoelectric transducer applications are based on perovskite-structured crystals. The particular form of the phenomenological theory as put forward by Devonshire for $BaTiO_3$ has therefore naturally been extended in a more or less straightforward way to the study of these other perovskite materials as well. The phenomenological theory is a useful tool for experimentalists because it provides a relatively simple framework within which the various physical properties of ferroelectric crystals of interest can be correlated. For example, the physical properties of a particular crystal as measured subject to one set of experimental conditions can be predicted subject to a different set of conditions, which may be inconvenient or otherwise difficult to access in the laboratory. Alternatively, the physical properties of a given crystal as measured by different types of experimental methods can be compared with the predictions of the theory and so checked against one another for consistency. In those instances where discrepancies between theory and experiment are observed, it is then possible to investigate the origins of the discrepancies, and to trace these to factors such as the quality of the specimens under study, to errors in the measurements themselves, or more importantly, to the possible occurrence of other physical phenomena in the crystal that may not have been anticipated at the outset of the experiments. Following this approach, a self-consistent set of physical properties characterizing the monodomain single-crystalline state can be established. For crystals for which such information is available, polycrystalline-averaging methods may then be used to estimate the properties of corresponding randomly axed ceramic material.

The perovskite crystals and ceramics of greatest interest today exhibit ferroelectric phenomena that are typically more diverse and complex than those first treated by Devonshire in the case of the perovskite compound $BaTiO_3$. More specifically that is to say, these latter materials are comprised of multicomponent oxides in pseudobinary or pseudoternary solid solution systems and tend to have complicated phase diagrams in which different types of phase transformation phenomena may occur. In connection with the electromechanical behavior of such materials, the most important feature of these phase diagrams is that they may exhibit, at fixed pressure, very narrow boundaries separating adjacent phases in the composition–temperature plane in proximity to which various physical properties show pronounced extrema in the magnitudes of their coefficients. These “morphotropic phase boundary (MPB)” solid solutions include $PbZrO_3$–$PbTiO_3$ (PZT) and $Pb(Mg_{1/3}Nb_{2/3})O_3$–$PbTiO_3$ (PMN–PT) and related systems and so comprise the most widely exploited materials currently in use for piezoelectric transducer applications.

Accordingly, this chapter will be primarily concerned with problems that are amenable to treatment in terms of general thermodynamic principles and that are specifically associated with the behavior of ferroelectric perovskites and their solid solutions. First, some essential elements of the phenomenological theory of ferroelectricity will be reviewed as a means of introducing the concepts and definitions related to phase transitions in ferroelectric perovskite crystals. Following this, problems related to the role of isomorphous ion substitutions on the order of the paraelectric to ferroelectric phase transition(s), to their expected influence on the monodomain single-crystalline properties, and to the mixing thermodynamics and topology of the generic morphotropic phase diagram will be considered. Insights as drawn from thermodynamic considerations will be used to suggest possible fruitful future areas of enquiry related to the processing and microstructures of piezoelectric materials based on ferroelectric solid solution systems.

22.2 Phenomenology

The phenomenological theory as applied by Devonshire to barium titanate crystals is a particular form of the Ginzburg–Landau theory of phase transitions. The use of the Devonshire theory in describing the thermal, dielectric, and piezoelectric properties of $BaTiO_3$ and related perovskite materials has been described by many authors, and is treated in detail, e.g., in the text of Jona and Shirane [6]. More general considerations associated with the foundations of the phenomenological theory and its application to a variety of other physical phenomena as observed in ferroelectric crystals are elaborated in considerable depth in the monograph of Strukov and Levanyuk [7] and the treatment given by them will be followed here. However, here, attention will be restricted to the specific case of the ferroelectric perovskites and focused only on those aspects that are especially relevant to describing the pressure–temperature or composition–temperature phase diagrams of these materials as the essential starting point for enabling an understanding of the physical properties that they may display.

For this purpose, it is sufficient to begin by asserting without further justification that the behavior of a ferroelectric crystal in the vicinity of a structural phase transition may be described by some thermodynamic potential function $\Phi = \Phi(\theta, p, \eta)$, where Φ is the scalar quantity of energy density (energy per unit volume), θ is the absolute temperature, and p is the pressure. The variable η represents the order parameter of the structural phase transition(s). In the high-symmetry or prototype phase $\eta = 0$, and in the lower-symmetry phase or phase(s) $\eta \neq 0$. The thermodynamic potential describes both the prototype phase and the lower-symmetry phase(s) derived from it and so the form of the dependence Φ on η is restricted by the condition that Φ must be invariant with respect to any transformations of the high-symmetry phase. Specifically, this requires that

$$\Phi(\eta) = \Phi(\eta^*), \tag{22.1}$$

where η^* is the order parameter transformed by the symmetry operations of the prototype phase. Close to the temperature θ_c at which the phase transition occurs, η is vanishingly small and the thermodynamic potential may be expressed as a series expansion in powers of the order parameter η:

$$\Phi(\eta) = \Phi(0) + \Phi'\eta + \frac{1}{2}\Phi''\eta^2 + \cdots, \tag{22.2}$$

where $\Phi(\eta)$ depends on the external parameters of temperature θ and pressure p. Consequently, the exact form of the thermodynamic potential describing a structural phase transition in particular crystal depends on the symmetry of the prototype phase, on the transformation properties of the order parameter, and on the temperature and/or pressure dependencies of the resulting expansion coefficients Φ', Φ'', and the corresponding higher-order terms.

For the sake of simplicity, it is convenient to first consider the specific case of a second-order transition that occurs in a crystal on going from a centrosymmetric prototype phase I to a lower-symmetry phase II characterized by the appearance of a single-order parameter η. In accordance with (22.1), the inversion operator $\bar{1}$ of the centrosymmetric paraelectric phase imposes the restriction that all odd-rank terms in (22.2) must vanish. At some particular pressure $p = p_c$, the second-order transition I $\leftrightarrow$ II occurs at the Curie temperature θ_c, where at temperatures $\theta > \theta_c$ the function $\Phi(\eta)$ is a minimum for $\eta = 0$ in phase I and at temperatures $\theta < \theta_c$ the equilibrium state of phase II is characterized by $\eta \neq 0$. For such a case, (22.2) evidently has the following form:

$$\Phi(p,\theta,\eta) = \Phi_0(p,\theta) + A(p,\theta)\eta^2 + B(p,\theta)\eta^4 + C(p,\theta)\eta^6 + \cdots, \tag{22.3}$$

where at the pressure $p = p_c$ the coefficients $A(\theta)$, $B(\theta)$, and $C(\theta)$ in the vicinity of θ_c are expressed as series expansions in powers of $(\theta - \theta_c)$:

$$\begin{aligned} A(\theta) &= A(\theta_c) + A'(\theta - \theta_c) + \frac{1}{2}A''(\theta - \theta_c)^2 + \cdots, \\ B(\theta) &= B(\theta_c) + B'(\theta - \theta_c) + \frac{1}{2}B''(\theta - \theta_c)^2 + \cdots, \\ C(\theta) &= C(\theta_c) + C'(\theta - \theta_c) + \frac{1}{2}C''(\theta - \theta_c)^2 + \cdots. \end{aligned} \tag{22.4}$$

For the situation where $B(\theta) > 0$ and $C(\theta) \geqslant 0$, the transition I$\leftrightarrow$II then occurs as $A(\theta)$ passes continuously through zero. In the lowest order approximation, (22.4) gives $A' > 0$, $B(\theta_c) > 0$, $C(\theta_c) \geqslant 0$, and $A(\theta_c) = B' = B'' = C' = C'' = 0$. The thermodynamic potential (22.3) becomes simply

$$\Phi(\theta,\eta) = \Phi_0(\theta) + \frac{1}{2}\alpha(\theta - \theta_c)\eta^2 + \frac{1}{4}\beta\eta^4 + \frac{1}{6}\gamma\eta^6, \tag{22.5}$$

where $\alpha = 2A'$, $\beta = 4B(\theta_c)$, and $\gamma = 6C(\theta_c)$. It follows that the condition $A(p_c,\ \theta_c) = 0$ defines the line of the second-order transitions in the $p - \theta$ plane between phases I and II and so describes the pressure–temperature phase diagram of the crystal. The transition between phases I and II of course need not be of second order. First-order transitions are described by (22.3) if for the coefficient of quartic term we have $B(p,\ \theta) < 0$ and for the sextic term $C(p,\ \theta) > 0$. Alternatively, it may happen that a line of first-order transitions crosses over to a line of second-order transitions. Such a situation occurs, for example, in crystals placed under hydrostatic pressure, or in some instances, when the chemical composition of the host crystal is altered by ion substitution. The point at which this crossover occurs is termed a *tricritical point* and corresponds the case where the coefficient of the quartic term changes sign, viz., when $B(p,\ \theta) = 0$.

Application of the equilibrium condition to (22.5) requires that

$$\Phi(\theta) = \Phi(\theta, \eta_0(\theta)) \tag{22.6}$$

with the equilibrium value of the order parameter $\eta_{\text{eq}} = \eta_0$ being determined from

$$\frac{\partial \Phi}{\partial \eta} = 0, \qquad \frac{\partial^2 \Phi}{\partial \eta^2} > 0. \tag{22.7}$$

Generalizing from the above considerations, the thermodynamic potential for a crystal undergoing a structural phase transition may be written in the general form

$$\Phi(p, \theta, \ldots, N, \eta_1, \ldots, \eta_n) = \Phi_0(p, \theta, N) + \Phi_1(p, \theta, \ldots, N, \eta_1, \ldots, \eta_n), \tag{22.8}$$

where N is some additional external parameter other than pressure or temperature, η_i are variables representing the order parameters of the system, and the function $\Phi_0(p, \theta, N)$ represents that part of the free energy density that does not depend on these variables. The form of the function of $\Phi_1(p, \theta, \ldots, N, \eta_1, \ldots, \eta_n)$ is determined by the transformation properties of the order parameter subject to the symmetry operations of the prototype phase and by the temperature, pressure, or other dependencies of its resulting nonzero coefficients. Minimizing $\Phi(p, \theta, \ldots, N, \eta_1, \ldots, \eta_n)$ with respect to η_i

$$\frac{\partial \Phi}{\partial \eta_1} = 0, \ldots, \frac{\partial \Phi}{\partial \eta_n} = 0 \tag{22.9}$$

gives a set of values $\eta_0(p, \theta, \ldots, N)$ which when substituted into (22.8) gives the equilibrium thermodynamic function:

$$\Phi(p, \theta, \ldots, N) = \Phi_0(p, \theta, \ldots, N) + \Phi_1(p, \theta, \ldots, N, \eta_{01}, \ldots, \eta_{0n}). \tag{22.10}$$

In the case of the perovskite crystals, the cubic prototype phase belongs to the centrosymmetric symmetry group $m3m$. As such, it is paraelectric and nonpiezoelectric. The order parameter η for the transition from the cubic prototype phase into one or more lower-symmetry ferroelectric phases is identified with the polarization vector $\boldsymbol{P}$ and is related to the amplitude of a transverse optic soft mode that condenses at θ_c. The Curie principle asserts that the "proper" ferroelectric phase(s) that may appear in the crystal are those that are subgroups of the prototype phase and so must contain all symmetry elements common to the symmetry group of the prototype and the symmetry group of the order parameter. The order parameter $\boldsymbol{P}$ is a polar vector quantity and its symmetry group is ∞m. Shuvalov [8] has determined all such possible ferroelectric phases and these are listed in Table 22.1. As shown in the table, the cubic $m3m$ prototype phase of the perovskite crystals admits seven proper lower-symmetry ferroelectric phases: one tetragonal ($4mm$), one rhombohedral ($3m$), one orthorhombic ($mm2$), three monoclinic (m), and one triclinic (1).

Table 22.1. Possible proper ferroelectric phase transitions from symmetry group $m3m$ (adapted from [8])

Prototype symmetry group $m3m$	Possible proper ferroelectric phases, F						
Point symmetry of domain state	$4mm$	$3m$	$mm2$	m	m	m	1
Number of orientation states	(3)	(4)	(6)	(12)	(12)	(12)	(24)
Orientation direction	$D4$	$D3$	$D2$	$A4$	$A'2$	$A''2$	$A1$
Direction of the ∞-axis	[001]	[111]	[110]	$[hk0]$	$[hkk]$	$[hhl]$	$[hkl]$
Order of the transition $m3m \to F$	II	II	I	I	I	I	I

The parentheses enclose the number of orientation states equal to one half the number of vector directions of the polar vector. The symbols D and A indicate whether the orientations are definite or arbitrary with respect to the following axis. The symbols I and II indicate whether the transition from the paraelectric phase to ferroelectric phase may be of second-order II or must be of first-order I. In all cases, the lattices of domains are parallel

The polarization vector $\boldsymbol{P}$ has three components proportional to the order parameter $\eta = \{\eta_1, \eta_2, \eta_3\}$ and so (22.8) may contain terms of the form

$$\eta_i, \quad \eta_i\eta_j, \quad \eta_i\eta_j\eta_k, \quad \eta_i\eta_j\eta_k\eta_l, \quad \eta_i\eta_j\eta_k\eta_l\eta_m, \ldots, \tag{22.11}$$

where the range of all letter suffices i, j, k is $1, 2, 3$. Since the thermodynamic potential must be invariant with respect to any transformations of the symmetry elements of the point group of the prototype phase, $\Phi(\eta_1, \eta_2, \eta_3) = \Phi(\eta_1^*, \eta_2^*, \eta_3^*)$, the operation of the generating elements $\bar{3} \parallel [111]$ and $4 \parallel [001]$ of the point group $m3m$ results in a thermodynamic potential of the form

$$\begin{aligned}\Phi(p, \theta, \eta) = \Phi_0(p, \theta) &+ \frac{1}{2}\alpha(\theta - \theta_c)(\eta_1^2 + \eta_2^2 + \eta_3^2) + \frac{1}{4}\beta_1(\eta_1^2 + \eta_2^2 + \eta_3^2)^2 \\ &+ \frac{1}{2}\beta_2(\eta_1^2\eta_2^2 + \eta_2^2\eta_3^2 + \eta_3^2\eta_1^2) + \frac{1}{6}\gamma_1(\eta_1^2 + \eta_2^2 + \eta_3^2)^3 \\ &+ \frac{1}{2}\gamma_2\left[\eta_1^4(\eta_2^2 + \eta_3^2) + \eta_2^4(\eta_1^2 + \eta_3^2) + \eta_3^4(\eta_1^2 + \eta_2^2)\right] + \frac{1}{2}\gamma_3\eta_1^2\eta_2^2\eta_3^2 .\end{aligned} \tag{22.12}$$

For a mechanically free crystal, and with temperature-independent coefficients β_i, and γ_i, (22.12) is in an alternative but generalized equivalent form to that given by Devonshire [1]. By minimizing this function with respect to components of the polarization vector

$$\frac{\partial \Phi}{\partial \eta_1} = 0, \quad \frac{\partial \Phi}{\partial \eta_2} = 0, \quad \frac{\partial \Phi}{\partial \eta_3} = 0, \tag{22.13}$$

there is one trivial solution

$$\eta_{01} = \eta_{02} = \eta_{03} = 0 \tag{22.14}$$

and three others of the form

$$\begin{aligned}
&(1)\ \eta_{01}^2 = \eta_{02}^2 = 0, \quad \eta_{03}^2 \neq 0, \\
&(2)\ \eta_{01}^2 = \eta_{02}^2 \neq 0, \quad \eta_{03}^2 = 0, \\
&(3)\ \eta_{01}^2 = \eta_{02}^2 = \eta_{03}^2 \neq 0,
\end{aligned} \tag{22.15}$$

or the equivalent ones obtained by interchanging η_{01}, η_{02}, η_{03}. Comparing these equations with the possible proper ferroelectric phases listed in Table 22.1 it is seen that the solution (22.14) corresponds to the paraelectric prototype phase and that the solutions (1)–(3) of (22.15) correspond to the tetragonal, orthorhombic, and rhombohedral ferroelectric phases having definite directions of the polar vector parallel to the crystal directions $\langle 001 \rangle$, $\langle 110 \rangle$, and $\langle 111 \rangle$, respectively. However, the lower-symmetry monoclinic and triclinic phases listed in the table may not be described by a function of the form (22.12). It was shown by Vanderbilt and Cohen [9] that the monoclinic phase is described only when eighth-order invariants are retained in the expansion, and for the triclinic phase, only if the expansion is carried out to invariant terms of the 12th order. The most general form of the thermodynamic potential that would include such invariants has so been called the "higher-order" Devonshire theory.

It is apparent that it is not possible to decide, either on the basis of symmetry considerations alone or by any general analysis of the thermodynamic potential function itself, which of the possible transitions indicated in Table 22.1 may in fact occur in a particular crystal as the external parameters of temperature θ and pressure p are varied. This is evident, for example, in considering the perovskite titanate compounds $SrTiO_3$, $PbTiO_3$, and $BaTiO_3$. Mechanically free crystals of $SrTiO_3$ display no ferroelectric transitions on cooling from high temperature, $PbTiO_3$ undergoes one transition $m3m \rightarrow 4mm$, and $BaTiO_3$ displays the sequence of ferroelectric transitions $m3m \rightarrow 4mm \rightarrow mm2 \rightarrow 3m$. The particular phase transition or sequence of transitions observed in the pressure–temperature phase diagram of a given crystal depends on the ratio of the coefficients of the thermodynamic potential α, β_i, $\gamma_i, \ldots$ and, following from (22.4), the possible dependencies of one or more of these coefficients on temperature and/or pressure.

It is possible, however, based on the above considerations, to make some predictions as to whether such transitions may take place continuously, or if they must take place discontinuously. For cubic crystals, the transition to the ferroelectric phase may be second order under the conditions that at the Curie temperature (1) there appears a polarization vector having only one component directed along a unique axis of the prototype phase, e.g., $\pm\eta_1$, $\pm\eta_2$, $\pm\eta_3$ or (2) there appears a polarization vector having three equal components $\eta_1 = \eta_2 = \eta_3$. Such a situation is realized in the perovskite crystals for the transitions $m3m \rightarrow 4mm$ and $m3m \rightarrow 3m$, all other possible transitions listed in Table 22.1 being first order [8]. Even though transitions to the tetragonal and rhombohedral phases *may be* second order, this possibility appears to be the exception rather than the rule. For example, in the perovskite compounds $KNbO_3$, $BaTiO_3$, and $PbTiO_3$, all of which exhibit the transition $m3m \rightarrow 4mm$, in each case the transition is found to be first order [6].

To illustrate how the parameters of the thermodynamic potential may be determined for some particular crystal of interest, we now consider the simple case of the single first-order transition in $PbTiO_3$ that occurs, at atmospheric pressure, at a temperature $\approx$760 K [10]. In analyzing the experimental data, it is convenient to define the order parameter $\boldsymbol{\eta} = \boldsymbol{n}\eta$, where $\boldsymbol{n}$ is a unit vector in the direction of the polarization and η is its absolute value. In the tetragonal $4mm$ phase, the polar vector appears along one of the cube axes of the prototype phase with $\boldsymbol{n} = \{0,\ 0,\ 1\}$. At atmospheric pressure, thermodynamic potential then assumes the form

$$\begin{aligned}\Phi(\theta, \eta, \boldsymbol{n}) = \Phi_0(\theta) + \frac{1}{2}\alpha(\theta - \theta_c)\eta^2 + \frac{1}{4}\left[\beta + \beta'(n_1^4 + n_2^4 + n_3^4)\right]\eta^4 \\ + \frac{1}{6}\left[\gamma + \gamma'(n_1^6 + n_2^6 + n_3^6) + \gamma''(n_1^2 n_2^2 n_3^2)\right]\eta^6.\end{aligned} \tag{22.16}$$

Because there is only one transition corresponding to the spontaneous polarization vector directed parallel to the directions $\langle 001 \rangle$, the basic features of the first-order phase transition will be captured even when the anisotropic contributions of the sextic invariants in (22.16) are neglected. Making this approximation, the expansion of the thermodynamic potential begins with

$$\Phi(\theta, \eta, \boldsymbol{n}) = \Phi_0(\theta) + \frac{1}{2}\alpha(\theta - \theta_c)\eta^2 + \frac{1}{4}b_0\left[1 - \lambda(n_1^4 + n_2^4 + n_3^4)\right]\eta^4 + \frac{1}{6}\gamma\eta^6, \tag{22.17}$$

where the constants α, b_0, $\gamma > 0$. Substituting $\boldsymbol{n} = \{0,\ 0,\ 1\}$, the entire quartic coefficient is $\beta_0 = b_0(1 - \lambda)$ and it is evident that the transition is first order for $\lambda > 1$. From (22.13),

$$\eta_0^2 = -\frac{\beta_0}{2\gamma}\left\{1 + \left[1 - \frac{4\alpha(\theta - \theta_c)\gamma}{\beta_0^2}\right]^{1/2}\right\}. \tag{22.18}$$

From this, it is seen that the highest temperature to which the ferroelectric phase may exist as a metastable phase in the absence of an applied external field is given by

$$\theta^* = \theta_c + \frac{\beta_0^2}{4\alpha\gamma}, \tag{22.19}$$

Differentiating (22.17) with respect to temperature, the entropy and specific heat at constant pressure are

$$\sigma = -\left(\frac{\partial \Phi}{\partial \theta}\right)_p, \quad C_p = \theta\left(\frac{\partial \sigma}{\partial \theta}\right)_p. \tag{22.20}$$

Following from (22.17) and (22.20), the temperature dependence of the "excess" entropy $\Delta\sigma$ of the phase change is

$$\Delta\sigma = \sigma - \sigma^0 = -\frac{1}{2}\alpha\eta_0^2 \tag{22.21}$$

and the temperature dependence of the "excess" specific heat ΔC_p of the phase change is

$$\left(\frac{\Delta C_\mathrm{p}}{\theta}\right)^{-2} = \left(\frac{C_\mathrm{p} - C^0}{\theta}\right)^{-2} = \frac{4[\beta_0^2 - 4\alpha(\theta_0 - \theta_\mathrm{c})]}{\alpha^4} + \frac{16\gamma}{\alpha^3}(\theta_0 - \theta), \tag{22.22}$$

where θ_0 is the temperature at which the cubic and tetragonal phases are in equilibrium and is defined by the condition $\Phi(\eta = 0) = \Phi(\eta = \eta_0)$.

Under mechanically free conditions, the strain associated with the distortion of the cubic crystal lattice is of the same order of smallness as the square of the order parameter, $S_{ij} \propto \eta^2$ with

$$S_{ij} = q_{ijkl}\boldsymbol{n}_k\boldsymbol{n}_l\eta^2, \tag{22.23}$$

where the fourth-rank tensor q_{ijkl} is symmetrical with respect to any permutations $i \leftrightarrow j$, $k \leftrightarrow l$, and $(ij) \leftrightarrow (kl)$. Since $\boldsymbol{n} = \{0,\ 0,\ 1\}$, the relation between strain and polarization in the tetragonal phase is then

$$S_{ij} = \begin{pmatrix} q_{12} & 0 & 0 \\ 0 & q_{12} & 0 \\ 0 & 0 & q_{11} \end{pmatrix} \eta^2. \tag{22.24}$$

From (22.24), the crystal lattice parameters c_t and a_t of the tetragonal ferroelectric phase are given by

$$\begin{aligned} a_1 = a_2 = a_\mathrm{c}(1 + S_{11}) = a_\mathrm{c}(1 + q_{12}\eta^2) = a_\mathrm{t}, \\ a_3 = a_\mathrm{c}(1 + S_{33}) = a_\mathrm{c}(1 + q_{11}\eta^2) = c_\mathrm{t}, \end{aligned} \tag{22.25}$$

where a_c is the crystal lattice parameter of the cubic prototype phase.

From measurements of the temperature dependence of the specific heat and the crystal lattice parameters and using (22.21), (22.22), and (22.25), we may readily examine if a thermodynamic potential function of the type (22.17) adequately describes the behavior of these quantities in the vicinity of the transition, and if so, to determine values for the parameters α, β_0, γ. Experimental data suitable for such an analysis may be obtained on single-crystal specimens or on well-crystallized and coarse-grained powder specimens. Representative results as measured over a wide temperature range [11, 12] are shown in Fig. 22.1. Figure 22.2 shows the variation with temperature of the crystal lattice parameters, together with that of the quantity $(\Delta C_\mathrm{p}/\theta)^{-2}$, close to the transition between cubic and tetragonal phases. It is seen in this figure that within the accuracy and resolution of the experimental data, all of the main features of a first-order phase change as predicted from the phenomenological theory are displayed. First, in accordance with (22.19), there is a temperature interval $(\theta - \theta^*) \approx 18\,\mathrm{K}$ over which the cubic and tetragonal

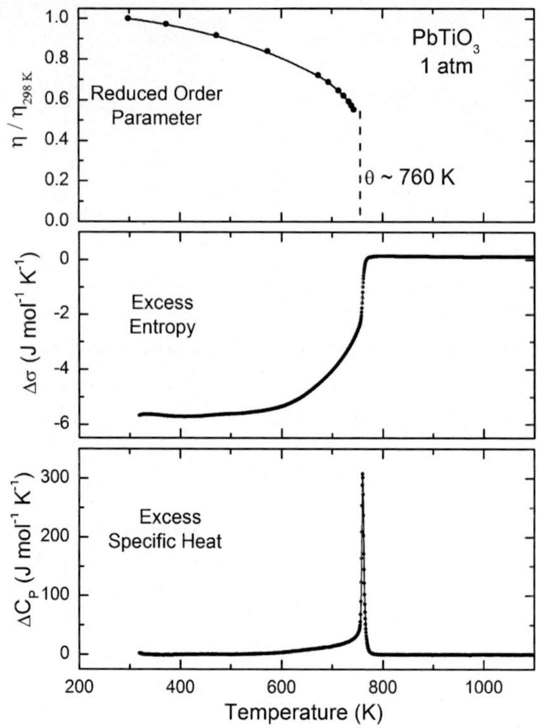

Fig. 22.1. Temperature dependence of the excess specific heat, excess entropy, and order parameter of $PbTiO_3$ at atmospheric pressure (adapted from [11, 12])

phases coexist. Second, as expected based on (22.22), we see that there is a linear dependence of $(\Delta C_{\mathrm{p}}/\theta)^{-2}$ on θ with a slope given by the prefactor of the second term in this equation and a discontinuity at θ_0 given by the first term. Following from (22.19), this linear dependence extrapolates to zero at θ^*, the temperature above which the metastable ferroelectric tetragonal phase is no longer observed. Finally, comparing the observed behavior of the crystal lattice parameters of the tetragonal phase with (22.25), it is evident that the implicit expectation that the coefficients q_{11} and q_{12} do not depend on temperature is fulfilled through the transition. To complete the analysis, we need only to find the value for the coefficient α. This can be done with the aid of (22.21) and the results are shown in Fig. 22.3.

The above analysis confirms that, at least in the case of the perovskites exhibiting a single proper ferroelectric phase transition, the thermodynamic potential function given by the phenomenological theory is well constrained by just three temperature-independent constants $\alpha > 0$, $\beta_0 < 0$, and $\gamma > 0$ over a rather wide temperature interval near the transition. As estimated from the experimental results in Figs. 22.2 and 22.3, this interval corresponds to a reduced temperature range $(\theta^* - \theta)/\theta^* \approx 0.1$ or $\approx$75 K. This observation

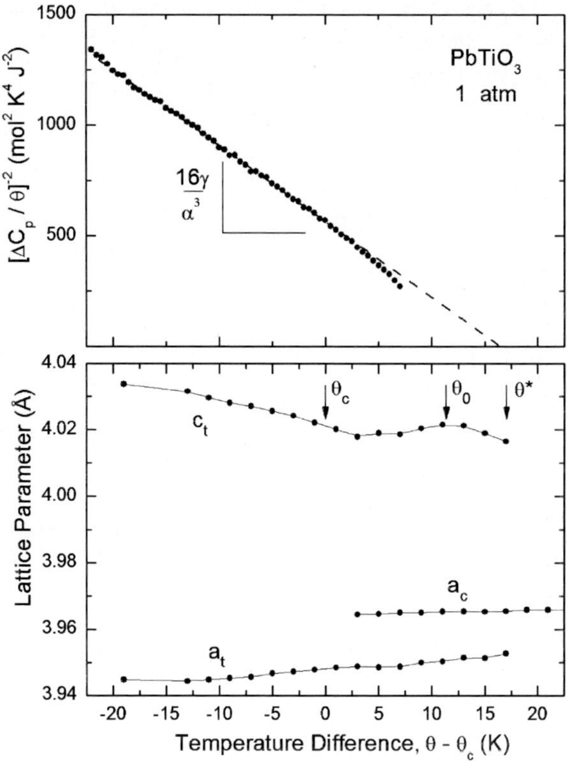

Fig. 22.2. Temperature dependence of the crystal lattice parameters (a_c of the cubic phase and a_t and c_t of the tetragonal phase) and the quantity $(\Delta C_p/\theta)^{-2}$ of $PbTiO_3$ at atmospheric pressure (adapted from [11, 12])

extends the utility of the analysis of such transitions as occur, e.g., in the $PbZrO_3$–$PbTiO_3$ MPB system considered below, where the single proper ferroelectric transitions $m3m \to 4mm$ and $m3m \to 3m$ occur over most of the composition–temperature diagram.

To take account of the effects of pressure on the phase transition, the elastic energy must be added to the thermodynamic potential. The linear stress-induced elastic strain is $S_{ij} = s_{ijkl}T_{kl}$, and for the cubic crystals there are three nonzero components of the elastic compliance tensor, s_{11}, s_{12}, and s_{44} in Voigt notation. If the applied pressure is hydrostatic, there are three nonzero components of the stress tensor $-T_{11} = -T_{22} = -T_{33} = -p$ and following from (22.25), elastic contributions to thermodynamic potential become

$$\Phi_{\text{elas}}(p, \eta) = -\frac{3}{2}(s_{11} + 2s_{12})p^2 - (q_{11} + 2q_{12})\eta^2 p. \qquad (22.26)$$

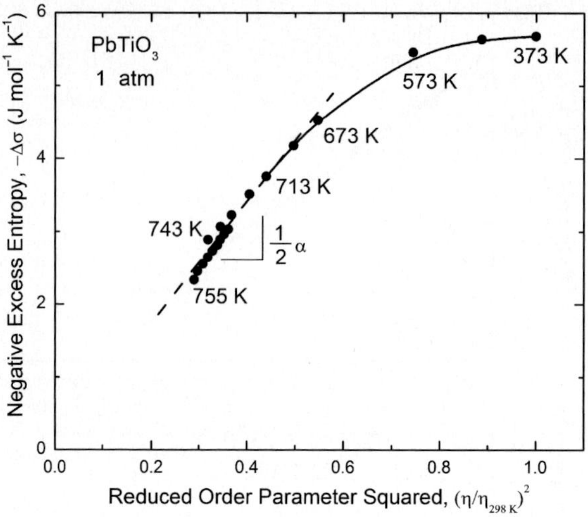

Fig. 22.3. Dependence of the excess entropy on the square of the order parameter for $PbTiO_3$ at atmospheric pressure (adapted from [11, 12])

Substituting (22.26) into (22.17) gives

$$\Phi(p,\theta,\eta,\boldsymbol{n}) = \Phi_0(\theta) - \frac{3}{2}(s_{11}+2s_{12})p^2 + \left[\frac{1}{2}\alpha(\theta-\theta_\mathrm{c}) - (q_{11}+2q_{12})p\right]\eta^2 + \frac{1}{4}b_0\left[1-\lambda(n_1^4+n_2^4+n_3^4)\right]\eta^4 + \frac{1}{6}\gamma\eta^6. \tag{22.27}$$

The pressure-dependent part of the above expression predicts that the application of a hydrostatic pressure will produce (1) a decrease in the unit cell volume Ω of the crystal and (2) a linear shift of the Curie temperature. The shift of the Curie temperature with pressure follows from the Clausius–Clapeyron relation and is in low-pressure limit ($p \rightarrow 0$):

$$\frac{\partial\theta_\mathrm{c}}{\partial p} = -\frac{2(q_{11}+2q_{12})}{\alpha}. \tag{22.28}$$

In the case of $PbTiO_3$, the value obtained from (22.28) is $\approx -85\,\mathrm{K\ GPa^{-1}}$, in reasonably good agreement with the value $-71\,\mathrm{K\ GPa^{-1}}$ as determined directly from dielectric measurements [13]. As mentioned earlier, however, a change in an external parameter such as hydrostatic pressure may also produce a crossover from first-order to second-order transition behavior, e.g., a tricritical point. Based on measurements made at relatively low pressures, Samara [13] suggested the presence of a tricritical point in the pressure–temperature diagram of $PbTiO_3$. Subsequent structural studies by X-ray and neutron diffraction carried out to higher pressures [14, 15] support this indication, and an approximate pressure–temperature diagram constructed using this data appears as shown in Fig. 22.4.

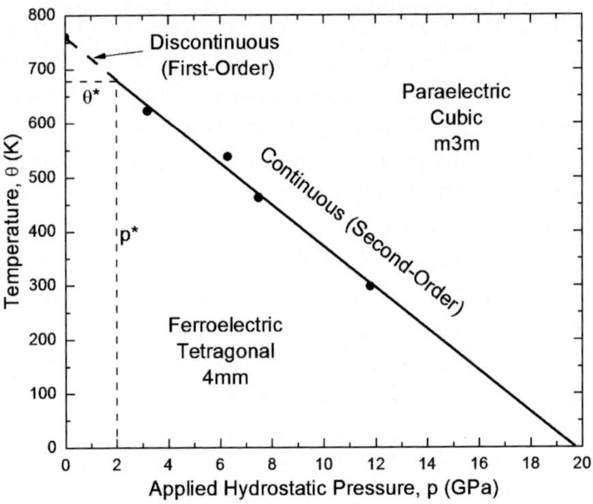

Fig. 22.4. Pressure–temperature phase diagram for $PbTiO_3$ as constructed using the data of Ramírez et al. [14] and Sani et al. [15]. The transition line was determined by the best linear fit to the experimental data over the full range of pressure and the location of the tricritical point (p^*, θ^*) is approximate

As shown in the figure, the transition temperature changes nearly linearly with pressure as expected based on (22.27), and at low pressures the transition is first order. The point (p^*, θ^*) corresponds to the tricritical point where in (22.4) the coefficients $A(p^*, \theta^*) = B(p^*, \theta^*) = 0$. The position of this point as determined from the variation of η with pressure is difficult to locate exactly but the available data suggest an approximate value $p^* \approx 2 \pm 0.5$ GPa and $\theta^* \approx 650 \pm 25$ K. The thermodynamic potential (22.27) will obviously not account for such a point since the quartic term does not depend on pressure. Rather, some of its coefficients may have pressure dependencies (which have not been explicitly taken into account) or alternatively there are higher-order contributions to the function $\Phi(p, \eta)$ that may become important at high pressures.

As mentioned earlier, the relative stabilities of the tetragonal, rhombohedral, and orthorhombic phases are determined by the ratio of the coefficients in (22.16). By imposing the solutions corresponding to $\boldsymbol{n} = \{0,\ 0,\ 1\}$ for the tetragonal phase, $\boldsymbol{n} = \{1/\sqrt{3},\ 1/\sqrt{3},\ 1/\sqrt{3}\}$ for the rhombohedral phase, and $\boldsymbol{n} = \{1/\sqrt{2},\ 1/\sqrt{2},\ 0\}$ for the orthorhombic phase, Devonshire [1] showed that if temperature-independent values of the quartic and sextic invariant terms are appropriately chosen, the minimum of the thermodynamic potential function becomes progressively lower with decreasing temperature as the polarization vector changes its direction from $\langle 001\rangle$ to $\langle 110\rangle$ to $\langle 111\rangle$, and it is possible to describe the transition sequence $m3m \to 4mm \to mm2 \to 3m$ as observed in $BaTiO_3$. Consequently, if these parameters are scaled to experimental data by some method similar to that described earlier, a single set of coefficients that best agree with all known measured properties can be determined. Using these coefficients, it is then possible to investigate the relative phase stabilities and physical properties of the crystal for some other particular set of thermal, mechanical, or electrical boundary conditions of interest.

Thus, one of the main practical advantages of the phenomenological approach is that when properly chosen and analyzed, a relatively small set of experimental data can be used to infer in a self-consistent way a great deal of information about the thermodynamic behavior and physical properties of the crystal of interest. Viewed from this perspective, a thermodynamic potential function properly constructed from the phenomenological theory then provides a compact description of the properties of the crystal. In the case of ferroelectric solid solutions, relative stabilities of the ferroelectric phases and the transitions among them will of course depend not only on temperature and pressure but also on the composition variable c. Following from the most general form of the thermodynamic potential given by (22.8), it is possible to extend the above analysis to describe some of the features of the temperature–composition ($\theta - c$) phase diagram. The application of such an analysis to the specific case of the $PbZrO_3$–$PbTiO_3$ solid solution is given in Sect. 22.3 and some thermodynamic considerations that arise in connection with their piezoelectric properties are discussed.

22.3 Aspects of the Thermodynamic Behavior of Ferroelectric Solid Solutions

The conventional composition–temperature phase diagram of the $PbZrO_3$ – $PbTiO_3$ system is shown in Fig. 22.5 [16]. This diagram was determined by a number of workers in a series of studies, and Glazer et al. [17] provide an excellent summary covering both the early literature and more recent developments. Perhaps, the most significant among these is the report by Noheda et al. [18] that an intermediate phase of monoclinic symmetry appears below room temperature in the vicinity of the composition $c \approx 0.5$. Following this report, there has been a resurgence of interest in the possible existence and range of stabilities of other intermediate phases. Consequently, the conventional diagram shown in the figure has undergone a series of revisions and these are discussed in the review article of Woodward et al. [19]. The possible relation of intermediate phases to the exceptional piezoelectric properties of MPB compositions has naturally received a great deal of attention from both the experimental and theoretical perspectives. Since the topic is one of ongoing research, it will not be covered in detail here and the reader is referred to the recent reviews of Noheda [20] and Bell [21] for further discussion. Here, attention will be focused on three related subjects that are directly amenable to treatment in terms of the phenomenological theory and which are generally relevant to an understanding of the structure-sensitive properties of ferroelectric solid solution systems.

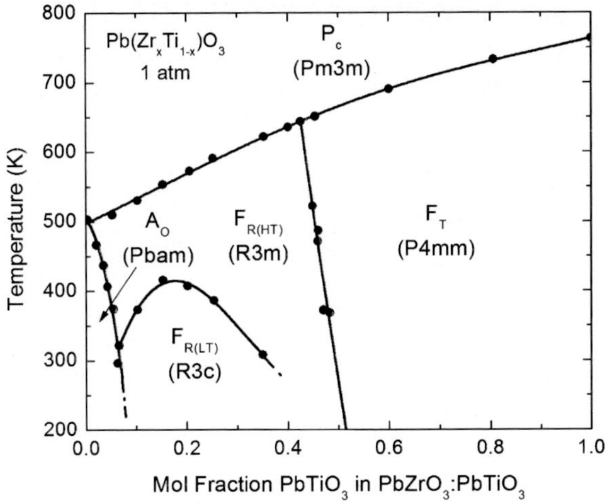

Fig. 22.5. Conventional composition–temperature phase diagram for $PbZrO_3$: $PbTiO_3$ at atmospheric pressure (adapted from [16])

These problems concern (1) the influence of isomorphous ion substitutions on the order of the paraelectric to ferroelectric phase change, (2) the relative contributions of the homogeneous, monodomain state to the dielectric and piezoelectric property responses, and (3) the mixing thermodynamics and topology of the morphotropic phase diagram.

As shown in Fig. 22.5, the paraelectric to ferroelectric transitions that occur over much of $c - \theta$ plane in the $PbZrO_3$–$PbTiO_3$ system are of the type $m3m \rightarrow 4mm$ and $m3m \rightarrow 3m$. Although the general expectation is that these transitions are of first order, based on symmetry principles (Table 22.1), this need not necessarily be the case. Just as pressure as an external variable influences the first-order nature of the phase change in $PbTiO_3$, it may occur that a change in the composition variable c has an analogous effect. Whatmore et al. [22] first investigated this possibility and their X-ray studies, performed on single-crystal specimens, indicated the presence of such a point located at $c^* \approx 0.06$ and $\theta^* \approx 515\,\mathrm{K}$. The fact that the phase change in $PbTiO_3$ is first order necessarily implies that if one such a point exists, there must be at least one other. A number of subsequent studies have been carried out to more precisely determine the location of such a point or points [23–27]. While there is presently no general agreement concerning the exact locations where these points occur, all of the available data do suggest that the first-order nature of the phase transition decreases significantly on approaching the center of the $c - \theta$ diagram from either end member. The studies of Eremkin et al. [23] provide a convincing demonstration of this and their data, showing the discontinuities in the unit cell volume $\Omega(\theta_c)$ and tetragonal distortion $\delta(\theta_c) = [(c_t/a_t) - 1](\theta_c)$ as functions of composition are plotted in Fig. 22.6.

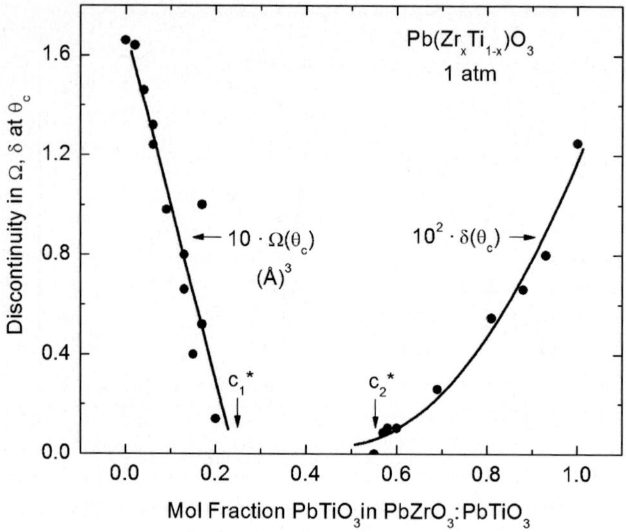

Fig. 22.6. Discontinuity in the unit cell volume $\Omega(\theta_c)$ and tetragonal lattice distortion $\delta(\theta_c)$ as a function of composition in the $PbZrO_3$:$PbTiO_3$ system at atmospheric pressure (adapted from [23])

As shown in the figure, the vanishing values of $\Omega(\theta_c)$ and $\delta(\theta_c)$ indicate the existence of two tricritical points: one located at $c_1^* \approx 0.22$ and another located at $c_2^* \approx 0.55$. The possible existence of such points in the $c-\theta$ diagram of ferroelectric solutions has important consequences that follow directly from the phenomenological theory. First, as discussed earlier, a tricritical point corresponds to the condition that the quartic term(s) in the thermodynamic potential change their sign(s). Consequently, there is a close connection between the location of a tricritical point and the topology of the $c-\theta$ diagram. Such a connection has in fact been observed. In the phase diagram corresponding to Fig. 22.6, Eremkin et al. [23] found that the morphotropic boundary separating tetragonal and rhombohedral phases was not vertical as shown in Fig. 22.5, but instead showed pronounced temperature dependence with the boundary "bending" sharply toward more Zr-rich compositions and intersecting the line of paraelectric to ferroelectric transitions at the point (c_1^*, θ^*) with $\theta^* \approx 610\,\mathrm{K}$. Second, the domain wall energy on either side of a tricritical transition is expected to display a different dependence on temperature below the θ_c. Salje [28] shows that the domain wall energy at $\theta < \theta_c$ for a tricritical transition increases linearly with $(\theta - \theta_c)$, while in a second-order transition the domain wall energy increases as $|(\theta - \theta_c)|^{3/2}$. As a result, the domain wall energy is always greater for the tricritical transition compared with that for the second-order transition and so domain wall formation in the latter case is much more strongly favored. On changing composition in a narrow region in the vicinity of a tricritical point, dramatic changes in the domain structure may occur and these can be expected to strongly affect the dielectric and piezoelectric response.

In this connection, it is of interest to interrogate the relative contribution of the monodomain single-crystalline state to the observed dielectric response of compositions located close to the MPB. Following the approach of Devonshire, Amin et al. [29] have estimated the ratios of all the coefficients in thermodynamic potential (22.12) for $PbZrO_3{-}PbTiO_3$ by fitting the phase boundaries of the $c{-}\theta$ diagram subject to the constraint that, as observed experimentally, the ferroelectric orthorhombic $mm2$ phase is everywhere metastable. Using a procedure not dissimilar from the one described above for the example case of $PbTiO_3$, measurements of the specific heat and crystal lattice parameters were then used to scale these parameters. In subsequent work, Haun et al. [30] accounted for the presence of two tricritical points in the $c{-}\theta$ diagram and further trimmed the values of the coefficients in the vicinity of the morphotropic boundary. To do so, dielectric data were acquired at liquid helium temperatures where the extrinsic contributions to the response are expected be frozen out. Because the Landau theory is generally not valid far from the Curie temperature, such an approach should be regarded as a model phenomenological theory with adjustable constants. In this capacity, it serves as a reasonable a first-order approximation to illustrate how the monodomain contributions to the dielectric and piezoelectric properties may vary in a narrow interval near the point where the expansion coefficients of the thermodynamic potential have been constrained. Using this approach, the dielectric susceptibility $\chi_{ij}^{\theta,\mathrm{T}}$ (and hence the relative dielectric permittivity tensor $K_{ij}^{\theta,\mathrm{T}}$) can be calculated from (22.12), viz.,

$$\frac{\partial^2\Phi}{\partial\eta_i\partial\eta_j} = \frac{1}{\chi_{ij}^{\theta,\mathrm{T}}}. \tag{22.29}$$

Figure 22.7 shows the temperature dependence of K^{T} as measured on an electromechanically soft (Nb-modified) morphotropic composition [31], having a Curie temperature approximately the same as that of the corresponding unmodified material. Also shown in the figure are the computed monodomain values of K_{11} and K_{22} at 4.2 K for the unmodified material and the temperature variation of the corresponding polycrystalline average value of these, $\langle K^{\mathrm{T}}\rangle$. At temperatures <50 K, the experimental and computed values closely agree as expected, but as the temperature is raised the increase that is observed in the measured data far exceeds the computed results, which would be predicted from the temperature variation given by the phenomenological theory (22.29). At room temperature, the value of the measured permittivity is greater than the predicted value by a factor of >2. It is apparent from the figure that over most of the temperature range, the monodomain contributions make up only a small portion of the total observed dielectric response and that the vast majority of the response observed in the experimental data

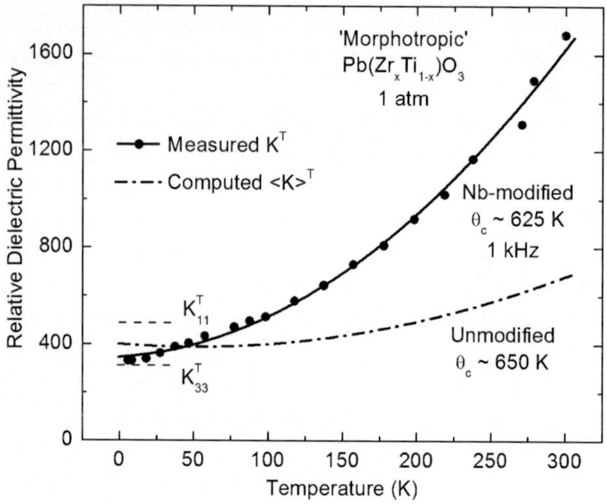

Fig. 22.7. Comparison of the measured relative dielectric permittivity of a Nb-modified morphotropic $Pb(Zr_xTi_{1-x}O_3)$ composition with the relative dielectric permittivity calculated by averaging the monodomain contributions as given by the phenomenological theory. The computational results are based on the work of Haun et al. [30] and the experimental data are taken from Zhang et al. [31]

at room temperature is not intrinsic to the homogeneous, monodomain state. The piezoelectric moduli d_{mkl} are given by an equation of the form

$$d_{mkl} = \varepsilon_0 \chi_{mj}^{\theta,\mathrm{T}} q_{ijkl} \eta_i, \tag{22.30}$$

so the response of $\langle d \rangle$ will show a trend similar to that of $\langle K^{\mathrm{T}} \rangle$.

Since twinning is unavoidable in ferroelectric perovskite crystals, much of the extrinsic contributions must be attributed to the thermal activation of domain wall motion and domain wall-defect interactions. In solid solution systems, however, more complex patterns of spatially inhomogeneous states may arise than those are found in stoichiometric compounds due to equilibration of the system with respect to the compositional degree of freedom. In fact, according to the Gibbs phase rule, the diagram shown in Fig. 22.5 cannot be an equilibrium diagram since this requires that a two-phase field must separate the single-phase regions. The conventional diagram shown above can only be observed under the special condition that diffusional processes providing equilibration of the system are fully frozen and that the transformations are diffusionless, i.e., that there are no atomic redistributions. In contradiction to this fact is the observation that, except under very specialized sample preparation conditions, some region of phase coexistence is almost always observed in $PbZrO_3$–$PbTiO_3$ materials and the width of this region varies substantially depending on both the specimen under study and the resolution and sensitivity of the measurement technique used to study it. Examples from the literature for data collected near room temperature give values that range from essentially 0 to as much as 15 mole % [32, 33].

The origins of phase coexistence in morphotropic phase systems have long been controversial and Isupov [34] has collected much the literature on this topic. Based on considerations of the thermodynamics of mixing in the approximation of the theory of regular solutions, it is possible, however, to easily establish some limiting cases. Following from (22.8), the thermodynamic potential can be written as

$$\Phi(c,\theta,\eta) = \Phi_0(c,\theta) + \Phi(c,\theta,\eta), \tag{22.31}$$

where the first term on the right-hand side of the equation represents the free energy of mixing and the second term is the Landau potential. Adopting the conventional regular solution model for the mixing free energy and using the form (22.17) for the η-dependent part of the thermodynamic potential gives

$$\begin{aligned} \Phi(c,\theta,\eta,\boldsymbol{n}) = & \frac{1}{2}Wc(1-c) + k_{\mathrm{b}}\theta\left[c\ln c + (1-c)\ln(1-c)\right] \\ & + \frac{1}{2}\alpha(c)[\theta-\theta_{\mathrm{c}}(c)]\eta^2 + \frac{1}{4}(c-c_{\mathrm{m}}) \\ & \times b_0\left[1-\lambda(n_1^4+n_2^4+n_3^4)\right]\eta^4 + \frac{1}{6}\gamma\eta^6, \end{aligned} \tag{22.32}$$

where k_b is a Boltzmann's constant, W is the atomic mixing parameter, and $c_{\mathrm{m}} = 0.5$ is the nominal composition of the diffusionless tetragonal–rhombohedral MPB. Based on measurements of the enthalpies of formation, Rane et al. [35] have shown that the regular solution model is approximately obeyed for $PbZrO_3$–$PbTiO_3$ and that the system displays a positive enthalpy of mixing, $W > 0$. To make some computations using (22.32), we may take b_0, γ, $\lambda > 0$ to be constants, independent of c, and give $\alpha(c)$ and $\theta(c)$ linear dependencies between the known values for $PbZrO_3$ and $PbTiO_3$. Noting that $\boldsymbol{n} = \{0,\ 0,\ 1\}$ for the tetragonal phase and $\boldsymbol{n} = \{1/\sqrt{3}, 1/\sqrt{3}, 1/\sqrt{3}\}$ for the rhombohedral phase and applying the minimum principle (22.9) gives the equilibrium phase diagram for different values of W [36]. The results of such an analysis are shown in Fig. 22.8.

The diagram in the panel at left was computed by finding the locus of points in the $c-\theta$ plane where the free energy density curves of the tetragonal and rhombohedral phases are equal. This diagram represents the diffusionless limit. The remaining two diagrams were computed in the conventional way by finding the points of common tangency on the free energy density vs. composition curves. The diagram at center represents the situation where the $PbZrO_3$–$PbTiO_3$ system behaves as an ideal solution ($W = 0$) and the diagram at right represents the situation where the systems behaves as regular solution with a positive enthalpy of mixing ($W > 0$) constrained to be sufficiently small such that no decomposition may occur in the cubic phase. If diffusional processes are operative, the diagrams in Fig. 22.8 indicate that the

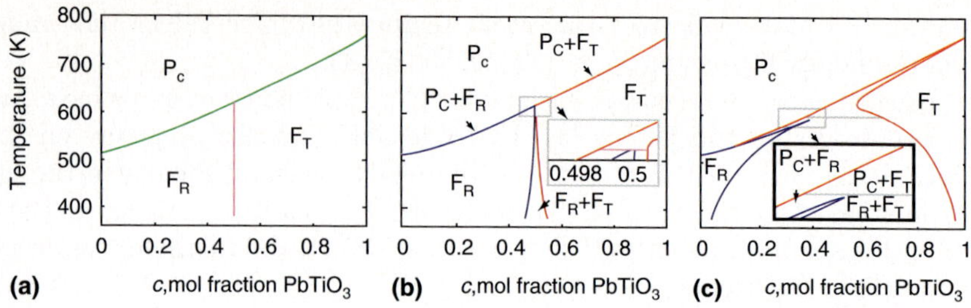

Fig. 22.8. Calculated composition–temperature diagrams of the $PbZrO_3$:$PbTiO_3$ system at atmospheric pressure: (**a**) diffusionless, (**b**) ideal solution, and (**c**) regular solution with a positive enthalpy of mixing (adapted from [36])

time–temperature–transformation (T–T–T) history will determine whether decomposition in fact occurs into a mixture of tetragonal and rhombohedral phases or if the cooling rate is sufficiently fast to provide the metastable diffusionless diagram. Given the large ranges of phase coexistence reported in the literature [32–34], it is reasonable to speculate that diffusional processes are in fact operative under normal processing conditions of $PbZrO_3-PbTiO_3$ materials. The region of phase coexistence that is typically observed at room temperature will then lie somewhere between diffusionless limit and that which is given by the equilibrium condition depending on the details of the (T–T–T) history. The fact that spontaneous equilibration processes result in the formation of two-phase states that are by definition compositionally, structurally, and electrically inhomogeneous is expected to have important implications for the precise control of the processing of $PbZrO_3-PbTiO_3$ materials and also in understanding the relations of their domain structures to properties at compositions close to morphotropic boundaries.

The mixing thermodynamics of ferroelectric solid solutions represents an area of research into these systems that has not yet been covered in detail. Takayama-Muromachi and Navrotsky [37] have shown that there are systematic correlations among the heats of formation ΔH_{f} of the perovskite oxide compounds and that the measured enthalpies scale almost linearly with the degree of deviation from the ideally close-packed cubic structure as characterized by the parameter:

$$\Delta a_{\mathrm{c}} = a_{\mathrm{c}}^{\mathrm{calc}} - a_{\mathrm{c}}^{\mathrm{meas}}, \tag{22.33}$$

where $a_{\mathrm{c}}^{\mathrm{meas}}$ is the measured (equivalent) cubic crystal lattice parameter and $a_{\mathrm{c}}^{\mathrm{calc}}$ is computed from the ionic radii. Further, they showed that when the changes in the Madelung energy contributions ΔE_{M} to the lattice energy ΔU were separated from the nonelectrostatic contributions ΔE_{N} by means of the approximation

$$\Delta H_{\mathrm{f}} \approx \Delta U = \Delta E_{\mathrm{M}} + \Delta E_{\mathrm{N}} \tag{22.34}$$

that these latter two quantities also scale linearly with Δa for a large number of perovskite compounds. Since the change in Madelung energy can be directly

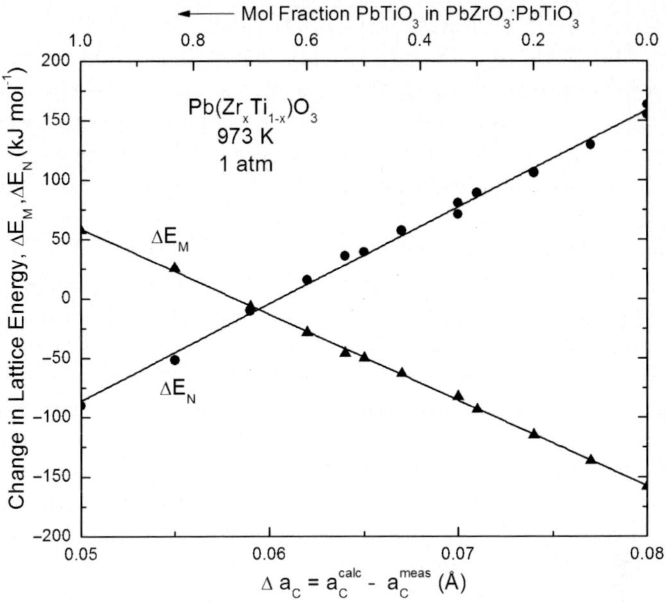

Fig. 22.9. Changes in Madelung energy (ΔE_{M}) and nonelectrostatic energy (ΔE_{N}) contributions to the change in lattice energy in the $PbZrO_3$:$PbTiO_3$ perovskite system as functions of the degree of deviation from the ideally close-packed cubic structure (Δa_{c}) (adapted from [27])

computed, the measured formation enthalpy data can be used to determine ΔE_{N} by difference. The expected trend is nicely displayed for the $PbZrO_3$–$PbTiO_3$ system as shown in Fig. 22.9.

Alternatively, Davies and Navrotsky [38] have made an analysis of the atomic interaction parameter W for both cation and anion mixing in large number of divalent oxide compounds. Their results show that analogous trends exist for these solid solutions wherein W scales with the difference in the molar volumes of the end-member compounds. Knowledge of the mixing properties for different ferroelectric solid solution systems therefore provides an empirical means to correlate the topologies of the phase diagrams with the crystallochemical parameters of ionic size, charge, and/or degree of covalence.

22.4 Summary

The phenomenological theory provides a convenient framework within which the phase transitions and physical properties of ferroelectric crystals can be interpreted. Because the parameters of the theory are directly related to macroscopically observed quantities, they can be determined in a self-consistent way from a small number of experimental measurements. The general observation that for ferroelectric perovskite crystals the range of validity of the theory away from the Curie temperature is wider than that would be expected extends its utility in describing the physical properties of these technologically important materials. Deviations of measured properties from those predicted by the parameters of theory thus provide the logical starting point for investigations into more complex phenomena and for inquiries into their possible microscopic origins.

Formal extension of the theory to include composition as a parameter of the thermodynamic potential can be used to compute possible topologies of the composition–temperature phase diagram and to explain the existence of spatially inhomogeneous states that may be more complex than those associated only with the conventional symmetry twin structures. Considerations predicated on the possible topologies of these diagrams at equilibrium may provide insights useful in constructing hypothesis-driven experiments, leading to better control of their piezoelectric and other structure-sensitive properties. Knowledge of the mixing properties of such systems further provides a possible empirical means to correlate the topologies of such phase diagrams with crystallochemical parameters.

References

1. A.F. Devonshire, Philos. Mag. **40**, 1040 (1949)
2. A.F. Devonshire, Philos. Mag. **42**, 1065 (1951)
3. J.C. Slater, Phys. Rev. **78**, 748 (1950)
4. C. Kittel, Phys. Rev. **82**, 729 (1951)
5. L.E. Cross, Philos. Mag. Ser. 8, **1**, 76 (1956)
6. F. Jona, G. Shirane, *Ferroelectric Crystals* (Pergamon, New York, 1962)
7. B.A. Strukov, A.P. Levanyuk, *Ferroelectric Phenomena in Crystals* (Springer, Berlin, 1998)
8. L.A. Shuvalov, J. Phys. Soc. Jpn **28**, 38 (1970)
9. D. Vanderbilt, M.H. Cohen, Phys. Rev. B **63**, 094108 (2001)
10. K. Wójcik, Ferroelectrics **99**, 5 (1989)
11. G.A. Rossetti Jr., J.P. Cline, A. Navrotsky, J. Mater. Res. **11**, 3179 (1998)
12. G.A. Rossetti Jr., N. Maffei, J. Phys.: Condens. Matter **17**, 3953 (2005)
13. G.A. Samara, Ferroelectrics **2**, 277 (1971)
14. R. Ramírez, M.F. Lapeña, J.A. Gonzalo, Phys. Rev. B **42**, 2604 (1990)
15. A. Sani, M. Hanfland, D. Levy, J. Solid State Chem. **167**, 446 (2002)
16. B. Jaffe, W.R. Cook, H. Jaffe, *Piezoelectric Ceramics* (Academic, London, 1971)
17. A.M. Glazer, P.A. Thomas, K.Z. Baba-Kishi, G.K.H. Pang, C.W. Tai, Phys. Rev. B **70**, 184123 (2004)
18. B. Noheda, D.E. Cox, G. Shirane, J.A. Gonzalo, L.E. Cross, S.-E. Park, Appl. Phys. Lett. **74**, 2059 (1999)
19. D.I. Woodward, J. Knudsen, I.M. Reaney, Phys. Rev. B **72**, 104110 (2005)
20. B. Noheda, Curr. Opin. Solid State Mater. Sci. **6**, 27 (2002)
21. A.J. Bell, J. Mater. Sci. **41**, 13 (2006)
22. R.W. Whatmore, R. Clarke, A.M. Glazer, J. Phys. C: Solid State Phys. **11**, 3089 (1978)

23. V.V. Eremkin, V.G. Smotrakov, E.G. Fesenko, Ferroelectrics **110**, 137 (1990)
24. B. Noheda, N. Cereceda, T. Iglesias, G. Lifante, J.A. Gonzalo, H.T. Chen, Y.L. Wang, Phys. Rev. B **51**, 16388 (1995)
25. S.K. Mishra, A.P. Singh, D. Pandey, Philos. Mag. B **76**, 213 (1997)
26. G.A. Rossetti Jr., A. Navrotsky, J. Solid State Chem. **144**, 188 (1999)
27. G.A. Rossetti Jr., J.P. Cline, A. Navrotsky, Y.-M. Chiang, J. Phys.: Condens. Matter **14**, 8131 (2002)
28. E.K.H. Salje, Phase Transitions **34**, 25 (1991)
29. A. Amin, M.J. Haun, B. Badger, H. McKinstry, L.E. Cross, Ferroelectrics **65**, 107 (1985)
30. M.J. Haun, Z.Q. Zhuang, E. Furman, S.J. Jang, L.E. Cross, Ferroelectrics **65**, 45 (1989)
31. X.L. Zhang, Z.X. Chen, L.E. Cross, W.A. Schluze, J. Mater. Sci. **18**, 968 (1983)
32. B. Noheda, D.E. Cox, G. Shirane, R. Guo, B. Jones, L.E. Cross, Phys. Rev. B **63**, 014103 (2000)
33. T. Yamazaki, A. Onodera, H. Fujishita, Y. Ishikawa, M. Tanaka, Ferroelectrics **266**, 139 (2002)
34. V.A. Isupov, Ferroelectrics **266**, 91 (2002)
35. M.V. Rane, A. Navrotsky, G.A. Rossetti Jr., J. Solid State Chem. **161**, 402 (2001)
36. G.A. Rossetti Jr., W. Zhang, A.G. Khachaturyan, Appl. Phys. Lett. **88**, 072912 (2006)
37. E. Takayama-Muromachi, A. Navrotsky, J. Solid State Chem. **72**, 244 (1988)
38. P.K. Davies, A. Navrotsky, J. Solid State Chem. **46**, 1 (1983)

23

Effective Medium Theories

W. Kreher

23.1 Introduction

Predicting macroscopic properties of materials involves two tasks:

1. Projection of macroscopic loading conditions onto the microstructural level with sufficient and adjusted details (resulting in fluctuating internal fields)
2. Averaging the local reaction of material to obtain the macroscopic response to the loadings

For piezoelectric/ferroelectric polycrystalline materials, the relevant microstructural level is presented by special domain configurations within grains which have different orientations. The local material response comprises the elastic, dielectric, and intrinsic piezoelectric effect as well as the restructuring of domain configuration by movement of domain walls or complete reorientation of the spontaneous polarization, i.e., domain switching. Other processes as moving free charges and/or defect dipoles are normally not considered in present micro–macro models. They appear in the form of already averaged local properties, e.g., as an effective frictional force or activation energy for domain wall movement.

Obviously, exact three-dimensional solutions for the internal fields and processes are impossible, so several simplifications are introduced in the models depending on the aim (i.e., whether linear, nonlinear, or complete repolarization behavior is considered). Typical approximations are:

- Simplification of domain configuration either by a particular deterministic domain arrangement (e.g., stack of plate-like domains) or by assumption of a set of noninteracting representative domains
- Simplification of the interactions between neighboring grains (e.g., assumption of homogeneous fields, smoothing by the effective medium approach)

The present chapter describes these modeling approaches in some detail, putting emphasis on partially exact solutions and practical applicability.

Section 23.2 describes the relevant modeling tools, namely the exact solutions for a lamellar domain stack and for an ellipsoidal inclusion in a homogeneous matrix. On this basis, the effective medium approach for a polycrystalline solid is introduced. Section 23.3 extends the theory to the extrinsic properties which are determined by the (reversible) movement of domain walls. Finally, Sect. 23.4 gives a short overview about modeling the nonlinear behavior which is connected with extensive domain reconfiguration.

To represent the theoretical background in a compact and comprehensible form, it is necessary to introduce a short form notation for the piezoelectric material laws. These so-called constitutive equations, relating mechanical stress and electric displacement to strain and electric field, are expressed by

$$\begin{aligned} T_p &= c^E_{pq} S_q - e_{lp} E_l + T^E_p, \\ D_k &= e_{kq} S_q + \varepsilon^S_{kl} E_l + D^S_k, \end{aligned} \qquad p,q = 1,2,\ldots,6,\ k,l = 1,2,3. \tag{23.1}$$

(We make use of the abbreviated Voigt notation and the Einstein summation convention.) Here, T^E_p is the stress state which arises if a solid, which is clamped ($S_q = 0$) and short circuited ($E_l = 0$), undergoes a ferroelectric/ferroelastic transformation, and D^S_k is the electric displacement observed under the same conditions. Of course, these remnant properties may be expressed by the corresponding quantities under stress-free conditions (i.e., transformation strain and spontaneous polarization); yet the representation adopted here is especially suited for the derivation of material properties within the effective medium approximation.

Though we do not explicitly consider pyroelectric and thermoelastic effects, it should be mentioned that these effects are comprised by T^E and D^S, too. Thus, any theoretical result shown below can be utilized to make predictions also for these thermal properties.

Combining mechanical and electrical field quantities into 9-dim vectors gives the compact notation:

$$j_i = L_{ij} f_j + j^{\mathrm{s}}_i, \quad i,j = 1,2,\ldots,9 \tag{23.2}$$

with

$$j = \begin{pmatrix} T_p \\ D_k \end{pmatrix}, \quad L = \begin{pmatrix} c^E_{pq} & e^{\mathrm{t}}_{pl} \\ e_{kq} & -\varepsilon^S_{kl} \end{pmatrix}, \quad f = \begin{pmatrix} S_q \\ -E_l \end{pmatrix}, \quad j^{\mathrm{s}} = \begin{pmatrix} T^E_p \\ D^S_k \end{pmatrix}. \tag{23.3}$$

Here, the superscript ^{t} means the transposition of the 3×6 matrix of piezoelectric constants. The matrix L represents the complete set of the linear elastic, piezoelectric, and dielectric constants; it is a symmetric 9×9 matrix, i.e.,

$$L = L^{\mathrm{t}}. \tag{23.4}$$

Note that this symmetry is a consequence of the special representation of f with a negative sign for the electric field.

All preceding equations fit into the standard vector–matrix calculus which is especially convenient for the numerical evaluation. Subsequently, uppercase letters denote 9×9 matrices and lowercase letters denote 9×1 vectors. So, in most cases, we can avoid writing any indices. In particular, we find for the inversed material law

$$f = Mj + f^{\mathrm{s}}. \tag{23.5}$$

Again, the matrix M of inverse linear material constants is a symmetric 9×9 matrix and f^{s} represents the remnant properties under fixed stress and charge conditions:

$$M = L^{-1} = \begin{pmatrix} s_{pq}^{D} & g_{pl}^{\mathrm{t}} \\ g_{kq} & -\beta_{kl}^{T} \end{pmatrix}, \quad f^{\mathrm{s}} = -Mj^{\mathrm{s}} = \begin{pmatrix} S_{p}^{D} \\ -E_{k}^{T} \end{pmatrix}. \tag{23.6}$$

Care must only be taken when a transformation to a local coordinate system (e.g., fixed to the crystal or domain axes) has to be performed. This must be done exclusively in the original tensor representation of the field quantities.

Other combined notations of mechanical and electrical field variables (e.g., the Barnett–Lothe notation) are also possible – yet the variant presented here is best suited for the compact description of effective medium models.

23.2 Linear Intrinsic Properties

We consider here a piezoelectric ceramic consisting of grains with a given random or textured orientation distribution. Each grain consists of a certain arrangement of domains forming a twinned and hierarchically ordered structure. In contrast to the extrinsic properties (see Sect. 23.3), here the domain structure is assumed to be fixed. The poling state is described by a particular orientation distribution of domains.

An exact solution for the internal fields and the effective (macroscopic) behavior of the ceramic is impossible. But a sufficiently satisfying solution is obtained if, as a first step, the domain configuration is approximated by a lamellar domain stack which can be exactly solved (see Sect. 23.2.1). On a second step, the interaction between neighboring grains is taken into account in a smoothed form by means of the effective medium approach (EMA). The EMA is an approximation well known for long in mechanical and electrical problems for polycrystals and composites (connected with the names of Bruggemann, Kröner, Hill). EMA means that the surroundings of a particular grain are replaced by a homogeneous matrix, described by material constants which are given by the still unknown effective properties. To evaluate the effective constants, one first has to solve the general inclusion problem. This is possible by means of the generalized Eshelby tensors (Sect. 23.2.2). The resulting equations for the generalized EMA are described in Sect. 23.2.3.

23.2.1 Exact Solution for a Lamellar Domain Stack

In general, grains have a complex hierarchically structured domain configuration. First, Arlt [1] has analyzed domain configurations with different complexity (see also [2]). Yet for application in effective medium theories, it is generally not necessary to consider the domain configuration in full detail since most subtleties of internal fields are averaged out when proceeding to the macroscopic level. Therefore, a good compromise between necessary details and practically tractable models is provided by a lamellar stack of domains with plane-parallel domain walls. Every lamella is described by a linear material law as given in Sect. 23.1 with known fixed material constants. The thickness of lamellas is assumed to be much smaller than the grain size.

If a homogeneous external field $\bar{f}$ is applied to such a lamellar stack, internal fields will result which are homogeneous in each lamella except for small end regions of the domains. Those disturbances, however, are negligible when we consider macroscopic properties (though possibly not for nonlinear processes as domain nucleation).

Let us assume that the normal of the domain stack is parallel to the 3-direction. Then, continuity of mechanical and electrical fields yields the following equations:

$$f_i = \bar{f}_i \quad \text{for} \quad i = 1, 2, 6, 7, 8, \tag{23.7}$$

$$j_i = \bar{j}_i \quad \text{for} \quad i = 3, 4, 5, 9. \tag{23.8}$$

These equations are valid for each lamella, i.e., the field components appearing in (23.7) and (23.8) must be homogeneous throughout the stack.

Now, we consider an individual lamella. Then, (23.7) and (23.8) together with the local material law (23.2) form nine conditions for nine unknowns $(f_1, \ldots, f_9)$. However, since we only know $\bar{f}$, which is prescribed, we do not have the field components $\bar{j}_3$, $\bar{j}_4$, $\bar{j}_5$, $\bar{j}_9$ necessary to utilize (23.8). The missing four additional global equations are obtained if the average field condition is taken into account:

$$\langle f_i \rangle = \bar{f}_i, \quad i = 3, 4, 5, 9. \tag{23.9}$$

Here, $\langle \cdot \rangle$ means averaging over all lamellas. Of course, (23.9) is valid also for $i = 1,\ 2,\ 6,\ 7,\ 8$, but since these field components do not fluctuate within the stack (cf. (23.7)), the equations are trivially fulfilled.

Thus, (23.7)–(23.9) represent a complete set of equations to determine each lamella field, which can be solved analytically. Due to the linearity of the problem, the stack behaves like a homogeneous material described by effective material constants:

$$\bar{j} = \langle j \rangle = L^{\text{stack}} \bar{f} + j^{\text{s,stack}}. \tag{23.10}$$

The effective constants L^{stack} and $j^{\text{s,stack}}$ depend on the properties of domains and their respective volume fraction. Typically, the lamella properties do not fluctuate arbitrarily; instead, one has a limited number of domain types (inclined at fixed angles to the stack orientation). Then, the internal fields are identical for domains belonging to a certain type, and the averaging is done by means of volume fractions of domain types. In Sect. 23.3, a domain stack consisting of two types of domains forming 90° domain walls will be considered.

23.2.2 Exact Solution of the Piezoelectric Inclusion Problem

When a certain inclusion (whether it is a homogeneous single-domain grain or a multidomain grain with certain homogenized properties) is embedded into a homogeneous matrix material, the general inclusion problem arises. In most cases, the shape of the inclusion can be approximated by an ellipsoid (including spherical, fiber-like, or plate-like shapes). Then, the general problem can be reduced to the elementary inclusion problem: Consider a homogeneous piezoelectric material where a small ellipsoidal subregion (much smaller than the sample volume) undergoes a homogeneous transformation f^{s} (cf. (23.6)). In this case, the resulting internal field is homogeneous inside the inclusion and given by

$$f = S f^{\mathrm{s}}. \tag{23.11}$$

The matrix S (which is not symmetric) is the generalization for the piezoelectric case of the depolarization factors for pure dielectric materials and of the Eshelby tensor for elastic materials. This matrix depends on the (homogeneous) piezoelectric properties of the material and on the shape of the inclusion. The calculation of this Eshelby matrix has been extensively considered in the literature (see, e.g., [6,9]).

For illustration, we present in Fig. 23.1 numerical results for a poled ceramic, which is transversely isotropic (hexagonal symmetry). The inclusion has spheroidal shape with the axis of rotation parallel to the poling direction (axes $a_1 = a_2$, a_3). The results were obtained by using the analytical expressions, which involve a numerical integration, given by [9] (after correcting a number of misprints).

Provided the Eshelby matrix is known, it is a relatively easy task to construct the solution to the general inclusion problem. Here, the piezoelectric material constants may differ (L, j^{s} for the inclusion and L^*, j^* for the surrounding matrix) and an additional far field f^0 is applied by prescribed external boundary conditions. Again, the field inside the inclusion is homogeneous and is found to be

$$j = Lf + j^{\mathrm{s}} \quad \text{with} \quad f = A(f^0 - S(L^*)^{-1}(j^{\mathrm{s}} - j^*)). \tag{23.12}$$

The matrix A, often called *concentration factor*, is defined by (cf. [5])

$$A = (I + S(L^*)^{-1}(L - L^*))^{-1}, \tag{23.13}$$

where I is the 9×9 unit matrix. The Eshelby matrix in (23.12) and (23.13) depends on the piezoelectric properties of the matrix and on the geometry of the inclusion, i.e.,

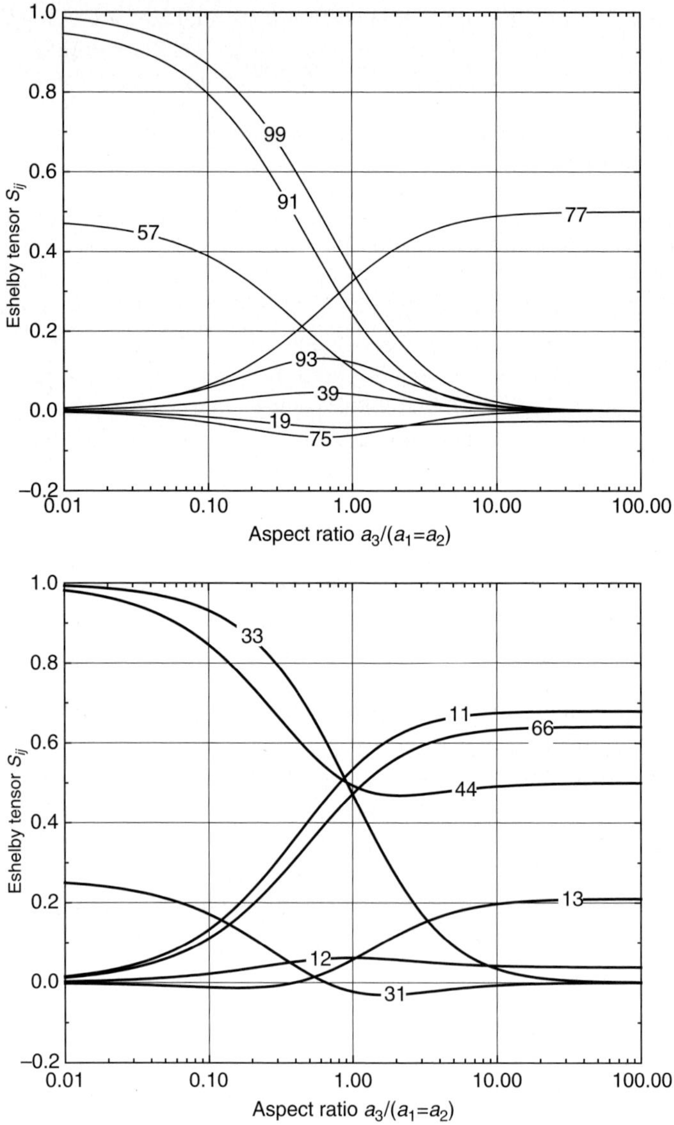

Fig. 23.1. Nonvanishing components of the Eshelby tensors. Piezoelectric material constants are given by $c_{11}^{\mathrm{E}} = 126\,\mathrm{GPa}$, $c_{12}^{\mathrm{E}} = 55\,\mathrm{GPa}$, $c_{13}^{\mathrm{E}} = 53\,\mathrm{GPa}$, $c_{33}^{\mathrm{E}} = 117\,\mathrm{GPa}$, $c_{44}^{\mathrm{E}} = 35.3\,\mathrm{GPa}$, $e_{31} = -6.5\,\mathrm{C\,m^{-2}}$, $e_{33} = 23.3\,\mathrm{C\,m^{-2}}$, $e_{15} = 17.0\,\mathrm{C\,m^{-2}}$, $\varepsilon_{11}^{\gamma} = 15.1\,\mathrm{nF\,m^{-1}}$, $\varepsilon_{33}^{\gamma} = 13.0\,\mathrm{nF\,m^{-1}}$. The components of S are dimensionless except $[S_{19}] = [S_{39}] = [S_{57}] = \mathrm{nC/N}$, $[S_{91}] = [S_{93}] = [S_{75}] = \mathrm{N/nC}$

$$S = S(L^*, \psi), \tag{23.14}$$

where ψ stands for the shape (and possibly orientation) of the inclusion considered.

The elementary inclusion problem is recovered if we put $L = L^*, j^* = 0$, $f^0 = 0$, $j^{\mathrm{s}} = -Lf^{\mathrm{s}}$ (cf. (23.6)).

23.2.3 Effective Medium Approach

The effective (macroscopic) piezoelectric properties of the polycrystalline ceramic relate macroscopic fields in the same form as the local constitutive equations:

$$\bar{j} = L^* \bar{f} + j^*. \tag{23.15}$$

The macroscopic fields can be obtained by averaging over all local fields:

$$\langle j \rangle = \bar{j} \quad \text{and} \quad \langle f \rangle = \bar{f}, \tag{23.16}$$

where the averaging procedure $\langle \cdot \rangle$ may be performed as a spatial average or by a statistical average.

Within the EMA, it is the essential assumption that the local fields inside the grains can be calculated using the solution of the inclusion problem (23.12) where the (still unknown) effective properties are set as matrix properties. Thus, the interaction of an individual grain with all neighboring grains can be taken into account in a smoothed form.

Considering a polycrystalline material, the averaging according to (23.16) with (23.12) must include all variations of inclusion properties, i.e., piezoelectric grain properties L, j^{s}, and grain shape and orientation statistics if present. Since (23.15) and (23.16) are valid for any applied external field $\bar{f}$, one obtains a sufficient number of equations to determine the unknown effective properties L^*, j^* and also the unknown far field f^0, which need not be equal to the average field since it may contain modifications due to the surrounding grains.

Often, only one of the conditions (23.16) is stated. As a substitute, the far field f^0 is replaced simply by the average field. However, in general, such an arbitrary assumption yields nonunique results depending on whether the first or the second condition in (23.16) is used. In our approach, the symmetric treatment of j and f yields unique results and, in addition, offers the possibility to summarize unknown fluctuating fields within the effective far field f^0.

Evaluating (23.15) and (23.16) with (23.12) and (23.13) gives after some manipulations (eliminating f^0) the following result for the effective constants according to EMA:

$$L^* = \langle A^{\mathrm{t}} \rangle \langle A^{\mathrm{t}} L \rangle, \tag{23.17}$$

$$j^* = \langle A^{\mathrm{t}} \rangle \langle A^{\mathrm{t}} j^{\mathrm{s}} \rangle. \tag{23.18}$$

Here, A^{t} is the transposed matrix A from (23.13). Since A depends also on the effective properties L^* (explicitly and via S), (23.17) has an implicit form which must be solved numerically. When L^* is known, the remnant properties j^* can be evaluated directly. Numerical results are shown in Sect. 23.3.2 in comparison with extrinsic properties.

23.3 Linear Extrinsic and Loss Properties

In Sect. 23.2, all material properties (volume fraction of domains, properties of grains) have been assumed to be constant, i.e., independent of the applied field. In most cases, however, domain walls are shifted by the application of an external field giving an additional contribution to the effective piezoelectric constants and the remnant properties. In first order, the material behavior is still linear but with modified extrinsic material constants.

To describe this effect within an effective medium model, the input properties must be adjusted according to the change in volume fraction of different domains. In this section, we restrict ourselves to small signal properties, i.e., complete reordering or creation of domains is excluded (nonlinear large signal behavior is considered in Sect. 23.4). Under this condition, we can determine the (reversible) shift of domain walls from the principle of minimum free enthalpy. In Sect. 23.3.1, model predictions are demonstrated for the typical case of the lamellar stack of domains. If these results are fed into the effective medium model, one obtains predictions for the extrinsic macroscopic properties which are shown in Sect. 23.3.2.

Even under small signal conditions, domain wall motion is connected with the dissipation of energy (i.e., in the thermodynamic sense, it is a nonreversible process). If one incorporates this effect into the models, one may relate the microscopic friction forces for domain wall movements to the macroscopically observed hysteretic loss. A typical result is discussed in Sect. 23.3.3.

23.3.1 Equilibrium Domain Configuration Under Applied Fields

We consider here a single grain embedded in a homogeneous matrix (the matrix volume is much larger than the inclusion volume). The grain consists of a lamellar stack of 90° domains, i.e., just two types of domains are present, differentiated only by the orientation of otherwise identical material. Then, we have just one microstructural state variable, namely the volume fraction ξ ($0 \leqslant \xi \leqslant 1$) of one type of domains.

Each grain is modeled according to (23.10) where the stack properties now serve as effective grain properties. Thus, the grain material constants become dependent on the domain volume fraction:

$$L^{\mathrm{grain}} = L^{\mathrm{grain}}(\xi), \quad j^{\mathrm{s,grain}} = j^{\mathrm{s,grain}}(\xi). \tag{23.19}$$

These properties can be fed into (23.17) and (23.18) in place of L and j^{s}, respectively. Together with matrix properties supposed to be known, (23.12) yields f which is now the average inclusion field. It is not necessary to determine the residual stress and electric field outside the inclusion since rigorous expressions are available which relate the potential energy of the solid to the field inside the inclusion (see [16, 17]). Here, potential energy means the internal field energy including the potential of the loading forces. For example, under isothermal conditions and for prescribed strain and electric field, the potential energy is a combination of free elastic energy and free electric enthalpy.

Figure 23.2 shows a typical result demonstrating the equilibrium position of domain walls. Without an external field, both types of domains have equal volume fractions, whereas under the action of the electric field the favorably oriented domains grow on the expense of the other domains. Such calculations take into account the complete mechanical and electrical interaction fields (except for the small end regions of the domains). They can be performed for any orientation of the grain and for any properties of the surrounding medium provided the matrix S of Eshelby tensors is known.

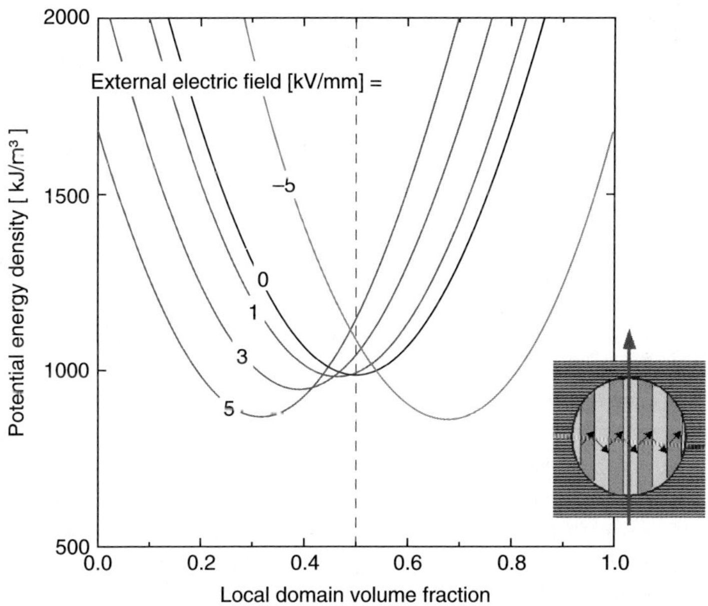

Fig. 23.2. Potential energy w_{p} per volume of a specific inclusion as a function of the volume fraction ξ of the nonfavorably oriented domain. The inclusion is embedded in a matrix which has isotropic properties (i.e., a nonpoled material). The far field f^{0} prescribes a varying electric field under the condition of a constant strain (after [21])

23.3.2 Effective Extrinsic Piezoelectric Constants

The results according to (23.19) may be used within the EMA, but now allowing for small reversible shifts of domain walls. Here, all grains are described by a lamellar stack of 90° domain walls where only the orientation Ω and the domain volume fraction ξ vary. The domain volume fraction is determined by the minimum of the potential energy as described in Sect. 23.3.1. Of course, the local equilibrium value depends on the applied field and on the local orientation:

$$\xi = \xi(\bar{f}, \Omega). \tag{23.20}$$

This together with (23.19) gives the grain properties which in turn can be fed into the effective medium formalism (23.17) and (23.18). The resulting effective properties now become dependent on the applied field. Though in general this relationship is nonlinear, it can be linearized for small fields. To this end, we consider the incremental form of (23.15):

$$\mathrm{d}\bar{j} = \mathrm{d}L^* \bar{f} + L^* \mathrm{d}\bar{f} + \mathrm{d}j^*. \tag{23.21}$$

In first order, the increments $\mathrm{d}L^*$ and $\mathrm{d}j^*$ will be proportional to the applied field increment $\mathrm{d}\bar{f}$. Therefore, we obtain

$$\mathrm{d}\bar{j} = \hat{L}^* \mathrm{d}\bar{f}, \tag{23.22}$$

where the new material property matrix $\hat{L}^*$ is the sum of the already known effective intrinsic properties and the effective extrinsic contributions due to the variations $\mathrm{d}L^*$ and $\mathrm{d}j^*$ caused by the moving domain walls:

$$\hat{L}^* = L^* + L^{*\ \mathrm{extrinsic}}. \tag{23.23}$$

Figure 23.3 shows numerical results vs. a scalar parameter θ_p serving to describe the degree of polarization of the piezoelectric polycrystalline material in a simple form. The approximate orientation distribution function (ODF) of grains is constructed in the following way. Starting from an isotropic ODF, every grain is completely switched by 180° if the angle between the local average polarization vector of the grain and the poling direction is greater than $\pi - \theta_\mathrm{p}$. Thus, the nonpoled ceramic has $\theta_\mathrm{p} = 0$, and the fully poled material is represented by $\theta_\mathrm{p} = \pi/2$. The results in Fig. 23.3 show the large difference between intrinsic single-domain properties, effective intrinsic properties of the polycrystal (which include the orientation effect and the coupling between the d_{31}, d_{33}, and d_{15} effect), and effective extrinsic properties (which include domain wall movements). The predictions for the extrinsic properties are in good accordance with experimental values if we suppose an effective poling angle $\theta_\mathrm{p} = 0.4\pi$.

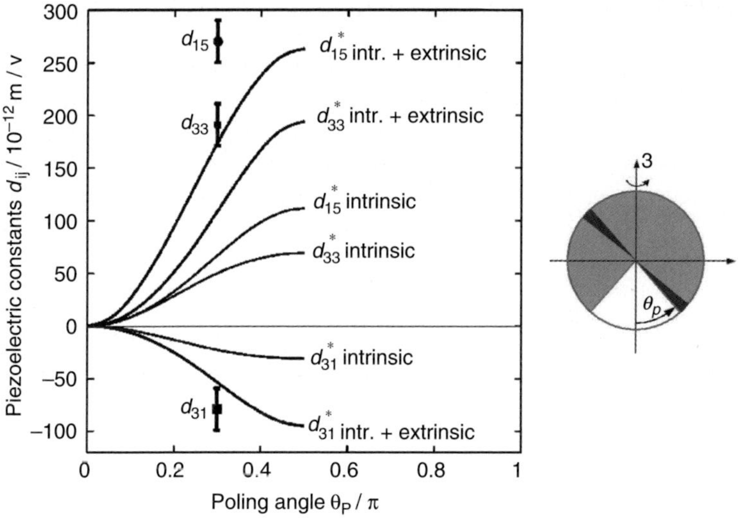

Fig. 23.3. Predictions for intrinsic and extrinsic piezoelectric constants vs. the polarization state for barium titanate (after [22]). The polycrystalline material consists of spherical grains with an orientation distribution according to the poling state described by the poling angle θ_p. Each grain is modeled by a lamellar stack of 90° domains. The piezoelectric constants of a single domain are given by $d_{31} = -33.4\,\mathrm{pm\,V^{-1}}$, $d_{33} = 90.0\,\mathrm{pm\,V^{-1}}$, $d_{15} = 540\,\mathrm{pm\,V^{-1}}$

23.3.3 Dissipative Domain Wall Motion

Any motion of domain walls is connected with a certain amount of dissipated work. The simplest model for this dissipative process is obtained if again the lamellar stack of twinned domains is considered. As discussed earlier, here the volume fraction of domains ξ can serve as a simple local state variable. Since the dissipated work is connected with the total amount of translation of domain walls, this energy is proportional to the accumulated change of volume fraction. So, we obtain for the dissipated work per volume the following expression:

$$w_\mathrm{d}(\xi) = D_0 \int_{\xi_0}^{\xi} |\mathrm{d}\xi|. \tag{23.24}$$

Here, ξ describes the final position of the domain walls and ξ_0 is the starting volume fraction. D_0 is a material property which represents the frictional forces acting on a moving domain wall (it is also possible to consider an aging process by assuming a dissipation constant which depends on the accumulated domain change instead of the constant value D_0).

Equation (23.24) must be applied to any grain orientation and can be combined with the effective medium approximation for the extrinsic properties. In essence, the equilibrium condition for the local domain volume fraction gets the form (cf. Fig. 23.2)

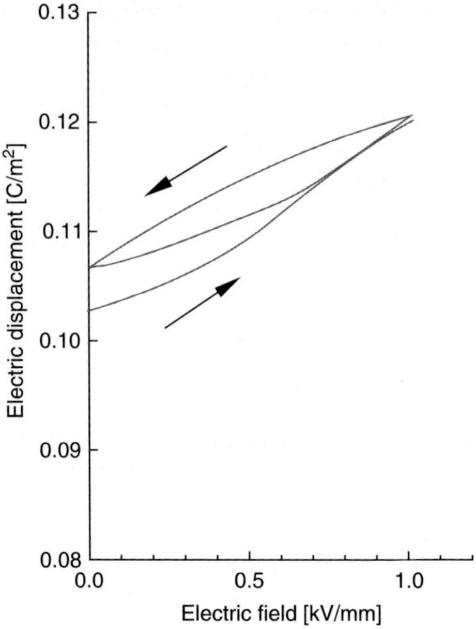

Fig. 23.4. Hysteretic behavior of electric displacement of $BaTiO_3$ vs. applied electric field (after [22]). Dissipation constant $D_0 = 100\,\mathrm{kJ\,m^{-3}}$

$$\frac{\mathrm{d}}{\mathrm{d}\xi} w_\mathrm{p} = \frac{\mathrm{d}}{\mathrm{d}\xi} w_\mathrm{d} = \pm D_0 \quad \text{for} \quad \dot{\xi} \gtrless 0. \tag{23.25}$$

This condition ensures that the dissipated work must be supplied by the release of potential energy.

If this scheme is invoked into the formalism, one obtains a threshold and a hysteresis describing the loss properties due to frictional movement of domains. An example is shown in Fig. 23.4. Thus, macroscopic loss measurements can be correlated to the microscopic dissipation constant D_0.

23.4 Nonlinear Large Signal Behavior

If a ferroelectric ceramic is subjected to large electrical and/or mechanical loadings, complete restructuring of the domain configuration occurs including domain wall motion, domain reorientation/switching, and formation/dissolution of domains. These processes are important during the poling process and also when the complete hysteresis curves are considered.

Theoretical modeling is much more complicated than for the linear behavior and no comprehensive model is feasible. In particular, it is not sufficient to model polydomain grains by the lamellar domain stack model with a simple fixed selection of domain orientation as for the linear model. Instead, one has to allow for a complete set of possible domain orientations (comparable with the complete set of glide systems in polycrystal plasticity). Such a model has been developed by Huber et al. [11] including the interaction of grains using a nonlinear effective medium approximation. Though from the theoretical point of view this model is very satisfying since both the local material behavior and the averaged interaction between grains are taken into account, for practical applications within a FEM code, e.g., to solve boundary value problems relevant to piezoelectric devices, these models require too much computational effort.

So, relatively simple switching models are frequently applied, where single-domain grains, usually without interaction, are considered. The models start with a certain assumption on the crystal/domain orientation. Under the action of external fields, the orientation may switch to another domain variant (in most cases, tetragonal symmetry is assumed so that there are six orientations possible). The condition for this transformation is based on an energy criterion: switching occurs if the release of free enthalpy exceeds a critical value. Sometimes, the analytical expression for the energy release is supplemented by a phenomenological weighting factor which differentiates the electrical and mechanical contribution. In most cases, the local stress and electric field are put equal to the macroscopically prescribed value (e.g., [13]), though there are some approaches where the local fields are calculated within a FEM model [12]. Also, the effective medium approach has been used [14] but for an isotropic effective medium only, i.e., the effect due to poling of the surroundings is neglected.

Pure phenomenological models as the Rayleigh model (see, e.g., [7]), the Preisach approximation (e.g., [3, 20]), and the models with switching surfaces [8, 15, 18] will not be considered here (see Sect. 24 for a more detailed discussion).

A new development is presented by the phase field models (e.g., [23]). But these models are presently restricted to 2D calculations, i.e., they are too complex for a realistic 3D description of polycrystalline ceramics.

Molecular field models consider the interaction between neighboring domains and grains by introducing an internal field. So, they may be regarded as being a phenomenological variant of the effective medium models described earlier where the internal field is calculated directly on the basis of the inclusion problem. Molecular field models may be important for relaxor ceramics where a local random fluctuation of the poling state independent of the crystal structure has to be considered [19].

Somewhere in between the simple switching models, the numerically troublesome nonlinear effective medium model of Huber et al. [11], and the phenomenological models, we find the models with a reduced set of microscopically motivated internal variables. For practical applications, these models appear most promising, so we describe this approximation in some detail.

The basic idea behind the model is to adopt the viscous concept known to be successfully applicable in modeling polycrystal plasticity. For ferroelectric/ ferroelastic materials, it means that all possible domain orientations may simultaneously transform into any other possible orientation. Only the transition rate differs as a function of the driving force which is given by the gain of free enthalpy. Thus, one can describe a gradual change of the polarization state of the individual crystals and avoids the problematic sharp switching transition, which is not realistic for a multidomain crystal and, additionally, causes problems in the numerical evaluation. Moreover, it is easily possible to include rate- and temperature-dependent behavior.

Let us denote the set of internal state variables by ξ_r with $r = 1, 2, \ldots, n$, where n is the number of states (one can find models with $n = 2, 6, 8, 20, 42, \ldots$). It is meaningful to restrict the variation range of ξ in the sense of weighting factors (so an interpretation as volume fractions of certain special domains is possible though not necessary for the model):

$$\sum_{r=1}^{n} \xi_r = 1, \quad 0 \leqslant \xi_r \leqslant 1. \tag{23.26}$$

The isotropic state is equivalent to $\xi_r = 1/n$. Under fixed loading conditions, the potential energy of the polycrystalline material is a function of these internal parameters:

$$w_{\mathrm{p}} = w_{\mathrm{p}}(\bar{f}, \xi_1, \ldots, \xi_n), \tag{23.27}$$

where $\bar{f}$ represents the macroscopically prescribed strain and electric field (often mechanical stress instead of strain is used as variable – but since only small variations of the internal state variables during a load increment are considered, any selection of loading variables will give identical results provided the potential energy is properly selected).

Given an expression for the potential energy, we recover the effective constitutive equation by

$$\bar{j} = \frac{\partial}{\partial \bar{f}} w_{\mathrm{p}}(\bar{f}, \xi_1, \ldots, \xi_n). \tag{23.28}$$

Furthermore, the driving force for an increasing internal state variable ξ_r is obtained as

$$\phi_r = -\frac{\partial}{\partial \xi_r} w_{\mathrm{p}}(\bar{f}, \xi_1, \ldots, \xi_n). \tag{23.29}$$

Since only transitions from one state to another are possible, the transition rate from state r into state s is a function of the difference of driving forces. This function can be approximated by a power law [10]:

$$v_{rs} = v_0 \left(\frac{\phi_s - \phi_r - w_{\mathrm{c}0}}{w_{\mathrm{c}} - w_{\mathrm{c}0}} \right)^m (n\xi_r)^\alpha \quad \text{for} \quad \phi_s - \phi_r \geqslant w_{\mathrm{c}0} \text{ (otherwise } v_{rs} = 0). \tag{23.30}$$

Here, v_0 is a characteristic frequency, $w_{\mathrm{c}0}$ denotes a threshold below which no transition will proceed (cf. (23.25)) and w_{c} is a characteristic energy above which the transition rates become large (in principal, these parameters may depend on the transition type r–s). Furthermore, two empirical constants m and α must be adjusted to experimental data. The second power term takes into account the saturation of the transition if the weight of the starting state becomes zero.

It is important to note that (23.30) is a unique function of state, so that in numerical simulations no iteration is necessary when a certain load increment is prescribed. Instead, one directly obtains the new domain configuration of internal state variables by summing all transition rates:

$$\dot{\xi}_r = \sum_{s=1}^{n} (+v_{sr} - v_{rs}). \tag{23.31}$$

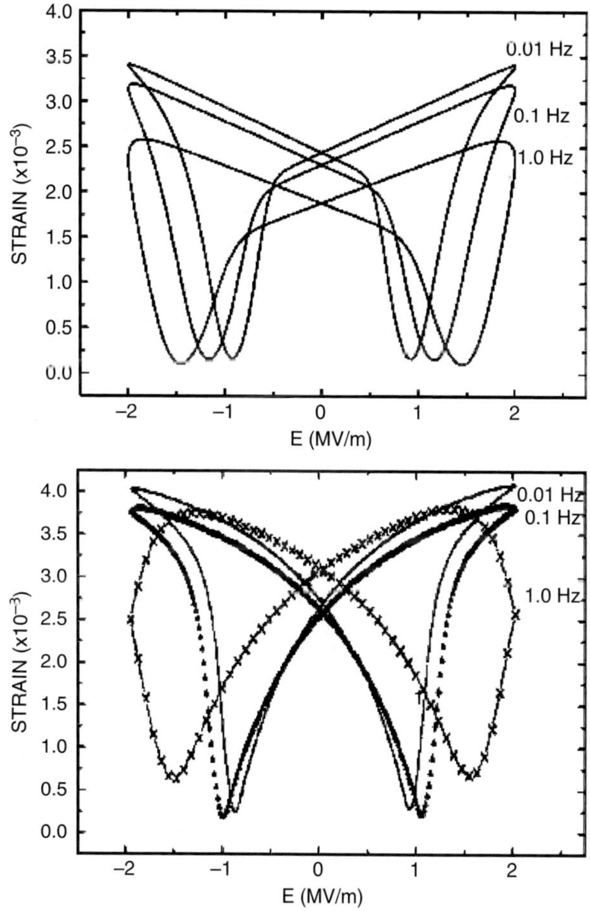

Fig. 23.5. Simulated strain hysteresis curves under sinusoidal loading for soft PZT (after [4]). Model parameters: $n = 6$, $m = 8$, $\alpha = 3$, $w_{c0} = 0$. Comparison with experimental results after [24]

What we still need is an expression for the potential energy. Here, the most simple assumption is provided by a linear combination of the potential energies $w_{\mathrm{p}}^{(r)}$ of certain domains, where r stands now for a particularly orientated domain:

$$w_{\mathrm{p}}(\bar{f}, \xi_1, \ldots, \xi_n) = \sum_{r=1}^{n} \xi_r w_{\mathrm{p}}^{(r)}(\bar{f}) \tag{23.32}$$

(this is often called a *Reuss approximation* according to a known model in the elastic theory of polycrystals). The various models differ only by the number and the type of domains considered.

Figure 23.5 brings an example for such a simulation. More details of the approach as well as comparison with other models and experiments for non-proportional loading paths can be found in Huber and Fleck [10].

23.5 Summary

Effective medium theories may be used to calculate internal fields on different structural levels (domains, grains) so that the linear and non-linear material behavior of piezoelectric materials can be modeled. Some applications have been presented in this chapter and more examples may be found in the literature. Below some open problems and possible applications are listed:

- Pyroelectric/thermoelastic properties
- Evaluation for time-dependent properties
- Conduction problems
- Effect of free charges (screening of polarizations, modification of interaction energy)
- Phase mixtures
- Relaxor behavior (diffuse domains)

References

1. G. Arlt, J. Mater. Sci. **25**, 2655 (1990)
2. G. Arlt, Ferroelectrics **189**, 91 (1996)
3. A.T. Bartic, D.J. Wouters, H.E. Maes, J.T. Rickes, R.M. Waser, J. Appl. Phys. **89**(6), 3420 (2001)
4. A.Y. Belov, W.S. Kreher, J. Europ. Ceram. Soc. **25**(12), 2567 (2005)
5. M.L. Dunn, M. Taya, J. Am. Ceram. Soc. **76**(7), 1697 (1993)
6. M.L. Dunn, M. Taya, Proc. R. Soc. Lond. A **443**, 265 (1993)
7. D.A. Hall, J. Mater. Sci. **36**(19), 4575 (2001)
8. J.H. Huang, Int. J. Solids Struct. **33**(20), 1439 (1996)
9. J.H. Huang, J.S. Yu, Compos. Eng. **4**, 1169 (1994)
10. J.E. Huber, N.A. Fleck, J. Mech. Phys. Solids **49**(4), 785 (2001)
11. J.E. Huber, N.A. Fleck, C.M. Landis, R.M. McMeeking, J. Mech. Phys. Solids **47**(8), 1663 (1999)

12. S.C. Hwang, G. Arlt, J. Appl. Phys. **87**(2), 869 (2000)
13. S.C. Hwang, C.S. Lynch, R.M. McMeeking, Acta Metall. Mater. **43**(5), 2073 (1995)
14. S.C. Hwang, J.E. Huber, R.M. McMeeking, N.A. Fleck, J. Appl. Phys. **84**(3), 1530 (1998)
15. M. Kamlah, C. Tsakmakis, Int. J. Solids Struct. **36**(5), 669 (1999)
16. H. Kessler, H. Balke, J. Mech. Phys. Solids **49**(5), 953 (2001)
17. R.M. McMeeking, S.C. Hwang, Ferroelectrics **200**, 151 (1997)
18. R.M. McMeeking, C.M. Landis, Int. J. Eng. Sci. **40**(14), 1553 (2002)
19. J.C. Piquette, E.A. McLaughlin, W. Ren, B.K. Mukherjee, J. Appl. Phys. **92**(1), 438 (2002)
20. G. Robert, D. Damjanovic, N. Setter, A.V. Turik, J. Appl. Phys. **89**(9), 5067 (2001)
21. J. Rödel, W.S. Kreher, Comp. Mater. Sci. **19**(1–4), 123 (2000)
22. J. Rödel, W.S. Kreher, J. Europ. Ceram. Soc. **23**(13), 2297 (2003)
23. J. Wang, S.-Q. Shi, L.Q. Chen, Y. Li, T.-Y. Zhang, Acta Mater. **52**(3), 749 (2004)
24. D. Zhou, M. Kamlah, D. Munz, in *Smart Structures and Materials 2001*, vol. 4332, ed. by C.S. Lynch. Proceedings of SPIE, SPIE – The International Society for Optical Engineering, Washington, 2001, pp. 64–70

24

Finite-Element Modelling of Piezoelectric Actuators: Linear and Nonlinear Analyses

T. Steinkopff

24.1 Requirements and Possibilities

Because of their excellent properties, piezoelectric actuators have became manifest in different applications like diesel injection, textile machines, and for nano-positioning. They are used in different designs: by means of linear actuators and bending actuators, forces up to several kilo Newtons and displacements up to several millimeters can be realised, respectively.

Piezoelectric actuators are mostly used under so-called large-signal driving conditions and in a wide range of temperature. For instance, in automotive applications like injection systems, typically conditions are forces up to 1 kN, voltages of 160 V, and temperatures from −40 to 150°C. As shown later, the electromechanical loading is often higher than the so-called coercive properties of the piezoelectric ceramic like PZT. As a consequence, the engineer who is interested in an optimal design and adaptation of the piezoelectric device to its electromechanical environment must take the nonlinear effects into consideration. The computer-aided design (CAD) and engineering (CAE) supply both macroscopic values like displacement and force and internal loading like mechanical stress and electric field distributions. The latter is the basic information for aspects of reliability and risk assessment.

The computer-aided engineering is carried out by means of the numerical method of finite elements. A lot of comprehensive and comfortable commercial software has been developed and is well established for solving the most different problems from easy elastostatics to multiphysics where complex coupling mechanisms are included.

The piezoelectric problems consist in a bi-directional electromechanical coupling. In sensor applications, external forces give rise to the induction of electrical energy. In the reverse case, electrical voltage can be used in actuators to generate mechanical energy from the electrical one. The coupling coefficients give answer about the maximum value of transformation which can be reached in the case of actuator applications only when the external loading stiffness does agree with the actuator stiffness.

Nowadays, the most used commercial Finite Element software for solving piezoelectric problems is based on the linear description of materials behaviour. The developers shrink from implementing the complex nonlinear behaviour which is essentially characterised by the ferroelectric and ferroelastic hystereses. At the other end, the user which is interested in nonlinear analyses is forced to program so-called user elements or to apply iterative algorithms. The two lead to disproportionately excessive costs.

With the FE software COMSOL an innovative approach is given to take the nonlinear behaviour into the numerical consideration. Examples of both linear and nonlinear piezoelectric analyses will also be given here.

24.2 The Piezoelectric Effect and Its Description

Piezoelectric actuators are based on the ferroelectric and ferroelastic ceramic PZT. The polycrystalline materials consist of multi-domain single grains with slightly different composition. The macroscopically observed piezoelectric effect in ceramics is the result of two different effects (1) the intrinsic piezoelectric effect of the tetragonal and rhombohedral crystal cells and (2) the extrinsic effect which is caused by domain wall movement within the single grains. The latter is based on the spontaneous properties of the elementary cell. Interaction with applied electrical fields E and mechanical stresses T can give rise to the spontaneous polarisation P_{spont} and deformation S_{spont} being redirected into other crystallographic equivalent directions. That means that favourably oriented domains grow at the expense of unfavourably oriented regions ("domain switching"). It is safe to assume that these processes are time dependent whereby the ferroelectric and ferroelastic hystereses can be understood.

The starting point of view is the electromechanically coupled materials behaviour:

$$\begin{aligned} S &= s^E \cdot T + d \cdot E = (c^E)^{-1} \cdot T + d \cdot E, \\ D &= d \cdot T + \varepsilon^T \cdot E \cong P, \end{aligned}$$

where c^E are elastic coefficients (for constant electric field), ε^T are dielectric coefficients (for constant mechanical stress), and d are piezoelectric coefficients (so-called charge constants).

Generally, the coefficients are functions of the loading conditions, loading direction, loading rate, and frequency:

$$c^E, d, \varepsilon^T = f\left(T, E, \frac{\partial T}{\partial t}, \frac{\partial E}{\partial t}, \ldots\right).$$

In cases of transient analyses, the variable t corresponds to the real time. In cases of static analyses, it should be understood as pseudo-time, first of all to describe the sequence and direction of load changes ΔE and ΔT.

For linear approximation, the materials coefficients are considered to be constant during loading which is sufficient for a first design evaluation. Thus, it is custom that for the purpose of comparison of different actuator designs, the properties like blocking force and free elongation are calculated by means of the so-called secant moduli (slope of the dotted line in Fig. 24.1).

A less well-known example for the nonlinear behaviour of piezoelectric ceramics should be given here. Considering the stress–strain behaviour of a piezoelectric ceramic under uniaxial compression, a peak of high compliance is observed. This compression value is designated as coercive pressure analogous to the coercive field strength. As mentioned earlier, the domain switching can be induced mechanically and electrically. For this reason, the coercive pressure should be dependent on the applied electric field. Experiments showed that there is a linear dependence. The electric field which is parallel to the direction of remanent polarisation stabilises the polarisation state and increases the pressure which is necessary for depolarisation (Figs. 24.2 and 24.3).

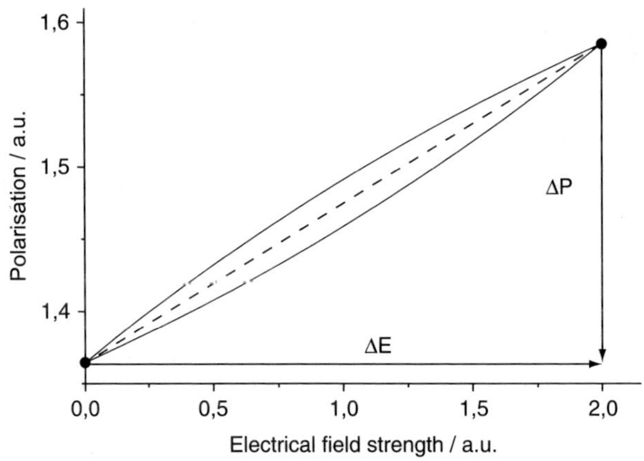

Fig. 24.1. Large-signal hysteresis and secant modulus

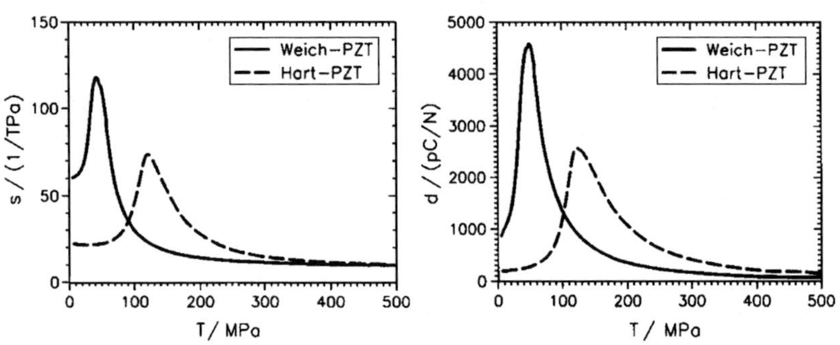

Fig. 24.2. Dependence of compliance s and of piezoelectric coefficient d on pressure [1]

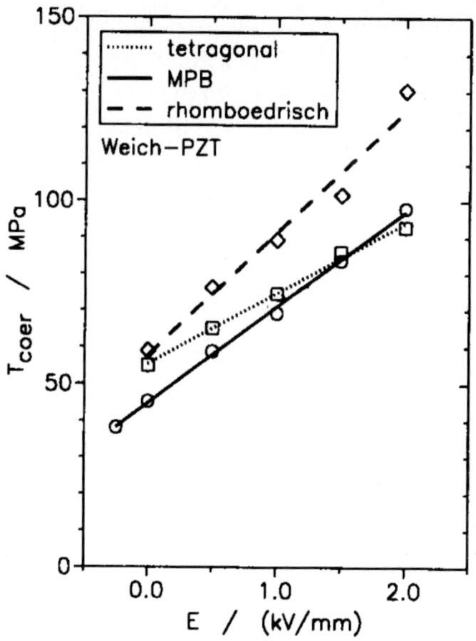

Fig. 24.3. Dependence of coercive pressure on the parallel applied electric field [1]

24.3 Models for the Piezoelectric Materials Behaviour

The electromechanical behaviour of piezoelectric ceramic is essentially determined by its polarisation state which can be described by a pole distribution function analogous to the classic orientation distribution function. The pole distribution function specifies the volume part for domains with polarisation vectors $(\varphi,\ \vartheta)$ within $0 \leqslant \varphi < 2\pi,\ -\pi/2 \leqslant \vartheta \leqslant \pi/2$. Changing the polarisation state by electrical or mechanical loading means that favourably oriented domains with $(\varphi,\ \vartheta)_1$ grow at the expense of unfavourably oriented ones with $(\varphi,\ \vartheta)_2$. Thus, the pole distribution function value at $(\varphi,\ \vartheta)_1$ increases in the same manner like the pole distribution function value at $(\varphi,\ \vartheta)_2$ decreases. Note that using the pole distribution function, the non-Euclidean metric $\sin\vartheta\,\mathrm{d}\varphi\mathrm{d}\vartheta$ must be taken into account.

Micromechanical models put the inner variables like pole distribution function to use. Domain switching processes are modelled on the base of so-called switching criteria like

$$\Delta W_{\text{total}} = \Delta W_{\text{mechan}} + \Delta W_{\text{electr}} = T \cdot \Delta S + E \cdot \Delta P \leqslant \Delta W_{\text{crit}}.$$

where T and E correspond to the actual loading conditions. Domain switching is defined to be observed if the work done by switching reaches a critical value. ΔS and ΔP are the changes of remnant strain and polarisation between states 1 and 2, and they represent the ferroelastic and ferroelectric contributions to total strain and polarisation, respectively, assumed that domain switching is realised.

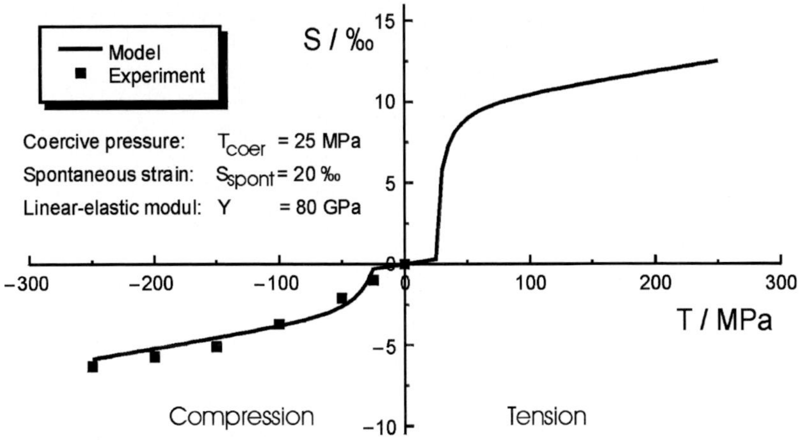

Fig. 24.4. Strain asymmetry between tension and compression for a ferroelastic material without grain-to-grain interaction. The total strain dependence on uniaxial stress is shown. Experimental data from [1]

In the recent few years, a lot of micromechanical models have been developed. Models without grain-to-grain interaction are worth to estimate the maximum values of strain and polarisation attainable by switching. Applying the ferroelectric considerations of [2] onto the ferroelastic behaviour, the strain asymmetry between tension and compression of originally untextured (unpoled) ceramics can be shown (Fig. 24.4). In the case of uniaxial tension and compression, maximum ferroelastic strain values of $4\pi \cdot \sqrt{2}^{-1} \cdot S_{\mathrm{spont}}$ and $-4\pi \cdot \left(1 - \sqrt{2}^{-1}\right) \cdot S_{\mathrm{spont}}$ can be reached, respectively.

Obviously, these maximum values for both polarisation and strain cannot be reached because of grain-to-grain, domain-to-domain interaction, and other deformation mechanisms. Finite-element methods recommend by their selves to take electromechanical interactions into account. Figure 24.5 shows results of finite-element simulations based on ferroelastic domain switching [3] which are compared to uniaxial data derived from bending experiments [4]. Because of mechanical interaction, the strain asymmetry occurs not until stress values of 40 MPa for this case of a soft PZT material with low coercive stress.

It is worth to mention that using FE methods for modelling domain switching effects, complex phenomena like auto-catalysis arise in a natural way [5]. From the point of engineering view, micromechanical models are only useful in cases of simplified considerations [6]. For more complex simulations of piezoelectric devices, the materials behaviour must be described phenomenologically [7]. Analogous to the classic elasto-plastic theory yield and hardening parameters can be well defined for piezoelectric ceramics according to experimental data [8]. Generalised yield stresses should be related to coercive stresses and fields. Kinematic hardening is useful to describe both the ferroelectric and ferroelastic hysteresis.

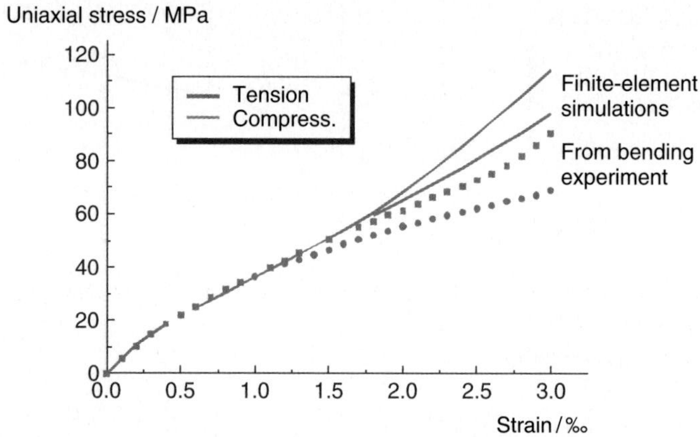

Fig. 24.5. Results of finite-element simulations based on ferroelastic domain switching compared to uniaxial data derived from bending experiments [4]

24.4 The Finite-Element Method

To describe the basic idea of the finite-element method, special emphasis is given to static analyses of piezoelectric devices. Thus, the starting point is the differential equation system for elastostatic and electrostatic problems, respectively:

$$\text{Elastostatics:} \quad \text{div}\, \overleftrightarrow{T} - \vec{f} = 0, \tag{24.1}$$

with

$$\overleftrightarrow{T} = \overleftrightarrow{T}(\overleftrightarrow{S}(\vec{u})), \tag{24.2}$$

$$\overleftrightarrow{T} \cdot \vec{n} = \vec{t}, \tag{24.3}$$

with $\vec{u}(\vec{r})$ being geometrical displacement, $\vec{f}(\vec{r})$ volume force density, and $\vec{t}(\vec{r})$ surface force density.

$$\text{Electrostatics:} \quad \text{div}\, \vec{D} - \rho = 0, \tag{24.4}$$

with

$$\vec{D} = \vec{D}(\vec{E}(\varphi)), \tag{24.5}$$

$$\vec{D} \cdot \vec{n} = \sigma, \tag{24.6}$$

with $\varphi(\vec{r})$ being electrical potential, $\rho(\vec{r})$ density of free charges, and $\sigma(\vec{r})$ surface charge density.

In addition to the essential boundary conditions (24.3) and (24.6), natural boundary conditions can be given. In the case of piezoelectric behaviour, the mechanical stress T and the electric displacement D are dependent on both the strain S and the electric field E:

$$
\begin{array}{ll}
T = c^E S - e_t E, & \textbf{with} \quad S = sym\ grad\ u \\
T = Y \cdot S - e \cdot E, & \textbf{and} \quad E = -grad\ \varphi \\
\text{div } T = \text{div}\{Y \cdot S - e \cdot E\} = f, & \text{div } P = \text{div}\{\varepsilon \cdot E + e \cdot S\} = \rho,
\end{array}
$$

with $S =$ sym grad u and $E = -$ grad φ:

$$
\begin{aligned}
\text{div } T &= \text{div}\{Y \cdot \text{sym grad } u + e \cdot \text{grad } \varphi\} = f, \\
\text{div } P &= \text{div}\{-\varepsilon \cdot \text{grad } \varphi + e \cdot \text{sym grad } u\} = \rho.
\end{aligned}
$$

Here, vector and tensor notations are neglected, and for reasons of simplicity $Y = c^E$ and $D = P$.

The differential equation system

$$
\begin{aligned}
&\nabla\{Y(T,E,\ldots) \cdot \text{sym grad } u\} + \nabla\{e(T,E,\ldots) \cdot \text{grad } \varphi\} = f, \\
&\nabla\{e(T,E,\ldots) \cdot \text{sym grad } u\} + \nabla\{-\varepsilon(T,E,\ldots) \cdot \text{grad } \varphi\} = \rho
\end{aligned}
$$

can be written in generalised matrix notation, here shown for the one-dimensional case:

$$
\frac{\mathrm{d}}{\mathrm{d}x}\left\{\begin{pmatrix} Y(T,E,\ldots) & e(T,E,\ldots) \\ e(T,E,\ldots) & -\varepsilon(T,E,\ldots) \end{pmatrix} \cdot \frac{\mathrm{d}}{\mathrm{d}x}\begin{pmatrix} u \\ \varphi \end{pmatrix}\right\} = \begin{pmatrix} f \\ \rho \end{pmatrix}.
$$

Generally, it is not possible to find the exact solution for boundary value problems like above because of complex domains of interest, loading and boundary conditions. Using the equivalence between strong and weak formulation of boundary problems, the finite-element method is the well-established numerical method for solving the variational problem related to the differential equation. The solution of variational problems consists in extreme values of so-called functionals Π which are often interpretable as potential or energy.

In detail, for the functions of interest (u, φ) piecewise approximations by means of elementary functions N_i are done:

$$
u \approx \widehat{u} = \sum_i a_i \cdot N_i.
$$

In this way, the functional Π keeps dependently on the expansion coefficients a_i only. In the common-used Ritz's method, an extreme functional value is realised if the derivatives of Π with respect to the expansion coefficients a_i are set to zero:

$$
\delta\Pi = 0 \Rightarrow \frac{\partial \Pi}{\partial \vec{a}} = 0.
$$

For cases of elastostatics and electrostatics, the functional is a quadratic term in the Ritz's coefficients. For elastostatic problems, a simple equation results between the nodal point displacements $\vec{a}_{u,i}$ and forces $\vec{F}_k$ with stiffnesses $\underline{K}_{ik}$. For electrostatic problems, an analogous relation consists between nodal point potentials $a_{\varphi,i}$ and charges Q_k with capacitances C_{ik}:

$$
\begin{aligned}
\Pi(\vec{a}) &= \tfrac{1}{2}\vec{a}_u^T \cdot \underline{K} \cdot \vec{a}_u + \vec{a}_u^T \cdot \vec{F} \Rightarrow \underline{K} \cdot \vec{a}_u = \vec{F}, \\
\Pi(\vec{\varphi}) &= \tfrac{1}{2}\vec{a}_\varphi^T \cdot \underline{C} \cdot \vec{a}_\varphi + \vec{a}_\varphi^T \cdot \vec{Q} \Rightarrow \underline{C} \cdot \vec{a}_\varphi = \vec{Q}.
\end{aligned}
$$

In cases of nonlinear behaviour, the stiffnesses and capacitances are dependent on the nodal point displacements and electric potentials, respectively.

For piezoelectric static problems, following system of algebraic equations can be derived:

$$\begin{pmatrix} \underline{K}(\vec{a}) & \underline{P}(\vec{a}) \\ \underline{P}(\vec{a}) & -\underline{C}(\vec{a}) \end{pmatrix} \cdot \begin{pmatrix} \vec{a}_u \\ \vec{a}_\varphi \end{pmatrix} = \begin{pmatrix} \vec{F} \\ \vec{Q} \end{pmatrix}$$

with piezoelectric coupling P_{ik}. Solving this system of nonlinear algebraic equations by means of Newton–Raphson algorithms, the loading history is divided into sufficiently small loading steps with linearised relations.

From the solution for the so-called primary variables $(a_{u,i}, a_{\varphi,i})$, the secondary variables like strain, stress, electric field strength, and electrical displacement are derived as follow:

$$\vec{T} = \underline{\underline{Y}} \cdot \vec{S} - \underline{e} \cdot \vec{E}, \quad \vec{S} = \frac{1}{2}\left(\frac{\partial \vec{u}}{\partial \vec{r}} + \left(\frac{\partial \vec{u}}{\partial \vec{r}}\right)^T\right), \quad \text{with} \quad \frac{\partial \vec{u}}{\partial \vec{r}} = \vec{a}_u^T \cdot \frac{\partial \vec{N}_u}{\partial \vec{r}},$$

$$\vec{D} = \varepsilon \cdot \vec{E} + \underline{e}^T \cdot \vec{S}, \; \vec{E} = -\frac{\partial \varphi}{\partial \vec{r}}, \quad \text{with} \quad \frac{\partial \varphi}{\partial \vec{r}} = \vec{a}_\varphi^T \cdot \frac{\partial \vec{N}_\varphi}{\partial \vec{r}}.$$

Note that these quantities are piecewise continuous only.

24.5 Linear and Nonlinear Piezoelectric Simulations of Bending Actuators

A first impression of the properties of piezoelectric bending actuators can be obtained by analytical solutions for layered structures [9], assuming there are no irregularities in geometry, material, and driving conditions. In other cases, the electromechanical behaviour must be described numerically first of all by means of finite-element methods. Examples for more complex investigations are:

- Geometrical inhomogeneities like non-active regions, for example, strips along the outer regions of the piezoelectric active layers which arise from hidden inner electrodes
- Strikes which are incorporated for limiting the deflection
- Residual stresses arising from the fabrication processes which influence the remanent polarisation state and, consequently, the deflection behaviour of the bending actuator
- Last but not least the nonlinear materials behaviour

Different effects which have an influence on the piezoelectric behaviour will be discussed here.

In applications like textile machines and pneumatic valves, bending actuators are used to realise high deflections of several 100 μm up to several millimeters. Bending actuators consist of at least two layers which are different in material, or in electrical loading. Typical examples are so-called

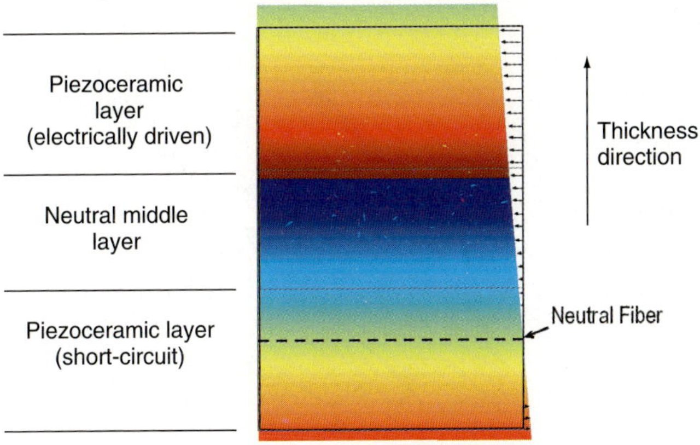

Fig. 24.6. Geometry, loading, and bending deformation of the trimorph actuator. *Colours* correspond to the axial stress levels (*red* for tension, *blue* for compression stresses)

trimorphs for which the deformation behaviour has been analysed by means of finite-element methods (Fig. 24.6).

24.5.1 Loading Conditions

In the following, the static and dynamic behaviour of trimorph bending actuators consisting of two active layers (top and bottom layer) and an inactive middle layer (Fa. Argillon) will be discussed. Aspects of reliability are in the foreground where piezoelectric actuators are used permanently for instance in textile machines. There, a lot of several thousand bending actuators are used under bi-directional operation conditions. Long-term investigations for different driving conditions (Fig. 24.7) [10] show a non-monotonic dependence of the mean-times-to-failure on the driving voltage: bending actuators driven with 275 V (magenta) show a higher lifetime than actuators driven with 225 and 250 V (orange, blue). By means of finite-element analyses, this unexpected behaviour can be understood.

Starting with static analysis, Fig. 24.8 shows typical deformation plots for both the free elongation and the blocking state of a trimorph actuator carried out with the FE software ANSYS© [11]. Note that there is no significant difference in the axial stress values between the free end (FE) and near to the clamped region (CR) although failure is observed to be a result of undercritical crack growth near the clamping region (CR).

A better understanding can be reached by considering the deformation and stress over time behaviour corresponding to Figs. 24.9 and 24.10. In Fig. 24.9, snapshots give an idea of the transient deformation behaviour. In textile applications, the deflection direction alternates with frequencies of about 50 Hz by changing the piezoelectrically active layer (top or bottom layer). In the left

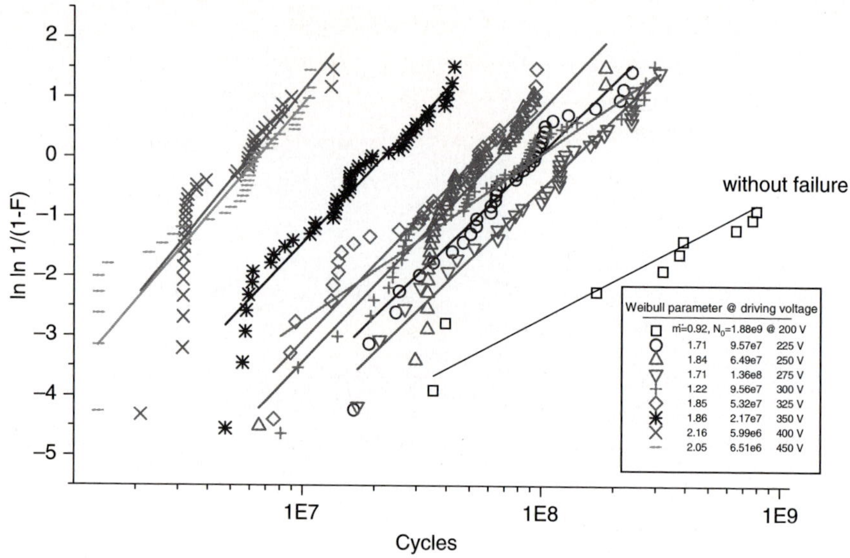

Fig. 24.7. Weibull plots for bending actuators which were driven with voltages between ±200 and ±450 V which corresponds to electric field strength up to 2 kV mm^{-1}

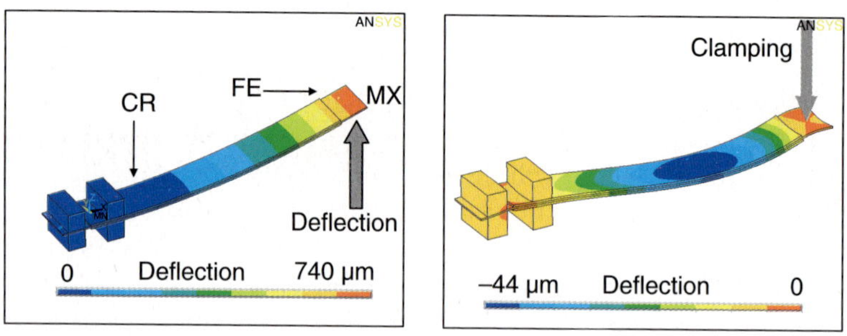

Fig. 24.8. Deflection behaviour of a bending actuator. *Left*: free actuator. *Right*: clamped actuator. Keep in mind the different scaling for deflection

and right column, the transient behaviour of the bending actuator is shown for the case that the piezoelectric top and bottom layer is driven, respectively. After switching (driving changes from the top to the bottom layer at $t = 20$ ms), inertial effects have a strong influence on the deformation behaviour and the internal loading conditions. Dependent on driving frequency, in-phase and out-of-phase deflection behaviour of the actuator can be observed.

For inappropriate switching moments, high dynamic tension stresses which are of the order of the static stresses are observed near the clamping region (Fig. 24.10). As a consequence for different operation conditions (voltage, frequency, strikes yes/no), totally different stress levels and mean-times-to-failure seem to be clear. Here, finite-element calculations are very helpful to analyse

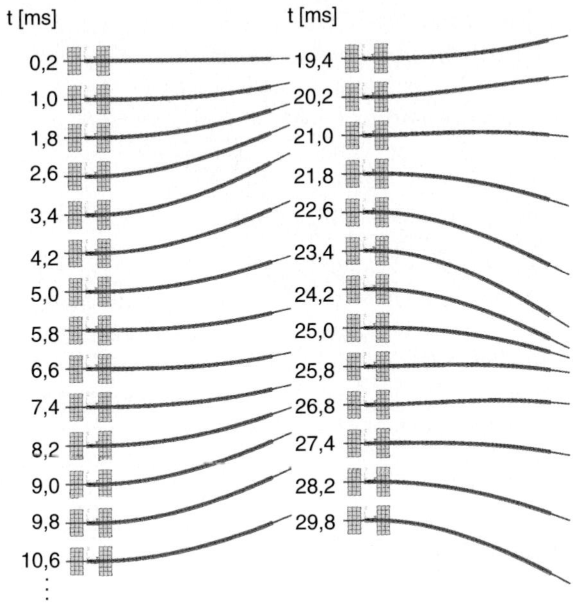

Fig. 24.9. Deformation over time behaviour of the bending actuator

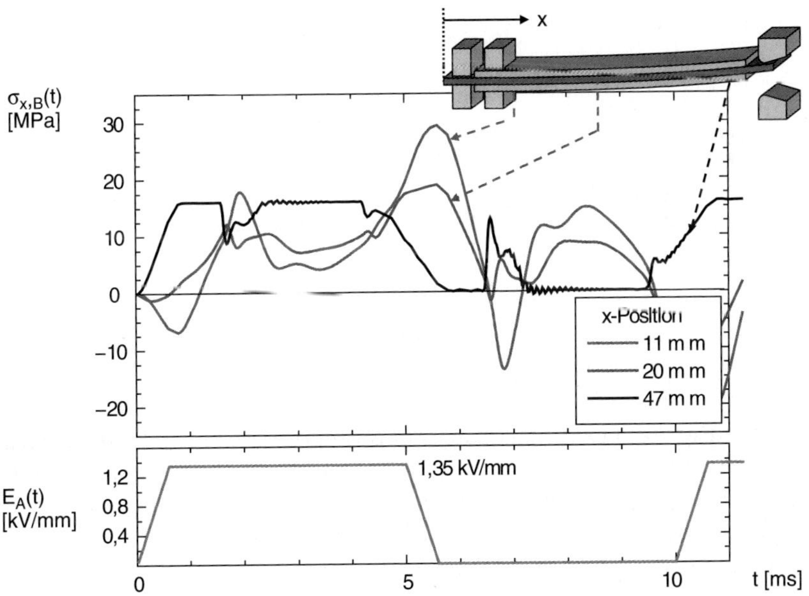

Fig. 24.10. Signal and stress over time for different x positions in the active layer of the trimorph actuator

the transient deformation behaviour in order to both optimise the driving conditions and reduce the tension stress levels within the actuator.

Note that even for linear materials behaviour, the electric potential within the piezoelectric layers shows a quadratic-like dependence on the thickness coordinate.

24.5.2 Nonlinear Materials Behaviour

For a more realistic design of bending actuators, different aspects of the nonlinear materials behaviour must be taken into account. From linear analyses, it is well known that the in-plane mechanical stresses are linear dependent on the coordinate in thickness direction. This is in contrast to the stress distribution in stack actuators where homogeneous tension stresses are observed in the so-called inactive regions and only parallel to the direction of elongation.

In large-signal applications of bending actuators, the tension stress values can reach the coercive stress. That means that domain switching is influenced by both the electric field and tension stresses. The former maintains the remnant polarisation state; the latter can cause the domains to switch into in-plane directions. Obviously, the higher the electric field strength is the higher is the in-plane tensile stress. It is dependent on actuator design and materials used for.

Finally, the ratio between the electric field strength and the resulting tension stress must be compared with the electric-field-dependent coercive stress value delivered from experimental data (Fig. 24.3). As a consequence, in almost cases, a higher piezoelectric strain should be reached by in-plane tensile stresses first of all realised by the fabrication processes. Nonlinear simulations are useful to determine the minimum value of tensile pre-stresses that are responsible for higher piezoelectric deflection during operation.

Up to now, nonlinear piezoelectric simulations are limited to only few finite-element codes developed in the framework of academic projects. Here, nonlinear simulations are presented which have been done by the commercial FE software package COMSOL© [12]. The basic idea of COMSOL consists in providing a universal partial differential equation (system):

$$e_a \frac{\partial^2 u}{\partial t^2} + d_a \frac{\partial u}{\partial t} + \nabla \cdot (-c\nabla u - \alpha u + \gamma) + \beta \cdot \nabla u + au = f.$$

It can be free formulated and thus can be adapted to the particular problem by the engineer. It is easy to see that the matrix c plays the central role for implementing nonlinear materials behaviour. This is realised into two steps. In the first step, the user defines the nonlinear behaviour

$$Y, d, \varepsilon = f\left(\frac{\partial u}{\partial x}, \frac{\partial \varphi}{\partial x}, \dots\right)$$

by means of the primary variables $(u,\ \varphi)$ and their derivatives. In the second step, COMSOL generates the weak formulation which corresponds to the nonlinear differential equation.

In the following, simple examples are given which demonstrates the influence of nonlinear elastic behaviour on the strain and stress distribution of a trimorph actuator.

Starting point is an elastic stiffness Y that may depend linearly from stress according to ferroelastic softening under tension stresses:

$$Y(T) = Y_0 - \Delta Y \cdot \frac{T}{T_0}.$$

To compare piezoelectric materials with different elastic behaviour, various scenarios can be considered because of the relationship between the piezoelectric charge coefficient e and voltage coefficient d:

$$e = Y \cdot d \Rightarrow e(T) = Y(T) \cdot d(T).$$

For reasons of simplicity, two cases are considered here: $d =$ constant (case A) and $e =$ constant (case B) materials. That means in case A, materials are compared which have the same piezoelectric d coefficient but different nonlinear elastic behaviour. In case B, the same procedure is applied for materials with the same piezoelectric e coefficient. Figure 24.11 shows the axial stress T_{xx} and electrical field strength E_y distributions in dependence of the coordinate in thickness direction (left: case A, right: case B).

In case A where materials with same d coefficient are compared, the influence of softening is much higher than that in case B ($e =$ constant).

In Fig. 24.12, the influence of the softening parameter $Y_{\text{soft}} = \Delta Y / T_0$ on the deformation behaviour is shown.

While the nonlinear materials behaviour considered here is very simplified, the results give a first impression of its influence on the complex electromechanical behaviour in bending actuators. For case B materials ($e =$ constant), the influence of the nonlinear elastic behaviour is much smaller than that for case A materials ($d =$ constant).

24.6 Conclusions

It could be shown that finite-element analyses are helpful for solving more complex piezoelectric problems. To get initial estimates about the maximum stress levels, linear finite-element calculations seem to be sufficient.

In the next years, one key aspect of finite-element activities in the field of piezoelectric applications will be concentrated on finding practicable formulations of the nonlinear materials behaviour not only in regard to the electromechanical loading conditions but also in regard to history, frequency, and time. From the author's point of view, there is no way to find a unified materials law that can incorporate the influence of all these parameters.

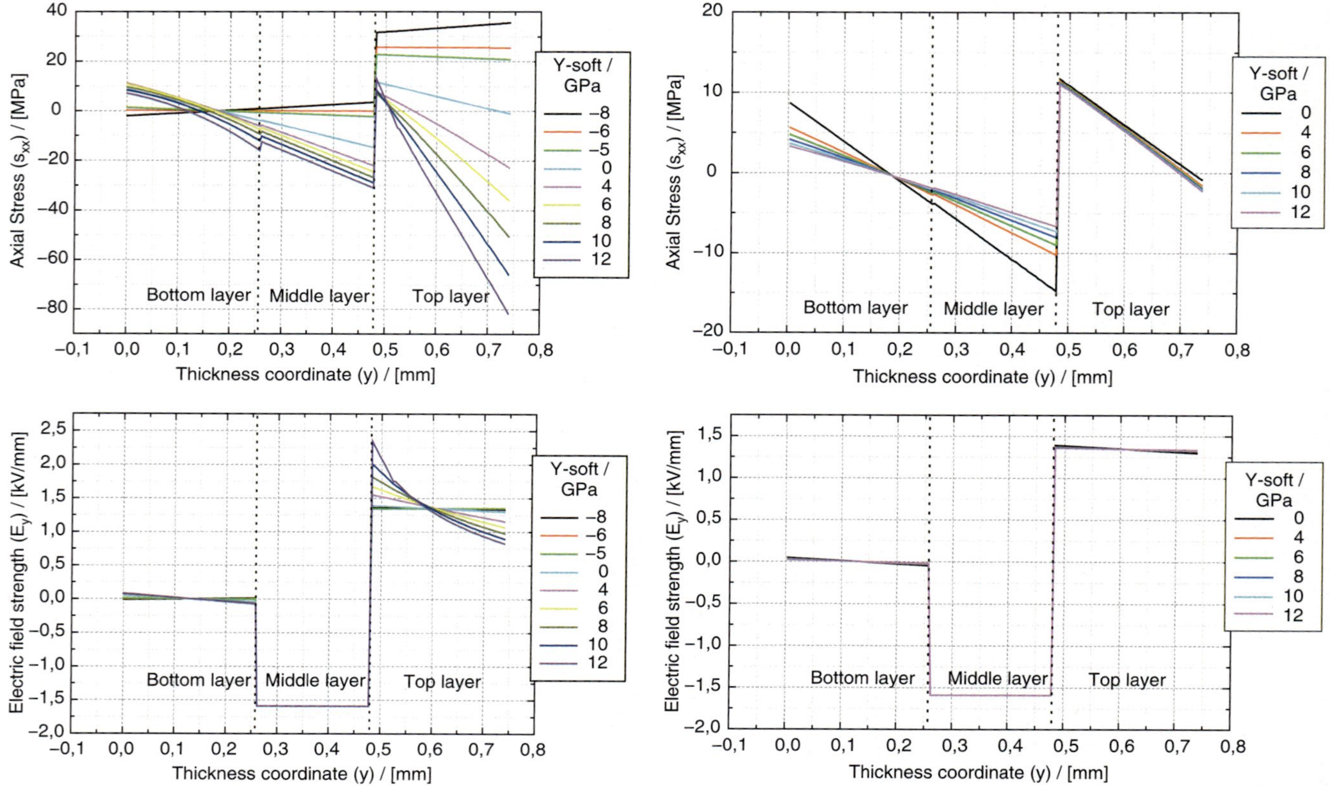

Fig. 24.11. Distributions of axial stress (*top*) and electric field strength (*bottom*) for cases A and B materials (*left and right*, respectively) with different nonlinear elastic behaviour

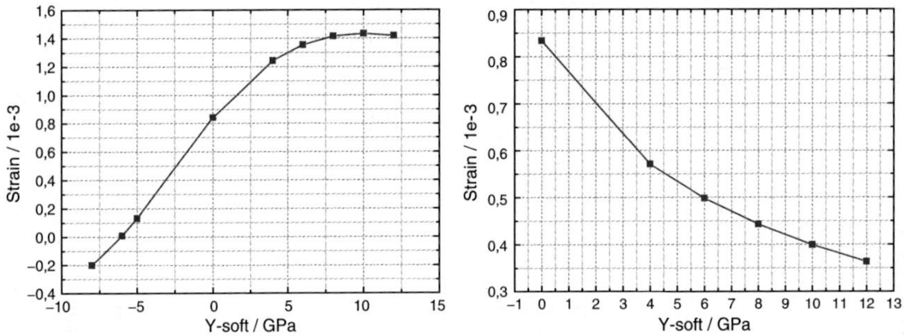

Fig. 24.12. Deformation behaviour of trimorph actuators with cases A and B materials behaviour (*left and right*, respectively). Keep in mind the different scaling for strain

References

1. A. Schäuffele, K.H. Härdtl, J. Am. Ceram. Soc. **79**, 2637 (1996)
2. T. Michelitsch, W.S. Kreher, Acta Mater. **46**, 5085 (1998)
3. T. Steinkopff, Ferroelectrics **222**, 383 (1999)
4. H. Weitzing, G.A. Schneider, Private communications, 1998
5. S.C. Hwang, J.E. Huber, R.M. McMeeking, N.A. Fleck, J. Appl. Phys. **84**, 1530 (1998)
6. M. Kamlah, Continuum Mech. Themodyn. **13**, 219 (2001)
7. C.M. Landis, J. Mech. Phys. Solids **50**, 127 (2002)
8. T. Fett, D. Munz, Adv. Eng. Mater. **5**, 718 (2003)
9. D.L. DeVoe, A.P. Pisano, J. Microelectromech. Syst. **6**, 266 (1997)
10. M.P. Denzler, Ph.D. Thesis, Cuvillier, Göttingen, 2004
11. ANSYS, Inc., Southpointe, 275 Technology Drive, Canonsburg, PA 15317
12. COMSOL AB, Tegnérgatan 23, SE-III 40 Stockholm

Part V

The Future

25

Trends in Ferroelectric/Piezoelectric Ceramics

N. Setter

25.1 Introduction

Ferroelectrics and piezoelectrics are bound to keep their place as important engineering ceramics. Capacitors and piezoelectric transducers are the most widely spread applications of these group of materials. In thin-film form, ferroelectrics and, more widely, polar materials have now been used for several years in rf devices and in non-volatile memories. Components based on piezoelectric films are also being developed and entered the market of various micro-sensor and micro-actuator applications.

Barium titanate ($BaTiO_3$–BTO) and lead–zirconate–titanate ($Pb(Zr,Ti)O_3$–PZT) continue to be the most important ferroelectric and piezoelectric materials in commercial devices. New materials have been introduced constantly. While during 1970s–1980s, those were the relaxors (with giant electrostriction), the 1990s saw the introduction of relaxor ferroelectrics such as $Pb(Zn_{1/3}Nb_{2/3})O_3$–$PbTiO_3$ (PZN–PT) increasing the coupling coefficient from ≈0.7 of PZT to ≈0.90 and over [1]. In the last decade, AIN became an important piezoelectric material due to its use in piezoelectric electronic components (bulk acoustic wave devices) for mobile communications [2]. $SrBi_2Ta_2O_9$ (SBT) competed with PZT in ferroelectric non-volatile memory (FRAM) applications with a superior fatigue resistance [3]. Barium–strontium titanate (BST) was investigated for DRAM applications and finally seems to have found market in tunable ferroelectric capacitors for microwave components. In 2004, Toyota published results showing that a lead-free material, modified $(K,Na)NbO_3$ (KNN) can potentially reach properties of the best lead-containing ferroelectrics [4].

The processing technology of piezoeoelectrics/ferroelectrics has been adapting itself to new needs too: finer powders became available [5], thick film technologies for piezoelectrics have been developed [6], and textured growth (the growth of oriented ceramics) gained importance [7]. Following the excellent properties of piezoelectric single crystals, a method of transforming piezoelectric ceramics into crystals has been reported with expected improvement in composition homogeneity control [8]. Processing of thin films by both physical and chemical methods reached a certain level of maturity with a rather tight control of composition (a challenge in particular in lead-containing compounds), stress, and orientation [9]. Recently, the production of one-dimensional (1D) nano-structures of ferroelectric and piezoelectric materials has been reported [10].

Trends that point to directions of future developments in the field of piezoelectrics and ferroelectrics are summarized below.

25.2 New Materials (Composition and Structure)

The pursuit of new materials is a constant challenge in the field of ferroelectrics and piezoelectrics. Recent advances that need further development include lead-free materials, high-temperature materials, and materials for microsystems and high-frequency applications.

25.2.1 Lead-Free Piezoelectrics

In 2004, researchers from Toyota reported a lead-free material, modified KNN, with excellent piezoelectric properties [4]. The reported performance of the material, a solid solution of KNN and $LiTaO_3$, approaches that of undoped PZT. When prepared in a textured fashion, (001) oriented, the piezoelectric properties even exceed those of commercial (doped) PZT4 composition (Table 25.1). It was shown later [11, 12] that similar compositions prepared in conventional and easily up-scalable methods have sufficient high properties to interest a further development. Part of the performance of the material is a result of the downward shift of tetragonal to orthorhombic phase transition temperature due to specific chemical modifications [13]. In addition,

Table 25.1. Comparison of properties of PZT4 and modified KNN [4]

Piezoelectric property		Units	LF4T (Toyota)	PZT4
Curie temperature	T_c	(°C)	253	250
Piezoelectric	K_p		0.61	0.60
Coupling constant				
Piezoelectric	d_{31}	$(\mathrm{pC\,N^{-1}})$	152	170
Charge constant	d_{33}	$(\mathrm{pC\,N^{-1}})$	416	410
Piezoelectric	g_{31}	$(10^{-3}\,\mathrm{Vm\,N^{-1}})$	11.0	8.3
Voltage constant	g_{33}	$(10^{-3}\,\mathrm{Vm\,N^{-1}})$	29.9	20.2
Dielectric constant	$\varepsilon_{33}/\varepsilon_0$		1570	2,300
Normalized strain	S_{max}/E_{max}	$(\mathrm{pm\,V^{-1}})$	750	700

the processing of the KNN-based materials is less straightforwards than that of PZT. As the properties in various functioning conditions (temperature range, electrical and mechanical fields, stability with time, etc.) are not yet fully explored, it is too early to estimate the utility of the material. It is certain, though, that the Toyota material shakes the so-far unbeaten paradigm that lead-free materials cannot deliver performance to compete with those containing lead. Since regulations in Europe and Asia require substitution of lead by environmentally clean materials, there is no doubt that lead-free piezoelectrics will continue to be explored fervently in the near future.

25.2.2 Morphotropic Boundary Compositions

The important role of the morphotropic boundary in enhancement of piezoelectric and dielectric properties of ferroelectrics was recognized soon after the invention of PZT, some 50 years ago [14]. Park and Shrout extended this idea to relaxor ferroelectrics, showing that excellent piezoelectric properties are obtained at the morphotropic boundary between the tetragonal PT and the relaxor (rhombohedral) PZN at given crystallographic directions [1]. Randall and Shrout proposed recently a number of piezoelectric ceramics, such as $(1-x)BiScO_3-(x)PbTiO_3$, which are attractive for high-temperature applications and are based too on morphotropic boundary compositions [15]. It seems that the KNN ceramics do not benefit from proximity to a morphotropic boundary but it seems that this effect can still be further explored [16].

25.2.3 Anisotropy of Properties

In addition to the enhancement of piezoelectric properties at the morphotropic boundary, attention has been given in recent years to the enhanced piezoelectric properties at directions that are inclined to the polar direction. This is the case in many perovskites, simple [17,18] and complex [19] alike. This enhancement originates from intrinsic phenomena, ease of polarization rotation [20,21] or/and polarization contraction under field, and from extrinsic phenomena, related to density and orientation of domain walls in the poled materials and their activity [22]. It has been emphasized recently [21] that the various origins of enhanced intrinsic piezoelectric activity, be it the morphotropic boundary or dependence on orientation, temperature, electric field or pressure variations, can all be generalized as being manifestations of flattening of the Gibbs free-energy profile of the material as a function of these various parameters. Namely, conditions that lead to isotropization of the free-energy profile result in enhanced instability and enhanced piezoelectric response (Fig. 25.1). The topic has been discussed thoroughly in Chaps. 5, 9, and 21.

These results are likely to have practical implications in the future. On one hand, it may be possible to prepare devices based on utilization of materials in

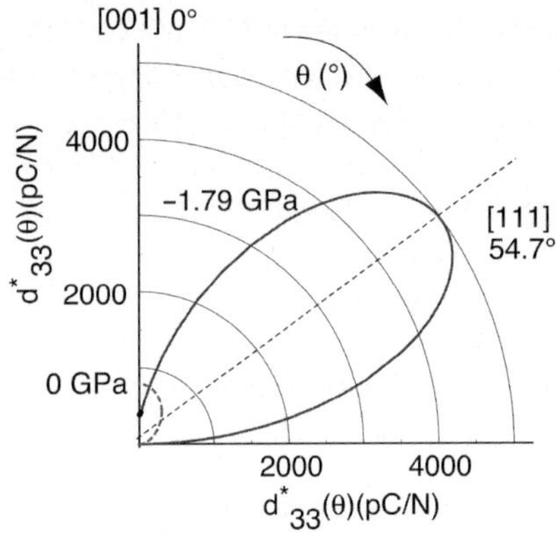

Fig. 25.1. Effect of compressive stress $\sigma_3 = 0$ (*dashed line*) and $-1.79\,\mathrm{GPa}$ at $T = 300\,\mathrm{K}$ on anisotropic piezoelectric enhancement in $PbTiO_3$. At $\sigma_3 = 0\,\mathrm{Pa}$, values for d^*_{33} (*dashed line*) are multiplied by 10. For σ_3 close to the coercive stress (approximately $-1.9\,\mathrm{GPa}$), values of d^*_{33} are strongly sensitive to input parameters and vary between 10^3 and 10^4 [21]

special orientations and utilization of the 'domain engineering' approach. On the other hand, it may be possible to use materials under special conditions (e.g. use of thin films under high DC field conditions, or use of films under special stress conditions by choice of appropriate substrates and processing conditions) that result in enhanced piezoelectric properties.

25.2.4 New Piezoelectrics in High-Frequency Electronics

The increase in operation frequency for wireless communication and the interest in miniaturization and integration motivated the recent replacement of many of the discrete surface acoustic wave components by bulk acoustic wave ones based on piezoelectric thin films. This has been discussed in Chap. 15. At present, the thin films are mostly AlN. In spite of the limited value of its coupling coefficient, AlN has advantages of a high sound velocity and ease of process integration. It is clear that search of new materials will continue: one possible direction is a further exploration of ZnO. This material 'lost' the competition with AlN due to its sensitivity to defects. Very recently, a spectacular improvement of its piezoelectric d_{33} coefficient was reported through appropriate doping [23]. If this will turn to influence also the coupling coefficient, it may be well an interesting material to pursue. At present the origin of this improvement is obscure. Another direction awaiting further work is the development of tunable filters by the use of paraelectric materials under bias field [24].

25.3 The Emergence of First-Principle Calculations as a Tool in Material and Process Design of Ferroelectrics and Piezoelectrics

First-principle calculations of ferroelectric and piezoelectric materials have been done since the early 1990s and contributed substantially to understanding of ferroelectricity and piezoelectricity – R. Cohen (Chap. 21) reviews these accomplishments. With intensification of the dialogue of experimentalists, theoreticians and computation scientists, a wealth of new information can be expected with a strong predictive nature towards further utilization of piezoelectrics.

25.3.1 Modelling of New Materials

The potential of computations in predicting highly performing piezoelectrics is exemplified by several recent studies. Spladin and co-workers [25] predict a strong piezoelectric activity in the hypothetical perovskites $BiAlO_3$ and $BiGaO_3$, although these materials can be prepared only under high pressure [26]. First-principle calculations may give directions for tailoring materials for desired properties. For example, Ballaiche and co-workers [27] predicted recently that a certain arrangement of Sc and Nb in the B-site of the perovskite lattice of $Pb(Sc_{1/2}Nb_{1/2})O_3$ (PSN) will give exceptionally high piezoelectric and dielectric properties. Such an arrangement can be made, in principle, by a layer-by-layer deposition of thin films. Such unusual properties have already been measured on bulk PSN ceramics, some 10 years ago, although, in that case the exact local distribution of the Sc and Nb throughout the material was not known. Superlattices [28] with enhanced spontaneous polarization have also been predicted, providing guidelines in producing new materials. First-principle calculations could well lead to the discovery of new ferroelectric materials.

25.3.2 Computations of Properties

Recently, this has been exemplified by a combined experimental and theoretical study of $BiFeO_3$. $BiFeO_3$ is a well-known high T_c ferroelectric material. Experimentally, in the bulk ceramic form, this high T_c ferroelectric shows a low remanent polarization, perhaps due to typically high leakage (due to Fe^{+2} and electron hoping) that prevents exploitation of the large polarization of the material. Following the finding that $BiFeO_3$ thin films possess a substantial higher spontaneous polarization than the bulk [29], Neaton, Rabe et al. [30] predicted using density functional theory that the material will have polarization of 90–100 $\mu C\,cm^{-2}$. This finding increases the technological interest in $BiFeO_3$ since it is a lead-free material of a high ferroelectric transition temperature ($T_c = 830°C$).

First-principle calculations have been used recently extensively for interpretation of applied phenomena in the field of ferroelectrics and piezoelectrics: enhancement of piezoelectric properties in relaxor ferroelectrics [31] orientation dependence of their piezoelectric properties [20] and pressure-induced giant piezoelectric effects in $PbTiO_3$ [32] are some examples.

Size effects have also been investigated by computations. The lower thickness limit to ferroelectricity is one of the issues that are finding theoretical [33] and experimental [34] answers simultaneously. New ferroelectric phase transitions are predicted in nano-discs of PZT with diameters as low as 3.2 nm [35]. It is interesting also to note that BN which has been synthesized in nano-tube forms for some years is predicted to have large electric polarization along the nano-tube axis [36, 37].

It will be interesting to see the power of first-principle calculations in resolving properties related to functioning problems such as domain wall contribution to the piezoelectric activity and piezoelectric non-linearity.

25.3.3 Calculations of Functioning of Ferroelectric/Piezoelectric Components

Seldom reported so far, explanation and further prediction of functioning of ferroelectric *components* can also be obtained by first-principle calculations. The problem of polarization fatigue can illustrate this. The problem of polarization fatigue in ferroelectric thin films is a well-known one: upon a large number of polarization switchings, the value of the remanent polarization of PZT is reduced – the material is fatigued. This is problematic for the use of PZT in memories, and the manufacturers overcome the problem by the use of oxide electrodes such as IrO_2 instead of the usual metallic electrodes (e.g. Pt), upon which the fatigue problem is eliminated. It has been shown recently [38] using first principles that the use of oxide electrodes instead of pure metallic ones, which is known to have beneficial impact on the switching and dielectric properties of ultra-thin capacitors, can be explained as arising due to the polarizability of the conductive metal oxide (Fig. 25.2). An extension of this study shows, by combining phenomenological theory and first-principles calculations, that interface asymmetry in thin films of realistic thickness leads in $BaTiO_3$ to smearing of the phase transition, an induced piezoelectric response above T_c [39]. At the same time Stengel and Spaldin [40] show by ab initio calculations that the permittivity of a ferroelectric thin layer is improved with the use of elemental metal electrode in comparison with the use of metal–oxide electrode.

The impact of first-principle calculations is bound to increase further: prediction of lead-free piezoelectrics and multiferroics is one needed direction to pursue, understanding and prediction of properties of 'real materials' such as domain wall contribution to piezoelectricity and role of defects, and above

Fig. 25.2. Penetration of the ionic polarization from the ferroelectric into the metal–oxide electrode. (The *gradient* in shading represents the concentration of free charge carriers, while the *solid line* represents the absolute value of polarization.) First-principles investigation explains this way the improved remanent polarization of ferroelectrics capacitors with oxide electrodes relative to metallic (non-polarizable) electrodes [38]

all, better insight into behaviour at surfaces and interfaces and behaviour of systems containing piezoelectrics could result in accelerated development of piezoelectrics and ferroelectrics.

25.4 Advances and Trends in Fabrication Technology

25.4.1 Bulk Technology

Fabrication technology of piezoelectric and ferroelectric materials continues to advance. Finer and more uniform powders became available [41]. In the last years, there has been a strong interest in preparation of textured ceramics [42]. This was motivated by the excellent properties observed at single crystals off the polarization direction in relaxor ferroelectrics [1]. Indeed, the recent breakthrough in lead-free piezoelectrics involved textured ceramics [4] too. The excellent properties of 'domain engineered' relaxor-ferroelectric single crystals and $KNbO_3$ [43] single crystals also motivates the interest to grow crystals from ceramic substrates (Fig. 25.3). These areas are not yet entirely developed and one can expect applicable results in the future.

25.4.2 Ceramic Thin Films

The fabrication technology of ferroelectric thin films has advanced substantially too, and in the last years a better understanding and control of homogeneity, orientation, texture and, for certain applications, epitaxy has been obtained. This in turn allows improved properties through the detailed studies of processing–structure–property relationship.

Few examples are given to illustrate this. (1) With a tight control of substrate–film relationship and subsequent strain in the films, Eom, Schlom

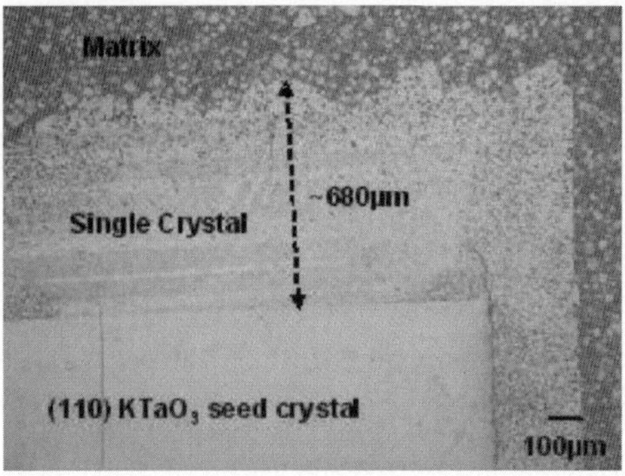

Fig. 25.3. KNN crystal grown from dense ceramics using template-assisted Solid State Single Crystal Growth (SSCG). (Courtesy of J. Fisher, A. Bencan and M. Kosec, Jozef Stefan Institute, Ljubljana)

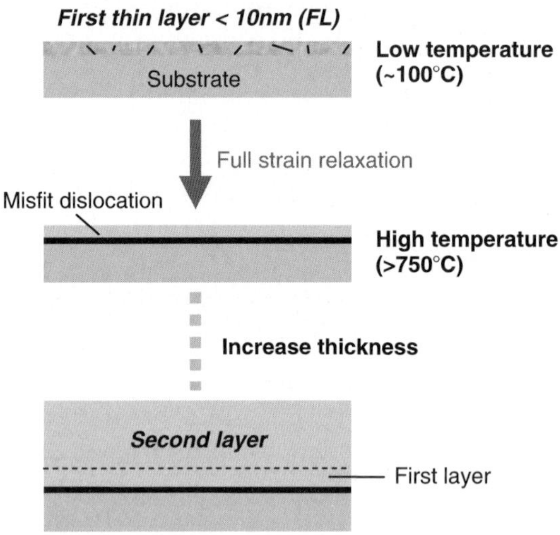

Fig. 25.4. Full strain relaxation and dislocation confinement during two-step growth of STO. The inset is a TEM micrograph of the film cross section showing the dislocation confinement to the first layer [46]

and co-workers obtained $BaTiO_3$ films with Curie temperature of the order of 400–600°C (while T_c of $BaTiO_3$ crystals is 120°C) which enhanced the room temperature ferroelectric properties [44]. The properties were similar to those previously predicted by LGD calculations [45]. (2) In another work, the strain in epitaxial $SrTiO_3$ (STO) films was controlled by a dislocation relaxation mechanism using a two-step growth process, where the crystallization of a pre-deposited very thin amorphous layer of identical composition acted as an absorption layer for the misfit strain (Fig. 25.4). Two-step grown films had a figure of merit for tunable ferroelectrics in microwave applications that was some 20 times higher than conventionally prepared films [46]. (3) With a tight control of oxygen partial pressure, Kingon and Srinivasan [47] processed PZT directly on Cu foils obtaining excellent properties, e.g. low polarization fatigue, and showing the possibility of use of low-cost high conductivity copper electrodes for a range of piezoelectric and ferroelectric thin-film applications.

25.4.3 Self-Assembly

There has been a considerable interest recently to obtain self-assembled ferroelectric structures on substrates. This interest has its roots in microelectronics and in the approaching lower limit to conventional lithography but captures interest in piezoelectrics as well.

Various strategies have been employed so far to obtain self-assembled nanocapacitors or piezoelectric arrays: Waser and co-workers [48] as well as Alexe and co-workers [49] obtained PZT nano-islands on STO due to the instability at high temperatures of the very thin spun-on layer. Islands of diameters as small as 20 nm and thickness of less than 10 nm were shown to be ferroelectric if they were free from misfit dislocations [50]. A second strategy to obtain island growth is to exploit lattice mismatch between the substrate and the material grown [51, 52]. While the former methods do not result in ordered arrays, patterned templates have been used to obtain lithography-modulated self-assembled ferroelectric ordered arrays [53]: lithography patterned ultrathin titania layers served as templates. The PbO that was captured by the titania template when PZT was sputtered helped in nucleation and growth of the piezoelectrically active PZT only on the pre-defined (down to 50-nm diameter) spots.

25.5 From Macroscopic to Nano-Composites

In the mid-1970s, Newnham developed the concept of piezoelectric composites, composites made of piezoelectric ceramics and non-piezoelectric polymers [54]. Due to the concentration of force on the stiff piezoelectric ceramic phase, the resulting overall piezoelectric coupling coefficient of the composite is comparable with that of pure piezoelectric ceramics. At the same time, the acoustic impedance of the composite is reduced due to the compliant polymer phase, cross-talk between individual piezo-elements in the composite is reduced, and the device is flexible and less brittle than a device based on ceramics alone. The properties of the composites can be optimized for various applications based on the connectivity of the two phases [55]. Indeed, the piezoelectric polymer–ceramic composite concept was adopted by many industries, in particular for ultrasonic medical imaging [56–58]. Later the fabrication technology of piezoelectric composites was further developed, in particular Safari and collaborators developed various techniques including relic composite technique, viscous suspension spinning, and fused deposition of multi-materials [59]. For actuator applications, the idea was extended to metal–ceramic composites in which flex-tensional configuration amplifies the strain (or the stress) of the element [60].

Recently, it was shown that the concept of composites is very attractive for high-frequency ferroelectric tunable applications. In these applications, the requirement is to use a ferroelectric material with a high tunability (a strong change of permittivity as a function of applied DC field) and a low permittivity. However, it is shown theoretically that the tunability is a strong function of the permittivity, so the two requirements contradict each other. Sherman et al. [61] modelled analytically dielectric–ferroelectric composites with low dielectric concentration and showed that the permittivity of the composite decreases relative to the permittivity of the ferroelectric phase but its tunability remains as high as that of the pure ferroelectric phase. The origin of this phenomenon is the field concentration on the ferroelectric phase. Following this model, Yamada et al. [62] fabricated a thin-film composite by means of a selective epitaxial growth process. The growth procedure resulted in a self-buildup mechanism that led to a columnar composite structure, where epitaxial (ferroelectric) and amorphous (dielectric) BST columns are electrically connected in parallel (Fig. 25.5a). In agreement with the model, the measured properties showed a linear decrease of the effective permittivity of the composite upon increasing of the concentration of the dielectric phase, while the tunability stayed fairly unchanged until the volume of the amorphous (dielectric) BST reached 70% (Fig. 25.5b). The growth procedure is down-scalable to nano-composite structures which are of interest considering the small size of the capacitors that are required for tunable applications.

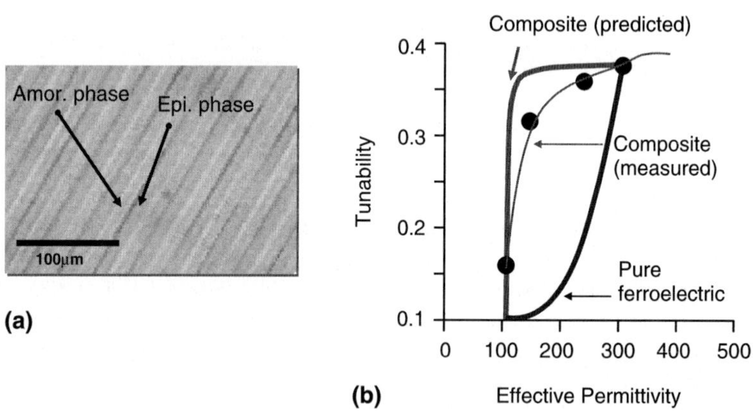

Fig. 25.5. (**a**) Optical micrograph of a fabricated composite structure. (**b**) Comparison between the relation between permittivity and tunability in pure ferroelectric and in parallel-connected composite (calculated and measured) [62]

A composite approach has been used recently to obtain a sensitive magnetic field sensor (of interest, e.g. in magnetoencephalogy) using magneto-electric coupling. Viehland et al. [63] used laminates of a strong ferromagnetic metal and piezoelectric ceramics to convert the magnetic field-induced strain in the magnetostrictive metal to voltage signal of the piezoelectric layer through the piezoelectric conversion of the coupled strain. Such composites can be prepared on the nano-scale as well, and here the magneto-electric coupling may be stronger. Ramesh and collaborators have shown recently an elegant method of growth of such nano-composites by a simultaneous deposition of ferroelectric and ferrimagnetic phases from a single target. Due to phase unmixing based on self-assembly, the ferromagnetic spinel and the ferroelectric perovskite phases formed columnar nano-structures. This was shown to be the case with $CoFe_2O_4$–$BaTiO_3$ [64], $CoFe_2O_4$–$PbTiO_3$ [65] and $CoFeO_3$–$BiFeO_3$ [66] systems. In the latter, coupled piezoelectric–magnetic switching was evidenced.

It seems that the use of nano-composites can be beneficial in piezoelectric and ferroelectric thin-film applications and merits to be further explored.

25.6 1D Nano-Structures and Nano-Devices

The interest in small anisotropic ferroelectric structures such as nano-discs, nano-rods, nano-wires and nano-tubes falls within the general interest in such structures following the discovery of carbon nano-tubes [67] and of the possibility to synthesize nano-tubes and nano-rods of inorganic materials including oxides [68, 69]. Advances have been made both in processing of such nano-materials and in their modelling. Very little is known about the actual properties of 1D nano-elements; demonstration of potential utilization is emerging.

Piezoelectric nano-rods of ZnO have been prepared extensively as free-standing particles and as arrays on various substrates. Other related configurations such as nano-belts of ZnO and nano-springs and nano-helics [70] were reported as well. The zinc-blend boron nitride (BN) was prepared in the form of nano-tubes [71].

An attractive way to prepare nano-rods is to use nano-porous silicon or alumina as templates and to fill them up with the suitable polymeric precursors which are then heat treated to form the ferroelectric oxide (BTO, PZT, etc.) and subsequently separated from the template by etching the latter away. This process was demonstrated on various ferroelectric materials and resulted also in ordered nano-rod arrays [72, 73]. This method is attractive for industrial up-scaling.

Various chemical processing routes show promise for ferroelectric nano-wire fabrication; solution phase decomposition of barium titanium isopropoxide to produce BTO nano-wires with diameters 5–60 nm and lengths up to 10 μm was shown [10]. A molten salt technique was used to produce BTO

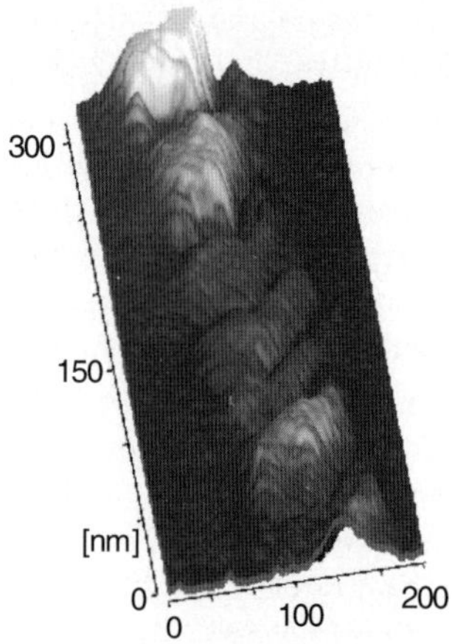

Fig. 25.6. 3D visualization of the piezoelectric activity of $KNbO_3$ nano-rod and the existence of ferroelectric domains with piezoelectric activity [84]

nano-wires in an up-scalable way [74]. BTO and STO nano-tubes were produced by reacting titania nano-tubes with barium and strontium hydroxides [75]. A hydrothermal growth of potassium niobate (KNO) nano-wires was demonstrated [76]. Hydrothermal method was used also for the fabrication of arrays of BTO nano-tubes on titanium substrate [77]. Very little has been reported so far on the growth mechanism of piezoelectric nano-tubes and nano-wires: while KNO seems to grow by a nucleation and growth mechanism in sub-saturated liquid [78].

The ferroelectric to paraelectric transition was recently reported on BTO nano-wires [79] but the crystalline structure of ferroelectric 1D structures has been only slightly explored [80,81]. Domain walls formation and disappearance upon cycling around the ferroelectric phase transition were demonstrated in PZT nano-wires [82]. Rare are yet the detailed reports on the piezoelectric properties of nano-tubes and nano-wires but switching of the ferroelectric polarization and local piezoelectric activity have been detected using force microscopy (Fig. 25.6) [83, 84]. The field of 1D small piezoelectrics is at its infancy.

25.6.1 Piezoelectric Nano-Devices

Will nano-devices offer new functionalities or other advantages over existing technology?

One application that is being explored is the use of nano-wires for energy harvesting: voltage generation by periodical mechanical actuation of a single $BaTiO_3$ nano-wire was demonstrated recently [85]. Several authors demonstrated energy harvesting using piezoelectric nano-wire arrays: ZnO nano-wire arrays [86] and GaN nano-wire arrays [87] were mechanically actuated by scanning an AFM tip over the wires and the generated mechanical energy was converted into electrical energy by the piezoelectric effect. The estimated electrical energy produced was $\approx 10\,\mathrm{pW}\,\mu\mathrm{m}^{-2}$. The collection of the energy is done in an elegant way taking advantage of the semiconductor nature of the nano-wires used: the nano-wire/conductor contact acts as a diode and the charge generated piezoelectrically is collected in a unidirectional way. The current leading idea is to convert mechanical energy that is available continuously such as body movement into electrical energy, so that self-powering can be used for wireless low power consumption devices. Development in this direction was shown recently by Zhong Lin Wang et al. [88] who created a textile structure that contains the ZnO nano-wires and metal wires for the current collection.

An inherent exploitation of nano-technology can be obtained when the phenomena at stake are coupled at the same scale. This was shown recently [89] in utilization of piezoelectric semiconductors to build cantilevers in which the width of the depletion region (hence the piezoelectric active region) was manipulated by an electric field, creating positive or negative deflection according to the extent of the field applied, which was translated into an increased vibration amplitude relative to a stationary piezoelectric cantilever. It is a similar effect to that obtained by bi-layer actuators, but on a nano-meter scale, tunable, and processed via standard integrated semiconductor technology, where the electrodes and the piezoelectric active regions are all based on variation of the doping in a GaAs cantilever. This mechanism functions only when the dimension of the cross section of the cantilever is of the same order of magnitude as that of the depletion region, namely, sub-μm. The authors further demonstrate various devices based on piezoelectric GaAs nano-structures such as nano-electromechanical XOR (exclusive or) logic gate and resonance-frequency control.

It is reasonable to expect that piezo-NEMS (nano-electromechanical systems) will develop further with possible applications in sensors, transducers and computing applications. The interest of piezoelectric semiconductors is also likely to intensify due to the small dimensions of the components.

25.7 Conclusion

Major trends in materials and processes in the area of ferroelectrics and piezoelectrics were outlined above. Emphasized directions were new materials, in particular lead-free and high-temperature ones as well as materials for communication technologies and MEMS–NEMS. The increased role of ab initio calculations in prediction of new material systems, properties and functioning of components and devices was pointed out. Trends in processing include 'domain engineering', nano-composites, self-assembly of small elements on substrates, and new 1D structures.

The question of new applications was not discussed although it is clear that the entire field is driven by applications. Intensification of use of piezoelectrics and ferroelectrics can be foreseen in several areas: with the present strong development of polymer electronics, it is likely that ferroelectric polymers will grow in importance. Increased investments into medical diagnostic tools may be projected as an opportunity for a further increased use of piezoelectrics in this domain. Certainly, highly performing lead-free materials will be welcome here. Increased automation and a more stringent control on efficient energy use and on clean environment are indicative for an intensified use of piezoelectric sensors and actuators for these applications.

Acknowledgement

This review is part of the activity of MIND – European Network on Piezoelectrics 6th FP. The European Commission is acknowledged for the financial support.

References

1. S.-E. Park, T.R. Shrout, J. Appl. Phys. **82**, 1804 (1997)
2. M.-A. Dubois, P. Muralt, Appl. Phys. Lett. **74**, 3032 (1999)
3. J.F. Scott, C. Araujo, Science **246**, 1400 (1992)
4. Y. Saito, H. Takao, T. Tani, T. Nonoyama, K. Takatori, T. Homma, T. Nagaya, M. Nakamura, Nature **432**, 84 (2004)
5. Y. Sakabe, Y. Yamashita, H. Yamamoto, J. Eur. Ceram. Soc. **25**, 2739 (2005)
6. P. Tran-Huu-Hue, F. Levassort, F. Vander Meulen, J. Holc, M. Kosec, M. Lethiecq, J. Eur. Ceram. Soc. **21**, 1445 (2001)
7. T. Takeuchi, T. Tani, Y. Saito, Jpn. J. Appl. Phys. 1, **9B**, 5553 (1999)
8. H.Y. Lee, in *Piezoelectric Single Crystals*, ed. by S. Trolier-McKinstry, L.E. Cross, Y. Yamashita p. 160, (2004)
9. S. Trolier-McKinstry, P. Muralt, in *Electroceramics Based MEMS*, ed. by N. Setter (Springer, Berlin, 2005), pp. 199–215, Chap. 10
10. J.J. Urban, W.S. Yun, Q. Gu, H. Park, J. Am. Chem. Soc. **124**, 1186 (2002)
11. E. Hollenstein, M. Davis, D. Damjanovic, N. Setter, Appl. Phys. Lett. **87**, 182905 (2005)
12. G.-Z. Zang, J.-F. Wang, H.-C. Chen, W.-B. Su, C.-M. Wang, P. Qi, B.-Q. Ming, J. Du, L.-M. Zheng, S. Zhang, T.R. Shrout, Appl. Phys. Lett. **88**, 212908 (2006)
13. T.R. Shrout, S.J. Zhang, J. Electroceram. **19**, 111 (2007)
14. B. Jaffe, R.S. Roth, S. Marzullo, J. Appl. Phys. **25**, 809 (1954)
15. T.R. Shrout, R. Eitel, C. Randall, in *Piezoelectric Materials in Devices*, ed. by N. Setter (2002), pp. 413–432
16. N. Klein, D. Damjanovic et al. J. Appl. Phys. **102**, 014112 (2007)
17. K. Nakamura, T. Tokiwa, Y. Kawamura, J. Appl. Phys. **91**, 9272 (2002)
18. S. Wada, S. Suzuki, T. Noma, T. Suzuki, M. Osada, M. Kakihana, S.-E. Park, L.E. Cross, T.R. Shrout, Jpn. J. Appl. Phys. 1 **38**, 5505 (1999)
19. X. Du, U. Belegundu, K. Uchino, Jpn. J. Appl. Phys. 1, **36**, 5580 (1997)

20. H. Fu, R.E. Cohen, Nature **403**, 281 (2000)
21. M. Budimir, D. Damjanovic, N. Setter, Phys. Rev. B **73**, 174106 (2006)
22. S. Wada, K. Yako, H. Kakemoto, T. Tsurumi, T. Kiguchi, J. Appl. Phys. **98**, 014109 (2005)
23. Y.C. Yang, C. Song, X.H. Wang, F. Zeng, F. Pana, Appl. Phys. Lett. **92**, 012907 (2008)
24. A. Noeth, T. Yamada, V.O. Sherman, P. Muralt, A.K. Tagantsev, N. Setter, J. Appl. Phys. **102**, 114110 (2007)
25. P. Baettig, Ch.F. Schelle, R. Lesar, U.V. Waghmare, N.A. Spaldin, Chem. Mater. **17**, 1376 (2005)
26. A.A. Belik, T. Wuernisha, T. Kamiyama, K. Mori, M. Maie, T. Nagai, Y. Matsui, E. Takayama-Muromachi, Chem. Mater. **18**, 133 (2006)
27. A.M. George, J. Iniguez, L. Bellaiche, Nature **413**, 54 (2001)
28. S.M. Nakhmanson, K.M. Rabe, D. Vanderbilt, Phys. Rev. B **73**, 060101R (2006)
29. J. Wang, J.B. Neaton, H. Zheng, V. Nagarajan, S.B. Ogale, B. Liu, D. Viehland, V. Vaithyanathan, D.G. Schlom, U.V. Waghmare, N.A. Spaldin, K.M. Rabe, M. Wuttig, R. Ramesh, Science **299**, 1719 (2003)
30. J.B. Neaton, C. Ederer, U.V. Waghmare, N.A. Spaldin, K.M. Rabe, Phys. Rev. B **71**, 014113 (2005)
31. L. Bellaiche, D. Vanderbilt, Phys. Rev. Lett. **83**, 1347 (1999)
32. Z. Wu, R.E. Cohen, Phys. Rev. Lett. **95**, 037601 (2005)
33. J. Junquera, Ph. Ghosez, Nature **422**, 506 (2003)
34. D.D. Fong, G.B. Stephenson, S.K. Streiffer, J.A. Eastman, O. Auciello, P.H. Fuoss, C. Thompson, Science **304**, 1650 (2004)
35. I.I. Naumov, L. Bellaiche, H. Fu, Nature **432**, 737 (2004)
36. E.J. Mele, P. Král, Phys. Rev. Lett. 056803 (2002)
37. S.M. Nakhmanson, A. Calzolari, V. Meunier, J. Bernholc, M. Buongiorno Nardelli, Phys. Rev. B **67**, 235406 (2003)
38. G. Gerra, A.K. Tagantsev, N. Setter, K. Parlinski, Phys. Rev. Lett. **96**, 107603 (2006)
39. G. Gerra, A.K. Tagantsev, N. Setter, Phys. Rev. Lett. **98**, 207601 (2007)
40. M. Stengel, N.A. Spaldin, Nature **443**, 679 (2006)
41. K.S. Seol, S. Tomita, K. Takeuchi, T. Miyagawa, T. Katagiri, Y. Ohki, Appl. Phys. Lett. **81**, 1893 (2002)
42. S. Trolier-McKinstry, E. Sabolsky, S. Kwon, C. Duran, T. Yoshimura, J.-H. Park, Z. Zhang, G.L. Messing, in *Piezoelectric Materials in Devices*, ed. by N. Setter (2002), pp. 497–512
43. K. Nakamura, Y. Kawamura, IEEE Trans. UFFC **47**, 750 (2000)
44. K.J. Choi, M. Biegalski, Y.L. Li, A. Sharan, J. Schubert, R. Uecker, P. Reiche, Y.B. Chen, X.Q. Pan, V. Gopalan, L.-Q. Chen, D.G. Schlom, C.B. Eom, Science **306**, 1005 (2004)
45. N.A. Pertsev, A.G. Zembilgotov, A.K. Tagantsev, Phys. Rev. Lett. **80**, 1988 (1998)
46. T. Yamada, K.F. Astafiev, V.O. Sherman, A.K. Tagantsev, P. Muralt, N. Setter, Appl. Phys. Lett. **86**, 142904 (2005)
47. A. Kingon, S. Srinivasan, Nat. Mater. **4**, 233 (2005)
48. A. Roelofs, T. Schneller, K. Szot, R. Waser, Appl. Phys. Lett. **81**, 5231 (2002)
49. I. Szafraniak, C. Harnagea, R. Scholz, S. Bhattacharyya, D. Hesse, M. Alexe, Appl. Phys. Lett. **83**, 2211 (2003)

50. M.W. Chu, I. Szafraniak, R. Scholz, C. Harnagea, D. Hesse, M. Alexe, U. Goesele, Nat. Mater. **3**, 87 (2004)
51. H. Fujisawa, M. Shimizu, H. Niu, K. Honda, S. Ohtani, Mater. Res. Soc. Symp. Proc. **596**, 321 (2000)
52. S. Karthauser, E. Vasco, R. Dittmann, R. Waser, Nanotechnology **15**, S122 (2004)
53. S. Bühlmann, P. Muralt, S. Von Allmen, Appl. Phys. Lett. **84**, 2614 (2004)
54. R.E. Newnham, D.P. Skinner, L.E. Cross, Meter. Res. Bull. **13**, 525 (1978)
55. J.F. Tressler, S.A. Robert, E. Newnham, Mater. J. Electroceram. **2**, 257 (1998)
56. T.R. Gururaja, W.A. Schulze, L.E. Cross, R.E. Newnham, IEEE Trans. US **32**, 499 (1985)
57. H. Takeuchi, C. Nakaya, Ferroelectrics **68**, 53 (1986)
58. A. Shaulov, W.A. Smith, Proc. 1985 IEEE Ultrasonics Symp. 648 (1985)
59. A. Safari, J. Cesarano III, P.G. Clem, in *Piezoelectric Materials in Devices*, ed. by N. Setter (2002), pp. 467–496
60. J.F. Tressler, A. Dogan, J.F. Fernandez, J.T. Fielding Jr., K. Uchino, R.E. Newnham, in *1995 IEEE Ultrasonics Symposium Proceedings*, ed. by M. Levy, S.C. Schneider, B.R. McAvoy (IEEE, Piscataway, NJ, 1995), p. 897
61. V.O. Sherman, A.K. Tagantsev, N. Setter, D. Iddles, T. Price, J. Appl. Phys. **99**, 074104 (2006)
62. T. Yamada, V.O. Sherman, A. Nöth, P. Muralt, A.K. Tagantsev, N. Setter, Appl. Phys. Lett. **89**, 032905 (2006)
63. S. Dong, J.F. Li, D. Viehland, J. Cheng, L.E. Cross, Appl. Phys. Lett. **85**, 3534 (2003)
64. H. Zheng et al, Science **303**, 661 (2004)
65. J.H. Li, I. Levin, J. Slutsker, V. Provenzano, P.K. Schenck, R. Ramesh, J. Ouyang, A.L. Roytburd, Appl. Phys. Lett. **87**, 072909 (2005)
66. F. Zavaliche, H. Zheng, L. Mohaddes-Ardabili, S.Y. Yang, Q. Zhan, P. Shafer, E. Reilly, R. Chopdekar, Y. Jia, P. Wright, D.G. Schlom, Y. Suzuki, R. Ramesh, Nanoletters **5**, 1793 (2005)
67. S. Iigima, Nature **354**, 56 (1991)
68. R. Tene, L. Margulis, M. Genud, Nature **360**, 444 (1992)
69. T. Kasuga, M. Hiramatsu, A. Hoson, T. Sekino, K. Niihara, Langmuir **14**, 3160 (1998)
70. Pu Xian Gao, Yong Ding, Wenjie Mai, William L. Hughes, Changshi Lao, Zhong Lin Wang, Science **309**, 1700 (2005)
71. N.G. Chopra, R.J. Luyken, K. Cherrey, V.H. Grespi, M.L. Cohen, S.G. Louie, A. Zettl. Science **269**, 966 (1995)
72. Y. Luo, I. Szafraniak, N.D. Zakharov, V. Nagarajan, M. Steinhart, R.B. Wehrspohna, J.H. Wendorff, R. Ramesh, M. Alexe, Appl. Phys. Lett. **83**, 440 (2003)
73. F.D. Morrison, L. Ramsay, J.F. Scott, J. Phys.: Condens. Matter **15**, L527 (2003)
74. Y. Mao, S. Banerjee, S.S. Wong, J. Am. Chem. Soc. **125**, 15718 (2003)
75. Y. Mao, S. Banerjee, S.S. Wong, Chem. Commun. 408 (2003)
76. J.-F. Liu, X.-L. Li, Y.-D. Li, J. Nanosci. Nanotechnol. **2**, 617 (2002)
77. N.P. Padture, X. Wei, J. Am. Ceram. Soc. **86**, 2215 (2003)
78. E. Vasco, A. Magrez, L. Forro, N. Setter, J. Phys. Chem. **109**, 14331 (2005)
79. J.E. Spanier, A.M. Kolpak, J.J. Urban, I. Grinberg, L. Ouyang, W.S. Yun, A.M. Rappe, H. Park, Nanoletters **6**, 735 (2006)

80. A. Magrez, E. Vasco, J.-W. Seo, C. Dieker, N. Setter, L. Forro, J. Phys. Chem. B **110**, 58 (2006)
81. G. Suyal, E. Colla, R. Gysel, N. Setter, Nanoletters **4**, 1339 (2004)
82. J. Wang et al., Appl. Phys. Lett. **90**, 133107 (2007)
83. W.S. Yun, J.J. Urban, Q. Gu, H. Park, Nanoletters **2**, 447 (2002)
84. G. Suyal, E. Colla, R. Gysel, N. Setter, Nanoletters **4**, 1339 (2004)
85. Z. Wang, J. Hu, A.P. Suryavanshi, K. Yum, M.-F. Yu, Nanoletters **7**, 2966 (2007)
86. Zhong Lin Wang, Jinhui Song, Science **312**, 242 (2006)
87. W.S. Su, Y.F. Chen, C.L. Hsiao, L.W. Tu, Appl. Phys. Lett. **90**, 063110 (2007)
88. Xudong Wang, Jinhui Song, Jin Liu, Zhong Lin Wang, Science **316**, 102 (2007)
89. S.C. Masmanidis, R.B. Karabalin, I. De Vlaminck, G. Borghs, M.-R. Freeman, M.L. Roukes, Science **317**, 780 (2007)

Index

Springer Series in

MATERIALS SCIENCE

Editors: R. Hull R. M. Osgood, Jr. J. Parisi H. Warlimont

50 **High-Resolution Imaging and Spectrometry of Materials**
Editors: F. Ernst and M. Rühle

51 **Point Defects in Semiconductors and Insulators**
Determination of Atomic and Electronic Structure from Paramagnetic Hyperfine Interactions
By J.-M. Spaeth and H. Overhof

52 **Polymer Films with Embedded Metal Nanoparticles**
By A. Heilmann

53 **Nanocrystalline Ceramics**
Synthesis and Structure
By M. Winterer

54 **Electronic Structure and Magnetism of Complex Materials**
Editors: D.J. Singh and D. A. Papaconstantopoulos

55 **Quasicrystals**
An Introduction to Structure, Physical Properties and Applications
Editors: J.-B. Suck, M. Schreiber, and P. Häussler

56 **SiO_2 in Si Microdevices**
By M. Itsumi

57 **Radiation Effects in Advanced Semiconductor Materials and Devices**
By C. Claeys and E. Simoen

58 **Functional Thin Films and Functional Materials**
New Concepts and Technologies
Editor: D. Shi

59 **Dielectric Properties of Porous Media**
By S.O. Gladkov

60 **Organic Photovoltaics**
Concepts and Realization
Editors: C. Brabec, V. Dyakonov, J. Parisi and N. Sariciftci

61 **Fatigue in Ferroelectric Ceramics and Related Issues**
By D.C. Lupascu

62 **Epitaxy**
Physical Principles and Technical Implementation
By M.A. Herman, W. Richter, and H. Sitter

63 **Fundamentals of Ion-Irradiated Polymers**
By D. Fink

64 **Morphology Control of Materials and Nanoparticles**
Advanced Materials Processing and Characterization
Editors: Y. Waseda and A. Muramatsu

65 **Transport Processes in Ion-Irradiated Polymers**
By D. Fink

66 **Multiphased Ceramic Materials**
Processing and Potential
Editors: W.-H. Tuan and J.-K. Guo

67 **Nondestructive Materials Characterization**
With Applications to Aerospace Materials
Editors: N.G.H. Meyendorf, P.B. Nagy, and S.I. Rokhlin

68 **Diffraction Analysis of the Microstructure of Materials**
Editors: E.J. Mittemeijer and P. Scardi

69 **Chemical-Mechanical Planarization of Semiconductor Materials**
Editor: M.R. Oliver

70 **Applications of the Isotopic Effect in Solids**
By V.G. Plekhanov

71 **Dissipative Phenomena in Condensed Matter**
Some Applications
By S. Dattagupta and S. Puri

72 **Predictive Simulation of Semiconductor Processing**
Status and Challenges
Editors: J. Dabrowski and E.R. Weber

73 **SiC Power Materials**
Devices and Applications
Editor: Z.C. Feng

Springer Series in
MATERIALS SCIENCE

Editors: R. Hull R. M. Osgood, Jr. J. Parisi H. Warlimont

74 **Plastic Deformation in Nanocrystalline Materials**
By M.Yu. Gutkin and I.A. Ovid'ko

75 **Wafer Bonding**
Applications and Technology
Editors: M. Alexe and U. Gösele

76 **Spirally Anisotropic Composites**
By G.E. Freger, V.N. Kestelman, and D.G. Freger

77 **Impurities Confined in Quantum Structures**
By P.O. Holtz and Q.X. Zhao

78 **Macromolecular Nanostructured Materials**
Editors: N. Ueyama and A. Harada

79 **Magnetism and Structure in Functional Materials**
Editors: A. Planes, L. Mañosa, and A. Saxena

80 **Micro- and Macro-Properties of Solids**
Thermal, Mechanical and Dielectric Properties
By D.B. Sirdeshmukh, L. Sirdeshmukh, and K.G. Subhadra

81 **Metallopolymer Nanocomposites**
By A.D. Pomogailo and V.N. Kestelman

82 **Plastics for Corrosion Inhibition**
By V.A. Goldade, L.S. Pinchuk, A.V. Makarevich and V.N. Kestelman

83 **Spectroscopic Properties of Rare Earths in Optical Materials**
Editors: G. Liu and B. Jacquier

84 **Hartree–Fock–Slater Method for Materials Science**
The DV–X Alpha Method for Design and Characterization of Materials
Editors: H. Adachi, T. Mukoyama, and J. Kawai

85 **Lifetime Spectroscopy**
A Method of Defect Characterization in Silicon for Photovoltaic Applications
By S. Rein

86 **Wide-Gap Chalcopyrites**
Editors: S. Siebentritt and U. Rau

87 **Micro- and Nanostructured Glasses**
By D. Hülsenberg and A. Harnisch

88 **Introduction to Wave Scattering, Localization and Mesoscopic Phenomena**
By P. Sheng

89 **Magneto-Science**
Magnetic Field Effects on Materials: Fundamentals and Applications
Editors: M. Yamaguchi and Y. Tanimoto

90 **Internal Friction in Metallic Materials**
A Handbook
By M.S. Blanter, I.S. Golovin, H. Neuhäuser, and H.-R. Sinning

91 **Time-dependent Mechanical Properties of Solid Bodies**
By W. Gräfe

92 **Solder Joint Technology**
Materials, Properties, and Reliability
By K.-N. Tu

93 **Materials for Tomorrow**
Theory, Experiments and Modelling
Editors: S. Gemming, M. Schreiber and J.-B. Suck

94 **Magnetic Nanostructures**
Editors: B. Aktas, L. Tagirov, and F. Mikailov

95 **Nanocrystals and Their Mesoscopic Organization**
By C.N.R. Rao, P.J. Thomas and G.U. Kulkarni

96 **Gallium Nitride Electronics**
By R. Quay

97 **Multifunctional Barriers for Flexible Structure**
Textile, Leather and Paper
Editors: S. Duquesne, C. Magniez, and G. Camino

98 **Physics of Negative Refraction and Negative Index Materials**
Optical and Electronic Aspects and Diversified Approaches
Editors: C.M. Krowne and Y. Zhang